MOVABLE BRIDGE ENGINEERING

MOVABLE BRIDGE ENGINEERING

Terry L. Koglin

John Wiley & Sons, Inc.

Published by John Wiley & Sons, Inc., Hoboken, New Jersey
Published simultaneously in Canada

For general information on our other products and services or for technical support, please contact our Customer Care Department within the United States at (800) 762-2974, outside the United States at (317) 572-3993 or fax (317) 572-4002.

Wiley also publishes its books in a variety of electronic formats. Some content that appears in print may not be available in electronic books. For more information about Wiley products, visit our web site at www.wiley.com.

Library of Congress Cataloging-in-Publication Data:

Koglin, Terry L.
Movable bridge engineering / Terry L. Koglin.
p. cm.
Includes index.
ISBN 0-471-41960-5
1. Bridges, Movable. I. Title.
TG420 .K64 2003
624′.8—dc21
2002155407

10 9 8 7 6 5 4 3 2 1

CONTENTS

PREFACE ix

ACKNOWLEDGMENTS xi

I SELECTION 1

1 To Bridge or Not to Bridge 3

2 Movable Bridges versus Fixed Bridges 10

3 Types of Movable Bridges 20

4 Bascule Bridges 33

5 Vertical Lift Bridges 55

6 Swing Bridges 81

7 Application of Types 100

8 Bridge Replacement 116

II DESIGN 123

9 Preliminary Aspects 125

10 Structural Issues 133

11 Detailing—Bascule Bridges 144

12 Detailing—Vertical Lift Bridges 170

13 Detailing—Swing Bridges 191

14 Machinery—Bascule Bridges 209

15 Machinery—Vertical Lift Bridges 219

16 Machinery—Swing Bridges 227
17 Machinery—General 235
18 Controls and Power 267
19 Rehabilitation 299
20 Operating and Maintenance Manuals and Bid Documents 305

III CONSTRUCTION 311

21 Construction Management 313
22 Structure 317
23 Fabrication and Installation of Machinery and Electrical Components 325
24 Balancing 335
25 Replacement and Repair of Existing Bridges 338

IV MAINTENANCE 341

26 Substructure and Superstructure 347
27 Machinery, Control, and Power Systems 351
28 Maintenance Particulars 357

V INSPECTION 363

29 Inspection Basics 365
30 Structure 373
31 Machinery—Bascule Bridges 380
32 Machinery—Vertical Lift Bridges 394
33 Machinery—Swing Bridges 409
34 Machinery—General 421
35 Electrical 452
36 Inspection Reporting 460

VI EVALUATION OF EXISTING BRIDGES 465

37 Bridge Management 467

38 Structure 472

39 Machinery 477

40 Electrical Systems, Traffic Control Systems, and Operators' Houses 487

41 Failure and Bridge Functionality Analysis 490

REFERENCES 493

BIBLIOGRAPHY 495

GLOSSARY 511

APPENDIX I UNITED STATES PATENTS ON MOVABLE BRIDGES 515

APPENDIX II NORTH AMERICAN MOVABLE BRIDGES 531

INDEX 679

PREFACE

This book is intended to be a description of movable bridges of the United States and a comprehensive source of engineering information for bridges of this type. It discusses selection, design, construction, maintenance, inspection, and evaluation of movable bridges. Movable bridges are basically very large machines, so that although this area does not dominate the book, there is a heavy emphasis on mechanical machinery. Structural and electrical issues are discussed, but because a great deal of information in those areas also pertains to many subjects other than movable bridges, their presentation here is limited. Space limitations prevent the more complete examination of several topics in movable bridge engineering that would be of interest to many readers. Other topics that some would have liked to see discussed have been left out altogether. No attempt has been made to present a "cookbook" method of solving problems, not so much due to space limitations as to the belief by the author that each movable bridge is unique and should be designed, constructed, maintained, inspected, and evaluated on that basis. Anyone with a specific movable bridge problem is invited to ask the author for specific assistance. Contact can be made via the Wiley web site (www.wiley.com/go/movablebridge). Any other comments, particularly updates of movable bridge information such as the listing of movable bridges which can also be found at the Wiley web site, are welcome.

Movable bridges located outside the United States are discussed only where their unique attributes require them to be mentioned, in instances in which the type of bridge or its use is rare or unknown in the United States, or to provide historical information on bridge construction predating usage in the United States. Details described are generally limited to those that are unique to bridges that are designed to be moved. The various chapters discuss machinery, applications, special structural considerations such as reversing loads, traffic, maintenance of movable bridges, and other aspects that do not pertain to fixed bridges. Structural issues that apply equally to fixed and movable bridges are not included in this book. Certain structural considerations, such as the actions of the various types of movable bridge spans under dead load, live load, and wind and other environmental loadings, are discussed so that structural analysis can be undertaken by whichever method is convenient to the engineer performing the work.

Architectural considerations have become more important in recent years, for railroad as well as highway bridges. The appearance of a bridge, whether movable or fixed, has an effect on the community surrounding it, and many communities have made the claim that they are entitled not to have an eyesore in their midsts. Architectural considerations are only briefly discussed here, beyond the very general notation as to which types of movable bridge may be functionally suitable in a particular situation and have the more pleasing forms.

A number of communities have decided that they wish to preserve their historic treasures, and sometimes preservation of an old bridge means maintaining restricted access to an area, for better or worse. Many communities have also decided that they want to cut the cost of government, which often means that funds for maintenance or replacement of aging structures are hard to find. Beyond suggesting ways to accomplish various tasks such as maintenance and inspection, this book does not address any of these issues.

Movable Bridge Engineering is organized into functional parts from an engineering standpoint, with conceptual, design, construction, maintenance, inspection, and evaluation work grouped separately. Within these sections, different parts or types of movable bridges are described in separate chapters. No attempt is made to provide for equal length of parts or chapters.

Citations of other works within the text may lead the reader to further information or greater detail on the subject under discussion. Footnotes indicate the sources of information presented, where additional related material may be available. Photographs are by the author, unless otherwise credited. Original drawings are by Ethan Fuster.

ACKNOWLEDGMENTS

The following people have contributed to this work by patient review and making valuable suggestions for its improvement. On bridge structural issues, Ulrich Gygax, Arthur Fishfeld, and Maria Grazia Bruschi offered much invaluable assistance. They also helped with grammar and punctuation. Dennis Marchetti offered guidance in the discussion of electrical systems. George Perkons provided his many years of experience with mechanical issues. John E. Nixon not only carefully reviewed the text for typographical and grammatical errors, but also added his invaluable comments on the issues dealing with wire ropes, hydraulics, and many other subjects of which he is a master. Brad Hollingsworth also took time from his busy schedule to review the sections of the manuscript dealing with machinery and offered many valuable suggestions. These people pointed out many errors, and it is hoped that all have been corrected. Many other persons also helped with suggestions and bits of information, including personnel of Union Pacific and Canadian Pacific, especially Donald J. Lewis of Canadian National/Illinois Central. Particularly, Christopher Hahin, MetE, PE, Engineer of Investigations, of Illinois Department of Transportation provided information on several movable bridge projects. The responsibility for the finished product rests solely with me.

To all who have provided assistance in any form, my gracious thanks.

I

SELECTION

1

TO BRIDGE OR NOT TO BRIDGE

HISTORY OF BRIDGES, REASONS FOR BRIDGING, AND WATER-VERSUS-LAND CONFLICTS

Movable bridges were first commonly used as protective devices over the moats at medieval castles (Figure 1-1); the "drawbridge" frequently seen in old Robin Hood and other swashbuckling movies was usually a bascule type that pivoted upward on trunnions. A typical bridge may or may not have been counterweighted, and it was usually operated via a windlass through chains or ropes. The floor of the bridge leaf acted as a strong door, impeding entry and providing resistance to projectiles fired from catapults.

Earlier than the Middle Ages, there is evidence of bascule-type structures built in Egypt as early as the fourteenth century B.C. These were movable bridges across moats that provided protection from marauding armies or robbers. According to Herodotus, Queen Nitocris of Babylon built a form of retractile bridge, for protective purposes, across the Euphrates at about 460 B.C. There are some reports of ancient movable spans, that allowed water traffic to pass through bridges that were built across estuaries or other waterways. These bridges were usually built for military purposes and were often temporary. A famous example is a pontoon bridge erected by Xerxes across the Hellespont at about 480 B.C. (Figure 1-2). It was rebuilt around 313 B.C., with three spans capable of being floated out of the way for water traffic, in the manner of some floating "pontoon" swing bridges in use today.

Almost all early bridges were built to allow land traffic to pass over a waterway. Bridges were difficult to finance in earlier times and were usually constructed for military purposes, such as those mentioned earlier. Some locations with substantial commercial traffic developed the need for bridges fairly early in history. The City of London, England, was located, at least from Roman times, on the main north–south

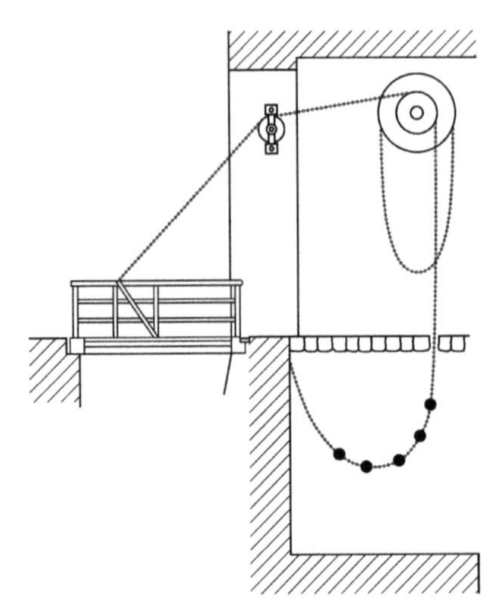

Figure 1-1 Drawbridge at Medieval Castle. Typical of such structures that were precursors of modern bascule bridges.

British trade route, at the lowest point on the Thames River that was easily crossable. The Romans built a fairly permanent wooden bridge there, perhaps largely out of military necessity, but this bridge disappeared in ruin during the Dark Ages. A wooden bridge was built, according to Richard Thompson,[1] in 994, with a draw span. This bridge was in place at the time of William the Conqueror (1066–1087), but it was in poor condition by the twelfth century.

At that time, it was agreed that a better bridge was needed, and the famous stone "London Bridge," of nursery rhyme fame, was built (Figure 1-3). This bridge originally had a short bascule draw span, which may have been intended for navigation rather than defensive purposes. But the bridge piers so disrupted the flow of water downstream that the passage of vessels through the draw span was difficult, and the movable span was seldom used. It was last opened for a boat in 1381, during the reign of Edward IV.[2] The bascule span had some value as a protective device and was raised during occasional rebellions and other crises such as in an attempt to stop the spread of the Great Fire of 1666. The limited utility of the draw span, combined with its requirement for extra maintenance, resulted in its eventual replacement by a fixed span, on which a house was built. One of the last instances of a drawbridge being used for defensive purposes was in 1775, at the beginning of the American Revolution, in Salem, Massachusetts.[3]

Many late medieval and Renaissance designers developed various types of movable bridges.[4] Leonardo Da Vinci developed plans for a bobtail (asymmetrical) swing

[1]Richard Thompson, *Chronicles of London Bridge* (London: Smith-Elder, 1827).
[2]Paul Murray Kendall, *Richard III* (New York: Norton, 1955), p. 279.
[3]Robert Leckie, *George Washington's War* (New York: HarperCollins, 1992), p. 93.
[4]O. E. Hovey, *Movable Bridges,* Vol. 1 (New York: John Wiley & Sons, 1926), p. 4.

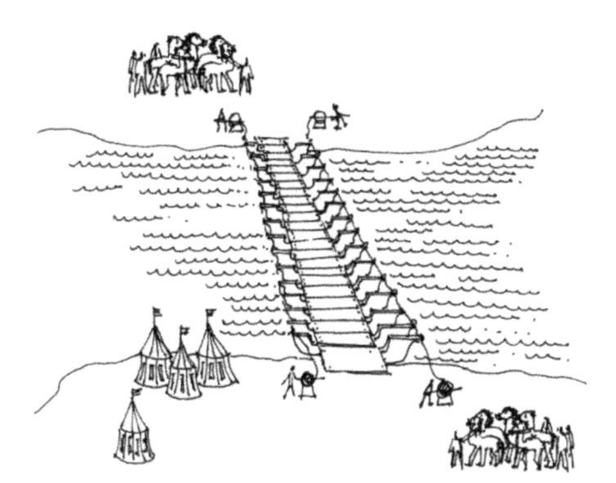

Figure 1-2 Xerxes' Pontoon Bridge at the Hellespont (Dardanelles). One of the earliest documented movable bridges, it was a temporary structure built for military purposes.

bridge and for a trunnion bascule and even designed a vertical lift span. The heel trunnion type of bascule bridge was developed at this time, and it eventually became so common in the Netherlands that it came to be known as the "Dutch draw." The rolling lift bridge was also developed at an early time as a means of providing a bascule bridge that, when opened, would not obstruct a canal towpath.

As commercial activity accelerated after the invention of the steam engine, conflicts arose between water and road traffic. Many types of movable bridges were developed, including bascule types that had elaborate balancing arrangements and very complicated operating mechanisms. Most movable bridges of the early industrial era were of the swing type, which rotates horizontally on a pivot located under its center of gravity, because they were simpler to design and generally thought to be more economical than other types.[5] With the pivot located under the center of gravity, the bridges were balanced in a horizontal plane in all directions about the pivot point, As power was expensive at that time, most people thought that swing bridges were cheaper because they did not have to be lifted and lowered. With the development of the main highway between New York and Philadelphia in the late eighteenth century, three of the main rivers crossed—the Hackensack, the Passaic, and the Raritan—were spanned with toll bridges built in the 1790s. As these rivers are all navigable, the bridges included draw spans. The Camden & Amboy Rail Road, built in the 1830s, required a draw span at its bridge over Rancocas Creek in New Jersey. The bridge, a shear pole swing span, was the site of an accident in 1853. A train went through the opened draw span. One of the first railroad draw bridges in the country was the Paterson & Hudson River Rail Road's bascule bridge over the Hackensack River, built about 1833.[6]

Swing bridges continued through the nineteenth century to be the most common type of movable bridge.[7] O.E. Hovey, writing in the 1920s, continued to extol swing

[5]Charles H. Wright, *The Designing of Draw Spans*, Vol. 2 (New York: John Wiley & Sons, 1898), p. 87.

[6]John T. Cunningham, *Railroads in New Jersey*, Vol. 2: *The Formative Years* (Andover, New Jersey: Afton, 1997), p. 52.

[7]Wright, p. 87.

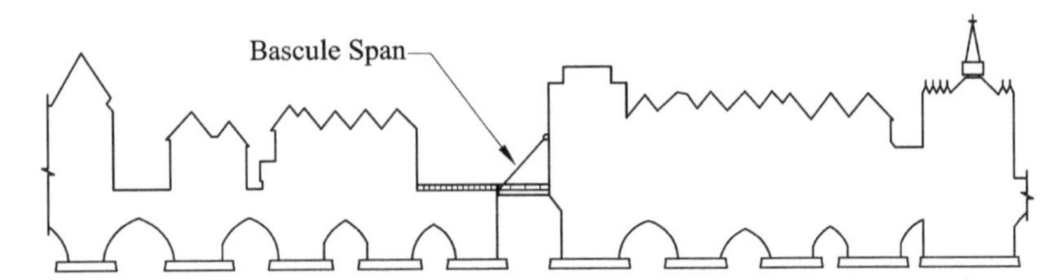

Figure 1-3 Old London Bridge. Built in 1116, during the reign of Henry I, it stood for 500 years. The bascule span was removed some years before the bridge was replaced, as navigation was impractical here because of swift currents under the bridge.

bridges because they did not require expensive counterweights. Near the end of the nineteenth century most movable bridges were of the swing type, which required little in the way of sophisticated engineering, provided that the span length was not too long and the bridge did not have to open or close very quickly. The rapid development of other types of movable bridges occurred around the end of the nineteenth century, with bascule variations, vertical lift bridges, retractile bridges, and other types coming into general use. A great many of these bridges were impractical and either did not last long or were not repeated. This was an age of complex mechanisms, and movable bridges were equipped with elaborate roller bearings (Figure 1-4) and complicated wedging and locking mechanisms. Most of this equipment proved to be not very durable and was prone to failure without constant attention, so these mechanical wonders eventually disappeared from the scene.

The city of Chicago, Illinois, developed rapidly during the nineteenth century, as it was located at the southwest corner of Lake Michigan and at the confluence point of the north and south branches of the Chicago River, one of the few water routes inland to the west. In the early nineteenth century this point became the eastern end of the preferred trade route between the Great Lakes and the Mississippi River. Commercial traffic was thick on the river, especially after the Illinois and Michigan, and then the Illinois and Mississippi, canals expedited water passage further inland. Canal development was far outstripped by local growth in the city, which forced the construction of many draw spans over the river despite the objections of the navigation interests. The conflicts multiplied when railroads came into use, as Chicago, because of its central location, developed into the railroad capital of the country.

Even today, more than 150 years since development began, Chicago has more movable bridges than any other locality in the world. The city was forced to find a way to eliminate the need for constant operation of its movable bridges. The ultimate solution was to move the main waterway to the south, away from the surface street congestion of downtown Chicago. The Calumet Sag Channel was built, in an area of relatively low property values, southwest of the city. It connects the main Chicago area Great Lakes harbor, Lake Calumet, with the Mississippi River system via the Illinois Waterway. Thanks to proper planning and engineering, every single bridge across the Calumet Sag Channel was built as a fixed bridge. Today many of the Chicago River bridges no longer operate, and those that do operate open almost exclu-

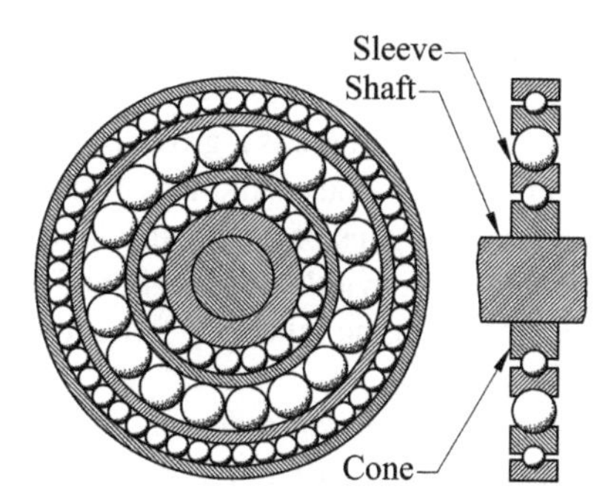

Figure 1-4 Old Bascule Bridge Roller Bearing. This example is typical of many complicated mechanical devices and systems that were tried on movable bridges at the beginning of the modern era. Most quite quickly fell into disuse.

sively for pleasure craft such as sightseeing tour boats, and do so on a regulated timetable. All local traffic still passes over the Chicago River and its navigable branches on movable bridges. Only certain expressways and interstate highways cross the navigable parts of the river on high level fixed bridges.

A notable U.S. Supreme Court case of 1857 gave land transportation companies the right to build bridges across navigable waterways and gave Abraham Lincoln national exposure as the counsel for the Rock Island Bridge Company. This company had built the first railroad bridge across the Mississippi River, at Rock Island, Illinois.[8] The bridge carried the tracks of the Chicago, Rock Island & Pacific Railroad, a main rail line west out of Chicago. The waterway interests contested this construction, even though the bridge was movable, saying they had a constitutional right to free and unrestricted passage up and down the river.

This particular case developed when a steamboat, the *Effie Alton,* hit the bridge and burst into flames. Lincoln argued that the boat was not destroyed because of the bridge, but because one of its paddle wheels failed, causing loss of control of the boat. The court ended up deadlocked, which amounted to a victory for the railroad and the bridge. The decision did not give land transportation, in the form of railroads, equality with waterway traffic in terms of rights of transit, but it did allow bridges to be built across the Mississippi River in spite of their being hindrances to the movement of steamboats both upstream and down.

Regulations were promulgated that required certain minimum clearances to be provided under bridges for navigation, resulting in limitations on the height of steamboats and their smokestacks. Any proposed bridge that would not provide the minimum clearance was required to open promptly on demand for water traffic. These regulations have remained largely in effect on all inland waterways in the United States, except for some artificial intrastate waterways built without federal assistance. The only significant example of such a situation today is the New York State Barge Canal system, the modern version of the old Erie Canal. Although commercial traf-

[8]David Herbert Donald, *Lincoln* (New York: Simon & Schuster, 1995), p. 157.

fic has almost disappeared from this system, it remains active for pleasure craft. New York State has its own rules for bridge clearances and regulations for movable bridges on the New York State Barge Canal. There is a group of what appear to be mass-produced movable bridges on the west end of the Erie Canal, between Rochester and Buffalo. These are 15 vertical lift spans of practically identical design and construction, each of almost identical length, width, and height and operating in the same way, crossing the canal at small towns in the relatively flat countryside in this region (see page 74).

In the United States, the federal government has traditionally been held to have constitutionally mandated authority over navigable waterways. Article II, Section 8, of the United States Constitution gives Congress authority over interstate commerce, which was construed to include control over navigable waters and the bridges and other impediments built on or over them. Over the years since the mid-nineteenth century, almost every river and stream big enough to float a canoe, as well as most larger lakes, have been declared navigable, and many have had significant improvements constructed on them at national expense. Many of the larger, more costly improvements have been made in the guise of flood control, initiated in the years after the Civil War by the Army Corps of Engineers, led by General A. A. Humphries. The primary beneficiaries of many of these projects have been commercial barge operators. In several locations, these projects have provided competition for the railroads, competition that otherwise would not exist, in the transportation of heavy bulk freight, such as coal and grain. The flood control value of the locks, dams, and levees has been in dispute since the early large construction projects, dating back to the Civil War era,[9] and recently environmentalists have found fault with such construction.

Budget cuts and the failure of the Mississippi River control system, and others, to protect adjacent communities during high water have resulted in a change in thought on the part of the Corps and Congress. One development, in California's Napa Valley and the Snake River in Idaho, is to return the waterways and their environs to their natural states. The effect of this change in thinking on movable bridges is not clear. The Corps of Engineers might concentrate its efforts on commercially viable waterways, but may find considerable opposition from trucking, pipeline, and railroad interests.

The utility of navigable waterways declined precipitously after railroads came into general use in the mid-nineteenth century. In fact, most states that had developed canal systems, including New York and Pennsylvania, levied special taxes on railroads operating in competition with their canals, as they recognized that the mule-pulled canal boats were poorly suited for competition with the railroads. Most of these early canals went out of business anyway, between the 1840s and the 1890s, as railroads took away almost all their business. By 1900, however, railroads had gained such a monopoly that canals were looked at as a way to provide competition for them. The City of New York completed the Harlem River Ship Canal in 1910 to provide waterway competition, in conjunction with the Erie Canal, for the New York Central Railroad, which had a monopoly on direct railroad service to Manhattan from the west. This resulted in several movable bridges being built or upgraded over the com-

[9]A. E. Morgan, *Dams and Other Disasters* (Boston: Porter-Sargent, 1971), pp. 78–80.

bined Harlem River and Spuyten Duyvil Creek between Manhattan and the Bronx. In the 1920s, the Erie Canal was enlarged for the last time in a further attempt to provide improved competition for the railroad.

Waterways again declined in popularity for commercial freight movement as the trucking industry developed after World War I; many movable bridges built at or before that time remain in place at locations where they open very seldom. Heavy commercial inland waterway traffic remains today only on a few waterways, such as the Great Lakes between Lake Superior and Lake Huron, the Mississippi River system, and the Gulf Intracoastal Waterway. Most navigable waterways now see only pleasure craft; traffic in recreational power- and sailboats has mushroomed since World War II. These boats also necessitate the opening of movable bridges, on demand at most locations, and have required thousands of movable bridges to be in operation almost continuously on weekends and holidays, particularly in coastal waters in the vicinity of popular recreation areas.

In some urban areas, former industrial dockyards have emerged from seedy neglect to become fashionable marinas, and the movable bridges near them have had to be rehabilitated to accommodate the increased boat traffic after many years of little or no use of the waterway. This trend may be expected to continue. The Newtown Creek area of New York City, which partially divides the boroughs of Brooklyn and Queens, has experienced a recent development of interest in recreational boating. The sight of weekend kayakers paddling past decayed industrial waterfronts is more than somewhat incongruous. In the not too distant future, further gentrification in the area, which comprises Maspeth, Greenpoint, and Long Island City—all still gritty industrial areas today—may include luxury marinas with large yachts moored within them. The dozen or so drawbridges in the area may eventually have more openings in a single weekend than in the entire last decade of commercial activity. In many urban locations movable bridges can become a serious impediment to traffic, particularly during rush hours. At many of these locations, restrictions have been imposed on movements of waterway traffic to minimize traffic jams on the streets. These restrictions can apply to commercial as well as pleasure boat traffic.

2

MOVABLE BRIDGES VERSUS FIXED BRIDGES

WHY MOVABLE BRIDGES ARE UNDESIRABLE

A movable bridge is a dangerous thing. Many lives have been lost and property has been destroyed because attempts were made to cross movable bridges that were open, vessels attempted to pass through movable bridges that were not open, or bridge operators attempted to open movable bridges when vehicles were on them or to close them when vessels were under them. There have been rare incidents of movable bridges collapsing under load (Figure 2-1) and of counterweights falling off vertical lift and bascule bridges. These events have usually been attributed to structural failure rather than failure of a movable bridge mechanism.

Counterweight ropes on a vertical lift bridge are structural members, serving as the primary means of support for the bridge counterweights, which are equal in weight to the lift span itself. These ropes are subject to mechanical wear, but catastrophic failures have been attributed to corrosion, which can affect any steel structural member. On the other hand, failure of a swing bridge wedge drive system can result in inadequate support of the bridge superstructure under live load, overstressing the main bridge members. Failures occurring in a part of the movable bridge mechanism most commonly happen when the bridge is operating, either opening or closing. It is unusual for injuries or damage to occur to persons, vehicles, or vessels under these conditions, but the bridge may become jammed in a partially open position, blocking both navigation and bridge traffic.

A movable bridge causes delays to vessels waiting to pass through its navigation opening, and delays to vehicles waiting for the bridge to be put back into position and readied for the acceptance of vehicular traffic. At locations such as Knapps Narrows, in Maryland, and Glen Island, New York, the movable bridge operates several hundred times a day in the peak season, and backups develop on both the highway and

Figure 2-1 Collapsed Swing Bridge over Chicago Sanitary and Ship Canal. This type of collapse, with the bridge failing under load, is an extremely rare occurrence. (From the *Chicago Daily News.*)

the waterway. Because they are large and heavy, the bridges cannot be moved and repositioned quickly. Many movable bridges, particularly highway bridges, have auxiliary devices for protective purposes, such as warning lights and gates. These must be actuated in sequence, adding to the time required to open a bridge for navigation and reopen it to traffic after a vessel has passed.

Movable bridges have many more complicated and delicate components than fixed bridges of comparable size. These include motors, gears, wiring, and control panels, which add to the cost of the structure, increase maintenance requirements, and are prone to failure, thus making the bridge less reliable. A movable bridge's environment is much harsher than most people realize, so that electrical and mechanical components that perform over a reasonable lifetime in an industrial environment can fail quickly on a movable bridge. Temperature variations, humidity, salt air at many bridge sites, poorly skilled operators, and indifferent maintenance all contribute to the overall poor reliability record of movable bridges. The design of movable bridges is much more difficult than that of fixed bridges, as all the complexities of their machinery must be kept in mind, as well as the structural aspects of the application.

WHY MOVABLE BRIDGES ARE DESIRABLE

At many locations, a movable bridge is the only practical means of crossing a waterway. The alternatives to movable bridges that allow traffic to cross a waterway include ferries, aerial transporters, tunnels, and high level bridges. Providing no crossing at all, so that a long detour is required, or closing the waterway to navigation and placing a simple, restrictive bridge across it are options that may be selected, sometimes unconsciously. Each of these alternatives has some drawbacks and should be studied before a decision on a crossing is made.

A movable bridge may be the least costly alternative to provide a means of crossing a navigable waterway. In a particular situation, such as a short span across a canal in an urban area, a movable bridge may be the only workable solution. It may be the most convenient option for the public, for instance, in an urban situation where a high level bridge would need very high piers and long approaches and require heavy vehicles to travel up steep grades. In many circumstances where steep grades are to be avoided, such as for railroads crossing navigable waterways between relatively low-lying embankments, a movable bridge can well be the only economically practical means of crossing the stream.

ALTERNATIVES TO MOVABLE BRIDGES

Today, in the United States, there are sufficient federal funds available to replace all the movable highway bridges in the country with "high level" fixed spans that would provide sufficient clearance for navigation to pass under them. At many bridge locations, this would seem to be ideal, as delays and dangers to highway and water traffic would be reduced or at least minimized. Unfortunately, the situation is not as simple as that. Local financial participation, required by law for most projects, could not be obtained for most of these bridge replacements. Many municipalities plead poverty for even the most necessary of infrastructure repairs. At many movable bridge locations a high level bridge cannot be constructed without causing some detrimental effect, so alternatives must be considered. Aesthetic opposition to high level bridges, which can be formidable, often develops.

Ferries were once the most common means of crossing a waterway. Parts of New York Harbor were jammed with passenger and vehicle ferries by the time the era of consolidation began in the late nineteenth century. East River ferry traffic ended after the Brooklyn Bridge and several later structures were built. Most of the ferry services across the lower Hudson River were operated by the railroads that approached New York from the west. Passengers, vehicles, and railroad cars were carried. Passengers and vehicles switched to rail and highway tunnels when they were constructed between 1904 and 1938. Some rail freight cars are still carried across New York Harbor on ferries, or car floats, to this day. New York City, in an effort to reduce truck traffic on city streets, is attempting to increase rail freight car ferry service between New Jersey and Long Island. The increased length of freight trains, extending to 150 or more cars today, as opposed to 30 or 40 cars 100 years ago, makes the railcar ferries impractical for through traffic. Most through rail freight to New England now bypasses New York on routes detouring 100 to 150 miles north of the harbor. Although a through rail freight tunnel has been contemplated for New

York Harbor for the past 100 years, it is yet to be built. The City of New York is still studying the construction of such a tunnel. The rail freight route the city is proposing would cross Arthur Kill, between New Jersey and Staten Island, on an existing vertical lift span (see Figure 5-17) and continue to Brooklyn via a tunnel to be bored under the Narrows, at the approximate location of the Verrazano-Narrows Bridge. Rail ferry service once operated across the Mackinac Straits in Michigan. Some efforts were made to accommodate rail freight cars on the Mackinac Bridge, but these were abandoned as the economic decline of Michigan's Upper Peninsula left very little demand for such traffic.

The ferries had a serious handicap in the days prior to the application of radar to navigation, as darkness, or worse yet, fog, made operations extremely hazardous. Delays due to weather or heavy traffic along the river made ferry travel time-consuming and unreliable. After an absence of several decades, passenger ferry traffic has been reinitiated in the Hudson River and East River portions of New York Harbor, as congestion of other modes of passenger transport made additional options desirable. Radar allows the services to operate in spite of the weather, and decreases in other shipping on the East River and Hudson River have reduced delays to tolerable levels. For some people, these new ferries actually provide the quickest means to cross the Hudson River.

Aerial transporters, or aerial ferries, were occasionally used at waterway crossings where traffic was not particularly heavy and high clearances over the navigation channel were required. A rigid structure similar to a gantry crane was constructed over the waterway, and a simple short platform was suspended below it. This platform shuttled between terminals at each side of the waterway, propelled by cables attached to the gantry (see page 29). The aerial ferries were relatively expensive, considering their small capacity. The last was built in Germany in the 1930s,[1] and it is believed there are none in service any longer.

Underwater tunnels have been used for almost 200 years to cross busy waterways in urban areas. The first "modern" subaqueous vehicular tunnel was promoted, designed, and built in the 1830s, under the Thames River by Marc Brunel, father of the famous I. K. Brunel, who built the Great Western Railway with the Saltash Bridge and many other engineering marvels of the early nineteenth century. Underwater vehicular tunnels usually have the advantage, over high level bridges, of depositing travelers close to the river's edge. The long approaches necessary for a high level bridge, which may have to clear the water by 200 ft in a major harbor, are not needed for tunnels. The typical tunnel depth under the water surface of an estuary may be only 50 ft or 60 ft, so that in the more common situations a tunnel would not have to go as far down as a bridge would have to go up to provide unrestricted passage for navigation.

The Baltimore Harbor Tunnel, the Boston harbor tunnels, and other underwater highways at major ports were constructed at a time when the construction of bridges to span these waterways was almost unthinkable. In smaller communities, such as Fort Lauderdale, Florida, tunnels have been used in the downtown areas to pass under waterways, to avoid the elevation required and resulting approach structures for a

[1]*Engineering News-Record* (October 16, 1930).

fixed bridge. Tunnels are not ideal solutions, however, as they are quite expensive in comparison to their very limited traffic capacity. Tunnels are psychologically restrictive as well: many people become disoriented when driving through tunnels, on bright sunny days visibility can be a problem when entering and exiting tunnels, and any kind of accident can be that much more difficult to clear because of the restricted space inside a tunnel.

SPECIFIC LOCATIONS WHERE MOVABLE BRIDGES ARE ALMOST INEVITABLE

"Overpasses" are the solution of choice for most authorities when confronted with conflicts between transportation modes. The interstate highway system was constructed with the intent of having a completely grade-separated system, so that all local streets and highways either passed over or under an interstate highway. At some locations construction of complete interchanges required three- or four-level intersections of roadways and accomplished what was desired at a very high cost. Every railroad track was also grade separated from the "Defense Interstate Highway Network," as it was officially called.

The limits of practicality were reached when it came to putting these limited-access superhighways over waterways. In Chicago, the Eisenhower Expressway, I-290, crosses the South Branch of the Chicago River on four bascule bridge leaves (see Figure 2-2). It would be impossible, with a high level bridge over the river, to provide the intersections with local streets that were an integral part of the planning for this highway. Putting I-290 into an under-river tunnel would also make it very difficult, if not impossible, to provide the local street connections required, even by building ramps in tunnels, which would get very expensive.

At certain other locations, the sheer expense of building a bridge at a sufficiently high fixed clearance over the waterway to satisfy navigation requirements was deemed to be too daunting. The Columbia River passes through a broad, low plain as it skirts the north side of Portland, Oregon. Large oceangoing ships occasionally navigate the river in this area, so a very high clearance was required for the I-5 bridge. Rather than construct the massive fixed bridge with high piers and extremely long approaches that would be needed to satisfy all requirements, twin mid-level vertical lift structures were provided. These structures open very seldom, minimizing inconveniences to highway traffic, by virtue of providing sufficient clearance over the water in the lowered position to allow most craft, even a considerable amount of commercial traffic, to pass under them. On the rare occasion that a large ship must pass under the twin bridges, traffic is inconvenienced, but the bridge openings are restricted to times of day with reduced highway traffic as much as possible.

On the other side of the continent, over the Hackensack River in northern New Jersey, it was deemed necessary to replace the Wittpenn Bridge, a 70-year-old vertical lift highway bridge, because of deterioration and the substandard roadway configuration. Every effort was made to develop a workable, practical solution that would

Figure 2-2 Interstate Highway 290 over the Chicago River, Chicago, Illinois. This is one of the few interstate highway bridges crossing a river on a movable bridge.

eliminate the movable bridge from the crossing. Because of dense development on the east side of the river and the necessity to maintain direct access to commercial sites on the west side of the river, development of an economical high level fixed bridge that would provide sufficient clearance for oceangoing vessels that navigate the stream was deemed impossible. As of June, 2002, after ten years of trying to find an alternative, the solution of choice was a movable bridge. The bridge will be located at such a height as to eliminate the majority of bridge openings, which are for smaller vessels such as tugs and barges, requiring the bridge to be opened only for the infrequent large oceangoing vessel. The Federal Highway Administration, the watchdog for the federal funding for the project, is by policy generally against the use of movable bridges and required reevaluation of a high level fixed bridge alternative in this case before agreeing to construction of a movable bridge.

A separate project in northern New Jersey, the Portway project, is barely into the planning stages (as of 2000). This project will provide a separate artery for trucking containers off ships and railcars from the port and rail yard area west of the Hackensack River and south of the Wittpenn Bridge, to other intermodal rail yards and to the tunnels to New York and the bridge to New England. It was contemplated to use the new Wittpenn Bridge to cross the Hackensack, as part of this project. Consideration was given to providing a rehabilitated or new movable Wittpenn Bridge for local traffic and a high level bridge, on a separate alignment, for the port- and rail-originated

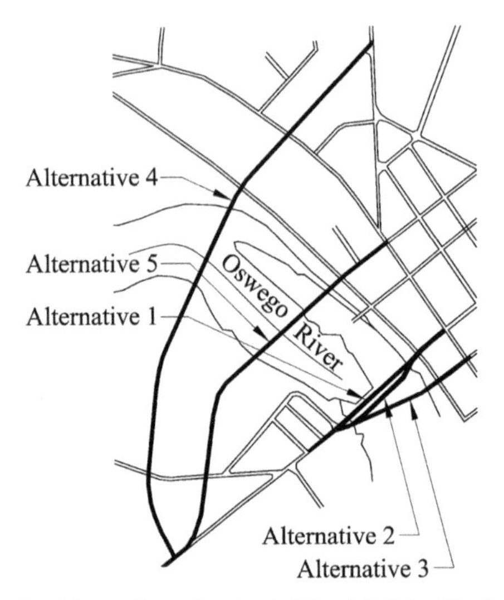

Figure 2-3 Alternatives for Lock Street Bridge Replacement

traffic. The new high level bridge would also allow vehicular traffic from Interstate 280, which currently terminates about 2 miles west of the Wittpenn Bridge, to continue east to a new grade-separated right-of-way to the Holland Tunnel, greatly easing traffic congestion in the area. It remains to be seen how the project will conclude; at last notice only one new bridge was to be built—another higher, larger vertical lift span.

At many locations people have decided, for aesthetic reasons, that new movable bridges should be built instead of high level fixed spans. For many years the standard solution to traffic congestion along coastal highways was to replace a low level movable bridge, which was most likely a bascule type, with a high level fixed bridge. It was assumed by all responsible parties that the public would prefer doing away with the inconvenience of the movable bridge. After many of the fixed bridges were built, people began to register complaints about unsightly obstructions to their views of the ocean, among other objections. For a period the construction of high level fixed bridges over narrow waterways in low-lying areas was the official policy of the Federal Highway Administration (FHWA), seconded by the transportation departments of many states with coastal highways; however, the practice is becoming less common. A number of projects that called for replacement of movable highway bridges with high level fixed bridges have been cancelled.

The FHWA still refuses to sanction movable bridges without show of just cause, so opposition must be raised to a high level fixed bridge at such locations before a new movable bridge will be considered for federal financial assistance. This attitude appears to be changing, as the dislike of high level fixed bridges has spread to include proposed bridges over inland waterways. Many recent projects have changed course, replacing fixed bridge with movable bridge construction, particularly in bridge replacements.

At Phoenix, New York, an old and badly deteriorated bascule bridge across the Oswego River and adjacent canal required replacement. The standard engineered solution called for a high level fixed bridge to replace the old movable bridge. Even though the high level bridge required a clearance of only 24 ft or so above the water, the local people refused to accept this solution, which they believed would put a "Chinese Wall" through the center of their town. Alternatives placing the new bridge outside the town were rejected as being too circuitous and inconvenient (Figure 2-3).

At a community meeting held in Phoenix on August 9, 1983,[2] the public expressed its desire, among other concerns, that the bridge would "be close to the center of Phoenix" and "be visually unobtrusive." The only solution that met these criteria, as well as the other requirements of the project, was a movable bridge. In spite of the fact that the old bridge was in precarious condition, community opposition forced a long delay in its replacement while an option including a movable bridge, which would avoid a high embankment through the town, was developed.

In the early 1990s a competition was held to provide a design for a replacement for the Woodrow Wilson Memorial Bridge, a bascule bridge that carries Interstate Highway I95 across the Potomac River below Washington D.C. The competition called for a fixed bridge at a sufficiently high elevation to allow all existing water traffic to pass under it. Several architecturally sophisticated designs were submitted, but all were finally rejected when residents of Alexandria, Virginia, on the south bank of the river, found them unacceptable, claiming that the new bridge would destroy their view of the river and bisect the town (Figure 2-4). This, with other similar objections, including the high cost of ramp connections to the high level bridge, was deemed sufficient cause to cancel the fixed bridge project. A few years later another competition was held for the bridge replacement, this time calling for a movable bridge design of the bascule type that would satisfy the local aesthetic requirements (Figure 2-5) and provide practical interchanges. This effort is leading to an acceptable replacement Woodrow Wilson Bridge, which is to be a bascule structure. The bridge and its approaches were under construction as of July 2002.

Figure 2-4 Woodrow Wilson Bridge High Level Alternate. This is one of the entrants in the earlier competition. (Courtesy of Maryland State Highway Administration.)

[2]New York State Department of Transportation, *Transportation Project Report–Lock Street Bridge over Oswego River and Canal,* 1984.

Figure 2-5 Woodrow Wilson Bridge Bascule Alternate. This bridge is currently under construction. (Figure courtesy of Potomac Crossing Consultants.)

Figure 2-6 Tower Bridge, London, England. This bridge is a combination suspension bridge and double-leaf bascule. The upper cross member above the bascule span is necessary to sustain the longitudinal tension of the suspension cables, which are actually iron bars.

APPLICATION OF MOVABLE BRIDGES

Movable bridges can be used in any bridge application. They can be built on a skew or on a grade. There are currently examples of swing, bascule, and vertical lift bridges used in these situations. Movable bridges can be built to any width and almost any length. A movable span can be placed in any part of a long causeway or other similar structure. Movable spans have even been built in combination with suspension and cable-stayed bridges (Figure 2-6).

3

TYPES OF MOVABLE BRIDGES

MODERN TYPES

Many different schemes have been developed over the centuries for providing a means to get a roadway, that crosses a waterway, out of the way of vessels, and then back in place for road traffic. With changes in available technology, new materials, increases in loads, and greater concern for safety, new types of movable bridges have occasionally been developed and older types have ceased to be used. In the late nineteenth and early twentieth centuries there was a proliferation of bridge types and variations of types produced. Many types of movable bridges that were developed with great expectations, at least by their inventors, have gone into history with only a single example, if any, built and put into in use.

Today there are only three types of movable bridge in common use: the bascule bridge, both single and double leaf types, the swing span, and the vertical lift bridge. The vertical lift is sometimes called a "lift bridge," but as the bascule bridge also lifts and is sometimes referred to as a lift bridge, the term is not exact. To be precise:

All movable bridges that are counterbalanced and open by pivoting about a horizontal axis should be called bascule bridges.

All movable bridges that open by pivoting about a vertical axis should be called swing bridges.

All movable bridges that open by lifting, without rotating or translating horizontally, should be called vertical lift bridges.

There are some bridges that appear to fit these definitions but actually do not, such as those that appear to be bascule bridges but do not have counterweights and are not otherwise balanced. For convenience, these bridges are lumped with their look-alike cohorts. Of the approximately 3000 movable bridges in existence in the United States today, about two-thirds (1900, plus or minus) continue to operate. Of these operable bridges, approximately 770 are bascule bridges, 750 are swing bridges, 270 are vertical lift bridges, and the rest are other types, including 80 removable spans, 6 retractile, and 30 pontoons, some of which are retractile and others of which are swing types. Of the new movable bridges built in the last two decades or so, about 60 percent are bascules, 35 percent vertical lift bridges, and the rest swing or other types.

The bascule bridge is commonly represented by double leaf types. There are also a few double swing bridges, in rare cases where they fill a special need. Paired bascules, with two separate bridges alongside each other, are not uncommon. Sometimes they are combined in such a way that those traveling over them have no idea that the structures are separate. Paired vertical lift spans are used occasionally as well. Triple or quadruple bascule and vertical lift bridges are also used, but rarely; these are almost exclusively railroad bridges in congested areas near terminals. Triple single leaf railroad bascules are in use at the approach to South Station in Boston, Massachusetts, although they have not been required to open for many years. Triple railroad vertical lift bridges are used at the east end of Pennsylvania Station in Newark, New Jersey, over the Passaic River (Figure 3-1). Quadruple single leaf two-track railroad rolling lift bridges were installed in the early twentieth century (1909 and 1910) over the Chicago Sanitary and Ship Canal, which connects the Chicago River with the Des Plaines River west of Chicago (Figure 3-2). These bridges provided eight tracks for

Figure 3-1 Dock Bridges, Newark, New Jersey. This is a combination of three semi-independent vertical lift bridges. The nearest has one track; the middle has two tracks at two different elevations, with one at a gradient; the far one has three tracks.

Figure 3-2 Western Avenue Bascule Bridges, Chicago, Illinois. This cluster consists of four independent single leaf Scherzer rolling lift bascule spans. The bridges were oriented in alternating opposing positions to allow them to be closer together. The bridges did not originally have machinery installed; although it was later put in, they no longer operate.

rail freight transfer traffic between the heavily used freight yards of the Chicago area. The Chicago Sanitary and Ship Canal, however, never developed any significant waterborne traffic. Most of the bridges over this canal that were intended to be movable did not have machinery installed when they were built. Some of those that did have machinery installed had it removed later (Figure 3-3). At present, one of the aforementioned quadruple railroad bascule bridges is out of service and another has only one track in use; none of these bridges operate.

Swing bridges, as mentioned earlier, were once the most common type of movable bridge, but very few have been built in the last 60 years. A few have recently been constructed to fill special needs. Other than bascule, vertical lift, and swing variations, the only other types of movable bridge that have been built in recent years are the retractile and pontoon types, with both types combined in single structures. A few of these have been built in Washington State, because of special conditions of wide, deep waterways requiring low-cost bridging and navigation by large vessels. One retractile pontoon has been built in Hawaii (see page 25). The swing bridge and other types are largely represented today by older structures. Most movable bridges now built are either bascule or vertical lift types, as these most readily answer the need for reliable, structurally sound, economically constructible, and economically operable movable bridges.

Figure 3-3 Western Avenue Vertical Lift Span, Chicago, Illinois. This bridge was constructed as a fixed bridge, then had machinery added; the machinery was then removed, and finally (after this photo was taken) the towers were removed.

The trend in the past few decades has been to simplify each movable bridge installation as much as possible. Fewer moving parts means less maintenance and less likelihood of breakdown or deterioration. Contradicting this trend is the increasing complexity of electrical control systems, which have become more complicated in an effort to reduce the skill needed by bridge operators. The application of solid-state electronics has resulted in the improvement of reliability and durability in these systems. More control functions are handled automatically in the typical newer bridge, as opposed to letting the human bridge operator perform most of the decision functions in an older bridge. A further reason for more complexity in movable bridges is the need to provide greater protection for the public in the form of traffic barrier gates, traffic lights, and so on. Unfortunately, these devices also cause a movable bridge to take longer to operate.

Structurally, new movable bridges follow the same trends as fixed bridges. For example, a cable-stayed bascule was proposed but rejected for the new Woodrow Wilson Bridge over the Potomac River (see page 17). Most new movable bridges are of plate steel girder construction rather than truss spans, except when there is an extremely long span. Truss spans, when built, are simplified as much as possible to reduce construction and maintenance costs. Latticed truss members have not been used since the 1950s. Concrete has been used as the main structural material in two recent movable bridges in Washington State, but is not likely to be used extensively until its long-term durability for movable bridge spans is demonstrated and complications due to the heavy moving masses are resolved. Today highway bridges are more often designed for permanent use rather than for 20- or 30-year lifespans, and the repairability of a steel structure is a big advantage.

Because of higher labor costs, the simpler designs are cheaper to fabricate even though they may use more material than a more complicated but structurally efficient design. This is just the opposite of the situation 100 years ago. The American inland waterway system, plus the very large cranes available today, allow many movable spans to be fabricated complete in an economical permanent yard, then shipped to the site and installed in a simple operation, saving field labor costs, which can be expensive.

Today there are many existing examples of bascule, swing, and vertical lift bridges that were designed and built 80 to 100 or more years ago. These movable bridges were designed to a standard that is obsolete today and have many more moving parts than a modern bridge. They include rim bearing swing bridges, rolling counterweight bascule bridges, multiple-counterweight (more than two) vertical lift bridges, and other types, of which there are currently very few examples.

ODD TYPES OF MOVABLE BRIDGES

There are some types of movable bridges, discussed in the following paragraphs, that are considered obsolete. A few examples still exist. There have been some new examples built in recent times of a few unusual bridge types.

Retractile

The retractile bridge, when mounted on land, is supported on some kind of roller system and rides on a set of horizontal tracks. The retractile bridge is usually constructed so that the deck of the movable span is at the same elevation as the deck on the approaches; the tracks are at an angle, 45° or so, from the longitudinal bridge axis. The bridge opens by riding back on its tracks at this angle until the channel is clear. Some retractile bridges were built or proposed back in the late nineteenth and early twentieth centuries, which retracted straight back along the bridge axis, but these required an auxiliary device to shift the approach roadway deck, usually upward, before retracting. This type is known to exist only on pontoon-type retractile bridges today, as discussed in the following section, "Pontoon Retractile."

Pontoon Retractile

The pontoon retractile bridge is a floating span that retracts parallel to an approach span when opened to clear a navigation channel for a vessel. This type was proposed for some railroad bridges across the Mississippi River in the nineteenth century and may actually have been used. The patent drawing for a pontoon movable railroad

Figure 3-4 Pontoon Movable Bridge over the Mississippi River, Built in 1874. This movable bridge was the precursor of modern versions of the type. It lasted until 1961, having to be extensively rebuilt only once or twice. (From Derleth.[1])

bridge built between Prairie du Chien, Wisconsin, and McGregor, Iowa, shortly after the Civil War indicated optional swing and retractile channel span openings for navigation.[1] The pontoon bridge that was built there in 1874 was rebuilt in the early twentieth century and lasted until 1961 (Figure 3-4). The Hood Canal Bridge, completed in 1961, off Puget Sound, is a larger example of this type, with a 600-ft draw opening. This bridge has reinforced concrete retractile spans, a deviation in the roadway for the east approach, and a hydraulically operated lift deck on the west approach span that raises to clear the retracting pontoon. The newest example of the type is the Ford Island Bridge at Pearl Harbor, Hawaii. This type of bridge is advantageous at locations where piers would be very expensive because of deep water, but vulnerability to storm damage is a serious problem. One of the Hood Canal draw pontoons sank in a storm and had to be replaced; this has also occurred at other pontoon bridges of

[1]August Derleth, *The Milwaukee Road,* (New York: Creative Age Press, 1948), p. 114.

similar construction. The draw pontoons are high-maintenance structures and complicated and time-consuming to operate. At locations where the bridge does not have to open often for navigation, the cheaper construction cost of a long reinforced concrete retractile pontoon span can be a great economic advantage over an expensive vertical lift truss span with its associated piers, towers, and counterweights, or a high level fixed bridge.

Pontoon Swing

A pontoon swing bridge is a floating span that rotates in a horizontal plane to provide an opening for navigation. These are usually rather small bridges, and most examples remaining in use today are located in the bayou country of Louisiana.

Shear Pole Swing

A shear pole swing bridge is pivoted on one end, approximating a bobtail-type or asymmetrical swing span (see page 93). The free end of the shear pole swing span is supported by tension rods or cables reaching down from an overhead structure. This structure usually consists of derrick-type posts, forming an A-frame around and above the pivot end of the bridge. This A-frame is guyed appropriately by cables stretching to the rear and side of the A-frame. The operation of the bridge may be by means of a cable drum arrangement that pulls the bridge open and closed, but some of these bridges had rack and pinion machinery drives, similar to conventional swing bridges. Typically, the bridge was a simple span, with the long end simply sliding into the closed position onto supporting masonry plates or strike plates. In some cases the railroad-type had simple open joints in the rails at the ends of the swing span. There were once many examples of this type of bridge in use, including one over the Delaware and Raritan canal at Princeton, New Jersey, and another, which lasted until the 1980s, on the Long Island Rail Road over Reynolds Channel on southern Long Island (Figure 3-5). This bridge had a full set of supporting end wedges and movable miter rails (see Figure 13-6), like a more standard type of swing bridge. There are only one or two examples of this type of bridge remaining. One is a single-track railroad span across Mantua Creek near Paulsboro, New Jersey. This bridge has been converted from manual operation to a hydraulically driven wire rope cable drive.

Folding

The folding, or "jackknife," bridge was an open deck structure, usually built to support a railroad track, that opened by swinging sideways. It differed from the conventional swing bridge in having a separate pivot for each line of stringers. The stringers were loosely linked together. This type of bridge lost favor quickly when it was discovered that any link failure could cause a train to fall into the water. There are not known to be any examples of this type of bridge remaining in use, although some were built into the twentieth century.

Figure 3-5 Shear Pole Swing Span. This bridge, on the south shore of Long Island, was replaced by a rolling lift bascule a few years ago.

Removable Spans

There are several "movable" bridges around the country that are simple spans fitted for temporary removal. These may have a span length of 20 ft or less, but some are quite large and would require a rather heavy crane or other substantial equipment to move them out of the way. Because of the difficulty in opening these bridges, they are practical only for spans that seldom open for navigation or for which the regulations allow for substantial advance notice—weeks or months—to the bridge owner prior to requiring an opening. Some of these installations may have been intended to satisfy the letter of the navigation regulations without incurring the expense of building a truly movable bridge.

Many such removable spans are quite small; the average length is probably in the 30- to 50-ft range. A particular example of a larger removable span installation is the 103-ft long single-track railroad span with 21 ft of vertical clearance above the level of the Wisconsin River at Prairie du Chien (Figure 3-6). This bridge, built in 1908, is just upstream from where the Wisconsin flows into the Mississippi. The original bridge, a swing span, was built for the Chicago, Burlington & Quincy Railroad (CB&Q) in 1886 as part of its extension to St. Paul. The present bridge has two identical single-track deck girder spans where the original arms of the swing bridge were, but only one is supposedly removable. It would take a great deal of effort and expense to open this bridge for navigation, as it is in a remote and relatively inaccessible location; it is possible that it has never been opened. It may have originally been intended to have

Figure 3-6 Removable Span over the Wisconsin River. This bridge, built in 1908, has probably never had to be displaced for navigation since it was erected. The removable span is one of the large girder spans supported at one end by the round pier.

another swing bridge at this location, as the center pier is configured for a swing bridge (it may have been built for the original), and the removable span may have been used as a temporary expedient. The river under the bridge was part of the Fox-Wisconsin Waterway, a canal linking the Great Lakes at Green Bay, Wisconsin, with the Mississippi River at Prairie du Chien.

Approximately 60 years after this bridge was built, the middle section of the waterway upstream from the bridge was declared an "advance approval area" by the United States Coast Guard. Many of the movable bridges downstream have since been replaced with fixed bridges with low vertical clearances over the water, so the railroad's selection of the "removable"-span-type bridge for the location has proved to be a correct one. No commercial traffic of any significance ever used the waterway, so that most of the movable bridges over it seldom opened. Dozens of movable bridges of various types were built over the section of this waterway that was later abandoned for all practical purposes. There was a considerable waste of expenditure in providing machinery that may have never been used. A few of these "movable" bridges still exist and carry traffic today, although some have had their machinery removed (Figure 3-7). One vertical lift bridge was cut back to become, in effect, a removable span. The only traffic under the "draw" at Prairie du Chien is the occasional small pleasure craft or fishing boat. As of 1984, the date of its last publication of *Bridges over Navigable Waters,* the U.S. Coast Guard listed five movable bridges on

Figure 3-7 Former Vertical Lift Span. This bridge, over the Fox-Wisconsin Waterway at Portage, Wisconsin, was deactivated in approximately 1970 by having the machinery and towers removed.

the Wisconsin River, including two other railroad bridges. None of the bridges, fixed or movable, above Spring Green on the Wisconsin River or above Oshkosh on the Fox River are listed by the U.S.C.G. This stretch of approximately 150 miles is the "advance approval area." The highway bridges on the Wisconsin River are no longer movable, so it is unlikely that the CB&Q bridge or the other movable railroad bridges on the Wisconsin River will need to be opened for navigation.

Transporter Bridges

Transporters were not really bridges, but actually had a more descriptive name: "aerial ferries." These structures consisted of tall towers on each side of the navigation channel, connected by a trussed gantry or similar sort of "bridge" high enough to clear navigation. A small platform was hung from the gantry by wire ropes, supported on a roller carriage. The carriage was propelled along the gantry, shuttling from one side of the waterway to the other. The platform was large enough to carry a few cars or trucks, and pedestrians could either ride the platform or, in some cases, climb the towers and walk across the waterway on the gantry. There was a transporter bridge in, Duluth, Minnesota, one in Liverpool, England, across the Mersey, and one at Newport, Wales, across the Usk (Figure 3-8). Their extremely limited capacity, while being almost as expensive to build as a full bridge, has rendered them obsolete; no examples are known to exist today.

Figure 3-8 Newport Transporter, Wales. This type of device was once fairly common for providing transport across busy waterways, particularly at harbors. This example has been removed, and none of the type are believed to be in existence any longer.

EARLIER TYPES OF MOVABLE BRIDGES

Many older-type movable bridges are still in existence, and those that are at locations where traffic does not demand improvements, such as wider roadways or higher load capacity, are likely to remain in place for a considerable length of time. These older bridges include the unusual types, such as the land-based retractile bridge, some examples of which still operate, plus a few that no longer open for marine traffic (Figure 3-9). Pontoon swing bridges also exist in small numbers, as do some archaic vertical lift spans that bear little or no resemblance to modern types.

Eighty-five years ago, J. A. L. Waddell described the movable bridges of the time:

> Movable bridges may be divided into the following classes: (1) Ordinary swing spans; (2) bobtailed swing spans; (3) horizontal folding draws; (4) shear-pole draws; (5) double rotating cantilever draws; (6) pullback draws; (7) trunnion bascule bridges; (8) rolling bascule bridges; (9) jack-knife or folding bridges; (10) vertical-lift bridges; (11) gyra-

Figure 3-9 Borden Avenue Retractile Span, New York City. This bridge, built in 1904, is no longer operable, and its replacement with a fixed span is being considered.

> tory lift bridges; (12) aerial ferries; transporter bridges, or transbordeurs; and (13) floating or pontoon bridges.
>
> Of these 13 types only Nos. 1, 7, 8 and 10 are in frequent use today. Nos. 2, 5, 12, and 13 are employed occasionally and Nos. 3, 4, 6, 9, and 11 have been relegated into oblivion."[2]

Waddell was a little premature in his pronouncement, as all of the bridge types mentioned, except the folding draw, or swing, and the jackknife draw survived into the 1980s. At least one variation of the gyratory lift bridge was built in the latter part of the twentieth century. This bridge is lifted slightly by a vertical hydraulic cylinder, which then acts as a center pivot, supporting the spans while hydraulic slewing cylinders provide rotation.[3] It appears that at least a few of these odd types will continue in use a little longer. Presumably Waddell included the heel trunnion bascule with trunnion bascules, as there were and are many examples, including very new bridges.

There are many retractile bridges, referred to by Waddell as No. 6, pullback draws, in use today, with two over commercial waterways in New York City alone. It is a fact,

[2]J. A. L. Waddell, quoted in "Fifty Year History of Movable Bridge Consruction," Part I, *ASCE Journal of the Construction Division* (September 1975), p. 513.

[3]Andrzej Studenny, "Hydraulic Lifting and Rotating System" Heavy Movable Structures, Inc., *4th Biennial Symposium,* Ft. Lauderdale, Florida, 1992.

however, that few of these bridges operate any longer. The shear pole, No. 4, is almost a thing of the past, with one or a few examples remaining. The shear pole railroad swing bridge over Mantua Creek, a small tributary of the Delaware River in southern New Jersey, is still active and has been rebuilt recently, but may see replacement by a more modern span if rail freight traffic continues its resurgence. The pontoon bridge, No. 13, is alive and well. At least one relatively new combination pontoon and retractile bridge exists in Washington State, with plans to partially replace another in kind, and one was recently constructed at Pearl Harbor, Hawaii.

4

BASCULE BRIDGES

The word *bascule* is French for "seesaw." The word was applied to a balanced bridge that pivoted on a horizontal axis at a right angle to its longitudinal centerline. *Bascule* strictly applies only to those bridges that consist of a single moving element, which pivots about a horizontal line near its center of gravity so that the weight on one side of the pivot axis nearly balances the weight on the other side. The balance is usually not exact. If the bias is towards keeping the bridge closed, it is referred to as *span heavy.* If the bias is towards keeping the bridge open, or causing it to open, it is called *counterweight heavy.* Many bridges pivot about a horizontal axis, but do not take the configuration of a seesaw; these bridges are generally all called bascule bridges as well, and accepted usage allows the term to encompass all the variable types that pivot in the same manner. The deck section or span of a bascule bridge that moves is referred to as a "leaf," as in the pages of a book, which move in a similar way. The leaf can also be called a "span," but that term applies to any movable bridge section, such as the full swinging length of a swing bridge or the lifting portion of a vertical lift bridge. The term *span* also applies to any length of a double leaf bascule or fixed bridge between supports, such as in "a three-span continuous bridge."

The outer end of the bascule span, or leaf, is called the toe of the leaf. The inner end, at the part of the leaf nearest the pivot point adjacent to the approach span or abutment, is called the heel of the leaf (Figure 4-1). The heel of the bridge leaf is supported on the pivot pier, also called the bascule pier, and the toe of a single leaf bascule bridge is supported on the rest pier. The first bascule bridges were probably intended for protective purposes, as indicated previously, for medieval castles and other fortifications in even earlier times. There are indications that some of the earliest ones were truly like seesaws, using equal sides rather than counterweights.

Many of these early bridges pivoted on a shaft, called a trunnion. Trunnions also supported medieval cannons so that they could be pivoted up or down to adjust the

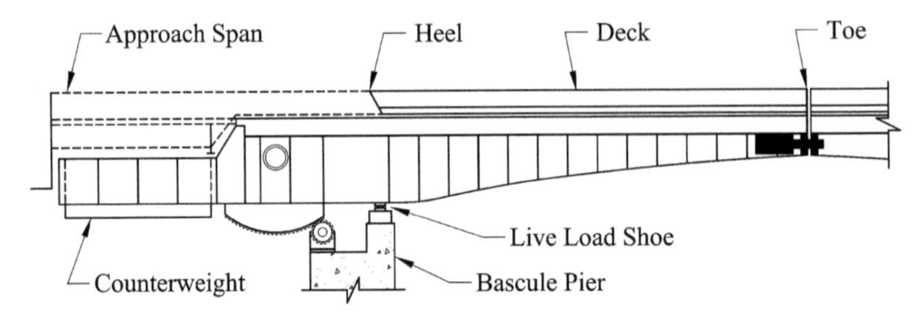

Figure 4-1 Elevation of Typical Bascule Bridge Leaf. The terms applied to the various parts of the bridge are generally considered standard.

range (Figure 4-2). The word *trunnion* is frequently used incorrectly to describe any heavy shaft that rotates while supporting a load. A true trunnion is merely a pivot and never rotates more than a fraction of a turn. The simplest type of bascule bridge is called a "simple trunnion" type and consists of a bridge leaf with a counterweight rigidly attached to the rear portion of the main support members, supported near its center of gravity by trunnions that rest on bearings. As stated previously, some experts consider that this is the only true bascule bridge, a balanced single unitary structure pivoting at a fixed point.

Almost all double leaf bascule bridges act as cantilevers, under dead and live load, with each leaf stabilized in the lowered position by resting on live load shoes that keep the leaf from rotating farther downward after it is closed, when traffic loading is

Figure 4-2 Ticonderoga Cannon. The cannon barrel has projections, called trunnions, on which it is supported in the carriage. The word "trunnion" is from the French meaning "trunk" or "stump."

on the bridge. The counterweights act as the anchor spans for dead load. The two mating leaves, when closed, are kept aligned with center locks, sometimes called "shear locks," acting to force the tips of the two leaves to deflect together under live load. This prevents the formulation of a step in the roadway as a vehicle, the weight of which has caused one leaf to deflect, approaches the other undeflected leaf.

VARIATIONS ON BASCULE BRIDGES

One difficulty with a true bascule bridge is that when one end goes up, the other end must come down. The roadway on many bridges is only a few feet above the water surface, so that means must be provided to prevent the counterweights of low level bascule bridges from dipping into the water. Most major variations on the bascule bridge type have been developed and implemented to solve this problem.

Some variations on the bascule type were developed at the end of the nineteenth and the beginning of the twentieth centuries: rolling lift bridges, which rock back when opening, like a rocking chair; heel trunnion bridges, which have the bridge and counterweight on separate pivots, but connected by links; and many variations developed and perhaps patented by a number of different inventors. Most of the odd types never had an actual prototype constructed, while many had only a single commercial example built.

Many bascule bridges are of the double leaf type, which consists of two leaves pointed toward each other and linked together at their ends where they join over the navigation channel. The double leaf bascule has an advantage over the single leaf bascule with the same span opening and height over water, etc, of opening more quickly. Double leaf bascules also provide greater navigation clearance when in the closed position, because the girders are usually tapered to a thinner section at the toes of the leaves. Moreover, double leaf bascules provide less wind resistance when open, use smaller counterweights, and require smaller machinery. Double leaf bascules, of course, require twice as many of every component, except perhaps an operating control station, as required for single leaf bascules. Some bascule bridges, single or double leaf, are "twinned" by having two parallel spans across the navigation channel. Some of these are connected so that the pair of leaves on one side of the navigation channel acts as one, and some are left independent so that each leaf acts separately.

A few double leaf bascules act as single-span truss bridges, notably the railroad bridge at Sault Sainte Marie over the Soo Canal (Figure 4-3). Some act as arch bridges, such as the Tacony-Palmyra bascule span over the Delaware River, which was designed to act as a three-hinge arch when carrying traffic (Figure 4-4). Very rare is the asymmetrical double leaf bascule, in which one leaf, usually the shorter, acts as a cantilever and the toe of the other leaf rests on the first, so that the second leaf acts, when carrying traffic, as a simply supported span. The railroad bridge at Boca de Rosario, Uruguay, is an example of this variation, with the short span "cantilevered" by being supported by cable stays, and the long span a simple span with its toe resting on the short span (Figure 4-5).

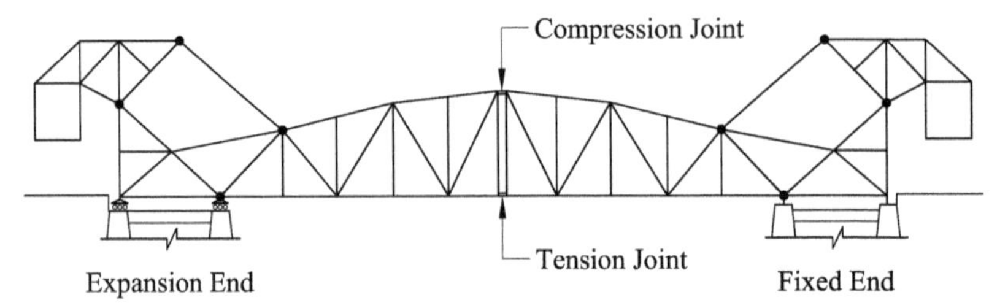

Figure 4-3 Double Leaf Heel Trunnion Bascule Bridge. This type is rather rare. It forms a simple truss by locking a tension connection at the bottom chords and forming a compression connection at the upper chord. It is supported on one end by expansion bearings.

Simple Trunnion

A simple trunnion bascule bridge consists of a unitary rigid displaceable structure supported on a horizontal pivot (Figure 4-6). Sometimes the pivot shaft is stationary, and the bridge pivots around it. More often, the pivot shaft is fixed to the bascule span, which is a true trunnion arrangement, and the ends of the trunnions are supported in sliding or antifriction bearings. If a simple trunnion bascule bridge is constructed at a low elevation above the water, a watertight pit, providing space for the counterweight end of the span as it opens, must be included inside the bascule pier.

Figure 4-4 Double Leaf Rolling Lift Bascule Bridge. The bascule span is to the right of the arch bridge. It acts as an arch to carry live load, with compression joints at the upper chords at the center of the span.

Figure 4-5 Combination Cable Stay—Simple Span Double Leaf Bascule Bridge. This bridge is located in Uruguay, on the north shore of the Rio de la Plata. It is about 70 years old, but is believed to still be in operation. The bridge is actually used as a loading dock for transloading sand from small railcars to the small ships or barges as shown in the photo. (Photo courtesy of Ron Ziel.)

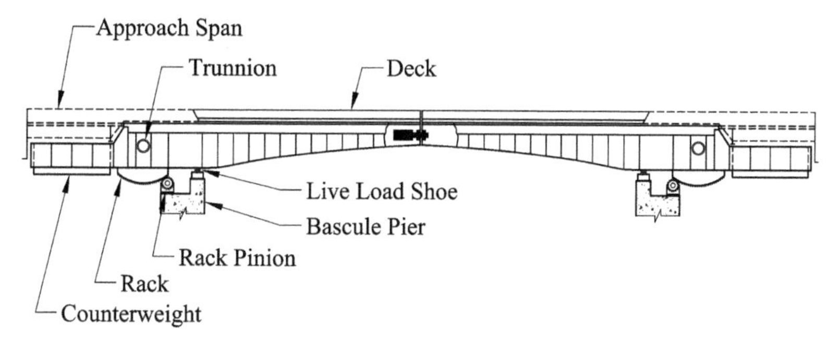

Figure 4-6 Double Leaf Simple Trunnion Bascule Bridge. This type of movable bridge is perhaps the most common.

This pit must be large enough to contain the rear, or counterweight, end of the bridge, and allow it to drop below the water level without dipping into the water as the bridge swings open. If the counterweight enters the water, even to a small extent, operating difficulties occur, as the bridge balance is changed because of the buoyancy effect of the displaced water. Occasional immersion can also accelerate deterioration of the steel and concrete components of the counterweight, particularly if the water is saline.

The simple trunnion type of bascule bridge is often referred to as the Chicago type, as great development of this style was made there around 1900.[1] Chicago type bascules can be very large; the Michigan Avenue bridge, a double-deck, double leaf bascule span, weighs about 6700 tons. Each leaf is actually two leaves, in twin fashion, forming—structurally, mechanically, and electrically—two parallel double leaf bascule bridges in intimate contact (Figure 4-7). On some twin simple trunnion bascules the two leaves next to each other, on both sides of the river if double leaf, are structurally tied together with diaphragms and so operate as single leaves, one on each side of the waterway. In some installations these connections can be disabled if necessary in an emergency.

Heel Trunnion

A heel trunnion bascule, sometimes called a Strauss bascule, is not a true bascule bridge in the strictest sense, because the movable leaf itself is a simple span and not balanced (Figure 4-8). The French, who developed the term *bascule,* call the heel trunnion bridge a "balance bridge." The heel trunnion type is considered a bascule bridge variation and is referred to as a type of bascule bridge in this book. The heel trunnion bascule allows all the advantages of a bascule bridge type of construction, at a low elevation above the water, without the need to construct and maintain a watertight pit for the counterweight. The heel trunnion also allows the pivot point for the bascule leaf to be placed considerably forward on the bascule pier, in comparison with a simple trunnion bascule, allowing a shorter bridge leaf to span the same width of navigation channel. This latter advantage may have been what drove Strauss and others to develop this type of bridge, as it would be cheaper to construct than a longer span and it would be competitive in price with the very economical Scherzer type rolling lift bridge. There are many variations of the bascule design that accomplish these objectives; the conventional or Strauss heel trunnion type of bascule is only one of several developed that minimize the movable span length and eliminate the counterweight pit.

Many bascule bridges referred to as heel trunnion types are really not of this type. The true heel trunnion has a simple bridge span that is hinged at one end, on a horizontal axis at right angles to the roadway, and supported by some means at the other end. None are known that are cantilevered on live load shoes, as the far forward position of the main trunnion makes this construction virtually impossible. The counterweight of a heel trunnion bascule is mounted separately, on its own pivot parallel to the pivot of the bridge span, but is located at a higher elevation so that as the counterweight swings down when the bridge span swings up, the lowest part of the counter-

[1]Henry Ecale, "New Chicago Type Bascule Bridge," *ASCE Journal of Structural Engineering* (October 1983): 2340–2354.

Figure 4-7 Michigan Avenue Bridge, Chicago, Illinois. This is a double leaf, double-deck twin bascule span. It has a total of eight deck sections. The western, or upstream, twin is under reconstruction here, so that the eastern twin is visible to the rear.

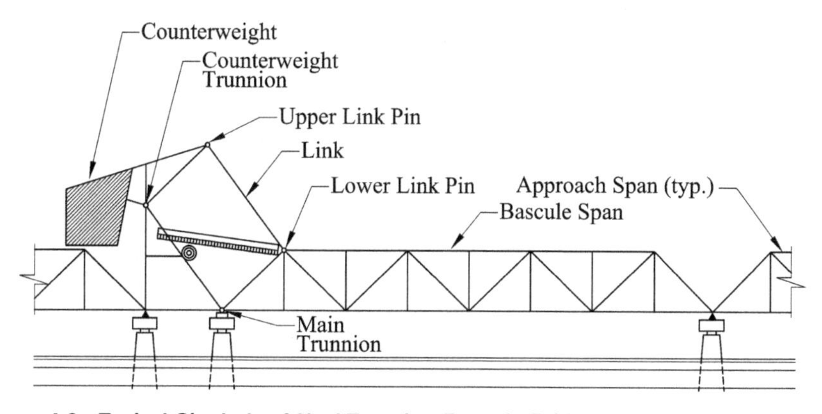

Figure 4-8 Typical Single Leaf Heel Trunnion Bascule Bridge. This type has been commonly used for railroad bridges. It minimizes span length by placing the pivot point very close to the navigation channel, and raises the counterweight on a separate pivot so that the bridge can be very low over the water. Not many are built today. Other types, such as vertical lift bridges, have become more popular for railroad bridges.

weight will be above the water surface. The counterweight is connected to the bascule leaf by links, so that the counterweight and bridge leaf operate together (see Figure 4-8). The connection is usually arranged so that the span and counterweight remain exactly parallel to each other and the links remain parallel to lines running from the centerline of the main trunnions to the centerline of the counterweight trunnions.

The typical Dutch draw is a double leaf bridge, but is common mainly in Europe. There are very few large double leaf heel trunnion bascule bridges in the U.S. known to the author. The heel trunnion type of bascule bridge is usually single leaf, with the heel end hinged on bearings mounted on the bascule pier and the toe end supported by simple bearings at the rest pier. The trunnions at the heel of the bascule leaf are referred to as the *main trunnions.* The counterweight of a true heel trunnion bascule is supported on a separate rigid structure and connected to the bascule leaf only by the links. The trunnions that support this counterweight are called *counterweight trunnions.* The connection points at the hinges between the bascule span and the connecting linkage are called *lower link pins.* The connection points at the hinges between the counterweight structure and the connecting linkage are called *upper link pins.*

Most heel trunnion bascule bridges have a single large block counterweight centered over the roadway. Many heel trunnion bascules have the counterweight split, with a section on either side of the roadway, to allow a smaller supporting tower to be used (Figure 4-9). The split type counterweight can pass alongside the roadway deck and extend somewhat below it when the bridge is fully opened. This variation is fairly common on railroad bridges.

The two-dimensional projection of a conventional or Strauss type heel trunnion bascule bridge forms a parallelogram in a vertical plane parallel to the longitudinal axis of the bridge. The four nodes of the parallelogram are the centerline of the main trunnions, the centerline of the counterweight trunnions, the centerline of the upper link pins, and the centerline of the lower link pins. The distance between the main trunnion centerline and the centerline of the lower link pins, measured on the moving span, is equal to the distance between the counterweight trunnion centerline and the upper link pin centerline, measured on the counterweight frame. The distance between the centerline of the main trunnions and the centerline of the counterweight trunnions, measured on the bridge support frame, is equal to the distance between the upper link pin centerline and the lower link pin centerline, measured along the connecting linkage (see Figure 4-8).

Many people assume, incorrectly, that all heel trunnion bascule bridges were designed by Joseph Strauss. Heel trunnion bridges of the type he designed are commonly called "Strauss bascules" in many parts of the world.[2] In fact, many other persons designed, and some patented, bascule bridges of the heel trunnion type. Some of these were developed a considerable time before the first Strauss bascule was built. The quaint Dutch draw is a heel trunnion bascule, frequently in double leaf form, that dates from the Middle Ages. Many of the earlier heel trunnion bridges did not have a rigid parallelogram type frame connecting the counterweight with the bascule leaf, but used various cable or rope arrangements. Although some of the other designers

[2]Henk de Jong and Nico Muyen, *Beweegbare Bruggen* (Rijswijk: Elmar, 1995), p. 162.

Figure 4-9 Bascule Split Counterweight. This arrangement allows a somewhat smaller superstructure to support the counterweight. It is frequently seen on heel trunnion bridges, as shown here; it is also used on rolling lift bridges.

had no success in getting their examples constructed, Thomas E. Brown, who was contemporaneous with Strauss, had a few examples of his variations on the heel trunnion type of bascule bridge constructed. A patent on a movable bridge type was considered a valuable marketing tool at the beginning of the twentieth century, a time when many railroads were investing in improvements to their tracks and structures and massive expansion in highway construction was beginning. Many variations on the heel trunnion type, as well as others, were developed at this time, some just to be different and to gain an opportunity for procuring a patent or to avoid infringing on a competitor's previously issued patent. Some of these variations were less than practical, but were built anyway; many were never constructed and exist only on paper.

The heel trunnion design, by partially freeing the bascule leaf from its counterweight, allows the construction of very large and long bascule spans. The St. Charles Air Line bridge over the South Branch of the Chicago River, a two-track railroad bridge built in 1919, is 260 ft from heel to toe, longer than the entire span of almost every double leaf bascule bridge. This bridge was constructed to such an extreme length to allow the river channel to be relocated, as part of a proposed navigation improvement project, after the bridge was completed. A single-track heel trunnion bascule railroad bridge of similar length was built in Cleveland, Ohio, in the 1950s, replacing an older bascule bridge of shorter length, increasing the 161-ft wide navigation opening to a clear span of more than 220 ft (Figure 4-10). The extreme length in this

Figure 4-10 River Street Bascule, Cleveland, Ohio. This single leaf, single track heel trunnion bascule bridge span length is more than 220 ft.

case allowed the bridge to span the river at one of its bends without placing a pier in the navigable part of the water, allowing easier transit by ore boats and other large vessels delivering materials to steel mills upstream.

Articulated Counterweight

Many bascule bridges with counterweights that pivot separately from the bascule leaf are incorrectly called heel trunnion bascules, but they should not be confused with that type of bascule bridge. This variation can perhaps best be described as the *articulated counterweight* type. Bridges of this type have the counterweight supported on pivots or hinges connected to rearward extensions of the main bascule girders, away from the navigation channel in relation to the main trunnions. The pins at these hinges are sometimes referred to as counterweight trunnions. This type of construction has more often led to serious difficulties and service failures in recent years than any other type of movable bridge. There are two basic variations on the articulated counterweight bascule, over deck and under deck. Each has its own particular shortcomings, which can be fatal (see following).

Joseph Strauss, in the early twentieth century, developed economical bascule bridges at low elevations above the water to compete with the Scherzer rolling lift bascule. Strauss was particularly instrumental in developing the articulated counterweight type of bascule bridge. Most, if not all, of these bridges were designed by his firm, and this bridge type is often called the "Strauss bascule." This sometimes leads to more confusion in identification, as true heel trunnion bascules, some of which his firm also designed, are also sometimes referred to as Strauss bascule bridges as mentioned earlier. Even O. E. Hovey had trouble with this distinction. See his discussion on page 116 of Volume I of *Movable Bridges*. The articulated counterweight design was a cheap non-rolling-lift bascule bridge for applications in which the bridge was at a very low elevation above the water. Strauss's belief, shared by many other movable bridge engineers, was that the bascule bridge was the best type of movable bridge overall, and that the problems encountered in trying to maintain a bascule bridge with a counterweight that dipped below the waterline when opened should be amenable to correction without having to resort to constructing watertight counterweight pits or to expensive dense metallic counterweights.[3] He developed several versions of the typical overhead counterweight type of pure heel trunnion bascule bridge, but then came up with the idea of pivoting the counterweight at the rear of the bascule girder. A clear advantage of this type of bridge is that the gravitational force of the counterweight acts at the center of the connection hinge point, which can be located higher or lower on the bascule girder than the center of gravity of the counterweight itself. This, in turn, allows the main trunnions to be located higher or lower, which can lead to cost savings in some instances.

The over deck type of articulated counterweight bascule bridge has the counterweight supported on posts, resting on pins or trunnions, supported by bearings mounted on the tail ends of the bascule girders, to the rear of the main bridge trunnions, on the opposite side of the main trunnions from the navigation channel (Figure 4-11). The counterweight is stabilized by being linked to rigid posts that are mounted on the pivot pier. The counterweight is connected at its top to pivoting struts, extending forward from the top of each side of the counterweight, connecting to the top of the posts on each side of the roadway (Figure 4-11).

A famous failure of a bridge of this type occurred when the counterweight stabilizing posts were inadequate and thus failed because of excessive secondary stresses. The particular bridge was a double leaf highway bridge over the Hackensack River in New Jersey, carrying what is now Truck Route 1 & 9. The posts were slender columns, which provided inadequate resistance to secondary forces. Wind loading and live load vibration appear to have been sufficient to cause one of the posts to buckle, toppling the counterweight. This happened shortly after the bridge was opened to traffic in December 1928, causing a stir in the journals of the day,[4] with many recriminations back and forth about the nature of what is today called the "state of the art." This type of bascule immediately fell from favor, in spite of retrofit work on many existing bridges of this type that stiffened the stabilizing columns and provided safety cages

[3]Joseph Strauss, U.S. Patent 738,954, September 15, 1903.

[4]*Engineering News-Record* (June 6, 1929; November 14, 1929).

Figure 4-11 Trunnion Detail—Strauss Overhead Articulated Counterweight Bascule. The near trunnion is the main trunnion, the counterweight trunnion is to the rear.

around the counterweights. Although some examples of the type have been built quite recently (Figure 4-12), most designers shun it.

The overhead or over deck counterweight version of this type of bridge has almost as much clearance over the water as the conventional heel trunnion, but the under deck type gained very little increase in clearance over that of a simple trunnion bascule bridge in the same situation. Nevertheless, many under deck articulated counterweight bascule bridges were built, in all areas of the country, from the early twentieth century up to the 1930s. This construction proceeded in spite of these bridges requiring extra moving parts and a counterweight pit almost as extensive as that required for a simple trunnion bascule built at the same elevation over the water. A few of these articulated counterweight type bridges have failed, and others have been replaced because of fear of failure or for other reasons, such as general age-related deterioration. In some of his patent documents, Strauss claimed that the under deck articulated counterweight bascule had an advantage in its compactness, enabling the counterweight to be tucked under the deck more snugly. This advantage is not always readily apparent.

The usual version of the under deck type of articulated counterweight bascule bridge has the counterweight suspended from hangers that are connected by pivot pins to projections extending to the rear of the bascule girders (Figure 4-13). This type of counterweight connection can allow the bridge to be built at a somewhat lower elevation above the water than a simple trunnion type. The bottom of the counterweight

Figure 4-12 New Strauss Overhead Counterweight Bascule. In spite of evidence that this type of bridge is prone to failure, this example was constructed in the 1980s.

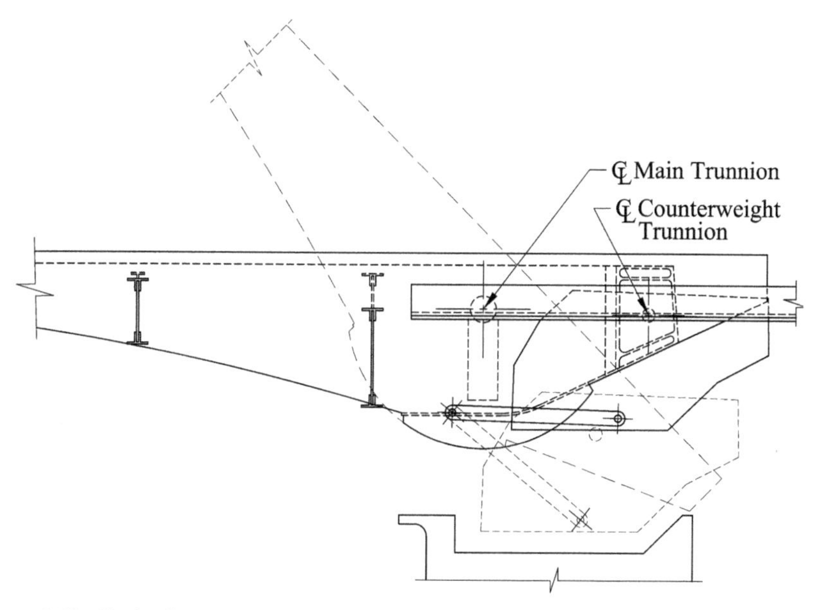

Figure 4-13 Under Deck Articulated Counterweight Bascule Bridge. This type, developed by Joseph Strauss, was quite popular. These bridges developed problems with the connections to the counterweights, resulting in some counterweights falling and others becoming difficult to operate. Very few remain in service today.

stays horizontal, so its center of gravity can drop lower without any of the counterweight entering the water. The pins supporting the counterweight are referred to as *counterweight trunnions,* similar to those in the overhead counterweight version of the articulated counterweight bascule bridge. These counterweight trunnions have very poor access, and many have suffered from lack of maintenance. Several have failed as a result of seizure of the counterweight trunnion in its bearing, causing the counterweight hanger to be wrenched loose from the trunnion as the bridge operated and allowing the counterweight to fall into the bascule pit. Typically, the counterweight trunnion seizes in the bronze bushing mounted at the rear end of the bascule girder, and the hangers separate from the trunnion as the bridge is being opened or closed. In one example of a failure, the means of attachment of the counterweight hangers to the trunnions was simplified by the fabricator, departing from the designer's detail. This modification certainly hastened the failure, although lack of lubricant in the bearing led to seizure of the trunnion in the bearing, which led directly to failure of the connection.

On certain other under deck counterweight versions of the articulated counterweight bascule, the link between the counterweight and the bridge support frame, which acts to maintain the frame-bascule-leaf-counterweight arrangement in a parallelogram relationship (see Figure 4-13), had been allowed, as a result of disuse and poor lubrication, to seize at one of its link pins and had been wrenched loose during bridge operation. Although this is not as immediately catastrophic as the failure of a counterweight trunnion connection, it can make the bridge difficult or impossible to operate.

Rolling Lift

William Scherzer is generally credited with having developed the rolling lift type of bascule bridge (Figure 4-14). His four-track bridge for the Metropolitan West Side Elevated Railroad, built in 1893 over the Chicago River near Van Buren Street in Chicago, is generally recognized as the first modern bridge of this type. His brother, Albert, took over the firm after William's untimely death and made many improvements to the rolling lift type over the years. Rolling lift bascule bridges are still being built in generally the same configuration as the Scherzer original. The rolling lift bridge has the distinct advantage over all other types of movable bridges in that it pulls itself out of the way of navigation as it rotates open. Thus, the navigation channel is free for boats more quickly with this type of movable bridge than with any other. The rolling lift bridge, because it translates away from the channel as it rotates open, does not have to open to as great an angle as other types of bascules to provide the same clearance over a navigable waterway. The rolling lift movable bridge type that rides on a curved track as it opens and closes is generally referred to as a "Scherzer type" wherever it is built.[5]

[5]*SNCF Bulletin Ponts Metalliques No. 17* (Paris: Office Technique pour l'Utilisation de l'Acier, 1994), p. 162.

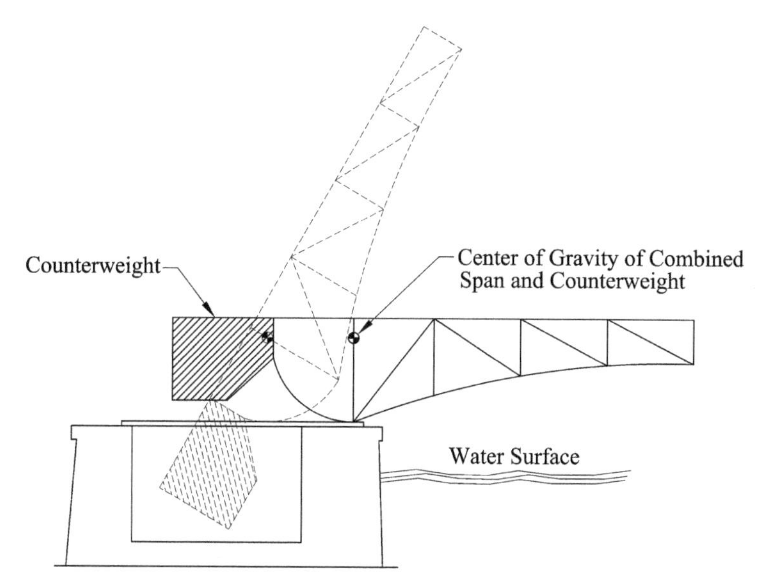

Figure 4-14 Scherzer Rolling Lift Bascule Bridge. This type was developed by William Scherzer in the 1890s. There are many examples existing today, some nearly 100 years old. It is still used for new bascule bridges. The chief advantage of the type is that the bridge rolls away from the navigation channel as the bridge opens, allowing a shorter span to be used and providing quicker opening and closing times.

The Scherzer type bridge has a curved tread plate fixed to the bottom of the each of the bascule girders, on which they rest when the bridge is closed and on which they rock backward as the bridge opens. One of the longest Scherzer type bascule bridges is the single leaf Hilton railroad bridge in Wilmington, North Carolina. The single leaf Scherzer rolling lift type has been favored for railroad bridges because the heavy live load is carried directly on the main girders, without relying on trunnions for support.

A variation on the rolling lift type bascule bridge was developed by Theodore Rall, in which the bridge leaf is supported on a pair of wheels that ride on tracks. By means of a linkage connecting the bascule girder to the pier, the movable span moves backward as it opens, in a controlled fashion, although not in the same manner and usually not quite as far as the Scherzer bascule. Several Rall type bascule bridges were built, and a few are still in operation. Examples of large Rall bascules include the Hanover Street Bridge over the Patapsco River in Baltimore, Maryland, and the Broadway Bridge over the Willamette River in Portland, Oregon. The Rall bascule bridge was advertised as having an advantage in that the support wheels were lifted off the tracks when the bridge was fully seated and in position to carry traffic (see page 165). This appears to have been a feature intended to generate sales over the Scherzer type bascule, which has difficulty accommodating wear at the supporting tread plates. The Hanover Street bascule bridge in Baltimore incorporates this feature, which was used to advantage in a rehabilitation of the bridge performed in 1990.[6] None of the highway versions of the Rall bascule investigated by the author had this feature. At least two, the large Broadway Bridge in Portland, Oregon (Figure 4-15), and the small Stone Harbor Bridge, over Great Channel on the Atlantic coast of southern New Jersey, have experienced extensive deterioration at the support wheels and the stabilizing linkages, which has proved difficult to correct. On both these bridges, the trunnions and rollers continue to support dead and live loads when the bridge is in the lowered position.

Scherzer type rolling lift bridges with overhead counterweights can extend quite high in the air. To reduce the size and the expense of the structure, many Scherzer rolling lift bridges incorporate counterweights made of cast iron or another dense material so that the counterweight is reduced in size. This allows smaller-diameter treads to be used, with smaller-radius segmental girders, and cuts down on wind resistance. Some Scherzer bascule bridges have split counterweights placed outside the roadway to cut down on the overall size of the structure and allow the counterweights to swing down alongside and below the roadway when the bridge opens. The split counterweights alongside the roadway are particularly prevalent on railroad bascules, which go to extreme lengths with single leaf spans, in some cases exceeding 200 ft. Placing the counterweights outside the superstructure allows them to reach below the roadway when the bridge is open, without requiring complicated roadway joints or long cantilevered rear decks on the bascule span.

Noncounterweighted

Many bridge engineers were shocked by the "bascule" bridge built in Sheboygan, Wisconsin, a few years ago. This bridge is a single leaf bridge span without a counterweight, which opens like a trap door being lifted. It should not have been such a surprise, as counterweightless bascule bridges had been built many years earlier in that area. In Milwaukee, Wisconsin, a few miles south of Sheboygan, the 16th Street bascule bridge was built in the 1890s, utilizing a design that called for no counterweight (Figure 4-16). The concept apparently did not succeed, as the bridge was replaced in 1929. There are some differences between the Milwaukee and the Sheboygan structures. The Milwaukee bascule bridge was a double leaf balanced type that required a latch at the rear of each leaf to support the live load. The Milwaukee bridge was balanced to minimize the forces required to open and close it, because power was relatively expensive in those days, as were large electric motors and other prime movers. By means of a pivoted strut arrangement, the moving leaves of this double leaf bascule were supported near their centers of gravity, so counterweights were unnecessary. As the bridge opened and closed, the struts traversed a small arc, resulting in the only change in elevation of the overall mass of the bridge. The disadvantage of the bridge was that traffic rode on a deck that could fall if the locking arrangement failed.

The Sheboygan single leaf bridge, opened to traffic in 1996, operates on a different principle. The designers of this bridge were confronted with the need for a bas-

[6]R. J. Slattery, *Rehabilitation of the Hanover Street Bridge,* Heavy Movable Structures, Fourth Biennial Symposium, Ft. Lauderdale, Florida, 1992.

Figure 4-15 Broadway Bridge, Portland, Oregon. This double leaf bascule span is one of the largest Rall bascules.

cule span to be built at a low elevation relative to the water and were puzzled at what to do about the counterweight. Their decision was to simply eliminate it. The Sheboygan bridge is completely noncounterbalanced and relies on brute force—provided by huge hydraulic cylinders—to open and close the span. The designers were so proud of their feat that they put large picture windows in the sea wall so that pass-

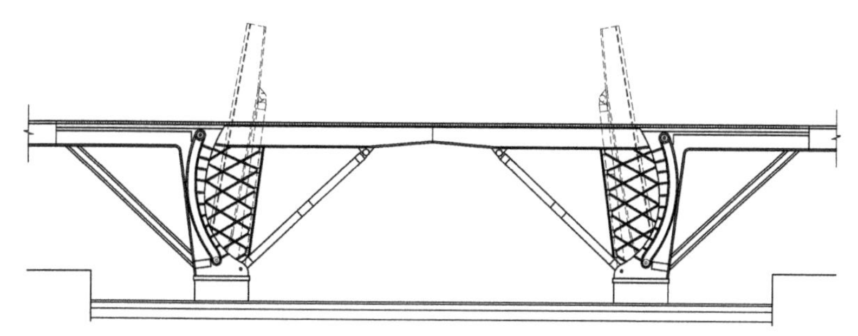

Figure 4-16 Counterweightless Double Leaf Bascule Bridge. This type was developed to allow elimination of the counterweight while providing for a balanced bascule bridge. It proved unpopular, as a defect in the latching mechanism allowed the weight of a vehicle on the rear of the leaf to cause it to drop.

ing boaters could observe the mechanism at work. The machinery is substantial. Operation of the bridge requires 1200 horsepower to operate the bridge in normal time; less horsepower can be used in an emergency, although it takes longer for the bridge to open. This bridge is not really a bascule bridge at all, because a bascule bridge is a "balance" bridge (see Chapter 1). It is classified as a bascule bridge because it opens and closes in the same manner as a bascule, by pivoting about a horizontal axis.

The designers of the Sheboygan counterweightless bridge claim that providing and paying for the extra power required to operate it is cheaper than having a counterweight, but this seems unlikely even if the cost of a watertight counterweight pit is included in the cost of the counterweight. The safety of such a design is also questionable, as a failure of the drive system in the fully open position may result in free fall of the leaf, causing a considerable impact.

Double Deck

There are a few double-deck bascule bridges in existence, in spite of the difficulties in accommodating a second deck on a bascule bridge, as opposed to the relative ease of doing so on a vertical lift span or swing bridge. There are a few notable examples of double-deck bascule bridges in Chicago. For instance, the Michigan Avenue Bridge has two levels of highway traffic, both including sidewalks (see page 39). This bridge had a serious accident during reconstruction a few years ago. There is no evidence that the accident was directly related to the nature of the bridge's double deck. Another double-deck bascule bridge, just a few blocks upriver at Wells Street, has a highway deck on the lower level and rapid transit tracks on the upper level (Figure 4-17). A third such example in Chicago, the Lake Street Bridge, is over the South Branch of the Chicago River.

Other Bascule Types

The Brown type of bascule bridge, as indicated earlier (page 41), had some success in the early twentieth century. One example, the Ohio Street Bridge in Buffalo, stood until fairly recently. An earlier version was proposed over Newtown Creek in New York City, using ropes to connect the bascule span to the counterweights, similar to many other early bascule bridges (Figure 4-18). Other unusual types of bascule bridges were also built, such as an example of the Waddell-Harrington type at Vancouver, British Columbia, in Canada.

ADVANTAGES OF BASCULE BRIDGES

A bascule bridge provides the greatest rapidity of operation of any commonly used movable bridge. It opens quickly and thus gets out of the way of an approaching vessel faster. It also has the ability to pass smaller vessels navigating a channel without opening fully; it is usually easier and safer for a vessel to pass a partially opened bas-

Figure 4-17 Double Deck Double Leaf Bascule. This bridge carries Wells Street and pedestrians on the lower level, and the CTA elevated rapid transit line on the upper level.

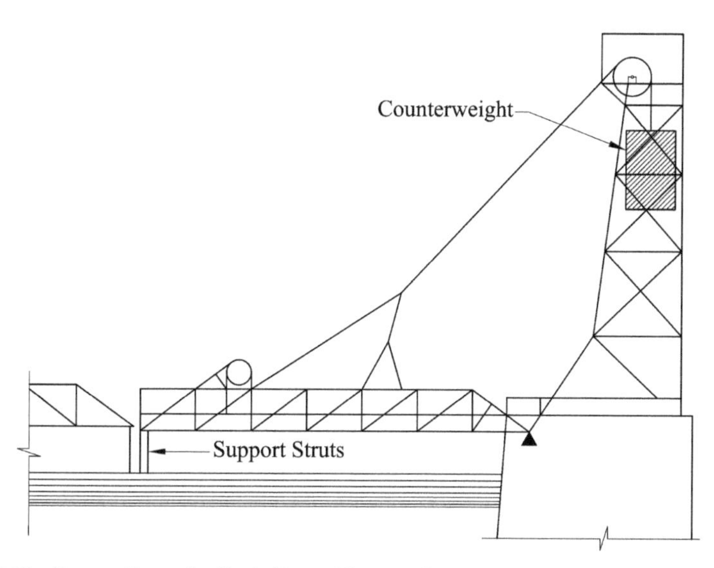

Figure 4-18 Brown Bascule, Early Type. Thomas Brown tried very hard to develop a marketable type of movable bridge in the late nineteenth and early twentieth centuries. He had little success, even with a great deal of further development work; he sold few of his bridges, which were somewhat more complicated than comparable contemporary designs.

cule draw than to navigate through a partially opened swing or vertical lift type. This advantage is even greater with double leaf bascules than with single leaf types. The bascule bridge has most of its superstructure outside the navigation channel when partially or fully opened, and is thus less likely to sustain severe damage in the event of a collision with a vessel than either a swing bridge or a vertical lift bridge. Ship captains are more likely to approach more closely to a bascule bridge than to a swing or vertical lift prior to the full opening; as a result, a bascule bridge can stay open a shorter time for the vessel to pass through than can the other types.

The bascule bridge leaf can provide an automatic barrier to vehicular traffic, preventing an automobile from driving into the waterway as the bridge opens. Double leaf bascules can provide this barrier on both sides of the navigation channel. By selective location of the rear break in the roadway joint of the leaf, the barrier function of the open leaf can be realized. Injury to a vehicle's occupants may occur when the partially or fully opened leaf is struck, but the occupants of vehicles that pass off the approaches of movable bridges into a waterway usually drown.

The bascule bridge, particularly the double leaf bascule bridge, is the most readily accepted, architecturally, of any type of movable bridge. The typical double leaf bascule span, being a symmetrically opposed cantilever, can be easily and relatively honestly configured to resemble an arch bridge. The only type of bridge considered for the eventual Woodrow Wilson Bridge replacement was a double leaf bascule; several variations were proposed (see page 18). The winning design was intended to look like an arch bridge, without fully imitating an arch, as was done with the Arlington Memorial Bridge. Single leaf bascules have also been treated in this manner, with a generally positive effect but with less structural truth in an architectural sense. Swing bridges and vertical lift bridges can be tidied up and frills can be added to them, but the fact that they are large machines is difficult to disguise, particularly in the case of a vertical lift bridge.

The double leaf bascule bridge, by virtue of the fact that it has no need for structural integrity at the center of the navigation channel, provides the greatest clearance under the span at the midpoint of the navigation channel of any commonly used type of movable bridge. The bascule girders, as they project from the piers, can be tapered down in depth until they are no deeper at the end of the leaf than the end floorbeam is required to be.

DISADVANTAGES OF BASCULE BRIDGES

Bascule bridges must resist wind loading to a greater extent than other common movable bridge types. Severe wind loading occurs only with the bridge in the open position; thus, it is not combined with live load and can be sustained at higher allowable stresses. Wind loading requires the bascule bridge machinery to be much more robust than it otherwise would have to be and is a bigger part of the machinery design for bascule bridges than for vertical lift or swing bridges.

Most double leaf bascule bridges, particularly those in which each leaf acts as a cantilever in supporting the live load, demonstrate a severe weakness in their need for a shear lock to connect the opposing ends of the main girders of the two moving leaves where they meet at the center of the navigation channel. These locks are prone to wear, as they are vulnerable to an accumulation of road dirt and other debris, which contaminates the lubricant. The locks are also subject to severe shocks when heavy traffic passes over the bridge. As wear increases, the shocks increase, accelerating the rate of wear until the center locks reach a state of almost complete uselessness.

Cantilevered double leaf bascule bridges are greatly dependent for stability on excellent alignment of the leaves, proper seating on their live load shoes, engagement of their live load anchors, if present, and on the two leaves being mated by their center locks. Differential distortion of the leaves, such as due to temperature differences, or wear on the aligning components, results in poor engagement of these components and poor seating of the bridge. If left uncorrected, the condition develops to the extent that the bridge leaves bounce noticeably as heavier traffic, such as buses, pass over the movable span.

Simple trunnion bascules and some articulated counterweight and rolling lift bascule bridges, when built at low elevations above the water, must have watertight counterweight pits, which are expensive to build and difficult to maintain.

Bascule bridges are usually quite rigid structures, particularly in the area of their pivot axes and where they are driven open and closed. Bascule leaves are usually driven from two or more points, so some form of equalization device is required to make the drive forces equal at the several places they are applied on a bascule leaf. This equalization can be achieved electrically, mechanically by means of a *differential* (see page 210), or by designed-in torsional flexibility of the drive train.

MODERN DEVELOPMENT IN BASCULE BRIDGES

Several decades of early development led to more complicated versions of bascule bridges to avoid shortcomings, addressing particularly the need of low-level simple trunnion bascules for watertight pits. The trend then turned toward the elimination of complicated mechanical devices, as it was seen that they required a certain minimum level of inspection, maintenance, and renewal to avoid problems due to irreversible deterioration. The conventional wisdom came to be that these more complicated types of bascule bridges should be avoided. Modern standard practice is to design and construct mainly simple trunnion bascule bridges. There are a few consulting firms, however, and a few owners who prefer to use Scherzer or the more complicated types where their advantages can be used to significant beneficial effect.

The articulated counterweight bascule, overhead or under deck, is almost a thing of the past. The heel trunnion bascule has had problems with fatigue of superstructure members that undergo stress reversal during operation, and with trunnion and other large bearings that have deteriorated and have been difficult to repair. Typically, these forms of deterioration occur only after 60 or 70 years or more of service, but have still given the heel trunnion bascule a bad reputation in the United States. Yet the heel trunnion bascule is still popular, particularly in Europe, where competent maintenance tends to be the rule rather than the exception, as it has proven to be a durable and versatile type of movable bridge when properly designed and cared for. For a few owners who want to minimize delays to traffic over a bridge, and who are willing to pay for quality design and construction to maximize the life span of such a bridge, the operational advantages of the double leaf Scherzer type rolling lift bridge are too great to be ignored. Single leaf Scherzer bascules, with their direct load bearing at the main girders rather than relying on trunnions, continue to be favored for railroad bridges. The Rall bascule, on the other hand, is considered obsolete, and it is unlikely that any more of this type will be built.

As a result of forcing the simple trunnion bascule into as many applications as possible, its shortcomings, recognized 100 years ago, are being revealed again as it is applied in situations where it is not ideal. A recently completed project, replacing deteriorated old paired Scherzer rolling lift spans with simple trunnion bascules, supposedly solved the lower elevation problem by building counterweight pits that would stay dry. The pits flooded at least once before the project was completed, and the owner of these new paired bridges is now saddled with an operational and maintenance problem that may be equal to or greater than that eliminated by replacing the worn-out old bridges. A new bridge, built to exactly the same plans used for the bridges that were replaced, would probably have lasted a minimum of 50 years without any structural or substructural problems. The old bridges were more than 90 years old when they were finally replaced.

It will presumably be recognized again in the future that the simple trunnion bascule is ideal only at locations where the counterweight of the open bridge will remain above the water level when the bridge is in the open position. For low-level bridges, where maintenance is irregular, a properly designed rolling lift bridge with overhead counterweight can be used, although it will eventually have to have its worn-out treads replaced if it opens frequently. Where maintenance can be expected to eliminate long-term deterioration, the heel trunnion bridge should be used and can be expected to have a very long life, provided fatigue-prone details are avoided.

5

VERTICAL LIFT BRIDGES

The vertical lift type of bridge consists of a simple span, usually of the truss type, that is raised straight up, without tilting, to provide sufficient clearance under the bottom of the span for vessels in the navigation channel to pass through. Almost all vertical lift bridges are supported on towers, either one at each end for the full width of the bridge or one at each corner of the vertical lift span. The towers have rotating counterweight sheaves mounted at their tops, with ropes on these sheaves that are connected at one end to the lift span. The other ends of the ropes are connected to counterweights, so that the weight of the span is balanced by means of the counterweights hanging from the ropes passing over the counterweight sheaves (Figure 5-1).

Small vertical lift bridges were built in the early nineteenth century over canals in England and the United States as well as in Europe. These were usually girder spans of 50 ft or less, with very short lifts. This may be the most economical configuration of a vertical lift bridge, but it is likely that if serious economic analysis had been done at the time, a bascule bridge would have been cheaper in every instance. It is difficult to conceive a situation calling for a small movable bridge in which a vertical lift span would have functional or configurational advantages over other common types of lift bridges. At least one small nineteenth-century English vertical lift span was built with cast iron towers, a material that would do wonders for their longevity. The towers would probably still be in excellent condition today if the bridge were still in existence. Several small nineteenth-century French vertical lift spans built to carry highway traffic over canals and other small waterways were hydraulically operated and may not have been counterweighted.[1]

There were some movable bridges erected over the Erie Canal earlier in the nineteenth century, which may be considered vertical lift spans, but their means of oper-

[1]O. E. Hovey, *Movable Bridges,* Vol. 1 (New York: John Wiley & Sons, 1926), p. 150.

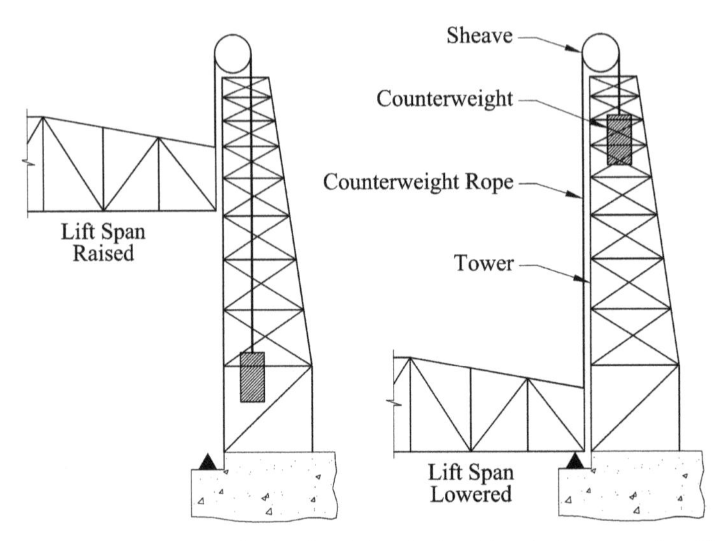

Figure 5-1 Vertical Lift Bridge Balancing. The dead weight of a typical vertical lift bridge is supported at the tops of towers by steel wire ropes connected to the vertical lift span on one end, and to the counterweights on the other. The ropes are looped over sheaves supported on shafts and bearings mounted at the tops of the towers.

ation was somewhat complicated and they did not appear to have been repeated elsewhere. These bridges were all replaced early in the twentieth century; they are discussed further on page 74.

The vertical lift bridge considered to be the forerunner of the modern type was erected in Chicago at South Halsted Street, over the Chicago River, and opened to traffic in 1894. This is the vertical lift bridge more or less as we know it today, the brainchild of J. A. L. Waddell, a prolific and colorful engineer and promoter of bridge projects. The Waddell design for the South Halsted Street vertical lift bridge consisted of two towers, a truss-type bridge span between them, counterweights and ropes hanging via sheaves on the towers, and an operating mechanism under one of the approaches that propelled the vertical lift span up and down by means of $\frac{7}{8}$ in.-diameter operating ropes that pulled the bridge up, and $\frac{7}{8}$ in.-diameter counterweight ropes that pulled the counterweights up, thereby lowering the span. Eight $1\frac{1}{2}$ in.-diameter ropes connected the span to the counterweight by means of 12-ft diameter built-up sheaves on the tops of the towers. Although it was later replaced by a double-leaf bascule bridge, this bridge was considered successful, and within a few years many roughly similar bridges were erected throughout the country.

Engineers at the time expressed concern about the cost of renewing the 14,000 ft of rope required for operation of the bridge.[2] Waddell soon modified his scheme so that on his later vertical lift bridges less rope was required for propulsion of the bridge and the number of bends in the ropes was reduced. Operating ropes on Waddell-type

[2]Charles H. Wright, *The Designing of Draw Spans,* Vol. 2 (New York: John Wiley & Sons, 1898), p. 216.

vertical lift bridges that open regularly have to be replaced fairly often, as they tend to wear and to break. The counterweight ropes, which are larger than the operating ropes but operate over larger-diameter sheaves and do not make as many bends during bridge operation, tend to last much longer. Later refinements in the span drive vertical lift bridge, by other engineers, have further reduced the number of bends transited by the operating ropes, improving rope life.

On most modern vertical lift bridges with counterbalancing, the movable bridge leaf is supported by counterweight ropes mounted on sheaves atop the bridge towers. These ropes are tensioned by counterweights so that the counterweight ropes exert an upward force on the span approximately equal to the span weight (see Figure 5-1). The dead weight of the vertical lift span thus exerts pressure on the pier, not at the live load shoes for the span, but at the base of the tower legs.

A vertical lift with drive machinery on the lift span is called a span drive vertical lift bridge. The vertical lift bridge with drive machinery at the center of the lift span, pulling the bridge up and down by means of ropes extending to each corner, is called a Waddell type span drive vertical lift bridge. This type corresponds to Waddell's revised design after South Halsted Street. In some later vertical lift bridges the machinery was removed from the span and one set of machinery was placed on the top of each tower, with the supporting counterweight ropes used to drive the bridge open and closed. This arrangement is called a tower drive vertical lift bridge (Figure 5-2). A further variation on the vertical lift bridge is the span-tower drive, or tower-span

Figure 5-2 Commodore Schuyler Heim Bridge, Long Beach, California. This bridge on the right, built in 1948, is a very wide tower drive vertical lift span. The adjacent bridge is the relatively new (1997) Alameda Corridor bridge to Terminal Island.

Figure 5-3 Span Tower Drive Vertical Lift Bridge. This example was built in the 1970s on the New Jersey shore. As this type is expensive to construct, it is less popular than other types.

drive, which has the drive machinery on a platform supported on fixed trusses or girders extending across the channel between the tops of the towers, above the lift span itself (Figure 5-3). Another vertical lift bridge variation is the tied-tower type. This is not a span-tower drive but can be either a span drive or tower drive bridge. The towers of this type are tied together for structural purposes only.

Most vertical lift bridges have a single counterweight at each end of the lift span, but some have been built with a counterweight at each corner of the span (Figure 5-4) and others have been built with multiple counterweights. An attempt is being made in Milwaukee to construct a vertical lift bridge without counterweights and without the associated ropes and machinery. The first such bridge, a relatively small pedestrian span, will be raised by brute force applied via hydraulic systems or other jacking devices. The bridge will lift only a few feet to provide clearance for navigation, similar to other vertical lift bridges over the Milwaukee River in Milwaukee.

Some other vertical lift bridges have the towers tied together by means of struts reaching across between the tower tops, but these struts are not related to the drive system as they merely stabilize the towers (Figure 5-5). These cross-span struts do not affect the designation of the type of bridge, which may be span drive or tower drive. There are several examples of both types in Cleveland, Ohio, where there was some concern that unstable subsoils might allow vertical lift bridge towers to tilt. The evidence is inconclusive. On page 60 is a listing of the vertical lift bridges on the Cuyahoga River in Cleveland, their type, and the apparent tilt of their towers:

Figure 5-4 Troy–Green Island Vertical Lift Bridge. This bridge has four separate counterweights, one in each tower.

Figure 5-5 Tower Drive Vertical Lift Bridge. This type of vertical lift bridge is driven from the tops of towers. It usually has no machinery on the vertical lift span.

Location	Name	Type	Apparent Tilt
Mile 0.29	Drawbridge	Span drive, tied towers	No tilt
Mile 0.57	Willow Avenue (over Old River)	Tower drive, tied towers	Tilt
Mile 1.51	Columbus Road (highway)	Span drive, no tower tie	No tilt
Mile 1.84	Columbus Road (railroad)	Span drive, tied towers	No tilt
Mile 2.02	Big 4	Span drive, no tower tie	No tilt
Mile 2.04	Carter Road	Span drive, no tower tie	No tilt
Mile 2.54	Eagle Avenue	Span drive, no tower tie	No tilt
Mile 3.05	NKP	Span drive, no tower tie	No tilt
Mile 3.42	West Third Street	Span drive, no tower tie	No tilt
Mile 5.23	W&LE	Span drive, no tower tie	No tilt

The Drawbridge and NKP (see Figure 5-6) are railroad bridges with heavy freight traffic; Columbus Road (railroad), Big 4, and W&LE are also railroad bridges, but Big 4 is not presently in use and Columbus Road (railroad) & W&LE see only light to moderate traffic. The others are all highway bridges. The Willow Avenue and West Third Street highway bridges have very heavy truck traffic. The tied-tower bridges tend to be closer to Lake Erie, where firm footings may be harder to find. See Figure 5-7 for locations of these and other movable bridges over the Cuyahoga River.

The ropes that hold up the movable spans and counterweights of vertical lift bridges are susceptible to wear, corrosion, and fatigue, but it is not uncommon for them to last 40 to 50 years before needing replacement (Figure 5-8). There is at least one vertical lift bridge, the Burlington-Bristol Bridge over the Delaware River between New Jersey and Pennsylvania, that operates fairly frequently (several hundred times a year) and, as of 1998, was still using its original counterweight ropes. All of the ropes were installed in 1930, except for two that were removed for testing and replaced with new ones. The testing indicated that the ropes were in good condition. On the other hand, the Dock Bridge in Newark, New Jersey, had its ropes replaced only 10 years after the bridge was built; the ropes had deteriorated for an undetermined reason. There seems to be no difference in deterioration rates between the counterweight ropes of span drive and tower drive bridges when their installations and situations are similar.

The vertical lift bridge is ideally suited for applications calling for a very large movable span, as it is inherently the most stable and the least complicated in design and construction. In addition to very long spans, some very heavy vertical lift spans have been constructed. The Union Pacific Railroad Steel Bridge over the Willamette River in Portland, Oregon, built in 1913, weighs about 2400 tons, for the lift span alone (Figure 5-9). The counterweights weigh as much again, so that the total load on the counterweight sheaves exceeds 4500 to 5000 tons, depending on the source cited. The Steel Bridge is a double-deck span drive bridge, constructed with a highway span above and a separately liftable span with two railroad tracks on the lower level. The

Figure 5-6 Waddell Type Span Drive Vertical Lift Bridge. This type of vertical lift bridge is driven via ropes from the machinery room on the vertical lift span.

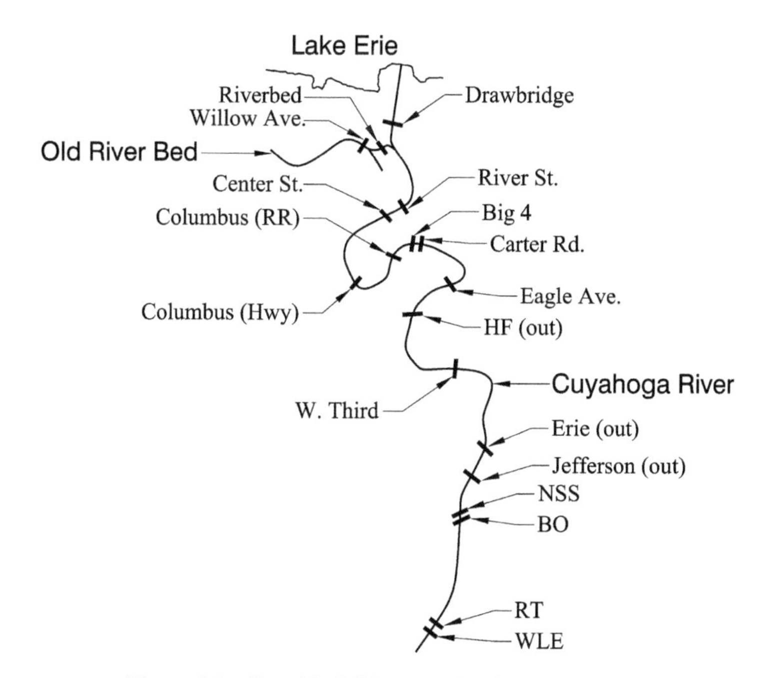

Figure 5-7 Movable Bridges on the Cuyahoga River

Figure 5-8 Old Counterweight Ropes. These ropes are the original on this bridge, more than 50 years old when this photograph was taken.

Figure 5-9 Double-Span Drive Vertical Lift Bridge. This is the Steel Bridge in Portland, Oregon. The lower deck can lift while the upper deck remains stationary, or both decks can lift.

highway deck has recently had a rapid transit line added. The Tomlinson Bridge in New Haven, Connecticut, now under construction (2002), a single-level combined rail and highway through-truss vertical lift bridge, weighs in excess of 6000 tons.

Another very heavy vertical lift span is the Broadway Bridge over the Harlem River in New York City. This bridge carries six highway traffic lanes on the lower decks, plus outboard sidewalks, all of concrete pavement, and three elevated subway tracks on the upper deck. The weight of this lift span when completed in 1963 was 2526 tons, so that with counterweights, ropes, and sheaves, the moving load exceeds 5000 tons. This bridge is a tower drive vertical lift span. A more recent large vertical lift span is the 320 ft long, 108 ft wide deck-girder-type Danziger Bridge in Louisiana, completed in 1987. This bridge is a tower drive type, with a vertical lift span weighing 2200 tons. The deck girder bridge may have been cheaper to fabricate than a truss, even at this length. It is likely that at least a few feet of depth under the deck could have been saved by using a truss design, even though heavy floorbeams would have been required for a bridge of such width. The deck girder concept can eliminate much concern for live load support from the design as multiple main girders can be used, each with its own live load supports at the piers. Dead load support can be arranged by using substantial end floorbeams that reach out to the counterweight rope connections.

VARIATIONS ON VERTICAL LIFT

Span Drive

A span drive vertical lift bridge has the bridge drive machinery placed on the moving leaf itself (Figure 5-10). Most older span drive vertical lift bridges are configured in the modified Waddell style (Figure 5-11). They have operating ropes, smaller in diameter than the counterweight ropes, that are used to pull the span up and down the towers. The operating ropes in this type of drive are driven from a machinery frame mounted in approximately the center of the span. The lift span itself is usually of the through-truss configuration, and the machinery frame is usually located on the top of the truss or, with some deeper trusses, on a special platform above the deck. Normally, there are 2 uphaul and 2 downhaul ropes operating at each corner of the bridge, for a total of 16 operating ropes. The ends of the operating ropes are attached to geared operating drums, with two pairs of ropes attached to each of four drums (Figure 5-12). The drums are geared together so that each drum plays rope in and out at the same rate. One pair of the ropes at each drum is the downhaul ropes, and the other pair at the drum is the uphaul ropes. These ropes extend from the drums to the corners of the lift span, where they run around deflector sheaves, so that the uphaul ropes extend to the top of the bridge towers and the downhaul ropes extend to the bases of the towers. Adjustable connections are made to the towers (see Figure 5-13).

A disadvantage of the typical Waddell vertical lift bridge drive configuration is the necessary slack in the operating ropes, to prevent them from becoming overtight and breaking prematurely. As the bridge raises or lowers, it passes through a state of zero

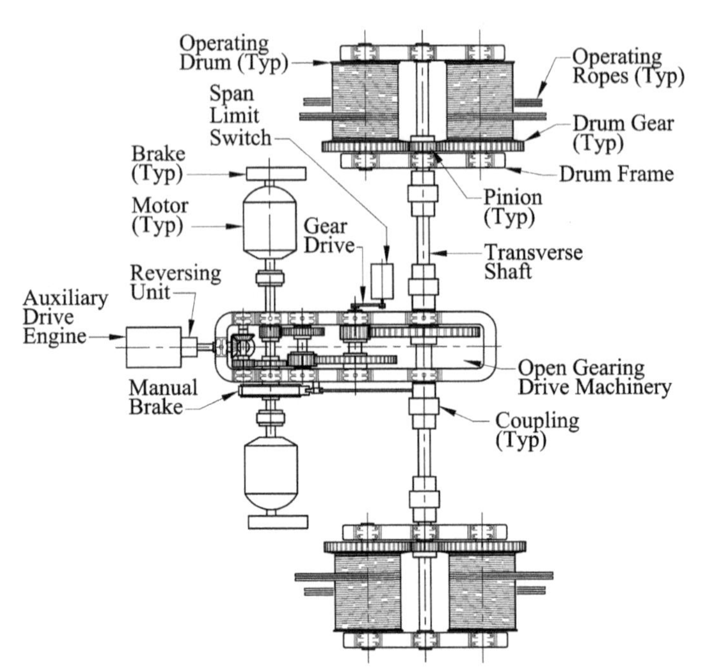

Figure 5-10 Waddell Type Span Drive Vertical Lift Bridge—Machinery Plan. This type has not been built in many years, but several older examples still exist. It uses a complicated system of haul ropes to pull the span open and closed (see Figure 5-11).

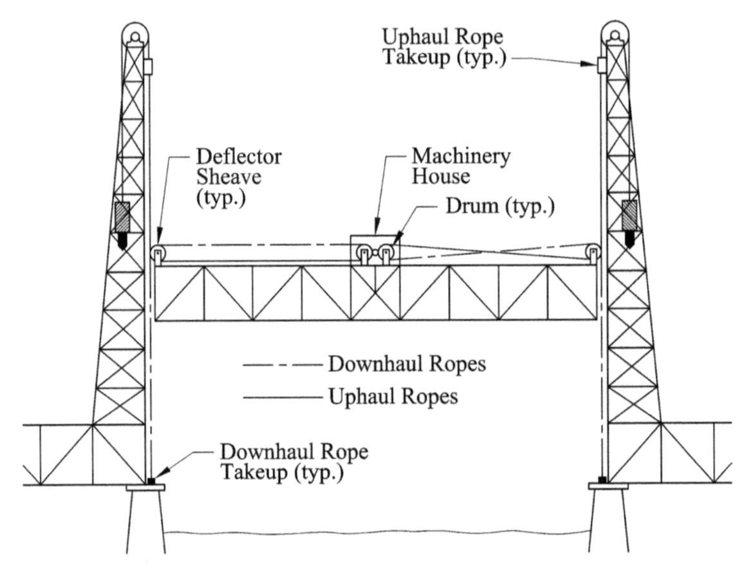

Figure 5-11 Waddell Type Span Drive Vertical Lift Bridge—Elevation. (See Figure 5-10) The haul ropes, or operating ropes, take a tortuous path from their connections at the tops and bottoms of the towers to the operating drums that pull them in and play them out. These ropes are prone to breakage due to fatigue because of the reverse bending they undergo around the drums and the deflector sheaves, as well as the intermediate sheaves or rollers between the drums and deflector sheaves.

Figure 5-12 Operating Rope Drums for Waddell Type Vertical Lift Span. Each of four mechanically synchronized drums pulls in and plays out pairs of ropes as the bridge opens and closes. The ropes of each drum connect to the corresponding corner of the moving span. One pair extending to fixed connections at the top of the tower, called the uphaul ropes, and one pair extending and fixed to the tower base, called the downhaul ropes.

imbalance as the weights of the counterweight ropes are equalized on their sheaves. The load on the bridge drive shifts at this point from driving to overhauling, and the slack is taken up, causing a shock load on the machinery, including the operating ropes. This inherent shock load makes it difficult to provide automatic control of this type of bridge, because of the free movement of the span and the sudden change in load. The shock load can also, if allowed to become too large, cause the sudden failure of the operating ropes. The flexing of the operating ropes around various sheaves leads to eventual fatigue failure of the wires in the ropes. This problem can be ameliorated somewhat by reducing the tension in the ropes when the bridge is seated, which produces a slack condition when they are not actually lifting but which exacerbates the aforementioned shock loading. Auxiliary counterweights can be applied, to reduce the imbalance load to nearly zero (see page 181).

The slackness in the operating ropes gives the Waddell type vertical lift bridge one advantage over other types of movable bridges. As long as the span balance is fairly even, with some bias toward span heaviness, the span will seat firmly at all four corners, as there are no pent-up machinery forces preventing the bridge from seating evenly. It is also almost impossible for this type of vertical lift bridge to develop a skew, as the haul rope connections are fixed and friction is not relied upon to drive the bridge up and down.

Figure 5-13 Span Drive Vertical Lift Bridge—Operating Rope Adjusters. The adjusters are attached to the towers. The ends of the operating ropes are tensioned and adjusted by means of the threaded rods.

A common modern variation on the span drive vertical lift bridge places the operating drums at the corner of the span (Figure 5-14). This configuration requires less operating rope to be used and because of fewer rope bends generally results in the operating ropes lasting longer with less breakage. This type of operating arrangement for vertical lift bridges is seen primarily on railroad vertical lift spans built in the last 30 years. The same advantages of a span drive are present in this type as in the Waddell type vertical lift bridge.

Another span drive vertical lift variation is a patented rack-driven "Strauss" vertical lift bridge. The same Joseph Strauss who was so prolific with bascule bridge designs (see Chapter 4) also developed some significant variations on the vertical lift bridge. There are two examples, over the Illinois River, of rack-driven vertical lift spans (Figure 5-15). These are both through-truss highway bridges, built in the late 1920s or early 1930s. The movable spans are supported in the conventional way for vertical lift spans, by means of counterweight, ropes, sheaves, shafts, and bearings. The drive for these bridges is unusual in that at each tower, facing the lift span, a gear rack is mounted vertically, with the axes of the teeth parallel to the longitudinal bridge axis (Figure 5-16). A motor and gear reduction located in a house atop the center of the span drives a long shaft reaching to the ends of the lift span. At each end of the lift span additional gear reduction drives a rack pinion, which is held in mesh with the vertical rack mounted in the tower. At the expansion end of the bridge, extra-wide

Figure 5-14 Span Drive Vertical Lift Bridge. This new example carries the Alameda Corridor onto Terminal Island. It has operating drums at each corner of the vertical lift span, powered via shafting from the central operator's and machinery house on the moving span.

Figure 5-15 Strauss Vertical Lift Bridge. This example, over the Illinois River, is now in the process of being removed and replaced with a high level fixed bridge. It is a fairly conventional vertical lift span, but is operated by racks and pinions instead of ropes (see Figure 5-16).

Figure 5-16 Strauss Vertical Lift Bridge Machinery. The racks are mounted vertically on the centerline of each tower. The pinions are driven by machinery on the span, coordinated by a longitudinal shaft the length of the span.

teeth are provided so that mesh is maintained during expansion and contraction of the bridge. This type of drive has some of the advantages of the Waddell type vertical lift bridge, without the nuisance of occasional broken operating ropes. It is as free, or freer, of skew problems as other span drive vertical lift bridges, but can develop failures in proper seating. One of these bridges is being replaced with a high level fixed span (2002). Another Strauss vertical lift bridge variation is discussed on page 72.

Tower Drive

A tower drive vertical lift bridge has no separate operating ropes, but requires a sophisticated device to maintain the two ends of the span at the same elevation, as they are driven separately at the two ends of the vertical lift span and are not mechanically connected or synchronized (Figure 5-17). A tower drive bridge has counterweights, counterweight ropes, sheaves, shafts, and bearings similar to those of any other typical vertical lift bridge. The counterweight sheaves have large-diameter ring gears, usually in sections, attached to one side of their rims, which provide the means of driving the bridge. Machinery is mounted on the floor at the top of the tower, consisting of a drive motor or motors, gearboxes or open gearing, and shafting connected to pinions, one of which is engaged with each ring gear mounted on the sheaves. Both of the counterweight sheaves at the top of the tower are driven thus, or if four sheaves

Figure 5-17 Tower Drive Vertical Lift Bridge. This particular vertical lift bridge is one of the world's longest. It spans Arthur Kill on a busy part of the New York harbor. This type of vertical lift bridge has no machinery on the moving span. Two separate drives, one on the top of each tower, operate the bridge. The drives must be coordinated to avoid tilting of the vertical lift span.

are present at the tower, they are driven by means of additional machinery that links all the main counterweight sheaves on a tower together mechanically. The lack of a mechanical connection between the drives at the two ends of the bridge prevents a direct engine auxiliary drive from being used (see page 288).

Each of the two towers of a tower drive vertical lift bridge has a similar arrangement. If the bridge has a tower at each corner, the two towers at each end have a structural cross connection supporting the machinery. The bridge drive machinery is controlled by two separate systems. The bridge operator controls the positioning and speed of the bridge in operation from the operator's station, usually located in a fixed control house on one of the towers. The drive motors are controlled dually, as in any other movable bridge using multiple drive motors. In addition, an automatic control system senses the relative speed at the drive of each tower, speeding up the slower drive and/or slowing the faster drive so that they operate at the same speed within the limits of the system. There are many types and variations of electrical control devices to keep the two ends of a tower drive bridge at the same elevation and moving at the same speed. The most common and least troublesome form of this system is sometimes called a power selsyn tie apparatus, in which a motor is installed at each tower, permanently connected to the mechanical drive at its tower. Power to these motors is not fed from the overall bridge control system. Rather, these motors are connected to each other, so that the slower is powered by the faster, which acts as a generator. Energy from the faster moving end of the lift span is thus transferred to the slower moving end, slowing down the fast end and speeding up the slow end. Another type of control feeds span position and speed to a Ward-Leonard or amplidyne main motor control system. Although the Ward-Leonard system is capable of very precise speed control, it is not foolproof and occasionally results in excessive skew. Some tower drive bridges are also equipped with skew limit switches, so that when skew exceeds a safe amount, the bridge drive shuts down. When excessive skew causes a shutdown, correction must usually be done manually before normal operation is resumed.

Modern solid-state devices allow another coordination system to be used. Tachometers are connected to the drive motors at each tower, sending a drive motor speed signal to the control house from each drive. The speed of the motors is compared with the desired speed selected by the control system, and voltage to the motors is automatically increased or decreased to speed up or slow down the motors so that they operate at the selected speed within the allowable range of error allowed by the system.

Another possible means of control is strictly digital. The towers of the bridge can be fitted with markers so that each end of the vertical lift span, or each corner, can have an equally spaced set of control points, ranging from the fully closed to the fully raised position. Sensors placed on the moving span can count the number of position markers passed between the lowered position and the bridge's present position, adding markers for upward movement and subtracting for downward movement, thus determining the height of the bridge above its fully seated condition. This information can then be fed to a logic circuit to correct positioning errors or variations in vertical position between ends, or even between corners, depending on how many independent drives operate the bridge. The information can be used for other control purposes, such as to determine the speed at which the bridge should be operating. This system can also be used to make other decisions dependent on the bridge's vertical position, such as what color the navigation lights should be—red or green.

Unfortunately, this is not all the control needed by a tower drive vertical lift bridge. Contrary to all other common types of movable bridges (except for a variation discussed in the following section), the tower drive vertical lift bridge relies on friction to drive the span open and closed. This friction is developed between the counterweight sheaves, directly driven by the drive machinery, and the counterweight ropes that are connected to the vertical lift span on one end and the counterweights on the other. Slippage occurs between these elements, which can be exacerbated by the need to lubricate the ropes to prevent abrasion and corrosion and by the flexing of the ropes as they pass over the sheaves.

Some tower drive vertical lift spans have no inherent means of correcting for counterweight rope slippage, relying on infrequent usage and the hope that all ropes slip over all sheaves at equal rates and in the same direction. Other bridges of this type use special controlled seating functions to correct for the rope–sheave slippage each time the bridge is lowered after a lift.

Many tower drive vertical lift bridges have clutch-actuated differential mechanisms in the drives at the towers. The clutch is normally locked, so that the left and right sheaves at a tower turn at exactly the same number of revolutions per minute while the bridge is lifting and lowering. On some of these bridges, when the bridge is being seated after a lift, the clutch automatically disengages and the differential mechanism is activated. The two sheaves may then move in a relationship similar to the wheels at the driving axle of an automobile. At each end of the span, the side of the span that seats first is stopped by the live load shoes, and the other end can continue to be driven down until it seats. The differential allows the sheave on the side that has not seated to continue rotating, and the side that has seated continues to receive an equal drive torque, but stalls against the live load shoes.

Span-Tower Drive

A further variation of the vertical lift type is the tower-span drive or span-tower drive vertical lift (Figure 5-18). This type drives the bridge via ring gears on the counterweight sheaves and through the counterweight ropes as in a tower drive bridge, but

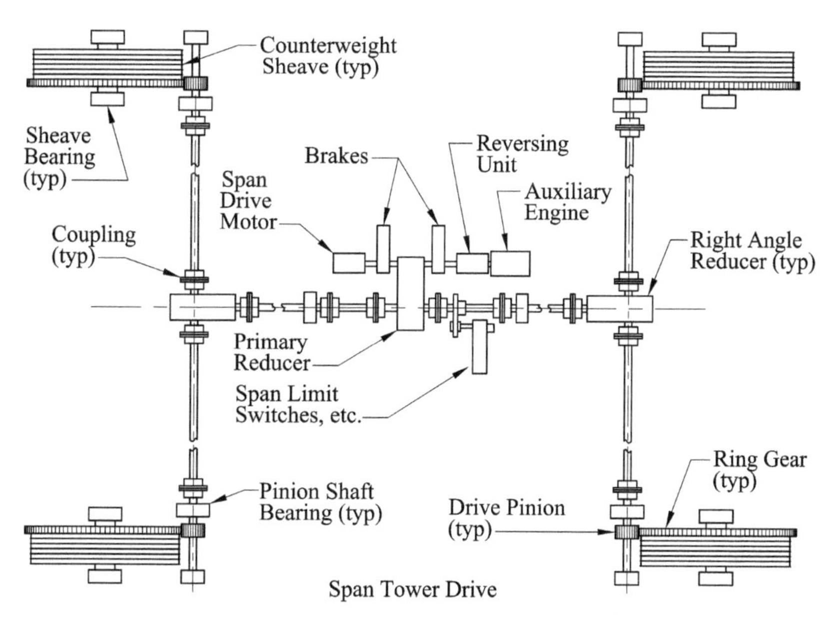

Figure 5-18 Span-Tower Drive Vertical Lift Bridge Machinery Plan. This type of movable bridge is relatively rare. The extra expense of the structure connecting the tops of the towers is justified by thus allowing provision of mechanical coordination of the lifting of each corner of the vertical lift span without requiring the use of operating ropes.

by only one set of drive machinery, which is mounted on a rigid span reaching across the waterway between the tops of the towers. This system provides mechanical synchronization of the raising and lowering of the bridge, almost as effectively as a span drive vertical lift bridge, without placing the machinery on the lift span. This type of vertical lift bridge is often a deck type structure; it has been fairly popular in New Jersey and Louisiana. It eliminates the operating ropes, which are a significant maintenance item for span drive vertical lift bridges. Most tower-span drive vertical lift bridges are rather small, with short spans of usually less than 200 ft, but may be as wide as any other type of movable bridge.

Slippage of the counterweight ropes at the counterweight sheaves affects the positioning of a tower-span drive vertical lift bridge in the same way it affects the positioning of a tower drive bridge. To correct for slippage, many tower-span drive bridges have a clutch arrangement at each tower, plus an additional clutch-actuated differential in the main drive at the center of the platform above the lift span. This differential is also normally locked out, so that the two ends of the lift span are driven at equal speed. When seating, the differential is actuated so that if one end of the span seats first, it stalls, and the other end continues to be driven down until it seats.

The clutches of the differentials are frequently automatically actuated at each seating of the bridge. Some of these vertical lift spans have manually actuated clutch differentials, which are actuated only after a seating problem develops. The differentials are engaged for a single operation to make a discrete correction, after which they are again manually locked out until the problem next arises.

Noncounterweighted

As indicated on page 58, a vertical lift bridge is to be built with no counterweights and no towers. This bridge, which is to be a pedestrian span, was about to be built when this book went to press.

Strauss/Rall/Strobel

Strauss and Rall (or Strobel) separately designed vertical lift variations that used the counterweight arrangement of the heel trunnion bascule to balance the lift span (Figure 5-19). Each tower supported a counterweight on a truss frame, linked to its end of the lift span. That did away with the counterweight ropes and sheaves by providing a pivoted counterweight arrangement. This introduced a sinusoidal function into the bridge balance. Rack and pinion machinery was used to operate the bridges. These were extremely ungainly structures, with severe complications in maintaining the moving parts. Only a few of these bridges were built. It is debatable whether, over a longer service life, the absence of need to maintain and eventually replace operating and counterweight ropes would justify the expense of the complicated drive and counterweighting devices on these bridges. A highway bridge of this type was built over the canal channel of the Ohio River at Louisville, Kentucky, in 1915. The "Strobel" type, which also eliminated the counterweight ropes and sheaves, was built for

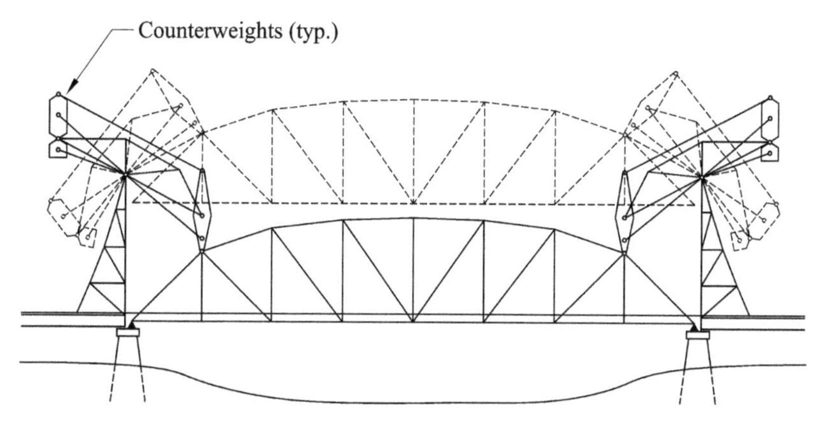

Figure 5-19 Strauss Vertical Lift Bridge Formerly at Louisville, Kentucky, over the Ohio River. The bridge no longer exists.

the Chicago, Burlington & Quincy Railroad over the Illinois River at La Salle, Illinois. The Strobel Steel Construction Company obtained control of the Rall patents, and the counterweights and their trusses on the Strobel variation were supported in a manner similar to the way in which those of a Rall bascule bridge were supported. Each counterweight and its truss rode on rollers on a track and moved toward and away from the lift span as it was raised and lowered. Both the Louisville and La Salle bridges are no longer in existence. There is an operating railroad bridge of the Strauss type south of Tacoma, Washington.

Double-Deck

Double-deck movable bridges are not unusual, but the vertical lift span allows some features of double decking that are not available on other types of movable bridges, inasmuch as no rotation occurs between fixed and movable spans as the bridge opens and closes. The Houghton-Hancock vertical lift span, in the Upper Peninsula of Michigan, has a lower deck that was originally equipped with a set of railroad tracks. The upper deck carried a conventional highway. The bridge was arranged so that it could be lifted part way, aligning the lower deck on the moving span with the upper level approach roadway. The lower deck has highway paving in addition to the railroad track, so that automobile traffic can drive across the bridge in the partially raised position while smaller vessels pass below. The bridge still carries highway traffic, but is no longer in use for rail traffic. In another variation, the Steel Bridge in Portland, Oregon, the lower, railway, deck could be partially raised while the upper highway deck remained in place, allowing highway traffic to continue to cross the partially raised bridge. The upper deck was lifted only for the passage of tall vessels. The bridge is still in use for railway traffic; the upper deck was rebuilt to carry a light rail transit line.

Erie Canal Bridges

When the Erie Canal was enlarged to accommodate larger freight vessels, it was found to be inconvenient to raise many of the roadways that crossed the canal, particularly at the many small towns that grew up along its length. The original fixed bridges and primitive movable bridges at these locations were replaced with various types of structures. Many of these were movable bridges of unusual designs not found anywhere else.

Squire Whipple designed a few movable spans that were built in the 1870s over the Erie Canal. These bridges, which could be called precursors of the vertical lift type, had an overhead rigid truss structure, supported on towers, with the roadway deck suspended from tension rods. The deck was lifted to clear the canal. These bridges were small, about 60 ft long by 18 ft wide. The bridges were manually operated by means of a tread wheel mounted in the overhead truss. The lift portions of the bridges were counterweighted like modern vertical lift spans, but actual operation of the lift was accomplished by falling weights, which were then lifted back into position by means of the tread wheel. The lifting deck arrangement of the earlier bridges was incorporated in a few later bridges, such as the ASB Bridge in Kansas City, Missouri, and was used to support the lower deck of the Steel Bridge in Portland, Oregon as described earlier.

Many of the Erie Canal bridges were replaced with slightly less unconventional vertical lift spans around 1915, about the time of another enlargement of the canal

Figure 5-20 Erie Canal Bridge at Medina, New York. This is one of many similar vertical lift bridges on the New York State Barge Canal between Rochester and Buffalo, New York.

Figure 5-21 Towerless Vertical Lift Bridge. Several of these modern towerless vertical lift bridges cross the Milwaukee River in downtown Milwaukee, Wisconsin.

system. Fifteen examples (14 operable) of this type of bridge are in use in western upstate New York, between Rochester and Tonawanda (Figure 5-20). They have no visible towers such as those that form the distinctive feature of a conventional vertical lift span. Each bridge is lifted and counterbalanced by mechanisms that are under the approach roadways, which push the bridge up less than 10 ft to provide clearance over the canal when needed. Many of these bridges have been rehabilitated in recent years, and it can be assumed that they have several years of useful life ahead of them. The present type of Erie Canal vertical lift span may be considered the model for the short lift vertical lift spans in Milwaukee, the nonhydraulic examples of which have a somewhat similar drive system. Over the past 35 years or so, several bridges of this later Erie Canal type have been built, and proved successful, in downtown Milwaukee, over the Milwaukee River (Figure 5-21).

ADVANTAGES OF VERTICAL LIFT SPANS

Vertical lift bridges have an advantage in that they can be made to almost any length required; they are limited only to the maximum length of a simple span bridge. There are several examples of vertical lift bridges with lengths in excess of 500 ft. For very long movable spans that do not have to lift very high to clear marine traffic, the vertical lift bridge is likely to be less expensive than the bascule or swing bridge. Typical replacement movable railroad bridges on inland waterways are 300 ft long vertical lift spans that lift only 40 or 50 ft. This is sufficient to clear navigation on such waterways as the Mississippi River system.

The vertical lift bridge acts as a simple span for both live load and dead load. Thus, as long as it is reasonably well balanced, it will seat stably on its bearings and will not move vertically at the bearings under live load. For these reasons, vertical lift bridges are simpler to design and construct than bascule and swing bridges. Because the vertical lift span so closely approximates a fixed span, it is ideal for heavily loaded structures such as railroad bridges.

The vertical lift span can be built at any desired width, with as many trusses or main girders as needed, if the roadway can be so divided, as the counterweights can be connected as necessary to balance the bridge and the resulting sheaves and ropes can be connected and supported as required. If the roadway must be continuous across the width of the bridge, large floorbeams, with intermediate live load bearings at the ends as necessary, can support the deck at the end of a bridge of extreme width to avoid deflections at the joints. This feature offers a contrast to multiple main girder bascule bridges, which require careful alignment to avoid interference or binding when operating—an effect that is difficult to achieve. Many two-girder bascule bridges that are very wide have difficulty with bearing alignment because of deflection changes when opening and closing. Vertical lift spans compare even more favorably to wide swing bridges, which require wide center piers in the navigation channel and are impossible or extremely impractical to fit into the channel opening when they are too wide. A deck girder vertical lift bridge can have multiple main girders, each with its own live load supports with "floorbeams" that really only support dead load by transferring it to the counterweight rope connections outside the girders.

Vertical lift bridges do not rotate relative to the roadway when opening or closing. This allows double-deck vertical lift bridges to be provided with a lower deck that lifts independently of the upper one. Thus, land traffic need be only partially disrupted to clear smaller vessels, as the upper deck can be left in place, open to roadway traffic, while the lower deck is lifted a few feet to allow the smaller vessels to pass under it. The entire bridge, including the upper deck, is lifted only to allow relatively large vessels to pass through the navigation channel. In a variation of this concept, the entire double-deck vertical lift bridge is lifted a small amount, so that the lower deck on the span is matched with the upper deck on the approaches, for the same effect.

DISADVANTAGES OF VERTICAL LIFT BRIDGES

The key disadvantage of a vertical lift bridge is that the overhead clearance is limited when the bridge is open, but this has seldom resulted in an accident. Most accidents involving collisions between vertical lift bridges and vessels navigating the channel have occurred as a result of a vessel's striking a partially opened bridge. Such an accident may be caused by a vessel entering the navigation channel prematurely, before the bridge is fully raised, or by a bridge beginning to lower in front of an approaching vessel. Rarely has a vertical lift bridge been lowered on top of a vessel in the channel, although it has been known to happen. The vertical lift bridge is the only common type of movable bridge that always obstructs the full width of the navigation channel, even when fully opened. Serious accidents have occurred because of the bridge's being struck by a vessel when nearly or partially opened. Such a mishap occurred at Portsmouth, Virginia; the ship departed the scene with the vertical lift span resting on it. Another incident occurred on the Chesapeake and Delaware Canal when the vertical lift span was lowered, in fog, in front of a ship, causing damage that tied up the waterway and closed the bridge to traffic for a considerable period of time.

A vertical lift bridge requires expensive towers to allow it to be raised to the open position. Because of machinery restraints and the counterweight rope connections, these towers must be as much as 60 ft taller than the navigation clearance required. As the maximum required clearance height increases, the height of the towers and their resultant cost also increase. The cost increases at a nonlinear rate, as a result of the increase in the wind moments of the towers, particularly with the span in the fully opened position, that must be resisted as the height increases.

Aesthetically, the vertical lift bridge has about the worst overall record of any type of movable bridge, although there have been some relatively attractive vertical lift bridges built. These include such bridges as the span over the Hudson River between Troy and Green Island, New York (see Figure 5-4), the pedestrian bridge across the East River at 103rd Street in New York City, and the Milwaukee bridges mentioned earlier. The chief advantage of the vertical lift span, its capability of long spans, results in its usually taking the truss form, which is usually an eyesore. There have been several attempts to rectify this aesthetic deficiency while retaining the truss form for longer spans, such as the Marine Parkway Bridge (see Figure 7-4) in New York City, the Schuyler Heim Bridge in Los Angeles (see Figure 5-2), and others, with what may be indifferent results. The tower drive type, often selected for aesthetic reasons, is an occasional cause of misalignment and jamming of the vertical lift bridges to which it is applied, due to failure of the devices synchronizing the operation of the two ends of the bridge. The truss type span drive vertical lift bridge is generally considered incapable of effective aesthetic treatment, but there are some relatively pleasing renditions of this type, such as the Tower Bridge in Sacramento, California (Figure 5-22). Recent changes in the perceptions of the avant-garde have resulted in an "industrial chic," which may bring back the trussed vertical lift bridge for "signature bridge" applications, in which efforts are made to produce bold architectural statements.

In a recent effort, the Danziger Bridge in New Orleans used the deck-girder type for a fairly long vertical lift span of more than 300 ft, which is artistically pleasing but may not be particularly efficient structurally. Most people interested in aesthetics, however, when thinking of vertical lift bridges, recall the lower Newark Bay bridge of the Central Railroad of New Jersey between Bayonne and Elizabethport (since demolished) or the "trio over the Hackensack River, Jersey City," which are referred to as just plain "ugly."[3] When the producers of the movie *The Blues Brothers* wanted to

[3]Martin Hayden, *The Book of Bridges* (New York: Galahad Books, 1976), p. 107.

Figure 5-22 Tower Bridge, Sacramento, California. This bridge is an unusual span drive type, in that efforts have been made to produce an architecturally pleasing effect with the bridge superstructure. This is one of the few span drive vertical lift bridges with any sheathing on the towers, for instance.

Figure 5-23 Cluster of Movable Bridges over the Calumet River, on Chicago's South Side. This grouping typifies the reason for the aesthetic sentiments of most persons regarding movable bridges.

capture the depressing squalor of South Chicago, they used the four vertical lift spans over the Calumet River as a backdrop. True, there was a heel trunnion bascule bridge and a rigid cantilever-truss span in the picture, but they were grossly overshadowed by the vertical lift spans (Figure 5-23).

THE FUTURE OF VERTICAL LIFT BRIDGES

After a period in which the vertical lift span was seen as the ideal movable bridge for most applications, the following spate of counterweight rope replacement projects in the past decade, the occasional need for replacement of badly deteriorated counterweight sheaves, and a few near-catastrophic counterweight sheave shaft failures, have caused a rethinking of this approach. The only ensured use for the vertical lift span is in the very long spans required at certain locations. Improvements in the stability of double leaf bascules, and the willingness of owners to accept longer single leaf bascules, will eliminate the vertical lift span from sole consideration for intermediate-length movable bridges.

Figure 5-24 Vertical Lift Bridge Counterweight Sheave with Static Shaft. The shaft does not turn, thus eliminating fatigue considerations from its design. The sheave is supported on the shaft on spherical roller bearings.

For the longest movable bridges, vertical lift spans will usually remain the only practical choice. Improvements must be made in their durability, particularly in the ropes and in the sheave shafts. The longevity of the ropes can be increased by finding a practical means of lubricating and preserving them. Corrosion of the ropes has been partly due to severe air pollution, which is largely a thing of the past in many parts of the United States. The wearing of the ropes, as they rub against the sheave grooves, can be reduced by providing at least minimal lubrication. Reliable automatic lubrication should be provided, to avoid the need to depend on normal maintenance. Fatigue failure of sheave shafts can be dealt with by making them stationary; this has been done with success at the Carter Road Bridge, in Cleveland, Ohio (Figure 5-24), and at one of the I-5 bridges at Portland, Oregon. Unfortunately, the use of a stationary sheave shaft is a technical violation of both AASHTO (2.9.7) and AREMA (6.5.36.8b), both of which require that counterweight sheaves are to be shrunk fit onto their shafts and secured by dowels. 2000 LRFD AASHTO continues this requirement, as well as the use of one-piece forgings for sheave hubs (presumably only for the hubs of welded sheaves). In lieu of stationary sheave shafts, the proper design of rotating sheave shafts can eliminate the likelihood of fatigue failure for all practical purposes.

To maintain the desirability of vertical lift bridges, their reliability must be improved. Tower drive bridges should be selected with caution, as they require a higher level of maintenance and operational competence for reliable operation than do span drive vertical lift bridges, and they require more sophisticated control systems. Counterweight rope life must be improved, if vertical lift bridges are to be an equal or better choice among the various types of movable bridges.

6

SWING BRIDGES

A swing bridge opens for navigation by pivoting in a horizontal plane about a vertical axis, on a bearing mounted on a central pier called the pivot pier. When the swing span is closed and carrying traffic, its ends are supported on rest piers, or if the total bridged length is short, on the abutments. The movable part of a typical swing bridge is referred to as a *swing span,* with two "arms" consisting of cantilevered structures extending from the pivot point of the bridge.

Swing bridges have been around in more or less their present configuration for 200 years or so. One was used on the Pennsylvania Main Line Canal, built in the 1820s.[1] The earlier examples were made of wood, most commonly in a trussed form. A few of the smaller examples of older timber-beamed swing bridges still exist, or did so until recently, in out-of-the way places, such as in Cape Cod, Delaware, or on the Eastern Shore of Maryland. These bridges have been or are rapidly being replaced, however, primarily because of their limited load capacity or the inadequate width of the deck for two-way automobile traffic.

Unlike the other common types of movable bridges, a swing bridge is not lifted out of the way to clear a channel for navigation. The swing bridge's props are removed, and it is pushed out of the way, without lifting. The implication is that, unlike bascule and vertical lift bridges, the swing bridge is not inherently stabilized by gravity. This is, in fact, the case. In addition to requiring some sort of stop device to position the swing bridge horizontally to carry traffic, it must be properly supported vertically to carry traffic; thus, the bridge is shimmed up at certain points after closing so that it is stable and overstress is avoided.

Many, probably about half, of the older swing bridges in use today are railroad bridges dating from the late nineteenth or early twentieth century. Most are fairly

[1]Robert McCullough and Walter Leuba, *The Pennsylvania Main Line Canal* (York, Pennsylvania: The American Canal and Transportation Center, 1973), p. 46.

large symmetrical truss bridges, up to the double-deck 500-ft long rim-bearing Government Bridge over the Mississippi River at Rock Island, Illinois, built in 1893, and another double-deck structure, the 524-ft long Burlington Northern Santa Fe (BNSF) railroad bridge at Fort Madison, Iowa, both with highway decks on the lower level, and the three-track center-pivot-type built in 1892 for the Long Island Rail Road (LIRR) at Long Island City, in Queens, New York. Until it was put out of commission by a ship collision, one of the longest swing bridges in the United States was a 521-ft railroad bridge built in 1908 over the Willamette River in Portland, Oregon. This bridge was replaced by a vertical lift span of the same overall length in 1989. Some large highway truss swing bridges still in existence date from the nineteenth century; an example from 1880, the 7th Avenue or Macombs Dam Bridge, crosses the Harlem River in New York City, near Yankee Stadium. This bridge is presently undergoing extensive renovations and is expected to remain in service for many more years. There are many other truss type swing bridges of various sizes—both highway and railroad—in use. There are also several smaller old girder types extant at various locations.

It was generally believed, in the era between the Civil War and the Depression of the 1930s, that a swing bridge was the most economical means of spanning a navigation channel. The reason for this belief was, mainly, that a symmetrical swing bridge did not require a counterweight or its supporting structure to balance the bridge. In most applications, an approach span of some length was required in addition to the span across the navigation channel, and the second arm of the symmetrical swing bridge could accommodate this need. In many cases in earlier days, the Army Corps of Engineers, which at that time had jurisdiction over obstructions to navigable waters, did not object to the placement of a pivot pier for a swing bridge in the middle of a waterway as long as the space on each side of the pivot pier was wide enough for the largest vessels using the waterway to fit through. Prior to the early twentieth century there was much commercial traffic on many of the nation's inland and coastal waterways, and the small vessels that carried this traffic could easily pass through a narrow channel on one side of a swing span. Most earlier movable bridges in the United States were railroad bridges, paid for by the private railroads whose tracks were carried on them. As long as a bridge could physically be moved out of the way for navigation, speed of operation and convenience of the mariner were not important considerations. As long as the bridge was bigger and more substantially constructed than the vessel passing through it, the bridge owner was not particularly concerned about damage to the bridge in a collision. Swing bridges were usually operated to swing away from an approaching vessel, however, to minimize the time to complete passage as well as to minimize the likelihood of a collision.

A swing bridge does not lift into the air as it opens, so that, as compared with a bascule bridge, it does not have to resist or overcome substantial wind forces to operate. Yet wind forces must be considered, particularly for center-bearing-type swing bridges, as the overturning moment due to wind can be significant. Swing bridges must be able to move against unbalanced wind forces, which can be substantial. Although horizontal wind speeds are usually lower near the ground, many swing bridges are sited in exposed locations where the wind can reach high speeds. The wind can act more on one arm of a swing bridge than on the other, so that large wind moments may have to be resisted to operate a swing bridge, just as in operating a bascule bridge. For a large through-truss swing bridge, these forces can be great, indeed.

Since the development of cheaper girder spans, trusses have fallen out of favor and longer spans than necessary are no longer desirable. The shortest movable bridge that can accommodate a navigation channel of a given size is never a swing bridge. In addition, development of commercial multiple-barge tow traffic on inland waterways has made center piers undesirable. The Truman-Hobbs Act has allowed, since the 1940s, for the replacement, partially at public expense, of railroad bridges that are "hazards to navigation." The bridges to be replaced are usually movable bridges. Typically, the federal government pays the cost of the improvement to navigation, and the railroad pays the cost of the increase, if any, in the capacity of the bridge or any other improvements. The federal share of the costs may range from 80 to 90 percent. Most bridges replaced under this program have been swing bridges, and most replacement bridges have been vertical lift types. The Act includes local and state owned highway bridges, and funding has been tightened so that a very strict interpretation of improvements has shifted more of the cost of a project to the bridge owner. Thus, a project is often resisted by the bridge owner, sometimes successfully, sometimes not.

In addition to truss versus plate girder construction, the basic structural distinction in swing bridge design is in the support of the swing bridge—rim bearing versus center bearing. Traditionally, swing bridges pivoted about some sort of a center bearing, but as railroad swing bridges got larger and heavier toward the end of the nineteenth century, the rim bearing type of swing bridge was developed, on the assumption that it offered the best way to carry heavier loads.

Swing bridges require considerably more complicated machinery for operation than bascule or vertical lift bridges. The ends of the swing bridge have to be free when the span swings open or closed, but must be rigidly supported for carrying traffic. Therefore, wedges, retractable rollers, or jack and shoe arrangements are provided at each end of the swing span to lift the ends of the swing span and support them with the roadway or railroad tracks properly aligned. Swing spans went out of favor a few decades ago, largely because of functional shortcomings as well as the increased cost of construction and maintenance resulting from the additional mechanical components. Railroad swing bridges provide additional difficulties, as the railway tracks must be continuous across the joints at each end of the swing bridge when carrying traffic, yet must be cleared free when the bridge is swinging. Mechanically actuated miter rails are usually required on swing bridges for this purpose, but they are fraught with difficulties, especially in heavy-duty high-speed service. Most other types of movable railroad bridges do not require mechanical actuation of the miter rails or end rails, but can use pieces solidly connected to the movable span that mate with pieces solidly mounted to the approach. Miter rails that are not mechanically actuated tend to be much more durable than the mechanically actuated type.

Occasionally, a new swing bridge is built where an in-kind replacement is desired, such as at a historically significant location when the objective is to replace an old inadequate swing bridge with a new bridge of the same type, yet eliminate some of the deficiencies of the existing structure. New swing bridges are sometimes built at locations where a high profile structure is undesirable.

Such a situation developed at Yorktown, Virginia, where it was found necessary to build a bridge across the York River near the site of Cornwallis's surrender to George Washington in 1781. The Department of the Interior refused to allow any structure to be built that would be visible from the more critical locations within the national historic park at this site, or that would create a general impression at the park that was not in keeping with its historical context. The U.S. Navy has an active depot on the river upstream from the park, which requires a bridge to be able to pass the largest naval vessels. These somewhat contradictory requirements resulted in construction of a low-level bridge that would allow passage of very large vessels, including aircraft carriers. The decision made at the time called for a double swing bridge to be built, that would provide unlimited vertical clearance and about 400 ft of horizontal clearance when open. When traffic on the bridge developed to such an extent that more lanes were necessary across the river, reconstruction was proposed to widen the spans, which were in good condition and able to take the additional load. It was decided to replace the bridge with a new structure, as this was shown to involve less delay to traffic. In the 1990s an in-kind but wider replacement was constructed.

VARIATIONS ON SWING BRIDGES

Center Bearing

The typical center bearing swing bridge is supported by a circular disc, which has a convex spherical surface, fixed to the bottom of the superstructure. Normally, a loading girder or pair of loading girders is connected to the top of this bearing by suitable connecting framing where necessary. The two main girders or trusses of the swing bridge are supported at the ends of the loading girder assembly. This provides a simple, primarily nonredundant load path, which ensures equalized loading of the members. The convex disc supports the bridge weight on a matching concave disc, firmly mounted on a base on the pivot pier. The bridge spins like a toy top, supported at what may be considered a single point.

Balance wheels, which are attached to brackets on the swing bridge superstructure, ride on a circular track at the perimeter of the pivot pier to keep the swinging bridge from being tipped over by the wind or by some slight imbalance. Wedges or some other sort of positionable support devices prop up and stabilize the bridge when it is in position to carry traffic, so that most of the live load is carried through these supports rather than through the center bearing. The center bearing type of swing span concentrates the dead load of the bridge at a point at the center of the pivot pier (Figure 6-1) but distributes the live load, usually through six support points: two center wedges, and four end wedges (Figure 6-1).

Many different ideas were tried during the latter part of the nineteenth century to provide for higher-capacity swing bridge pivots. Several complicated center roller bearing designs were developed; of those that were built few are known to have survived. This type of bearing worked quite well when it was properly fabricated and installed, but it was very expensive. One such center bearing in 1884, was removed

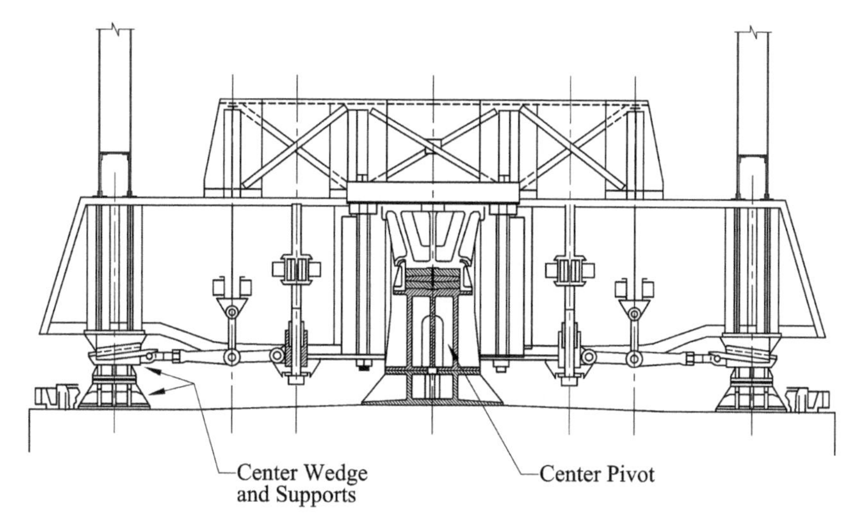

Figure 6-1 Center Pivot Swing Bridge—Center Support Arrangement. The loading girder transfers the dead weight of the swing span to the center pivot. The center wedges, when engaged, provide a firm support under the ends of the loading girder for live load on the bridge.

from a Chicago, Milwaukee & St. Paul Railroad bridge in Chicago that was replaced; the bearing was stored, then sold to the Chicago & Northwestern Railroad (C&NW), and used on another bridge in Chicago. This bearing was later removed again and installed, in 1909, in a C&NW bridge in Milwaukee.[2] Recently, roller bearing center pivots have again come into use (see page 92). By the late 1890s, the rim bearing type of bridge swing support (see page 89) was settled upon by most parties as the most desirable for very large swing bridges, but this did not continue for long (see page 87).

Some center pivot swing bridges were constructed with a double bearing surface at the center pivot. A loose bronze bearing piece, called a "melon seed" or "pumpkinseed," with top and bottom spherical surfaces forming a lenticular shape, was placed on a forged steel disc with a depressed spherical surface at its top. This steel disc was rigidly connected to a support that was placed on the top of the center pier. A second forged steel disc, similar to the first, was placed on top of the pumpkinseed, concave side down, and rigidly connected to the bottom of the swing bridge superstructure (Figure 6-2).

This redundant arrangement was apparently intended to accommodate poor fabrication practices, as a seizure at one bearing surface would not cause a bridge failure, but the bridge would continue to pivot at the other surface. Hovey[3] expressed concern that this double bearing arrangement could allow the bridge to swing off-center, and there have been reports of such occurrences, but the system has been used to support

[2]O. E. Hovey, *Movable Bridges,* Vol. 2 (New York: John Wiley & Sons, 1927), p. 55.
[3]Hovey, Vol. 1, pp. 297–299.

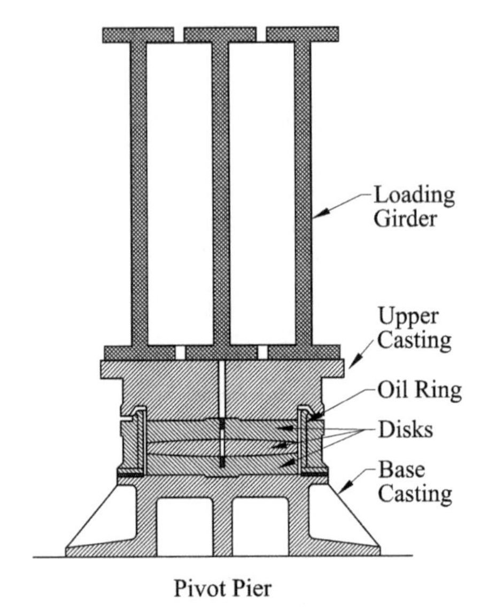

Figure 6-2 Swing Bridge Three-Piece Center Pivot Bearing. This type of bearing is found on many railroad swing bridges.

some very large truss-type swing bridges and has worked quite well (Figure 6-3). It was a rather expensive design and is not recommended by either AASHTO or AREMA. Many of the bridges that used this type of pivot-support system are still in use today, ranging from small 100-ft long deck girder swing spans to 300-plus-ft long double-track railroad truss bridges. The LIRR swing bridge over Dutch Kills in Long Island City, New York, which is more than 100 years old, was still operating a few years ago with its original support system of this type. Although the bridge is a three-track-wide through-truss span, it is only about 150 ft long and at present there is only one track on the bridge. Other larger bridges, such as the Point-No-Point bridge, the Newark Draw, and the Portal Bridge, all in northern New Jersey, are approaching or have reached 100 years of age with their original center bearing of the pumpkinseed or lenticular type, without any serious pivot problems.

There are several bridges with the pumpkinseed center bearing type of support still in existence, and most are functioning, opening and closing, without any pivot problems. An exception is the 450-ft long swing bridge of the New York and Long Branch Railroad, built in 1908, over the Raritan River. This bridge has had center pivot problems throughout its existence. The entire center bearing assembly was replaced with a new one of similar design a few years ago, but the difficulties persist. The trouble may be due to impact damage and pier instability, rather than attributable to the center pivot (see page 135). The problem with this bridge does not appear to have been with any deficiency in the design or construction of the bridge or with its center pivot bearing, but lies in the difficulty with which navigation passes through the shipping channel. The bridge is occasionally struck by fairly large commercial vessels, result-

Figure 6-3 Large Center Bearing Swing Bridge. This bridge was one of the first successful large swing bridges not to be supported on a rim bearing.

ing in a poor connection between the upper part of the pivot assembly and the bridge superstructure. The bridge has been hit many times in its life, some of these collisions resulting in dislocation of the center pivot assembly and damage to the center pier. It is doubtful that a different center support would have eliminated the damage experienced. To the contrary, the original design, which is still in use even though most of the components have been replaced, is possibly the most economical from the standpoint of minimizing collision damage and reducing the cost of repairs.

In the late nineteenth and early twentieth centuries center bearing construction was thought to be unsuitable for the largest, heaviest swing bridges. The Dutch Kills Draw, built in 1892 at Long Island City for the Long Island Rail Road, and the NY&LB bridge at Perth Amboy, built in 1908, suggested (inconclusively) that this concept was not true. Modern—that is, twentieth century—construction methods have allowed suitable loading girders to be fabricated and connected to the bridge trusses, and modern machining methods are capable of producing the suitable large spherical surfaces necessary for a center bearing of appropriate capacity. Almost all swing bridges built in the last several decades have been center pivot type bridges. Modern higher-quality bronze has allowed this simpler type of bearing to be used in larger bridges with little difficulty. A double-deck railroad and highway swing bridge built for the Southern Pacific Railroad at Sacramento, California, which weighs 3374 tons (see Figure 6-4), supported on a center bearing. The bridge was completed in 1911, and very few rim bearing swing bridges have been built since.

Figure 6-4 Very Large Center Bearing Swing Bridge. This bridge was one of the first successful very large swing bridges not to be supported on a rim bearing. It is even heavier than the bridge shown in Figure 6-3.

One of the more common problems with the typical center bearing swing bridges is failure by seizing of the bearing itself. This occurs occasionally with two-disc type center bearings when the bearing surfaces are poorly fabricated or when lubricant has been lost or was never properly installed, so that high friction develops at the bearing surfaces. The dowel pins holding the bronze disc shear when the bridge is operated, as the bronze disc has stuck or welded itself to the supporting steel disc. The bridge then spins, not between the bearing pieces, but about the surface between the upper bronze bearing piece and the bracket connecting it to the lower part of the superstructure.

Something similar to this occurred at the Southern Pacific Railroad bridge at Sacramento, mentioned earlier, in 1996, 85 years after the bridge was put in service. As far as can be ascertained, the bearing received no attention other than regular lubrication for its first $8\frac{1}{2}$ decades of existence. As the bearing was buried in concrete because of pier height restrictions, it was almost impossible to inspect. In 1996 the bearing seized and was dismantled for inspection. The center bearing was the three-disc pumpkinseed variety, directly supporting the bridge loading girders, and it was found that the bearing surfaces had seized and that the lower disc was turning on the bottom support casting when the bridge was rotated. The center bronze disc was remachined, an extra filler shim piece was put in to make up for the loss of height, and the assembly was reinstalled. The bridge was found to work perfectly after reassembly.[4]

[4]Jeffrey S. Mancuso and Jeffrey W. Newman, "Emergency Repair of a 7 Million Pound Swing Span," Heavy Movable Structures, Inc., *6th Biennial Symposium,* Clearwater Beach, Florida, 1996.

Rim Bearing

The rim bearing type of swing bridge is supported on a set of tapered rollers arrayed around the perimeter of the pivot pier of the bridge, spreading the load around the edge of the pier. The most obvious visible characteristic of a rim bearing swing bridge is a structural device known as a drum girder, or rim girder, that rests on top of the roller assembly. The rim bearing swing bridge superstructure proper sits on this substantial piece of steelwork, which is not required for a center bearing swing bridge. For the typical rim bearing swing bridge, a heavy track is laid on the pier to match tread plates mounted on the bottom of the rim girder. Rollers are then placed between the tread plates and the tracks and arranged so that the rollers move in a circle around the track on the pier as the bridge is rotated. A vertical post or heavy casting with cylindrical sections is firmly anchored to the center of the pier, and the bridge rotates around it, aligned by radial struts. The rollers are separately connected to the post by an arrangement often referred to as a "spider," consisting of radial rods and connection pieces, or of a fabricated framework.

Generally, the axes of the rollers are horizontal, with the tread plate above and the track below conical, as are the rollers. The apexes of all the conic sections are at a common point at the center of rotation of the span. Some rim bearing swing bridges were built with a flat tread plate attached to the bottom of the rim girder. This eliminated the tendency of the rollers to center the bridge and required the center post to perform all the work of keeping the bridge centered while swinging. Wind loads can sometimes be considerable, so because of its capability in resisting lateral loads, the symmetrical conical arrangement of rollers and tread plates has been considered the better choice.

The normal arrangement of the rim bearing, rollers, and track results in a situation that many bridge inspectors find startling: as the bridge rotates on the center pier, the rollers move around the pier as a group at half the rate at which the bridge moves. The rollers are interconnected and held in position by rods or structural members so that the axes of all the rollers are on radii from the center of rotation of the bridge. The framework that keeps the rollers in alignment rotates about the bridge with the rollers. The roller support system in this configuration acts as a giant tapered roller thrust bearing. As the rolling surfaces of the rollers and the loaded surfaces of the tread plate and the track are sections of cones, all of which must meet at a point at the centerline of the bridge, only rolling contact occurs. There is no sliding or resultant high friction and wear.

It is difficult to transfer the weight of the separate trusses or girders to the several rollers of a rim bearing swing bridge—usually in excess of 50—equitably. Many engineers spent a great deal of time developing various schemes to distribute the bridge load as evenly as possible around the rim girder. As the rim bearing was usually specified because of a need to support a very heavy truss-type swing bridge, the equitable distribution of these loads was an important design goal.

An example of a pure rim bearing swing bridge, the Madison Avenue Bridge over the Harlem River in New York City, has an intermediate cross-girder grid set between the three-truss superstructure and the cylindrical drum or rim girder. The bridge is a true rim girder type, with no load supported by the center pivot, which acts only as a centering device to keep the center of the bridge in the middle of the pivot pier. The cross-girder arrangement takes the load from the three trusses and distributes it directly to 12 points equally spaced along the top flange of the circular drum. At the time of the bridge's construction the scheme was thought to be an elegant one, and O. E. Hovey, writing in 1926, 16 years after the bridge was constructed, praised the design because it distributed the load to 12 points on the drum, a 50 percent improvement over the usual maximum of 8.[5] However, analysis of this bridge suggests that the 12 bridge load points on the rim girder may vary in load by a factor of as much as 3:1. If this is the case, then whether it has always been the case or whether changing conditions over the years have caused a redistribution of dead loads on the bridge, is not known.

Many different schemes have been developed to distribute the dead loads on a rim girder swing bridge, but these rely, for their accuracy, on careful construction and a sufficiently rigid finished product so that the load patterns do not change in service. Unfortunately, it is difficult to monitor the placement of steel during construction to see that all the members in a multiply redundant load path are carrying their intended share of the load. Equal loading is then made more difficult by the fact that the live loading, and the deflections of the main members, are changed when the bridge end supports are released or activated during opening and closing cycles. Some hysteresis, or plastic deflection, is bound to be present in a large structure such as a swing bridge, due to imperfectly connected joints and various unreconciled internal stresses. Moreover, temperature differentials cause distortion that changes the load distribution.

In addition to ensuring equitable distribution of the dead load onto the drum girder, a means must be provided to distribute the load equally to the rollers supporting the rim girder bridge. This can be accomplished by making the drum girder very deep, but there are usually limits to depth because of geometric constraints. Many railroad bridges have track elevations that are at a very low elevation above the water, and in addition to the need to provide some depth for the bridge floor system, it is desirable to keep the top of the pier as high above the water as possible. This leaves a limited amount of space for the rim bearing rollers and supports. The solution has usually been to rely on extreme precision in the manufacture and erection of the bridge. The rolling surface on the circular track must be perfect, the track must be evenly supported on the pier, the tread plate on the bottom of the rim girder drum must be perfect, and each roller must be perfectly placed and have the same conic section. The only adjustment that may be available is to position the rollers slightly inward or outward on their radial axes.

The 7th Avenue, or Macombs Dam, Bridge over the Harlem River in New York City, built in 1880, more than doubled the difficulty by having two sets of rim bearing rollers, one located inward and one outward on the same elevation, with separate tracks, tread plates, and drum girders for each. The drums are rigidly connected, so no equalization occurs between the sets of rollers. There is liberal allowance for adjustment of the radial position of each roller, but Wright stated, in 1898, "And it is evident that the most perfect workmanship and the greatest care must have been used

[5]Hovey, Vol. 1, p. 59.

throughout to insure the action of all the wheels."[6] This bridge is still standing, a testament to the quality of the work that went into its construction, but this type of rim bearing bridge was, and is, rare indeed.

The typical roller array of a rim bearing swing bridge, acting as a giant conical antifriction thrust bearing, prevents horizontal movement of the swing span except around the vertical axis through the center, stabilizing the bridge against any unbalanced forces. It also stabilizes and adequately supports the center of the span for carrying traffic, so that no auxiliary center supports are required. As indicated earlier, a great deal of precision is required in the manufacture and installation of the set of rollers and treads that support a rim bearing swing bridge. This difficulty has practically eliminated the rim bearing type from what little new swing bridge construction is performed today. At one time, as indicated previously, many designers believed that very large and heavy swing bridges could be supported only on rim bearings, but this was shown not to be the case. The Pennsylvania Railroad, noted for its engineering prowess, and which seemed to go out of its way to construct expensive high-quality civil works projects in the late nineteenth and early twentieth centuries, has several existing swing bridges dating from that era. Not one of them is a rim bearing bridge.

Rim bearing bridges continued to be built, particularly for the larger size swing bridges, through the 1920s. One of the largest swing bridges, a 524-ft long, 5000-ton double-deck rim girder span across the Mississippi River at Fort Madison, Iowa, was completed for the Atchison, Topeka & Santa Fe Railway in 1924 (see page 82). It was becoming apparent that their higher fabrication costs, and more difficult erection, were not warranted. Both the American Association of State Highway and Transportation Officials (AASHTO) and the American Railway Engineering and Maintenance of Way Association (AREMA) now recommend that the rim bearing type not be used for new construction.

In spite of their higher cost and the doubtfulness of their increased capacity or durability, smaller rim bearing bridges are not unusual. An example is the Toledo Terminal Railway single-track through-truss bridge over the Maumee River in Toledo, Ohio, a little more than 200 ft long, built in 1903. This bridge has not been used for many years and has recently had the railroad tracks removed from its deck. Most, if not all, of the smaller rim bearing swing bridges are railroad spans.

The rim bearing may regain acceptability because of seismic considerations and the availability of prefabricated rim bearing assemblies. A rim bearing can provide greater lateral support than a center-pivot-type bearing, and uplift restraint can be easily included. A few swing bridges have been constructed recently or retrofitted with large, prefabricated precision roller thrust bearings of equivalent overall size to a rim bearing roller assembly. These prefabricated bearings can incorporate a lateral capacity and uplift resistance impossible to achieve with the conventional array of rollers on tracks. Because the swing bridge is the only common type of movable bridge that does not require a counterweight that concentrates a great deal of mass in a very small space, designing them to resist earthquakes can be a simpler task than designing bascule or vertical lift bridges for such resistance. In addition to the above, recently,

[6]Charles H. Wright, *Design of Draw Spans,* Vol. 2 (New York: John Wiley & Sons), p. 193.

spherical roller bearing type of center pivot bearings designed to resist significant seismic forces have been specified for new swing bridges.

Combined Bearing

Many bridges that appear externally to be rim girder bridges are actually combined rim- and center bearing bridges, with part of the load carried on the rim bearing and part carried at a center bearing connected to the rim bearing by radial struts. These struts are present on a pure rim bearing bridge, but serve only to maintain the position of the rim girder about the bridge center. On the combined bearing bridge, the struts are primary load-carrying members. The trusses or main girders rest directly on them, while the outer end of each strut is supported by the rim girder and the inner end is supported on the center pivot. The distribution of load between the rim and center bearing varies, but on most bridges of this type the major part of the load is carried on the rim bearing. An example of this type of support is the New York and Lake Erie Railroad Greenwood Lake Branch bridge over the Passaic River, north of Newark, New Jersey. This bridge is a two-track deck-girder type, only about 200 ft long, built in 1908. Another example on the same rail line is the NY&GL bridge "DB" over the Hackensack River in New Jersey (see Figure 6-5).

Figure 6-5 Swing Bridge Combined Bearing Center Pivot. This type of bearing supports part of the dead and live load; the rim bearing rollers support the rest.

Bobtail Swing

There are some examples of bobtail or asymmetrical swing bridges, trussed type and girder types, from the early twentieth century, but they seem to be concentrated in the upper Midwest. The Chicago, Milwaukee & St. Paul Railroad was a particularly vigorous proponent of this type of bridge, having several trussed examples in the Chicago and Milwaukee areas, and deck girder types both in urban areas and at rural locations, such as on the Wisconsin River. Most of these bridges are still in use for railway traffic, but some swing bridges in the rural areas no longer open for navigation. The bobtail swing was generally used at narrow waterways, as was the LIRR span at English Kills in New York (Figure 6-6), but some examples, such as the one over the Wisconsin River at Sauk City, Wisconsin, is combined with several truss-type approach spans. The necessity of adding an "expensive," as O. E. Hovey put it, counterweight to balance this bridge, when a symmetrical second arm could have easily been added and the approach spans merely shortened, seems incongruous. It appears that the bridge was rope-operated from the shore, so that the short shoreward span provided some advantage in the shorter length of operating rope required and a better angle of incidence of the tensioned operating rope at the span. Occasionally, more modern versions of this type of swing bridge have been built where they fit the circumstances.

Some of the very short and lightweight bobtail swing spans are not cantilevered at their long arms when in the closed position, but act as simple spans, with a very short

Figure 6-6 Bobtail Swing Bridge. Typical of such structures of older construction.

cantilevered span extending from the other side of the pivot point, not reaching beyond the extremity of the pivot pier. A particular variation of this type, called the shear-pole-swing bridge, once quite common, was not balanced and had no counterweight at all. This type was supported and stabilized when swinging, and in the open position, by an overhead cable arrangement (see Figure 3-5).

Double Swing

There were a few double swing bridges erected in the nineteenth century, such as a highway bridge over the Cuyahoga River in Cleveland, Ohio, that was removed many years ago. These bridges quite possibly would not have been built in such a form if long spans, such as the vertical lift and double leaf bascule, had been fully developed at the time. Modern double swings in this country have been used exclusively where overhead restraints make vertical lift or bascule bridges undesirable. In addition to the G. P. Coleman double swing bridge at Yorktown, Virginia, there are a few other double-swing bridges in existence, including most remarkably a concrete version in Seattle Washington (see below). The longest double swing was a railroad bridge over the Suez Canal, with two 520-ft long swing spans, that replaced an earlier, smaller bridge of similar configuration built in 1941. The bridge was built in 1965 and destroyed in the 1967 war and is now being replaced in kind. The lack of moment continuity at the joints at the ends of the swing spans would present a problem for an American railroad with heavy live loading, but is apparently not a concern in Egypt. A vertical lift bridge could achieve an equal or greater clear span length without the troublesome, particularly for heavy railroad traffic, truss joints at center span, at equal or less cost, but the overhead clearance for navigation would be restricted.

One of the heaviest movable bridges in existence is the concrete double-swing bridge, constructed in 1985 in Seattle, Washington. This bridge has a center-pivot type of support, but the bearing is cylindrical rather than spherical. There are extra stabilizer bearings at the center, so that balance wheels are not needed for operation, but a bearing ring around the rim of the pivot pier is used to support live loads. The entire swing span superstructure is lifted hydraulically to clear the supports and allow the bridge to swing. Releasing the lift allows the bridge to rest on the supports at the perimeter of the pivot pier, which carry dead and live loads. This bridge is similar to some hydraulically operated swing bridges constructed in the late nineteenth century, which were not very successful because of limitations in hydraulic machinery technology and precision at the time. A single-span hydraulic swing bridge constructed at Bowling, England, prior to the twentieth century was lifted hydraulically and turned by means of hydraulic cylinders. According to Wright, "Rollers are provided in case the ram fails to work, and in this case the turning is similar to an ordinary American bridge."[7] A railroad bridge was built near Glasgow, Scotland, over the Forth and Clyde Canal, in the late nineteenth or early twentieth century, with a hydraulic lifting arrangement similar to that of the concrete Seattle bridge.

[7]Wright, Vol. 2, p. 214.

Double Deck

The swing bridge, like the vertical lift span, can easily accommodate a second deck, because there is no hinge at either end of the moving span such as there is on a bascule bridge. Many double deck swing bridges are in existence, primarily because several railroads built swing bridges and added highway decks so that the costs of the bridges could partially be recouped via tolls on highway vehicles. These are usually very large through truss structures. Such bridges exist at Rock Island, Illinois, and Fort Madison, Iowa, over the Mississippi River (see page 91), at Sacramento, California, (see page 87), over the Sacramento River, and at other locations. At most locations the railroad is on the upper deck because of the typical topographic conditions at a river crossing, necessitating the railroad to approach from a higher level at grade. The Sacramento bridge is an exception to the rule, as the railroad is at the lower elevation, running along the river bank. This bridge has the highway deck at the upper level, below the top chord of the swing span. Approach ramps bring the highway up to the required elevation.

ADVANTAGES OF SWING BRIDGES

The swing bridge has the lowest-profile envelope of any of the commonly used movable bridge types. Its dimensions do not exceed, along any axis, the full extent of the functioning bridge during any part of the operating cycle. The swing bridge thus presents the least area to the wind of any ordinary type of movable bridge of comparable moving roadway dimensions and has the least wind load on it. The swing bridge does not, as compared with the bascule or vertical lift bridge, project into the air when opening, so the structural moment, particularly on the foundations, due to wind is much less for a swing bridge than for other common movable bridge types.

Symmetrical swing bridges can provide two movable spans in one moving structure. On some waterways, two separate channels can be an advantage in keeping the waterway traffic orderly, or can be convenient at paired waterways such as the locks of a busy canal. In the past, this was a marked advantage at many locations with a multitude of small commercial craft, which required a bridge to be opened often. Upstream-bound traffic could pass on one side of the center pier, and downstream-bound traffic on the other at the same time.

At locations where long approach spans are required, the second balanced arm of the symmetrical swing may substitute for a fixed approach span. The second arm can reach across navigable water or nonnavigable water. In some cases the second arm reaches across dry land, such as at the Canso Causeway Bridge in Nova Scotia, eliminating the need for an approach span or fill. Whether a second arm is more economical than building a counterweight and finding another means to bridge the approach span gap is a matter of labor and material costs. The correct solution depends on time and place.

Swing bridges do not lift to open, so they are less noticeable in operation. They can also be, and usually are, symmetrical, so that providing a pleasing appearance is

Figure 6-7 Swing Bridge under Fixed Bridge. This application of swing bridges, to fit under a high-level fixed bridge, has been repeated in congested areas.

not as difficult as it is with some other common types of movable bridges. Thus, swing bridges have often been selected at locations where a movable bridge is necessary but where aesthetics play an important part in design.

Because a swing bridge does not lift up in the air to open, and a symmetrical swing bridge normally has no massive counterweight placing a large concentrated load on the piers, the swing bridge has been successfully built with less massive piers than a bascule or vertical lift bridge.

The low profile of a swing bridge can allow it to fit under a high level bridge at the same location, whether the swing bridge is open or closed (Figure 6-7). The Center Street asymmetrical swing bridge in Cleveland, Ohio, is an example. This was also, at least in part, the reason for building the concrete double swing bridge in Seattle in 1985.

DISADVANTAGES OF SWING BRIDGES

A swing bridge relies, to a greater extent than a bascule or vertical lift bridge, directly on mechanical components to support it and its live load. As the swing bridge has many more moving parts than the typical bascule or vertical lift span, it requires more maintenance than those types. In times of tight labor markets, which is usually the case in the United States, this is a serious drawback to the selection of a swing bridge. Because it has three or more major mechanical functions to go through in opening or closing, a swing bridge takes longer to operate than a vertical lift or bascule span. Moreover, once it has been prepared to swing, a swing bridge usually takes longer to move from closed to open and back again, than a bascule or vertical lift span of the same size. It generally takes longer for a vessel to navigate the draw opening of a swing bridge than the openings of other common types of movable bridges, because the draw opening is restricted at water level for a longer period until the swing bridge is fully opened.

Swing bridges, among all types of movable bridges, are assumed by some designers to be least affected in operation by winds. Unfortunately, this is something of a myth, as substantial rotational wind moments can be developed on a swing span. Swing bridges are sometimes constructed at locations that experience high winds, under the assumption that the swing bridge will be less susceptible to damage as a result of its low profile and typically symmetrical construction. The leading edge of a swing span, opening into the wind, produces a greater resistance to wind than the trailing edge. Apparent symmetry does not translate to zero wind resistance in opening and closing the bridge.

As a result, the wind resists rotation of the swing bridge as long as the leading edge is moving upwind, but if the bridge swings far enough to bring the longitudinal axis of the swing span parallel to the flow of wind and then continues to rotate, the net wind moment shifts from resisting the operation of the bridge to assisting it. If the wind is strong enough to produce a rotational moment greater than the sum of the other bridge resistances, the rotating span will "overhaul" on the drive machinery, with the span driving the machinery and motor instead of vice versa. A shock load will then be experienced by the drive machinery as the backlash is transferred from one side of the various gear teeth in the drive train to the other. If the difference in moments is great enough, the shock can cause the machinery to fail. This type of failure can cause a break in the drive train, resulting in a freely rotating bridge with no effective drive or control system. The swing bridge, in practice, usually has a lower maximum allowable wind speed for operation than bascule or vertical lift spans.

A swing bridge requires much more machinery than a bascule or vertical lift bridge. The usual swing bridge, which can open in either direction, does not automatically align itself when closing, but requires an active centering device to quickly align it in the closed position, something that is not required for bascule or vertical lift bridges. The typical swing bridge also requires mechanical end-lifting devices to hold up the ends of the bridge when it is carrying traffic. The end lifts must develop a positive dead load reaction as required by AASHTO and AREMA, so that deflection of the opposite arm of the swing span, due to live load, will not cause the end of the span to lift off its supports.

A swing bridge is normally placed so that its pivot pier is in the center of the navigation channel. This may be convenient in providing separate directional channels for two-way marine traffic, but it puts the bridge in a vulnerable position for ship impact and reduces the maximum size of vessel that can fit through the navigation channel. Although it would reduce the likelihood of impact somewhat if the bridge swung away from the vessel entering the draw opening, this is not always done. Many swing bridges are usually opened in the same direction, regardless of the travel of the vessel. It is not unusual for a swing bridge to be struck many times throughout its life by vessels having difficulty navigating the channel. The swing bridge in such a position is normally protected by an expensive fender system to reduce the damage from such impacts. The swing bridge with its additional piers and fendering becomes a greater obstruction to water flow, as well as navigational flow, so that undesirable scouring can occur in one part of the river bed and undesirable silting may occur in another.

Swing bridges require a lateral clearance equal to the length of the swing span leaf for operation. A great deal of space can be wasted if parallel railroads or highways on swing spans cross a waterway in a congested area, in allowing each swing bridge to have sufficient space to open. In the event that expansion of the width of the roadway or railroad is required without decreasing the width of the navigation channel, a swing bridge must usually be replaced with a longer swing span or another type of bridge, whereas a vertical lift or bascule bridge could simply have a similar bridge built alongside, close to the original bridge.

Swing railroad bridges are the only type of movable bridges in common use that require expensive and fragile mechanical rail disconnect devices to allow the bridge to swing open and closed and provide a properly aligned surface for carrying traffic. Serious accidents have occurred because this mechanism failed in service and the failure was not detected by the safety devices in place on the bridge. This problem has been a source of considerable inventive activity since railroad trains began operating at higher speeds more than a century ago, and completely satisfactory solutions have not yet been developed. Many types of miter rails have been developed to separate the railway rails so that the bridge can swing open, and are in use, but none have been proven fully foolproof, maintenance free, and durable. The Pennsylvania Railroad (PRR) (see page 91) built several swing railroad bridges the late nineteenth and early twentieth centuries. The railroad was always very concerned about the safety of its trains and passengers and spent a great deal of effort to develop an acceptable miter rail system when it built its Portal Bridge,[8] on its much heralded line into Penn Station in New York City, which opened in 1910. These attempts largely failed. The majority of the movable bridges built by the PRR after 1911 are either bascule or vertical lift spans, and none of its new high speed main line bridges after that date were swing bridges.

THE FUTURE OF SWING BRIDGES

Swing bridges have been considered by some designers to be functionally obsolete for some time—for many decades, in fact. On inland waterways, the United States Coast Guard prefers to replace swing bridges with vertical lift spans wherever the opportunity arises. There are still occasional applications where their disadvantages are

[8]Frederic Westing, *Penn Station: Its Tunnels and Side Rudders* (Seattle, Washington: Superior, 1978), p. 33. Reprint of *History of the Engineering Construction and Equipment of the Pennsylvania Railroad Company's New York Terminal and Approaches* (New York: Isaac H. Blanchard, 1912) (from *Transactions of the American Society of Civil Engineers,* Paper No. 1153).

outweighed by the particular applicability of a swing bridge for certain situations. There is also some preference for the swing bridge for aesthetic reasons, so the occasional swing span will continue to be built. There are occasional special situations for which a swing bridge is perhaps the most practical solution. In the past 20 years new highway swing spans have been built at Yorktown, Virginia, and Seattle, Washington, and new railroad swing bridges have been built at Atlantic City, New Jersey, and at Mystic and New London, Connecticut. At least three of these were selected for primarily aesthetic purposes; only one or two were chosen for mainly practical considerations.

7

APPLICATION OF TYPES

HIGHWAY

Quick-Acting—Double Leaf Bascule, Particularly the Scherzer Type Rolling Lift

Many movable bridges must open for small vessels hundreds of times a day during certain seasons. At the same time, the bridge tender must keep the roadway open to highway traffic for the longest periods possible. This situation usually arises where a small island in a resort area is popular for visitors coming by both boat and automobile, and the island is connected to the mainland by a movable bridge. At locations where a constant, heavy flow of vehicular traffic must be interrupted frequently for marine traffic that consists largely of smaller vessels, the double leaf bascule bridge can significantly reduce the delay for both traffic modes. The double leaf bascule provides the quickest opening at the center of a navigation channel. A partial opening at the center of the channel, of unlimited vertical clearance, is available almost immediately with a double leaf bascule, and is frequently sufficient for smaller vessels, such as fishing boats and pleasure craft, particularly sailboats with tall masts. The reason is that the toes of the two leaves move apart as the bridge is raised, so that vertical clearance at the center of the navigation channel becomes unlimited almost as soon as the bridge begins to open. Although U.S. Coast Guard regulations require that a movable bridge be fully opened for marine traffic, a bridge operator cannot prevent one of these smaller vessels from passing through the channel opening before the bridge is fully opened. The bridge can then quickly start coming down again after the vessel has passed, allowing highway traffic to resume passage over the bridge with minimal delay.

The Scherzer type rolling lift bridge, because of its inherent translational movement away from the navigation channel as it is rolling open, provides an extra advantage in reduced opening time as compared with other bascule bridges. When the

Scherzer bridge is of the double leaf type, the advantages of the rolling lift are more pronounced. The Rall rolling lift, because it does not translate as far back when opening as the Scherzer, is not as advantageous, but still clears the channel more rapidly than a similarly proportioned simple trunnion or heel trunnion bascule. The advantage of all types of rolling lift bridges when opening—the reduction of operating time—also applies when the bridge is closing again, doubling the savings in highway closure time. The Scherzer firm made great use of this facet of their bridge type's operation in its advertising in the early twentieth century.

The extent to which the leaves of a Scherzer bridge move back is a function of the diameter of the rolling lift treads; the larger the diameter, the farther the bridge moves back as it opens. The State of Wisconsin has made use of this fact at several locations in the northern part of the state. Both bridges connecting part of Door County with the rest of the state, across the Sturgeon Bay Ship Canal, are double leaf Scherzer type bascules. This type is also used for one of the newest bridges in the area, the Main Street Bridge (Ray Nitschke Bridge) in Green Bay.

The Rall type rolling lift bridge also moves away from the navigation channel as it opens, but usually to a lesser degree, depending on the dimensions of the linkage connecting the lift span to the pier. The double leaf version of this type of bascule was used for the Stone Harbor Bridge over Great Channel on the southern Atlantic coast of New Jersey. To some extent the advantage of a rolling lift bridge can be negated by the application of safety appliances, such as barrier gates, on a highway bridge, so that the bridge opening takes a long time anyway. It is also of little advantage when the bridge must open for a long period of time for a larger vessel.

It is more difficult, particularly for amateur mariners, to maneuver through partially opened single leaf bascules, swing spans or vertical lift spans, than through double leaf bascule bridges. It is more difficult for weekend mariners to judge the overhead clearance when passing under a vertical lift or partially opened single leaf bascule bridge. Swing bridges, because they stay at the same elevation throughout their movement, require caution on the part of mariners to avoid getting too close to the swinging end of the bridge. The double leaf bascule, which presents a clear opening with unlimited vertical clearance at the center of the navigation channel as soon as it starts to lift, is easier for a boat to approach more quickly without fear of a collision.

Stability—Single Leaf Bascule and Vertical Lift

The vertical lift bridge and the single leaf bascule bridge require the fewest mechanical components to support live loads (Figures 7-1 and 7-2). The center shear locks of double leaf bascule bridges are very difficult to maintain with the desired snug fit. They tend to loosen over time, with the result that the leaves bounce when heavier live loads traverse the structure. The swing bridge has many components, and these must be kept in proper order for the bridge to safely carry traffic (Figure 7-3). The end lifts usually end up out of order, as it is difficult to maintain them so that they hold the ends at the proper elevation. Center wedges for center pivot swing bridges also tend to be poorly adjusted. These defects can and sometimes do result in damage to main bridge supports, particularly when a center pivot type bearing is used.

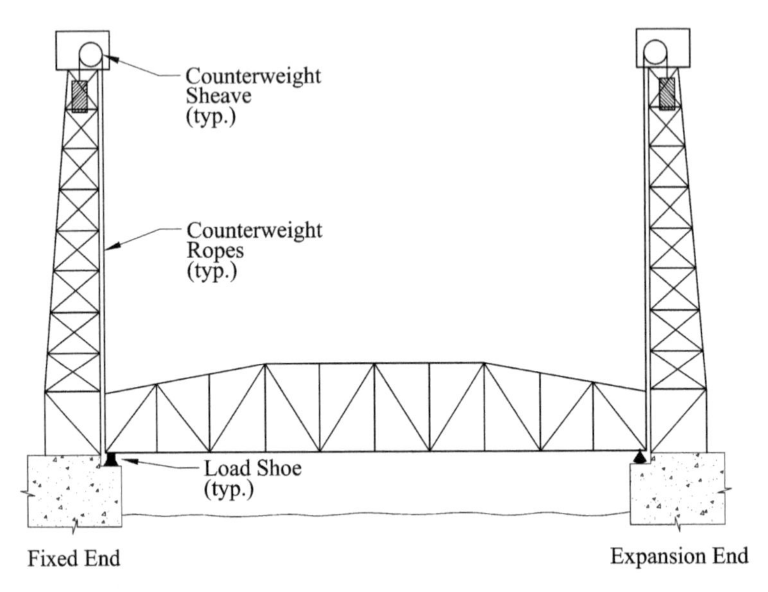

Figure 7-1 Vertical Lift Bridge—Seated. This type acts as a simple span when seated. No bridge operating machinery supports live loads.

The single leaf bascule bridge and the vertical lift span act as simple bridge spans in support of the live load and have no need for span locks for stability in support of live loads. There is no machinery of any type for these bridges on which the live load depends for its support, except for trunnions or rolling lift tracks for one end of the single leaf bascule. Any live load instability in these bridge types is usually found to result from improper balancing, so that the bridge does not seat firmly, or deteriora-

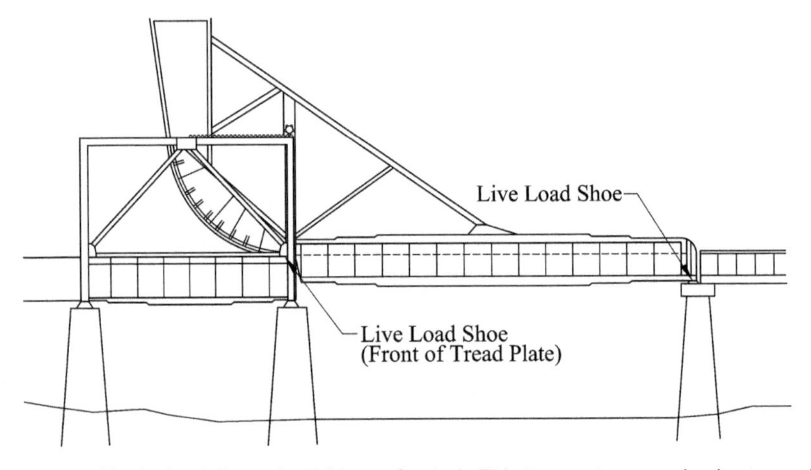

Figure 7-2 Single Leaf Bascule Bridge—Seated. This type acts as a simple span when seated. For a Scherzer rolling lift bridge, as shown, no bridge operating machinery supports live loads, unless one considers the forward end of the tread plates to be machinery.

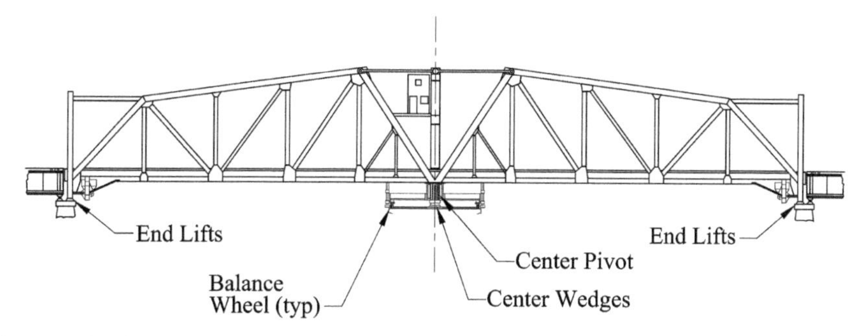

Figure 7-3 Swing Bridge—Closed and Ready to Support Traffic. This type of movable bridge incorporates many machinery components in the support of live loads.

tion of the live load supports. Large single leaf bascule spans for highway use have been constructed recently, such as the Erasmus Bridge in Rotterdam, the Netherlands, with a 59-m (193-ft 5-in) span and 33-m (108-ft 2-in) width.

Very Long Span—Vertical Lift

The longest possible ordinary movable bridge spans can most readily and economically be built of the vertical lift type. The vertical lift bridge, when seated, acts as a simple span. It is limited in length only by the structural limitations of simple span length. The vertical lift bridge can be built to any span length for which a simple truss span can be built. The longest vertical lift spans known to date are in excess of 500 ft, but less than 600 ft. Most of these longest spans are railroad bridges, such as the Buzzards Bay Canal bridge on Cape Cod, Massachusetts, and the Arthur Kill bridge between New Jersey and New York. There are vertical lift highway bridges in excess of 500 ft long, including the Burlington Bristol Bridge mentioned earlier. Some of these bridges are tower-drive spans, such as the 540-ft long Marine Parkway Bridge in New York (Figure 7-4). The Burlington Bristol Bridge is a Waddell type span drive bridge. All of the very long span vertical lift bridges, whether built for railroad or for highway use, are of the truss type. A vertical lift span could be built, if necessary, with a length equal to that of the longest simple truss nonmovable spans, up to a length in excess of 700 ft, such as the through-truss highway bridge over the Ohio River at Chester, West Virginia, built in 1976. There are simple truss fixed railroad bridges that have been built in excess of 700 ft in length.

Size and weight do not pose a problem for drive systems for vertical lift bridges. Extremely heavy vertical lift spans have been built, such as the Tomlinson Bridge in Connecticut, Danziger bridge over the Industrial Canal, connecting Lake Pontchartrain with the Intracoastal Waterway and the Mississippi River in New Orleans, Louisiana, the Broadway Bridge in New York City, and the Steel Bridge in Portland, Oregon. Higher-strength steels available today are coming into more general use for bridge building, and the use of high-tech composite materials can increase the practical maximum span length and reduce the weight.

Figure 7-4 Marine Parkway Bridge. Efforts were made to produce an attractive superstructure on this through-truss vertical lift bridge.

A single leaf bascule span could be built to a length nearly equal to that of a vertical lift bridge, but severe operating difficulties would develop because the bridge is lifted and balanced from only one end, considerably magnifying the forces involved as the entire leaf is cantilevered for dead load. Although the longest cantilever fixed spans are, in fact, nearly twice as long as the longest fixed simple spans, they are notoriously weak and inefficient structurally. Several of these that were built "economically" either collapsed or have had to endure load restrictions. The most successful bridge of this type, the Forth Railway Bridge, drew severe criticism for being overbuilt until it was found, through the examples of other bridges and more rigorous analysis, that such judgment was not necessarily valid. The vertical lift bridge, being lifted by essentially equal forces at each corner, is lifted and balanced by forces in a 1:1 ratio, not multiplied by the leverage ratio inherent in a bascule span. As a bascule bridge is lifted, its machinery is subject to increasing wind moments as it reaches the fully opened position. The vertical lift bridge machinery experiences no such moments opposing its machinery, but must simply withstand the direct wind forces pushing the span against the towers, plus small vertical wind components. The vertical lift towers and the bascule span structure, when open, must withstand the direct wind forces, regardless of direction, which tend to be greater at higher elevations.

In some cases a swing bridge can be advantageous for wide navigation openings. At Rock Island, Illinois, paired locks for the Mississippi River navigation system pass under a double-deck railway and highway swing bridge, with one set of locks on either side of the center pier. Although two separate movable bridges of another type, or one long vertical lift span, could be used in this situation, the swing span is convenient and probably the most economical alternative.

Very High Wind (Hurricane) Locations—Vertical Lift

The vertical lift bridge does not change its angular relationship to the wind as it raises and lowers. The vertical lift bridge and the swing bridge do not project angularly into the air as they open, so the effect of high winds on them is reduced as compared with the effect of such winds on bascule bridges. The swing bridge experiences variable wind moments as it opens, so operation can be difficult in a windstorm. Swing bridges located in high-wind areas must be adequately powered, with robust drive trains, to avoid breakdown when operated in such winds. As discussed earlier, the vertical lift bridge machinery does not experience direct wind forces of great magnitude, or wind moments, in operation, as there is no rotation about any axis of the span as it moves. The significant wind forces are horizontal, and the vertical lift bridge moves vertically, at right angles to the wind. The wind resistance parallel to the direction of movement of the span translates directly to operating force. Horizontal winds, which are the most prevalent, produce indirect operating forces due to frictional resistances at vertical lift bridge span guides and counterweight guides. Some vertical lift bridges, particularly those constructed in areas of fairly regular high winds, have roller-type guides on the span and, in some cases, on the counterweight as well, to reduce this friction. The towers of vertical lift bridges are subject to wind moments in proportion to their height, and the vertical lift span, as it opens, produces increasingly large moments in the towers during high winds, which must be resisted by the towers and foundations. The towers and foundations must be designed for such loads when the bridge is built in an area subject to high winds, particularly if the bridge is normally left open or opened in emergencies such as hurricanes. Several vertical lift railroad bridges are in this situation.

Winds usually have some vertical components, and winds are seldom constant, so there is some direct effect of the wind on the operation of a vertical lift span. Wind speed usually increases with elevation; thus, ground wind speed is not a good indication of wind velocity. The resultant pressure at the upper portions of a bridge, as much as 200 ft up in the air may be considerably higher. A difference of only 100 ft in elevation may show a tripling of wind speed. Swing bridges and bascule bridges are more directly affected by the wind than vertical lifts, to the point that it can be dangerous to operate them.

As discussed on page 97, the arm of a swing bridge that is upwind produces greater wind resistance than the downwind arm, so that there is a net wind moment on the span. What is worse, as the swing span opens into the wind, the wind moment will be reversed as the bridge axis comes parallel to the wind, resulting in a shock to the machinery. If the shock is great enough, which can be the case if the machinery is old and worn, a failure can occur, resulting in a free-spinning swing bridge. The greatest direct effect of the wind is experienced by bascule bridges, particularly single leaf bascule bridges, which project up into the air and may be required to operate directly against the force of the wind. Because the only part of the bascule that is exposed to the wind is on one side of the pivot point, the entire wind moment may be unbalanced. A retractile bridge is affected by the wind to a lesser degree than a swing bridge or bascule, because the bridge is not rotating into the wind, but only translating across it, so that there is no great leverage applied to the wind force to be resisted by the bridge machinery. The cross-sectional area of a retractile bridge that is closing into the wind is rather small, so the resultant increase in bridge drive force is not great.

Wind forces must be resisted by the foundations and machinery of a movable bridge as well as the superstructure, and this can add considerably to the cost of the bridge, particularly vertical lift and bascule types. The retractile and swing bridges stay at the same elevation open and closed, so higher wind velocities at higher elevations are not a factor.

Safety of Highway Traffic—Double Leaf Bascule

Safety is considered here only in respect to protecting the public from danger in situations where a bridge, being a movable one, which is displaced to accommodate marine vessels. The double leaf bascule bridge can be designed to provide automatic barrier protection for highway traffic, to prevent automobiles from plunging into the water. By careful location of the pivot axes of the leaves, combined with a particular layout of the heel joints, each leaf of the partly opened bascule becomes an upward-projecting ramp from its approach. The ramp becomes steeper as the bridge opens, so that it is impossible to climb when the bridge is fully raised. This is not an impact-attenuating-type barrier, however, and damage can be sustained by a vehicle hitting a fully raised bridge. Survivability is generally conceded to be much higher in an impact with such an object than in plunging off a bridge into a waterway. When the bridge just begins to open, an automobile can pass onto the bridge and run off the toe of the partly lifted leaf. Many old television "cops and robbers" programs have scenes of just such an event, staged on an actual bridge. An automobile piloted by a stunt driver leaps across the partially opened span from one side of the bridge to the other and continues on its way. Some double leaf bascule bridges have not been designed with consideration for the safety of motorists approaching or impacting the open or partially opened bridge—with tragic consequences.

Unlimited Vertical Clearance—Bascule or Swing

At some locations it is a great advantage to have no concern about a vessel's striking an open bridge. In these cases a bridge that completely clears the waterway is desirable. A bascule bridge, single or double leaf, is usually the most practical choice, but in some applications a swing bridge may be advantageous.

Of the bridge types that are considered preferable by AASHTO (section 2.1.2 in its *Standard Specifications for Movable Highway Bridges,* 1988) only the bascule and swing bridges completely remove themselves from the navigation channel to provide clearance for ships. The swing bridge, however, relies on an extensive and sometimes flimsy extended fender system (if it has one at all) to protect it from vessels, whereas the bascule bridge can remove itself completely from harm's way. The vertical lift span raises straight up and depends on its raised height to clear vessels—occasionally failing to do so. Other types, such as the pontoon retractile, achieve unlimited clearance when open, but have other drawbacks that make them undesirable, so these types should be avoided when possible. In waterways that are trafficked by oceangoing vessels, or vessels from distant ports, whose captains may be unfamiliar with local conditions, a movable bridge that completely evacuates the navigation channel is a safer bet for long-term survival.

Poor Foundations, Poor or Unknown Soil Conditions—Swing

For a given size of movable bridge, a swing bridge exerts much less force on the foundations than a bascule or vertical lift bridge. A swing bridge weighs less, because, at least in the symmetrical configuration, which is more common, it has no counterweight. It exerts less overturning moment on the foundations than a bascule or vertical lift span because it does not project into the air, either when open or when closed.

Not Recommended—Odd Types

There are many types of movable bridges constructed in the past of which only a few, or no, examples exist today. These types are inherently unstable, or take up a great deal of space to operate, or are more difficult to maintain or operate than the more common types of movable bridges. These include pontoon bridges, retractile bridges, folding bridges, double-swing bridges, and others, which should not be used unless extremely unusual site conditions make them desirable.

Some authorities would include ordinary swing bridges in this category, as they take up a large space in the navigation channel, are slow to operate, are vulnerable to collision, and rely on mechanisms to support traffic, but the large number of swing bridges in successful service contradicts this position. Most of the anti-swing-bridge literature was promulgated by "patent bridge" interests, such as the Scherzer company, that were trying to increase sales of their proprietary designs. Their arguments were and still are valid, although there are also legitimate points in favor of swing bridge use. AASHTO, in its *Standard Specifications for Movable Highway Bridges,* states:

> Movable bridges shall be of the following types:
>
> Swing Bridges
>
> Bascule Bridges
>
> Vertical Lift Bridges

No preference is indicated by AASHTO among these types. AASHTO makes no comment about the specific desirability or undesirability of double leaf bascules.

The pontoon retractile bridge type has found modern application at several locations on the Pacific Coast, due to extreme water depths and channel widths, such as the Puget Sound area. These bridges have not been without their difficulties, as a few have sunk in storms or in merely heavy weather. They take a long time to operate and require extensive maintenance and care. The Hood Canal Bridge, one of the largest, has an on-site staff of almost a dozen persons on weekdays, and maintenance personnel must be available for call on off-hours. A new pontoon retractile bridge was built in 1998 at Pearl Harbor, Hawaii. The type was selected, in part, because it could be designed so that it would minimally intrude on the atmosphere of the adjacent USS *Arizona* memorial. Opening this bridge is a major operation, requiring substantial advance notice. It has been opened only a half dozen times in its first three years of service.

At least two double swing bridges were built in the last decade of the twentieth century. Both were selected because of restrictions on the height of the structure. The Seattle bridge has another structure overhead, and the Coleman Bridge replacement is restricted by the National Park Service because of its proximity to the Yorktown National Battlefield.

RAILROAD

There has been negative opinion about the use of certain so-called "patent" types of movable bridges. There is some evidence that bridges of these types are less durable than the traditional types. The movable bridge types covered by patents include the following:

- Rolling lift: Scherzer and Rall bascules
- Heel trunnion bascules of almost all modern types
- Articulated counterweight bascules of all types
- Vertical lift bridges of all types
- Dozens of other odd, seldom constructed types, many of which had only one, unique, example constructed

The traditional types include the following:

- Simple trunnion bascules
- Center-bearing swings
- Rim-bearing swings, but these are considered undesirable by some as well

Some may argue with the list of "patent" types, but there is some validity to this position, as it has been demonstrated that some of the "patent" types have shortcomings. The rolling lift bridge is sure to wear out after a certain number of operations and will eventually require either replacement or expensive major rehabilitation. A Scherzer rolling lift bridge operating often in a harsh environment may require tread replacements every 20 years or less and can develop fatal disintegration of its segmental and support girders if tread wear is allowed to continue too far. The linkages, rollers, and treads of Rall rolling lift bridges also wear out after a certain number of operations. Many heel trunnion bridges (Figure 7-5) have developed defects at their hinges, and although these bridges may operate satisfactorily with such defects for many decades, many owners have thought it necessary to repair the defects to minimize the possibility of a bridge failure. Some heel trunnion bascules have experienced structural steel fatigue cracking due to the fact that they open to a very large angle, resulting in high degrees of stress reversal in operation. The articulated counterweight bascule presents a threat of catastrophic failure, but there is no reason that movable bridges of many of the patent types, if properly designed, conscientiously built, and scrupulously maintained, cannot last indefinitely.

The vertical lift bridge is also a patent type and almost always requires the replacement of counterweight ropes after a period of time. There are examples of these ropes lasting as long as 70 years, but these are exceptional cases. Fifty years seems to be an optimistic expectation for the life of the counterweight ropes of a vertical lift bridge. The operating ropes of a Waddell type vertical lift bridge do not last nearly as long, and replacement of these ropes should be considered part of the bridge maintenance expense.

(a)

(b)

(c)

Figure 7-5 Heel Trunnion Bascule Bridge. (a)–(c). In the opening process, the action of the span, counterweight, and operating strut can be seen.

AREMA, in its *Manual for Railway Engineering,* Chapter 15, Part 6, "Movable Bridges," states:

> Movable bridges preferably shall be of the following types:
>
> Swing
>
> Single leaf bascule
>
> Vertical lift

No preference is indicated by AREMA, other than in its commentary, between these types.

Double leaf bascule bridges were once used fairly often as railroad bridges, but serious accidents have happened because of failure of the center locks. A notable incident occurred on the Soo Line bridge over the Soo Canals at Sault Sainte Marie, Michigan. The center locks failed, causing a train to fall into the canal as the two bridge leaves separated.

The design engineer, when selecting a movable bridge type for a particular installation, must be aware of the advantages and limitations of the "patent" types while not ignoring the shortcomings of the traditional types. The simple trunnion bascule requires a large vertical clearance under its counterweight to allow it to open. Many railroads cross waterways at low elevations, so this type of bascule bridge cannot be used without providing for a waterproof counterweight pit, which is expensive to build and difficult to maintain. Counterweight pits should always be avoided where possible, as they entail an unnecessarily large capital and maintenance expense. If a bridge can be built at a sufficient elevation to allow the simple trunnion counterweight to clear the high-water line with the bridge open, then this is a very satisfactory form of movable bridge.

In the old days when long, heavy freight trains were pulled by steam locomotives with limited tractive effort, an uphill grade encountered in crossing a navigable waterway was a serious operating handicap. Such grades were avoided whenever possible, so many railroad bridges cross waterways at very low elevations, sometimes with the lower parts of the superstructure lapped by waves or even submerged at high tide (Figure 7-6). With modern diesel-electric locomotives, much greater tractive effort is available, so that raising the elevation of a waterway crossing to the extent that a simple trunnion bascule bridge can be installed without a pit is feasible in many situations. This has been done in some recent bridge replacements. At one location twin single leaf simple trunnion bascule spans at a higher elevation have replaced twin rolling lift spans. The resultant slight upgrade approaching the bascule spans from either direction was determined not to be an operating handicap for the train traffic, primarily passenger trains, crossing on this bridge. Unfortunately, counterweight pits were still required for this installation.

The swing bridge, after having been ignored as "obsolete" for several decades, has made a comeback as the shortcomings of the patent types have been revealed over the years. The utility of the swing bridge has also improved. As fewer multiple-track railroad bridges and more single-track bridges are built, the swing bridge provides a

Figure 7-6 Swing Bridge with High Water. This movable railroad bridge, constructed at a low elevation, has its rack pinion, rack, and track partly submerged at each high tide.

lesser obstacle to the navigation channel when opened. Increases in traffic can be handled by sophisticated signal systems, so the number of tracks on a line can be reduced at a bridge with little or no obstruction to traffic flow. The swing bridge has mechanical limitations, however, which puts it in a category of desirability that some believe to be no better than that of the "patent" types.

As discussed on page 98, these mechanical limitations of swing bridges, except for some of the very small bobtail types, including shear pole swings, include the necessity of mechanically actuated miter rails at the track joints at the ends of the swing span. Swing bridges also require very complicated machinery for end lifting and locking. All of this mechanical equipment requires competent maintenance attention for reliable operation.

The high load-carrying capability, due to its being a simple span type, and the simplicity of design of the vertical lift bridge outweighs, for most designers, the limitations of rope life and the possibility of fatigue failure in a counterweight sheave shaft. It has in recent decades become by far the preferred type of movable bridge for railroad applications. This is particularly true for medium to very long-span movable bridges. The last long-span heel trunnion bascule in the United States was built in 1956. The last long-span single leaf rolling lift bridge in the country, with a clear span of more than 200 ft, was built in 1973. The U.S. Coast Guard, when mandating movable bridge replacement over inland waterways under the Truman-Hobbs Act, invariably specifies the vertical lift type for new movable bridges and usually requires a channel width, where applicable, of 300 ft.

Short and Medium Spans—Single Leaf Bascule

The single leaf bascule is generally the most desirable form of movable bridge. It typically has the least first cost, highest reliability, and greatest longevity. It also requires the least maintenance and takes the least time overall to open or close. The span length limitation is not very restrictive—spans of more than 250 ft have been built with the technology of 1917. There is no reason that single leaf bascule spans crossing 300-ft or wider navigation channels cannot be built. The chief drawback is wind loading, which increases machinery loads and foundation forces in proportion to the square of the span length. For movable bridges of up to 200 ft in length, the single leaf bascule is most often the best choice. This type of bridge is uncomplicated and durable. The length of the span does not result in excessive wind-related stresses in the open position, and a truss-type span can easily be designed with sufficient rigidity to support the heaviest traffic on a longer span. If the track is at a sufficient elevation above high water, the simple trunnion type can be used, but if the bridge is at a low elevation, the heel trunnion type has been employed to avoid the need for a counterweight pit. The majority of larger railroad bascules in this situation have been of the heel trunnion type, and many are in successful operation today, with more than a few in excess of 50 years of age. There are also a few rolling-lift-type bascule bridges in railroad service in the 200-ft range, although this type has been most frequently used for shorter spans of 150 ft or less. Many Scherzer-type rolling lift bridges have been designed with overhead counterweights for low-elevation installation without the need for counterweight pits.

Long Spans—Single Leaf Bascule

For span lengths up to 250 ft, there are several single leaf bascule bridges in use, including both rolling lift and heel trunnion types. As long as the bridge is designed for all wind loads to which it is likely to be subjected and the other elements, including substructure and machinery, are properly designed for the length of the span, there is no reason to exclude the single leaf bascule bridge from a type study simply because of the length of the movable span. As compared with highway bridges, the railroad bascule is narrower, assuming it is built for only one or two tracks, and thus the maximum wind moment is limited. Railroad bascule bridges are invariably open-deck types, and longer bridges are typically the open-truss type, further reducing the effect of wind loads on the structure and reducing the maximum load requirements on the machinery.

Very Long Spans—Vertical Lift

For movable spans of extreme length, the vertical lift span is about the only practical choice. Several of over 250-ft clear span have been installed to replace swing bridges, replacing two short-width navigation spans separated by a pivot pier in the middle, by one long span without obstruction. Such changes have been mostly to the advantage of navigation interests, easing waterway congestion and facilitating navigation through the draw. The U.S. Coast Guard usually mandates a 300-ft clear channel for movable bridge alterations under the Truman-Hobbs Act, and has preferred vertical lift spans. Several vertical lift spans have been built to a length of 500 ft or slightly more. The longest railroad span, at 557 ft $11\frac{1}{2}$ in. center to center of bearings, is the Arthur Kill Bridge, between Staten Island and New Jersey. This bridge provides a 500-ft wide navigation channel, with a vertical clearance in the open position of 135 ft. It replaced a swing bridge with horizontal clearances of 202 ft on one side of the center pier and 212.5 ft on the other. This was a big improvement for tankers navigating the waterway. A vertical lift span can be built, if necessary, to a length equal to that of the longest simple-truss non movable spans, up to an excess of 700 ft, such as the bridge built by the Chicago/Burlington & Quincy and the Louisville & Nashville railroads, and used also by the Illinois Central Railroad, over the Ohio River at Metropolis, Illinois, an eyebar truss type designed by Ralph Modjeski and built in 1917. This bridge, in spite of its great length, has a very high live load capacity. Extremely heavy vertical lift spans have been built, including the 6000-plus ton moving load Tomlinson Bridge in New Haven, Connecticut, a combined rail-highway span, and others (see page 63).

Short and Medium Spans with Approach Spans—Swings Are Best Choice under Some Conditions

When the length of a span crossing a navigation channel is not great, say, up to 150 ft, and one or more sizeable approach spans are required to clear the edges of the waterway or to cross a valley alongside the waterway, a swing span may be practical (Figure 7-7). Both sides of the center pivot pier can be used as navigation channels if so desired. In many cases only one navigation channel is necessary or possible. Asymmetrical or bobtail swing spans have often been built in such situations, with the short arm of the swing sometimes extending over dry land.

Unlimited Vertical Clearance—Bascule or Swing

At some locations it is a great advantage to have no concern about a vessel striking an open bridge. In these cases a bridge that completely clears the waterway when open is desirable. The bascule bridge, single or double leaf, is usually the most practical choice, but in some applications a swing bridge may be advantageous.

Of the bridge types that are considered preferable by AREMA (Chapter 15, Article 6.2.2), only the bascule and swing bridges completely remove themselves from the navigation channel to provide clearance for ships.

The railroad bridge across the Suez Canal in Egypt was destroyed in one of the Arab-Israeli wars. Ships from all around the world, their sizes limited only by the capacity of the canal, pass through, so it was necessary to provide a bridge that would minimize the likelihood of collision. The structure was a double-swing bridge with a very wide clearance. It was decided to replace that bridge with a new structure virtually identical to the previous one, which would be the third double swing span on the site (see page 94). Double leaf railroad bridges are regarded with disfavor by

Figure 7-7 Swing Bridge over Delaware and Raritan Canal. This bridge, built in 1905, has probably not opened since the canal was closed in the 1930s. The canal is in a small valley, and rather than build a single movable span with additional approach spans or fill, a symmetrical swing span was constructed, although only one arm is over water. The canal towpath passes under the second arm of the swing span.

American railroads because of the difficulty in providing moment continuity at the juncture of the two movable leaves. The movable bridge was a much less expensive option than a tunnel or high-level fixed bridge. The relative infrequency of train operations prevents this bridge from being an obstacle to navigation.

High Wind, Such As Hurricane Locations—Vertical Lift

A bascule bridge is highly sensitive to wind loading, and the need for resistance to high winds when operating adds proportionately more to the power requirements for a bascule bridge than for a swing or vertical lift bridge. It may be just as well to build a bascule bridge so that its many advantages can be employed, making sure to provide sufficient driving and braking strength and power to withstand the wind loads expected. In many cases, however, a vertical lift or swing bridge is selected to reduce the effect of wind loading on the construction cost. The Louisville & Nashville Railroad constructed new swing bridges along the coast of the Gulf of Mexico in the 1960s to "lower . . . hurricane hazard."[1] New spans were installed near Gulfport, Mississippi, in 1967 and at Biloxi, Mississippi, in 1978. This approach has sometimes

[1] *Trains* (December 1967) p. 9.

failed to achieve the desired result—the new, not yet completed, swing bridge at Biloxi was blown off its bearings by Hurricane Bob.

A properly designed swing or vertical lift bridge, taking into consideration the wind loads expected at the site, can be less expensive over the long term than a bascule bridge. This is especially likely to be the case for a railroad bridge on a lightly traveled line, where the bridge is normally left in the open position and lowered only for oncoming train traffic. Some bascule bridges in this situation are equipped with special locking devices that transfer the wind moments directly from the open bascule leaf to the fixed superstructure or pier, bypassing the bridge drive machinery.

Tracks at Low Elevation above Water—Vertical Lift, Rolling Lift Bascule, or Heel Trunnion Bascule

Many railroad lines cross bodies of water at very low elevations. It is difficult for heavy freight trains to climb steep grades, so rail bridges over water are kept at the same elevation as the approaches, or nearly so. At many of these crossings, the tracks are only a few feet above the water. To avoid damage to the machinery, movable bridges with all their machinery as high as possible in relation to the rail level are preferable. A vertical lift span has all its machinery above the deck, and this may be part of the reason that so many vertical lift bridges are used to carry railroad tracks. Rolling lift bascules with counterweights above the track and heel trunnion bascules have all their drive machinery above the track level, and the lowest moving parts can be at or very near to track level. Swing bridges, on the other hand, have their main supports and at least part of their drive machinery below track level and can experience difficulties if the tracks are only a few feet above high water.

8

BRIDGE REPLACEMENT

A great deal of time is usually required—from ground breaking to opening to traffic—to erect a movable bridge and get it into proper operation. Replacement highway bridges are usually placed on adjacent alignments, allowing the greater part of the new structure to be installed without disturbing traffic on the existing span being replaced. This generally does not permanently affect traffic, as the transition curvature required for the deviation from the original alignment can usually be designed to avoid a speed restriction. Often, the highway is widened as part of the project, so that half of the new bridge can be erected adjacent to the old one, and the old bridge is then removed and the rest of the new bridge is built in its place. Many railroad bridges have been replaced by new bridges on adjacent alignments, but this is usually done only on secondary lines or in situations where the railroad does not consider the additional curvature introduced to be an impediment to operations. Replacement railroad bridges built on adjacent alignments can result in speed restrictions for high-speed traffic when space is not available for a suitable transition.

Sometimes, for railroad or highway bridges, right of way is not available to build an adjacent span and the new bridge must be built at the same alignment as the old. A fixed bridge can usually be replaced in a matter of hours. The Union Pacific Railway changed out its bridge across the Missouri River, between Council Bluffs, Iowa, and Omaha, Nebraska, in $1\frac{1}{2}$ hours in 1903.[1] There are many examples of fixed bridge replacements on busy rail lines, performed in very little time. Similar work has been done on movable bridges, but the minimum replacement time is usually somewhat longer because of the greater complexity of a movable bridge. Unless the situation is such that the new bridge can be completely constructed around the old, or placed adjacent to it so that a "roll out–roll in" operation is feasible, it is very difficult to com-

[1]August Derleth, *The Milwaukee Road* (New York: Creative Age Press, 1948), p. 202.

plete the movable bridge changeover in less than a 24- to 48-hour railway closure (see page 119).

Careful study should be undertaken before a decision is made to replace an existing movable bridge. Many difficulties can arise in areas not directly related to the condition or use of the structure. These include historical and ecological considerations, which may increase the cost of structure replacement significantly. Such factors can have an even greater effect on the extent of time required from the first indications that a major investment must be made in the structure, until completion of a bridge replacement. Without sufficient effort to push through and expedite a bridge replacement, obstacles and opposition can result in a very long, drawn-out, expensive project.

Certain kinds of deterioration or shortcomings in an existing structure, however, make it almost imperative to seriously consider the implementation of a replacement project. These problems can include gross substructure deterioration, extreme traffic congestion that can be resolved by replacement of the bridge with a structure of greater traffic capacity, difficulties with navigation due to the configuration or size of the structure such that it is a serious impediment to the movement of vessels on the waterway (see the following discussion on the Truman-Hobbs Act), substandard or obsolete construction resulting in lack of capacity so that loads or speeds on the bridge must be restricted, or structural deterioration so severe that massive amounts of in-kind superstructure replacement would be required to restore the bridge to adequacy. The possibility of hidden defects should also be considered, in that some deterioration may not be discovered until after the reconstruction has commenced, thus increasing the cost of rehabilitation. Not only can increased costs be incurred because of the extra work involved, but it may also be difficult or impossible to obtain the extra work for a competitive price.

If a movable bridge is to be replaced in its entirety, demolition of the existing structure may be a difficult and expensive part of the project. If possible, such as in the case of constructing a new bridge on an adjacent alignment, consideration should be given to leaving all or part of the existing structure in place, as a historical monument, park, or fishing pier.

Some movable bridges are replaced, not because of any lack of adequacy in carrying their designated loads, but because they are a nuisance or inconvenience to boat operators. The Truman-Hobbs Act, mentioned elsewhere in this text, is a United States federal law that mandates the removal of bridges that are impediments to navigation. Specifically, paragraph 512 of Title 33 states, "No bridge shall at any time unreasonably obstruct the free navigation of any navigable waters of the United States." Paragraph 513 says: "If the Secretary determines that any alterations of such bridge are necessary . . . he shall . . . issue . . . an order requiring such alterations of such bridge as he finds to be reasonably necessary for the purposes of navigation" (June 21, 1940, ch. 409, ~3, 54 Stat. 498.) Further in Title 33, paragraph 516 states, "The Secretary shall determine and issue an order specifying the proportionate shares of the total cost of the project to be borne by the United States and by the bridge owner. The bridge owner shall bear such part of the cost as is attributable to direct and special benefits which will accrue to the bridge owner as a result of the alteration, including the expectable savings in repair or maintenance costs." This Act was originally "pork barrel" legislation, sponsored by then-senator Harry Truman of Missouri in the 1940s. The first project under the law was the replacement of the Chicago, Milwaukee, St. Paul & Pacific (CMStP&P) Railroad bridge over the Missouri River at Kansas City, Senator Truman's base of operations.

The law, which originally provided for payment for practically the entire cost of a bridge replacement, was restricted to railroad bridges, as other programs allowed federal funds to be spent on highway bridge work. The law is now being administered to cover any bridge and to require larger percentages of the total cost to be paid by the bridge owner. Since these changes were promulgated, the law has become somewhat controversial. The target of most of the effort under the Act is those railroad bridges that are inconveniently sized or located in the navigation channels of inland waterways, rather than only derelict structures. The major force behind implementation of the law is mounted by inland marine interests, direct competitors of the railroads. Pressure has been exerted to get railroads to replace bridges that are perfectly functional, for the primary purpose of allowing larger barge tows to navigate more efficiently on the rivers of the United States.

These efforts are in direct contradiction to the railroads' self-interests, and they have frequently objected to implementation of the Act. In specific instances bridges have been replaced where the railroad has not had the political or legal resources to prevent such action, and bridge replacements have been delayed or postponed at locations where the owning railroad has been able to marshal adequate support for its position. The Peoria & Pekin Union Railroad single leaf heel trunnion bascule bridge over the Illinois River at Peoria, Illinois, was replaced, in spite of its being in excellent condition and having served for more than 50 years, in order to widen and straighten the navigation channel through Peoria. The replacement of the Union Pacific (originally Chicago & Northwestern) Railroad swing bridge over the Mississippi River between Fulton, Illinois, and Clinton, Iowa, with a high level fixed bridge, as the river operators desired, has not been implemented, at least not at the time (2002) of this writing. The high concentration of freight train traffic over the bridge results in the delays for openings for barge tows, but the bridge is more than adequate structurally. The Burlington Northern Santa Fe bridge at Fort Madison, Iowa, was ordered to be "altered,"and work is in progress.

If a project for replacement of an existing movable bridge by another movable structure proceeds, it should be kept in mind that certain types of movable bridges lend themselves to replacement by particular types.

REPLACING A SWING BRIDGE—USE A VERTICAL LIFT

Many railroad swing bridges, as they aged, have been replaced with vertical lift spans. The replacement spans were built immediately adjacent to the old bridge or, in some cases, on the same alignment. The advantage of using a vertical lift span to replace an existing swing bridge is especially apparent when it is desired not to change the alignment; building a new vertical lift span directly around and over an existing swing bridge is a relatively simple matter. The swing bridge can continue to carry traffic and be opened for navigation while the new vertical lift bridge is built. This is especially true in dealing with a symmetrical swing bridge, which has no counterweight to demolish. The towers for the new bridge can be built intermeshed with the existing span without affecting traffic, and if placed outside the limits of the swing span, the towers and lift span can be installed without affecting operations of the bridge or requiring modifications to it.

In some cases, such as the replacement of the CMStP&P swing bridge at Hastings, Minnesota, the old swing span was modified prior to replacement by being shortened slightly so that the new vertical lift span could be built with the towers at the desired locations. The new towers were erected in place and the vertical lift span was built off-site and floated into position after the old swing span had been removed. Alternatively, if overhead clearances permit, the new span can be built in the raised position and merely lowered for use after the old bridge has been taken out. This operation can be done very quickly, so that the interruption to train service can be limited to only a few hours. At Hastings the rail line was shut down for 36 hours during the changeout, or replacement of the old bridge with the new, as a convenient detour was available. The bridge could not be opened for marine traffic for a somewhat longer time. In 1990, the Union Pacific swing bridge at Alexandria, Louisiana, was replaced with a vertical lift span in 12 hours.

Longer track outages are usually scheduled so that ample time is available to make the changeover when the new span is floated in to replace the old. The replacement of a swing span by a vertical lift bridge, accomplished in 1989 for the Burlington Northern Railroad at Portland, Oregon, is a typical example (see page 82). The bridge is fairly busy, carrying considerable freight and some passenger trains. It was decided by the railroad that a 72-hour closure could be tolerated. This decision was aided by the fact that the bridge had been damaged by a collision and bypass routes had been put in operation immediately after the collision, so that some rail traffic was already getting past the bridge without using it. When time came to perform the changeout, the work was easily accomplished in the time allotted, and the operation was considered to be quite successful.[2]

REPLACING A BASCULE—USE ANOTHER BASCULE

When it is desired to replace a single leaf bascule span with another bridge on the same alignment, it may be feasible to erect a new bascule bridge on the opposite side of the channel, in the open position, facing the old span, without affecting bridge operation or traffic. This has been done in railroad bridge replacements, but it may not be possible for highway bridge replacements and the extra coordination and planning effort may not be worthwhile for a highway bridge unless additional right-of-way is exceptionally difficult to obtain, so that a new bridge on an adjacent alignment is not feasible.

A type of bascule bridge was developed with replacement of existing bridges in mind. The Abt, or American Bridge type of bascule, was designed by Hugo Abt to avoid interference with traffic while building a new bridge on an existing alignment (see Figures 8-1 and 8-2). Only a few bridges of this type are known to have been

[2] *Engineering News Record* (September 7, 1989).

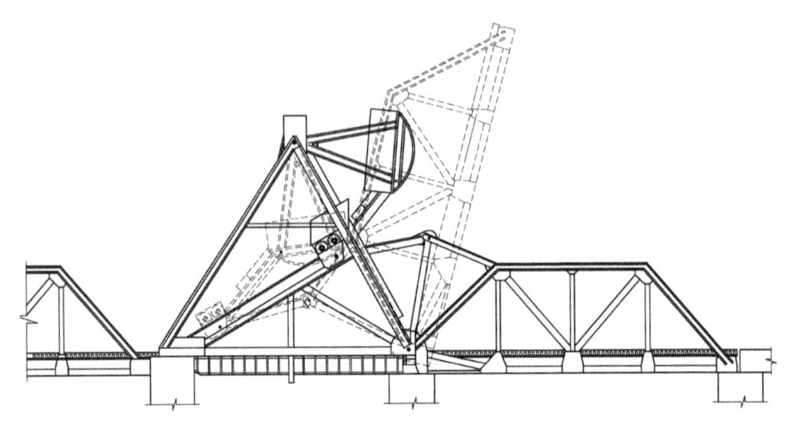

Figure 8-1 Abt Bascule Bridge Construction. This type of movable bridge was developed to minimize disruption of traffic during bridge replacement.

(a)

(b)

(c)

Figure 8-2 (a)–(c) Abt Bascule Bridge in Operation. This is one of the more unusual types of movable bridges still in fairly common use. There are several in existence.

built. One that is still in heavy use today is a two-track through-truss railroad bridge over the River Rouge near Detroit, Michigan. There is also a single-track through-truss version over the Manitowoc River at Manitowoc, Wisconsin. There are possibly one or two others extant, such as the Southern Pacific Railroad bridge in Beaumont, Texas. The Abt bridge counterweight moves in the opposite direction to a normal bascule counterweight and does not foul the approach when in the open position as a bascule with an overhead counterweight usually does.

REPLACING SUPERSTRUCTURE ONLY—USE SUPERSTRUCTURE OF SAME TYPE, SIMILAR TO EXISTING

When it is deemed justifiable to replace a movable span that does not have extensive overall deterioration, yet has defects that make span replacement economically feasible, it may be prudent to perform an in-kind replacement, changing as few basic dimensions of the existing bridge as possible. This can allow the piers and abutments of the bridge to be reused without change, which is a definite plus in this age of environmental considerations.

For highway bridges, this approach is seldom a viable option, as older bridges are usually functionally obsolete, with insufficient roadway widths or clearances, which require the existing bridge to be replaced with a wider span. An exception is the Carter Road Bridge, in Cleveland, Ohio, which had sufficient traffic capacity. The lift span was severely corroded because of advanced age in a harmful environment over the Cuyahoga River (so polluted that it had been known to catch fire in the area). Most of the damage to the bridge appeared to be due to salt. Winters in that area can be severe, and salt was (and is) plentiful locally for use on roads, being available from the salt mines on the Lake Erie waterfront. Some of the salt causing damage at this and other bridges in the area may, in fact, have been spillage from trucks hauling salt from the mines. The new structure consisted of the vertical lift span itself, sized identically to the original bridge, and moving parts and other machinery required to operate the bridge. The existing foundations and main towers were reused, because they were in good condition, at considerable savings in construction time and cost. The small approach spans were completely replaced, as they had suffered damage similar to that at the vertical lift span.

REPLACING ELECTRICAL OR MECHANICAL EQUIPMENT ONLY—DO A CLEAN SWEEP

In rehabilitating a deteriorated older movable bridge, it is generally much simpler, quicker, and perhaps cheaper to replace all electrical and mechanical drive equipment in one operation. If the bridge is to be taken out of service for the rehabilitation, the closure is more efficient because two largely independent crews can work on the bridge at the same time. If the bridge is to remain in service, a temporary operating system can usually take the place of the electrical and mechanical machinery while these are being replaced. This procedure usually involves fewer delays to traffic than trying to replace the components piecemeal.

Electrical equipment is usually not difficult to replace completely; it may, in fact, be necessary or desirable. Such equipment becomes obsolete quickly, and in-kind replacements for a bridge's electrical equipment may be unavailable after the bridge has been in service for 30 years, or even less. Age deterioration of electrical equipment may not be obvious, so the replacement of 100 percent of the electrical equipment may be easier than an attempt to coordinate a new subsystem with a basic arrangement that is 30 or more years old.

Much mechanical equipment in a movable bridge drive is extremely durable. Unless gears, shafts, and bearings are routinely overloaded, poorly lubricated, or poorly protected from the weather, they should last a very long time even on a bridge that operates frequently. It may be feasible to replace only selected components of an open gear drive system, but it is usually better to take the opportunity to provide an enclosed gear reducer drive system and thus reduce maintenance requirements for the bridge. Such subsystems as span lock drives and end lifts can also be replaced, in many instances, in such a way that high-maintenance items such as open gearing can be eliminated. The primary difficulty with general replacement of bridge drive machinery is in locating the optimum point of interface between new equipment and that which is to remain. Fitting new machinery to existing machinery or mounting new machinery onto existing steelwork or piers can be difficult.

II

DESIGN

References to specific sections of American Association of State Highway and Transportation Officials (AASHTO) or American Railway Engineering and Maintenance-of-Way Association (AREMA) publications, unless otherwise identified, are attributable to *Standard Specifications for Movable Highway Bridges* (1988) and *Manual for Railway Engineering,* Chapter 15, "Steel Structures" (2000), respectively.

In addition to the aforementioned standard references for movable bridge design, other current works on movable bridges are *AASHTO's LRFD Movable Highway Bridge Design Specifications* (2000) (2000 LRFD AASHTO) and the *AASHTO Movable Bridge Inspection, Evaluation, and Maintenance Manual* (1998). The value of both of these volumes is compromised by extensive commentaries, which may or may not contradict the text. The commentaries frequently add detail to statements in the texts, but it is unclear whether statements in the commentaries are mandatory, in spite of the occasional use of the word *shall.* The 2000 LRFD AASHTO is also hampered by incomplete editing, such that it has many errors and ambiguously worded statements. The 1988 *Standard Specifications for Movable Highway Bridges,* which was largely a revision of the previous edition of the same manual, has very few errors. It did have to be reissued in a revised version shortly after publication to eliminate some mistakes. Supplements were issued later to cover still other uncaught errors. AREMA's Chapter 15, "Steel Structures," which undergoes a constant review and revision process, is the clearest and most concise design specification, although it occasionally gains a questionable entry that may be deleted or revised a few years later. The standing AREMA committee responsible for the content of Chapter 15, "Steel Structures," has fought a continuing, and thus far successful, battle with the AREMA editing and publishing staff to be allowed to review the to-print copy of the manual and has minimized the number of typographical errors in that chapter of the published manual.

2000 LRFD AASHTO is a metric or ISO document. There is very little electrical or mechanical hardware manufactured in the United States for movable bridge use that is detailed in metric or SI (Système International, French for International System) dimensions, even though most units for electrical quantities are metric and the United States has been officially metric since 1867! The dimensions of most structural steel shapes available in the United States have been converted to metric by the expedient publishing of tables of properties of the various existing shapes in metric units, so at least that aspect of movable bridge design can be performed in metric units without great difficulty. A few states have issued proclamations that all bridge design must use metric units, but in specific cases they can usually be convinced that the machinery should be designed in traditional U.S. units. Many states that initially converted to metric under pressure from the Federal Highway Administration (FHWA) have since returned to traditional units, and others will probably do so eventually. There is presently no tangible benefit to building an American bridge in metric units, but for precision components there are many disadvantages. Proper tolerances are not available initially and were not developed until after many entities returned to traditional units. In addition to the unavailability of some components in metric units, there are the constant errors by people who fail to properly convert from one system to the other. Some of these errors are caught at little or no cost, and some are not. LRFD (Load and Resistance Factor Design) is a relatively recent phenomenon, presumably intended to provide for the design of more efficient structures by allowing designers to select factors for capacities of members rather than adhere to a rigid list of allowable stresses. This has resulted in design errors from having peoples unfamiliar with this process select the factors as well as errors in various LRFD manuals that were not recognized until after publication.

9

PRELIMINARY ASPECTS

There are several general principles that apply to the design of movable bridges. Movable bridges put bridge designers in the position of having to consider many points that are not relevant to the design of fixed bridges. Movable bridges are, in reality, large, complex machines and must be considered as such while still made to perform their primary function, which is to carry land traffic over waterways. Movable bridges may experience fatigue in their structural steel members, due to large variations in stresses, without accompanying live load, as a result of their movements when opening and closing. A few key points must be kept in mind:

1. A movable bridge should conform as nearly as practical to a fixed span when in the closed position and carrying traffic. It is impossible to achieve complete fixity, in any practical way, for many of the common types of movable bridges.

2. The structural and machinery parts should be independent of each other. Accomplishing this to a significant degree of completeness is impossible with most common types of movable bridges, although some machinery components can be isolated from the traffic-load-carrying function.

3. The operating machinery should be simple, easily installed, and easily maintained. The terms *simple* and *easily* are, unfortunately, open to interpretation. The skilled operating, maintenance, and construction personnel needed to make these terms apply are in short supply, so the typical movable bridge will not be sufficiently simple to build, maintain, or operate, to avoid difficulties and installation and maintenance will not be easy.

4. The control apparatus should be conveniently arranged so that the bridge operator can observe roadway traffic, the bridge movements, and navigation, and keep an eye on the moving span, particularly the ends, while maintaining physical and visual contact with the bridge controls. This is not easy to achieve and usually is not achieved.

On many movable bridges, no attempt has been made to provide for the operator to view bridge or waterway traffic. Some railroad movable bridges provide almost no such visibility.

Different common types of movable bridges adhere to the preceding principles to varying degrees. When these principles are, of necessity, violated, care should be taken to minimize the likelihood of future difficulties when maintenance, operation, and inspection tasks may not be ideally performed. Any modification to an existing system should be a fail-safe measure, so that in the worst circumstances the system will be no worse than it was without the modification.

SITE EVALUATION

The general location of a bridge is determined by traffic, but there may be wide variations within the general area, ranging from spots that are ideal for a movable bridge to locations that will be more expensive for construction and cost more in maintenance or other difficulties later. Careful evaluation of the specific site may result in considerable savings in construction cost, by reducing substructure costs or avoiding the need for extensive collision protection.

Find Narrowest Waterway with Highest Banks

The worst location for a movable bridge is the middle of a low-level causeway over a wide estuary, at a low elevation above the water, with many large vessels passing through the draw opening. Proper site evaluation would show that a movable bridge in such a location is especially susceptible to damage from being hit by vessels, is difficult to maintain, expensive to operate, and may be the cause of major traffic problems if put out of service. At this type of location, a practical solution may be elimination of the movable bridge by means of a long, high level fixed bridge or an underwater tunnel linking low level, long-distance approach spans, such as the Chesapeake Bay Bridge-Tunnel between the Norfolk, Virginia, area and the Eastern Shore (see Part I).

A long, low-level causeway with one or more movable spans is highly undesirable at such locations unless the waterway traffic is light, particularly in regard to tonnage. Many movable bridges in situations of this type, such as the New York & Long Branch Railroad bridge over Raritan Bay at Perth Amboy, New Jersey, a large swing span mounted only a few feet over the water, have received several hard hits from vessels such as barges and lighters (Figure 9-1). These vessels typically have insufficient impact power to completely destroy a bridge, so it would be necessary to construct a new span that could be designed to be better suited to its location. But these vessels are heavy enough that they more often inflict sufficient damage, particularly to movable bridge machinery and swing span pivot bearings and truss lower chords, that a bridge must be repaired repeatedly at considerable expense and inconvenience. Some of the bridges that fit this description have been replaced as "hazards to navigation," with the new bridges having larger navigation clearances (see page 117).

Figure 9-1 Raritan Bay. This location is typical of many in placing a movable bridge in danger of being struck by vessels, particularly if high winds are prevalent.

Avoid Locations Where Waterway Bends

Bends in a waterway, such as in a river or a tidal area, require careful handling by mariners. Many such areas have heavy barge traffic, particularly on the Mississippi River and its navigable tributaries, such as the Illinois River. On these waterways multiple barge tows of several thousand tons may be negotiating narrow navigation channels. These tows have been known to use the movable bridge piers as pivot points to make their turns in the river, so bridges that must be built in such locations should be particularly strong or protected by substantial fender and dolphin systems. The U.S. Coast Guard has revised its regulations in an attempt to prevent damage to bridge piers caused by heavy barge or vessel traffic using bridge piers as fulcrums to accommodate turns. Even so, there is the possibility of serious damage to a bridge by errant vessels. Even a fairly distant change in the course of a waterway can cause difficulties in navigating through the draw opening. This is an especially difficult and dangerous situation if the bridge passes over a river with strong spring flood currents. Even small pleasure craft can have difficulty in negotiating a sharp bend in a channel at the immediate vicinity of an obstruction such as a movable bridge. A large, heavy barge tow can easily destroy a weak or inadequately protected bridge. If it is necessary to position a bridge at such a location, the pier and protection system should not only be substantial, but it should be kept in good repair to prevent its being snagged by passing vessels. Site evaluation may be unable to avoid these situations, but it should be able to minimize their detrimental affects.

Avoid Highway Movable Bridge Locations with Adjacent Streets Parallel to Waterway

Site evaluation should always include the adjacent roadway network. Where there are cross streets on a roadway with a nearby draw span, general traffic tieups can develop if road traffic is heavy when the bridge has to be opened for navigation. Coordinated traffic signals can help prevent blockages of cross traffic, but only to a limited extent. The better solution is to have grade-separated approaches to the bridge. If possible, a movable bridge should have its roadway elevated, as this will accommodate an increase in the clearance for navigation under the closed bridge and reduce the frequency of required openings and the resultant traffic disruptions. In some cases, grade separation at the approaches may be impossible or extremely difficult. In these situations, if there is heavy highway traffic, extra effort should be made to minimize traffic problems at the approaches by providing coordinated signals, additional lanes, bypass routes, or other relief measures (see Figure 9-2). In severe cases, it may be possible to restrict bridge openings to off-peak highway traffic hours.

TRAFFIC STUDIES—HIGHWAY

Assume Maximum Future Traffic Projections

Movable bridges are substantially more difficult to widen than fixed bridges. In fact, some types, such as swing bridges, cannot be widened or have an adjacent parallel span added to increase capacity unless the two spans can be widely separated or it is acceptable to reduce the width of the navigation channel. The cost of machinery and control equipment increases at a lesser rate than the physical size of the bridge, so that some additional economies of size are available for movable bridges as opposed to fixed bridges. For instance, the electrical system for a six-lane movable bridge, all other things being equal, is a major item in overall cost, but is about the same price as for a four-lane movable bridge. Superstructure and substructure costs are largely a function of the length of the span and the width of the roadway, for both a fixed and a movable bridge.

Assume Truck Traffic Will Exceed Present Allowable Loads

There is a consistent long-term history of increases in the legal maximum vehicle load and axle load on the nation's highways. The truck lobby is strong, and in spite of phenomenal growth in railroad traffic in recent years, the trucking industry is an extremely important part of the country's economy. Truckers compete better against railroads by increasing the maximum loads they can carry, so there is constant pressure for increases in allowable truck weight and size. Conversely, because railroads do not have legal limits on the weight of their vehicles, but are limited by the ca-

Figure 9-2 Culvert Street Bridge, Phoenix, New York. This is one of the few "European"-type heel trunnion bascule bridges constructed recently in the United States. It has a few unique features, including a multigirder redundant superstructure, isolation of live load from the operating machinery, and machinery mounted on the counterweight. The warning gates are adjacent to the cross-street intersection.

pacity of the equipment, highway trailers and containers that ride "piggyback" on rail cars tend to be very heavily loaded. In areas near rail yards where these trailers and containers are transloaded between rail and highway, the average truck on the road may be grossly overweight. Local highway bridges can thus be overstressed on a regular basis. When a movable bridge is located on a roadway with considerable truck traffic, the bridge should be designed with substantial excess capacity.

AASHTO, in its *Standard Specifications for Highway Bridges,* specifies HS-20 loading. This results in designing for five-axle 80,000-lb trucks. The costs of counterweights, supports, and machinery all increase with the weight of the movable span, so that there is a larger increase in the overall cost for a movable bridge under these conditions, compared to lower weight limits, than for a fixed bridge. In spite of the increased cost for movable bridges, which are more difficult to replace than highway bridges and more complicated to repair, extra strength beyond that specified by AASHTO can be beneficial in the long run. As evidence shows that many large trucks are substantially overloaded, the more conservative approach appears to be prudent. Some states, such as Michigan, have no maximum vehicle weight but specify only maximum weight on axles. The result is very heavy vehicles that place a large concentrated load on a movable bridge. There has been pressure to allow similar "truck trains" on all the nation's highways.

TRAFFIC STUDIES—RAILROAD

Railroad traffic, being controlled to a greater degree than highway traffic, can tolerate a restriction due to a smaller number of tracks at a bridge than on the rail lines leading to the bridge. Many two-track rail lines have single-track bridges at major rail crossings (see Figure 3-6). This is particularly true of freight traffic. Passengers are less willing to be delayed because of such restrictions than is freight, so railroads carrying passenger traffic should be more concerned about future growth in traffic when designing a new bridge. There is also the danger of a train failing to heed signals at a point where the number of main tracks is reduced. This apparently is what caused the fatal rail collision in Maryland in 1987, on the approach to a bridge on Amtrak's Northeast Corridor.

Long-term projections of traffic levels are extremely difficult, and not very accurate, so it makes little economic sense to commit a substantial amount of present capital to satisfy conjectural future needs. For this reason, it is best to provide railroad movable bridges that can be expanded in the future as needed.

Bascule and vertical lift bridges are the only common movable bridge types to which more tracks can be added by building additional adjacent bridges, or by having the bridge itself widened, without reducing the width of the navigation channel. These types should be used on railroads that anticipate traffic increases. This practice was followed on many previous installations; in fact some "patent" bascule bridge advertising of the early twentieth century extolled this advantage. Many bridge piers and abutments on Amtrak's Northeast Corridor between Boston and New York were constructed to allow additional bridge spans to be installed. At some locations the second bridge span has been installed, and at some it has not. Many of the abutments and piers on the line between New Haven and Providence have vacant space for an additional two-track span. None of these, even after 80 years or more, have had the additional bridges constructed. A bascule bridge in Ohio was built for two tracks, expanded to carry four tracks, and now only carries two active tracks.

In spite of great increases in the tonnage carried by railroads, and in some locations the increases in passenger traffic, advances in other areas of technology have so far generally allowed the increased traffic to be carried on most rail lines without the need to add more tracks. Technological improvements, such as in signaling and communications, that allowed rail traffic increases without the installation of more tracks, may not continue in the future. Most projections indicate substantial increases in future railroad traffic, so the ability to increase the number of tracks should be kept in mind in building new bridges. The Association of American Railroads (AAR) has recently increased its maximum loading for four-axle freight cars. The AAR has found that this higher loading does not substantially increase the rate of deterioration of well-maintained heavy-duty track. Yet the increase in maximum load may require the strengthening of many railroad bridges across the country, both fixed and movable, as very few meet the AREMA standard E-80 loading criterion. Fatigue of bridge materials becomes an important factor when loading and train frequency increase.

OPTIMIZATION OF LENGTH

Consider Future Channel Widening or Realignment

It is difficult to predict the future needs of inland marine navigation, as the decisions concerning investment in locks, dams, dredging, and harbors are largely political in nature. On waterways heavily used by commercial vessels, where a particular bridge site is an effective restriction to navigation, it should be assumed that there will be interest in improving the channel in this area. As part of the site evaluation, various responsible authorities should be consulted to determine the likelihood of pressure developing for channel changes before a final design is initiated for a new bridge, or even for significant rehabilitation of an existing bridge. It may be found that a plan for significant change at the location in question, including work on nearby structures upstream or downstream, is in the offing and that outside funding will be available for a new span. Studies of future navigation needs should be performed before committing to a detailed design for a new movable bridge. One railroad bridge that had deteriorated was replaced with a new one of identical type and size, but 20 years later the U.S. Coast Guard decided the new bridge was a "hazard to navigation," and it is now to be replaced with a longer span.

A longer movable span, or one in a different location on the waterway, may be a significant and desirable improvement to navigation from the standpoint of the marine interests. As such, the cost of a bridge replacement, less betterments that provide bridge improvements from the bridge owner's point of view, is rightly borne by the federal government, which would pay the cost under the Truman-Hobbs Act when the proper documentary and other requirements have been satisfied (see page 117).

Additional span length usually makes a project more expensive, although environmental and other considerations may make fills and/or retaining walls undesirable. Movable highway bridges frequently develop creep problems, which would shorten span length, as the approach spans move toward the movable span. Many movable highway bridges have developed "growth" problems so severe that annual trimming must be done at the ends of the moving leaf or leaves to avoid structural interference in operation during hot weather. Minimizing the overall length of the superstructure, including approach spans, and building very substantial abutments can relieve this problem.

Insufficient research has been done on abutment movements to determine completely the cause and economical solutions of bridge "growth." Many bridges are built with minimal study of the soil conditions prior to construction. Such problems that are confronted during construction are sometimes dealt with in the most expeditious, rather than the most durable, manner. Filling of roadway expansion joints, so that the roadway itself produces expansion forces against the movable span, is likely to be a contributory factor. Unstable piers provide less resistance to longitudinal forces developed at the approach spans. Unstable abutments can also easily contribute to the problem. The difficulty appears to be more common with highway movable bridges than with railroad movable bridges, lending credence to the expansion joint theory.

There are many cases of railroad bridges with multiple approach spans, built on soil known to be unstable, that have survived for nearly 100 years with no expansion or "growth" difficulties.

OPTIMIZATION OF HEIGHT

A movable bridge should avoid openings as much as practical. Some railroad bridges can be built with some gradient at each approach to maximize the clearance under the navigation span. Diesel locomotives of today can pull longer trains up moderate grades that would cause the steam locomotives of yesteryear to stall. Vessels are required to lower any appurtenances such as masts when it can be easily done, but a few more feet of clearance will allow many smaller vessels, even some barge tows, to pass under a bridge without requiring an opening. Studies can be made to determine the actual clearance required for traffic under a bridge. Some movable bridges were built to clear commercial sailing vessels that no longer exist. At other locations, pleasure sailboats now pass through the draw. Many movable bridge replacements in recent years have been built at higher elevations than the bridges they replaced. Not only can this reduce the number of openings, but it also keeps more of the machinery and structural steelwork away from the water, thus reducing corrosion.

10

STRUCTURAL ISSUES

SUBSTRUCTURE

Soil conditions are very important in designing a movable bridge, and accommodation to them should not be left to chance. If subsoil conditions are not precisely known when a bridge construction project is put out to bid, provision should be made for adaptation of the substructure design to conditions found after construction begins.

There has been a great deal of effort expended since 1989 on seismic analysis and production of bridge designs compatible with the seismic conditions of a site. Movable bridges are more complex in this regard than fixed bridges, because of their large, heavy moving parts and the impossibility of fixing the moving part of the superstructure rigidly and permanently to the foundations. The problem may not be as severe as it sometimes appears to be, as movable bridges in areas of higher seismic activity have not generally shown any greater likelihood to be damaged in an earthquake than fixed bridges in those locations. Generally, movable bridges actually tend to suffer less damage than fixed bridges in earthquakes. This characteristic may be due to the extra care that goes into a movable bridge design because the bridge is movable. It may be also be due to a lack of real knowledge of earthquake phenomena, resulting in many fixed bridges being built with inadequate seismic resistance capability. It may be that the design of a movable bridge span, which permits it to move, allows it to absorb seismic shocks without damage.

To ensure the greatest strength and homogeneity in a movable bridge substructure, massive concrete placements should be avoided if possible. A hollow box structure, provided with suitable strengthening webs, allows the greatest strength of the concrete structure. This box can be ballasted with heavy fill, if necessary, except where space for the counterweight must be provided.

Bascule Bridges

Rolling lift type bascule bridges present special substructure design problems, because the entire dead load of each bascule leaf travels forward and backward on the pier during bridge operation. On a large bridge, this presents a shifting load of several thousand kips on the pier. For bridges founded on poor soil, it can present a serious stability problem; rolling lift bridges should not be used in such situations unless a means can be provided to ensure substructure stability.

Simple trunnion bascule bridges are frequently built at elevations that require the rear of the counterweight to be below the water level when the bridge is open. A suitable pit must then be formed in the bascule pier or abutment that will provide a dry space for the counterweight to be lowered into when the bridge is opened. Extra care must be taken in the design and construction of such structures to avoid the development of cracks, which can lead to leaks.

Wind loading must be taken into particular consideration in the substructure design for bascule bridges, especially long span single leaf bascules. These bridges may extend as much as 300 ft above water when open, where wind loads can be quite high. Heavy winds blowing parallel to the alignment of the bridge, combined with acceleration loads resulting from starting or stopping the bridge movement, can produce very large short-term loads on the foundations of a long single leaf bascule bridge. It pays to be conservative in estimating these loads so as to avoid shortening a bridge's life span.

Vertical Lift Bridges

Substructure design for a vertical lift bridge is not much more complicated than the design for a fixed bridge. It is important to consider the wind loading on the bridge in the open position, as a considerable overturning moment can be developed, particularly for truss-type bridges. It is also important to avoid settlement or other instability, as the height of the towers magnifies angular misalignment, which can result in severe operating problems. Many vertical lift bridges have interferences at the centering devices, span guides, or counterweight guides that are attributable to some out-of-plumbness of the towers.

Vertical lift bridges present the greatest potential risk for earthquake damage, because of the large mass of either the counterweight or the lift span itself in the raised position during a seismic event. In seismically active geographic areas, precautions must be taken; however, the evidence suggests that damage will not be as great as may be expected (see page 185).

Swing Bridges

The design of the substructure for a swing bridge must consider the effect on the center pier of the swing span superstructure supported on it without any connection to other piers. This can be an unwieldy arrangement. Round center piers for swing bridges have proved to be more durable than center piers of square cross section in plan. This probably has more to do with masonry mechanics, stream hydraulics, and ship impact than the direct effects of supporting the swing span itself. One bridge, the New York and Long Branch Railroad center pivot swing over the Raritan River at South Amboy, New Jersey, has had center pier difficulties throughout its existence. The top of the pivot pier, which is square in plan, is just barely at the water level, in the center of a navigation channel that sees a considerable amount of commercial traffic. The river is wide at this point, and the weather is usually windy. The bridge has been struck several times, and the pivot pier shows significant deterioration, including uneven settlement and less than ideal support of the center pivot as well as the circular track supporting the balance wheels.

Center piers for rim bearing swing bridges must be of more substantial construction that those for center bearing swing bridges. The rim bearing places all dead and live loads at the edge of the pier, but must still have a substantial center post at the middle of the pier to help resist horizontal loads and keep the bridge centered. A center pivot swing bridge concentrates all major vertical and horizontal loads at the center of the pier. The pivot pier of a center bearing swing bridge must carry dead loads at the center of the pier and need resist only live loads and other relatively minor vertical loads, due to wind or imbalance reactions at the balance wheels, at the edge of the pier, as well as horizontal operating loads at the rack. For both types of swing bridges, the bridge rotational drive forces are relatively minor as compared with the primary support loading. Maximum structural wind loads are less severe for typical swing bridges than for bascule or vertical lift bridges.

FENDERS

Any bridge, movable or not, that spans a navigable waterway that has traffic in large vessels needs protection from damage as a result of being struck by vessels. Rolling lift bridges and vertical lift bridges may need only a strong pier structure. Most swing bridges and some bascule bridges require substantial crib or timber fender systems to keep vessels from reaching the superstructure of the open bridge (see Figure 10-1).

Most movable bridges provide a narrower opening at the navigation channel than a fixed bridge at the same location could easily provide. In the past bridge owners and designers have usually attempted to provide the smallest possible movable bridge that the War Department, the U.S. Army Corps of Engineers, or the U.S. Coast Guard would allow to span the navigation channel. The primary reason was to avoid the great increase in cost: as movable bridges increase in size, machinery must be bigger to carry the heavier superstructure load; larger, more expensive bearings must be provided; and heavier power supply apparatus is needed to meet the much greater power requirements of a larger bridge—all resulting in higher total construction cost. Today, with a greater awareness of the potential for ship impacts, and with the larger and heavier commercial marine vessels, survivability in an impact is of greater concern. Some movable bridges can be built so that it is almost impossible for them to be hit, by keeping the opened structure a fair distance away from the navigation channel. When this cannot be done, as is in most instances, sufficient protection must be pro-

Figure 10-1 Bascule Bridge Remains after Collision with Ship. This bridge was not fully protected from impact as a ship passed in the narrow channel. The bascule span was completely torn free of its supports.

vided, perhaps in the form of dolphins, in addition to fenders. Extra functional sheathing, such as resilient rubber bumpers, for the piers themselves may also be justified. This is particularly relevant where the narrowness of the space between the piers does not allow for large, impact-absorbing fender systems. The fender system should be configured to guide larger vessels so that they cannot make direct impacts on the piers, as it requires much lower maximum forces to deflect a vessel along its path than to resist a direct impact.

Ideally, a bridge and its supporting piers should be placed out of the reach of any vessel navigating the waterway. This is not always possible, as a structure cannot always be provided that is high enough and wide enough to eliminate the possibility of impact by any vessel on the waterway. Even some supposedly "high level" bridges are occasionally struck by vessels. The Brooklyn Bridge, in spite of its having been designed and built with adequate clearance per all responsible authorities, has taken a hit at its superstructure now and then.

There are two ways to provide active protection for a bridge and its piers. One is to make the superstructure and substructure so strong and massive that no vessel that can navigate the waterway can hit the structure hard enough to cause serious damage. The other solution is to provide protection that absorbs or deflects the kinetic energy of a moving vessel that comes in contact with the structure, and dissipates this energy without damage to the bridge's superstructure or substructure. The Federal Highway

Administration (FHWA), following the Sunshine Skyway disaster, has prepared guidelines for ship impact studies.[1] Following these guidelines will allow a bridge to be protected from any reasonably likely ship impact.

Timber fender systems should be constructed of pressure-treated or otherwise protected timber to maximize life. Corrosion-resistant hardware should be used throughout the fender system, particularly in saltwater areas. Galvanizing does not add appreciably to the life of steel components, particularly in a saline environment, so stainless steel or bronze bolts and fittings should be considered.

SUPERSTRUCTURE

Bascule Bridges

Bascule bridge superstructures rotate from a horizontal, or nearly horizontal, plane in the closed position to an approximately vertical plane when open. The force of gravity causes a bascule bridge to sag between its main support girders or trusses. The amount of sag depends on the stiffness of the superstructure, which is greater in the plane of the superstructure than perpendicular to it. The deflection of the bridge due to gravity accordingly decreases as the bridge opens and increases as it closes. This change in deflection can cause difficulties with trunnion bearings or rolling lift treads unless it can be accommodated or reduced to miniscule proportions. Some bascule bridges have been built that deflected more than their supports could accommodate, so that difficulties have arisen at the supports. A bascule bridge should be designed so that, at its supports, it is nearly as rigid in a vertical plane across the bridge as it is in a horizontal plane. For a simple trunnion bascule, the massive counterweight at the rear of the bascule girders, combined with a heavy, deep floorbeam in the area of the trunnions, usually suffices to provide more than adequate stiffness. The same applies to a deck-type rolling lift bascule. Other bascule configurations, such as through-truss heel trunnion bascules, which may not provide such rigidity in the superstructure, should have main trunnion bearings or supports that will not be damaged by changes in deflection from the open to the closed position and vice versa. Spherical friction main trunnion bearings that are free to oscillate laterally as the bridge opens and closes can resolve this problem in cases where these deflections are large. 2000 LRFD AASHTO (6.8.1.3.2) explicitly suggests this. AASHTO (2.6.4) and AREMA (6.5.11) state that antifriction bearings mounted in separate pillow blocks shall be self aligning. Spherical antifriction bearings, if used, must be adequately sized to accommodate the transverse rotation due to superstructure deflection.

Simple trunnion single leaf bascule bridges act as cantilevers for dead load, in all positions, because they are balanced at the trunnion axis. Live load is carried on these bridges as on a simple span. Stresses at the top chord of the truss, or at the top flange of the girder, are thus tensile, and bottom chord or flange stresses are compressive with the bridge lowered and not carrying live load. Live load superimposes compres-

[1]Federal Highway Administration, *Bridge Ship Impact Study Manual* (Washington, D.C.: FHWA, 1993).

sive stresses in the top members and tensile stresses in the lower ones. A single leaf trunnion bascule that is "efficiently" designed and has a high ratio of live load to dead load can thus be susceptible to structural fatigue under live load without operating, because of the large change in stress resulting from application of live load.

Heel trunnion bascules provide balancing of the span (see page 38) through a connection of a tensile link to the leaf at some point forward of the main trunnions. If this link is placed far enough forward, the bascule leaf acts as a simple span for dead load (see Figure 10-2). If the link connection to the bascule leaf is at midpoint on the span or thereabouts, the bascule leaf acts as a two-span continuous bridge for live load, with both spans cantilevered for dead load (Figure 10-3). The maximum stress with dead and live load for a bascule leaf acting as a simple span for dead load is usually greater than for a simple trunnion bascule, but no stress reversal under live load occurs in the lowered position.

Most double leaf bascule bridges are cantilevers for dead load and for most live load conditions. The center connections between the leaves are usually designed to carry shear but no moment. On a double leaf bascule bridge the two bascule leaves are usually cantilevered out from the piers, and in the interest of economy of construction these leaves are made as light as possible. The bascule girders or trusses are usually made of a tapering section so that the truss or girder has maximum depth at the pivot point, which is the location of maximum moment, for strength. Depth is a minimum at the toes of the leaves, maximizing navigation clearance under the span when it is closed. This configuration results in leaves that can deflect visibly at their tips under live load. For this reason, the toes of the two mating leaves must be tied together in shear to make the live load deflection uniform, so that vehicles do not experience a vertical dislocation as they pass from one leaf to the other. On a leaf carrying live load while its mating leaf is unloaded, some stress reversal will occur, but the bascule girders should be designed so that the reversal is small so as to avoid fa-

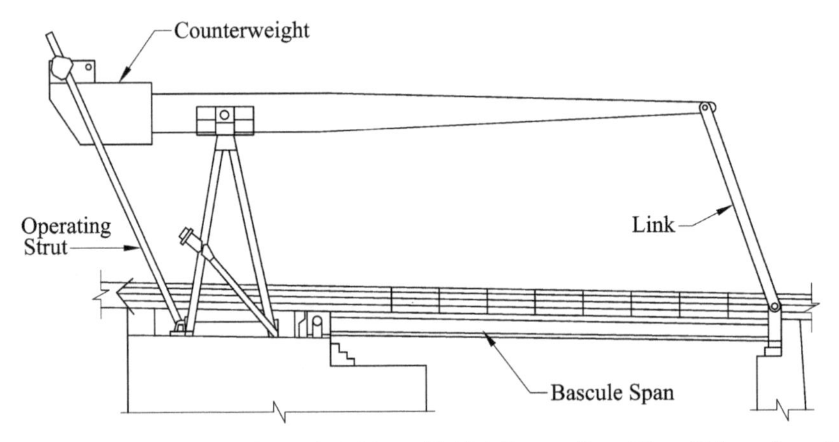

Figure 10-2 Heel Trunnion Bascule Bridge with Link Connection at Toe. This configuration allows the bascule leaf to act as a simple span for dead and live loads. It also eliminates live load stresses in the operating machinery.

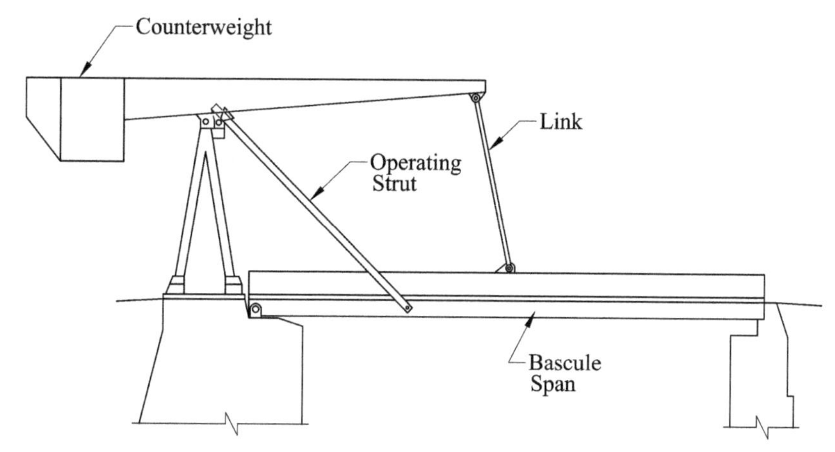

Figure 10-3 Heel Trunnion Bascule Bridge with Centered Link. This is a typical configuration for this type of bridge. The bascule span is cantilevered. Live load deflection of the span causes motion of the counterweight boom structure and of the operating machinery.

tigue problems. A double leaf bascule bridge with very high structural stiffness would not require center shear locks, as the live load deflection of the cantilevers is negligible. About the only economical way to obtain such a structure is to build a concrete bascule span, similar to the concrete swing bridges built in Washington State over the last few years (see page 142).

Many bascule bridges are designed to open to a very wide angle to provide a clear navigation channel while minimizing the size of the span and the width of the fender system. The structural members of a bascule span undergo a significant change in dead load stress as the bridge is opened. These changes are a direct function of the opening angle and can become radical if the opening angle is large. Such stress changes have resulted in fatigue failure of structural members of some older heel trunnion type bridges, particularly those designed for extreme economy of construction. An opening angle for a single leaf heel trunnion bascule bridge in excess of 80° or 85°, which is not uncommon, allowed it to be competitive in first cost with a Scherzer rolling lift bridge, which may open only 70° to 75° in the same application. A large opening angle for one of these bridges results in almost complete stress reversal in some of the main bridge members, and in the counterweight truss members, because of the complex load changes in the link connections, support members, and operating struts. Load at the main trunnions can become negative for some heel trunnion bascule bridges, resulting in uplift at the bearings. The main trunnion bearing caps on such bridges are usually mounted at an angle to reduce the load on the cap bolts.

The effect of wind loading on bascule bridges in the open position must be carefully analyzed, as this can produce the highest load case and stresses not only for the superstructure but also for the substructure, as well as for the machinery. The wind can blow in any direction and can increase the range of reversal of stresses in the main structural members. AASHTO (2.1.13) and AREMA (6.3.5) permit increases in allowable stress levels for wind loading on an open leaf; these can in some cases eliminate wind loading as the determining load case for structural design of the bascule leaf.

The deck of a bascule leaf must be sufficiently anchored to prevent disengagement due to wind loading with the leaf open. This is especially of concern with railroad bascules that open to very steep angles, particularly those with older, deteriorated decks.

The term *simple trunnion* suggests that this type of bascule bridge is the least complicated. This may be true in a mechanical sense, but simple trunnion bascules often pose taxing problems in detailing the support for the trunnions, and in finding space for the counterweights. Even placement of the machinery and its supports can be difficult. These issues were the subject of several court cases in the early twentieth century, one of which involved a patent infringement suit by Joseph Strauss against the City of Chicago. Strauss won the case, but he never sold any bascule bridge designs to that city afterward—so much for the marketing value of patents.

Vertical Lift Bridges

A vertical lift movable span has the least difficult superstructure to design of any common movable bridge. Whether it is a girder or a truss span, it acts as a simple span for both dead and live load. The only difference between a vertical lift span and a fixed span is the means of support of the dead load. The entire dead weight of a fixed simple span rests on the bearings at each corner of the bridge span. These bearings also support the live load on the bridge. The reactions at the masonry plates for a properly adjusted vertical lift span consists of the live load and some very small dead load reaction. Most of the dead load is carried by the counterweight ropes, which transfer this load to the towers. Connection of the counterweight ropes to the lift span is fairly straightforward. For deck girder spans, an extension at each end of each end floorbeam can accommodate the rope connections. For a truss span, a vertical hanger arrangement in the form of a secondary portal is generally used at each end, with rope connections at its top. Vertical lift spans do not rotate as they open or close, and no change in their dead load support occurs. They thus do not encounter difficulties in the superstructure deflection changes that can plague bascule and swing bridges.

Vertical lift bridge towers are rather unusual structures in that they may be quite tall, up to 200 ft, but most of the mass effect of the structure is at the top of the towers. Regardless of the position of the bridge—open, closed, or in between—almost all of the dead weight of the movable span, plus the mass of the counterweight and supporting sheaves and ropes, is supported by the bearings at the tops of the towers and transmitted by the tower superstructure to the ground. Any span-heavy imbalance (see page 33) that may exist will result in some dead load taken by the live load shoes of the seated bridge. A span drive bridge can carry imbalance loads in the operating ropes when the bridge is seated as well as when it is operating.

Swing Bridges

Most swing bridges are truss-type spans, as swing bridges tend to be longer spans, but there are many short- and not-so-short-span plate girder swing bridges. There have

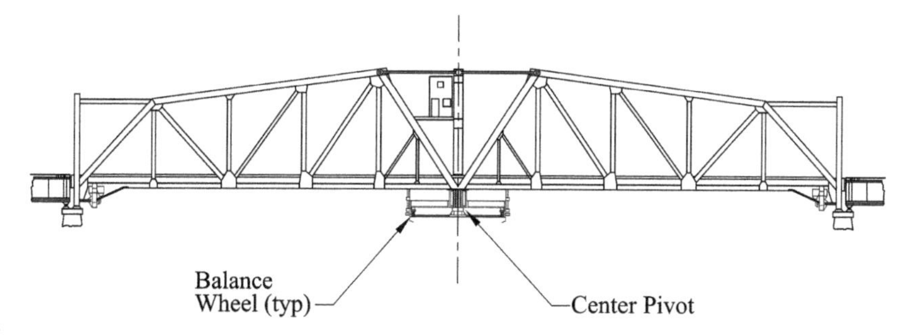

Figure 10-4 Swing Bridge Supported on Center Pivot. The swing span is supported at three points at each truss: each end and the center.

also been a few concrete segmental type swing bridges built (see page 94). Trusses are more economical of material than plate girders for longer spans, but higher fabrication costs for trusses require careful economic analysis before a decision is made as to the type of main members. Most swing bridges, whether they are of truss or girder construction, are symmetrical. These are double cantilevers for dead loads, particularly in the open position, but multiple span continuous bridges for live loads. Center bearing swing bridges are two span continuous for live load, as they invariably have a single pair of center wedges, one under each main member at the transverse centerline of the lift span for live load support at the center pier (see Figure 10-4). Rim girder swing bridges usually have an additional continuous span between the extremes of the center supports over the drum (see Figure 10-5).

The end lifts for swing bridges support not only the live load at each end of the bridge, but also carry some dead load so that uplift will not occur at the end of the span when the live load is only on the other side of the center pier. End wedges invariably become misadjusted on swing bridges after a period of time in service, unless competent inspectors and maintenance personnel keep careful watch over them, so a certain amount of overdesign of the superstructure is warranted to avoid damage from overstress and impact at the ends of the swing span.

Basic design analysis of the swing span superstructure must be performed for all conditions, as any one can prove to be critical. Center pivot bridges are by far the easiest to analyze and design, because of the simpler center support. Rim bearing swing bridges are difficult to design and construct. Equal distribution of loads to avoid large overstresses at the supports over the drum can be difficult. Analysis requires the rationalization of multiple redundancies. These are sufficient reasons to avoid specifying rim bearing swing bridges. In consideration of the difficulties of designing, detailing, fabricating, and installing the roller support arrangement, it is certainly sensible for AASHTO and AREMA to recommend that the rim bearing type swing bridge not be used.[2]

[2]American Railway Engineering and Maintenance-of-Way Association, Chapter 15 Commentary, (9.6.5.34.1), (Landover, Maryland: AREMA, 2000).

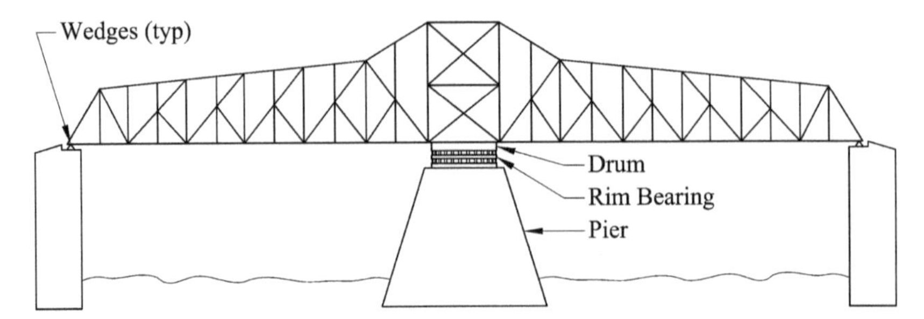

Figure 10-5 Swing Bridge Supported on Rim Bearing. Each truss is supported at four points: each end and at each side of the drum at the center.

Regardless of the type of main member or the type of support of a swing bridge, it is critical for the designer to be aware that stress reversals will occur at some points in certain of the main members, as a function of live load as well as the operation of the bridge. Although it is possible that bridge operation cycles will not achieve significant numbers for fatigue consideration purposes over the life of the bridge, live load cycles can easily do so. The critical locations for live load fatigue are most likely to be at and near the points along the main members that have no dead load stress when the end lifts are driven. Areas near these points cycle between compressive and tensile stress as live loads cross the span. The worst-case stress ranges for live load should be computed and an estimate made of the frequency of application of these loads. The members and connections should then be detailed to easily accommodate these stresses without the likelihood of suffering fatigue damage. The increase in cost of the structure will be small, as compared with the price of future repairs.

The few double swing bridges in existence span wide navigation channels and were selected for use because of overhead restraints or for other compelling reasons that required the use of this type of bridge instead of a single or double leaf bascule or a vertical lift span. The double swing bridge presents a stability problem, particularly under live loads, as there is no fixed structure available to prevent the closed bridge from collapsing into the navigation channel. Substantial restraints must be provided at each swing span, at the ends away from the navigation channel, to prevent the span from tilting into the channel when heavy live loads are on the center of the bridge, at or near the joint between the two swing spans. Some older bridges provided an arrangement so that the rear of each leaf, as it was swung into the closed position, came under an overhanging projection on its respective approach span.

The double swing bridges now in existence use different means to achieve stability. A double swing in Seattle, Washington, is constructed of concrete and is so heavy that the swing spans cannot be toppled from their pivot piers by live loads or any other expected loading. This bridge is extremely rigid, because of its massive concrete construction, so that it deflects negligibly under live loads, needs no end lifts to provide stability, and requires no center shear locks between the mating ends. This is truly a special-case bridge. Another double swing, built in 1997 to replace a narrower structure built in 1953 over the York River at Yorktown, Virginia, uses a more conventional approach to stabilize its two steel swing spans of deck-truss configuration. After the bridge is swung to the closed position, mechanical shear locks are driven to connect the rear ends of the swing spans with the approach spans, which are heavy enough to provide stability. Center locks, similar to those provided at double leaf bascule spans, are driven to connect the mating ends of the two swing spans in shear. Thus two two-span continuous bridges are linked, over four piers, with three suspended joints. The vertical shear connection to the approach spans is relied on to maintain the stability of the swing spans under live load.

11

DETAILING—BASCULE BRIDGES

Smaller simple trunnion bascule bridges can be supported on single-bearing trunnions mounted so that the bearings are outboard of the main bascule girders, with suitable bracing between the girders. A substantial moment connection is required for this type of trunnion. This is usually accomplished by placing a trunnion girder inboard of each main girder and attaching the inner end of the trunnion to it. A simple column is all that is necessary to support the trunnion bearing, and no structural interference need occur.

On larger simple trunnion bascules, double supported trunnions are frequently used, with a bearing inboard and outboard of each main girder or truss. This eliminates the need to resist the bending moment in the trunnion by means of a bracket or cross frame or truss, but finding a means of support for the inboard trunnion bearing may be difficult. Support of the bascule leaf is further complicated when a third or fourth truss or main girder is added to the bridge. If a simple column is used to support each bearing, the moving leaf and its counterweight must be detailed to clear this column in all positions of the bascule span. Usually, simple columns are not sufficient to stabilize the span, especially when opened, and various cross members, struts and braces must be added, all of which must still stay clear of all components of the moving leaf in every position it can take.

The machinery to operate the bridge usually is located near the counterweight area and must come in contact with the main bridge members in order to drive the leaf. The machinery is all that holds the bridge stable against wind loading in the open position, and so it can be forced to withstand very high loads. In a bascule pier or counterweight pit typical of the simple trunnion bascule type, proper support of the machinery can be difficult to achieve.

Most bascule bridges are equipped with some sort of bumper blocks that a part of the leaf will hit if it overtravels when opening. The bumper arrests movement of the bascule leaf and prevents interference elsewhere between the moving leaf and its supports or the pier.

TRUNNIONS

Simple trunnion bascule bridges, heel trunnion bascule bridges, and several other types of bascules require trunnions to act as hinge pins for the pivoting leaf and other components. Trunnions are a frequent source of difficulty, because they require skill in fabrication and installation and can degrade as a result of poor maintenance. The usual trunnion is a solid steel shaft, often an alloy steel forging. As a large piece of steel, sometimes more than 2 ft in diameter, it can have internal flaws. AASHTO (3.3.4) and AREMA (6.8.3) require trunnions or shafts more than 8 in. in diameter to be furnished with holes bored through along their longitudinal axes, as an attempt to detect flaws before further expensive fabrication and erection are performed. The trunnion for a simple trunnion bascule bridge is machined to fit precisely and very tightly into a collar or hub attached to a main bascule girder. The full dead load of the span is supported on the trunnion while the bridge is operated, and wind and other environmental forces must be resisted, as well as the bridge weight, at the trunnion-to-span connection without the assembly coming loose. There are three configurations of trunnions commonly used on bascule bridges. Large, heavy bascule bridges that may have moving leaves weighing in excess of 1000 tons usually have each trunnion supported on two bearings, one on each side of the bascule girder or truss (Figure 11-1). Smaller bascule bridges sometimes use a trunnion with only one bearing, on the outboard side. The inboard end of this trunnion is usually longer and tapered to a smaller diameter at its end, where it fits tightly into a separate girder parallel to the main girder, or into a diaphragm in a cross girder or beam extending between the main girders (Figures 11-2 and 11-3). Some bascule bridge trunnions are fixed, mounted on the pier, with the journal extending into a bearing mounted in the bascule girder (Figure 11-4).

When the bascule bridge is closed, most trunnions must support, and transfer to the bearings, the live load on the bridge, with impacts, as well as the dead load. The trunnion has to be precisely aligned with another trunnion mounted in another bascule girder for the same bridge leaf, or serious degradation of the bearings, and possibly the trunnions as well, will result. The axial holes bored through the trunnions are sometimes used to assist in aligning the trunnions when the bridge is erected.

Some bascule bridge leaves have multiple girders, with each girder having a trunnion that must be aligned with the others. It is very difficult to achieve perfect alignment with two trunnions per bascule leaf, and alignment becomes much more difficult with each additional trunnion added to a leaf. Static alignment can theoretically be attained over any number of points, but bascule bridges must rotate to as much as nearly 90° to open and close, and gravity and temperature differences can cause distortions of the leaf as it changes position. These changes can be semipermanent, as imperfect connections between major components can cause nonreversing distortions. A few attempts have been made at true multiple-girder simple trunnion bascule bridges, such as the State Street Bridge in Chicago, with three main girders per leaf, and the Mason Street Bridge, in Green Bay, Wisconsin, with four. These bridges have presented varying degrees of difficulty in erection and maintenance. Several more recent multiple-girder simple trunnion bascules have been designed and built as separate two-girder spans that have been tied together after erection of the leaves, in such

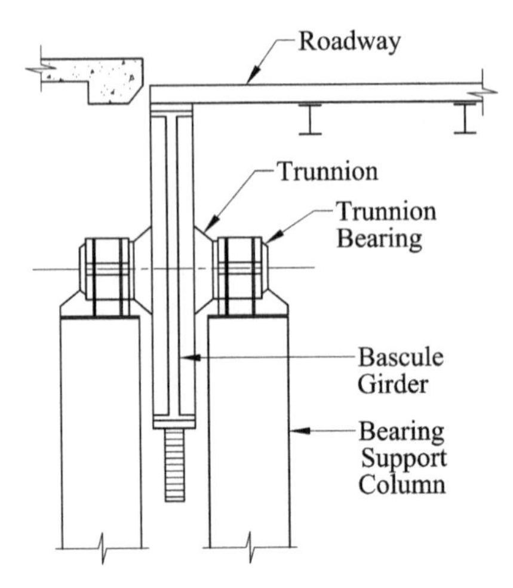

Figure 11-1 Two-Bearing Main Trunnion. This type of support is used on heavier trunnion bascule bridges.

a way that the roadways can be continuous across each "leaf," but the basic structures retain a limited degree of independence.

There are several trunnion and trunnion attachment details that reduce the amount of precision work necessary at bridge erection. A common method is to use a spherical housing for the trunnion bearing. This allows some misalignment of the bearing seat supporting the trunnion, without causing binding or point loading of the trunnion in the journal. Spherical roller bearings can be specified to eliminate not only the misaligned base from consideration, but also the misaligned trunnion. As the bridge opens, the trunnion can oscillate without binding, as the spherical roller bearing os-

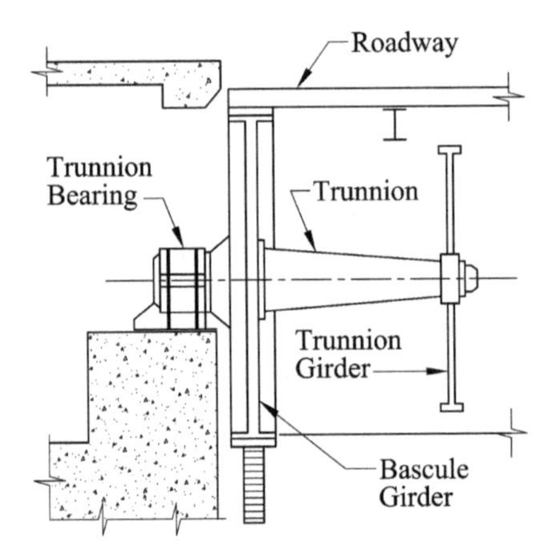

Figure 11-2 Single-Bearing Main Trunnion. This type of support is used on smaller trunnion bascule bridges. This example uses a trunnion girder to support the inboard end of the trunnion.

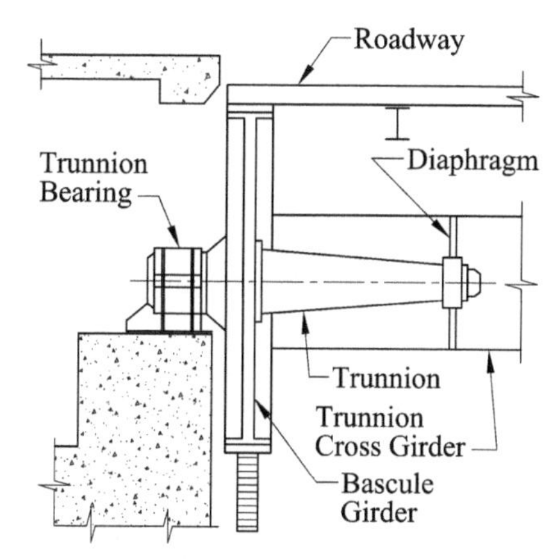

Figure 11-3 Single-Bearing Main Trunnion. This type of support is used on smaller trunnion bascule bridges. This example uses a cross girder to support the inboard end of the trunnion.

cillates within its fixed outer race to accommodate the movement of the trunnion. Some additional wear will be experienced at the roller bearing, but this will not be substantial, and the bearing can be specified to be of a large enough size to have sufficient reserve capacity to counteract the effect of this wear.

There has been no known fatigue failure of a bascule trunnion, but when vertical lift bridge sheave shafts began developing fatigue cracks in the late 1970s and early 1980s, AASHTO and AREMA lowered the allowable stress levels for all heavily loaded rotating shafts. Bascule trunnions are thus designed somewhat more conservatively than needed. This should not be of great concern, as the trunnions are a critical part of a bascule bridge and a slight increase in the size of the trunnions will not add a great deal to the cost of a bridge.

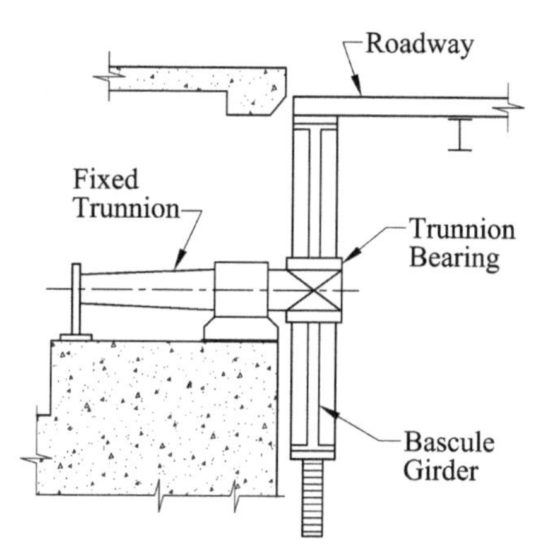

Figure 11-4 Fixed Trunnion. This type of trunnion bascule bridge support is unusual. It avoids cyclical fatigue stresses from bridge openings and closings.

Bascule bridge trunnions must be analyzed for the full range of stresses that they experience. All bascule bridge trunnions experience some stress variation as they open and close. The greater the opening angle, the closer the trunnion approaches full stress reversal. Fatigue is a real possibility for a poorly designed bascule bridge trunnion. The dead load stress variation depends on the angle of opening, but the number of cycles is limited to the number of openings. Live load stress variations can easily approach fatigue-sensitive numbers on a bridge carrying heavy traffic, and the stress range can be large on a small bridge with small trunnions (see page 138).

Most older bascule bridges have bronze-bushing type friction bearings for their trunnions. These are generally quite long lasting and serviceable. Such bearings tend to be rather large, and because they are always loaded, it is difficult to lubricate them fully. Moreover, because bascule bridge trunnions do not make full revolutions during operation, lubrication is generally difficult. Various configurations of lubricating grooves and channels have been used, with largely indifferent success. Yet even when the bearings have failed to receive maintenance and run dry of lubricant, they tend to continue to perform, although at the cost of much higher power requirements to operate the bridge and more rapid deterioration of the journal and bearing. Some of these bearings have a separate bushing in the lower half only. A solid cap is used in these cases, usually made of bronze, with a small clearance between the cap and the shaft journal. This arrangement, with the cap firmly bolted to the bearing base, assuredly prevents rotation of the bearing bushing in the base in the event that the bushing "freezes" to the journal. In such an event, the journal will break free from the bushing when the bascule leaf is rotated, and the bridge can continue to operate.

Many newer bascule bridges use antifriction bearings. These bearings eliminate wear at the journals, reduce maintenance requirements, lower the horsepower required to operate the bridge, and can help eliminate shaft fatigue failures. They can present a control problem, as operating loads are so low with antifriction bearings under normal conditions that there can be difficulty controlling the speed of the drive motors (see Chapter 18). The antifriction bearings reduce friction considerably, but there can be difficulty with galling or shelling of the bearings due to the bearings making only partial rotations. Bearing selection is critical; the bearing manufacturer should be consulted when specifying an antifriction trunnion bearing for a bascule bridge.

With rotating shafts supported in fixed bearings, lubrication is fairly simple, as the bottom of the bearing, where the lubricant tends to accumulate, is under load. With stationary shafts supporting the bridge on rotating bearings, lubrication is more difficult, as the top of the bearing is loaded. Extra caution must be taken with this configuration, and a larger antifriction bearing is usually required. It is also prudent to increase the size of a friction bearing used in this arrangement.

TREAD PLATES

Tread plates are truly the Achilles' heel of Scherzer rolling lift bridges. In the interest of construction economy in an age when materials were relatively expensive, thin plates, usually between 1 in. and 2 in. thick, were specified on most early bascule bridges built to the Scherzer patents. It was eventually found that these thin plates would fail in fatigue after a finite number of bridge operations. In some cases, the plates would fail without a great number of bridge operations, as a result of the fatigue effect of cyclical live loading such as on railroad bridges. It was also found that the deterioration of the tread plates, when left unchecked, would lead to fatigue cracking or other deterioration of the segmental girder connecting the bascule span to the curved tread plate. Fatigue-type cracking has also been found in the horizontal girders supporting the flat tread plates on which the rolling lift bascule moves back and forth during operation. A few multiple-main-girder rolling lift bascules have been built, such as the Pelham Parkway bridge in New York City. Unequal loading of the redundant tread plates can increase the likelihood of cracking.

It was eventually determined that the tread plate stresses could be reduced by increasing the thickness of the tread plates, making the plates of higher-strength steel, and increasing the operating radius of the tread. Some installations were made with very heavy, thick cast tread plates, in the hope that this would eliminate the problem, but most of these over a certain age have developed serious fatigue cracks as well. The castings are difficult to align and must be made up of segments, resulting in joints along the tread. These joints are frequently aligned imperfectly, resulting in high cyclical stresses when the bridge is operating and the formation of cracks in the material adjacent to the joints. The heavier tread plates are more expensive, negating to some extent the cost advantage of Scherzer-type bascule bridges. A minimum tread plate thickness of 3 in. is now specified by AASHTO (2.3.5) and AREMA (6.5.35.5c).

For these reasons, the Scherzer rolling lift bascule has been used less and less in new movable bridge construction over the last few decades, as longer life spans are being required for bridges from the outset, and it is understood that future tread plate replacement must be included in the life cycle cost of these bridges. When a large number of openings are expected over the life of the bridge, this type of bridge should be built only when its operational or load carrying advantages are considered essential. An example is the Scherzer single leaf bascule built by the Galveston, Houston & Henderson Railroad over the Intracoastal Waterway at Galveston, Texas, in 1919. Railroad traffic is heavy over the bridge, as it provides the only rail access to the port of Galveston. This bridge is at a low elevation, close to the water. Alternatives such as counterweight pits, a heel trunnion type bascule, and raising the bridge elevation were considered undesirable. This bridge had its tread plates replaced several times over its life, and when the bridge was completely replaced in the 1980s, the railroad specified that another Scherzer type bridge must be used.

To provide the best possible life for the tread plates of a rolling lift bridge, the following rules should be followed:

1. Use the highest possible strength material for the tread plates, curved and flat.
2. Avoid joints in the plates, welded or otherwise, for the full length of operating contact.
3. Use the thickest possible plate, within the limits of the steel manufacturing industry's capabilities of producing a homogeneous piece of rolled steel of the desired length, width, and thickness without flaws.

4. Use a wide tread, adequately connected throughout its width to the adjacent steelwork.
5. Provide good contact between tread plate and girder web.
6. Provide the largest practical radius of operation for the curved tread plates.
7. Provide support for live loads independent of the tread plates.
8. Machine all components precisely, and align all components accurately during erection.
9. Carry the dead load of the bridge when closed on the live load supports, so that the curved tread plates are lifted off the horizontal plates, or at least the load is reduced.

Following all these rules to their maximum extent can result in a very expensive addition to the cost of the bridge. Each installation has its own characteristics, which make some of the rules more economical to follow than the others. The engineer should make a careful study of the particular bridge design situation to determine which of the rules can be compromised to least disadvantage in durability of the structure while offering the greatest savings in construction cost, so that the arrangement is produced that is most economical over the projected life of the bridge. Items 7 and 9 can have an adverse effect on the stability of the bridge: When the bridge rocks forward to seat on the live load shoes, the support may move forward of the center of gravity, so that seating of the bridge becomes unstable.

DECK OPENINGS

Bascule bridges must rotate open and closed and provide a continuous roadway surface for their traffic, whether railroad or highway vehicles, without a large gap between the movable bridge deck and the deck on the approach spans. The deck of a highway bascule bridge may show a slight gap between the ends of the bascule span and the adjoining span, but it is desirable to minimize this gap. On railroad bridges, a continuous line of rail at each side of the track is a necessity for safety except for particularly crude structures carrying very slow, light traffic.

Because a bascule bridge rotates as it opens and closes, the relationship between the deck sections on the approach and bascule span, particularly at the heel of the bascule span, is complicated. Each point on the bascule bridge represents a three-dimensional question of location before, during, and after a bridge opening and closing. Clearances must be correctly provided where needed by careful design to avoid interference when the bridge opens and closes.

Railroad bridges can be fitted with more or less standard joint components that can be arranged as required by adjusting the structural steel or the deck timbers if necessary. It is impossible, of course, to build a bascule railway span with a conventional ballasted deck unless the maximum opening angle of the bridge is very small (and it would still be impractical).

Highway bascule bridges must have the deck joints painstakingly detailed in the design phase, as the proper relationship of the roadway surface to the main bridge members depends on workable joint details. It is desirable to provide a cross slope at the roadway, even with open grating decks, to facilitate drainage. This further complicates the geometry of these joints.

The toe joint is usually not a difficult detail, as the bridge members move almost purely vertically relative to each other. The heel joint at the deck of a bascule bridge can be very complicated. The bascule deck may travel back over or under the approach deck at the heel, depending on the location of the center of rotation of the moving leaf. On railroad bascule bridges, the end rails at the heel joint must be arranged to avoid interference and allow the bridge to open and close. For trunnion bascule bridges, the following relationships apply during bridge operation:

If the center of rotation of the moving leaf is below and ahead of the heel approach joint, the heel of the opening deck moves back and under the approach as the bridge is opened (Figure 11-5). This arrangement is an advantage for a highway bascule, as the moving leaf presents an effective barrier to automobile traffic. There is also no possibility of a vehicle impacting a bridge floorbeam (see below).

If the center of rotation of the moving leaf is below and behind the heel approach joint, the heel of the opening deck moves back and over the approach as the bridge is opened (Figure 11-6). This arrangement presents a barrier to automobile traffic on a highway bascule, but the rear edge of the deck of the moving leaf is in the path of oncoming cars, and a high-speed impact can be, and in cases has been, fatal to the vehicle's occupants.

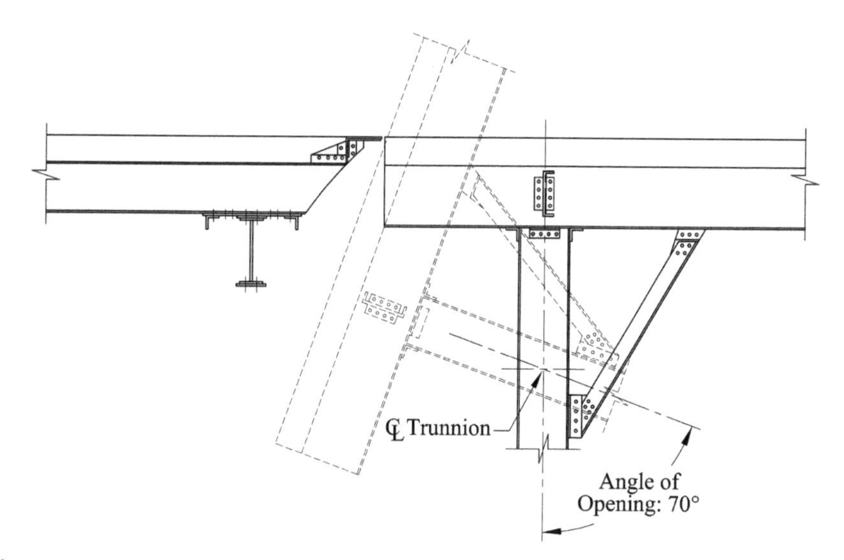

Figure 11-5 Bascule Bridge Heel Joint. With the pivot axis located below and forward of the deck joint, the moving edge of the joint drops below the fixed edge as the bridge opens.

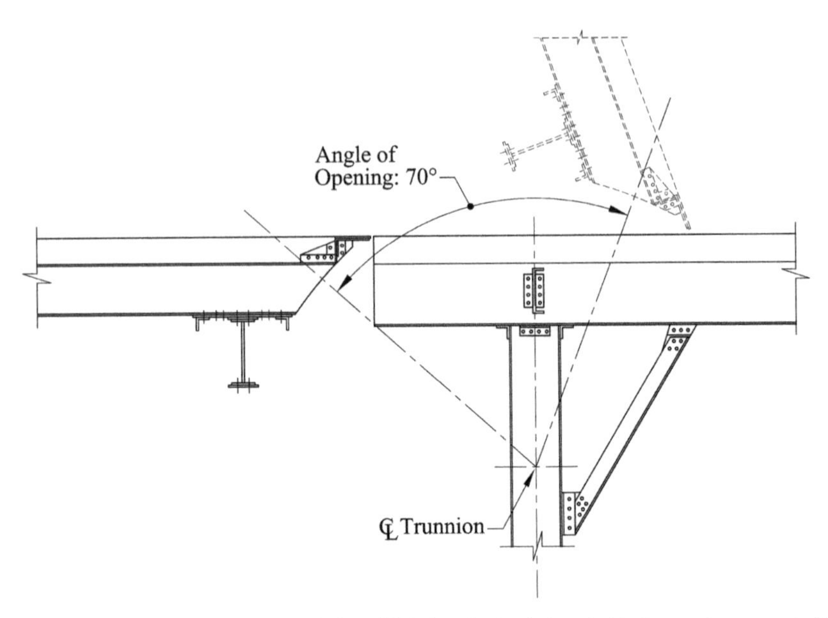

Figure 11-6 Bascule Bridge Heel Joint. With the pivot axis located below and rearward of the deck joint, the moving edge of the joint rises above the fixed edge as the bridge opens.

If the center of rotation of the moving leaf is above and forward of the heel approach joint, the heel of the opening deck moves down and away from the approach as the bridge is opened (Figure 11-7). This arrangement is uncommon for highway bridges as it necessitates the positioning of the trunnions and the main girders above the roadway, an aesthetically undesirable location, and it results in an undesirable opening in the roadway of the open bridge. Such an arrangement can be used on railroad bridges, where it is desirable to keep the trunnions as far as practical above high water, and the deck opening offers no disadvantages to railroad traffic. There will be a negative live load moment as a train enters or exits the moving leaf from the heel; this negative moment is considered undesirable.

If the center of rotation of the moving leaf is above and behind the heel approach joint, the heel of the opening deck moves up and away from the approach as the bridge is opened (Figure 11-8). This is a common arrangement on truss-type heel trunnion bascules. It produces an opening in the roadway and places the heel edge of the bascule span in the path of traffic, so it is undesirable for highway bridges. It is, however, a convenient arrangement for railroad bridges.

While some of the above configurations provide a barrier to traffic in the form of the need of the moving leaf, an impact-attenuating type barrier gate is more appropriate for roadways carrying higher-speed vehicular traffic (see page 278).

The amounts of movement at a roadway joint with the aforementioned arrangements are dependent on the distance of the joint from the center of rotation of the leaf.

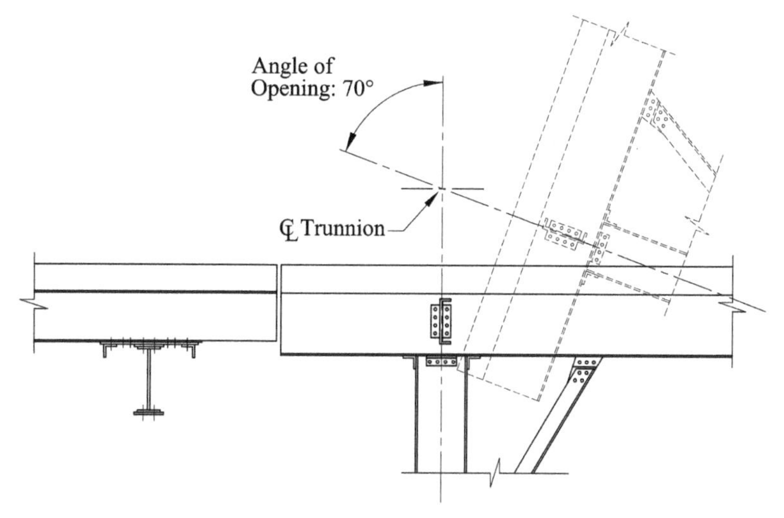

Figure 11-7 Bascule Bridge Heel Joint. With the pivot axis located above and forward of the deck joint, the moving edge of the joint drops below and forward of the fixed edge as the bridge opens.

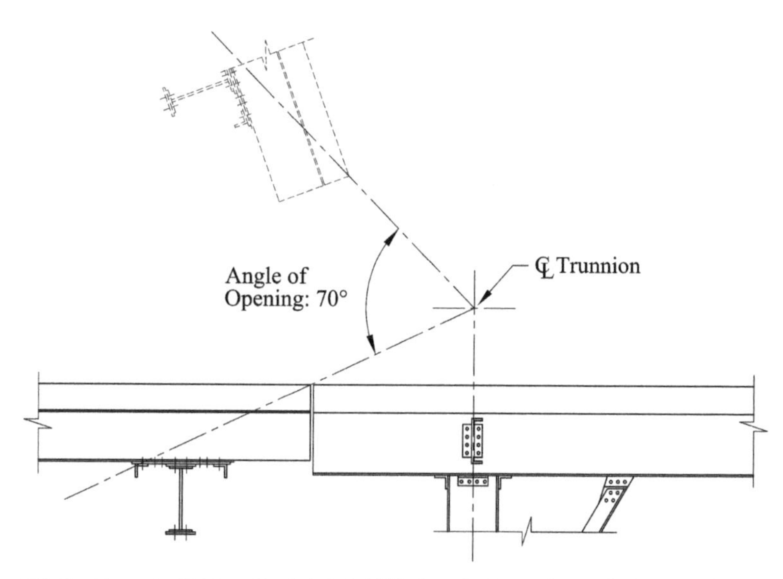

Figure 11-8 Bascule Bridge Heel Joint. With the pivot axis located above and rearward of the deck joint, the moving edge of the joint swings above and forward of the fixed edge as the bridge opens. This type is common on heel trunnion bascule bridges.

The movements are more complicated for rolling lift bridges, as the heel edge of the moving leaf has an additional component of movement toward the approach as the bridge opens, furthering the possibilities for interference.

COUNTERWEIGHTS

Almost all movable bridges are designed to optimize their unbalanced condition so that the amount of power required to operate them is as small as practical. This is particularly important for bridges that must be lifted out of the way to clear a navigation channel to allow a vessel to pass, such as bascule and vertical lift bridges. Swing bridges, retractile bridges, and some other types may require counterbalancing to stabilize them, so that toppling is prevented, and to equalize the load on their supports. Counterweights are usually designed to be as dense and as cheap as practical. Most bascule and vertical lift bridge counterweights are composed largely of concrete. Some concrete counterweights are made with a high-density aggregate to reduce their overall size. Other counterweights, in special cases, are made of cast iron or steel, and there have been very rare instances of counterweights containing lead.

Many counterweights were constructed by mixing steel punchings or scrap in the concrete to increase the density. A good many of these have deteriorated because of corrosion of the steel chunks within the concrete. The steel expands as it corrodes, putting pressure on the concrete and eventually causing it to spall. When a very dense concrete counterweight is desired, its formulation should be developed with care. On the Arlington Memorial Bridge, built in 1932 over the Potomac River, counterweight concrete with a density of 265 lb/ft^3 was required, with a strength of 2000 psi. Steel punchings and special Swedish iron ore were used, with the ore crushed and carefully graded to produce the desired proportion of fines.

The most practical and durable counterweight is a steel box filled with concrete, with the thickness of the steel box walls adjusted depending on the overall density required. The minimum thickness of the box walls should not be less than $\frac{3}{8}$ in., and should preferably be more than $\frac{1}{2}$ in. The concrete filler mixture can be adjusted in the field after the actual weight and center of gravity of the structural portion of the moving span are known, so that the span is brought as closely as possible to the desired final balance state. The counterweight's concrete composition and strength become less critical when it is enclosed in a box, as the concrete is then protected from deterioration and fully supported.

The cheapest counterweight is a concrete block cast around some kind of framework connecting it to the lift span. Counterweights of this kind have tended not to be as durable as steel-faced counterweights, but an exposed concrete counterweight may last longer as air pollution is reduced. Reductions in pollution will also, of course, lengthen the life span of exposed steel. A more serious source of deterioration in concrete counterweights on highway bridges was the increased use of deicing salts on roadways in the latter half of the twentieth century. In the case of underdeck counterweights, the salt passed down to the counterweights with the melted ice and soaked into them, causing rapid deterioration. Steel punchings within the counterweights would then be attacked, and their swelling, as they corroded, caused rapid crumbling of the counterweights. The reduced use of road salts should result in less such deterioration in the future.

Pockets, or empty spaces, are designed and built into the counterweights to allow adjustment of the weight of the counterweight after it has been completed with the concrete, if used, poured. To make precise adjustments to the balance of a bascule or vertical lift bridge, for which the balance state is a critical figure in the successful operation of the span, the pockets usually encompass 10 percent of the volume of the entire counterweight. Normally, there are several pockets in a single counterweight, so that the position of the weight added or subtracted can be varied to help provide the correct balance state. This feature can be used to make balance adjustments long after the bridge is completed, such as to compensate for repainting or a minor structural modification. Ordinarily, the pocket volume is designed to provide correct balance when it is half full, so that either positive adjustments, by adding weights, or negative adjustments, by removing weights, can be made as necessary.

The pockets are filled to the desired level with small portable balance blocks or weights. These portable weights can be made of concrete, iron, steel, or any other suitable material. It is usually best to limit the weight of each separate balance block to the amount one person can safely lift, no more than about 75 lb. Larger weights may be used, as it is very rarely that a bridge needs the precision in balancing that would require adjusting weights of less than 75 lb. In confined quarters, however, especially after a bridge has been in service for a considerable period of time, moving of the weights will be difficult enough and could be impossible for one person to comfortably remove or insert a heavy block in a pocket. For ease of movement during periodic adjustments, if any are expected to be necessary, at least a percentage of smaller weights should be used, preferably less than 50 lb.

Sometimes rigging, falsework, and enclosures must be temporarily installed on a span for repairs; if this material is light enough, the number of balance blocks can be adjusted sufficiently to allow the span to remain operable. More extensive repairs or reconstruction, such as redecking, usually requires a modification of the counterweight structure. Many older highway bascule spans developed severe operating difficulties after trolley tracks, along with the structure to carry the overhead wire, were removed from the bridge. Extensive, and in some cases ingenious, rebalancing efforts were required to get the span center of gravity to a location acceptable for operation. In many instances dead weight was added on top of the deck of the cantilevered portion of a rebuilt bascule leaf simply to act as ballast for proper balance.

Bascule and vertical lift bridges rely to a great extent on their balance state to provide stability under live loads. Adjustment blocks can help provide for continued proper seating after minor weight changes in the movable structure. For bridges in some locations, a major shift in adjustment blocks may be necessary on a seasonal basis. Sometimes snow and ice loading may accumulate on a bascule or vertical lift bridge to the extent that the bridge cannot be raised. In such cases, more weight can be added to the counterweight to maintain balance, by the addition of a few adjustment blocks. Many movable bridges have a store of such blocks nearby to accommodate addition to or removal from the counterweight(s) as needed.

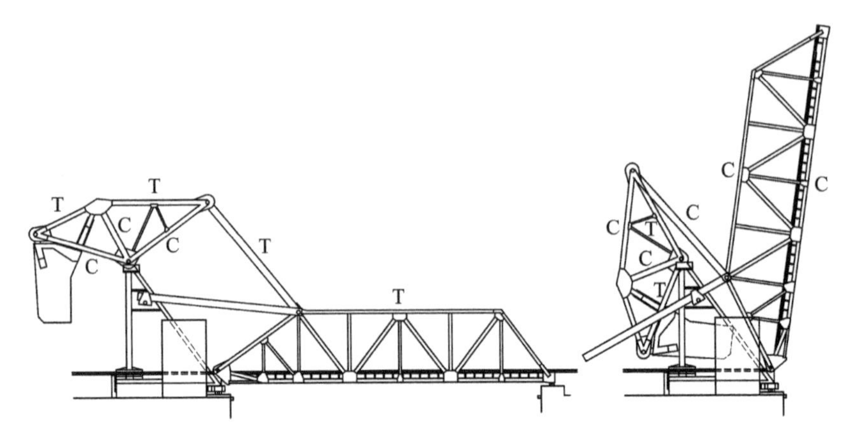

Figure 11-9 Heel Trunnion Bascule Bridge Open and Closed. This arrangement is typical for this type of bridge. The operating strut has a significant influence on the balance state of the bridge.

The external dimensions of a counterweight must be carefully detailed so that there will be no interference with fixed structural members as the bridge opens and closes. Space is at a premium, particularly in the confined, watertight counterweight pit that is necessary for some simple trunnion bascule bridges, particularly the Chicago type.

The counterweights for heel trunnion bascule bridges must be designed with care (Figure 11-9). The area where the counterweight swings is usually less restricted for this bridge type than for a simple trunnion bascule. Many single leaf through truss heel trunnion bascule bridges open to a very wide angle, almost 90°. The counterweights for these bridges are supported on large trussed frames, some of the main members of which undergo complete stress reversal during opening and closing operations. Some of these bridges have developed fatigue cracks in their structural members, resulting merely from opening and closing (see page 139).

SPAN LOCKS

On a single leaf bascule bridge, span locks are used as a safety device in the span operation interlocking procedure, to prevent, or at least make difficult, the opening of the span without withdrawing the locks. An ordinary single leaf bascule bridge that is reasonably well balanced does not require a span lock for stability unless its deck configuration is such as to provide an extremely large live load moment behind the center of rotation. On many single leaf bascule highway bridges that have span locks, the locks are found to be inactive, as they have been disconnected by maintenance crews. Single leaf railroad bascule bridges usually have the span locks interlocked with the railroad signal system, so the span lock must always be retracted before the bridge is opened, and driven after the bridge is seated. The span lock that is so interlocked acts as a safety device, linking the bridge position to the signal system, whether the span lock directly performs any useful function or not. To be an effective interlocking device, the span lock must be arranged so that it cannot be driven unless the bridge is seated.

The span locks on single leaf bascule bridges are usually mounted on the toe of the leaf. They typically take the form of a remotely actuated sliding lock bar, which engages a pocket or socket. Usually, the lock bar is mounted on the span and the pocket is mounted on the rest pier, but sometimes these locations are reversed. Some single leaf span locks are in the form of a hook, mounted on the rest pier, which is remotely actuated to engage a tab, eye, or other device on the toe of the bascule leaf. The positioning of this type of device can also be reversed, with the hook on the span and the tab on the pier. Unless the span locks on a single leaf bascule are used for live load stabilization, they can be very crude and cheap in design, with liberal tolerances at all moving parts.

On an ordinary double leaf bascule, span locks play an inherent and necessary role in stabilizing the bridge to carry traffic. Center locks are used to structurally connect the toe ends of the two mating bascule leaves when carrying traffic, so that the toes of the leaves deflect together when a live load is on the span. This prevents a vertical discontinuity and resulting bump from occurring when a vehicle, on one leaf of the span and pushing it down by its weight, contacts the other leaf.

Center span locks for double leaf bascule bridges have been developed in many different forms, but almost all are intended to act as shear connections and are often referred to as "shear locks." There is usually one shear lock at each mating pair of girders or trusses for a double leaf bascule bridge. The most common type of center shear lock consists of a simple sliding lock bar, mounted parallel to the axis of the bridge on one leaf, which is mechanically actuated to engage a socket on the toe end of the other leaf. The lock bar has two supports on the leaf on which it is mounted, providing vertical fixity when the lock bar is loaded. These supports are generally called *lock bar guides*. The lock bar guides and sockets are usually adjustable, to allow for reducing the clearance that develops in them as a result of wear. Some span lock bars are round, but rectangular cross sections are preferred to provide greater bending strength against the primarily vertical loading of the assembly. The rectangular section of the lock bar also makes the provision for shimming, at the guides and sockets, easier, so that clearances can be better adjusted at installation and as wear occurs. This type of center lock tends to experience excessive wear, as the extending and retracting bars are exposed to the dirt that is pulled into the bearing portions of the guides and sockets, contaminating the lubricant and accelerating abrasion. The deflection of the bascule leaves under live load also causes deterioration of the span locks, as the lock bars pinch in the sockets as the ends of the leaves rotate as they deflect downward. Several attempts have been made to eliminate some or all of these deficiencies. Cragg (U.S. Patent #5,327,605) cushions the guides and sockets to reduce the effects of pinching. Gilbert (U.S. Patent #2,610,341) provides a rotary socket to avoid pinching. Koglin (U.S. Patent #6,453,494) provides a rotating socket in a sliding block to avoid pinching and abrasion.

Other types of mechanically actuated center locks take the form of mechanically actuated jaw devices mounted on the toe of one leaf, which grip rigid tongues mounted on the other leaf (Figure 11-10). In some cases this arrangement is reversed, with the

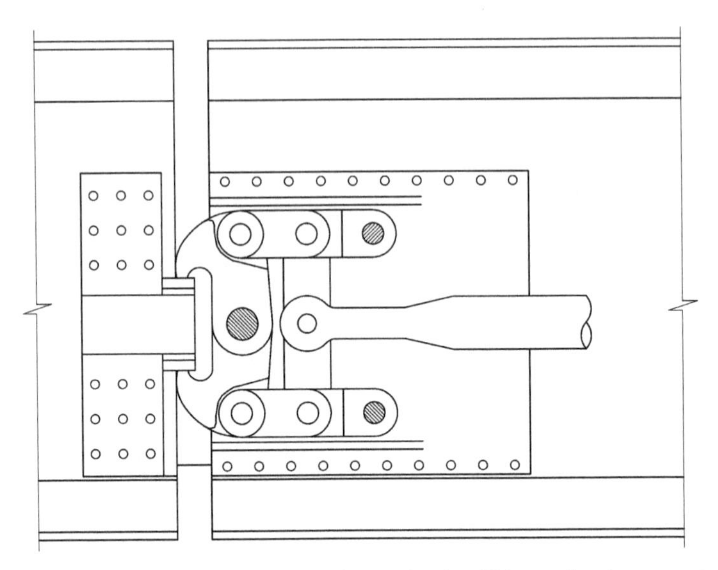

Figure 11-10 Bascule Bridge Pincer Type Center Locks. This type has been used on double leaf bascule bridges. It has been used for many decades and is still occasionally specified for new structures.

jaws rigid and mechanically actuated tongues extended from one leaf and expanded to grip in the jaws mounted on the other leaf. These devices have a theoretical advantage over other types of center locks in that actuation can be adjusted so that the mechanically actuated jaws or tongues can be set to rigidly grip the mating parts, eliminating vertical free play in the connection. This can result in greater wear, however, as the components slide against each other when the bridge leaves expand or contract as a result of temperature changes or live load deflection. Because of their greater complexity, with many more moving parts to maintain, these types of center locks are rarely specified for a bridge today. Although they can allow adjustment of the clearance at the locks without disassembly, this feature is rarely taken advantage of, and most jaw type center locks have excessive clearances similar to those of the bar-type locks.

Scherzer type double leaf rolling lift bridges, because they retract from the center of the navigation channel when they open, frequently use rigid male-female jaw and tongue fabrications, mounted as extensions of the main bascule girders or trusses, to act as center shear locks (Figure 11-11). These locks can be used on rolling lift bridges because the combination of longitudinal and rotational movement when opening and closing allows the locks to engage and disengage without the aid of any mechanical components. The disadvantage of this arrangement is that the two leaves must open and close in unison. This has been accomplished on some bridges by automatically controlling the movement of the two leaves at the nearly closed position, so that the rigid center locks are aligned without any manipulation by the bridge operator. This type of center lock tends to wear rapidly, because of the difficulty in main-

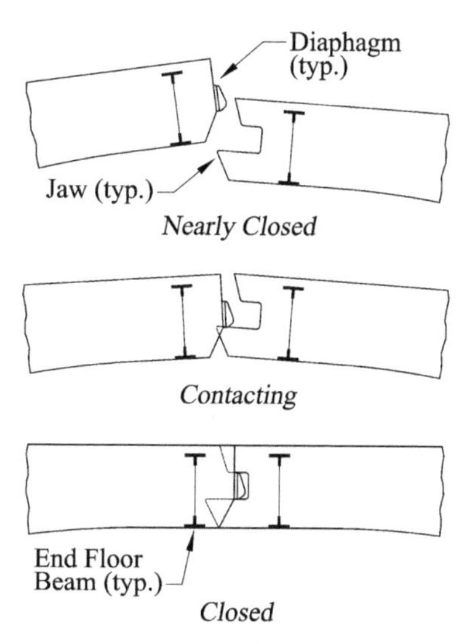

Figure 11-11 Scherzer Jaw Type Center Locks. This type of center lock can be used on Scherzer type rolling lift bascule bridges, because the translation of the leaves when opening and closing assists in disengagement and engagement.

taining lubrication. The locks can also experience high impact forces during engagement of the leaves in the event of imperfect synchronization of the mating leaves during closure, but their ability to be robustly constructed makes up, to some extent, for these disadvantages.

The vertical shear forces that are taken by the center locks can be quite large, depending on the size and type of construction of the bascule leaves. The lock components and the structural steel on which they are mounted must be designed to withstand these loads without overstress.

Some double leaf bascule bridges have been equipped with tail locks to help stabilize the leaves under live loads. These locks are usually installed only when the bascule leaves are arranged in such a way that the weight of vehicles on them can result in a moment tending to open the leaf. The tail locks may be in a form similar to the center locks that use a sliding bar to carry the load in shear, or they may consist of column arrangements that are mechanically moved to prop up the rear end of the leaf in the closed position.

CENTERING DEVICES

A bascule bridge is kept aligned by the rigid connection of its pivot end to the support on the pivot pier. For trunnion bascule bridges of all types, the alignment thus achieved is adequate for the life of the bridge, barring some damage or shortcoming in construction that has caused an inherent misalignment. Scherzer rolling lift bridges, which rely on the lugs and pockets on their tread plates to stay in alignment, lose precision of seating after the tread plate, lugs, and pockets have been worn down by operation and corrosion. In the Rall rolling lift, which relies on linkages for alignment, there is a relationship between wear and span alignment somewhat midway between that of the Scherzer rolling lift and the trunnion bascule, when the bridge is relatively new. Older Rall bascules have increasing difficulties with misalignment as severe deterioration of the aligning linkages occurs, and they can become quite unstable and a source of serious operating difficulties when linkage wear allows large positioning errors.

Both AASHTO (2.8.1) and AREMA (6.5.35.2) specify that centering devices are required to align the free end of a bascule leaf with the adjoining span or abutment. This is a necessary function for a railroad bridge to avoid derailment of a train due to misalignment of the rails. The centering device allows the bridge to close fully with the rails aligned; interlocking devices are usually activated after closure to make sure the rails are aligned before trains are allowed to proceed. The need for centering devices is questionable on highway bridges, except perhaps to avoid such gross misalignment that the guide rails or curbs along the sides of the bridge become an impact hazard. For the bascule types that are inherently resistant to misalignment, such as the trunnion bascule bridge, there appears to be no value in providing separate centering devices. Even on other types of bascule bridges, particularly single leaf spans, a fairly large degree of misalignment at the toe does no harm. Double leaf highway bascule bridges, if provided with sufficient lateral capacity at their center locks, also would not need centering devices.

BUFFER CYLINDERS

Buffer cylinders are frequently installed on bascule bridges. The purpose of these devices is to act as shock absorbers in the event that the control system for the bridge fails and a leaf comes crashing down onto the live load bearings. Buffer cylinders can be particularly useful if the bridge is substantially imbalanced so that it is very span heavy. Heavy impacts have been known to happen, and have been known to happen many times, as a result of brake failure, coupling failure, operator error, or for some other reason. It is assumed, in installing a buffer cylinder, that the action of the cylinder in such an instance will work to reduce the maximum stress in the components affected and perhaps prevent a failure. For this reason, buffer cylinders are required by AASHTO (1.1.14) and AREMA (6.5.3) on bascule and vertical lift bridges, although exceptions are allowed.

The typical buffer cylinder is a piston-cylinder arrangement that acts as an air compressor every time the bridge on which it is mounted is lowered. The cylinder itself is mounted rigidly on the span or connected to the bascule leaf by means of a pinned connection to allow for pivoting. The piston is actuated by a rod that extends from the bottom of the assembly. When the bridge is in the open position, the piston is at the bottom of its travel in the cylinder and the rod is fully extended. As the bridge is lowered, the rod contacts a strike plate mounted in the correct position on the pier, and as the bridge continues to close, the cylinder is brought down around the piston and rod, which are stationary after contact with the strike plate.

As the bridge continues to close, the air in the cylinder is compressed in an approximation to an adiabatic process; the volume of a specific amount (number of molecules) of ideally compressible fluid (air) is reduced without the addition or removal of heat. Continuation of the motion of the cylinder vis-à-vis the piston and rod causes the compressed gas in the cylinder to increase in pressure, according to the adiabatic relationship,

$$Pv^n = k$$

where

P = the pressure within the cylinder
v = the volume contained in the cylinder at any given time
n = the ideal gas constant, equal to about 1.4
k = a specific constant for the particular operation being studied

The adiabatic relationship assumes that there is no heat transfer into or out of the fluid during compression, which is essentially true with a buffer cylinder. The air heats up and exhausts out the needle valve before any substantial heat transfer to the cylinder body occurs. The increase in pressure is limited by installing a needle valve on the cylinder so that some of the quantity of air in the cylinder bleeds off as the volume inside the cylinder decreases and the pressure rises above atmospheric. The valve is adjusted to limit the pressure buildup in normal operation to a practical value.

Some movable bridges have buffers mounted to cushion the end of travel when they are opening as well as closing. These installations on bascule bridges usually have the cylinders mounted near the tail end of the leaf.

The process of operation of a buffer cylinder is highly variable. Various changes over time, such as wear, corrosion, and contamination with dirt or oil, can change the rate of air compression as the cylinder operates, or even eliminate compression completely. Unequal action of buffer cylinders, in a severe closing event, may do more harm than good, perhaps permanently warping the bridge superstructure or damaging the pier. Elimination of buffer cylinders should be considered in designing a new bascule bridge or rehabilitating an existing one, especially when a modern control system is being installed with automatically controlled seating of the bridge. There is also little need for buffer cylinders if the moving leaf is automatically stopped at the nearly closed position before seating. The advantages of compressed-air type buffer cylinders, when they work properly, are that they build up from practically zero force at contact to a maximum force after having traveled some distance and the maximum force increases with the speed of the span at impact. If it is believed that compressed-air-type buffer cylinders must be used, they should be combined by piping so that all the buffers mounted on a single span or leaf operate under nearly the same pressure.

RAIL JOINTS

Movable bridges carrying railroad traffic must provide guidance so that the wheels of the train equipment can pass onto and off the movable bridge without derailing. For single leaf railroad bascule bridges, the rail joints can be built so that they are rigid relative to the movable span and the approaches, with no mechanical movement of the rails required (Figure 11-12). Careful design of the joint details can allow the bridge to open and close without interference at the joints. Some small lateral movement of the joint components on a bascule span will occur relative to their mating components on the rest and bascule piers, but these can be accommodated within fixed joint designs by providing tapered cross sections on the moving components that drop into corresponding receivers on the fixed sections. More precise alignment can be accomplished with the aid of span centering devices. Longitudinal freedom of movement should be allowed within the assembled joints so that expansion or contraction of the bascule span due to temperature or live load effects does not cause binding at the joints. For these applications, several companies fabricate standard details that are readily available to mate with the various common sizes of rail. Many railroad companies have their own particular favorite joint details for movable bridge applications, which may be fabricated to the railroad's own specifications from rolled rail sections, cast steel, or other materials.

These joints are of two basic forms; sometimes both are used on a single railroad bridge. The occurrence of both forms can result from a partial changeout, or one type

Figure 11-12 Miter Rails on Bascule Bridge. These rails are firmly fastened down onto the superstructure of the approach, in the foreground, and the bascule leaf, at the far side of the joint.

may be used on the heel end of the bascule bridge, where the size of the gap is fairly constant, and another type on the toe, to allow for greater longitudinal expansion capability. The first type, and the most common, is achieved by the end of the rail section being cut on a vertical plane at an angle to the axis of the rail, with a suitable transition radius at each end of the cut. The rail end is usually bent, sometimes at two points, so that the web of the rail is left intact to the end of the rail and only the base and head are actually cut, in a tapering section, as in switch points. This is referred to as a "mitered" or "miter" rail. Often the rail cut is not straight, allowing the joint to be elongated, so that there is a more gradual transition from one rail to the next as the wheel passes over the joint.

The other common form of rail joint has the rails cut straight across at the rail ends, 90° to the rail axis. A third rail section or a heavy piece of steel of rectangular cross section is used as a bridging piece, attached to the outside of one of the rails and formed to overlap the other. This form is usually used only in low-speed service, because of the problem of worn wheels impacting the bridging piece. Some successful applications of this type have been used in higher speed service in which a bridge is primarily used for passenger trains, where the incidence of worn wheels is not great enough to cause a problem.

Double leaf bascule bridges are rare in railroading, because of the potential for catastrophe (see page 110) and the difficulty with deflection at the center when no moment connection is provided. There are a few double leaf railroad bascules in use, and the rail joints at the mating ends of the two leaves are more critical than those bridging the gap between the heel of the bascule leaf and the approach span. This type of construction was once common, in the late nineteenth and early twentieth centuries, when railroad trolley cars traversed many miles of track in the neighborhood of navigable waterways and double leaf highway bascule bridges were available for use by these railcars. After several decades of almost no such use, recent developments in mass transit have called for these details to reappear in urban areas, sometimes on the same double leaf bascule bridges that had their original trolley car tracks removed 20, 30, or more years earlier. The rail joints at the center of the bridge are difficult to design. Either the bridge must be opened and closed in sequence, with rigid joints, or mechanically actuated rail connections must be used to provide secure rail joints. Fixed rail ends that project beyond the toe of the leaves of the bascule span restrict the bridge so that one leaf must always close before the other.

LIVE LOAD SHOES

Bascule bridges require special supports to stabilize them under live loads. Single leaf bascules have live load shoes under the toe ends of the main girders. These live load shoes are similar in form to the sole plates and masonry plates used to support simple bridge spans. Generally, no attempt is made to combine the live load support with any other function, and the sole plate is free to slide around on the masonry or strike plate within the limits allowed by the bascule span itself. Some bridges combine centering devices with the live load shoes so that the masonry plates will have vertical projections along their outer sides, which mate with wearing surfaces on the outer edges of the girder ends so as to align the bridge leaf laterally as it is seated. The sole plates are firmly bolted to the bottoms of the main bridge members, usually with provision for shimming to adjust the end of the bascule span to the proper elevation. The masonry or strike plate is also adjusted to elevation when installed, usually by application of grout under the plate. The masonry plate is rigidly attached to the pier by means of anchor bolts.

Double leaf bascule bridges, particularly those in which each leaf acts as a cantilever span for dead and live loads, usually have live load shoes mounted on the main girder or truss so that they mate with masonry plates on the pivot or bascule pier ahead of the pivot point (Figure 11-13). On trunnion bascules these frequently consist of a set of shoes for each main girder or truss for each leaf, somewhat similar in appearance to the sole plate and masonry plate arrangement common on fixed spans. The live load shoes are most effectively located close to the forward edge of the bascule pier, to maximize their leverage, relative to the application point of live load. The sole plate and masonry plate on a double leaf bascule are both generally of very heavy construction, as compared with those for fixed spans of similar size, because they must be capable of receiving substantial impacts during the seating of the leaf without sustaining damage; they also receive magnified live loads due to the cantilever effect and are subject to possible impacts due to poor seating of the bridge.

Some double leaf trunnion bascules have rear anchor type live load supports instead of or supplementary to the forward live load supports. These are usually built into the approach span or abutment and may consist of vertical tension columns with cross members that span across the rear of the leaf or its counterweight. The anchors are difficult to adjust for proper contact, especially when acting redundantly with forward live load shoes. Generally, the forward live load shoes are brought into firm contact, and then the rear live load shoes or anchors are adjusted to very small clearances. Under heavy live loads, the bascule span and the pier deflect to some extent, as the live load shoes and the local area of the pier compress or deflect to some extent, bring-

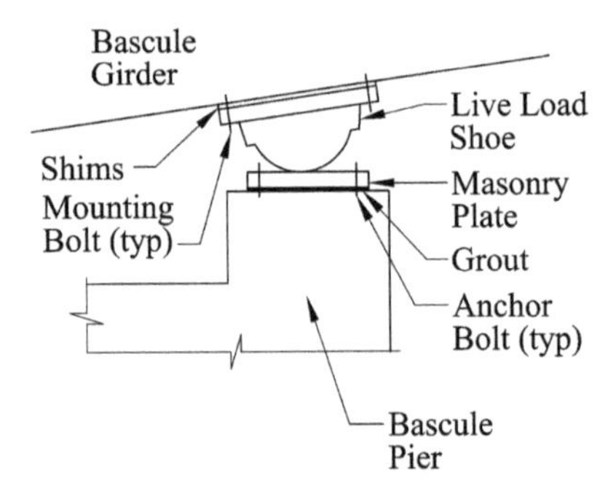

Figure 11-13 Live Load Shoes for Double Leaf Bascule. This type of live load support is most common for trunnion type double leaf bascule bridges. The shoes are mounted forward of the trunnions, so that load is taken off the trunnions as live loads pass.

ing all the live load supports into contact. This loading and deflection partially unloads the trunnions, lifting the rear of the bascule leaf slightly. It should be made certain that this action will not be great enough for the bridge to break contact at its trunnion bearings, as the impact on the bearing after the live load leaves the span can cause deterioration of the bearing.

On Scherzer rolling lift spans, the live load support at the bascule pier is frequently an extension of the rolling lift treads; sometimes the curved tread plates are flattened to allow a larger bearing area. Some older rolling lifts have no special provision for live load support in this area, simply relying on the curved tread plates on the bascule span and the flat tread plates mounted on the pier to do the job. This arrangement provides only line contact and in some cases proves to be insufficiently durable, and the tread plates wear excessively or fail at this point. Repeated loading has been known to cause breakage of the tread plates due to the propagation of cracks. Double leaf Scherzer bascules may have separate forward and/or rear live load supports added, as in trunnion bascule bridges. A common arrangement provides rear anchors with forward live load support directly at the rolling lift treads. The live load supports for Rall rolling lift bridges are similar to those for trunnion bascule bridges. Some Rall bascules were supposedly designed to have the bascule leaf tilt forward when seating so that the entire dead load and live load are supported on the live load shoe, with the rear anchor stabilizing the leaf. The support wheel of the Rall bascule would then be lifted off the support track, allowing maintenance, repair, or replacement of the wheel or its bearings without taking the bridge roadway out of service.[1]

When anchors are used at the rear of a bascule leaf, they must be adequately restrained to carry their intended loads. If a heavy approach span is used, its weight can provide the necessary uplift resistance for the anchor, provided that an adequate connection can be made. If it is necessary to attach the anchor to the masonry of the bascule pier, there must be adequate embedment to prevent the anchor from pulling out. If concrete is merely poured around the anchor where it exits the substructure, incremental release may occur, resulting in loss of sealing between the anchor bolt and the concrete. This can lead to continuing corrosion, significantly weakening the anchor until it eventually fails. If the anchor bolts can be substantially preloaded before grouting, the problem can be minimized.

A very few double leaf trunnion bascule bridges have rear live load anchors as the only live load stabilizing devices between the bascule leaves and the piers (Figure 11-14). This type of support can work effectively with Scherzer rolling lift bascules, but with trunnion bascules it results in trunnion stresses that increase substantially with live loads rather than decrease. On Scherzer rolling lift bridges the stress on the forward end of the tread plate, in contact when the bridge is closed, increases. An advantage is gained, with a rolling lift bridge, by having the pivot center immediately adjacent to the edge of the pier with the bridge closed, shortening the effective span length. No such advantage is gained with a simple trunnion bascule, as the pivot point

[1]Theodor Rall, U.S. Patent 669,348, March 5, 1901.

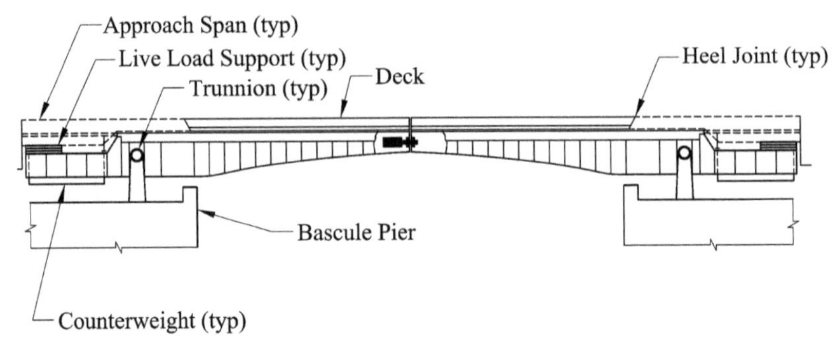

Figure 11-14 Trunnion Bascule Bridge with Rear Live Load Support. This arrangement is rarely used as it increases substantially the live load on the trunnions, which can lead to fatigue failure due to the large bending stresses that can be developed.

must be at least as far back from the edge of the pier as the depth of the bascule girder is below the centerline of the trunnion.

Single leaf heel trunnion bascules almost always carry the live load at the bascule pier on the main trunnions only. This is also commonly the case with single leaf simple trunnion bascules.

BALANCING

Almost all bascule bridges are balanced structures. A large amount of energy is required to lift the span when opening if it is not provided with a counterweight, and very heavy machinery is required to lift and hold the load against gravity. When a non-counterweighted bridge leaf is lowered, the difference in potential energy of the bridge between the raised and lowered positions of the center of gravity of the leaf must be dissipated in a controlled drop. Safety must also be a consideration, as a failure of the support system for an unbalanced span in the open position would be catastrophic, with possible loss of life as well as severe damage or destruction of the bridge. If the unbalanced span should happen to fall on a vessel navigating the draw opening, the damage will be compounded. Extreme caution must be exercised in designing and constructing a "bascule" bridge that is not counterweighted so as to eliminate the likelihood of such a failure. The bridge itself should be fail-safe and not dependent on good maintenance to prevent an accident.

Most early bridge designers were primarily concerned about cost of construction and operation. In the days when machinery, particularly motors and engines, were relatively expensive to build and maintain, it made sense to minimize the power requirements of a movable bridge by eliminating gravitational forces from the drive loads. As indicated previously, designers either tried to balance their bascule bridges exactly or used a slight negative imbalance to help accelerate a bridge when starting to open or close. The negative imbalance would also reduce the likelihood of large

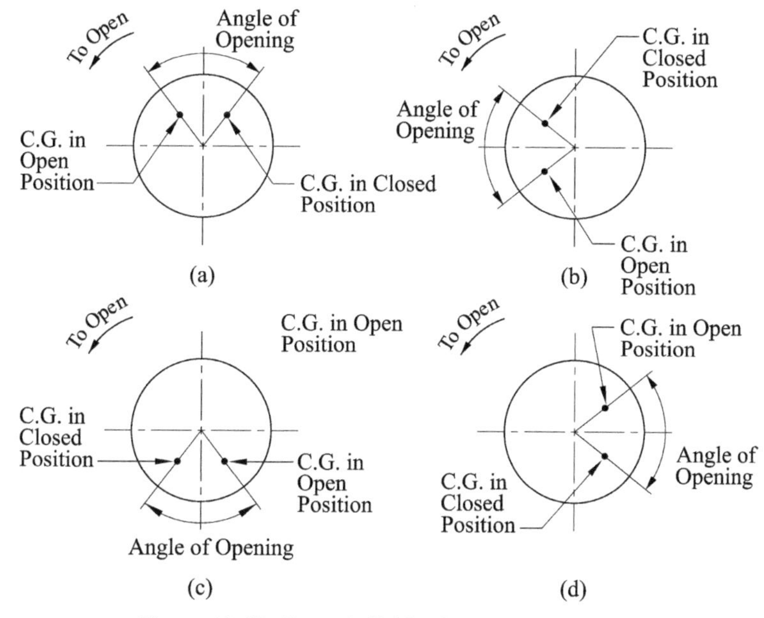

Figure 11-15 Bascule Bridge Balance Quadrants.

impacts on the bridge seats when closing, or overtravel when opening. Various types of span locks, sometimes several types on one structure, and large machinery brakes were used to supply the desired level of stability and safety.

Today, the damaging effects with faster traffic of unstable seating of movable bridges, and the increased liability in the case of an accident, have forced most designers to be more concerned about finding inherent means of stabilizing their movable spans in the closed position. The most readily available means of stabilizing a bridge is in the treatment of the balance state (see Figure 11-15). Gravity can do an excellent job of holding things down if given a little guidance. It is not necessary to remove all or even a large part of the counterbalancing weight of a movable span to stabilize it. By making small adjustments in the position of the overall center of gravity of the moving leaf, the desired amount of imbalance in the open and closed position can be achieved.

The best location of the center of gravity of a bascule bridge is generally agreed to be slightly above and forward of the center of rotation, measured when the bridge is in the closed position, making the bridge's span heavy when closed. This results in the moving leafs being exactly balanced when partly open, and going through a state of zero imbalance when the bridge opens, keeping the amount of force required to overcome gravity small. It places the center of gravity above and behind the center of rotation in the span open position so that the bridge will be counterweight heavy, held in the open position by a gravitational moment. More important, this produces a seating force due to gravity on the closed bridge, which tends to push the span down on

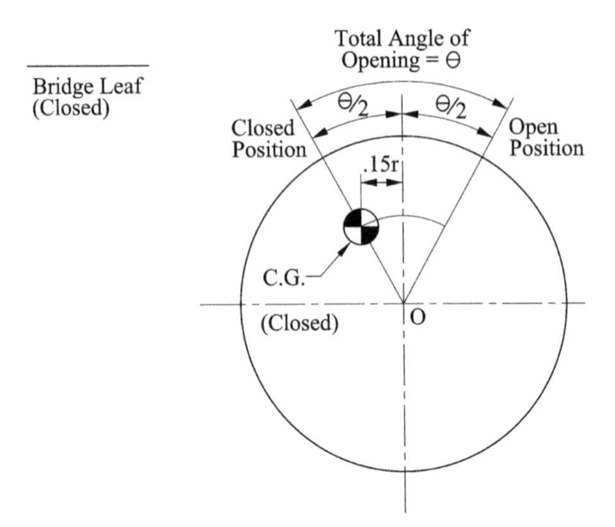

Figure 11-16 Ideal Bascule Bridge Balance.

its live load shoes. This will help keep the bridge from lifting up when no live loads are on the span and help to counteract the effect of impacts caused by heavy live loads on the span, which tend to bounce the bridge off the masonry or strike plates at its live load shoes.

The ideal amount of imbalance for a bascule span is still a matter of some debate. Various methods of determining the proper amount of imbalance have been suggested. Hool and Kinne[2] suggest that for a simple trunnion bascule on bronze bearings, the distance of the center of gravity from the center of rotation should be equal to the radius of the main trunnions, with the angle of this line forward of the vertical through the trunnion center by an amount approximately equal to one-half the full angle of opening (Figure 11-16). The intent is to produce a net gravitational moment in the closed position that will just overcome friction in the trunnion bearings, allowing the bridge to drop freely to firm seating at the live load shoes. Another approach is to provide for a certain gravitational reaction at the live load shoes, say, 2 kips, at each, with the balance of the bridge becoming negative at some point after it is half way open.

Whatever the desired balance, it must be carefully designed into the bascule span, as it is nearly impossible to make large shifts in the location of the center of gravity after the bridge and counterweight have been constructed (see Figure 11-17).

A bascule bridge rotates about a center as it opens and closes, so adjustment of the balance is not a simple matter. The position, as well as the amount of weight added or subtracted, is important. Some older bascule bridges, after undergoing major re-

[2] G. Hool and W. Kinne, *Long Span and Movable Steel Bridges* (New York: McGraw-Hill, 1923), p. 55.

BASCULE BALANCE SHEET

ITEM	WEIGHT	X	Y	Z	MX	MY	MZ
	LBS	FT	FT	FT	FT-LBS	FT-LBS	FT-LBS
B1	20000	40	3	15	800000	60000	300000
B2	14000	45	4	10	630000	56000	140000
B3	12000	50	5	5	600000	60000	60000
B4	12000	50	5	-5	600000	60000	-60000
B5	14000	45	4	-10	630000	56000	-140000
B6	20000	40	3	-15	800000	60000	-300000
F1	8000	100	5	0	800000	40000	0
F2	9000	80	4.5	0	720000	40500	0
F3	9500	60	4	0	570000	38000	0
F4	9800	40	3.5	0	392000	34300	0
F5	16000	20	0	0	320000	0	0
F6	18000	-20	-3	0	-360000	-54000	0
F7	10000	-40	-5	0	-400000	-50000	0
S1	6000	50	4	15	300000	24000	90000
S2	6000	50	4	8	300000	24000	48000
S3	6000	50	4	0	300000	24000	0
S4	6000	50	4	-8	300000	24000	-48000
S5	6000	50	4	-15	300000	24000	-90000
D	150000	50	5.5	0	7500000	825000	0
LL	8000	95	4	15	760000	32000	120000
LR	8000	95	4	-15	760000	32000	-120000

W TOT	368300 LBS	MX TOT	MY TOT	MZ TOT
		FT-LBS	FT-LBS	FT-LBS
		16622000	1409800	0

X CWT -30 FT - PER DESIGN

W CWT 554066.6667 LBS V CWT 3648 CU FT - PER DESIGN
INCLUDES POCKETS
VP= 960 CU FT

Y CWT -2.544459151 '=-MY/W CWT D CWT 151.8823099 LB/CU FT

Z CWT 0 '=-MZ/W CWT

Figure 11-17 Spreadsheet for Balance Computations. All components on the moving span are entered into the sheet, with coordinates.

construction, have been so far out of balance that sufficient weight could not be added to or removed from the counterweight to correct it, and weight had to be added to another part of the leaf to obtain an acceptable balance state. Other bascule bridges, as a result of reconstruction, had the concrete adjusting blocks in the pockets of their counterweights replaced with steel plates to increase the mass of the counterweights sufficiently to balance them. In some applications heavier material has been required.

12

DETAILING—VERTICAL LIFT BRIDGES

COUNTERWEIGHTS

The counterweights for vertical lift bridges are relatively simple to design, because they are merely heavy weights that must be of sufficient mass to balance the weight of the lift span. The counterweight produces a tension, due to the action of gravity on its mass, on the wire ropes from which it is suspended. These ropes pass over sheaves in bearings at the tops of the bridge towers and continue down from the other side of the sheaves to the vertical lift span, to which the ends of the ropes are attached. The sum of all the counterweight rope tensions on the vertical lift span is approximately equal and opposite to the gravitational force on the span. Typically, a vertical lift bridge has two counterweights, one at each end of the span. There is generally one set of ropes for each corner of the lift span, although many bridges have two sets of ropes at each corner, each set supported on its own sheave. For good operation and maximum overall durability, each corner of the lift span must be correctly balanced by having the right mass of counterweight connected to the ropes supporting that corner of the span.

Most vertical lift spans are symmetrical, so the counterweights must be likewise. If the lift span has a sidewalk on one side only, or an operator's house on one side, the counterweight detailing must take this into consideration. For adjustment of balance, pockets should be located in the counterweights so that the weight hanging from each group of ropes, connecting to one corner of the lift span, can be adjusted as independently as possible by removing or installing separate removable weights from the pockets. Thus, there should be two pockets at each counterweight, one located as close as practical to each end of the counterweight. The counterweights should work to exactly balance the weight of the lift span, ignoring the mass of the counterweight ropes.

Some vertical lift bridges have a separate counterweight for each corner of the lift span. Care must be taken to prevent these counterweights from binding at the towers because of rope twist. Counterweight guides can be provided with roller type bearings to bear against the guide rails, reducing friction due to the contact between the counterweight guides and rails as the ropes tend to unwind or lose their twist. Alternately, nonrotating ropes or equal numbers of left-hand and right-hand lay ropes can be used at each counterweight.

Counterweight Ropes

Counterweight ropes are main structural support members for a vertical lift span, but they are also wearing parts that unavoidably deteriorate as the bridge ages. The ropes corrode because they are very difficult to protect from the atmosphere, and they wear because they are not very well lubricated as they ride over unlined steel sheaves as the bridge is raised and lowered. The effect of rope wear is reduced by making the wires in the ropes larger; larger rope wires also corrode at a slower rate, all else being equal. Larger rope wires, however, fatigue more rapidly than smaller ones. Such fatigue occurs because the ropes bend or flex as they pass over the sheaves during bridge operation. The flex decreases as the sheave diameter increases. The bending stress in the wire increases as the wire diameter increases. The bending stress is cyclical, but fatigue is rarely the cause of failure of counterweight ropes in a vertical lift bridge. The bending stresses in a given wire in a rope tend to equalize out, due to the helical or spiral configuration of a wire in a strand, and of a strand in a rope. As these stresses equalize out, however, rubbing occurs between the wires as they make small longitudinal movements within the rope while bending over a sheave. These rubbing movements eventually result in what is called intrastrand or interstrand nicking (see page 400).

Usually 6 × 19 fiber core ropes, which consist of 6 strands made up of 19 wires each, twisted around a central rope or core made of Manila hemp or some other fiber material, are used as counterweight ropes (Figure 12-1). The most common counterweight ropes are of Seale construction, which provides for all outer wires in each strand being the same size. This type of construction produces a rope that actually has 25 wires per strand, and is sometimes called 6 × 25, but it is still considered a 6 × 19 type because the 6 extra wires are only fillers and are so small that they are not considered in computing the loaded cross-sectional area of the rope. The 6 small wires merely assist the outer perimeter of the strand in taking a more nearly circular shape. Both AASHTO (3.2.3) and AREMA (6.6.3) specify that wire ropes shall be made from improved plow steel. This is a cold-drawn carbon steel wire with great strength. Actually, most wire ropes are made from wire that is much stronger than improved plow steel, with strengths up to 260,000 psi, although the wire is still basically carbon steel with 0.83 percent carbon, the ideal percentage for producing hard, high-strength steel. The cold drawing process that is used to produce the wire is what gives it its great tensile strength. Such wires become pearlite filaments, which are very strong pure tension members. Cold drawing works best with plain carbon steel, so alloy steel or stainless steel wire ropes are relatively rare.

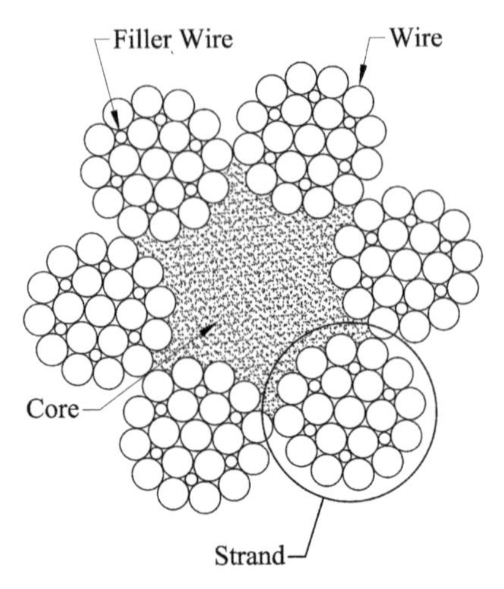

Figure 12-1 Cross Section of Vertical Lift Bridge Counterweight Rope. This is a typical rope, 6 × 19, with fiber core construction.

AASHTO (2.5.19) and AREMA (6.3.15.3) use what may be considered very conservative formulas for calculating the capacity of vertical lift bridge counterweight ropes. Both AASHTO and AREMA require that the total tension in counterweight ropes not exceed $\frac{2}{9}$, or 22.2 percent, of the specified ultimate strength of the rope, and that the direct tension load not exceed $\frac{1}{8}$, or 12.5 percent, of the rope's ultimate strength. The 2000 LRFD AASHTO has the same requirement. The total tension not only includes the load from support of the counterweight, the drag on guides, wind loads, and so on, but also the bending stress of the wire rope going around the supporting sheave. Both AASHTO and AREMA calculate this bending stress as follows:

$$K = \frac{0.8Ed \cos^2 L \cos^2 B}{D}$$

where

K = bending stress in the wire
E = modulus of elasticity of the wire
d = diameter of the wire
D = diameter of the sheave
L = helix angle of wire in strand
B = helix angle of strand in rope

The 2000 LRFD AASHTO simplifies this in an equation that it explicitly states is conservative:

$$\delta_b = E(d/D),$$

where δ_b is the same as *K*. The *E, d,* and *D* are practically the same in the two equations. Depending on the construction of the rope, the more complicated equation may produce a bending stress value 30 percent or more less than the simple (2000 LRFD AASHTO) equation.

The 2000 LRFD AASHTO equation is simply the formula for the bending of a wire on a diameter equal to the sheave diameter. The AREMA and AASHTO 1988 formulas reduce this bending stress value somewhat in proportion to the effect the helix angles have in reducing the bending stress. The actual bending stress in the wires depends on further factors, such as the amount of wire twist, the degree of lubrication of the rope (both internally and in the sheave groove), and the condition of the core of the rope. To gain the most from the preceding fatigue equations, AASHTO (3.2.2) and AREMA (6.6.2) specify counterweight ropes to be maximum of $2\frac{1}{2}$ in. in diameter, with a corresponding preferred minimum diameter of the sheave of 200 in., or 16 ft 8 in. In fact, fatigue failure of wires has been almost unheard of in vertical lift bridge counterweight rope, without the wires having become excessively worn, so that the 2000 LRFD AASHTO formula may be excessively conservative.

Avoiding the possibility of fatigue in the rope wires by using a formula that is definitely conservative, thus "overdesigning" the ropes, is not necessarily a bad idea. Fatigue breaks can occur inside a rope, where they are not visible. A failure may thus become imminent without warning. Corrosion and abrasive wear, on the other hand, can provide visible signs and normally produce a gradual deterioration over time. Larger ropes require larger sheaves to hold down bending stress, increasing the cost of the installation. Oversized ropes will tolerate a more corrosion and abrasive wear before requiring replacement.

Very low allowable unit loads for counterweight ropes requires many large-diameter ropes for a heavy bridge. This makes the ropes last as long as possible, postponing or avoiding the difficulty of replacing the counterweight ropes. Ideally, the ropes should last as long as the bridge, but the longevity of vertical lift bridges is increasing, as it is for most other bridges, because of the difficulties in finding the capital for bridge replacement and for environmental and other reasons. Thus, many vertical lift spans exist that are 70 or more years old, which are expected to last considerably longer in spite of general deterioration and perhaps a degree of functional obsolescence. Many of these bridges have been forced to undergo replacement of their counterweight ropes, and it appears to be wise to provide for ease of counterweight rope replacement and readjustment in designing a new bridge.

Failure of counterweight ropes is usually due to wear resulting from contact with the sheave grooves, which progresses until the wires become so thin at worn spots that they break (see Figure 12-2). Such wear can be reduced by enclosing the sheaves in weather tight housings and providing adequate lubrication and corrosion protection during use. The ropes are provided with lubricant preservative when fabricated, which usually protects them from corrosion and internal wear for several years. Many types of lubricant have been used on, and in, wire ropes. As long as a lubricant provides good corrosion resistance, offers some lubricating effect between the rope wires and between the rope and the sheave, and does not readily wash off or evaporate away, it should be adequate for a wire rope. Relubricating the ropes in service reduces exter-

Figure 12-2 Worn Counterweight Ropes for Vertical Lift Bridge. The rope wires have large wear spots. There are many broken wires that can be seen. These ropes were replaced as a result of these findings.

nal wear and can significantly reduce the rate of corrosion. If the lubricant penetrates to the interior of the rope, it can also help prevent internal wear, but this penetration is difficult to achieve. Some experts say it is impossible to add internal lubricant with ropes permanently mounted and loaded in such a way as are vertical lift bridges.[1]

As indicated earlier, most movable bridge counterweight ropes have fiber cores. Many, if not most, wire ropes are made with wire rope cores, which results in a higher breaking strength for the rope when new. These wire rope cores tend to break up over time, negating the increase in rope strength. They tend to abrade the inner surfaces of the wire rope strands. In addition, they do not hold as much lubricant as a new fiber core, when the fiber core is made of the proper material such as Manila hemp. Other types of fibers, such as sisal, have been used for the cores of fiber core ropes. Jute is often mentioned as a typical material for fiber cores, but it is a very poor material as it deteriorates rapidly because of abrasion and lack of rot resistance. Most newer fiber core ropes use plastic fibers of a material such as polyester or polypropylene. Plastic cores do not hold nearly as much lubricant as Manila hemp, but they hold up fairly well. Most wire rope salespersons promote the use of polyester cores, as they are cheap and readily available, by comparing them with jute cores. Manila hemp cores have proven themselves by successfully serving on many movable bridges for 70 years or more without allowing internal damage to the rope strands. Manila hemp should be specified when fiber core wire ropes are to be supplied as vertical lift bridge counterweight ropes.

[1]Frank Haas, "Frank J. Haas, A Personal Profile," *Wire Rope News and Sling Technology* (February 1999), p. 11.

Many different types of end connections have been used for counterweight ropes, but the most common is the standard open socket, connecting the rope to the bridge span and to the counterweight by means of a steel pin. The ropes are usually connected directly to the counterweight, and connected to the lift span by means of eyebolts or turnbuckles to allow for ease of individually adjusting the rope tensions. This situation is sometimes reversed, with the eyebolts or turnbuckles on the counterweight ends of the ropes instead of the span ends.

Sometimes lever-type equalizers are provided at the counterweight connections or at the eyebolt connections to the span in an attempt to ensure equal loading of the ropes, but their value is questionable. The levers are made triangular to conserve space in the confined area of the connections, and this results in a shift of load from the longer rope to the shorter one as a rope stretches. This transfer of load should eventually cause a return to equilibrium, as the shorter rope, now more heavily loaded, should stretch to approach the longer rope in length, tending to equalize the loads. Other than making the connections of the ropes to the span more convenient, equalizers have very little value on a vertical lift bridge. Several vertical lift bridges have been designed and built with no equalizers and have not suffered as a result. Greater redundancy is achieved if the ropes are not equalized. Proper adjustment at installation is usually necessary in any case to minimize the inequality of the rope loadings. Rope lengths must be very carefully measured during attachment of the sockets, but the rope lengths will still not be perfectly equal, and handling them prior to installation can result in a change of the effective length. Ropes are usually prestretched at the factory, as a part of the socketing process, and then marked with a stripe to indicate the proper relationship of the ends of the ropes to each other.

Some vertical lift bridges use block type or "button" sockets, similar to those used in other structural rope applications such as suspension bridges, for a slight saving in initial construction cost. Adjustment of rope tensions during installation is usually done by shimming at the load-bearing surfaces of the sockets. This makes later rope replacement much more expensive and time-consuming, as individual connection and adjustment of ropes with end connections of this kind is a great deal more difficult, particularly in the confined quarters present at counterweight rope span and counterweight connections. If only one or two ropes in a group are replaced, they will stretch and become slack, requiring them to be reshimmed later. If all the ropes are replaced at the same time, they are likely to stretch nearly equally, so reasonably equal tension is maintained.

A very few vertical lift spans have counterweight ropes attached with no adjustment possible, relying on the following to provide equal loading on the ropes:

1. Exact correct length to connections when manufactured
2. Small adjustment by twisting or untwisting ropes
3. Equalization of loads on ropes by working-in (see below)

Of these measures, the first is probably the most reliable, as it depends only on proper quality control throughout the bridge building process. The second can accommodate only very small length adjustments and is hardly worth doing. The third eventually ends up doing its job, intentionally or not. Any ropes initially more highly loaded will stretch more than the others as the bridge operates, so that they relax and shed load to the other ropes. This can also accommodate future changes to some extent. If some wires are broken on one rope in a group, for instance, the rest of the wires, which are unbroken, will stretch to some extent as they take up the load. When these wires stretch, they transfer load from the damaged rope to the other ropes in the array. Because of the rope's stretching, as indicated earlier, it is practically impossible to replace individual ropes with this arrangement so that the new rope will share load equally with the others.

To be able to replace damaged ropes in the future, or to replace a whole set of ropes after they have worn, it is a good idea to make the attachments easy to undo. Almost every vertical lift bridge in the United States has the counterweight ropes socketed on each end, with one rope per sheave groove. The sockets can take any of the forms indicated earlier. The turnbuckle or eyebolt tensioning arrangement is very good idea for the permanent connection at one end of a rope, as it can be reused to advantage later when replacing the rope. The eyebolt is simply loosened, the pins are removed from the open sockets at each end of the rope, and the old rope is pulled out and a new one pulled in. One end of the rope is attached, and the other is connected at the eyebolt and put in initial tension. The bridge is operated a few times, and the rope is retensioned. After a few iterations of the retensioning process, the new rope is at the same tension as the old ones. Ropes not having adjustable connections at one end are very difficult, if not impossible, to retension. If only one or a few ropes in such a group are replaced, they will soon lose their tension and become useless if not readjusted.

COUNTERWEIGHT SHEAVES

Counterweight rope sheaves must be made to a sufficiently large diameter to reduce the fatigue experienced by the ropes to a minimum as the bridge operates. As indicated earlier, AASHTO and AREMA have standard factors for sheave diameter as a function of rope diameter; these factors do not appear to be excessively conservative, but fatigue failures of counterweight ropes are practically nonexistent. AASHTO states, in section 2.9.7: "For main counterweight ropes, the pitch diameter of the sheave, center to center of ropes, shall not be less than 72 times the diameter of the rope, and preferably not less than 80 times."

The AREMA specifications contain an almost identical statement (6.5.36.8). The sheaves should be as wide as necessary to accommodate all the counterweight ropes connected to one corner of the lift span. Fabrication of such large sheaves, which may be up to 16 ft or more in diameter, is difficult. In the past they were usually cast of steel in one piece, but today welding is prevalent and economical for general industrial applications, so that there is little need for large castings for general use. Today it is difficult to locate foundries that can produce cast counterweight sheaves and to obtain competitive price quotes for them.

Because of the difficulty in finding a fabricator capable of supplying good-quality cast steel sheaves of the size required for a large vertical lift span, designers have been commonly specifying welded sheaves. These counterweight sheaves, whether cast or welded, are subjected to very high and complicated loadings, so that proper design and high-quality fabrication are essential for longevity. Proper joint detail is fundamental in welded sheaves, as in any highly and irregularly loaded weldment. This is especially the case with cyclical reversing loading, such as in vertical lift bridge counterweight sheaves. AASHTO (2.9.9) reduces the allowable stresses significantly for welded sheaves. The designer should not be overzealous about reducing the weight of the finished weldment, but should produce a sound design that is as easy to fabricate as possible. The reversing loads experienced by counterweight sheaves as they rotate are particularly brutal punishment for weldments, which tend to have many defects in their weld areas. It is important for welded counterweight sheaves, perhaps even more than cast steel sheaves, to be thoroughly and competently inspected after fabrication to minimize the likelihood of the presence of weakening defects in sheaves installed in the field. The track record for durability appears to be better for cast sheaves than for welded ones, but the cast sheaves have a longer history, and early developmental failures may be buried in time. With the use of welded sheaves, it is difficult to achieve a sufficiently thick hub with proper proportions to provide the desired shrink fit between the hub and the sheave shaft, while avoiding major sectional discontinuities at the hub-web interfaces; this is relatively simple to accomplish with cast sheaves. AASHTO (2.9.9) and AREMA (6.5.36.10) have similar requirements for welded sheaves, including the use of a one-piece forging for the hub of a sheave.

There have been isolated cases of failure in cast steel counterweight sheaves on vertical lift bridges (see Figure 12-3). This type of failure has occurred mainly on highway bridges that have gone through many opening and closing cycles and may have had significant numbers of flaws within the sheaves dating from the time of fabrication. These flaws may stabilize over time or may continue to the point that the sheaves begin to disintegrate.

Hovey, in the 1920s, recommended multiple counterweight sheaves for each corner of a vertical lift span, so that a single sheave could be taken out of service for repair without preventing operation of the bridge.[2] This supposedly redundant feature may be theoretically desirable, but the difficulty of mounting, supporting, maintaining, and inspecting paired sheaves in service has been substantial because of their necessary close proximity. Paired sheaves, unless designed so that each sheave can carry the whole load without overstress at the sheave ropes, shaft, or bearings, should not be considered redundant, except perhaps in the static load case only. Operation with only one of the paired sheaves in service may allow moving the span to the raised or lowered position in an emergency. Counterweight rope deterioration caused by operating the bridge during periods of double the normal load on the ropes may be severe and irreversible. Providing three or more counterweight sheaves per corner of the span, to reduce the overload when one sheave is removed, is probably impossible for all practical purposes. Most modern vertical lift spans are built with single

[2]O. E. Hovey, *Movable Bridges*, Vol. 2 (New York: John Wiley & Sons, 1927), p. 187.

Figure 12-3 Vertical Lift Bridge Cracked Counterweight Sheave. The cracks in these sheaves appeared shortly after the bridge was put into service. The cracks were likely the result of manufacturing defects rather than any service conditions. The repairs shown have held up for about 20 years; it is questionable whether the repairs were needed at all.

sheaves supporting each corner of the span. The maximum practical number of counterweight ropes on a single sheave may not be much more than 16, but even this results in a very large sheave to be fabricated. With the increasing weights of large vertical lift spans, paired sheaves should be used if necessary to limit the width of each sheave to about $3\frac{1}{2}$ ft maximum. Larger sheaves require exceptional care in design and fabrication.

The proper analysis of counterweight sheaves for design purposes can be accomplished by means of finite element analysis. There are many computer programs available for this purpose; performing a rigorous analysis is not difficult, provided the proper setup is made. Hovey, in Appendix B in Volume 2 of *Movable Bridges,* provides an acceptable manual method of designing and determining stresses for cast sheaves; this guide can also be used for welded sheave design. For cast sheaves, there are many old existing examples that have successfully stood the test of time, and these can be used as guides when identical or similar sheaves are desired. The fabrication specifications for these older examples are usually obsolete or unavailable, so there is some risk of not obtaining the desired finished product without considerable present-day input. Control of the production process is essential, regardless of the means of fabrication of the sheave, as a cast or welded sheave can easily contain fabrication flaws that may severely curtail its useful life. Internal flaws and stresses in counterweight sheaves must be avoided to the greatest extent possible.

COUNTERWEIGHT SHEAVE SHAFTS

Ideally, a sheave should rotate about a fixed supporting shaft so that the possibility of cyclical fatigue in the shaft is eliminated, but see page 80. The load transfer to the shaft is then taken through the top of the bearing, rather than the bottom. This requires a larger bearing to be used, because of the difficulty in lubricating the upper loaded parts of the bearing. Most counterweight rope sheaves are fixed to their shafts, which are supported on and turn in their bearings, so that the loaded area is below the shaft. In either case, the sheave hub should be of substantial construction. The hub should be of one piece with welded sheaves (see page 177). Hubs of welded or cast sheaves should have more than the minimum thickness around the bearing or shaft, and have the proper fit. The tightest shrink fit possible, FN-2 (per USAS B4.1[3]) (per AASHTO 2.5.17) (FN-2, per AREMA 6.5.1), should be performed on a hub-to-shaft fit when the shaft is to rotate with the sheave. There is some concern that the tight FN-3 fit overstresses the components being attached and can result in failure, particularly with the required dowels or keyways added. This can occur if the components are insufficiently substantial. Analysis of both hubs and shafts should be made to determine that overstress will not be a result of the tight fit. FN-2 fits produce a maximum of 0.07% interference; FN-3 goes to 0.1% interference. 2000 LRFD AASHTO requires an H7/s6 fit, which can produce 0.12% interference. If the hole and shaft are distorted equally with this fit, approximately 18,000-psi tension or compression is produced at the interface. This is normal (at a right angle to) the primary bending stress in the shaft or hub. The finishes inside the sheave hubs, whether mating to the bearing or to the shaft, should be smooth and true, per AASHTO and AREMA standards as cited, to ensure a durable fit.

The shafts supporting the counterweight sheaves of vertical lift bridges are frequently called “trunnions,” although this usage is not correct. When these shafts are fixed to the sheaves in the usual arrangement, they differ in a very significant way from true trunnions (see Figure 4-2) in that the shafts and the sheaves they support make complete revolutions under load. A typical vertical lift bridge sheave makes up to three or more rotations as the bridge lifts to its full height, and the same number in the opposite direction when lowering. Each of these rotations results in complete bending stress reversal in the shaft when it is fixed in the sheave and supported on bearings. Many older vertical lift bridges were designed prior to the development of a good understanding of metal fatigue, and the sheave shafts have failed on some of these bridges, which have spent many years raising and lowering several times a day for navigation.

To prevent, or at least minimize, the likelihood of fatigue failure on vertical lift bridge counterweight sheave shafts that rotate with the sheaves they support, these basic engineering principles should always be followed:

1. Do not provide large changes in diameter in the shaft, particularly at the locations of large bending stresses.

[3]USAS B4.1-1967 (1974), reaffirmed 1979, *Preferred Limits and Fits for Cylindrical Parts* (New York: American Society of Mechanical Engineers, 1974).

2. Where changes in diameter do occur, provide the largest fillet radius possible at the junction of the smaller diameter with the portion of the shaft that is of larger diameter.
3. Make the shaft of nickel steel or another alloy that is resistant to fatigue.
4. Do not add other stress risers, such as keyways, dowel pin holes, or lubrication passages, at points at or near maximum stress.

The 2000 LRFD AASHTO contains considerable information on fatigue considerations for sheave shaft design, including charts of stress concentration factors. Some vertical lift spans have been fitted with stationary sheave shafts, which eliminate fatigue from consideration (see page 80). Unfortunately, this arrangement does not provide for the most ideal situation for lubrication of the shaft bearings. Proper specification of the bearing and detailing of the mounting, with the assistance of the bearing manufacturer, can produce a satisfactory result that will give long years of service (see Figure 5-24). This can be provided in new construction or in a retrofit to an existing bridge.

Most vertical lift bridge counterweight sheave shafts that developed destructive cracks due to fatigue were supported in friction journal bearings. Wear, and sometimes corrosion, resulted in small stress concentration points that assisted in the development of fatigue cracks. Roller bearings, by eliminating the wear factor at the shaft, can help prevent fatigue cracks from developing.

COUNTERWEIGHT SHEAVE SHAFT BEARINGS

Most older vertical lift spans have bronze bushing type friction bearings for their sheave shafts. These are generally quite long lasting and serviceable. The bearings tend to be rather large, and because they are always loaded it is difficult to lubricate them fully. Various configurations of lubricating grooves and channels have been used, with largely indifferent success. Even when they have failed to receive adequate maintenance and run dry of lubricant, these bearings tend to continue to perform, although at the cost of much greater power required to operate the bridge. Many of these bearings have a separate bushing in the lower half only. A solid cap is used in these cases, made of either bronze or steel, with a small clearance between the cap and the shaft journal. This arrangement, with the cap firmly bolted to the bearing base, prevents rotation of the bearing bushing in the base in the event that the bushing "freezes" to the journal. In such an event, the journal breaks free of the bushing when the sheave is rotated, and the bridge can continue to operate as long as the bridge drive can provide sufficient torque to keep the shaft rotating. Wear of the journal and bearing will be substantial under these conditions.

Most newer vertical lift spans use antifriction bearings. These are an improvement over bronze bearings, as they eliminate wear at the journals, reduce maintenance requirements, lower the horsepower needed to operate the bridge, and can help eliminate shaft fatigue failures. Most counterweight sheave shafts that have failed in fatigue were supported by friction bearings; the fatigue cracks originated in the journal fillets at areas that were scored because of wear (see page 180). No sliding contact and resulting abrasion occur at a shaft with an antifriction bearing supporting it, and the shaft is also usually better protected from corrosion by the housing for the antifriction bearing. Antifriction bearings reduce friction considerably, but this is not necessarily a plus. Some vertical lift bridges with roller bearings on the counterweight sheaves have had braking difficulties because of the extreme ease of movement. This can be a particularly vexing problem when sudden snow loading causes an excessive imbalance of the span. Low operating friction can also cause difficulties with a control system that is incapable of controlling speeds at low, zero, or negative loads, particularly if the load pattern quickly changes from one to another of these (see Chapter 18).

With rotating shafts supported in fixed bearings, lubrication is fairly simple, as the bottom of the bearing, where the lubricant tends to accumulate, is under load. With stationary shafts supporting rotating sheaves, lubrication is more difficult, as the top of the bearing is loaded. Extra caution must be exercised with this configuration, and a larger antifriction bearing is usually required. It can also be wise to increase the size of a friction bearing used in this arrangement.

AUXILIARY COUNTERWEIGHTS

Most vertical lift bridges are equipped with auxiliary counterweights, which are intended to at least partially compensate for the change in effective balance of the span due to the counterweight ropes moving over their sheaves as the bridge opens and closes. These auxiliary balancing devices have been developed in many forms. Most serve to pull up on the lift span when it is in the lowered position, and pull down on it when it is in the raised position. Other types automatically vary the amount of mass at the main counterweights as the bridge raises and lowers.

One of the most common types of auxiliary counterweight uses a fairly large weight suspended on wire ropes (Figure 12-4). The ropes pass over sheaves on the bridge towers and then reach out and attach to the span. The assembly is arranged in such a way that the ropes are nearly horizontal between the span and the sheave when the span is lifted halfway. When the span is in the lowered position, the tension of the ropes has a significant upward vertical component at the span connection. When the bridge is fully lifted, the vertical component of the rope tension is downward at the span connection. The vertical component varies continuously between the maximum upward and maximum downward force, but the changing angle of the connection between the ropes and the span introduce a sinusoidal function into the vertical component. The counterbalancing effect thus does not exactly equal the change in counterweight rope load on the span, which is a linear function of span lift, but it is close enough for operational purposes.

Some other, usually older, vertical lift bridges use the same auxiliary rope and auxiliary counterweight concept, but connect the ropes to the lift span by means of a pivoting tower mounted at the top of the lift span, in the center of the span (Figure 12-5). This works similarly to the system with the ropes connected directly to the span, but adds a substantial amount of dead load to the bridge. With either of these types, it is

Figure 12-4 Vertical Lift Bridge Auxiliary Counterweight. The diagonal ropes extending from the center of the vertical lift span to the towers reach down and connect to weights hanging alongside the tower legs. These ropes pull up on the span when it is down, and down when it is up, compensating for the weight of the counterweight ropes.

typical to take advantage of the auxiliary counterweight ropes to provide a means of connecting electric power and control cables to the movable span, particularly handy for a span drive vertical lift.

Some vertical lift spans are equipped with auxiliary counterweights—a variation of that described earlier, located at each tower, connected to the ends of the span rather than at the center and providing no means of carrying electrical cables between the tower and the lift span. This type is used primarily on tower-drive vertical lift bridges, but has also been used on span-drive bridges.

Many vertical lift bridges use balance chains to counter the effect of the counterweight ropes (Figure 12-6), typically two chains per counterweight. The balance chains have one end attached to the bottom of a main counterweight and the other end connected to a fixed point on the tower. The amount of balance change with this arrangement is a linear function of the vertical movement of the bridge, so the counterweight ropes can be exactly balanced in all positions of the span. Balance chains are very heavy and add to the load on the counterweight sheave shafts. The

Figure 12-5 Vertical Lift Bridge Auxiliary Counterweight. The diagonal ropes extending from the mast reaching up from the center of the vertical lift span to the tower tops reach down and connect to weights hanging alongside the tower legs. These ropes pull up on the span when it is down, and down when it is up, compensating for the weight of the counterweight ropes.

chains are difficult to lubricate, and sometimes links seize; thus, permanent kinks develop in the chains, compromising the balancing effect. If kinking is allowed to continue, the operation of the bridge is impeded and can be prevented completely.

Some vertical lift bridges do not use auxiliary counterweight devices, but rely on the bridge drive power to overcome the imbalance due to the position of the counterweight ropes. Most of these bridges lift only a short distance, so the effect of the rope imbalance on power requirements is not great. The Hawthorne Bridge in Portland, Oregon, lifts about 110 ft to its full raised position without the assistance of auxiliary counterweighting. This bridge has had operating difficulties, but now opens typically only a short distance.

SPAN GUIDES

Span guides for vertical lift bridges have been the subject of some creative work by designers of movable bridges. It is not desirable to have a vertical lift span completely free to float around, or "blow in the wind," when lifted off its live load bearings, as the counterweight or operating ropes may abrade on the edges of their grooves, or even slip out of their respective grooves, should the bridge translate too far laterally. Longitudinal and lateral restraint must also be provided to avoid having the lift span

Figure 12-6 Balance Chains for Vertical Lift Bridge. These devices were once commonly used to compensate for the weight of counterweight ropes passing over their sheaves as the bridge operated.

impact the towers excessively when raising or lowering. In the closed position, the bridge must be aligned accurately so that the roadway is correctly located. This is especially important for railroad bridges, as the rails must be properly aligned at the joints at the ends of the lift span so as to avoid interference when the span closes and to prevent derailment of trains crossing the bridge. The span must also be located longitudinally when seated, firmly enough to transfer the longitudinal tractive forces from vehicles crossing the bridge. This is particularly true for railroad bridges carrying heavy freight trains (see page 185).

The most expeditious means of providing span guidance is to provide extensions at the ends of the lift span girders or trusses and have these extensions envelop the forward columns of the main towers of the lift span (see Figure 12-7). The extensions are equipped with rollers or friction shoes, and the towers are provided with vertical rails that mate with the contact points on the extensions. The rails are frequently designed with nonconstant thicknesses so that the clearance to the contact points on the extensions is reduced at the lower end of span travel. This produces accurate align-

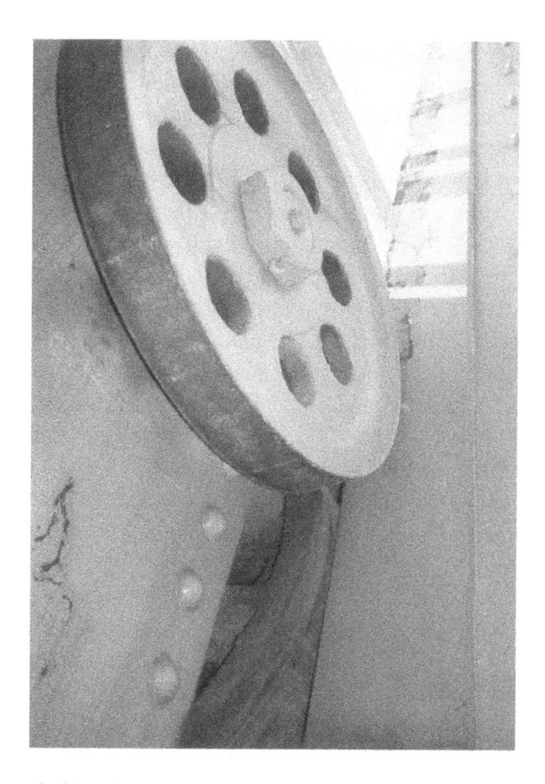

Figure 12-7 Span Guides for Vertical Lift Bridge. These devices keep the moving span aligned with the towers as the span raises and lowers. The guide shown incorporates a wheel for lower friction; many span guides use friction shoes.

ment of the bridge in the closed position. There has been some concern in recent years about the seismic stability and safety of vertical lift bridges. This can be a valid concern for vertical lift spans left normally in the raised position, but no damage is known to have occurred to any existing vertical lift bridge, in either the raised or the lowered position, that has experienced a sizeable earthquake, except at the counterweights.

Additional lateral centering devices are in some cases located at the underside of the end floorbeams of the vertical lift span. These engage mating devices mounted on the piers so that when the bridge is fully closed, the centering of the ends is as accurate as possible. These devices may be necessary for railroad bridges, but sufficient accuracy of positioning can just as easily be obtained in most cases by proper provision of lateral clearance at the main span guides. Wide spans that may require significant lateral clearance at the main guides, to allow for temperature variations, may require the auxiliary lower centering devices, particularly in railroad applications where track rail alignment is critical. Such accurate centering of the span is of doubtful utility for highway spans and should be dispensed with unless careful consideration of the particular application indicates that lateral centering is required. The additional centering devices are not only an added expense, but require maintenance to avoid operating problems. These devices sometimes interfere with the proper seating of the span after the main live load bearings on the vertical lift span experience significant wear. The centering devices may assist in resisting lateral forces during an earthquake, but their location substantially below the center of gravity of the span, at least for a truss bridge, will minimize their effectiveness under this loading.

A vertical lift bridge is free to float longitudinally when lifted because its only support is the ropes from which it is suspended. In addition to lateral guides to control the side-to-side movement of the lift span, longitudinal guides are installed on one end, designated the "fixed" end of the lift span. These guides allow only limited free longitudinal movement of the vertical lift span, just enough to avoid binding during normal operation. The guides can also act as longitudinal live load traction restraints, especially important for railroad bridges, but this function is usually performed by the live load shoes, which typically have a special detail at the fixed end of the span that prevents longitudinal movement at that end when seated. The other end of the vertical lift span, the "free" end, has no longitudinal span guides. On railroad bridges and most highway bridges the expansion end of the vertical lift span usually has a rocker or other expansion type of live load shoe under each main bridge member.

COUNTERWEIGHT GUIDES

Counterweight guides must be provided at vertical lift bridges to prevent impact of the counterweights against the tower superstructures. The counterweights can be energized laterally by the wind or by seismic forces. Wind forces are fairly small, but can cause substantial lateral displacement of an unrestrained counterweight when the bridge is in the raised position. Seismic forces can be quite large, but are so unlikely to occur that at most locations there is little point in designing for them. Where earthquakes have occurred, the most frequent damage to a movable bridge has been to vertical lift bridge counterweight guides. Such damage typically occurs at a vertical lift bridge that has been left in the lowered position during the earthquake. Rigid counterweight guides capable of withstanding earthquake forces without damage transmit the acceleration force of the counterweight to the tower. This may be considered in the design of a vertical lift bridge in a seismically active area, but the guides and towers must then be designed to withstand these loads.

Most vertical lift bridges are likely to be in the lowered position during an earthquake. If a span were to be in the raised position, the span guides would be forced to withstand forces developed during the earthquake. Counterweight guides must withstand much lower wind forces than span guides, although the seismic forces would be about the same at the span and at the counterweights. Counterweight guides designed solely for wind forces can thus be inadequate to fully restrain the counterweights in an earthquake. As the counterweights can typically swing some distance before striking and damaging critical structural steel, such as the main tower members, a counterweight guide that breaks free at a certain prescribed level or flexes elastically during an earthquake can prevent damage during a seismic event, up to a certain magnitude.

A counterweight guide that deforms permanently in an earthquake may prevent more serious damage, at the cost of making minor repairs to the guide later.

SPAN LOCKS

Vertical lift bridges are usually equipped with span locks to hold them in the closed position. The locks serve mainly as safety devices to prevent accidental opening of the span before traffic is halted. They may be interlocked with other systems. A vertical lift bridge is inherently imbalanced so that it is likely to remain seated when seated, so span locks are usually not needed for stabilizing purposes. The span locks on a highway vertical lift bridge have in many cases been deactivated after the bridge was put in service, as they were usually located in a spot that made them difficult to maintain and they offered no perceptible advantage in bridge operation or load carrying capacity. Some railroad vertical lift bridges are equipped with span locks to hold them in the raised, or open, position.

Span drive vertical lift bridge usually take the form of longitudinally acting lock bars that drive from the end of the lift span into sockets mounted on the tower. They may be activated by a linkage from a central device, located at the machinery house of a span drive vertical lift span, or they can be independently operated by an electromechanical or hydraulically operated device connected to each span lock. They are usually located at the top of the truss of this type of bridge. On a tower drive vertical lift bridge, the span locks are usually located on the pier, with one under each end of the lift span. The drive mechanism is on the pier, and the socket with which the lock bar engages is on the underside of the lift span. Some span drive vertical lift bridges also have this arrangement.

BUFFER CYLINDERS

Many vertical lift spans have buffer cylinders located on the ends of the span to cushion impact when seating. Some vertical lift spans also have buffer cylinders mounted on the tops of the towers to cushion impact in the event of an overtravel of the span when opening. Almost all buffers are of the compressed air piston-cylinder arrangement. See page 160 for further discussion of buffer cylinders.

RAIL JOINTS

Railroad vertical lift spans require an uncomplicated arrangement for the rail joints at the ends of the lift span (Figure 12-8). The ends of the rails on the lifting structure are simply extended beyond the ends of the lift span a sufficient distance to bridge the gap to the approach spans and provide a solid seating in chairs at the ends of the rails on the approaches, while avoiding interference, when the vertical lift span is opening, with any overhead structure on the approach span.

The forms of the rail joints for vertical lift bridges are similar to those used on rail-

Figure 12-8 Miter Rails on Vertical Lift Bridge. These rails are firmly fastened down onto the approach structure in the foreground, and to the vertical lift span, at the far side of the joint.

road bascule bridges and are discussed in Chapter 11. The clearance needed when the bridge is opening is not difficult to determine, as the deck does not rotate when the bridge opens. The joints may be of a different form on each end of the bridge, as normally one end of a vertical lift span is "fixed," while the other end moves longitudinally as the bridge expands and contracts because of temperature changes. At the expansion end of very long vertical lift spans, substantial longitudinal expansion capability, without allowing a gap in the rails at these joints, is required. Failure to provide a continuous support for the rail wheels over the joint will result in impacts that can shorten the life of the rail at the joint and may result in structural damage to the bridge.

LIVE LOAD SHOES

The most common live load shoes for vertical lift bridges consist of segmental rockers hung from the bottom of the expansion ends of the main members, and longitudinally restraining load-carrying devices at the fixed, or nonexpansion, ends of the trusses or girders. Some highway vertical lift spans that have no sizeable longitudinal tractive forces applied to them have simple masonry plate and sole plate combinations at each end of the lift span. These bridges rely on the span guides to hold the vertical lift span in the proper longitudinal position.

BALANCE

A vertical lift span, like a bascule bridge, is generally assumed to require counterbalancing of the weight of the moving leaf to provide for safe and economical operation. The only proposed noncounterweighted vertical lift span is very small, and lifts a short distance, so that the detrimental effects of not having counterweights are minimized. The costs of construction and maintenance for counterweights, sheaves, ropes, shafts, and bearings are large. There is no additional cost for larger bridge bearings, as both AASHTO (2.4.1) and AREMA (6.3.15) require that the live load bearings for a vertical lift span be capable of also supporting the full dead load of the movable span without overstress, to allow for replacement of the counterweight ropes and other repairs. The replacement of worn-out counterweight ropes causes an extreme disruption of bridge operations and is expensive. If a vertical lift bridge can be safely provided without counterbalancing, a significant advantage in saving capital costs and maintenance can be realized.

The total weight in the counterweights of a vertical lifting span is usually designed to exactly equal the weight of the lifting span, so when the bridge is lifted halfway, it is exactly balanced. The counterweight ropes are large and heavy, so the work against gravity in lifting them up over the counterweight sheaves and letting them back down again is significant. For this reason, auxiliary counterbalancing devices are typically provided, as discussed earlier (on page 181). On the other hand, stability in the closed and open positions of the bridge is enhanced by having some imbalance, which can be provided free, so to speak, by the counterweight ropes. Some vertical lift spans that do not lift to great heights do not have auxiliary counterweights. The Burlington Bristol Bridge, crossing the Delaware River between Pennsylvania and New Jersey, has been operating successfully for 70 years without auxiliary counterbalancing devices. It is a large vertical lift span, more than 500 ft long, but lifts only a maximum of 70 ft. It has 16 counterweight ropes at each corner of the span, which weigh about 6 lb per foot each. If the total weight of the two counterweights exactly equals the weight of the vertical lift span, the maximum imbalance of the span can be reduced to:

$$6 \times 16 \times 4 \times \frac{70}{2} = 28{,}000 \text{ lb}$$

Thus, there are 7000 lb of dead load per corner of the seated lift span. This weight is not greatly in excess of the desired dead load reaction at the live load shoes. Gravity produces the same retarding force, opposing closing of the bridge, when it is in the fully open position. Sufficient additional drive capacity must be provided to counteract this gravitational resistance when the span is started in motion from the seated or raised position, but this can be considered a good investment in enhanced bridge stability and the extra capital cost for the needed power is likely to be matched by the capital cost savings realized by not having to install auxiliary counterweights. The disadvantage of having no auxiliary counterweights is the tendency of the bridge to "run away" when approaching the nearly closed or nearly open position. A large impact can typically be taken safely by the live load shoes when the bridge is closing, but a vertical lift bridge without auxiliary counterweights should also have substantial stops at the tops of the towers to safely stop the bridge in the event that the control system fails when the span is opening.

There are additional variable masses on a rope-operated span-drive vertical lift bridge, as the operating ropes also transfer their mass to and from the lift span. The uphaul ropes are wound onto drums mounted on the lift span as the bridge raises. As they unwind during lowering of the bridge, their weight it transferred to the towers. The downhaul ropes are wound onto the lift span as the bridge lowers, but they are always supported by the lift span. Because there are few operating ropes and their weight per foot is small, this effect can usually be ignored. Eight uphaul ropes weighing 2 lb per foot produce a weight shift over a 70-ft lift of only 1120 lb for the entire bridge. The operating ropes partially counteract the effect of the counterweight ropes, as the weight of the uphaul operating ropes is transferred to the span as it lifts.

OPTIMUM DRIVE SYSTEM

The reliability of operation of long vertical lift bridges requires careful consideration. To guarantee avoidance of skew failures, a tower-drive system should not be used. To minimize control difficulties, the bridge should be span driven, with direct mechanical drive links from the point of power application to each end of the span. Rope drives should be avoided, as they wear out rather rapidly; they also stretch and cause misalignment. Vertical lift bridge machinery is discussed in more detail in Chapter 15. Controls are discussed in Chapter 18.

13

DETAILING—SWING BRIDGES

CENTER PIVOT BEARINGS

The center bearing for a swing bridge carries the entire dead load of the movable span, transmitting it to the center pier. This can be quite a heavy load, exceeding 3000 tons on some of the largest swing bridges. The bearing must be carefully designed, fabricated, and erected to sustain these loads. Usually, the bearing itself consists of a bronze disc, with the lower surface convex, resting on a steel disc of slightly larger diameter, with the upper surface concave. The convex and concave surfaces are usually spherical with a radius of several feet. The upper, convex, bronze surface is usually 2 in. or 3 in. smaller in radius than the lower concave disc. This concentrates the bearing pressure in the center of the bearing assembly so that the turning friction is less and allows for some slight tilt of the bridge while turning, without binding. Center pivot bearings are invariably lubricated by oil, such that the bearing contact surface is always flooded.

A type of center bearing frequently used for railway bridges has the weight of the swing span suspended from rods that hang from a frame resting on top of the upper disc. This arrangement was supposedly originally developed to allow leveling of the bearing after installation, but the utility of this detail for this purpose is not quite clear. The bearing itself is usually a 3-piece arrangement (see page 192). The center part of the bearing is usually spherical on its top and bottom (see Figure 6-2) and should be self-leveling without the suspending rod arrangement. The suspension arrangement is an expensive way to accomplish leveling, as shimming would be cheaper if, in fact, leveling is required. The extra expense is probably why it has not been seen in highway bridges, which are usually more sensitive to first cost considerations. The true value of this suspended arrangement is realized in later repairs to the center pivot, which may not be needed for several decades after the original construction. The suspender rod arrangement allows the pivot bearing assembly itself to be unloaded and dismantled for inspection or repairs without the need to jack or lift the swing bridge. No other common swing bridge center pivot detail allows this to be done.

Some center pivot swing bridges have been equipped with three-piece center bearings consisting of upper and lower concave steel discs and a center double-convex bronze disc, as shown in Figure 6-2. This is the "pumpkinseed" arrangement. These bearings supposedly allowed for occasional sticking of the bearing by providing a reserve sliding surface. It may have been effective, as several two-disc types have failed by seizing and required expensive repairs, whereas no similar failure of the pumpkinseed type has been reported. Because of the extra expense incurred for fabrication of the triple-piece bearing, plus the opinion of some engineers that it was prone to misalignment, this arrangement has not been used in new swing bridge construction for several decades. The "misalignment," followed by its jumping back into place, may be what keeps the pumpkinseed type of bearing from seizing. The 3-piece configuration also avoids the difficulty of fitting a piece of bronze to the structural steel. The structural connections are instead made via the stronger steel forgings. AASHTO (2.7.3) and AREMA (6.5.34.7a) require that center pivots be of the 2-piece type, with one disc of steel and one of bronze, excluding 3-piece center pivot bearings from new construction. This also excludes the use of antifriction type center pivot bearings; an exception to these requirements would be required in order to use such a bearing. 2000 LRFD AASHTO (6.8.2.3) continues the requirement of 2-piece bronze and steel center pivot bearings, but also allows antifriction type center pivot bearings.

The center pivot is always lubricated with oil, as grease tends to work out of the bearing surface and not return. Oil always tends to flow into the lowest portion of the lower concave surface, where the load is. This suggests that it is likely that the upper bearing surface of the pumpkinseed arrangement quickly becomes a fixed connection in service because of its being less well lubricated than the lower bearing surface, and this type of bearing probably acts at many installations simply as a two-piece bearing. The location, shape, and detailing of the oil passages for center bearings are very important. Oil must always be allowed to pass to the proximity of the center of the bearing with as little restriction as possible. At the same time, the strength of the bearing cannot be compromised by an excessive number of lubrication channels. Hovey provides a detailed discussion of several variations in oil grooves for center pivots.[1]

A few new center pivot details have been developed in recent years. The concrete double swing bridge in Seattle, Washington, uses a vertical hydraulic cylinder of extremely large diameter to lift the swing span slightly off its live load bearings and then allow the bridge to pivot on the cylinder. This design has not worked out very well, as the giant cylinder has proved difficult to manufacture and the bridge has been plagued with cracks in the cylinder. The problem has still not been resolved, 17 years after the bridge was built. Another new center pivot design calls for a very large spherical roller bearing to be installed, with its axis vertical, acting as the center bearing so that the bridge load is taken as thrust load on the bearing. It has been applied in several recently constructed swing bridges. This arrangement is intended to provide for lat-

[1]O. E. Hovey, *Movable Bridges*, Vol. 1 (New York: John Wiley & Sons, 1926), p. 300.

eral load capacity at the center bearing to satisfy seismic loading requirements. It remains to be seen whether this bearing will prove successful over an extended length of time. Precision antifriction bearings have been used as swing bridge center pivots in the distant past and were found to be lacking in durability, as discussed in Chapter 6.

Both AASHTO (2.7.3) and AREMA (6.5.34.7) specify that center bearing swing bridges must have center pivot discs that can be replaced without closing the bridge to traffic. It is explicit in both specifications that this means no closure to highway or rail traffic, as a bridge cannot turn without its pivot bearing in place. A few railroad swing bridges follow this rule strictly. Many railroad bridges of the late nineteenth and early twentieth centuries were constructed to allow the dead load of the swing span to be eased off the center pivot and placed on the center wedges. The center pivot assembly was configured as described in Figure 6-1, with the weight of the bridge suspended on threaded rods. The Pennsylvania Railroad, and a few others, built their swing bridges this way with a center bearing assembly that could be dismantled, configured in such a way that the dead load and live load rests on the center wedges while the center bearing assembly is removed for repair or replacement. The oldest known existing example of this arrangement is on a Long Island Rail Road swing bridge in New York City that was built in the early 1890s. Most center bearing swing bridges have the loading girder or girders resting directly on top of the center pivot bearing assembly. These bridges must have the center of the swing span jacked up a small amount, at least an inch or two, and then the bridge must be supported on the jacks or on falsework to allow the center bearing to be removed. This reduces the dead load reaction at the end lifts, but traffic can usually be carried across the span at reduced speed during the work.

Most highway swing bridges are of this arrangement. These bridges generally may be able to support live loads with the center pivot removed. To this end, a bridge is lowered to its normal height and provided with temporary supports, or it is kept lifted and traffic is allowed to cross the bridge at low speed.

BALANCE WHEELS

Balance wheels are required only for center bearing swing bridges. The main support rollers provide center pier stability for rim bearing swing spans. The balance wheels are not intended to carry live loads or dead loads, but to resist overturning moments developed when the bridge is not supported and stabilized on its end lifts and center wedges. The balance wheels carry dead load if the bridge is improperly balanced, when the end lifts and center wedges are retracted. The balance wheels also ensure that the swing span is not excessively tilted as it approaches the closed position, so that the end lifts, centering latches, or other parts of the swinging span do not foul at a rest pier. On some very long swing bridges the balance wheel clearance cannot be sufficiently reduced to eliminate the possibility of such fouling. In these cases special guides are placed on the pier adjacent to the end lift seats so that the end lifts or other devices are gradually lifted, taking the rest of the end of the bridge with them as the bridge closes, to prevent fouling.

The balance wheels are attached to the swing span, around a circle that is usually slightly smaller in diameter than the bridge operating rack. The wheels are adjusted to hang just above a circular track mounted on the center pier. The circular track may be integral with the circular rack that engages the rack pinion or pinions to drive the bridge open or closed, or the rack may consist of separate components that are bolted to the track. On some swing bridges the circular track and circular rack are totally independent of each other. On small swing bridges the circular track may be composed of simple railroad-type rails mounted in a circle on the pier.

The primary load that is intended to be resisted by the balance wheels is overturning moment due to wind loading while the bridge is not in the closed position and supported on its end lifts and center wedges. For a through-truss swing bridge, this moment is at its maximum when the wind is blowing at a right angle, normal to the longitudinal axis of the span. The balance wheels at the sides of the swing span, under the main support members, are thus most heavily loaded, and the magnitude of the maximum load may require multiple balance wheels at these locations, firmly mounted to the swinging structure (Figure 13-1). Depending on the configuration of the center pivot, the center for pivoting in the wind may be some distance above the pier. The configuration, radius, and the elevation of the bearing discs determine the location of this center. The overturning wind moment depends not only on the size and construction of the swing span, but also on the location of the pivoting center of the bridge. Provision must be made, for each angle of the horizontal wind to the span, for the balance wheels to be capable of resisting the overturning moment without overstressing any component whether the bridge is in the process of opening or closing, turning or stationary.

Figure 13-1 Swing Bridge Paired Balance Wheels. This center pivot swing span has pairs of balance wheels at the sides to help resist wind loading.

The amount of free vertical play of the balance wheels closest to the longitudinal axis of the swing span is critical, as the ends of the swing span, when not supported by the end lifts, must clear all obstructions on the rest pier so that the bridge can swing freely. Normally, only $\frac{1}{16}$-in. vertical clearance at the balance wheels is allowed in design, to minimize the free tilt of the span and thus minimize the amount of vertical movement required at the end lifts. A minimal but sufficient amount of extra clearance should be provided at the ends of the swing span, over the rest piers, so that as wear causes additional clearance at the balance wheels, fouling at the rest piers does not occur when the bridge tilts because of imbalance or wind loading. Special precautions must be taken if this clearance cannot be maintained, such as by providing rails on the rest piers to support the end wedges as the bridge swings. Improperly adjusted balance wheels, with little or no clearance, can result in live loads and possibly dead loads being supported at the balance wheels and the probable deterioration of the balance wheels or their bearing brackets.

RIM BEARINGS

The rim bearing type of swing bridge may generally be considered obsolete today, but many examples still exist, which are not likely to be replaced in the near future. The rim bearing for support of a swing bridge is, in reality, a giant tapered roller thrust bearing. As such, its proper design, fabrication, and installation are critical for its successful use. It is extremely difficult to accomplish these three tasks, and for this reason alone swing bridges should be very cautiously specified with rim bearings.

A rim bearing assembly consists of a multiplicity of truncated cones, each of which has its apex at a point on the exact center of rotation of the bridge. Most of these cones are small rollers, usually about 2 ft in diameter, arranged to travel in a circle around the edge of the center pier. These small cones are cast steel or cast iron rollers; up to 70 or more, as many as can be fitted in, are spaced equally about the circumference of the pier. The small cones rest on a larger one, consisting of forged tread plates arranged in a continuous circle, machined to mate with the rollers. The large cone is supported on some cast steel chair arrangement that is also continuous, and firmly attached to the pier. The small cones support another larger one, the inverse of the first, that is attached to the bottom of the rim girder supporting the swing span. Upper forged steel track pieces similar to the lower ones rest on top of the rollers and support the entire moving, and almost all the stationary, bridge dead load as well as most of the live load. The rim bearing assembly also resists most or all of the wind loading on the span.

All of these cones have the same face width, and all must be very accurately made and positioned for the bridge to work properly. The rollers are usually kept in alignment by a framework, not unlike the cage for the balls or rollers of an antifriction bearing, that holds the rollers at the proper spacing and keeps their axes pointed at the center of rotation. The framework, or "cage," consists of a large ring that contains the rollers and keeps them equally spaced, and radial struts or rods that keep the framework, and the rollers, at the proper distance from the center of rotation. This framework is usually referred to as a *spider* assembly. The radial rods or struts pivot about a center post via attachment to a collar that is loosely fitted to the post and rotates around it as the bridge is turned. This post also mates with radial members that are attached to the swing span via the rim girder itself and help to hold the bridge in its centered position. The swing span with its connections and the rollers with their connections rotate independently about the center post. The roller-frame-rod/strut-collar assembly rotates around the center at half the rate of the bridge itself, because of the rolling action of the rollers on the cone below attached to the pier and the cone above attached to the swing span.

The loads on the rollers must be as nearly to equal as possible to avoid unequal wear or damage to heavily loaded rollers. This is accomplished by accurate fabrication and assembly and by making the rim girder, which transfers the dead load of the swing span to the rollers, as deep as necessary to spread the load from a few points of bearing on the top of the girder, where the superstructure rests on it, to many points under the girder where it rests on the rollers. As long as the roadway is substantially higher in elevation than the top of the pivot pier, a suitably deep rim girder or drum can be provided.

Some old rim bearing assemblies have been replaced with a series of simple pillow block bearings supporting wheels on axles on the pier. The bridge is then supported on a circular track resting on the wheels or inversely, with the pillow blocks mounted on the span and the rollers riding on a fixed circular track mounted on the pier. This type of support places the full live and dead load of the swing bridge on the pillow block bearings and is prone to failure due to successive bearings being overloaded because of misalignment and unequal load distribution, but applications that have been in service for as long as 20 years are still functioning. Other rim bearing assemblies have been rebuilt more or less in kind, an alteration that can result in a substantial increase in useful life if it is done fairly carefully.

If an existing rim bearing assembly is badly deteriorated but the bridge itself does not warrant replacement, it is may be feasible to replace the bearing assembly with a prefabricated giant roller thrust bearing, as was done at the Center Street bridge in Cleveland, Ohio (Figure 13-2). The high cost of ordinary rim bearings, due to a great deal of expensive precision fabrication and complicated assembly, makes them undesirable. They are difficult to install properly and are sensitive to the quality of maintenance provided to such a degree that their service life may be short. Prefabricated rim bearing assemblies provide a more unified mechanism, which requires less labor to install.

The prefabricated assemblies can, when properly designed, provide substantial resistance to seismic events. It may be prudent to consider the use of these assemblies on swing bridges located in areas where there is a high likelihood of severe earthquakes, depending on the type of loads to be expected in a seismic event. If no appreciable uplift forces are expected, the traditional rim bearing assembly can provide substantial lateral resistance, because of the conical shape of the rollers and tread plates. The roller cage or spider itself must be able to transfer this horizontal force in shear. The center post of a rim bearing bridge can also be detailed to provide considerable strength to resist lateral forces.

Figure 13-2 Swing Bridge Rim Girder. This swing span has had its girder and rollers replaced with a new unit assembly.

END LIFTS

The end lifts of the swing bridge, when driven, become primary means of support for dead loads and live loads. Thus, the lifts must be rigid and stable. Ideally, a pedestal that is self-stabilizing should be used to eliminate any reliance on mechanical devices to stabilize the bridge. There have been a few such bridges designed. One example, built at the U.S. Marine Corps facility at Camp Le Jeune, North Carolina, uses blocks of prismatic cross section thrust between bearing pads, at the bottom of the end of each main bridge support member, and masonry plates on the rest piers. The ends are lifted, to provide the required dead load reaction at the ends, by jacks that lift the ends sufficiently to insert the blocks and then release the load onto the blocks, so that the jacks carry no load when the ends of the swing bridge are resting on the blocks and carrying traffic. This application is a special case, as the traffic over the bridge includes military tanks, which are extremely heavy live loads and have very little springing in their tread mechanisms to cushion impact.

One sure method of support for the ends of a swing bridge has been used, but to only a limited extent because of the large amount of work required in operating it. This method pulled the top chord at each truss back toward the center of the swing bridge, by means of screw drives, when the bridge was to be swung open. When the bridge was returned to the closed position, the screws were reversed and the ends of the swing span settled onto simple sole plate and masonry plate end supports. No mechanism was involved in any way in the support of the two arms of the swing bridge at the ends, so complete stability was ensured. The method required the complete lifting of each arm of the span, however, rather than only the few end panels, and required the lower chord near the center to be hinged. This was an expensive means of raising the ends and time-consuming for operation. The system was used a few times in the nineteenth century, but no modern examples are known. One of the few examples of this method was used was on the CMStP&P railroad bridge over the Mississippi at Hastings, Minnesota, built in 1871 and not replaced until 1981.

Another relatively stable system of end support is a roller arrangement set up so that the end of the swing span simply rides up a ramp as the bridge is closed, providing the desired lift at the ends. The rollers are arranged, one under each main truss or girder, with the roller axes parallel to the girders. The amount of lift available is limited via this system and it has been used only in small highway bridges or in bobtail swing spans that do not require actual deflection of the main girders. One bridge known to have remained in service until recently (and still in existence) with this type of end lift system is the CMStP&P bridge over the Wisconsin River at Sauk City, Wisconsin, a 63-ft span bobtail single-track deck girder railroad swing bridge.

The most common end lift device for swing bridges is a wedge system. This usually provides sufficient lift and reasonable stability under live load with a simple mechanism (see Figure 13-3). The wedges can be locked in the driven position by a

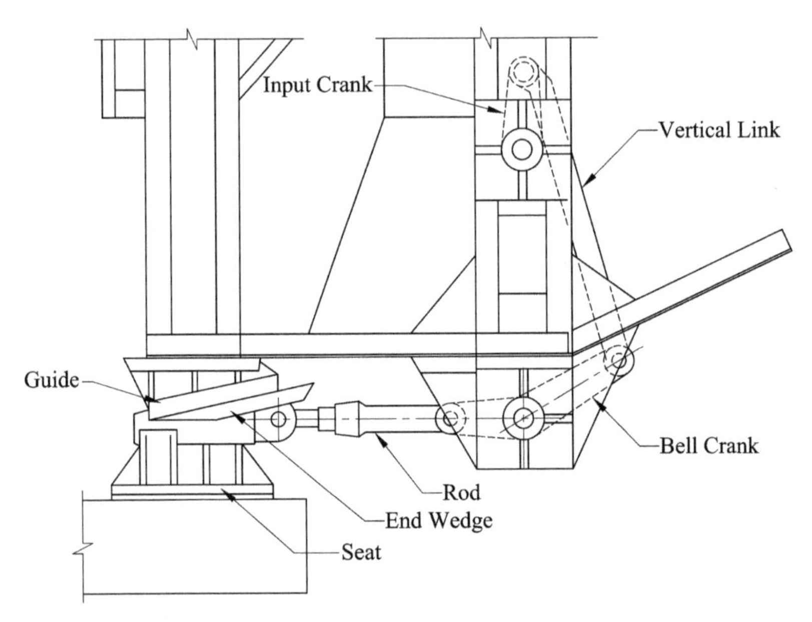

Figure 13-3 Swing Bridge End Wedges. This is the most common type of end support for swing bridges. The wedge is driven between a shoe, called a guide, on the bottom of the end of the swing span main truss or girder, and a seat mounted on the pier. As the wedges are driven, in unison or nearly so, the bridge main structural members are deflected so that the ends are lifted and there is a significant dead load reaction on the seats on the piers.

toggle mechanism incorporated in the drive. The wedges typically have a 1:6 slope on their top surfaces, which mate with similarly sloped guides attached to the bottoms of the ends of the main support members. The bottoms of the wedges are horizontal and rest directly on specially designed masonry plates, called wedge seats, attached to the rest piers. This type of system has been used in the majority of swing bridges with a high degree of success and is a reliable and economical means of support for highway and railroad swing bridges of short and medium span lengths. The end wedge lifting device has the advantage of being adaptable; it can also act as a centering device, by having ears added to the sides of the wedge seats, with tapered engagement. As the end wedges are driven, the sides of the wedges can act against the ears on the wedge seat to push the end of the bridge laterally to the aligned position (see page 204).

Very long swing spans can develop a substantial angular deflection when the end lifts are retracted, so that end wedge type lifts can tend to drive the leading edge of the wedge into the wedge seat, causing jamming, damage to the components, gouging of the wedge seat as the leading edge of the wedge digs into it, or displacement of the wedge seat or the entire rest pier. In the late nineteenth century, when very large swing bridges were first being designed and built, roller-type combination end lifts and supports were developed to overcome this difficulty (Figure 13-4). The more common version provided a swinging vertical link with a roller at its bottom, which engaged a bearing plate mounted on the rest pier as the link was rotated by a drive mechanism. The link and roller were built to be sufficiently sturdy to resist any vertical loads encountered in raising the bridge or supporting live loads. Many swing bridges built more than 100 years ago with this type of end lift are still in use today, with no substantive change. Other types of lifts developed to solve the problem include wheels mounted on mechanically driven eccentric cranks (Figure 13-5), vertical toggle and shoe arrangements, and variations on the jack and block method. See page 229 for further discussion of end lift details.

Very wide swing bridges may have a third end lift at each end. This is usually accompanied by a third, center, truss or main girder. Wedges or wheels on eccentric cranks are generally used for end lifts when a lift at the center of the ends is required.

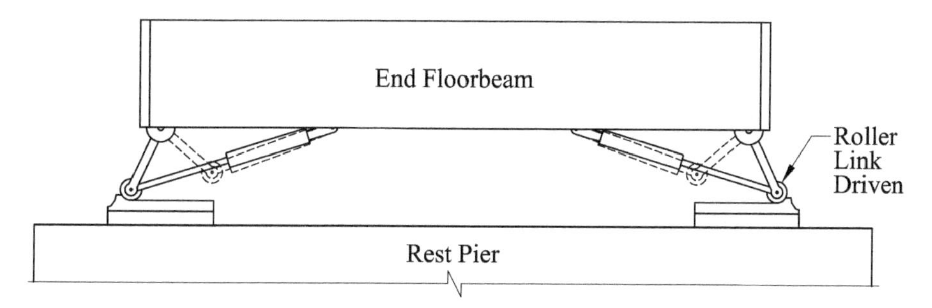

Figure 13-4 Swing Bridge Roller End Lifts. This type of lift was generally used on very large swing bridges. The roller mechanism drives outward from the center of the end floorbeam, by a toggle-type action capable of exerting large force at the end of travel.

Figure 13-5 Swing Bridge Eccentric Roller End Lift. This type of lift is not very common. The mechanism rotates an eccentric shaft, causing the roller to lift or lower, lowering or raising the end of the swing span.

RAIL JOINTS

Railroad swing spans require a means of separating the track rails at joints at the ends of the swing span (Figures 13-6 and 13-7). The ends of the rails on the swinging structure are usually extended beyond the ends of the swing span a sufficient distance to bridge the gap to the approach spans and provide a solid seating in chairs at the ends of the rails on the approaches, yet avoiding interference when the swing span is opening. The forms of the rail joints for swing spans are similar to those used on railroad bascule and vertical lift bridges and are discussed in Chapters 11 and 12, respectively. Failure to provide a continuous support for the rail wheels over the joint results in impacts that shorten the life of the rail at the joint and possible structural damage to the bridge. The unique aspect of swing bridge rail joints is that they usually must be mechanically actuated. Formerly, the actuation was typically performed by a device that simply lifted the end rails and then allowed them to drop via gravity. Because of a serious accident a few years ago, it is now required, at least on some swing bridges, that the end rails be forcibly held in position for rail traffic. This is especially helpful for high-speed rail traffic. See Chapter 16 for a further discussion of end rails and actuation devices.

CENTERING DEVICES

Swing bridges require two types of centering functions. These functions may both be performed by a single device, or they may be performed by devices that also perform

Figure 13-6 Miter Rails on Swing Bridge. These rails are firmly fastened down onto the superstructure of the approach, on this side of the arrows in the foreground, but are lifted mechanically on the end of the swing span, at the far side of the joint, to allow the bridge to be opened for navigation.

Figure 13-7 End Rails on Swing Bridge. These rails are firmly fastened down onto the superstructure of the approach, in the foreground, but are lifted mechanically on the end of the swing span, at the far side of the joint, to allow the bridge to be opened for navigation. This type of end rail, rather than being mitered, uses a bridge rail to align the rails and allow the wheels to pass over the joint.

other functions. Rarely are two separate devices installed on the same bridge to perform each of the two functions. Function #1 is to bring the swing bridge to a halt, when closing, in a position approximating the fully closed position. Function #2 is to align the swing bridge, when it is in the approximately closed position, so that other ensuing functions that may be required can be performed and the bridge is properly aligned in the horizontal plane to carry traffic.

For many years, the most common type of centering device on swing bridges was a gravity-powered vertical latch, usually one at each end of the swing span, that engaged a pocket on the rest pier when the bridge reached the fully closed position, performing function #1 quite adequately. This type of device was required on swing bridges by AASHTO until deleted from the 1978 Standard Specifications for Movable Highway Bridges. Without such a device, it is very difficult to align the bridge when it is closing. Many swing bridge designers have done away with the automatic latch-type centering device, which is considered by some maintenance personnel as a nuisance because of the necessity of its being free to operate by the light force of gravity. Lack of lubrication, fouling by debris, or slight damage can prevent the device from operating properly. Movable bridge designers have tried other means, such as proximity switches, laser beams, and position indicators, to control the position of the bridge through the drive machinery, but they are not as effective. A new swing bridge, with small amounts of free play in the drive mechanism and little play elsewhere, can be positioned by means of a limit switch actuated by span position, but the accuracy of this type of system deteriorates as the bridge ages. A swing bridge can theoretically be designed and built so that the bridge drive machinery can accurately control the position of the ends of the swing span, allowing the bridge to be centered with sufficient accuracy that a separate power-driven centering device can accurately position it, or the end lifts can be driven without further centering. The difficulty lies in the performance of the bridge in service: (1) It is difficult to construct a bridge to the precision required to provide this level of accuracy; (2) wear of the drive components results in loss of precision in positioning the bridge; (3) the bridge position may shift so that the closed position for which the machinery was adjusted may no longer be the correct position; and (4) temperature differences can cause a change in the shape of the bridge, making alignment at the ends more difficult.

Some of the sophisticated centering systems have performed extremely poorly in executing function #1. A number of rebuilt highway bridges have what appears to be no system at all, so that the bridge operator must jockey the bridge back and forth, almost at random, until the bridge stops near enough to the fully closed position that the end lifts can be driven. The final alignment of these bridges, when carrying traffic, may be off by several inches.

The many varieties of swing bridges are unique among movable bridges in that almost all can open in either direction and have no discrete end of travel where the span will be forced to stop in the open or closed position. The typical swing bridge thus requires a means of stopping and holding it in the proper position, both when open and when closed. The open position is usually not a problem, as positioning need not be accurate; the skill of the operator and a reliable set of brakes are all that is needed. Some railway swing bridges that are normally left in the open position have latches on the protective fender, but this is a rarity. Function #2, the positioning of a bridge when closed is more critical, especially for railroad bridges that have miter rails that must be aligned. The old gravity-actuated latch has usually been considered adequate to perform function #1 and, function #2 for highway bridges, as the accuracy required in aligning a highway span when closed is much less than that required for a railroad bridge, unless the highway bridge happens to have trolley or other railroad tracks on it. For railroad swing bridges, function #2 has been handled adequately by the end wedges or rollers. Some bridges have been built, or rebuilt, with more sophisticated systems, such as thrusting lock bar type devices, intended to perform function #2 (see page 204).

Every swing bridge that does not have a bumper that allows it to open in only one direction should have some sort of automatic latch that can stop and hold the bridge as it swings closed, satisfying the requirements of function #1. If the bridge is a railroad swing and the latch cannot position it accurately enough for miter rails to actuate, then an additional means should be used to provide accurate closure and satisfy function #2. The most effective and economical method is to use end wedges as centering devices, as described on page 204.

For the purpose of stopping a swing bridge in the approximately closed position required of function #1, the old gravity-actuated center latch is quickest and most reliable (Figure 13-8). This device consists of a vertical sliding bar with a roller mounted on its bottom and the axis of the roller parallel to the longitudinal axis of the swing span. The bar is mounted in guides on the end floorbeam of the swing span and counterbalanced by means of a linkage with a counterweighted arm so that a small force will lift the bar. The bar has a pawl hanging from it, near the top of the bar. The pawl engages a lever actuated by the end lift mechanism. When the bridge is centered, the bar rests in a pocket on the pier. The pocket has sloping external sides, which the roller rides up as the bridge approaches the centered position (see Figure 13-9).

When the bridge is to be opened, the end lift mechanism rotates the lever, raising the bar by means of the pawl. The bar is then lifted sufficiently so that the swing span can rotate. As the bridge begins to open, the roller on the bar contacts the edge of the top of the pocket, lifting the bar as the bridge continues to swing, until the pawl disengages from the lever. As the bridge continues to rotate, the roller rides down the sloped outer side of the pocket and the bar drops in its guides to the limit of its movement.

When the bridge closes, the roller contacts the sloped outer side of the pocket and rides up the slope, raising the bar. When the bridge is close to being fully centered, the bar drops into the pocket, stopping the bridge. The counterweight of the bar prevents it from dropping too quickly, so that if the bridge is traveling too fast for a safe stop, the bar will not engage the pocket and the bridge will continue past the fully closed position. Another attempt at closing the bridge is then made from the opposite direction.

Some versions of this arrangement have a pocket mounted on the pier with sprung sides so that the latchbar will engage and the moving span will cause the springs in the pocket to compress, slowing the span to a stop. The springs then

Figure 13-8 Swing Bridge Automatic End Latch.

expand back to their normal length, centering the bridge. Once the bridge has been latched, the end lifts and other devices can be actuated and the bridge can resume carrying traffic. Activating the end lifts resets the lever and pawl so that the latch is ready to be lifted for the next bridge opening.

For accurate centering of a swing bridge in the closed position satisfying function #2, many railroad bridges use a centering device similar in appearance to an enlarged bascule bridge center lock bar assembly. This device may be mounted vertically or horizontally. There is usually one device at each end of the swing span, mounted at the center with its pocket or socket at pier level. The utility of these devices is somewhat limited, as the miter rails still frequently foul when these devices are used. The centering devices usually activate after the end lifts, when opening the bridge, and before the end lifts when closing to avoid high loading due to sliding the end of the bridge laterally on the end lifts. AREMA requires these devices to be located as close as practical to the rails, but they are usually required to be mounted on the pier for rigidity, and it may be a substantial distance from the pier top to the rail level.

End lifts can be, and frequently are, adapted to also provide for rigid centering of a swing bridge in the closed position. AREMA (6.5.34.5) specifies that the end wedges should also center the closed span. AASHTO (2.7.2) has a similar statement. This is

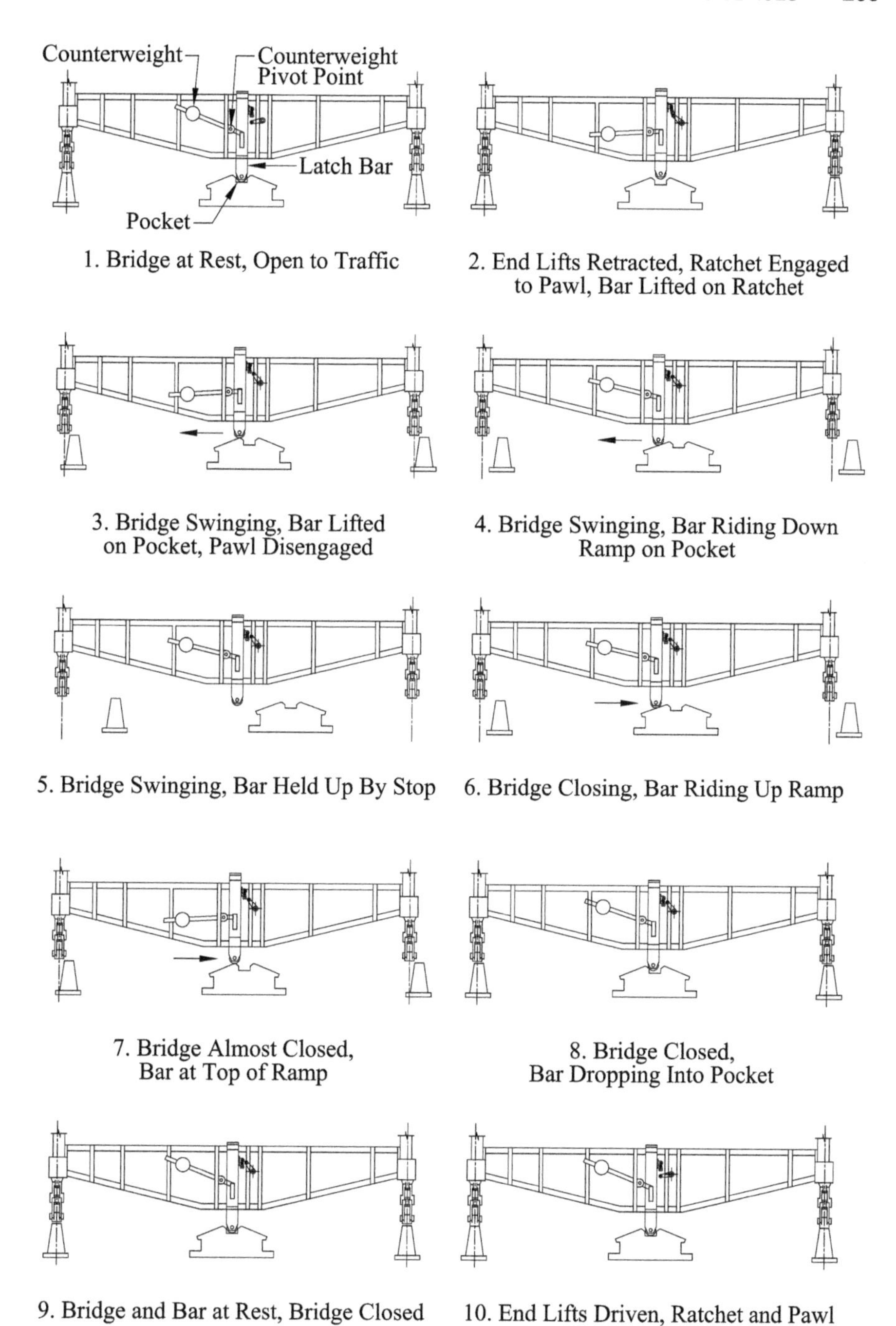

Figure 13-9 Swing Bridge Automatic Latch Operation.

accomplished with end wedges by providing "ears" on the sides of the end wedge seats, with enough space between them to allow the wedge to fit with a small clearance (see Figure 33-5). The end wedge is normally of a constant width through the greater part of its length, but has the leading edge tapered in plan so that there is ample clearance as the wedge enters the space between the ears to allow for span misalignment. As the wedge continues to be driven, contact between an ear and the corresponding tapered side of the wedge results in the development of a lateral force that pushes the end of the bridge to the aligned position. Laterally acting roller-type end lifts are typically provided with upwardly curved edges of the seat, which the roller approaches with a slight clearance when fully driven. This provides some centering of the bridge. It is more difficult to accomplish centering with jack-and-block devices, so a separate centering device is usually provided. Regardless of the type of centering device used, it is helpful to have a center latch to hold the bridge in approximate alignment while the centering device operates.

For high-speed rail systems, the rails must be precisely aligned at the ends of the swing span. Devices aligning each rail should be used in addition to special devices aligning the ends of the superstructure. Care must be taken to avoid pushing the track out of alignment when making the joint. The maximum permissible train speed over the bridge should be carefully determined, considering the alignment and stability provided by the rail system in use (see page 231). The swing bridge is not inherently desirable for carrying high speed rail traffic (see page 233).

CENTER WEDGES

Center wedges to support live load are needed only for center bearing type swing bridges, as a rim bearing swing bridge fulfills the live load support function at the center pier by means of the rim bearing assembly itself. The center live load supports for a center bearing swing bridge are invariably in the form of relatively flat steel wedges, usually with a 1:9 slope at the top surface. The mechanisms used to insert these wedges into position and to retract them are quite varied among different bridges.

The center wedges are never intended to carry any dead load of the bridge, as the extreme rigidity of the bridge structure at the cross section containing the center wedges does not allow a suitably small percentage of the dead load to be repeatably transferred to these wedges. Temperature and other subtle effects can cause large variations, from one operating cycle to another, in the load carried by the center wedges. Firmly driven center wedges, supporting live and dead loads, may eventually have most of the lubricant squeezed out from the bridge supporting surfaces. A very high force may then be required to withdraw the wedges. Worn wedge drives have been known to be unable to deliver such forces. Instead, the center wedges are usually driven to a predetermined position between the wedge seat on the pier and the wedge guide mounted on the bottom of the swing span. Adjustment is made in this position so that the center wedge comes as close as possible to making firm contact between the guide and seat without binding. For railroad bridges, AREMA (6.5.34.1) states that the main girders or trusses of a bridge shall rest on the driven center wedges, but stops short of requiring that the bridge dead load be supported on the center wedges. It states (6.3.13.2.d) that center wedges shall be designed for maximum live load plus impact. AASHTO (2.7.1) and (2.7.7.4) contain similar statements.

The force required to drive the center wedges is small, but they must be held firmly in place when driven. The 1:9 slope of typical center wedges is practically self-locking, but cannot be guaranteed to be so, because of variable lubrication and the uncertain load on the wedges. Many swing bridges use a form of toggle linkage acting directly against the wedge, with power supplied in the form of a nut-and-screw drive that acts vertically normal to the toggle. Sliding guide bars are used to restrain the drive assembly against live load and wedge driving forces, avoiding heavy loads on the screw and nut. The screw threads are difficult to lubricate and keep free of contaminants in the dirty environment of the usual swing bridge center pier, but as the load is light, this is not of great consequence.

To avoid screw drives, some swing bridges have a worm gear drive acting on the far link of the toggle. The starting and stopping of the drive is carefully controlled so that the wedge is properly positioned. On some older swing bridges, as originally constructed, a single drive point operates all the wedges and other secondary devices, with the position of the center wedges being the controlling one and all other devices adjusted so that they are in approximately their ideal position when the center wedges are fully driven and fully pulled. On newer swing bridges and most rebuilt older ones, the maintenance of the extensive shafting is eliminated by having separate, independent drives at each center wedge and at other components, controlled by limit switches that sense the position of the wedges and start and stop the drive motors accordingly. These have proven not to be as reliable as the mechanically linked drives, require longer times for bridge operating cycles, and have difficulty with synchronizing the positions of all devices.

A few center pivot swing bridges have been built with center wedge drives incorporating gravity, springs, or other nonpositive movement devices. These are intended to help keep the center wedges in firm contact with the seats and guides in spite of variations in positioning. With these devices in the drive train, the wedge drives until it meets a certain level of resistance. These devices have generally not proven satisfactory, as the wedges sometimes do not drive as far as they should, lubricant on the wedges sometimes allows them to be driven out when live load impacts the span, and they are occasionally driven too far so that they stick in the driven position and are difficult to retract in preparing to open the bridge.

BALANCING

Balancing need not be particularly precise for symmetrical swing bridges. Errors in the delivered weights of material tend to be spread out fairly evenly over the structure, so the center of gravity of the moving span tends to remain where intended. A simple concrete block can be poured in the approximately correct position, to provide balance about the pivot point, on a swing bridge that is imbalanced because of having an operator's house located on one side, or having an asymmetric deck, such as a bridge having a railway track and a highway deck of different weights alongside each other on the swing span, or having a sidewalk on only one side of the bridge roadway.

For an asymmetrical bobtail swing bridge, the effect of delivered weights of material not being equal to designed weights must be more carefully considered. A very large unanticipated imbalance can be developed in such structures, because of differences between designed and constructed weights, which cannot be corrected without allowance for adjustment. This type of structure should go through a balancing process during construction, similar to that required for a bascule bridge except that the vertical position of imbalance is not important, but lateral position may be critical. As weights may change during minor repairs or reconstruction later, or as a result of ice or snow loading, it may be desirable to include pockets of removable weights in the counterweight of such a swing bridge, as are provided for bascule and vertical lift bridges.

14

MACHINERY—BASCULE BRIDGES

Single leaf bascule bridges have less machinery than other types of commonly used movable bridges, and the layout of their machinery is simpler. Double leaf bascules have the same machinery arrangement as single leaf, on each leaf, plus span lock machinery where the leaves join, and may have additional equipment.

Bascule bridge machinery is subject to higher wind loading than that of other types of movable bridges, because the bascule leaf pivots up into the air when open so that a wind load parallel to the structure axis has an impact on one side of the pivot point and gains in leverage as the length of the bascule leaf increases. This results in a large moment that is resisted by the machinery. The machinery must be designed to move the leaf against the greatest wind load that is expected to occur when the bridge is required to operate. The usual result is that the motor size and machinery dimensions are determined almost entirely by the wind resistance with the bridge in the open position. At some installations, where the bridge is normally left in the open position, the machinery must hold the span open against a higher wind load, although it may not be necessary to operate the bridge against the higher wind. The wind load is opposed directly by the machinery and produces an opposite reaction at the span leaf bearing points. Some bascule bridges normally left open, or expected to remain open during storms, particularly hurricanes, have special span locks to hold the bridge in the open position during heavy winds.

Most bascule bridge leaves have two main girders or trusses, with the drive machinery operating off these members. Bascule bridges are usually quite rigid in the area of the final machinery drive. The superstructure framing is usually quite stiff in the area of the leaf supports. Rigid counterweight connections can add greatly to the stiffness. Usually, some means of equalizing the loading in these two main members is provided. In the rare cases with multiple main girders on a bascule span, such as at Mason Street in Green Bay, Wisconsin, several main girders, but not necessarily all of them, have drive machinery acting on them. Some means must be provided on bascule bridges to equalize the driving load at each drive point on a leaf. Exceptionally high loading at some machinery and structural components can result if this does not occur.

SIMPLE TRUNNION MACHINERY

Most simple trunnion bascule bridges have the span operating machinery mounted on the bascule pier, where power and control connections are relatively simple to make. This, or at least "the stationary part of the bridge" is the preferred mounting position for bridge drive machinery, per AASHTO (2.5.2) and for bascule bridge drive machinery per AREMA (6.5.35.6), although the simple trunnion bascule is one of the few types of movable bridges for which this is completely practical. The typical gear drive machinery normally engages circular rack segments, connected one to each main girder. On most bascule bridges, this results in two drive points per leaf, although a few bascules have a single drive at a rack centered on the leaf and some leaves have four drive points. When more than one pinion drives a leaf, a differential mechanism is usually placed at some point in the geartrain to provide equalization of the torque to each rack pinion that meshes with its mating rack segment. With four rack pinions, there may be two additional differential units per leaf. The differential or differentials provide equal drive loading on the main bridge members, in spite of unequal deflections of the bascule leaf or drive train elements that may occur because of temperature variations and other factors. On some bascule bridges the load equalization is accomplished by having separate drives with separate motors for each rack and relying on the installing electrician to equalize the output torque of the motors. This may be acceptable theoretically, but the torques are seldom initially equalized at the span drive motors, and even if they are, they tend to eventually come out of adjustment so that the load is not shared equally. Conversely, the differential gearing arrangement of load sharing is fairly precise, automatic and virtually maintenance free. Some newer electrical drive systems claim to be able to provide this quality of load sharing. Their utility remains to be proven in longer terms of service.

The typical movable bridge differential mechanism is similar in appearance and arrangement to that in the drive train of a motor vehicle. In an automobile, the purpose of the differential is to allow the two drive wheels to turn at different speeds while remaining equally powered when the car is going around a curve. The load equalization feature of the differential can be undesirable in this application, and some automotive drives have features that partly negate it. When one drive wheel of the car is on ice, for instance, the differential limits the drive force to that sustainable at the wheel with the least resistance, so that wheel spins while the other does virtually nothing.

On a movable bridge drive, load equalization is the primary purpose and freedom of movement of the two outputs is a minor advantage. On a typical older bascule bridge leaf, two bevel gears are mounted, one on the inboard end of each shaft driving one of the two rack pinions. The bevel gears mate with and are driven by bevel pinions mounted with their axes at right angles to the axes of the shafts on which the bevel gears are mounted, so that each pinion meshes with both bevel gears. There are usually two bevel pinions in the differential mechanism, but sometimes three or four are used. The bevel pinions are, in turn, driven by a spur gear, referred to as the spider gear, on which they are supported, which turns about the same axis as the bevel gears and the rack pinions. The spider gear is supported on inboard extensions of the bevel gear shafts, which are supported on bearings outboard of the bevel gears. The spider gear is driven by the machinery between it and the drive motor. Typically, there are several gear reductions involved. An output spur gear pinion meshes with the spider gear. During operation, the bevel pinions do not rotate on their shafts except to take up unequal slack in the drive, such as may be present if the drive machinery is more heavily worn on one side of the bridge leaf than on the other side. Another possibility for differential movement might be if there is a loose coupling or gear hub somewhere in the drive train between one of the bevel gears and its associated rack pinion. This feature also allows for equalized pinion loading if the leaf is warped because of temperature difference or for other reasons. In each of these cases the differential allows a slightly different amount of movement of the machinery on both sides of the bridge leaf, so that each pinion ends up bearing equally against its rack as movement of the leaf begins. On many, particularly larger, bascule bridges, there is additional reduction gearing between the differential unit and the rack pinions. The reduction to each rack pinion is the same.

There are other differential types that are used to divide the load equally between the two rack pinions. One type uses a planetary spur gear arrangement instead of bevel gears in the differential. Most newer bascule bridge drives have the differential mechanism, of whatever configuration it may be, mounted inside a gearbox, which usually also includes additional reduction gearing.

Some older deck-type simple trunnion bascule bridges have the drive machinery on the counterweight, with the main pinion or pinions driving against an internal rack segment mounted in the counterweight pit. This has the advantage of maximizing the rack pitch radius and thus minimizing the tooth load that must be developed at the pinions to operate the bridge. This advantage is probably more than outweighed by the difficulty in maintaining such inconveniently located machinery. Most maintenance cannot be performed except with the bridge raised and traffic blocked. It is also difficult to provide power to the machinery for moving the span. Most bridges of this type have severely corroded rack segments, due to excessive moisture in the counterweight pit, which is usually damp if not actually partially submerged a good part of the time.

A few simple trunnion bascule bridges are operated hydraulically by means of hydraulic cylinders connected between the pier and the moving leaf. This type of drive can have the advantage of great simplicity, but many cheaply built systems suffered from very poor quality design and components and had to be repaired or rebuilt shortly after going into service.

The most common arrangement of operating machinery for a simple trunnion bascule bridge is what is referred to as the "conventional" arrangement. This consists, for each leaf, of a centrally located motor, brakes, and differential gearing mounted on the pier. Additional reduction gearing, also mounted on the pier, is located between the differential and each rack pinion. The rack pinions are also mounted on the pier, and a circular rack segment mating with the rack pinion is mounted on each main bascule girder. Many older bridges have assemblies of open gearing, but newer and rebuilt bridges usually have the gear reductions and differential mounted in enclosed speed reducers.

HOPKINS DRIVE FRAMES

The State of Florida, which has a large number of bascule bridges, is the home of many with an unusual drive system for trunnion bascule bridges, the Hopkins frame. The Hopkins frame forms a connection between the bascule leaf and the pier, on which are mounted the drive motor, brakes, and reduction gearing. The frame is connected to the bascule leaf in such a way that the rack pinions, which are mounted on the Hopkins frame, are constantly in mesh with the rack segments, which are fixed to special rack girders on the bascule leaf, inboard of the two main girders. The Hopkins frame provides a modular drive system that is assembled in the shop and then brought to the site as a completed unit and erected on the bascule bridge. It would be easy to repair by simply changing out the entire unit, replacing it with a spare, and doing the detailed repairs in a shop rather than on the bridge. This did not turn out to be as good an idea as originally thought; the Hopkins frames proved to have a tendency to develop cracks at their points of attachment to the pier. The frames have a tendency to flex as the bridge opens and closes, because of changes in the shape of the bascule leaf in the open, as opposed to the closed, position. Some efforts were made to redesign Hopkins frames to avoid this problem, but to little avail. The popularity of these drives did not spread very far, and their use has diminished in recent years as some of the drives on existing bridges have been replaced by hydraulic operating machinery, or the bridges on which these drives were mounted have been replaced with new bascule bridges with more standard drives, or with fixed bridges.

Some Hopkins frames drive a single rack and pinion on a bascule bridge leaf, but most drives have two pinions driving two racks. The single rack is on the longitudinal centerline of the bridge leaf, whereas the two-rack system has rack girders located symmetrically about the bridge's longitudinal centerline, a few feet from it. The Hopkins drive frame working against two racks and pinions usually has a parallel shaft differential reducer mounted on it and, usually, additional exposed, or open, reduction gearing between the differential reducer output shafts and the rack pinions.

HEEL TRUNNION MACHINERY

The heel trunnion bascule bridge generally uses more complicated machinery than the simple trunnion bascule. The most striking feature of the heel trunnion bascule is its parallelogram arrangement connecting the counterweight to the bascule leaf. This feature is shared to some extent with the articulated counterweight bascule, and people sometimes confuse the two types. The heel trunnion bascule has the counterweight pivoted on a rigid A-frame mounted on the bascule pier.

It is very rare for a heel trunnion bascule to be operated by means of a circular rack segment engaged by a pinion. Most heel trunnion bascules are operated by means of an "operating strut," which is driven back and forth to push and pull the bridge open and closed (Figure 14-1). The operating strut usually has a straight gear rack attached to its bottom side. The mechanism that holds the operating strut against the rack pinion is called a *carrier* (see Figure 31-3). It usually has three sets of rollers arranged to keep the rack on the strut aligned with and at the proper distance from the pinion, but to allow the strut to translate freely when driven by the pinion. The carrier supports the weight of the operating strut, which can be very heavy. Pinion shaft bearings are mounted on the sides of the carrier frames, and the whole assembly is supported by additional pinion shaft bearings. These additional bearings are mounted on the counterweight support tower, in those installations that have the drive machinery located there (Figure 14-2). Most heel trunnion bascule bridges have the machinery oriented in this manner, but some are different. The HX bascule over the Hackensack River at Secaucus, New Jersey, has the drive machinery mounted on the bascule leaf and the

Figure 14-1 Operating Strut for Heel Trunnion Bascule Bridge. This is the most typical operating strut arrangement. The strut is pinned to the heel portal of the moving span, and connected to drive machinery resting on the A-frame tower supporting the counterweight.

Figure 14-2 Operating Machinery for Heel Trunnion Bascule Bridge. The machinery, mounted on the A-frame supporting the counterweight, connects to the operating struts, one on each side of the bridge, via the carriers that engage the pinions to racks mounted on the underside of the struts.

operating struts pinned to the upper part of the counterweight support tower (see Figure 14-4). The Culvert Street Bridge at Phoenix, New York, over the Oswego canal has the operating machinery mounted on the counterweight and the operating strut pinned to the base of the counterweight support tower, or pylon (see Figure 14-5).

The operating strut on a large heel trunnion bascule is so heavy that it can have a significant effect on the balance state of the movable span as it is opened and closed. The balancing effect is dependent on the arrangement of the strut connections. As indicated earlier, most heel trunnion bascules have the operating strut connected to the bascule leaf by means of a hinge pin, with the drive mechanism mounted on the frame that supports the counterweight (Figure 14-3). With the bridge in the lowered position, the strut is extended out toward the bascule span and a good portion of its weight is supported on the span connection, resulting in a shift in the span balance toward being more span heavy. When the bridge is in the opened position, the strut is extended back toward the counterweight, pivoting on the carrier assembly so that an upward reaction is exerted on the span at the strut connection, shifting the center of gravity of the moving bridge toward the counterweight. As mentioned earlier (page 210), this arrangement is preferred by AASHTO and AREMA, as it places the drive motors, brakes, and so forth, on a rigid nonmoving platform, making electrical connections easier and facilitating maintenance, inspection, and repair. This arrangement also provides a desirable shift in bridge balance during operation, stabilizing the closed and the open bridge.

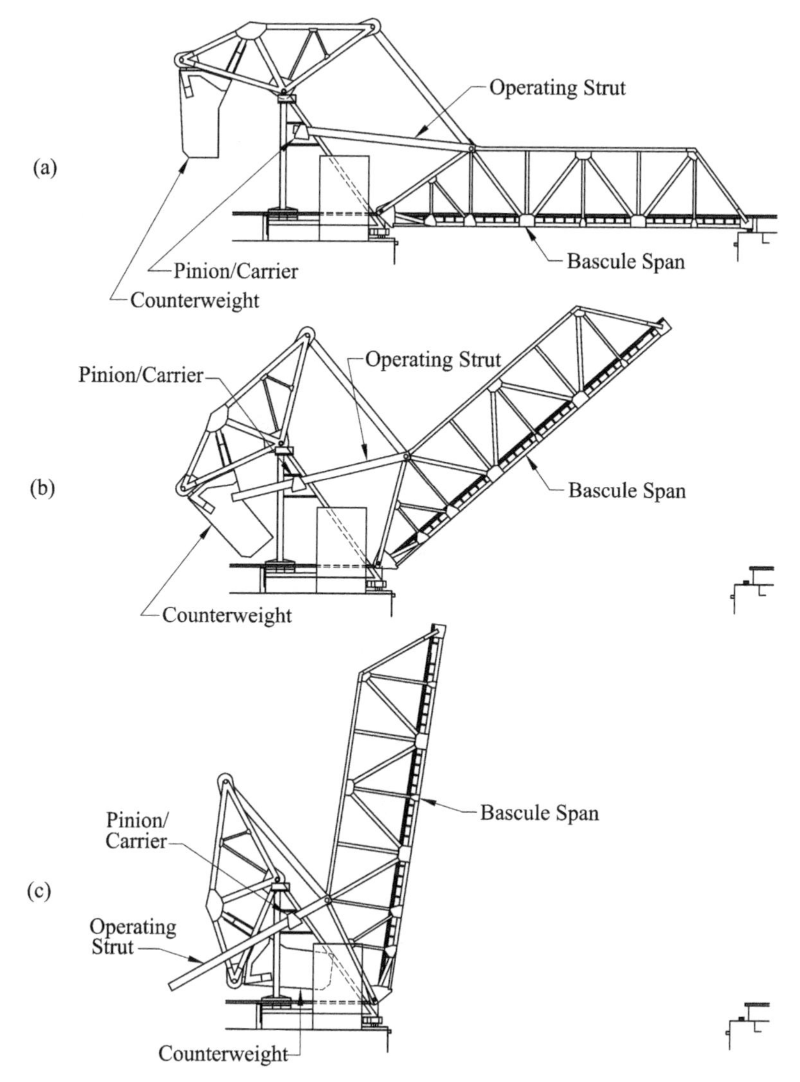

Figure 14-3 Heel Trunnion Bascule with Operating Machinery on Frame That Supports the Counterweight.

In those variations of heel trunnion bascules that have the struts pinned to the upper part of the counterweight support frame and the drive mechanism mounted on the bascule span (Figure 14-4), the strut weight shifts during operation toward making the bridge more counterweight heavy in the closed, or down, position and more span heavy in the open, or raised, position. This arrangement was favored by some early movable bridge designers, as it reduced the amount of power required to operate the bridge. Gravity assisted in opening the bridge when it was closed, and helped to start closing the bridge when it was opened. This arrangement also assisted in de-

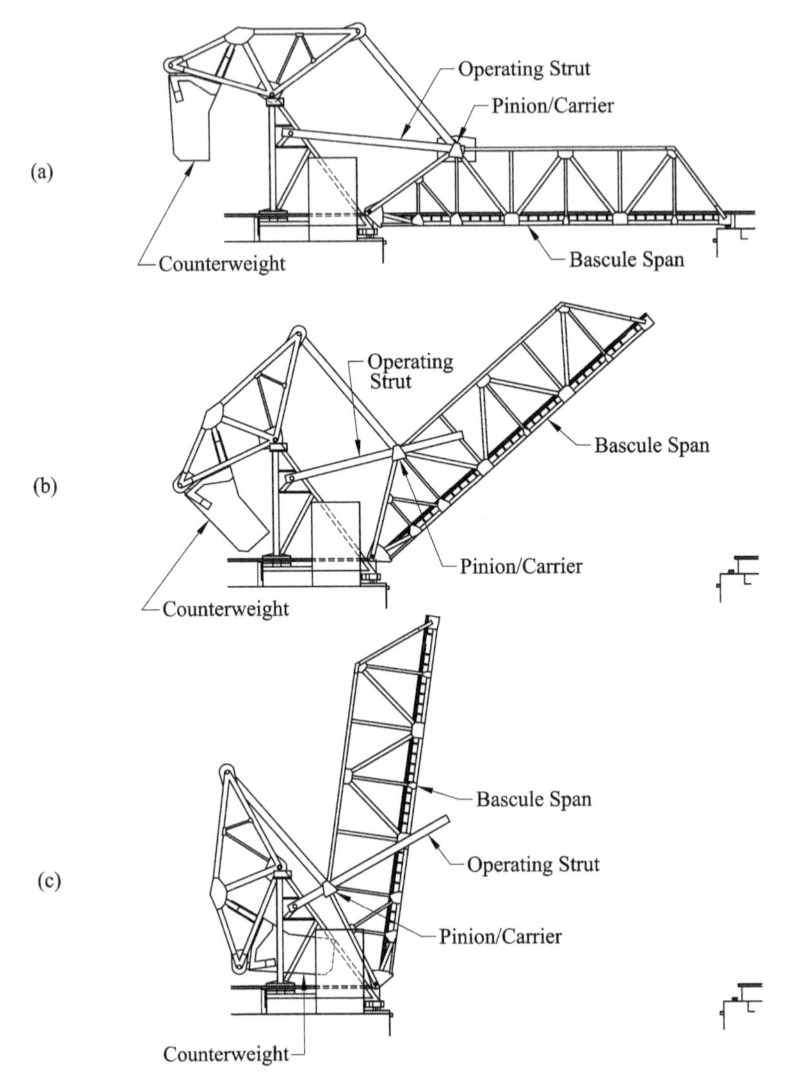

Figure 14-4 Heel Trunnion Bascule with Operating Machinery on Moving Span. As the bridge opens, more of the weight of the strut is carried by the bascule span, making it increasingly span heavy.

celerating the moving span as it reached full open or full closed position. This is an inherently unstable condition, except when the bascule leaf is in some partly raised position, and is generally not favored as a balance condition today.

For the few heel trunnion bascules that have the drive mechanism mounted on the counterweight and the strut pinned to the base of the counterweight support frame (Figure 14-5), the operating strut weight support is shifted toward the counterweight as this variation is opened; when it is closed, the weight is shifted back off the counter-

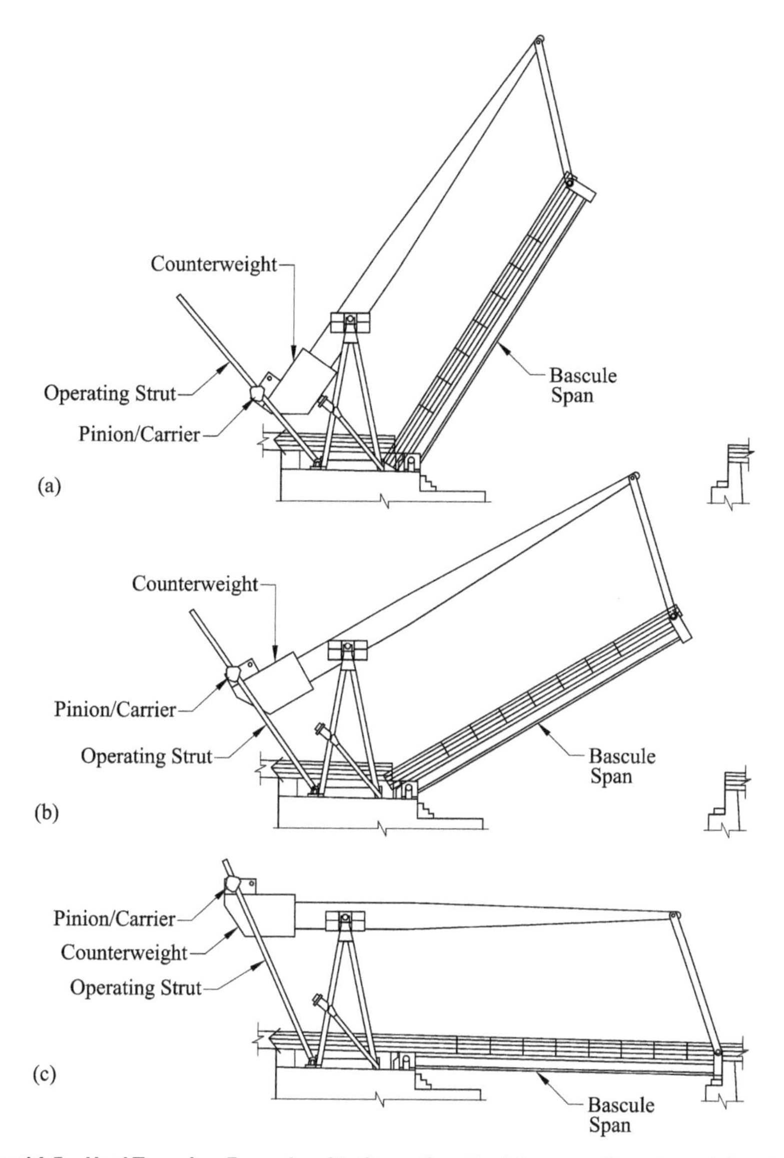

Figure 14-5 Heel Trunnion Bascule with Operating Machinery on Counterweight. As the bridge opens, more of the weight of the strut is carried by the counterweight, making it increasingly counterweight heavy.

weight. As with the arrangement of strut pinned to bascule span and machinery mounted on the rigid platform, this arrangement results in the moving bridge tending to be span heavy when closed and counterweight heavy when open. As discussed in Chapter 11, this is an inherently stable balance condition, but requires slightly more power to operate the bridge than when the condition is reversed.

ROLLING LIFT MACHINERY

Some older rolling lift bridges of the Scherzer type were propelled open and closed by gear-driven operating struts. The original William Scherzer rolling lift bridge patent describes a single powered tension-compression strut operating mechanism, mounted along the centerline of the bridge, connecting to the bridge superstructure below the deck, with the drive mechanism mounted on the pier behind the lift span. Most Rall type rolling lift bridges were also operated with a strut, but the struts were paired, one on each side of the bascule leaf. The drive mechanisms were mounted on the piers, and the struts were pinned to the moving leaf. Almost all curved-tread-plate-type rolling lift bridges less than 100 years old, whether of Scherzer design or otherwise, are driven by gear mechanisms mounted on the bascule leaf, engaging straight or slightly curved rack segments mounted on the pier adjacent to the rolling lift tracks. The slightly curved operating racks seem to have been intended for the purpose of sidestepping the Scherzer patents for rolling lift bridges. The Tacony-Palmyra bascule over the Delaware River, designed by the Ralph Modjeski firm, and the original Venetian Causeway bascules in Biscayne Bay, designed by Strauss, incorporate curved racks but are otherwise very similar to the Scherzer bascule bridges. To mate the drive pinions with the curved racks throughout their travel, the pinions are mounted so that their shafts are slightly offset from the rolling center of the bascule leaf. A few older rolling lift bridges have been converted to hydraulic operation by attaching hydraulic cylinders mounted horizontally at the sides of the bridge leaf, connecting to it where the pinions were formerly located. This arrangement has not proved entirely satisfactory, as a result of secondary forces on the hydraulic cylinders and the end connections caused by the horizontal orientation of the hydraulic cylinder. More recent new rolling lift bridges that have been designed to be hydraulically powered have used hydraulic motors operating through gear drives.

15

MACHINERY—VERTICAL LIFT BRIDGES

The machinery for a vertical lift bridge must raise and lower the span equally at all points to avoid binding between the vertical lift span and its supporting towers. The machinery must negate, or accommodate, the change in balance that occurs with a vertical lift span as the counterweight ropes pass over the sheaves on the tower tops, as this gradually changes the bridge from being span heavy in the closed position to being counterweight heavy in the open position. On most vertical lift bridges the machinery must also overcome friction in the counterweight ropes. This friction can be substantial as the ropes bend over the sheaves during bridge operation, and it should not be ignored.

SPAN DRIVE MACHINERY

The operating machinery of a span drive vertical lift bridge is mounted on the vertical lift span. The motors, brakes, and reduction gearing are located at the center of the vertical lift span, usually above the roadway. Most span drive bridges are of the through-truss type, so it is relatively simple to provide a support platform above the roadway or railroad tracks for the drive machinery. The machinery provides a direct mechanical connection to move each corner of the lift span up or down the same amount when the drive is actuated. This arrangement makes it feasible to provide a direct engine drive, if desired, for auxiliary or emergency purposes.

As indicated previously, there are several variations on span drive vertical lift bridge machinery. The older Waddell type may still be common, but the variation placing the operating rope drums at the corners of the lift span has been recently built in increasing numbers, being almost the only type of span drive bridge built in the last 30 to 40 years.

The Waddell type vertical lift bridge is driven by means of machinery located on the span. As mentioned earlier, these bridges are almost always through-truss type lift spans, as the through truss provides the most practical way to support the lifting machinery on a bridge of this type. An enclosure mounted above the roadway, at the center of the lift span, contains the operator's house and machinery room. Inside the machinery room (see Figure 17-17) are the drive motors and some reduction gearing in a frame, the brakes, and an emergency or auxiliary engine drive if the bridge is provided with one. The reduction gearing converts the high-speed, low-torque output of the drive motors or engine to low-speed, high-torque in horizontal shafting that lies transverse to the bridge axis. This shafting is extended from each side of the machinery room by means of floating shafts and couplings. Alongside the machinery room, on each side, near the main trusses, are additional machinery frames, one at each truss. These frames each contain two large rope drums, each with a large ring gear attached to it. The ring gears mesh with pinions mounted on the shafting that extends from the machinery room (see Figure 5-10).

Each drum of the Waddell type vertical lift bridge has four operating ropes attached to it, arranged so that as two ropes play off the drum, the other two are pulled in. One of these pairs of ropes is called the *uphaul ropes,* and the other pair is called the *downhaul ropes.* The ropes are arranged so that when the machinery is operated, all the downhaul ropes pull in at the same time, and when the machinery is reversed, all the uphaul ropes pull in at the same time (see Figure 5-11). All the operating ropes extend out to the ends of the lift span, so that two uphaul ropes and two downhaul ropes reach each corner of the lift span, adjacent to the towers. The operating ropes then make 90° turns around deflector sheaves, so that the uphaul ropes are headed up and extend to the tops of the towers, where they are secured, and the downhaul ropes extend down to the bases of the towers, where they are firmly attached. Adjustment devices are usually mounted at these connection points so that the lengths of the operating ropes can be adjusted to maintain equal tensions in the slack ropes and keep the span level when being lifted (see Figure 5-13).

Each drum has a single cylindrical surface on which the operating ropes wind and unwind. The cylindrical surface has a pair of rope grooves machined into it, in a continuous double helix, from one end of the cylinder to the other. The pair of uphaul ropes is attached to one end of the drum, and the downhaul ropes are attached to the other. The ropes are arranged so that they fill all the grooves at all times, except for a half turn where the uphaul and downhaul ropes meet (see Figure 5-12). Only one layer of ropes is placed on the drum. During operation, the uphaul ropes unwind as the downhaul ropes wind on to the drum, and vice versa, so that the same amount of rope is always on the drum. Typically, intermediate sheaves or rollers support the operating ropes as they extend from the drums to the deflector sheaves. Rollers are used only to support slack ropes, and intermediate sheaves are used to support the ropes under load.

A fairly common variation on the span drive places the four operating drums one at each corner of the vertical lift span and connects them via gearing and shafting from the power unit at the center of the top of the span (see Figure 5-14). This eliminates the long horizontal lengths of operating rope and the extra turns of the rope over the deflector sheaves, reducing the wear on the ropes and making them easier to adjust. This scheme is more common on newer span drive vertical lift bridges than any other drive method. It still retains the troublesome operating ropes, a primary reason why many designers and owners prefer tower drive vertical lift bridges.

Another variation, less common, eliminates the operating ropes and drums completely while keeping the drive power at the center of the top of the span. Longitudinal shafting connects the power unit to additional gear reductions at the center of each end of the top of the lift span. This machinery drives pinions meshing with geared racks that are mounted to each tower and extend vertically up each one. This is a variation of the "Strauss vertical lift bridge" design (see Chapter 5); two examples exist in Illinois, as highway bridges over the Illinois River (see Figures 5-15 and 5-16). Both are more than 60 years old (one is in the process of being replaced in 2002), but this type of drive system has worked well and has undoubtedly paid for its additional initial construction cost many times over in reducing breakdowns and eliminating operating rope replacement and maintenance expenses. The Shippingsport bridge, about 100 miles west of Chicago, had its racks and pinions replaced in the late 1980s, after more than 50 years and many thousands of bridge operations, but it is unlikely that the racks were actually worn out at that time. In the same time frame, with the same number of operations, a similar but rope-driven span-drive vertical lift bridge would be likely to have had a dozen or more of its operating ropes replaced. With long geared racks in common use today in such applications as elevators, this type of drive may become an economical alternative for vertical lift bridges.

TOWER DRIVE MACHINERY

A tower drive vertical lift span has all its operating machinery, that drives the bridge up and down, located at the tops of the towers. Usually, the operator's house is also on one of the towers. Thus, there is no mechanical equipment on the vertical lift span except for the counterweight rope connections, some secondary equipment such as span lock sockets and buffer cylinders, and the span guides. Almost all machinery maintenance for this type of bridge is centered at the tower tops, so tower drive bridges usually have an elevator at each tower to provide access. The tower drive machinery powers the counterweight sheaves, via circular rack gears mounted on their rims, to drive the bridge to the raised and lowered positions (Figure 15-1). The movement at each corner of the lift span must be coordinated with the movement at the other corners to avoid tilting. The operation of the bridge relies on friction between the counterweight sheave rope grooves and the counterweight ropes acting to drive the bridge up and down. This is the only common type of movable bridge to rely on friction for operation.

Tower drive machinery provides for equal vertical motion at each corner of the vertical lift span by

1. Controlling the drive motor(s) at each tower so that all the motors operate at the same speed

2. Providing gears and shafting at the top of each tower so that the counterweight sheaves at an end of the lift span rotate the same amount when the drive motor(s) at that tower rotates

There are two weaknesses in this system:

1. The drive motors at the towers do not always operate at the same speed; sometimes they become violently out of sync, and the span skews to such an extent that it may actually jam against the towers.
2. The span is driven by means of friction of the counterweight sheaves against the counterweight ropes. This friction is far from infinite, and the ropes flex as they work against the sheaves so that irregular, unpredictable slippage occurs, which can result in a skew of the lift span.

This misplacement or slippage usually causes a slow differential movement of the counterweight ropes supporting the span, which is not by any means necessarily equal between the groups of ropes at the different sheaves. Eventually, one end, or one corner, of the tower drive lift span seats substantially before or after the others. When this condition develops, without any other corrective action being taken, a great deal of force must be exerted to get the span to seat fully. The counterweight rope slippage must be periodically corrected to avoid failure of the bridge to seat. Several different means have been developed to accomplish such correction. For end-to-end differences, the drive machinery is electrically equalized at seating by disabling the synchronizing device so that the two ends of the span can be driven down to seat without regard to the relative positions of the two ends of the lift span. After the limit switches or sensors at all four corners of the lift span indicate that the span

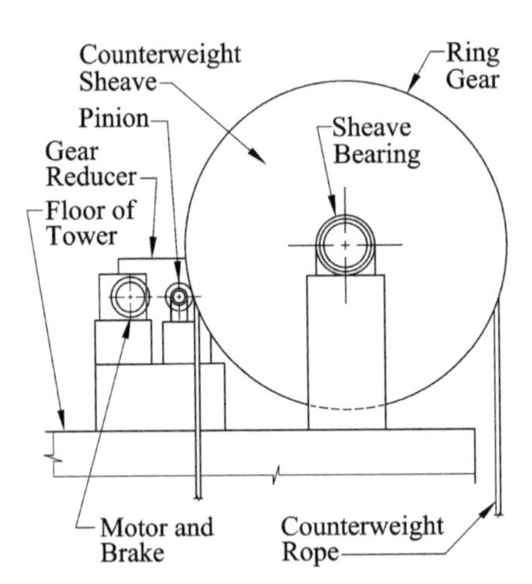

Figure 15-1 Tower Drive Vertical Lift Bridge—Operating Machinery Elevation. The drive machinery engages ring gears at the counterweight sheaves. The drive forces are transferred to the vertical lift span via traction of the counterweight sheaves against the counterweight ropes.

is fully seated, the electrical synchronizing device is reactivated. See Chapter 18 for a further discussion of the application of electrical controls to skew control.

Clutch differentials and adjustable couplings are the most common means of allowing the two sides at each end of a tower drive vertical lift span to be equally seated. Clutch differentials are usually combined with gear reduction in the form of an enclosed reducer, referred to as a *clutch reducer.* When the bridge is in the nearly closed position, the differentials are activated, so that the two counterweight sheaves on each end are driven with equal force instead of equal positioning, until the span is seated. Prior to the next lift, the differentials are again deactivated so that each of the counterweight sheaves is turning at the same number of revolutions per minute. See page 210 for a discussion of differential mechanisms.

Adjustable couplings, placed on the cross shafts between the center drive and the final drive at a sheave, can be used to correct seating problems with tower drive bridges, but there is a danger in that the coupling, when disconnected, separates the span machinery brakes and other controls from the lift point of the bridge. If the bridge is counterweight heavy, for instance, and an adjustable coupling is disconnected in an attempt to correct poor seating, one corner of the span can lift off at its live load shoes. This type of adjustment should not be done when the bridge is open to traffic, unless a safe means of securing the bridge in position is in place (see Chapter 28). Different types of couplings can be used to accomplish this repositioning on tower drive vertical lift bridges. An older type, specially made for the purpose, consists of a rigid flange type coupling with a multiplicity of holes for the connecting bolts, so that an almost infinite angular positionability is available between the two segments of shafting connected by the coupling. An ordinary gear type coupling can generally be used to make these adjustments, as the interchangeability of the several gear teeth on the hub, in their mesh with the internal teeth of the outer sleeve of the coupling, should allow sufficient accuracy in adjusting the angular position of the two shafts. Other couplings, such as friction types, can be used for this application, but this is quite uncommon (see Figure 15-2).

Another method of correcting for rope slippage and allowing firm seating at a tower-drive vertical lift bridge is to have a separate drive for each corner of the vertical lift span (Figure 15-3). The two drives at one tower are connected by a cross shaft at the axis of the motor shafts or at an available lower-speed shaft at each drive. A disengaging coupling that can be remotely controlled is placed on this connecting cross shaft. When the bridge is being seated, the coupling is activated so that each drive turns independently. When the bridge is to be lifted, the coupling is engaged so that the two drives must operate together. This requires the coordination of the operating speeds of the four drive motors during normal operation; there are many means available to accomplish this, for tower drive and other types of movable bridges.

Some very large vertical lift bridges have two counterweight sheaves at each corner of the vertical lift span. If the bridge is a tower drive type, each sheave must be driven. To avoid complications in the drive machinery, the two sheaves are usually on the same axis. The two sheaves are driven via a differential unit so that each sheave will receive the same drive force and relative slippage of the ropes on the two sheaves will be minimized and inconsequential.

Figure 15-2 Vertical Lift Bridge Friction Coupling. This coupling is intended to provide for firm seating at both sides of one end of a tower drive vertical lift span, by slipping the drive on one side that seats before the other side has reached the seated position.

SPAN-TOWER DRIVE MACHINERY

Span-tower drive machinery provides a tower drive vertical lift with mechanical coordination between the two towers. A longitudinal shaft reaches across the span, from end to end, connecting the drive gearing on one tower with the drive gearing on the other. The motors and brakes are located at the midpoint between the towers. A fixed platform reaches across between the tower tops, supporting all the connecting and drive machinery (see Figure 5-18). The operator's house is usually mounted on one of the towers. Most span-tower drive bridges have an enclosed right angle gear reducer at each tower and a parallel shaft reducer at the center of the overhead span. These provide the gear reduction necessary to develop the required bridge operating torque. The motors and brakes are connected to the input shaft of the parallel shaft reducer. An auxiliary direct-drive engine can, if desired, be connected to the input shaft of the primary reducer. Pinions connected to extensions of the output shafts of the right angle reducers engage ring gears mounted on the counterweight sheaves for the final reduction.

The counterweight ropes of span-tower drive vertical lift bridges have a tendency to creep on the sheaves in a similar fashion to the sliding movement of the counter-

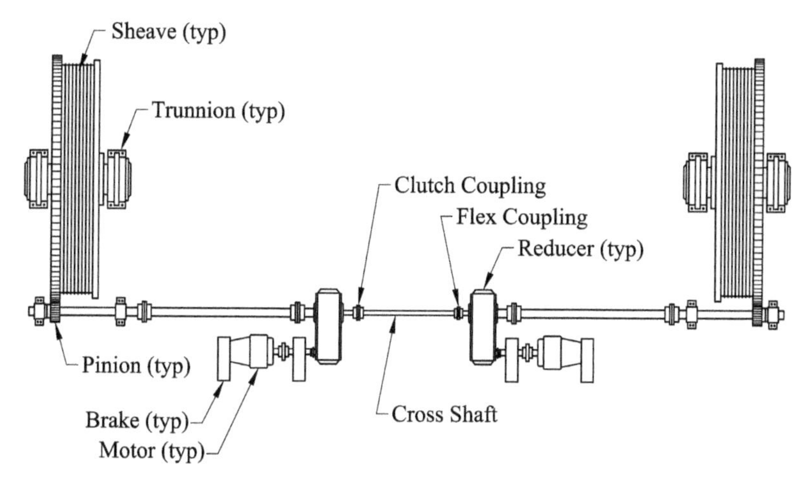

Figure 15-3 Tower Drive Vertical Lift Bridge with Clutch Coupling Seating Adjustment. As the bridge reaches the fully seated position, the couplings are deactivated so that the machinery on each side of the tower can operate semi-independently. When the bridge is at rest, fully seated, and to be lifted, the couplings are activated so that the two sides are held in mechanical synchronization.

weight ropes on the sheaves of tower drive vertical lift bridges. This creep, or incremental sliding of the ropes over the sheaves, may or may not produce unequal displacement of the counterweight ropes at the four corners of the vertical lift span. Some of these bridges have a mechanical means of adjusting the machinery to correct this creep, and others do not. One means of adjustment is a coupling in the longitudinal shafting, which can be disconnected and then reconnected after rotating one shaft sufficiently so that both ends of the span seat. A similar coupling at the cross shafts at each end of the bridge will allow an adjustment to be made so that all four corners of the lift span seat in unison. These couplings are similar in form to those used for the same purpose on a tower drive vertical lift bridge. This method requires considerable time and labor to disconnect and reconnect the couplings. It also requires the vertical lift span to be span heavy on all four corners when seated, or that an auxiliary means be available for holding the span down when the couplings are disconnected and the adjustment is being made.

Another way to correct seating in a span-tower drive vertical lift span is to use clutch reducers. These devices would be installed in a similar fashion to the installation of the same type of equipment on a tower drive vertical lift bridge, with one at each tower. In addition, a third clutch differential reducer would be installed at the drive point on the longitudinal shafting.

The clutches of these differential reducers may be activated every time the bridge is seated, or they may be used only when a problem with seating develops. This clutch, unlike the clutch of an automobile, does not slip while working, so wear does not normally become a problem. Some older tower-span drive vertical lift bridges have clutch differentials combined with exposed open gear reduction sets. These have a tendency to deteriorate because of corrosion due to exposure to the elements, but otherwise function similarly to enclosed clutch reducers. A span-tower drive vertical lift bridge can also use paired or quad drives and clutch couplings, as discussed on page 223.

Span-tower drive vertical lift bridges may be equipped with other machinery, such as auxiliary counterweights, span locks, and so forth, but they usually are not. These devices are described in other parts of this book.

16

MACHINERY—SWING BRIDGES

Machinery is more critical for swing bridges than for bascule or vertical lift spans, because mechanical drive components serve to hold live load supports in place, as is the case for all the end lifts and center wedges for a typical swing span. Swing bridges, unless they open in only one direction and come up against a stop when closed, must be mechanically aligned in the closed position, whereas bascule and vertical lift bridges come to rest in the closed position on live load shoes and masonry plates as they seat. Swing railroad bridges require mechanically activated rail lifts, which are an especially sensitive live-load-carrying component. Because there are so many mechanical operations associated with a swing span, the proper sequence of their operation may come into question. It may be necessary to perform as many as six separate operations to open a railroad swing bridge, after traffic has been properly stopped and the bridge is safe to open:

1. Separate catenary (if trains are electrically powered by overhead wire).
2. Separate end rail connections.
3. Disengage centering device.
4. Pull center wedges (if bridge is center bearing type).
5. Release end lifts.
6. Swing bridge open.

Except for additional interlocking functions that can apply to any type of movable bridge, and are not included in the preceding list, this sequence is usually considered preferable to any other. Some designers call for the centering devices to not be disengaged until after the end lifts are released. Not all of these functions, of course, may be present. The reverse procedure is used in closing, except that an automatic latch may be used to align the bridge in the closed position and the centering device, if present, is usually activated before the end lifts.

TURNING MACHINERY

The most critical part of the turning machinery for a swing bridge, other than the center support bearing, is the rack and pinion arrangement. In most swing bridges there has been inadequate attention given to this area, with a consequent failure of the rack pinion shaft bearing, the rack connections to the pier, or both.

Most swing bridges have a full circular rack mounted horizontally on the pier, and two diametrically opposite rack pinions mounted on the swing span, meshing with the rack and driving the swing. A few very large swing bridges have four rack pinions, such as the BNSF bridge over the Mississippi River at Fort Madison, Iowa. Other small swing bridges have only one rack pinion, such as the Milford Haven Bridge in Virginia. Some swing bridges, those with only one pinion or built on a skew so that they do not open 90°, may be built with less than a full 360° rack. This may save a little in initial construction cost, but can more easily result in a rack that is not properly aligned to the center of rotation of the bridge. Swing bridges require a great deal of effort to accelerate them and may have high starting friction, so the rack pinion loads are large. In addition to having broken pinion bearings or bearing brackets, many are found with cracks or broken teeth in their rack sections near the location where a pinion meshes with the bridge in the closed or open position.

A swing bridge should not suffer degradation at its drive machinery as a result of live load action, because the live load deflection of the span should be at right angles to the load direction of the rack and pinion teeth. Damage and wear to rack and pinion teeth that are in contact with each other, when the bridge is closed and carrying traffic, are mainly due to large forces developed when starting, and from acceleration forces developed in the drive resulting primarily from inaccuracy in gear tooth manufacture as well as wear in the drive train.

The rack pinions driving a swing bridge open and closed are usually mounted on shafts supported in bearings that are mounted on brackets. These brackets are usually cantilevered downward from the superstructure of the bridge. The extension is necessary for the rack pinions to meet the racks; the cantilevered extensions extend below the balance wheels or support rollers. There is limited space to provide for structurally sound brackets. Many swing bridges have undersized brackets that are overstressed at the loads they carry. The loads tend to increase as the bridge machinery wears. As center pivots wear, the diameter of the loaded area at the center of the bearing increases, increasing the friction torque. As rim bearings wear, their friction also increases, particularly if the rollers are allowed to become misaligned to even a small degree. Increased bearing clearance due to a worn friction bearing at the rack pinion shaft allows the rack pinion to push away and partly separate from the rack, decreasing the efficiency of the tooth mesh. This action also increases the bending moment on the gear tooth and the separating force between the rack and pinion. As the separating force increases, bearing wear increases. All this increases the torque required to turn the pinion to move the bridge. The pinion drive torque and the separating force have a direct relationship to the load on the bearing and its support bracket. The use of 24° pressure angle gear teeth can allow rack pinions with fewer teeth to be used, but the increased pressure angle also increases the separating force. On swing bridges, any increase in separating force puts the lower part of the pinion bearing bracket in further jeopardy.

Most swing bridges have two rack pinions, and these are usually driven from a single drive point, with bevel gears or right angle reducers providing the necessary right-angle bends in the drive train. A differential mechanism is usually employed to equalize the load at the two rack pinions (see page 210).

Many older swing bridges have only a single drive motor or two motors working together or alternately. The motors are selectively engaged with the turning machinery or with the end lift and other machinery by means of hand-operated clutches. These clutches are complicated and may have complicated linkages connecting them to actuating levers or wheels in the operator's house.

END LIFT AND CENTER WEDGE MACHINERY

Many swing bridges use worm gear drives to operate the end lifts and center wedges, as the load is small for the greater part of the travel of these devices and efficiency of power transmission is not a primary consideration. A large gear reduction is needed between the drive motor and the end lift or wedge linkage because of the large maximum forces developed during the driving of the end lifts or wedges, so worm gear drives are very nearly ideal for the application. Worm gears can also be self-locking, which helps to prevent the wedges from backing out after they have been driven, but this feature cannot be relied on absolutely to hold the end wedges in place after they have been driven. AASHTO (2.6.13) specifies that wedge-drive worm gear units "shall be self-locking." This feature can be achieved only in a theoretical sense, and it should not be depended upon to hold the end wedges in place. All swing bridge end lift drives that hold the ends of the bridge up, directly or indirectly, should have brakes or other positive locks included in the drive train that hold the drive stationary when the ends of the bridge are to be held in the raised position. These brakes or locks should be located so to be as close as practical to the end lifts. This is especially true for independent drive type systems. The large amount of shafting and machines tied together in the central drive system, with inherent resistance to motion in many function bearings, acts as a "sort of positive lock" for older swing bridge end lift drives.

Some older swing bridges use shafting to connect all the end lifts and center wedges, driving them all in mechanically synchronized fashion from a single center point. A few smaller swing bridges use tension links to connect the end lift driving and withdrawing machinery with the center machinery. These systems have the advantage of mechanical coordination of all the lifts, but maintenance and inspection of the shafting extending the length of the bridge is difficult because of lack of access. Most newer or rebuilt swing bridges have separate drives at each end and at the center to drive this machinery, dispensing with longitudinal shafting or tension links. Some mechanical coordination of movements is still performed with separate drives, such as in combined driving of both wedges at one end of the bridge by the same device. Electrical interlocks are used to coordinate the operation of the separate mechanical systems as necessary, so that the center wedges are retracted before the end lifts, for instance. Interlocking is discussed in detail in Chapter 18.

Other equipment, such as centering devices, may be separately driven or operated off the end lift drive. As with vertical lift bridges, the most reliable synchronization method is mechanical. The older type drive, with one power input near the center of the swing span, driving machinery at the center and both ends simultaneously, has functioned reliably on practically every bridge in which it was installed for decade after decade. Electrical and hydraulic systems, which do not have a fixed linkage such as shafting or rod linkages connecting the machinery at both ends of a swing bridge in one synchronized mechanism, have tended to experience constant difficulties. Typically, one or two end lifts drive before the others. On longer center pivot swing spans this can result in an inability to drive the last end lift, as the first lift to actuate (or the first two if they operate at the same end) causes the bridge to tilt down at the opposite end so that more lift is required at that end than is available from the drive. This problem can be minimized by keeping the free vertical play at the balance wheels to a minimum so that there is little free tilt capability in the entire swing span.

The chief difficulty with the single drive point longitudinal shafting arrangement is lack of access for maintenance. This type of drive can also be more expensive than having a separate drive, or drives, at each end of the swing span, connected to the control system and the power supply only by electrical wiring. The shafts extending from the center of the swing bridge to each end must have support bearings, as well as flexible couplings breaking their length, to accommodate misalignment at erection as well as the deflection of the swing bridge itself due to temperature, live load, and the operation of the end lifts. These shafts are invariably mounted under the roadway, making access to them for maintenance or inspection very difficult. Modern antifriction bearing pillow blocks and gear-type flexible couplings can substantially reduce maintenance requirements, but do not completely eliminate the need for maintenance and inspection.

Provision must be made for the lubrication of end wedges, as they tend to stick in the engaged position if the bridge is not operated often. On many swing bridges, maintenance personnel have adjusted the end wedges to minimal lift because of this "sticking problem." Extreme pressure (EP) lubricants with maximum environmental resistance usually work well, but they must be reapplied regularly to avoid the sticking problem. Other ways to prevent wedges from sticking include using harder materials for the wedges, guides, and seats or using dissimilar materials, such as a bronze wedge with steel guide and seat. On some newer swing bridges, particularly longer spans using higher-strength structural steels that do not require a structure as massive as older bridges, as they can be stressed to higher levels, it has been found that when end wedges are used to lift the ends of the swing span, the leading edges of the wedges can dig into the wedge seats, sometimes stalling so that failure to complete the lift results. This is due primarily to excessive dead load deflection of the bridge, as a result of insufficient stiffness, which is a function of the modulus of elasticity and cross-sectional area of the bridge steel and other geometric factors, not the steel's tensile strength. The solution, if wedges are desired or already installed, is to coordinate lift at both ends of the swing span (see above), provide a smooth leading edge to the wedges, use harder materials for the wedge seats, or increase the stiffness of the ends of the swing span so that angular change is reduced. In extreme cases, dissimilar materials can be used to reduce the friction or auxiliary end lifts can be provided (see above). For extra-high lift requirements, a jacking-type device should be used.

CENTERING DEVICES

As discussed in Chapter 13, centering devices on swing bridges that can open in either direction take two forms. One type, performing function #2, forcibly aligns the bridge in the closed position after it has been stopped in the nearly closed position by the bridge's operating machinery. The other type, performing function #1, acts to "snag" the bridge as it reaches the closed position and holds it there. A swing bridge may incorporate either type, both, or neither of these devices. The snag type, or centering latch, that performs function #1, is usually activated by a cam or lever attached to an end lift drive shaft, then engages automatically as the bridge closes. The mechanical type of centering device that performs function #2 invariably has its own drive system, including a motor, so that its operation can be controlled in proper sequence.

END RAILS AND ACTUATORS

All railroad bridges must provide continuous or at least well-aligned track surfaces for trains to ride on when crossing the bridge. Unlike bascule or vertical lift bridges, the rails on a typical swing bridge forming the surface on which a train rides cannot be automatically separated by merely having joints in them near the ends of the span. There are two common types of rail joints for the ends of the more ordinary swing railroad bridges. The most common type has the ends of the mating rails bevelled at the joints near the end of the span, so that they overlap when lowered into position to carry trains. These are frequently termed *miter rails.* AREMA (6.5.2 c) states: "Where the connections are of the miter type, the two sections shall be held positively in a transverse direction by guides, to prevent spreading at the miter joint." (See Figure 16-1.) In another form of lifting swing bridge end rail joints the rail ends are not mitered but cut square, and a short section of standard rail, machined so that the head fits against the head of the running rail, is fitted to the outboard side of the lifting rail. This short section of rail acts to align the running rail and form a bridge across the joint. AREMA (6.2.12) calls this an *easer rail* arrangement. The operating mechanism for this type of joint can provide positive seating. This joint has an advantage over the sliding block (see below) in that traction forces are accommodated, but it does not work well in freight service as worn wheels produce severe impacts at the joints when the outer edge of the wheel strikes the short piece of rail. This type of joint is used on some

Figure 16-1 Miter Rail Locks on Swing Bridge. These miter rails are of the cast "Conley" type. Each movable miter rail has a separate electrical lock that trips when the moving rail is fully engaged with its mate on the approach span.

commuter passenger railroads, at their swing and other movable bridges, and appears to be successful.

In some applications the easer rail is separate from the running rails and is mechanically positioned, extending and retracting from the swing span to engage and disengage. This allows the running rails to remain fixed to the bridge and approach track supports. The easer rail is a longitudinally sliding block of heavy steel which carries the wheels of the train over the gap in the rails at the end of the swing span. The block also acts to align the rails to some extent, but this system is not generally considered capable of carrying trains operating at high speed. This system also suffers because the traction administered by locomotive wheels sends the sliding block back, in the opposite direction of the train movement, while the rolling resistance of unpowered wheels tends to pull the block along in the same direction of movement as the train. Because of this buffeting action, the drive linkages for the sliding blocks tend not to last very long before needing repair or replacement. Some railroads call for reducing locomotive tractive effort when crossing this type of joint, but this does not help much because even the normal drag of freight car wheels causes the easer rail to be "kicked." This type of joint is favored by many railroads operating in cold weather, as it is less likely to be fouled by snow or ice.

On swing bridges, one rail at each joint must be mechanically lifted or retracted to clear the other rail when the bridge is to be swung open. The lifting rail or retracting component is almost always placed on the swing span so that the drive machinery is readily powered by the machinery on the bridge. The mechanism to lift or retract the rails is usually powered by means of links to the end wedge drive, to simplify the machinery. In the past, the lifting rails were usually held in the lowered position by gravity alone. More recently, connections have been developed to both lift the rail up and pull it down to achieve greater security in rail connections under train traffic (see the following discussion). Special electrical switches are used to make certain that the rails were fully lowered before trains are allowed to cross the bridge. These switches are often connected to small lock bars, which do not necessarily hold the rails down but assure that the rails must be fully lowered before the switches can be tripped.

A particular type of joint for swing bridges was developed some time ago and marketed by the Conley Company of Tennessee, called the "Conley Joint." The joint is made up of cast steel pieces, fabricated somewhat in the manner of cast steel turnout frog or crossing pieces. One piece is hinged so that it can tilt up, clearing another piece. The tilting piece and a third piece, from which it pivots, are mounted on the swing span, and the second piece is mounted on the approach. The mated pieces form a relatively smooth continuous rail section across the joint between the swing span and the approach, while allowing some thermal expansion capability. These have proven quite successful in freight and relatively slow-speed passenger service; they have a maximum practical speed limit of about 60 mph. The Conley-type joint is in use on several swing bridges, and imitations have been developed. It has the advantage of allowing positive control of the moving rail. It also forces the pivoting rail section to be properly aligned with its mating components, as the pivoting section cannot be completely lowered unless the rails are aligned.

Several swing bridges carrying passenger traffic use a miter rail type lift linkage that pulls the lift rails down when the bridge is closed, reducing the likelihood of misalignment at the miter rails. This development provides a means of allowing high-speed passenger trains to cross a swing bridge without excessively burdensome speed restrictions. Some attempts were made in the course of the Northeast Corridor Improvement Project, during the 1970s and early 1980s, to provide a swing bridge miter rail system that would allow the design speed for the project, 106 mph, to be achieved by passenger trains crossing the several swing bridges on Amtrak's rail line between Boston and Washington. A joint of the type produced by the Conley Company was determined to be the best available and was installed at a few bridges, but after evaluation it was determined that the joint was not sufficiently durable, as discussed earlier, at high speeds.

Of the five swing bridges on Amtrak's Northeast Corridor line between Boston and Washington, only two are on potentially high-speed sections of track. The southernmost swing bridge, over the Susquehanna River at Havre de Grace, Maryland, seldom opens for water traffic, so removable continuous rail is used across the bridge, eliminating the miter rail joint completely. At the Portal Bridge over the Hackensack River in the New Jersey Meadows, several different types of miter rails have been used during the more than 90 years that the bridge has been in existence, with varying degrees of less-than-overwhelming success. Miter rails made from modified milled standard rails are now in use on this bridge, with lift mechanisms that positively pull the lifting sections of the miter rails down into their mating fixed chair sections. After a passenger train derailment that was attributed at least in part to a failure of miter rails with the older type of lifting mechanism to seat properly, this positive type of mechanism has been mandated for swing bridges carrying passenger traffic. As of 2002, a study was under way to replace the Portal Bridge with a high-level fixed crossing over the Hackensack River, to completely eliminate concerns about having a movable bridge, particularly a swing bridge, supporting high-speed railroad track at this location.

17

MACHINERY—GENERAL

Friction values in machinery components can vary considerably and cannot be precisely predicted. In calculating the frictional resistance of a machinery drive train, the most conservative values should be chosen from the various test and hypothetical data available. Machinery resistance should be assumed to be high in starting a bridge in motion and propelling it, and should be assumed to be low in stopping the bridge or holding it in position. The machinery resistances presented by AREMA and AASHTO in their specifications and manuals for movable bridges are generally considered acceptable and are, without exceptions taken, appropriate to provide for an adequate prime mover. These resistances do not always provide for successful operation of a movable bridge in the various conditions under which a movable bridge may be required to operate. The 2000 LRFD AASHTO *Movable Bridge* Design Specifications (Section 5-4) makes an attempt to be more realistic in sizing the prime movers for movable bridges to match actual resistances.

Exception to the aforementioned statements applies when specially manufactured enclosed gear drives and enclosed antifriction bearing units are used on a bridge. These have small friction losses that approach the infinitesimal as far as the bearings are concerned and can almost be ignored during design. When these assemblies are properly designed and constructed so that they are veritable sealed units, impervious to external conditions on a bridge, they can operate for many years with high efficiencies and little or no maintenance attention. If they are grossly overloaded or otherwise abused, or suffer from submersion in floods or from other unexpected external factors, their lifetimes will be cut short unless some remedial action is taken. Reputable manufacturers of such components provide reliable units to suit the bridge application when provided with all necessary correct data as to loading, duty cycles, and environmental conditions of the particular application. Many manufacturers of mass-produced machinery components have categorized various service applications so that the designer can readily select the correct component for a particular application. Unfortunately, movable bridge applications are relatively uncommon in comparison to general industry applications, and no two movable bridges are the same. It is thus quite difficult to develop suitable blanket application specifications for movable bridge equipment.

An important element in the design of machinery for movable bridges that is frequently overlooked, or poorly addressed, is provision for maintenance. The machinery for a movable bridge is generally slow moving and carries heavy loads. Lubrication and other maintenance are not reliable, as very few bridge owners can afford to have full-time maintenance personnel on hand at a bridge. Thus, the machinery must be rugged, and ruggedness must take precedence over efficiency in power transmission or economy in fabrication or erection. Some components of bridge machinery are expected to perform satisfactorily for years without any attention whatsoever. On many movable bridges, "maintenance" personnel are called to the scene only when a failure occurs, to make repairs (see Part IV).

Many older highway bridges were designed solely for minimum construction cost, so that the machinery was cramped and access to lubricate or inspect it was practically nonexistent. Labor was cheap and plentiful, so efficiency in performing maintenance was not a consideration in bridge design. Many newer highway bridges are not laid out much better, and it appears in many cases that the machinery was added to the structure as an afterthought, with no consideration whatsoever for maintenance and inspection. Newer structures are expected to last a long time, and in many cases are assumed to be capable of infinite life. Maintenance and inspection of the machinery, on a regular basis, is essential not only to prolong the life of the bridge and its machinery, but to minimize the likelihood of breakdowns. No existing movable bridge has been designed to eliminate maintenance (see page 344).

In all cases, there should be access to every machinery component for maintenance. A person carrying a heavy load of tools or material should be able to walk directly to the point of attention without having to stoop, climb ladders, or crawl into a pit. Generous walkways to all areas, with proper safety appliances, should be provided. Pipe handrails and stanchions should never be used on any bridge, movable or otherwise. These components can and often do corrode from the inside out, becoming dangerously weak with little or no external indication of being so—on some "well-maintained" bridges these appliances were kept brightly and neatly painted until the day one of them snapped off. The tower top areas of vertical lift spans should be accessible via stairways with landings or, preferably, by powered elevators. Maintenance personnel tend to use the vertical lift span itself as an elevator if one is not provided, so an elevator definitely helps to keep a bridge in service during maintenance work. Overhead traveling cranes can be provided in machinery rooms and other areas where heavy components may require removal or replacement. If these lifting devices are not properly designed with access to the areas needing them, the money spent on them is wasted. Workers usually erect temporary rigging to accomplish their tasks rather than use poorly situated or poorly suited built-in equipment.

As labor is not cheap nor readily available, and adequate supervision is difficult to achieve, efficiency and ease of maintenance are essential to provide for a durable and economical structure. Ideally, necessary maintenance should be reduced to a minimum. In many cases maintenance functions can be reduced or eliminated by providing for automatic maintenance within a unit, such as an enclosed gear drive. Periodic maintenance inspection, however, is still essential to monitor performance of maintenance functions whether they are manual or automated. Certain types of machinery need less maintenance than others. Hydraulic machinery is inherently maintenance intensive, as fluids must be changed, fluid filters must be cleaned or changed regularly, and hydraulic cylinders, joints, and flexible hoses must be inspected and maintained regularly to avoid leakage. Conversely, enclosed gear drives, when properly specified, designed, built, and installed, can last the life of the bridge without maintenance.

Providing for maintenance is a point of good design, but it should not be the primary purpose of the designer. The primary function, the utility, safety, and functionality of the bridge, should be ensured first and the means of operation next; then provisions can be made for maintenance. If necessary, the primary function should be revised to accommodate the lesser, but not to the extent that the primary function is compromised. Maintenance itself is discussed specifically in Part IV.

MAIN BEARINGS

The main bearings for a swing, vertical lift, or bascule bridge must carry extremely heavy loads and must last for a very long time, but they must also sit motionless for long periods between operations at very slow speeds. Because of their specialized nature, main bearings for swing bridges are discussed in Chapter 13. The main bearings for bascule bridges and for the counterweight sheaves of vertical lift bridges, which are very similar to the bearings used in general industrial applications, are discussed in this chapter. Most older vertical lift and bascule bridges were constructed, with grease-lubricated bronze bushings supporting steel journals, to accomplish the requirements of durability and load capacity. Many bridges today still use this type of bearing. Its main drawback is the need for normal maintenance.

A few successful attempts at using antifriction bearings for the main supports of movable bridges were tried in early times, notably on the Tower Bridge in London, but the accuracy required in a large antifriction bearing, with the required durability and load capacity to be successfully installed in a movable bridge, was problematic at that time. One of the earliest successful attempts in the United States at applying roller bearings for heavy loads to the main bearings of a movable bridge was at the Borden Avenue bridge, a retractile structure over Dutch Kills in Long Island City, New York, built in 1908. The original bearings are almost 100 years old and are still in place, supporting the bridge and its traffic loads. Enclosed roller bearings have since been applied very successfully on vertical lift and bascule bridges of all sizes. Standardized large antifriction bearings are not designed specifically for such low-speed intermittent service, and the design of such bearings is a highly specialized field of mechanical engineering. Both AASHTO (2.5.14, 2.6.4, and 2.6.5) and AREMA (6.2.11.6, 6.4.5c, 6.5.10, and 6.5.11) have guidelines for configuration of the large roller bearings used in these applications, which have proven successful. A failure in these applications due to poor bearing selection is very expensive to rectify. Selection of bearings for such service should be left to the application engineering staffs of reputable bearing manufacturing companies with experience in such applications. For these applications, the small increase in initial cost due to providing a slightly oversized bearing is nothing as compared with the cost of having to replace a prematurely failed bearing later.

Antifriction trunnion and counterweight sheave bearings can last the life of a bridge without requiring maintenance functions, such as relubrication, to be performed. These bearings are sensitive to contamination and maladjustment, so any disassembly for inspection or maintenance purposes should be done with extreme caution and is, in all prudence, best left to a technician provided by, or at least trained by, the bearing manufacturer. Some noninvasive techniques can be used by a bridge machinery inspector to get a fair idea of the condition of the bearings (see Chapter 34).

SHAFT BEARINGS

At one time, babbitted bearings were used almost exclusively for shaft support on bridge machinery. These bearings were easy to fabricate but could support only small loads, tended to overheat, and wore out rapidly. They had the advantage of being easily repairable by having new babbitt poured in place. Most bridges now use either bronze bushed bearings or antifriction bearings to support gear shafts and the like. Antifriction bearings have the advantage of not requiring much in the way of normal maintenance, but if they fail, they can do so abruptly and completely, possibly taking the bridge out of service. Bronze bearings need regular lubrication, but they wear slowly so that there can be obvious advance warning of their need for replacement. Repairs or replacements can usually be made when needed without seriously disrupting bridge operation.

Whenever possible, intermediate shaft bearings should be avoided in bridge machinery applications. They are a source of misalignment if the bridge superstructure should shift and can be difficult to install correctly in the first place. Any shaft can be made stiff enough to support itself over a reasonable given length between couplings. If a very long shaft is required, it may be better to break it up into shorter segments. Areas where intermediate shaft bearings may be required include the following:

- Shafts connecting rack pinion drives for very wide bascule and tower drive vertical lift bridges
- Shafts connecting the end machinery to central drives on swing bridges
- Shafts connecting disparate drive machinery elements on span-tower drive and non-Waddell types of span drive vertical lift bridges

Any rotating shaft should be fixed axially at no more than one bearing point. Collars can be placed around one friction type bearing, or one fixed and one expansion antifriction bearing can be used.

Bearings adjacent to gears can be subject to very heavy loads. Even when well lubricated, heavily loaded babbitted bearings on older bridges require renewal, adjustment, or repair on a regular basis. For this reason, babbitt was infrequently used to support rack pinion shafts, which usually were of bronze on older bridges. Today these bearings are invariably bronze or of the antifriction roller type. The cost difference may be small, so that the choice is one of increased maintenance cost versus the possibility of a debilitating sudden bearing failure. Generally, antifriction bearings are chosen over the bronze bushing type, primarily because of reduced friction and reduced maintenance requirements. The same cautions apply to these antifriction bearings as to larger trunnion or sheave shaft types. They should be dismantled for inspection with caution. Internal inspection is best left to factory-trained technicians.

Oilless-type bearings should be used with caution on movable bridges. Teflon or other nonmetallic bearings are subject to being ruined by abrasion caused by foreign particles entering the bearing. This is not at all unlikely in the typical dirty environment of a movable bridge. Metal bearings with graphite inserts or impregnated with oil provide their own lubrication by wearing and are not suitable for heavily used machinery shafts, as the bearings will require replacement sooner than other machinery components. Oilless bearings are frequently used on hand-drive shafts, which are seldom operated. In these applications the bearings are likely to seize up, as the bearing material becomes bonded to the shaft material over time, because of even slight amounts of corrosion, when a protective coating of grease or oil is not present and the shaft does not rotate for a very long period of time.

COUPLINGS

There are many types of couplings that could be used in bridge machinery. Couplings can be rigid or flexible, and the degree of flexibility can vary. Couplings can be all metal or can have nonmetallic components. AASHTO and AREMA both require all-metal couplings for all primary drive trains. There are actually only a few types of couplings commonly used in bridge machinery.

For connecting electric motors to bridge machinery, grid type couplings are almost always used. These couplings consist of two hubs, one for the motor shaft and one for the machinery shaft, with matching longitudinal radial slots around the perimeters of the hubs. Bent pieces of flat metal wire, called grids, are fitted into the slots, connecting the two hubs and transmitting power between the shafts. A cover with seals is fitted over the hubs, helping to keep the grids in place and retaining lubricant. Grid type couplings allow for angular, parallel, and axial misalignment, as well as providing torsional cushioning between the shafts. The torsional cushioning helps to prevent the deterioration of electric motor commutators and windings due to impacts from the machinery.

To connect the drive shafts, gear type couplings are generally used. This type of coupling consists of two hubs. The hub mounted onto one shaft has external gear teeth, and the hub on the other shaft has internal gear teeth; when the two hubs are put together, the internal and external gear teeth mate. The gear teeth are cut with slightly curved faces so that the internal and external teeth can tolerate angular misalignment. This type of coupling, referred to as single engagement, allows for angular and axial misalignment between shafts. A double engagement gear type coupling allows for parallel misalignment. This coupling has two hubs with external gear teeth and a sleeve containing internal gear teeth that fits over the two hubs. A gear type coupling usually needs no separate cover, but typically has seals that allow lubricant to be retained.

Small jaw type couplings with inserts between the jaws are used in many cases to connect machinery parts that do not transmit heavy loads, such as driving position indicators, selsyns, and rotary limit switches.

Other types of couplings are, in some cases, found on older movable bridges. These can include plain jaw type couplings, sometimes of very large sizes, Oldham couplings, and chain type couplings. These are all difficult to maintain in a bridge environment and have limited load capacity, and thus they are now seldom specified on new bridges.

GEAR DRIVES

The power for most movable bridges is transmitted by gear drives. A gear is simply a wheel that is specially adapted for efficiently transmitting torque while providing positive (nonslip) engagement. A gear always mates with another gear. The gear drives for a movable bridge are of two kinds: open gearing—consisting of mating gears mounted on shafts, supported by bearings in various forms mounted on structural members, which may all be exposed to the environment—and enclosed gearing—consisting of gearboxes enclosing and supporting gearsets mounted on shafting, supported in bearings mounted in and part of the gearbox itself. Enclosed drives allow precise alignment of the gears, which will not change later because of loose bearing mounts. Almost all older movable bridges have reduction gear trains of the open type between the prime mover and the final drive points, whereas most newer bridges have all or most of the reduction gearing in the form of enclosed gears. The final drive points, for all types of gear-driven movable bridges, new or old, are invariably open gearing, typically pinions driving straight or curved racks or ring gears.

A few movable bridges have what could be called *fully enclosed open gearing.* This apparent misnomer describes a system in which the gears, shafts, and bearings are of the open type, but the assembly, consisting of one or more gear reductions, is enclosed within an oil-tight pan underneath and a cover above. The gears enclosed in such a manner can be of any type, but on movable bridges they are usually straight spur gears. The enclosure keeps the machinery clean and allows the use of an oil bath instead of grease for lubrication. An oil bath is, for most purposes, a higher-quality lubrication system than grease, as it is more reliable in keeping the gear teeth coated with lubricant, does not require much attention, is affected less by temperature changes, and can act as a heat sink. On some older bridges, open gearing is mounted on frames, usually made of cast steel, on which all the bearings are integral. These frames keep the gears in alignment as long as the bearings and journals are not excessively worn. Grease is generally used for lubricating gears on these frames, but an oil bath can be used if a frame is enclosed as described above.

The design criteria for open gearing for highway and railroad bridges are explicitly defined by AASHTO (2.6.12) and AREMA (6.5.19). The gearing is selected primarily on the basis of the breaking strength of the gear teeth. Resistance to wear is secondary and was not considered by these authorities until recently, as the duty cycle is ordinarily quite low for bridge machinery as compared with typical industrial applications. As the open gearing of a movable bridge moves at relatively low speed and is of crude construction by general machinery standards, a great deal of wear can usually occur before the functionality of the gearing is seriously compromised. On the other hand, a broken gear tooth can put a bridge out of commission immediately. Some specifications call for case-hardened or surface-hardened gears in bridge drives. These have good initial strength, fatigue strength, and durability, but as soon as the hardened surface wears away, the gears wear rapidly and can then fail. If bridge drive gears are hardened, they should be through hardened, but it is generally better to use a larger size gear tooth to provide the strength needed. AASHTO (2.6.12) requires bridge machinery drive gears to be designed for surface durability as well as for breaking strength, but does not give detailed guidance except in the 2000 LRFD AASHTO.

Gears are very specialized pieces of machinery, which have undergone centuries of development into the high-precision, reliable, and durable components in use today. The different aspects of a gear's geometry have special names and relationships (see Figure 17-1).

Specialized gear manufacturing companies have a cumulative total of billions of dollars invested in the equipment used to make gears. Unfortunately, most of this equipment is designed to make gears in the diametral pitch system, whereas AASHTO and AREMA still use obsolete circular pitch and stub tooth gear forms. To make matters worse, the United States federal government made a strong pitch for metrication a few years ago, with the result that in many states and local government entities re-

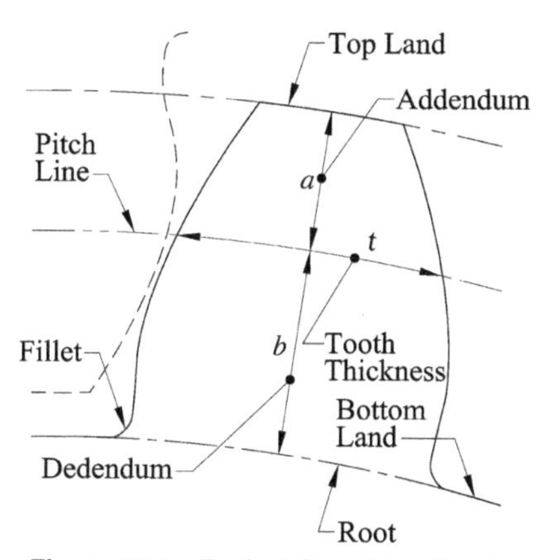

Figure 17-1 Typical Spur Gear Tooth.

quired that all new bridges and some rehabilitations be built completely to metric standards. These systems differ as follows:

Circular Pitch—p The size of gear teeth in inches, measured along the pitch line (see Figure 17-2). Typical larger circular pitch sizes used in movable bridge machinery are 1 in., $1\frac{1}{2}$ in., 2 in., $2\frac{1}{2}$ in., 3 in., 4 in., etc., up to a maximum pitch of $6\frac{1}{2}$ in. or thereabouts. This sizing system is based on the manufacture of cast tooth gears and is not used extensively today.

Diametral Pitch—P A gear tooth sizing system in which the diametral pitch number is the number of teeth in a gear of 1 in. pitch diameter. Typical diametral pitch sizes for movable bridge machinery are 20, 16, 12, 8, 4, 2, 1, and $\frac{1}{2}$. The smaller teeth are used on the gears in enclosed drive units. This gear size system is based on machine cutting of gear teeth, by a process called "hobbing," and is the most common tooth size system used in the United States today.

Module—m A gear tooth sizing system in which the module times the number of teeth equals the pitch diameter. This is the common metric gear tooth size designation, with the size measured in millimeters. A four module gear with 12 teeth has a pitch diameter of 48 mm. Modules are usually "metric," so 5, 10, 25, and 100 are typically used. This sizing system is not generally available from U.S. gear manufacturers, but is becoming more so. It can be considered the metric equivalent of diametrical pitch for manufacturing purposes.

The following are definitions of other gearing terms:

Addendum The radial distance from a gear's pitch diameter to its outside diameter (see Figure 17-1).

Backlash Space between gear teeth at the pitch line.

Clearance The space between the top land of the tooth on one gear and the bottom land of a mating gear.

Conjugate action The mesh of gears during operation so that the ratio of the rotating speeds of the two mating gears is constant during tooth engagement and disengagement—this is a basic requirement for power transmission gearing, to avoid or at least minimize acceleration and impact forces when gears are operating.

Contact ratio The number of teeth in contact in a gear set at a given time. Contact ratios should be more than 1. Ratios less than 1 produce impacts. The strict defi-

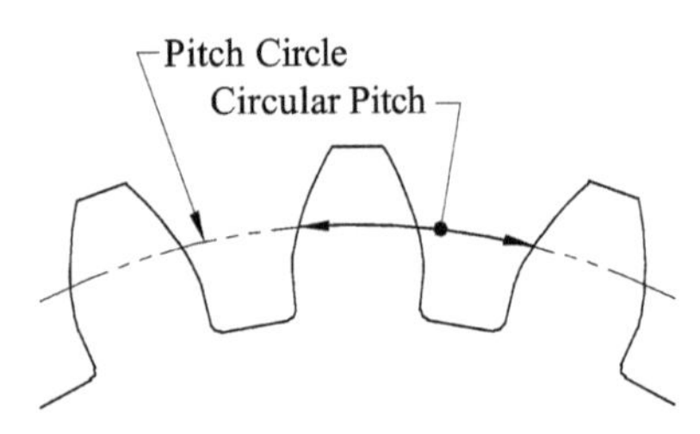

Figure 17-2 Circular Pitch.

nition of the contact ratio for a set of spur or helical gears is the length of the zone of action divided by the base pitch. What it really means is as stated above.

Dedendum The radial distance from a gear's root diameter to its pitch diameter (see Figure 17-1).

Involute A particular gear tooth shape that is notable for allowing errors in the center distances between gears and still providing conjugate action, and thus the same velocity ratio, and avoiding impact forces on the gear teeth.

Pitch diameter (Figure 17-3) The diameter of a circle formed by the pitch line of a gear. A straight gear rack such as on a Scherzer rolling lift or Strauss heel trunnion bascule bridge would have an infinite pitch diameter.

Pitch line A circle, the diameter of which is equal to the pitch diameter, or to the overall size of a gear with infinitely small teeth, that is, a friction wheel. The pitch line is also the line across the flank of the gear teeth at the pitch diameter (see Figure 17-4).

Standard teeth Gear teeth that have normal proportions, which would be obtained in ordering gears without specifying any special details. A standard gear tooth has an addendum that has a length in inches equal to

$$1/P$$

which is equal to

$$0.3183p$$

which is equal to

$$m/25.4$$

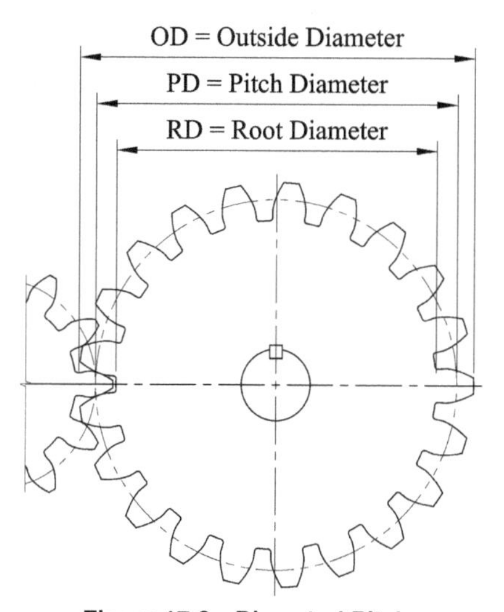

Figure 17-3 Diametral Pitch.

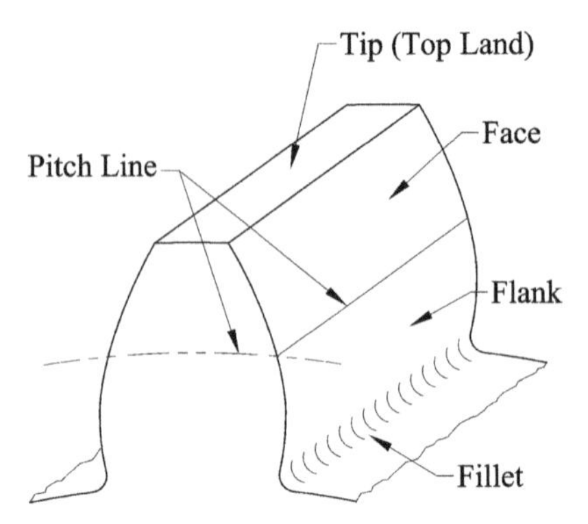

Figure 17-4 Gear Tooth.

In each of these systems, the dedendum of a gear tooth of standard proportions is equal to 1.25 times the addendum. However, most gears for movable bridges are made to the obsolete AGMA of 20° $b = 1.157/P$.

Stub teeth Gear teeth that are shortened, as compared with so-called full-depth teeth, by reducing the addendum, which is the distance from the pitch line to the tip of the tooth, and reducing the dedendum proportionately. It is desirable for the rack pinion to have as few teeth as possible while maintaining sufficient strength. Stub teeth produce more impact during operation, because of a lower contact ratio, but allow fewer teeth in the rack pinion without the tooth form being *undercut* (see page 246). There are several systems of stub gear teeth in use. The most common are the Nuttal stub (Figure 17-5) and the American Gear Manufacturers' Association (AGMA) stub (Figure 17-6). Both of these are considered obsolete by the gear industry. The AGMA stub tooth form is cited by AASHTO and AREMA. It is convenient, for bridge rack and pinion detailing purposes, to have a standardized special tooth shape to provide additional strength under certain conditions. For a movable bridge, this allows the rack and pinion to be defined without the necessity of preserving a detail drawing or keeping track of dimensions specified to four decimal places (Figure 17-7).

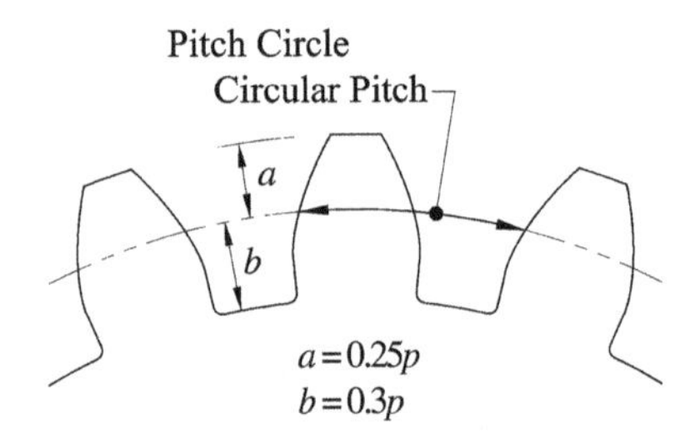

Figure 17-5 Nuttal Stub Teeth. The addendum $a = 0.25p$.

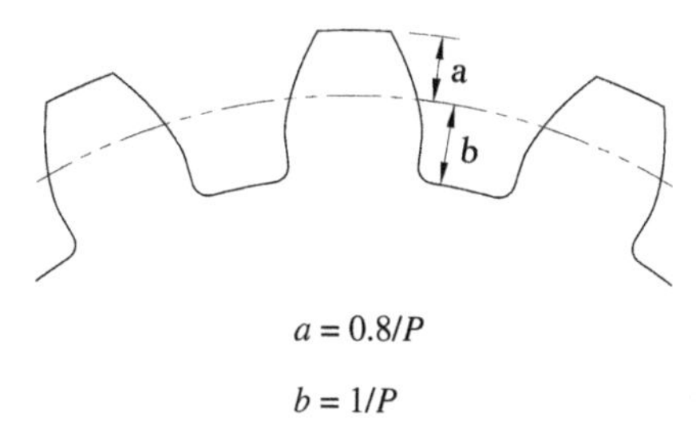

Figure 17-6 AGMA Stub Teeth. The addendum $a = 0.8/P$.

Tooth thickness The distance along a gear's pitch diameter from one side of a tooth to the other side of the tooth, measured through the tooth (see Figure 17-1). For any circular gear, this distance is measured along a curved line.

None of the aforementioned tooth size systems are compatible with each other, and each requires its own special tooling to fabricate gears. Stub tooth gears cannot mesh with full-depth gears, or at least, it cannot be done without interference or partial engagement and irregular operation (see Figure 17-8). Another variation is in pressure angle, which is the slope of the working sides, or faces, of the gear teeth. There are several different "standard" pressure angles, none of which are compatible

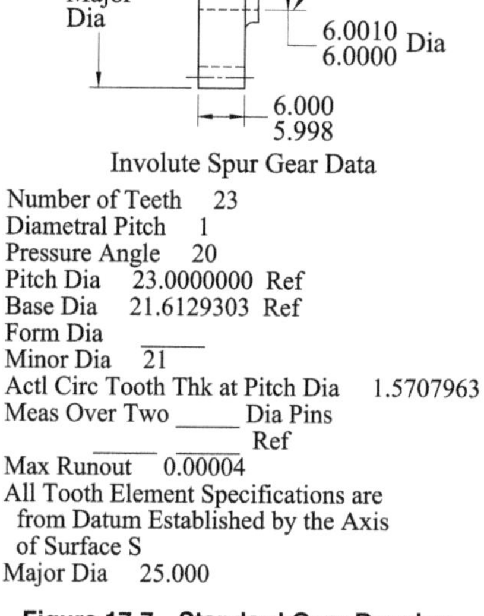

Figure 17-7 Standard Gear Drawing.

Figure 17-8 Improperly Mated Gears. The pinion on the left was installed as a replacement for the worn original. The fact that the original was a stub tooth gear was not noted at the time. The mesh of a full depth pinion with a stub tooth gear is the result.

with any other. Most movable bridge gearing today is made to a 20° pressure angle, but many older gears were made to the $14\frac{1}{2}$° pressure angle system. Some older movable bridges had the gears specified to 15° pressure angles. Gears made with different pressure angles will not mate properly and will not operate smoothly—they may not operate together at all. Much gearing outside the movable bridge industry is made to $22\frac{1}{2}$° or 24° pressure angle. This allows pinions with fewer teeth to be cut without undercutting (see Figure 17-9), but increases separating forces and thus produces larger bearing loads. It also can operate less smoothly than 20° gears. The high separating force is not as critical a disadvantage as it once was, as antifriction-type bearings can usually be easily provided to withstand these larger forces. The size of a sleeve bearing automatically becomes larger in proportion to the load it is designed to carry unless a higher-capacity material is used.

Naturally, the larger the teeth in a gear, all else being equal, the larger the load the gear can carry. Gears with larger teeth are allowed to have wider faces, which further increase the capacity of the gearset. AASHTO and AREMA allow the effective face width of open spur gears to be as much as three times the circular pitch. The pressure angle of the gear determines its thickness at its base; larger pressure angles result in

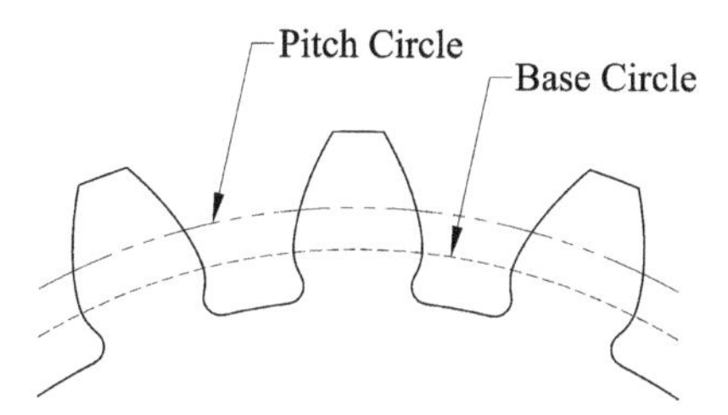

Figure 17-9 Undercut Gear Teeth.

larger bases and stronger teeth. Early gears were made with small pressure angles to reduce the load on the bearings supporting the gears, but modern gears use larger pressure angles to produce stronger teeth. As the number of teeth in a gear increases, along with its pitch diameter, the space available for the root of the gear increases and the teeth are stronger.

A straight rack has the greatest strength of any external gear in a given geometry system. Internal gears, with the teeth facing inward, are stronger yet, but have little application in bridge machinery drives, except in the occasional planetary reducer. Internal racks have also been used on a very few bascule and swing bridges (see page 211). The internal rack is very strong, but the mating pinion is subject to the same weaknesses as other pinions. The gear teeth of a particular internal-external gearset may be modified in a manner similar to other gearsets (see below). A gear with the smallest number of teeth possible has the least strength. If the gear has fewer teeth than the theoretical minimum for a given geometry series, it will be even weaker, because the roots of the teeth are undercut, but it will still mesh with another involute gear of the same pitch, depth, and pressure angle. Undercutting occurs naturally when a standard tool is used to cut (by hobbing) a gear with too few teeth. Undercutting is usually necessary for a gear with a small number of teeth, to provide sufficient space for the tips of the teeth of the mating gear. For full-depth gears of a given pressure angle, there is a certain minimum number of teeth in the gear to avoid undercut teeth. For a 20° pressure angle, this is 19 teeth. The larger the pressure angle of a gear tooth system, the smaller the minimum number of teeth in the gear without undercutting the teeth. In typical applications, the addenda of the mating gears are modified to eliminate undercutting and to increase the strength of the smaller gear of the pair. The result is a gearset with unique geometry details, particularly tailored for a certain application. This is sometimes done for open gearing used on bridge racks and pinions. It is often done with the gears used in enclosed speed reducers. Most often on movable bridges, gears with small numbers of teeth are given special profiles, such as radial, in the root area. This can result in rough operation, but is not very noticeable in slow speed applications such as racks and pinions.

Gear strength is, of course, also a function of the material of which the gear is made; stronger materials result in gears of greater capacity. The quality of a gear's fabrication (see page 249), and the speed at which it runs also affect its load capacity. Poor quality and high speeds produce larger impact forces and greater wear on the gear teeth in operation so that the load must be reduced to avoid failure.

The basic mathematical equation to determine the breaking strength of gear teeth is the Lewis equation, developed by Wilfred Lewis in the late nineteenth century. Lewis did a great deal of research on gearing. He proposed in theory, that a gear, would fail in bending by breaking a tooth at its root. He considered the tooth to be a cantilever beam and developed a form factor for the tooth to establish its beam strength. The form factor takes into consideration the size and shape of the gear tooth, which is a function of the pitch of the gear, the pressure angle of the gearset, and the number of teeth in the gear. As the number of teeth in a gear increase, the base of the tooth gets wider, starting with an almost barrel-shaped cross section (Figure 17-10) in a gear with few teeth and evolving to a broad-based trapezoid in a gear with many teeth (Figure 17-11).

The basic Lewis equation is as follows:

$$W = SpfY$$

where

$W =$ the allowable load on the gear tooth
$S =$ the permissible bending stress at the root of the tooth
$p =$ the circular pitch of the tooth
$f =$ the face width of the tooth
$Y =$ the form factor of the tooth

This equation has been expanded for movable bridge gearing to include a strength reduction factor for the speed of the gearset at the pitch line, and modified to develop the form factor as a function of the number of teeth in the gear and the pressure angle of the gearset. For 20° full-depth teeth, the equation per AREMA and AASHTO takes the following form:

$$W = fsp\left(0.154 - \frac{0.912}{n}\right)\frac{600}{600 + V}$$

where

$W =$ the allowable tooth load in pounds
$f =$ the effective face width in inches
$s =$ the allowable tensile fiber stress in pounds per square inch
$p =$ the circular pitch in inches
$n =$ the number of teeth in the gear
$V =$ the pitch line velocity in feet per minute

This equation allows gears in movable bridge service to satisfy the breaking strength requirements, with further specified restrictions as follows:

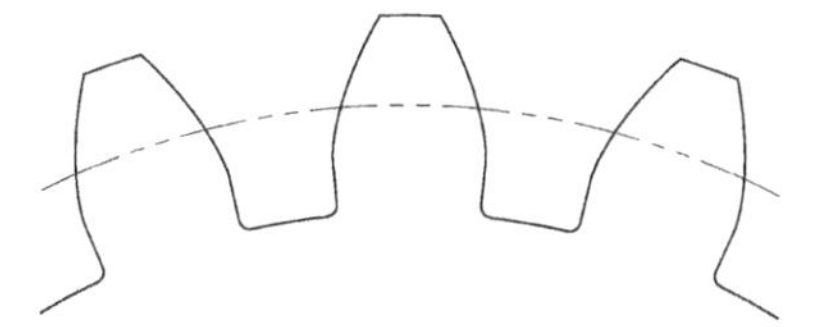

Figure 17-10 Pinion.

The effective face width must not exceed three times the circular pitch. This restriction reflects the difficulty in adequately aligning gears of wide face widths, and the bending potential along the gear axis. The effective face width is the narrowest face of the two mating gears. This is somewhat conservative for pinions, which usually have wider faces that fully participate in breaking strength.

Gearsets not shop assembled in common frames are to have their allowable stresses reduced by 20 percent. This restriction takes into account the variations of field assembly of machinery and possible later misalignment. It is assumed only one tooth in a gear takes the whole load.

The standards cited here presently call for the quality levels of gears to meet AGMA standards. Most bridge construction documents call for AGMA quality level 6 as the minimum. This is about the lowest level of quality of manufacture commensurate with cut gear teeth capable of transmitting any significant amount of power. As most open gearsets on bridges operate at low speed and are expected to continue to operate reliably for years with considerable wear, the expense entailed in achieving a higher level of quality in open gearsets is not justified.

Enclosed gear drives for bridges can be constructed with higher quality gearing, as the accurate alignment, automatic lubrication, and prevention of contamination can keep these gears in like-new condition for the life of the bridge. The chief advantages of using higher-quality gears for bridge drives are that they allow higher speeds and reduce the size of the gearboxes required to transmit the same loads. Because smaller size is usually not an advantage for bridge drives, and because bridge drives operate at lower speeds and for shorter times than gear drives in other applications, the higher quality of gearing available is not usually required in bridge gear drives, and it is more susceptible to degradation resulting from contamination, corrosion, or poor lubrication. Higher gear quality than absolutely needed is not recommended. There are situations in movable bridge engineering, such as on swing bridges, where space can be at a premium, when it will be advantageous to make the machin-

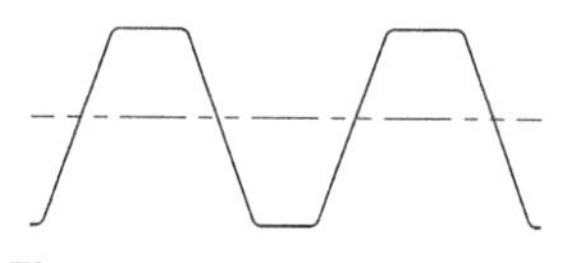

Figure 17-11 Gear or Rack.

ery package fit into as small a space as possible. In these situations, high quality, more compact gear reducers may be advantageous, and the higher fabrication cost of such units can be worth the benefits gained.

AASHTO has recently called for durability to be evaluated in designing gears for bridge drives. When a gear manufacturer designs gears, two calculations are usually made, one for strength, as discussed earlier, and another for durability. The durability calculation determines the allowable maximum load, considering local stresses in the areas of contact between the teeth of mating gears, allowing the proper-sized gear to be used, or the material properties to be specified adequately. Generally, the hardness of a gear can be increased to allow for acceptable levels of durability. This is preferably done in a bridge drive gear by increasing the strength of the material, as the ultimate strength and hardness bear a fairly constant relationship. Some gear teeth are surface hardened, as optimum strength and durability are often achieved by such a process. As discussed previously (page 241), surface hardening is not recommended for bridge drive gears, as long periods with lack of proper maintenance and inspection may result in a gear's deterioration and failure, due to the hardened surface wearing away before such wear is detected.

A special field of gearbox manufacture has arisen for bridge drives, but because the number of new units required annually for movable bridge use is small, and each gearbox must be specially designed for the application as to external shaft diameters and lengths, as well as the gearbox heights and center distances between input and output shafts, the number of manufacturers who participate in production of specialized bridge drive gearboxes is limited. The specific design of an enclosed gearbox drive unit, or gear reducer, is probably best left to a manufacturer. The manufacturer should be one who produces a standard line of gearboxes and guarantees their performance. The bridge designer should make an effort to fit the desired gearbox ratio, input speed, and output torque to the manufacturers' standard lines. Special features like lengths of external shafts are easily produced without changing the internal configuration of the gearbox. Extending the gearbox input and output shaft center distances to provide clearances for motors and brakes can be done by some manufacturers, but it is risky business. It is much simpler, and may not be any more expensive, to specify a gearbox of larger capacity than that needed, but which fits the physical layout of the machinery area. Some areas of standardized gearboxes, such as mounting feet and breathers, are underdesigned for the rigors of movable bridge use. It should be made certain that these areas are covered in the design drawings or specifications for a movable bridge project. It also should be recalled that most industrial applications have the load applied in one direction only, whereas movable bridges have reversing loads (see page 326). Some manufacturers have a standard factor by which the service factor is modified to account for reversing loads, and some do not.

Enclosed gear drives for movable bridges are of many forms, but most are adaptations of standard units produced by manufacturers for general industrial applications. The most common type of enclosed reducer on a movable bridge is the parallel shaft type (Figure 17-12). This reducer has the input and output shafts parallel to each other, hence the name. These shafts, and others that may be included inside the reducer, are usually in the same horizontal plane, or they may be arranged in a vertical

Figure 17-12 Parallel Shaft Differential Reducer. The gearbox in the center of the picture, behind the motor that powers it, delivers drive torque equally to the two secondary reducers, one to the right in front, and the other behind.

plane, depending on the needs of the bridge. If there are several gear reductions internal to the reducer, one or more shafts may be located outside the plane of the others, but remaining parallel to them, to decrease the overall length of the reducer. Another common type of enclosed drive used on movable bridges is the right angle reducer (Figure 17-13). This type has all the shafts in the same plane, but the input and output shafts are at right angles to each other. The vertical-type right angle reducer is frequently used on swing bridges. This application has an input shaft on a horizontal axis and the output shaft extending down and out from the bottom of the reducer. This allows easy connection to the rack pinion shafts on a swing bridge (Figure 17-14). Many swing bridges, even older ones, make use of the worm gear reducer. This type, because it is capable of very large gear reduction in a small space, is used to drive the end lift linkages on many swing bridges (Figure 17-15). A worm gear reducer is not very efficient mechanically, but this is not a great handicap as the end lift drives are usually heavily loaded for only a short portion of their travel.

Most enclosed drive gearboxes for main bridge drives use helical gearing. Some also use herringbone gearing or straight spur gears for parallel shaft gear sets. Helical gears operate smoothly but develop axial thrusts that must be resisted by the bearings and the gearbox frame. Herringbone gears are more difficult to fabricate, requiring special machinery, but their use eliminates axial thrust. For shafts that turn corners, bevel gearing, spiral bevel gearing, or worm gearing is generally used. Bevel

Figure 17-13 Right-Angle Shaft Reducer. The gearbox has its input and output shafts in a horizontal plane, but at right angles to each other.

gearing is somewhat less efficient than spur gearing, because of the axial shaft thrusts produced on the bearings holding the gears in place. Spiral bevel gears are smoother running than straight bevel gears. Worm gearing is very smooth running, but is also quite inefficient. Worm gearsets can be produced in very high ratios, as the worm gear is the only type that can operate with pinions with only one tooth, or pitch. Usually on bridge applications single-pitch worms are not used, as they tend to wear rather rapidly. Multiple-pitch worms, with two, three, or four pitches, are generally used. These also improve the mechanical efficiency somewhat over single-pitch worm drives. Worm gears are typically used only for end wedge drives for swing bridges and similar applications that do not require transmitting a large amount of power. Many older enclosed gear drives use bronze sleeve bearings to support the shafts, but almost all newer enclosed drives on movable bridges use antifriction bearings exclusively. These may be straight roller, spherical roller, tapered roller, or ball bearing types. Selection usually depends on axial load conditions.

AASHTO and AREMA provide for the use of stub tooth gearing in bridge drives (see page 244), as this allows for stronger pinions of fewer teeth. Wherever possible, full-depth standard-form straight spur gear teeth should be used for spur gear drives on movable bridges. As the stub tooth system is obsolete in general gear manufacturing, it should not be used in specifying new gearing. Where a pinion with a small number of teeth is required, such as a final drive rack pinion for a swing or bascule bridge, a larger pressure angle or a special tooth shape can be used. Larger pressure angles allow pinions to be made of fewer teeth without deviating from standard

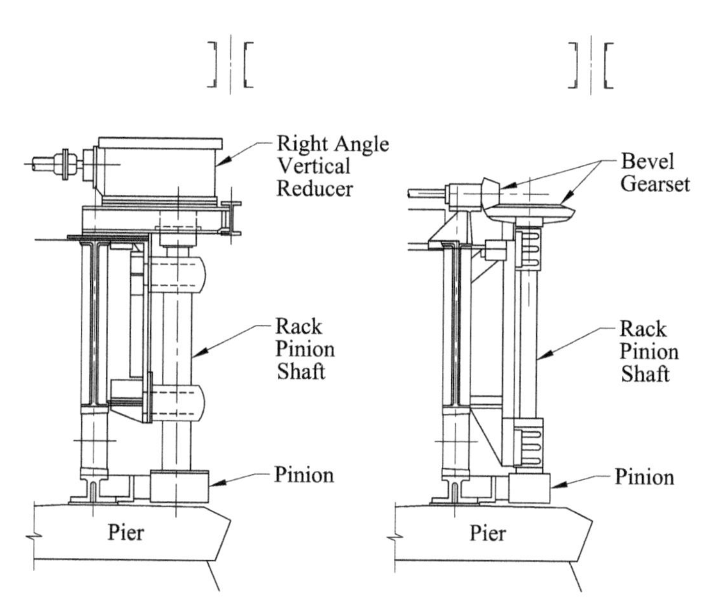

Figure 17-14 Swing Bridge Rack Pinion Shaft and Bearings. The assembly on the left was designed as a replacement for the original, on the right. The assembly on the right, a more typical arrangement, suffered many cracks and failures over the years. The new assembly eliminates stress concentrations and has robust components throughout. It is expected to prove considerably more durable.

Figure 17-15 Worm Gear Unit. This type of semi-enclosed gear drive is seen on many older swing bridge end lift drives. Newer bridges use fully enclosed drives.

forms, but increase the separating forces and require stronger bearings and supports. These components are a source of difficulty in movable bridge design even without aggravating them by increasing their loads, so it is better to use a special tooth shape for the rack and pinion.

In almost all cases, the rack and pinions must be specially designed and fabricated for the particular bridge anyway, so the only benefits in using standardized tooth shapes are (1) having the tooling readily available in gear manufacturers' shops to produce the gears and (2) being able to easily define the gears in a specification or on a drawing. The most efficient way to achieve the desired ratio between rack and pinion when using a pinion with a small number of teeth is to use modified addendum gearing, which can be fabricated with standard tooling. A standard pinion with a small number of teeth is relatively weak in tooth strength, whereas a mating standard rack with a large pitch radius is quite strong. If the gears have a 20° pressure angle and the pinion has less than 19 teeth, it will either be undercut or have special geometry.

Modifying the addenda of such a gearset reduces the rack strength while increasing the pinion strength, so that the two become more equal in load capacity. The pinion is almost always made of a stronger, harder material than the rack, to further equalize strength and wear. The number of teeth in the pinion can, with modified addenda, be reduced below the minimum number for a standard tooth shape without having a special shape or undercutting, a valuable bonus. It is virtually impossible to determine the fabrication details of a modified addendum gearset in the field, especially after substantial wear has occurred. For this reason, if modified addendum gearing is used, the pinion and rack tooth fabrication details should be recorded and stored for future access, to calculate data for inspection purposes and to supply replacement parts if needed. It may be better, however, to have a spare pinion and rack segment provided with the new bridge, or new machinery installation, and keep them on-site where they cannot be lost, preferably protected so that they will not corrode excessively.

Some rack and pinion applications call for a great increase of the pinion tooth strength so that it is able to match the rack. A few bridges have special racks and pinions with greatly reduced rack tooth thickness and greatly enlarged pinion tooth thickness to match (Figure 17-16).

The gear drive portions of bridge drive machinery are detailed after the prime mover has been selected. AASHTO (2.5.4) and AREMA (6.3.10) stipulate that machinery must be dimensioned to withstand 150 percent of the full load torque of the drive electric motor or motors without exceeding 100 percent of the allowable stresses as defined in their specifications. The machinery is also required to be sized to accept the starting or breakdown torque of the motor without exceeding a 50 percent overstress. The same authorities allow electric motors to operate delivering 180 percent of their rated full load motor torques for short periods of time while accelerating the span from the stationary position. A slight margin of liberalism in design is thus officially sanctioned, as the gearset can be loaded during operation to 180/150, or 120 percent of its allowable stress. In adverse conditions such as high winds, this acceleration can be occurring throughout the bridge opening or closing cycle. This require-

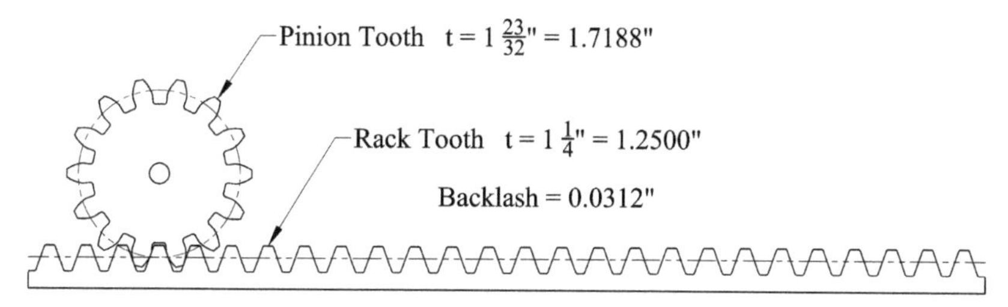

Figure 17-16 Rack and Pinion with Special Teeth. On a few movable bridges, racks and pinions with special tooth forms have been specified. A larger pinion tooth, and a smaller rack tooth to mate with it, are provided. Both the rack and pinion still have the same pitch and pressure angle as if they were standard teeth. Special information is required to produce such gearing, and to produce replacements.

ment has been changed in the 2000 LRFD AASHTO as the 180 percent figure is reduced to 150 percent (5.4.1).

In actual practice many bridge drives overload their electric drive motors by a higher margin during starting, as the starting friction at bearings combines with inertia to produce extremely high starting resistances that are usually limited only by the overload setting of the main breakers for the motor(s). Hovey recommended using 200 percent of motor full load torque for machinery design,[1] and this extra margin of strength is helpful if durability is desired. Internal combustion engine drives are not capable of producing such overloads without greatly sacrificing durability, and normal practice as sanctioned by AASHTO (2.5.4) and AREMA (6.3.10) is to size machinery components for 100 percent of the full rated output of these types of engines in multiple cylinder versions, about the only kind used today in these applications.

The sizes of all the components making up the drive train are determined by the power to be transmitted (see Chapter 18) and the forces required at the racks. The bridge drive force is determined at the point of contact between the rack and pinion, and machinery forces are developed from there to the prime mover. It is generally considered economical in the long run to have these components conservatively designed and to specify components that require as little maintenance as possible.

ELECTRIC MOTORS AND BRAKES

Most movable bridges are powered by electric motors, and retardation is provided by electrically actuated brakes. Motors are discussed more thoroughly in Chapter 18. These machines, however, form the interface between the electrical and mechanical parts of a movable bridge installation, so some discussion of their mechanical aspects is presented here. An electric motor produces output torque that causes a gear train or

[1]O. E. Hovey, *Movable Bridges,* Vol. 2 (New York: John Wiley & Sons, 1927), pp. 102–103.

other piece of machinery to move. This torque is directly proportional to the rated horsepower of the motor and inversely proportional to the motor's design speed. The rated horsepower and speed are used to calculate an electric motor's full load output torque. In regard to the aforementioned relationships, the rated torque of an electric motor is equal to

$$T = \frac{k\mathrm{HP}}{\mathrm{RPM}}$$

where

T = the rated full load output torque
HP = the rated full load output horsepower
RPM = the rated full load speed in revolutions per minute
k = a constant depending on the units being used to measure the output torque. For output torque in inch-pound units,
k = 63,025 inch-pounds RPM/HP.

One must be careful to specify rated full load speeds, horsepower, and so forth, because electric motors are capable of producing much higher outputs than their ratings indicate, limited only by the source of input electrical power and the ability of the motor to conduct current and withstand the heating effects of the power being transmitted. An electric motor not otherwise restricted can provide up to $7\frac{1}{2}$ times its rated full load output torque. Electric motors for movable bridge applications are usually rated on the basis of their one-hour ratings, which is the rate of power transmission that can be withstood for one hour at a time without causing damage to the motor. The motor must be allowed to cool between these one-hour applications, so they are only valid for intermittent use such as is experienced on the typical movable bridge. A one-hour rating, as opposed to a continuous rating, can save perhaps 10 to 20 percent in a motor's physical size, weight, and cost. If the intermittent use does not apply, such as for a movable bridge in a busy seaside port that may open hundreds of times a day, the continuous rating of the motor should be used unless it can be shown that the motor will be operating at only a fraction of its full load for the greatest part of the operating cycle. In many applications the motor is somewhat oversized, so a reduced load factor may be justified. It is usually safer to err on the side of conservatism in sizing motors, as the increased cost for a motor frame of a slightly larger size is usually not great. Larger motors will, however, require larger machinery to meet AASHTO and AREMA requirements (see page 254).

Large output torques must be resisted and transmitted to the motor support. The mounting of a motor must be done with the same care as the mounting of other components of the drive machinery. The motor loads are reversing, so any play in the motor mounts can quickly result in a loose motor. The motor should be mounted with turned bolts in finished holes, using the largest-size bolts that can be fitted into the motor mounting feet. Unfortunately, most industrial applications are non-reversing, so motor mounts can be too small for movable bridge use. Specifying TENV motors with long duty times usually eliminates this difficulty by providing a larger motor frame for the same rated torque and speed as an open motor with a smaller frame.

Electrically driven movable bridges are required by AASHTO (2.10.4) to have motor brakes that act in conjunction with the bridge drive motor, and machinery brakes that are separately controlled. Machinery brakes and motor brakes do not go through quite as rigorous a duty cycle as the main bridge drive motors, but they do receive reversing loads that can be substantial. The brakes must be able to stop the moving span and hold it in position under the most adverse environmental conditions.

Almost all movable bridge brakes are spring set and electrically released, so their capacity is almost purely a mechanical matter. Most movable bridge brakes are of the thruster type (Figure 17-17), as the thruster mechanism allows some engineered delay in the setting of the brakes so that shock loading of the machinery is minimized. The thruster is a self-contained hydraulic unit with an electrically driven pump, reservoir, and "hydraulic cylinder" within the assembly. Some older bridge drives use solenoid brakes, which instantly set and release, and a few bridges have hand brakes or air brakes. Most bridge brakes are of the drum type, with a cylindrical drum or wheel mounted on a shaft, acted upon by brake shoes pressing against the outside surface. The pressure of the brake shoe on the drum is produced by a coil spring, acting through levers and linkages of the brake mechanism. There are some disc-type brakes in movable bridge main drive applications, but they are relatively rare. Almost all

Figure 17-17 Thruster Brake. The brake wheel is mounted directly to the reducer input shaft. The thruster unit, the large vertical cylinder, is mounted in a frame to the floor.

main drive bridge brakes are mounted in a two-step process. The brake drum is mounted on a machinery shaft, usually in the shop where the machinery is manufactured or assembled. The brake mechanism, with linkage, brake shoes, spring, and actuator, is mounted in the field, or in the shop if the machinery is to be installed entirely as a unit, such as with a Hopkins frame. Sometimes the motor brake is mounted directly on the drive motor, in which case the motor manufacturer may mount both the brake drum and the mechanism. Some specifications require the mounting of coupling halves and brake wheels to the motor by the motor manufacturer; this generally gives a better fit and avoids problems of shaft or motor damage.

HYDRAULIC DRIVES

Hydraulic power is nearly as old as civilization. There may have been, in ancient times, movable bridges that were powered by being linked mechanically to waterwheels, which are simply hydraulic motors. Hydraulic power, in the contemporary sense, has been used to operate movable bridges since the beginning of the modern era, and even earlier to a considerable extent. As mentioned previously (page 55), some movable bridges in France were hydraulically powered in the early nineteenth century. All the early hydraulic systems used water as the working fluid, hence the name "hydraulic" (Greek *hydros* = water). The Tower Bridge in London, England, was equipped with hydraulic motor drives powered by giant accumulators and steam-powered pumps when it was constructed in 1894. The bridge was rebuilt in the 1970s with a modern hydraulic drive system, still using motors for the main bridge drive. There are still a few other hydraulically powered movable bridges of about the same age. Hydraulic drives had disappeared from new movable bridge construction by the early twentieth century, when electric motors and controls were developed that proved reliable and economical and required little or no maintenance. Wright, in the 1890s, provided an extensive discussion of hydraulic drives for movable bridges and provided information on several examples that had been built or were under construction at that time, when hydraulic power was comparatively common. Hool and Kinne, in 1923, does not mention hydraulic drives at all. Hovey, in 1927, gave hydraulic drives little more than passing mention, as by that time electrical controls and motor drives had reached an advanced state that made the hydraulics of the time obsolete, so that they were not seriously considered for new movable bridge drives. Hovey applies his discussion to hydraulic jacks as well as to drive systems.[2]

Solid-state electrical controls first began to appear on movable bridges in the 1970s, concurrently with the abandonment (for all practical purposes) of the movable bridge field by the large electrical manufacturers who had developed the earlier reliable electrical power and control systems. A massive decline in serviceability of new bridges occurred, as these new control systems proved to be extremely unreliable, new manufacturers failed to stand behind their products, and technology advanced so as to make

[2]Hovey, Vol. 2, pp. 227–241.

new designs obsolete, sometimes before the installations were completed. This trend resulted in a demand for an alternative to the electromechanical movable bridge drive.

Since World War II, hydraulics had made great strides as a reliable, precise means of controlling large machinery. People began to apply hydraulic systems to the operation of movable bridges with considerable success. Hydraulics, by nature, allow the installation of a very powerful drive unit in a very small space. Although this compactness is usually not a great advantage in movable bridge design, in some applications it can be very helpful in fitting machinery to a movable bridge. Many swing bridges, for instance, provide very little space for machinery, so the use of a compact hydraulic drive can leave more room, allowing access for maintenance and inspection.

The way in which the speed of a hydraulic motor is controlled makes it appropriate for use in movable bridges. A hydraulic drive can be controlled by means of a valve that is moved from the closed to the open position, gradually increasing the rate of flow, and gradually cutting off the flow. Careful design of such a valve and its actuation can result in automatic, reliable control of the rate of acceleration and deceleration of the prime mover. Other hydraulic systems use pumps that can be controlled to produce the same gradual increase or decrease in output. This can be convenient in starting and stopping the motion of a massive structure such as a movable bridge, which has a very large amount of inertia. Hydraulic systems can provide load equalization as the hydraulic fluid, at least under static conditions, is at the same pressure throughout the system. Proper design of piping and controls can minimize differences in pressure drop, so the load equalization capability of a hydraulic drive can equal that of a mechanical system. Hydraulic systems can also, by use of flow dividers or other means, provide equal movement at several different force application points. The Hood Canal pontoon retractile bridge (see page 25) has a noncounterweighted lifting approach span that is hydraulically operated, using four cylinders operating in unison on each section of the lifting approach span. Hydraulic systems are sensitive to large changes in ambient temperature. In locations where summer temperatures may reach 100°F and winter temperatures may plunge to below zero, hydraulic systems can have severe adjustment problems and damage may actually occur to such components as the hydraulic cylinders. When in an acceptable environment, properly installed, and properly maintained, hydraulic systems can function well.

Caution must be used in designing a hydraulic drive system for a movable bridge. On a large swing bridge that had been powered by a steam engine since it was built in 1908, a new hydraulic motor drive was installed. The hydraulic motors were mounted directly on the old steam engine crank shafts, which were geared to a clutch shaft that alternately drove the bridge turning drive and the end wedge–center wedge drive. The first time the bridge was operated, the bridge operator attempted to stop the bridge rotation in the same way he had for years—by reversing the engine. The large torque capacity of the hydraulic system, plus the incompressibility of the hydraulic fluid, resulted in a sheared drive shaft and a free-spinning bridge. Following repairs, the operating procedures were altered to avoid repetition of that incident. After 20 years of use, that hydraulic system was, as of 1999, in the process of being replaced with an electromechanical drive.

Hydraulic drives are not very efficient mechanically. A feature of a typical hydraulic power system is some type of heat exchanger, sometimes in the form of what appears to be excessive lengths of piping in the neighborhood of the pumping unit. A typical hydraulic system has a large fluid reservoir, which can act as an effective heat sink. The cost of the actual amount of energy used to open and close a bridge over the course of a year is usually trivial as compared with other operating expenses, so hydraulic systems do not really suffer because of being burdened with higher power costs than other types of systems incur. The slightly larger prime mover required because of extra power losses as a result of the comparative inherent inefficiency of hydraulics is likewise a relatively small capital expense as compared with overall capital costs of a movable bridge.

Many hydraulic systems for movable bridges have been designed utilizing "off the shelf" hydraulic components designed for other industries. This can result in a drive system with very low first cost, but the reliability and durability of some of these systems in use have been poor, as the components cannot tolerate the low level of maintenance of a typical movable bridge. In addition, the movable bridge environment can be, in many ways, harsher than the environment to which most machinery is subjected. There are hydraulic components available as standard manufactured items that are perfectly capable of withstanding the rigors of movable bridge use, but many of them come at a substantial premium in price, so that the initial cost advantage of hydraulic systems becomes small or nonexistent when these components are used.

Hydraulic systems are, by nature, maintenance-intensive power systems. The hydraulic fluid levels should be checked regularly and replenished as needed, and the fluid must generally be replaced after certain intervals. Other routine but critical preventive maintenance, which is unnecessary in electromechanical drives, should also be performed on hydraulic systems, such as checking the tightness of joints and inspecting for leaks and replacing or cleaning the filters. Yet, hydraulic systems on movable bridges usually receive about the same amount of maintenance as nonhydraulic bridges, so that low fluid levels, coatings of hydraulic fluid over all surfaces in the vicinity of the hydraulic units and their piping, overused filters, damaged seals, and other defects develop until a bridge breaks down and repairs must be made.

Most of the inexpensive hydraulic drive components are manufactured for more maintenance-intensive environments, such as those in which construction equipment operates, where the machine operator carefully checks out his machine each day before starting work, or for the factory, where a full-time maintenance staff prowls the work area and fixes problems before they shut down expensive assembly lines. The expected life span of machinery in these applications is usually much less than it is for machinery on a bridge. On hydraulic bridges, many breakdowns have occurred in hydraulic cylinder drives when rod ends without the minimal bronze bushing thickness have seized and snapped off, and high pressure hoses have become entangled and ripped off, because of the lack of constant surveillance that hydraulic systems need for best performance.

The general tendency is for bridge owners to be enthusiastic about hydraulic drives when they are first installed, because a functioning hydraulic system can usually be installed on a movable bridge more quickly and cheaply than an electromechanical system. In time, the enthusiasm wanes as higher operating and maintenance expenses and more frequent breakdowns are noticed.

The operational advantages of hydraulic systems over electromechanical drives have largely disappeared with the development of solid-state electronic control systems capable of reliably operating electromechanically powered movable bridges. Some owners who have trumpeted their impending conversion to hydraulic drives for all their movable bridges, have, a few years later, quietly resumed specifying electromechanical systems for their movable bridge drives.

There are two basic means of converting hydraulic fluid flow to bridge drive power. The simplest is the hydraulic cylinder, which is assuredly the most common form of linear actuator in industry. The hydraulic cylinder can be built long enough to provide almost any amount of extension and can be built of any diameter to produce the required operating force. Many functions, such as lock bar actuation, wedge driving and withdrawing, and centering device operation, seem to be naturally adaptable to hydraulic cylinder operation and can be done with very simple drives. Hydraulic cylinders are used fairly frequently for bascule bridge main drives and occasionally for swing bridge rotation. Hydraulic motors are perhaps more commonly used for main bridge drives, particularly for bascule and swing bridges. Motors capable of producing very high output torques are available as standard items from manufacturers of hydraulic equipment. Such motors can dispense with some or all of the gear reduction required with electric motor bridge drives. The rack and pinion final drive is invariably still necessary, although some hydraulic motor drive schemes have attempted to dispense with it as well.

MACHINERY SUPPORTS

Machinery support appears to be an area in which the standard specifications of AREMA and AASHTO may not be sufficiently conservative, as many bridges have experienced failures in machinery supports. The chief cause appears to be in the value of maximum bridge starting torque (MBST), as calculated according to AASHTO and AREMA requirements. This value is computed using a prescribed static friction value for all bearings and ignoring inertia. The given static friction value is likely to be too low, as the primary bridge support bearings for bascule and swing bridges support live loads, resulting in poor initial lubrication at movement. Live load impacts tend to push the lubricant from the bearing contact surfaces; thus, values for unlubricated surfaces should be used, raising the friction coefficient by a factor of 2 or 3 or more. It is also nonconservative to ignore acceleration in the starting process, because unless the bridge is "babied" when starting, full power will be applied to get the bridge moving (see pages 255 and 282). This power is limited by the setting of the circuit breakers for an electric drive or by the physical capacity of an engine or hydraulic system. As designers and erectors do not want the bridge to fail to operate because of lack of power, these limitations are usually quite liberal. As a bridge ages, its machinery tends to become less efficient, and maintenance personnel may further increase the trip current settings of the motor protection to ensure the delivery of adequate starting torque.

Instead of the standard value of 150 percent of full load torque of an electric motor, the design value for the machinery supports and final drive components of a movable bridge should more prudently be established at 200 or 250 percent, which is closer to the maximum possible output. Hovey (1927), in his discussion of machinery design for movable bridges, suggests these higher values (see page 255).

Many movable bridges have required repeated repairs of the machinery supports over their lifetimes, as the basic weakness of these components could not be corrected by in-kind retrofitting. The most common area of failure is at the support of the primary drive pinions for gear-driven bridges. Many swing bridges that must operate with any regularity have experienced failures in this area, as have a few bascule spans. AASHTO (2.5.8) and AREMA (6.3.11) provide for greater stiffness at machinery supports than at other structural steel components, but increased strength needs are not explicitly considered.

Just as in the early days of the industrial revolution, when engineers mounted gearing and other machinery on wooden frames to save the cost of expensive metal, many bridges have the drive machinery mounted on concrete platforms. The machinery bases are usually attached to the concrete with anchor bolts. It is difficult to align machinery properly with this arrangement, and difficult to maintain the alignment during use. Mating gears, including racks and pinions, should be mounted in a frame that is as strong and rigid as possible, with adequate attention given during erection to proper alignment and mounting. For best results, if a fully exposed steel framework cannot be used, a steel grillage can be imbedded in a concrete pier. The machinery can then be attached to the grillage, either directly or with supporting frameworks.

FASTENERS

Both AREMA (6.5.25) and AASHTO (2.6.18) call for machinery components to be fastened with turned bolts (Figure 17-18), but AREMA also allows the use of finished bolts with fitted holes, so that the fit between bolt and hole is similar to that achieved with a turned bolt. Turned bolts provide a locating fit, to avoid misalignment of the machinery, without relying on a friction connection, so that alignment is retained if the bolts come loose. AASHTO and AREMA require the turned bolt to have a body diameter $\frac{1}{16}$ in. larger than the major diameter of the threads. This allows the bolt to be inserted in a slightly misaligned hole without damaging the threads. AASHTO specifies an LC-6 fit between the body, or shank, of the turned bolt and the holes into which it fits. This fit is a "Locational Clearance" fit, specified in the American Society of Mechanical Engineers' specification USAS B4.1, "Preferred Limits and Fits for Cylindrical Parts," originally published in 1967. The specification dates from the 1920s and might be considered obsolete today. It has no metric application or equivalents, but other standards are available for metric fits. American National Standards Institute (ANSI) B4.2-1978 "Preferred Metric Limits and Fits" is a metric counterpart to USAS B4.1. The 2000 LRFD AASHTO specifications references a metric tolerance specification, but up to this time almost all precision fits on a movable bridge

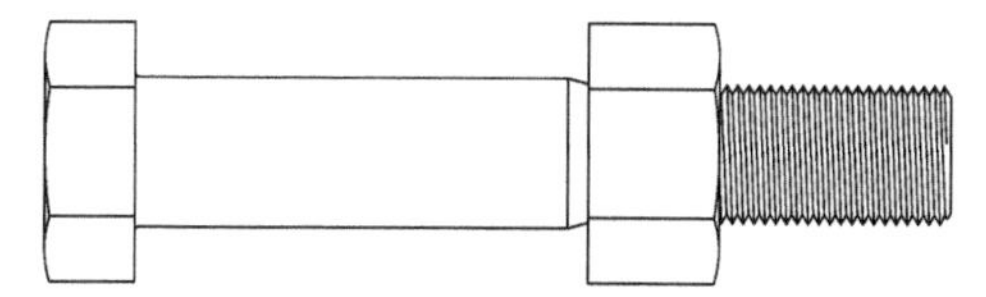

Figure 17-18 Turned Bolt. The turned bolt, as shown, has traditionally been used for machinery and some structural applications. It has been superseded largely by the high-strength bolt, but still has application where stress reversals occur.

are in English units; some of the fits have been defined within the particular project specifications.

The LC-6 fit provides a slight clearance between the turned bolt shank and the holes, about 0.003 in. for a 1-in. diameter bolt. Such clearance allows for occasional removal of the bolt for repairs without destroying the fit or finish of the bolt and the holes. This is particularly applicable when American Society for Testing and Materials (ASTM) A-307 material, a "regular strength" bolt material, is used for the bolt. AREMA (6.5.25) specifies an LT1 fit between turned bolts and finished holes, which is a "transition" fit per USAS B4.1, a slightly tighter fit than that provided with an LC-6 fit. Some specifications, particularly older ones, require "drive fits" of the turned bolts (usually not high strength) in their holes. This provides a very solid connection, but is also difficult to disconnect. This requirement also has the advantage (some would call it a disadvantage) of requiring very precise alignment at assembly.

With higher-strength turned bolts, a tighter fit can be used with less likelihood of galling or other damage to the bolt if it is frequently removed and reinserted. For this reason, ASTM A-449 or A-325 turned bolts may be provided with tighter fits than those specified by AASHTO. The high strength of the bolts also, of course, ensures greater tensile strength in the bolts when needed. Machinery is sometimes held in place with high-strength bolts without fitted holes, on the assumption that the high-strength bolts will maintain their grip and produce a friction connection and the machinery will not become loose. This theory works in some applications, but where high reversing loads (see page 326) are experienced, such as at rack pinion bearing supports, the bolts often come loose and misalignment or an operating failure is the result. Some applications call for locating pins to be used. These tend to deteriorate rapidly with disassembly and reassembly.

Some bridge construction specifications call for standard manufactured shoulder bolts in place of turned bolts. Shoulder bolts have precisely machined bodies, but the body diameter is much larger than the thread diameter in these bolts as compared with AASHTO or AREMA turned bolts. This makes little difference in a low-strength bolt relying on shear for the connection, but results in high stress concentration at the neck between the base of the threads and the turned body. Where these bolts are used in applications producing high cyclical tensile stresses in the bolts, such as in the attachment of rolling lift tread plates to girders, fatigue failure of the bolts has resulted.

MACHINERY ENCLOSURES

By taking adequate precautions, movable bridge machinery can be made almost impervious to the effects of weather. Stainless steels, bronzes, some types of plastics, and even glass, can satisfactorily resist weathering for long periods of time. It is difficult to make all components of movable bridge machinery, electrical and mechanical, sufficiently resistant to weathering to have a good probability of surviving 20 years or more without replacement. This is particularly true of mechanical moving parts, such as hydraulic cylinders, gearing, and related components. For a bridge to stand a chance of surviving the design life period, most movable bridge designs provide for enclosures of some type to protect the machinery. If the machinery enclosure is a house (Figure 17-19), the house should provide for installation and removal of the largest components mounted within it, by the use of double doors or trap doors if necessary. The lifting and moving of heavy components should be facilitated, such as by providing an overhead crane of sufficient capacity to lift the heaviest component in the house.

Some machinery components, like racks and pinions or swing bridge end wedges, usually cannot be protected from direct exposure to the environment. It is especially important for these components to be fabricated of corrosion-resistant materials or to be formed so that they are sufficiently robust to survive a great deal of weathering.

Surprisingly, the use of machinery enclosures for movable bridges is a controversial topic. The argument goes something like this:

1. "If I have good machinery enclosures, I won't need as much maintenance because the machinery, being better protected, will stay in good condition longer."
2. "If I don't have any machinery enclosures, I will get better maintenance, because the machinery will be more accessible and, being better maintained, will stay in good condition longer."

Both these premises may be true. The best machinery life will be attained if there are adequate machinery enclosures in place that allow easy access for maintenance, and if the machinery is covered by an effective maintenance program.

The most cost-effective bridge maintenance program is another matter. Enclosures cost money. Maintenance costs money. Removing the enclosures to perform maintenance costs money. Performing maintenance, such as cleaning bird debris from gear teeth, that would not have to be done if enclosures were in place, costs money.

Experience suggests that machinery enclosures pay for themselves by prolonging the life of open gearing and other machinery. Properly protected gears rarely wear out, because even an old thin coating of lubricant can remain effective for a long time if it is not washed or driven off or made ineffective by rain, wind, sun, or snow. The best machinery enclosure is a machinery room, secure but with easy access via a doorway, with plenty of headroom inside. The room should be secure from entry by unauthorized persons or by animals that could do harm, such as by causing fires. Openings into the machinery room should be carefully designed to minimize the ease

Figure 17-19 Machinery House. This house protects the main drive machinery of a Waddell vertical lift bridge. The bridge is operated from elsewhere.

with which birds can enter the room. This can be difficult at areas where the bascule girders pivot on the main trunnion bearings, where the counterweight ropes of a vertical lift span exit the tower top enclosures, or at the underside of a swing bridge. Effective measures can be taken at each of these locations, although there will be some maintenance effort required to sustain the effectiveness of the details.

A machinery room should be readily accessible to authorized personnel via a properly protected catwalk, stairway, or other entrance. An elevator should be used if the machinery room is located at a considerable distance above or below the ground or roadway. A machinery room provides protection for the machinery without the need to resort to removable covers, which in many cases are not effective, as they are often misplaced or simply not reinstalled after removal for maintenance access.

Sometimes it is impossible to protect all the machinery within a room, however, and separate covers must be provided, if practical, for all machinery components that should be protected. There are some types of bridges, such as rolling lift bascules, that have the drive machinery dispersed about the superstructure, making complete coverage impossible. It is also impossible to fully enclose swing bridge components such as end lifts and centering devices. In some cases, machinery covers that are provided by design make maintenance difficult, and maintenance personnel remove these covers permanently or modify them so that their effectiveness is reduced. It may be more beneficial in the long run to provide partial covers that provide some protection for the machinery while allowing easy access for maintenance.

All machinery, electrical and mechanical, should be readily and easily accessible. It is important to install ladders and walkways to all machinery not readily accessible within the machinery room. All ladders and walkways should be designed to provide ready and safe access for inspection, maintenance, and minor repairs. They should be strong and durable and meet all safety codes. Steel ladders and walkways should be galvanized or coated with some other protective finish, although this usually does not provide acceptable long-term protection in harsh environments. The use of fiberglass or stainless steel ladders and walkways should be considered where corrosion can be a problem, but fiberglass can be short on durability, and both are expensive.

Totally enclosed motors and fully enclosed gear reducers may not necessarily be weather proof. These components and others will either require additional enclosure or special fabrication and materials to be able to sustain themselves when exposed to the environment.

18

CONTROLS AND POWER

CONTROLS

There are many types of control systems in existence for movable bridges. Some installations are more than 100 years old, some are the latest in the state of the art, and some are examples of experiments in cutting-edge technology. Different control systems are required for different types of prime movers: direct engine drive, either internal combustion (of the gas or diesel variety) or steam; hydraulic motors or cylinders; or electric motors with gear drives. Control systems differ in the amount of skill required on the part of the bridge operator for successful operation, ranging from fully automatic to fully manual bridge control operation. All systems, even totally manual hard drives, must perform the same functions:

1. A system must provide sufficient power at the prime mover to operate the bridge under any expected conditions, and operate the bridge at normal speed under normal conditions, but must not provide excessive power or force or torque so that a driveline component is damaged, thus rendering the system inoperable.
2. The system must control the speed of the bridge in operation so that it does not operate too fast, thus damaging a component as a result of overspeed.
3. The system must control the rate of change in speed when starting and ending an operation, avoiding overstress of components during acceleration or deceleration, and bringing the bridge to a stop without excessive impact.
4. The system must control the positioning of the bridge so that it can be stopped at the desired point when opening, and stopped when closing at the correct position, for a swing bridge without overtravel, and without causing excessive impacts to the bridge seats for vertical lift and bascule bridges.

5. The system must integrate the various ancillary devices, such as end lifts for swing bridges, span locks, and other devices required for the particular installation, so that mechanisms are operated in their proper sequence.

The simpler control systems require the person operating the bridge to perform all the interlocking functions, except perhaps for some mechanical devices that may prevent certain incompatible functions from being performed simultaneously. Most bridges built since the mid-twentieth century have an electrical interlocking system, regardless of whether the bridge machinery is powered by a gasoline, steam, diesel, hydraulic, or electric prime mover. This interlocking apparatus will fulfill the requirements of function 5 in the preceding list, and perhaps function 4 as well. AASHTO requires that function 4 be performed automatically by having limit switches to stop motors and set brakes at the end of travel of the bridge. Some older systems perform function 3 in a crude fashion, and functions 1 and 2 are usually inherent in the system, with insufficient power available to require discrete devices to perform these functions. Many newer control systems perform functions 1, 2, and 3 by means of specific hardware and programming, taking most of the skills required out of the hands of the person operating the bridge, and many control systems of the past 25 years perform all functions, 1 through 5, so that at these locations, an on-site human operator can be dispensed with completely. There has perhaps been more innovation in the area of control systems than in any other area of movable bridge engineering in the last 20 years. *State of the art* is a term often used in respect to movable bridge control systems. There is some confusion as to its meaning. Some consider it to mean (a) "the most up-to-date." Others think it means (b) "the best available." Those who subscribe to (a) have had bridges built that never worked correctly, until they were rebuilt. Those who follow (b) have had bridges built that cost more than the cheapest available designs, but have ended up with solid, reliable bridges. Not all arguments about state of the art, as it applies to movable bridges, pertain to controls, but many that persist today do.

A few railroad movable bridges have been converted to completely automated remote operation. Some highway movable bridges have been built so that they could be remotely controlled. The overall track record is a mixed success, with some remote operations having been converted back to on-site supervision. A number of municipalities and other public movable bridge owners have expressed interest in remote operation as a means to cut operating costs. The potential for vandalism and liability has been sufficient to prevent any movement toward remote control highway bridges. An alternative means to control operating costs that has been implemented at many movable bridges has been to put the bridges on an advance-notice schedule. This step requires U.S. Coast Guard approval, which is usually obtainable for bridges that open very seldom. It is especially convenient for bridges that open only for commercial traffic, as commercial vessels are almost always equipped with marine radio for advance communication to the bridge operators. Some railroad movable bridges, on lines with little traffic, are normally left in the open position. The train crew approaches the bridge, stops, closes the bridge, crosses it with their train, reopens the bridge and continues on their way.

Waterways with a significant amount of pleasure craft traffic are usually forced to provide full-time bridge tenders. Some locations, such as on the Chicago River, have been able to provide scheduled openings for pleasure craft so that roadway traffic is interrupted only at stated times. This accomplishes about the same saving in operating expense as advance notice. By grouping the vessel passages on the Chicago River to a single time each day, the number of openings on busy streets in the Chicago Loop area has been minimized.

Electrical

The heart of the control system for an electromechanically driven movable bridge is the apparatus that feeds power to the electric span drive motors. This apparatus in the last half century has undergone the most radical evolution of any part of a movable bridge. The first phase of this development was gradual, as the means of varying the output torque and speed of the drive motors changed very little in overall function, but new components were developed to increase the capacity of the equipment without changing the overall method of converting the line power to the controlled flow to the motors. After the development of the transistor and subsequent solid-state technology, and their application to motor control in general industry, specific devices were developed that were capable of controlling electric motors for bridge applications. The result has been tremendous increases in the flexibility of bridge control systems.

From the beginning of the application of electric power to the movable bridge drive, both AC and DC drive motors were used. DC has always been simpler to control, as merely changing the applied voltage changed the speed at which a loaded motor would run. The DC motor could also easily provide dynamic or regenerative braking to retard movement and hold speed to a desired level if the load was overhauling the motor. Unfortunately, DC motors required extra maintenance because of wear at the commutators and brushes. DC drives operating from utility power required an expensive converter to change the line current from AC to DC, so AC drives were used whenever possible. The earlier AC variable speed drive was limited in power and in its capacity to control speed, so DC systems, including the converter apparatus, were used, particularly on larger bridges. The development of high power wound rotor AC motors with variable external resistances inserted by means of slip rings, offered sufficient motor speed control for movable bridges. The later development of solid-state control systems has allowed the elimination of DC drive motors on movable bridges. The AC wound rotor motor with secondary resistance[1] control and the typical squirrel cage motor are not normally capable of providing significant retarding effort for decelerating or against overhauling loads, so the use of motor brakes is necessary in some situations to limit a bridge's speed. The solid-state drives have the additional advantage of being able to positively control speed of an AC motor when the motor is being driven, or backdriven or overhauled, such as when a

[1]General Electric Company, "Electrical Equipment for Movable Bridges," Document GEC-1029 17 MB (Schenectady, New York, Jan. 3, 1955), p. 2.

span heavy bascule or vertical lift bridge is being lowered. Through the 1960s many of the largest movable bridges continued to use DC drive motors and controls, with attendant rotary converter apparatus. Today, all types of systems are in operation in the field, as their inherent durability has kept many of the oldest types of controls in use. The Government Bridge over the Mississippi River at Rock Island, Illinois, still had, at least until recently, its original DC drive system that was installed in 1893.

The manual, semimagnetic, and magnetic control systems, as they are called, are the antithesis of solid-state control systems. They have many moving parts and thus they require frequent maintenance, which must be performed to some extent, even if the bridge does not operate frequently, in order to avoid failures. For this reason, among others, many bridge designers rushed to implement solid-state controls as soon as components became available in the late 1960s and early 1970s. The need for additional ruggedness and reliability in these systems was immediately obvious, as many of these new systems failed almost as soon as a bridge was put into service, and some solid-state systems functioned only as long as the factory technicians who installed the systems maintained a constant vigil over them. This is, in all probability, a primary reason for the sudden popularity of hydraulic movable bridge operating systems in the 1970s and 1980s.

Today, most newer movable bridges operated by electromechanical means are controlled by solid-state systems using silicon controlled rectifier (SCR) technology. Many older bridges have had their control systems replaced with SCR drives.

Most other bridge mechanisms, such as span lock drives, end lift machinery, center wedges, and traffic gates, have much simpler controls, as the motor is simply turned on or off, and reversed. Relatively simple standard motor starter units accomplish this function, differing only in size according to the horsepower of the motor being controlled. Most of these motors are of the simple squirrel cage induction type.

Communications

The U.S. Coast Guard has a special set of regulations that movable bridges must follow, including a specific configuration of navigation light for each type of movable bridge.[2] The regulations include distinct requirements for pontoon, retractile, swing, bascule, and vertical lift bridges. Navigation lights passively communicate to boat operators the fact that there is a movable bridge ahead and tell them the location, width, and height of the navigation channel. Most of these navigation light systems also indicate whether the movable bridge is open or closed. This is sometimes accomplished by having a light on the side of the bridge change from red to green when the moving span is completely in the clear for navigation.

Many secondary passive communication systems have been used to assist mariners in navigating past a movable bridge. A device commonly found in tidal waters, or where water levels can be affected by flooding, is a clearance indicator. Such devices are usually mechanical, activated by a buoyant component that raises and lowers with

[2] United States Coast Guard, "Bridge Lighting and Other Signals," Enclosure 6 COMDTINST M16590.5, *Bridge Administration Manual,* (Washington, D.C., 1986), pp. 15, 17–20 (unnumbered).

the water level, causing a pointer to move relative to a board marked with clearance measurement numbers. Sometimes the process is reversed, and the board moves while the pointer is stationary. These devices are usually not a high-priority item for maintenance personnel, but they are normally located in highly exposed positions, subject to the weather and possibly corrosive salt water. Their components should be made as durable as possible by using stainless steel or bronze instead of galvanized or plain steels and by protecting them as much as possible.

The bridge operator must be able to communicate with the captains of vessels navigating the draw opening, with operators of vehicles crossing the bridge, and with pedestrians in the neighborhood. The universal tool for such communication is the whistle or horn, which is blown to respond to a request for an opening and to warn that an opening of the bridge is imminent. Other tools include marine radios, bells, warning lights, and public address systems. Generally, for crossings where most vessels are pleasure craft, the whistle or horn is satisfactory for communication. On commercial waterways, radio is, if not a necessity, a great convenience, as it is desirable to have advance warning of the approach of large vessels, which may have difficulty navigating the draw opening. On some Midwestern navigable rivers that carry large barge tows of several thousand tons, advance warning of the need for an opening during spring floods can add considerably to the life expectancy of a bridge. The U.S. Coast Guard dictates requirements for operations and communications at all movable bridges in the United States that obstruct federal navigable waters. The Coast Guard has taken over this responsibility from the U.S. Army Corps of Engineers, which is now concerned mainly with dredging needs. Most operating and communication requirements are specifically tailored to a particular waterway or a particular bridge.

Interlocking

One of the most important control functions (#5 on page 268) for a movable bridge is the policing of proper sequential operation of all functions. AASHTO (2.10.34) and AREMA (6.2.7) require a form of automated interlocking. There are three common means of performing this function without relying on the expertise or training of a human bridge operator:

Mechanical interlocking
Permissive circuit electrical interlocking
Electronic programmable digital interlocking

Very few movable bridges have completely eliminated the judgment of the person operating the bridge in the interlocking process. Many movable bridges with electrical interlocking systems have parts of the systems bypassed because of deterioration or malfunction of devices, so the human operator must perform more functions of judgment than were intended in the original design. It is very important, in designing a bridge control system, to consider the capabilities of the persons expected to be operating the bridge. It is also important, in designing an interlocking system, to make certain that it will be able to function at the level of reliability and durability anticipated. Many failures have occurred, for instance, in electronic programmable logic controller (PLC) digital interlocking; it has been found that the hardware, in service, is not as capable of withstanding the rigors of movable bridge service as was anticipated. These systems are sensitive to many kinds of external interference, and lightning strikes of any magnitude can totally destroy a PLC system. For this reason, heavy-duty lightning protection apparatus and line surge protectors must be installed with any of these systems. PLC systems do not require heavy current flows through the interlocking circuitry, so much smaller limit switches can be used, including small proximity switches, which can last longer when not abused, as compared with mechanically actuated lever arm or plunger-type limit switches.

Permissive circuit electrical interlocking is based on the use of relays to open circuits to prevent, and close circuits to allow, the various bridge functions to occur. The relays are controlled by limit switches on the various pieces of apparatus, so that a power circuit for a double leaf bascule span lock, for instance, cannot be closed until limit switches on each leaf have tripped to indicate that the leaves have reached and held the fully closed position. The relays and limit switches are prone to wear and external damage, but when proper heavy-duty units are specified, installed, and maintained correctly, they are likely to have a long and useful life.

Some bridges that operate automatically, or partially automatically, use relays to direct sequential operations in what is called a *relay logic system.* Such systems have, at some locations, been replaced on movable bridges by PLC systems. Some new bridges are built with PLC-based logic systems, but many are built with relay logic systems to avoid the problems associated with PLCs.

Each type of movable bridge has its own interlocking logic sequence that is largely invariable regardless of the size or configuration of the bridge. Some of the logic steps are the same for every type of movable bridge, but there are some variations between railroad and highway bridges.

Basic Sequence for Highway Bridges

1. *Stop traffic,* usually with traffic lights.
2. *Lower warning gates,* if present; if both on-bound and off-bound gates are used, the on-bound are lowered first, then the off-bound.
3. *Lower barrier gates,* if present.
4. *Open bridge for marine traffic.*
5. Stopping the bridge in the fully opened position will allow the navigation lights to change to green, signaling vessels that they may pass through the draw opening.
6. *Close bridge to marine traffic.*
7. Moving the bridge away from the fully opened position will cause the navigation lights to revert to red.
8. *Raise barrier gates,* after bridge has been closed.
9. *Raise warning gates.*
10. *Return traffic lights to green.*

Basic Sequence for Railroad Bridges (on Heavy Traffic Lines)

1. *Bridge operator requests permission of train dispatcher to open bridge.*
2. *Train dispatcher stops train traffic,* by setting approach signals to STOP or communicating directly with train crews.
3. *Train dispatcher closes switch allowing power to bridge,* for operation; this is done remotely, after it is confirmed that all trains approaching the bridge have stopped. Dispatcher also releases rail locks.
4. *Bridge is opened for marine traffic.*
5. Stopping the bridge in the fully opened position will allow the navigation lights to change to green, signaling vessels that they may pass through the draw opening.
6. *Bridge is closed to marine traffic.*
7. Moving the bridge away from the fully opened position will cause the navigation lights to revert to red.
8. *Train dispatcher receives signal that bridge is closed, and ready for train traffic;* this is done automatically via limit switches on the bridge.
9. *Train dispatcher sets rail locks and opens switch cutting off power for bridge.*
10. *Train dispatcher allows train traffic to resume,* either by changing STOP signal to APPROACH or CLEAR, or by direct communication with train crews.

The preceding restrictive interlocking method for railroad bridges was adopted after many years of experience with accidents at drawbridges, with trains plunging into open draw spans or striking the counterweights of bascule or vertical lift bridges (see Figure 18-1). It was found in many of these accidents that the bridge operator, who had control of the bridge and the adjacent train signals, opened a movable bridge directly in front of an approaching train. Adding a second person to the control process and restricting power to the bridge have eliminated this type of accident.

For each type of movable bridge, the interlocking system can be as follows:

Single Leaf Bascule Bridge

1. Lowering traffic gates allows power to span locks, if present.
2. Withdrawing span locks, if present, allows power to bridge drive.
3. Release of machinery brakes allows power to drive motors and motor brakes.
4. Release of motor brakes allows power to drive motor(s).
5. Approaching the nearly open position causes drive motor(s) to operate at reduced speed.
6. Reaching the nearly open position causes power to be cut to drive motor(s), and motor brakes to set.
7. Stopping at the nearly open position allows the bridge operator to continue to open the bridge, at reduced speed.

Figure 18-1 Accident at Railroad Vertical Lift Bridge. The bridge was opened as the train was approaching, with insufficient time or distance to stop. (Photo courtesy of John McCown.)

8. Reaching the fully open position cuts power to drive motor(s) and sets motor brakes.
9. Stopping the bridge in the fully opened position and setting machinery brakes will change the navigation lights to green, signaling vessels that they may pass through the draw opening.

After the vessel has passed

10. Release of machinery brakes allows power to release motor brakes.
11. Release of motor brakes allows power to drive motor(s).
12. Moving the bridge away from the fully opened position will cause the navigation lights to revert to red.
13. Approaching the nearly closed position causes drive motor(s) to operate at reduced speed.
14. Reaching the nearly closed position causes power to be cut to drive motor(s) and motor brakes to set.
15. Stopping at the nearly closed position allows the bridge operator to continue to close the bridge, at reduced speed.

16. Reaching the fully closed position cuts power to drive motor(s) and sets machinery and motor brakes.
17. Staying at the fully closed position allows span locks, if present, to be driven.
18. Driving of span locks, if present, allows traffic gates to be raised.

Double Leaf Bascule Bridge The system is similar to that for a single leaf bascule, but span locks are always present. If the center shear locks are of the mechanically actuated type, they are actuated in the sequence presented in the preceding list. If the center shear locks are of the rigid structural type, the bridge leaves must be coordinated at opening and closure. On some bridges, this is accomplished by mechanical sensors that control the positioning of the leaves as they approach the closed position.

Span Drive or Span-Tower Drive Vertical Lift Bridge The system is similar to that for a single leaf bascule; span locks may or may not be present. The bridge may have two or four fully seated limit switches, which must all be tripped to obtain a fully seated indication.

Tower Drive Vertical Lift Bridge The system is similar to that for a span drive or span-tower drive vertical lift bridge, but additional limit switches or other means of control are used to monitor and correct for longitudinal skewing of the bridge. Excessive skewing will cause the span drive motors to stop and the brakes to set.

Swing Bridge Some older swing bridges have two drive systems, one to turn the bridge and one to operate the end lifts, center wedges, and other devices. This type of system can have the following interlocking sequence:

1. Withdrawal of the end lift system with associated devices to the fully retracted position allows the span drive motors and brakes to be actuated.
2. Stopping the bridge in the fully opened position changes the navigation lights to green, signaling vessels that they may pass through the draw opening.

After the vessel has passed

3. Moving the bridge away from the fully opened position causes the navigation lights to revert to red.
4. Reaching and holding at the fully closed position allows the end lifts and associated devices to be driven.
5. Reaching the fully driven position at all the end lifts will allow the traffic control devices to be actuated to open the bridge to traffic.

Most newer and many rebuilt swing bridges have each device separately actuated, or a few devices, such as both end lifts at one end of the swing span, operated by a separate mechanism, necessitating more interlocking circuitry. For such a center bearing highway swing bridge, the sequence for actual bridge operation could be as follows:

1. Withdrawing both centering devices to the fully retracted position allows the center wedge drives to be actuated.
2. Withdrawing both center wedges to the fully retracted position allows the end lifts to be actuated.
3. Withdrawing all end lifts to the fully retracted position allows the bridge's turning machinery to be actuated.
4. Stopping the bridge in the fully opened position changes the navigation lights to green, signaling vessels that they may pass through the draw opening.

After the vessel has passed

5. Moving the bridge away from the fully opened position causes the navigation lights to revert to red.
6. Reaching and holding at the fully closed position (or very close to it) allows the centering devices to be driven.
7. Reaching the fully driven position at both the centering devices allows the end lifts to be driven.
8. Reaching the fully driven position at all the end lifts allows the center wedges to be driven.
9. Reaching and holding the fully driven position at the center wedges, while the end lifts and centering devices remain in the fully driven position, allows the bridge to be reopened to traffic.

Not all bridges have all of the interlocking circuitry indicated in the preceding lists. Railroad swing bridges have additional limit switches to monitor the action of miter rail actuators, and some may have separate limit switches to monitor the position of the miter rails themselves, directly. Many railroad bridges also have separate switches, connected to the track rails at the joints at the ends of the movable span, that can be closed only when the bridge is fully closed and the rails seated properly, to allow the railroad signal to display a clear aspect. Some movable bridges, particularly poorly balanced bascule and vertical lift bridges, require more complicated manipulation of the motor and/or machinery brakes to hold the movable span seated until the span locks are engaged.

Wiring

Most movable bridges are "hard wired." All switches, contactors, motors, brakes, and other pieces of apparatus are connected by insulated copper wiring to form the necessary circuits to control and operate the bridge. There has been some effort to make electrical connections for movable bridges without connecting everything by wiring, in situations where making the wiring connection is difficult. These situations include tower drive vertical lift spans and double leaf bascules that would require submarine or aerial cables to connect parts of the bridge separated by the navigation channel.

This can also be the case for swing bridges. Microwave and radio signals have been used, but their reliability is not absolute. For permanent connections of power and control components, hard wiring is usually the only sensible choice. The Hood Canal Bridge is an exception, where a very wide and very deep waterway is crossed by a very large movable bridge with a very wide navigation opening. Replacing a submarine cable is an expensive proposition in such a case. An aerial cable is out of the question because of the size of the vessels that navigate the draw and the difficult environmental conditions. Microwave is used to carry some signals in this application, backed up by a telephone modem connection.

On older movable bridges, with crude control systems, earth grounding may be used with satisfactory results. Modern bridges with more sensitive control systems usually require direct system grounding, and four-wire connections are almost universal for system reliability as well as safety. All wiring is usually enclosed in metal conduit, doubly protected from corrosion by being galvanized and plastic coated. The present codes require safety devices such as ground fault interrupters, which further complicate modern wiring systems.

Traffic Control Systems for Highway Bridges

Unlike the situation with railroad movable bridges, for which a train dispatcher typically orders trains to stop and then gives a bridge operator permission to open a bridge, there is no direct control of vehicular traffic on highways available to the operator of a movable bridge. When a movable highway bridge is to be opened for a marine vessel, the highway traffic must be stopped safely and prevented from entering the area of the movable bridge while the bridge is out of its lowered position. Several means are available for a highway bridge operator to interrupt traffic so that a bridge opening can be made, but many accidents occur because of the lack of direct communication to vehicle operators. Many motorists make every effort to avoid being delayed by a bridge opening, even to the point of driving around or through lowered gates. Little, poor, or no training is provided in the typical state driving manuals for motorists' encounters with movable bridges. Thus, clear warnings must be provided to motorists when a movable bridge is to be opened, and it may be necessary or desirable to provide an absolute means of preventing entry to the bridge's draw opening. The *Manual on Uniform Traffic Control Devices*[3] specifies the arrangement of traffic protection devices at drawbridges. In addition to bells and horns or whistles, the devices that can be employed for the safety of motorists include the following:

1. Traffic signals
2. Warning gates
3. Impact barriers

[3]Federal Highway Administration, *Manual on Uniform Traffic Control Devices* (Washington, D.C.: Department of Transportation, U.S. Government Printing Office).

Almost any combination of these devices may be present at a movable highway bridge installation, but warning gates are usually present as a minimum. On new installations, AASHTO requires traffic lights (2.1.7) and suggests warning gates (2.1.6). Red-yellow-green traffic lights are preferred, but some bridges have used stacked red lights. Normally, the traffic control devices are interlocked so that they must be operated in sequence before the bridge can be opened. The traffic lights are actuated so that they turn from green to yellow to red. The bridge operator waits until traffic clears and then actuates the warning gates, usually lowering the onbound warning gates and then the offbound warning gates, if present. After the warning gates are down and traffic has stopped, the bridge operator closes the barrier gates, if the bridge is equipped with them. At most bridges, each of the steps after the first one cannot be initiated until an interlock has been tripped indicating that the prior step has been completed. Most bridges are equipped with additional interlocks, preventing initiation of a bridge opening until the final traffic control step has been completed.

Traffic lights for a bridge may consist of a simple set of signals at the bridge, actuated by the operator. In more heavily trafficked areas, particularly with other main roads intersecting the bridge roadway at grade, the bridge signals may be interconnected with other traffic signals so that the traffic not passing over the bridge is allowed to continue, minimizing tie-ups resulting from the bridge opening. At some locations, traffic exiting the bridge can be given the green light to help clear the area for bridge operation.

The warning gates for movable bridges are usually similar to the crossing gates used at railroad-highway intersections. Many small, older movable bridges have only a single warning gate at each approach, protecting only the on-bound lane or lanes. A movable highway bridge without any warning gates is very rare indeed. In areas where motorists are more aggressive, it is necessary to add a second set of gates to protect the offbound lanes so that it is more difficult for motorists to drive around them in an attempt to cross the bridge before it raises. The warning gate is placed a suitable distance from the joint between the approach and movable span roadways. The *Manual on Uniform Traffic Control Devices* prescribes distances and other details for such devices. In some situations, particularly in urban areas, the available distance from the movable span, in which a warning device can be placed, is limited. In these cases, warning devices must be placed very carefully to maintain a safe roadway.

The use of impact or crash barriers on highway movable bridges is a controversial issue. There are no rigid standards for the use of these devices and no consensus on whether impact-attenuating barriers are better than rigid barriers. Several types of impact-attenuating barriers are available, all produced in proprietary fashion by specialized manufacturers. These are described in the following paragraphs.

The "Buda" or "Sawyer" or "Mollenberg-Betz" (BSMB) Net This is the only common type of impact-attenuating movable bridge barrier that is theoretically completely reusable (see Figure 18-2). When a vehicle impacts it, a friction device allows a wire rope, attached to a net that has trapped a vehicle, to play out, absorbing the energy of the vehicle. After the vehicle stops and is removed from the net, the wire rope can theoretically be rereeled and the net is ready for another impact. Reuse of the rope and net are not guaranteed. Care must be

Figure 18-2 BSMB Impact-Absorbing Barrier Gate. This gate, shown in the raised position, drops to block the roadway when the bridge is to be opened. If a vehicle hits the gate, the gate extends, slowing the vehicle, and the impact causes wire ropes to be drawn through a braking mechanism, absorbing the kinetic energy of the vehicle.

taken to ensure that the "reusable" wire rope and net have not been damaged in the incident. This type of device has also been difficult to get back into working order after more than usually severe impacts. The gate is raised and lowered vertically, without tilting, by means of mechanisms located in small towers on each side of the roadway. Lokran Industries provides an updated, improved version of this gate that meets FHWA requirements for this type of installation.

The Entwistle Metal Bender This device (Figure 18-3) uses a net similar to that used by the BSMB gate, but the energy of the vehicle is absorbed by metal tapes that are drawn through a series of rollers as the net is pulled by the vehicle. The tapes are plastically deformed when used and must be replaced with new ones after each impact. This device has the advantage of a carefully engineered resistance to impact. The gate usually operates in a similar fashion to the BSMB gate. The metal bender itself is a sole-source product.

The B&B Barrier This device is housed in a hollow metal beam that is placed across the roadway (see Figure 18-4). An annealed stainless steel rope is stretched plastically when the gate is hit by a vehicle. The complete gate must be replaced after a collision. This gate opens and closes in an angular fashion, like a warning gate, but the free end latches onto a socket on the opposite side of the roadway or on the opposing mating gate arm. There are at least two manufacturers that produce this type of gate.

Figure 18-3 Entwistle Metal Bender Barrier Gate. This gate, shown in the raised position, drops to block the roadway when the bridge is to be opened. If a vehicle hits the gate, the gate extends, slowing the vehicle, and the impact causes stainless steel strips or bars to be drawn through a convolution of rollers, absorbing the kinetic energy of the vehicle.

Some movable bridges that support low speed roadways have so-called rigid barrier gates installed (see Figure 18-5). These gates may provide some shock absorption, depending on their construction and on the speed, mass, and rigidity of the vehicles hitting them. They are intended, however, only to prevent vehicles from departing the roadway. The gates may be constructed of metal or concrete; some roll on and off the roadway, others lift up to clear and lower to block traffic. Some rigid gates have been built that lift up out of the roadway, but these are difficult to maintain, especially on roadways subject to icing.

Care should be taken in specifying warning and barrier gates for use on movable bridges so that units with the desired reliability and durability are provided. There are many manufacturers of these devices, and the market is competitive. Part of the reason for such an active market is that many of these devices wear out, or are destroyed in use, and must be replaced with new. This is an area in which "or equal" should be used with caution in a specification or set of bid documents, as it can be very difficult to prove that a particular manufacturer's product is not as reliable and durable as another's and the functionality of the two products may not be identical. If a particular manufacturer's product has given good service and has had a satisfactory life, consideration should be given to having that manufacturer's product be sole sourced for the application (see page 308).

Figure 18-4 B&B Type Barrier Gate. The gate contains an annealed stainless steel rope, which, when the gate is engaged, blocking the roadway, stretches across from side to side. A vehicle impacting the gate causes the wire rope to stretch, elastically and plastically, absorbing the kinetic energy of the vehicle.

Bridge Operators

As suggested earlier, the person to whom the operation of the bridge is entrusted is a significant element in the design process for a movable bridge. The level of skill expected of this person by the bridge owner is an important consideration in the design of the movable bridge's control system. Although safety and liability concerns suggest that many decision-making processes should be removed from the control of a human being, other aspects of movable bridge control may be either within the domain of a human being or left to the control apparatus. Yet some functions of bridge operation may best be handled by a human being.

Fully automated movable bridges, with no human intervention in the processes of opening and closing a bridge, usually work best when the bridge can be left in the open position, clear for navigation. This leaves the railroad bridge as about the only practical application for fully automated control. There are still various concerns about safety that cannot be ignored. It is difficult to completely bar unauthorized entry into an unmanned installation such as a bridge. There are many instances in which trespassers on railroad bridges have been hit and killed by trains. Railroads have usually avoided liability in such incidents by showing that all that was practical was done to avoid these occurrences. In the case of a remotely operated movable

Figure 18-5 Rigid Barrier Gate. This gate is a solid box member. It is driven down the poles on the side of the roadway to block traffic. It has little impact-absorbing capacity and is thus best deployed on local streets with slow traffic, as shown.

bridge, it is difficult to show that everything practical had been done to avoid the injury or death of a trespasser. The cost of maintaining remote cameras, motion detectors, annunciators, and other communication devices, security fences, or other protective equipment for prevention of injury to stray persons may outweigh the cost of having a competent human operator on-site at the bridge.

Push-button control, sometimes referred to as automated control, requires the human bridge operator only to initiate the bridge opening process when a vessel has requested passage. All operations following until the vessel has passed, are then sequenced, initiated, and completed by machine. In some highway bridge applications, the bridge operator still controls the working gates. After the vessel has passed and cleared the draw opening, the human bridge operator initiates closure in the same way as the opening was initiated and the bridge does the rest. This somewhat reduces the operator skill level required.

POWER

The first step in designing the drive system for a movable bridge, the mechanism that moves it in and out of position for carrying traffic, is to determine the power requirements. AASHTO (2.5.3) and AREMA (6.3.6) provide detailed specifications for the environmental loads on the bridge due to wind and ice and provide standard values for friction in the various moving components. The power required is first determined by establishing the maximum force that must be developed on the span to rotate a bascule or swing span on its respective axis, or to lift or lower a vertical lift bridge. The maximum drive force is established at the moving span, ignoring the machinery drive train itself, whatever its configuration will be. This maximum force is the sum of several components, including the following:

1. *The friction to move the span,* by translation or rotation, or some intended combination of both. This friction has two values. The starting friction is determined by using the static coefficient of friction at the main bearings. The moving friction is determined by using the dynamic friction coefficient at the main bearings. Friction type bearings have substantially different coefficients of friction for these two conditions. There is no appreciable difference between the dynamic and static coefficients of friction for antifriction type bearings such as roller bearings. The load on the bearings is not constant, but is the sum of the dead load of the span, including any variables such as a partially loaded fuel tank, any wind loading that will add to the load on the bearings, or subtract from it, snow or ice loading, and any possible reaction due to the action of the prime mover on the span. Any or all of these component loads may vary during bridge operation as a function of a number of independent variables including the opening angle of the bridge, and this variation should be considered in determining the maximum power required. The span friction also includes such secondary effects as the bending of counterweight wire ropes around the sheaves of a vertical lift span, the friction at counterweight trunnions and link pins of a heel trunnion bascule bridge, and the link and roller friction of a Rall bascule bridge. Some of these effects are trivial, whereas others, such as trunnion friction, may be quite significant.

2. *The imbalance of the bascule or vertical lift span.* At one time many engineers believed that a movable bridge span should be perfectly balanced, to minimize the net power required to operate the bridge. Some engineers thought that it was prudent to adjust the bridge's balance so that the bridge would have a net gravitational force tending to lower it when it was open, and tending to open it when it was closed, as this would at least partly counteract the effect of inertia when starting the bridge moving and reduce the required size of the prime mover and machinery. Today it is generally accepted that bascule and vertical lift bridges should be slightly imbalanced so that gravity tends to stabilize the bridge in the lowered position, by pulling it down, and stabilize it in the raised position by pulling down on the counterweight, holding the span up. This approach forces a slight increase in the size of the machinery, the size of the prime mover, and the amount of power required for operation, but the added costs are not significant today. The extra safety inherent in having a stable bridge is generally considered to be worth the price of a slightly greater initial investment in machinery. Swing bridges normally have no gravitational imbalance effects directly influencing power requirements.

3. *Acceleration of the span at the beginning of movement, and deceleration at the end of travel.* This load is relatively minor in computing the maximum power required, but will be a significant part of the long-term power costs of operating a bridge. Acceleration is usually the greatest single component of the bridge operating

load during normal operation, as high winds are seldom blowing when the bridge must operate and ice or snow loading may never occur. The combination of acceleration and overcoming static friction at the beginning of movement of a bridge causes the greatest machinery wear, and racks and pinions invariably show the greatest wear at the teeth in contact during the performance of this work. In calculating the stresses in the machinery components and supports, it is perhaps best to assume greater overloading of the prime mover during acceleration of the movable span, when the prime mover is an electric motor supplied by utility line power, than the overloading described by AASHTO and AREMA. This is particularly significant at the racks and pinions of swing bridges, for which failures due to overloading are not unknown. Designing for 250 percent of full load torque instead of the specified 150 percent may be warranted for a frequently operated span (see page 255). Many bridges have recorded 250 to 300 percent of full load motor torque during starting, as opposed to the AASHTO or AREMA 180 percent. The reasons for these high torques may or may not be somewhat spurious, but nevertheless they frequently exist (see page 261). In making a final determination on the size of the bridge drive motor or engine, the inertia of the machinery drive train must be included. The higher rotational speed of these components during operation increases their inertial effect considerably in proportion to the inertia of the span itself. This inertia can act to reduce power requirements, as the momentum of the machinery can assist in starting the bridge movement, provided that there is sufficient backlash in the system to allow the drive motor to come to speed before the rack pinions are fully loaded. In this scenario, however, impacts can be induced in the machinery that may result in failure of worn components.

4. *The wind load on the span and counterweights.* The wind can act against bascule and swing bridges to produce a moment retarding or assisting motion. This load is obvious on bascule bridges and easily computed. For swing bridges, somewhat unrealistic assumptions are made by AASHTO and AREMA in order to develop a wind moment value, but the net result is usually acceptable. The wind can act against a vertical lift span, pushing the span up or down. Lateral wind can add to the lifting or lowering resistance of the vertical lift span by increasing friction at the span guides.

5. *Snow or ice loading on the span and counterweights.* Such loading adds to the weight and the resultant friction load and may also increase the wind load. It can also affect the balance of the span and cause brake holding difficulties.

The net sum of these loads is resolved to a force at each operating strut, rack, or rope. This then translates to a force at the pinion or other main component and is carried back to the prime mover, incorporating friction losses in the drive train along the way. A prime mover is thus selected that can produce the required resultant rated torque or force to ensure that it will be possible to start the bridge moving whenever necessary. The resultant prime mover is then checked to see that it produces enough torque or force at full speed to drive the bridge open or closed after movement has been started, allowing for machinery resistances. If necessary, the prime mover's rated torque at rated full speed is increased to meet this requirement. AASHTO and AREMA have developed standard means of applying the aforementioned loadings to determine power requirements. The formulations of the two authorities are similar and include various load cases, with corresponding allowances for variations in operating times of the span to accommodate variable loads without unduly increasing the amount of power that must be installed. The loading to be applied under various environmental conditions depends on the type of span. The bascule bridge is more sensitive to wind loading, and additional power must be provided for this bridge type to ensure that it is capable of operation in adverse conditions. Among the common movable bridge types, the operation of a vertical lift bridge is the least affected by wind.

Three levels of power requirements are recognized:

Maximum bridge starting torque (MBST)—the power needed to develop the force necessary to break the bridge loose from a static condition and get it moving

Acceleration—the power needed to bring the moving bridge up to full speed

Full speed—The power required to maintain the bridge at a constant speed while opening or closing

The 2000 LRFD AASHTO attempts to provide a more realistic rating system for motors and machinery for these conditions, with less room for interpretation or misinterpretation by designers.

After the power required at the span is determined, the losses in the machinery drive train must be determined so that the prime mover can be sized (Figure 18-6). Overall efficiency of a drive train to move a bridge can be quite low, as little as 60 percent, so that the size of the prime mover is increased considerably. Efficiencies of auxiliary drives, such as for the end lifts on a swing bridge, can be even lower, because of the use of worm gear sets or drive screws, so losses there can be even greater.

If an electromechanical drive is used, then once the amount of power required at the prime mover for a bridge drive system is established, the span drive motor can be selected. AASHTO (2.10.17) specifies 900 rpm maximum driving open gearing, and 1800 rpm max for enclosed reducers. AREMA (6.7.5.8) requires that, for a main bridge drive, the full speed of an electric motor should not exceed 900 rpm (can be 900 or 1200 rpm per 2000 LRFD AASHTO). For durability of the entire system, 900 rpm maximum is recommended. For an electric motor gear drive bridge operating system, motors are readily available that operate at the desired speed in almost any horsepower rating. About the smallest size needed for any bridge is 10 hp, but motors ranging in size up to 200 hp, even 500 hp and more, are available from standard catalogs of many manufacturers. Because of the nature of movable bridges, the motors usually must include special features, such as corrosion resistance and moisture resistance, which prevents their being purchased from stock.

Electric motors should be totally enclosed, nonventilated (TENV), which minimizes contamination by the dirty environment usually surrounding a bridge drive (see Figure 18-7). A TENV motor also avoids reliance on air circulation for cooling in its confined environment, which can also be hampered by poor maintenance conditions and the controlled low speeds at which the motors sometimes operate. The use of special windings and insulation is a good idea, to avoid water damage and ensure long life. The National Electrical Manufacturers Association's (NEMA) Design D-type motors have usually been specified because of their high available starting torque.

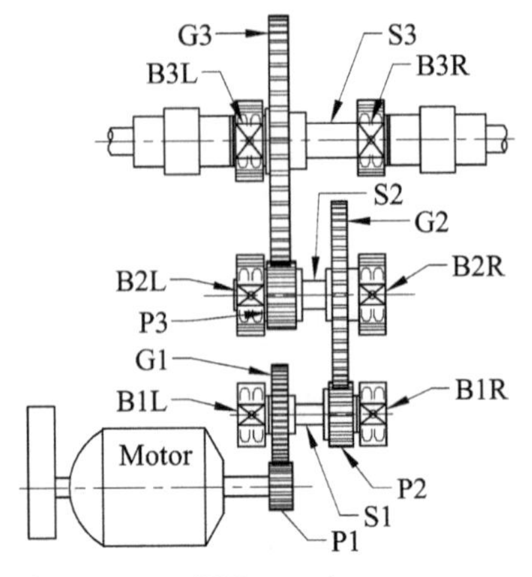

fP1 = Tmotor/RP1, pounds
fG1 = 0.98xfP1, pounds
Ts1 = fG1xRG1-tB1L, inch-pounds
fP2 = (Ts1 - tB1R)/RP2, pounds
fG2 = 0.98xfP2, pounds
Ts2 = fG2xRG2-tB2R, inch-pounds
fP3 = (Ts2-tB2L)/RP3, pounds
fG3 = 0.98xfP3, pounds
Ts3 = fG3-tB3L-tB3R, inch-pounds

Figure 18-6 Friction Losses in Gearing Drive Train. The resulting loads at gears and shaft, after friction losses are deducted, are shown.

Newer drive systems can provide the specified operating characteristics with the more common Design C or B motors. AASHTO (2.10.18) specifies that span drive motors shall be totally enclosed fan cooled. AREMA (6.7.5.9a) specifies that span drive motors shall be totally enclosed crane, hoist, or mill type, except (6.7.5.9c) alternating current motors shall be squirrel cage type if variable voltage or secondary resistance control is not used. 2000 LRFD AASHTO (8.5.1) specifies crane and hoist or mill duty motors for span drives, and totally enclosed motors when rated at 30-minute duty or of the wound rotor type. 2000 LRFD AASHTO (8.5.2.2.1) specifies that squirrel cage AC motors shall be NEMA Design B unless specifically determined otherwise. It may not be surprising if an occasional competent electrical engineer goes his own way in specifying electrical motors for a movable bridge. Heavy cast iron frames are better at withstanding shock loads and occasional removal and reinstallation than cast aluminum or sheet metal housings. Some motors are specified with corrosion-resistant fittings and stainless steel shafts to prevent deterioration in the harsh environment most drives have to withstand. DC motors were once preferred because of their better speed control capabilities, in spite of the difficulties in supplying DC power. Wound rotor AC crane-type motors were developed with sufficient speed control available to be used for operating bridges, so most bridge drives installed after World War II were AC motors. Today, reliable, fully functional, and economical solid-state drives are available to operate either AC or DC motors. Modern solid-state rectification allows DC power to be supplied in almost any quantity, but

Figure 18-7 TENV Motor. The totally enclosed nonventilated motor was a standard requirement on movable bridge applications for many years.

difficulties in commutator maintenance remain. In many applications, AC motors can now be the much cheaper squirrel cage type instead of the wound rotor type.

Bridge drives are unusual in comparison with most motor applications in that they must frequently run at reduced speed while delivering full torque. Most bridge control systems and operating procedures call for the bridge to be operated, while it is being seated, at "creep speed," which may be 10 percent or less of full motor speed. The acceleration and deceleration rates are controlled to limit torque output from the motor while bringing the bridge up to full speed, because of the high inertia of the moving bridge. This is usually especially true of swing bridges. Reduced-speed operation can result in extreme overheating of electric motors.

All else being equal, among bridges of equal span size, without large imbalances or external forces such as wind applied to them, the swing span requires the least power to operate and the vertical lift requires the most. The difference is in the amount of friction retarding movement of the bridge. A vertical lift bridge drive must overcome the resistance of counterweight ropes operating over sheaves. A swing bridge has a point bearing or a set of rollers supporting its moving weight, both of which are low-friction devices. The swing bridge has no gravitational forces to overcome during opening and closing, except some minor vertical movement at the end lifts. A bascule bridge, its weight supported on shaft-type bearings as are the sheaves of vertical lift spans, has an intermediate friction value. Including wind and all other resistances to motion, per the standard power requirements calculations, the bascule bridge requires the greatest amount of power because of its higher wind resistance, and the vertical lift bridge has the least, because of its minimal wind resistance. Of the other movable bridge types, the pontoon has a higher power requirement as a result of having to move through water, and the roller-supported retractile has a low, perhaps the lowest, power requirement because of slight wind resistance, low friction, relatively low inertia, and no gravitational forces to overcome.

In times of cheap energy, the actual power required to operate a bridge is not a major factor in its cost and is of no concern in selecting the bridge type. This was not the case prior to the mid-twentieth century. In many, if not most, instances, the selection of a movable bridge type at that time was heavily weighted by power considerations. Hovey, writing in the mid-1920s, preferred the swing bridge, partly because of its lower overall power requirements. This is not to say that the amount of power required to operate a movable bridge is trivial. Power supply may be expensive, particularly if the bridge is located in a remote area, as the nearest source of three-phase electric power may be miles away. Larger power requirements also require larger emergency generators, if such are to be installed on the bridge. The noncounterweighted bascule bridge in Sheboygan, Wisconsin, required a great deal of special construction to supply the substantial amount of electric current it needed, and a special arrangement was made with the local utility for the bridge's power supply. If a "balanced" bascule or vertical lift bridge has a large imbalance, its power requirements for normal operation can be quite high.

Many early electrically powered swing bridges, in order to cut construction cost, had a single motor to drive all the bridge machinery, with clutches used to connect the motor to turning or wedge drives as needed. Today the cost of fabricating, installing, and maintaining such clutches would exceed the cost of the additional motor or motors by a substantial margin.

The vast majority of movable bridges in the United States are powered by electromechanical drives, as discussed earlier. There are other power sources in use, some of which have become quite popular. The most common of these are hydraulic drives, which are discussed in more detail in Chapter 17. Almost all hydraulic drives rely on electric motors as their primary power source, but the electric motor for such a drive is operated with only a simple on-off control. Until recently, many older movable bridges were still powered by steam engines. These invariably had coal-fired boilers and required constant attention from a bridge operator, and the fuel usually required manual handling. The reliability and durability of these systems could hardly be equaled by any other form of power operation.[4] With increasing difficulties in utility power generation and delivery, and other problems with oil and gas supplies, such low-tech systems may bear a second look.

Auxiliary Power

Many bridges that must be operated very reliably, because they carry important highway or rail traffic and span critical, heavily used waterways, are equipped with auxiliary power or drive systems. AREMA (6.2.5) requires that power-operated movable railroad bridges be provided with a means of emergency operation. AASHTO

[4] O. E. Hovey, *Movable Bridges,* Vol. 2 (New York, John Wiley & Sons, 1927), p. 99.

(1.1.13) suggests that movable bridges be equipped with emergency power sources. Sometimes these auxiliary power systems consist of engine generator sets or secondary electric service installations, but these are useless if the bridge control system fails. For this reason, most critical bridges (except tower drive vertical lifts) are equipped with a manual or direct engine-drive auxiliary power source, whether they do or do not have an emergency or auxiliary generator or another alternative electric power source to provide backup power. Many auxiliary power sources, whether electric or direct mechanical drive, provide reduced power, so that the bridge operates at reduced speed when the auxiliary source is in use. If an electric auxiliary source provides reduced power, an electric motor smaller than the normal drive is generally provided. This usually drives through additional reduction gearing so that the full maximum drive torque is available at the bridge's final drive, allowing operation under any of the conditions prescribed by AASHTO and AREMA. Such an arrangement generally requires that a disconnect clutch be provided to isolate the auxiliary drive motor when not in use, to avoid overspeeding of the auxiliary motor through backdriving. The maximum allowable speed of an electric motor depends primarily on the amount of centrifugal force it can withstand, which is a function of its rotating speed, not its load. If a direct engine or manual drive is employed, there must also be a means of operating secondary devices, such as span locks for bascule bridges and end lifts for swing bridges. These devices are usually operable manually with the use of removable hand cranks, electrically interlocked so that the main power cannot be applied when a hand crank is engaged. Some swing bridge wedge drives are engine powered, usually by the same engine that drives the turning machinery, with selective engagement by means of clutches.

Secondary Electric Service Many movable bridges are conveniently located where two sources of utility electric line power are available at or near the bridge. This power may come from two separate power companies, or both sources may be from the same company, but reach the bridge through separate and relatively independent distribution systems. This condition can occur in urban as well as rural locations, as it is often the case that the waterway a bridge crosses is a natural barrier that has kept development separate on each side.

Once the second source of line power has been established at the bridge, it is a relatively simple matter to provide for automatic transfer of power from one source to the other in case of a supply failure. Automatic transfer switches are common and readily available pieces of electric apparatus that merely need to be installed and wired into the bridge's electrical circuitry. It is generally best to have a system of automatically switching over from one source to the other on a regular basis, as occasional exercising of the equipment tends to keep it functional. Occasional use is also a handy in-service testing mechanism, allowing a failure to be detected in the course of ordinary operation, rather than possibly having it occur in an emergency when service personnel may be heavily occupied with other duties.

Engine Generator Sets Movable bridges at major waterway crossings have, for many years, been routinely equipped with engine generator sets when the owners have been concerned about reliability of operation (see Figure 18-8). There was a flurry of

Figure 18-8 Emergency Generator. Many movable bridges are equipped with such a device, in some cases connected so as to automatically start up and provide power in case of a failure of utility power. Some emergency generators, such as this one, supply full power to a bridge for operation as normal. Others provide reduced power, so a bridge must operate at reduced speed while power is supplied by the generator.

activity to add installations of this type to many other movable bridges when the supposed Y2K crisis became imminent, as this episode highlighted the possible sensitivity of power supplies to various sources of interruption. The U.S. Coast Guard has also become more concerned in recent years about the reliability of movable bridge operation and has put pressure on bridge owners to provide for backup operating systems, particularly at installations where the main power source is known to be unreliable.

An engine generator set must supply sufficient power to operate the bridge plus whatever amount of power is required for other functions, such as navigation lights and traffic control equipment, if a second generator is not provided for these other functions. The generator must supply sufficient power to satisfy the maximum needs of the bridge drive during operation. AASHTO (2.10.14) and AREMA (6.7.5.5) allow the drive motors to be overloaded to 180 percent of their rated power output during periods of acceleration of the span, but 2000 LRFD AASHTO (see page 255) reduces this allowance somewhat. Regardless of whether the main drive motors are used when operating with generator power, or a smaller auxiliary drive motor is used with additional gear reduction, the generator should supply enough power to satisfy this requirement, plus whatever other power is required, at the rated output of the generator. This may result in a generator with an engine of 100 hp or more driving a bridge drive with a 50 hp prime mover.

Engine generator sets can be connected to a bridge's electrical system in the same manner as a second line power source. In conjunction with automatic connection, the engine generator set also requires an automatic starting system. This apparatus is all readily available, as many public and private institutions, such as hospitals, have found a need to protect their systems from damage due to power source interruption and have installed engine generator sets. The market for such equipment has been a thriving one in recent years, with the advent of extensive electronic data storage systems in many commercial and public areas making the loss of power for any period of more than a few seconds devastating for many end users. Manual start-up and connection can be employed on a movable bridge, but this can require some skill and effort on the part of the bridge's operating personnel. It may also require groping in the dark to locate the starting panel and get the generator running. If an engine generator set is supplied with sufficient power for unrestricted operation, it can be considered the primary source and utility line power dispersed with. This could provide more reliable operation at some locations, and might cost less. A different system could provide auxiliary power (see following).

Many installations of engine generator sets at movable bridges call for the system to be installed in the operator's house or the machinery room, but it is generally better, when space is available, to place the engine generator set in a separate enclosure. Isolating the engine generator set from the bridge control and operation areas minimizes the likelihood of damage due to a fuel fire and reduces the probability of ear damage to personnel resulting from the noise of the system when operating. An engine generator set should be installed with adequate ventilation so that the air supply for combustion within the engine and for the cooling of the engine is not impaired. A manufacturer-supplied unit complete with all weather housing, mounted outdoors on its own pad, can satisfy these requirements. If protection must be provided for security or other reasons, and a house is required, cooling and combustion air should be provided directly from outside if possible. Any ducts or louvers must be carefully designed to avoid overly restricting the air flow. Engine exhaust should be directed outside by the shortest possible route. A separate means of exhausting the cooling air, with minimal or no restriction, should be provided.

Several types of fuel can be used to power an auxiliary engine generator set. Gasoline is definitely not recommended, as it is dangerous for it to be stored at a facility and it deteriorates when stored for prolonged periods. Moreover, gasoline engines require maintenance, such as cleaning the carburetor and spark plugs and adjustment and occasional repair of the ignition system. Natural gas or propane eliminates some of the difficulties associated with gasoline-fueled engines, but not all of them. Propane and natural gas fuels present a fire hazard and are also vulnerable to leakage. Explosions are possible. Although natural gas sources are generally quite reliable, interruption can occur in a significant event like an earthquake. Diesel fuel is the best for the purpose, by almost any measure of consideration. Diesel engines are extremely reliable, and diesel fuel is quite stable and fairly safe from fire. The most common complaint about diesel engines is their difficulty in starting, but this can be overcome by selecting the right engine design, providing the proper starting equipment, and providing for exercising the generator regularly. Turbochargers or superchargers are often specified with auxiliary engines (particularly diesels) to reduce the size and cost of the installation for a given power output. Turbochargers can deteriorate or fail, and require their own maintenance, so consideration should be given to doing without them.

Direct Engine Drives Direct engine drives are used as primary power on some movable bridges, and as emergency power on many others (see Figure 18-9). A direct engine emergency drive is advantageous, as it completely bypasses the electrical system, so any electrical fault will not prevent bridge operation. Direct engine drives cannot be used on most tower drive vertical lift bridges because of the lack of a mechanical connection between the towers. Most direct engine drives require the presence of the bridge operator to start the engine, manipulate a clutch engaging the drive, and control the speed of the bridge when it is moving. Some remote drives have been installed so that the operator remains in the operator's house while using the engine drive, but it is probably best that the operation of the engine and machinery can be observed directly by the operator. Occasionally, the engine of an engine generator set is used as the prime mover for a direct engine drive auxiliary, providing what might be considered triple redundancy, with some economy of first cost realized by furnishing and installing only one engine and requiring less space overall. If the generator is used to supply house electric power while the engine is operating the bridge, there may be some fluctuation in lighting intensity in the operator's house or at the

Figure 18-9 Auxiliary Engine Drive for Vertical Lift Bridge. This drive is a combination—it provides electric power for emergency operation, and can also provide direct mechanical drive to the vertical lift span. A second engine-generator, for lighting only, is to the left rear in the picture.

navigation lights, or other areas as the bridge is driven open and closed. The same considerations apply for exercising, fuel, and turbocharging for direct engine drives as for engine generator sets (see page 291).

An internal combustion engine is not capable of being overloaded in the short term as much as an electric motor being fed from a central utility power station. By adjustment of fuel settings, operating speed, and other details internal combustion engines can be overloaded in the short term to very high power levels, at an extreme increase in maintenance cost and severe shortening of engine life. Both of these results are undesirable in the environment of a movable bridge. The engine used for a direct drive must be capable of producing sufficient power, without being excessively overloaded, to meet the maximum needs of the bridge drive during operation. For most bridges, this is under maximum bridge starting torque (MBST) conditions (see page 285), such as in attempting to lower a bascule bridge against a heavy wind.

Direct engine drives require more maintenance than engine generator sets, as the gears, shafts, couplings, clutches, and bearings must be lubricated occasionally. More effort is required for test operations and for actual bridge operation, and two or three people may be required to properly coordinate operation by means of a direct engine drive.

Hand Drives Hand drives are very unpopular with bridge maintenance people because of the physical effort needed to use them and the long time required to open or close a bridge with hand power. Hand drives can also be unsafe, as the lack of an interlocking device, or its failure, can cause the normal power to be applied to a bridge drive while the hand drive is in use. Injuries, and even death, have occurred to maintenance or other personnel using a hand drive when normal power has been suddenly and unexpectedly applied to the machinery. Hand drives are available on many movable bridges for emergency use, but the actual procedure in the event of a main drive failure is usually to wait until a fault is repaired and then continue operation in the normal mode. Some bridges that operate seldom, if ever, have only hand drives available. These may be found rusted and inoperable when the time comes to use them. Many bridges that do operate often, using a form of nonhuman power, have hand drive mechanisms that are ignored. These may also be found rusted and inoperable when they are needed. Hand drive modes include chain (Figure 18-10) operation, capstans, and cranks. Operation may be performed by one person, or may require four or more people to provide the necessary bridge drive force depending on the type of manual operation. Most hand drives can be informally adapted to operation with an air-powered impact wrench or a similar power tool so that their use is a real practicality.

Hand drive systems are very simple and inexpensive. In some areas, the U.S. Coast Guard has mandated that movable bridges have auxiliary operation modes available in the event of failure of normal power sources. New hand drive systems have been installed, or existing ones have been rehabilitated, on many movable bridges to satisfy this requirement. The power requirements for operating a movable bridge are determined on the basis of operations under the more adverse conditions expected. Even so, operation of a typical bridge under more quiescent conditions, at normal speed, may require the expenditure of 10 or 15 hp. To be able to operate a larger

Figure 18-10 Chain Type Hand Drive. This bascule bridge is manually operated. The pulleys on top of the counterweight are fitted with chain loops, operated by persons on the ground, to raise and lower the bridge.

bridge with only one or two, or even three or four persons, it is necessary that operation be slowed down considerably by large gear reductions or some other form of torque or force multiplier, so as to produce the necessary driving force or torque on the moving span. The result is manual operation that may take several hours to raise or lower the bridge. In some cases this kind of operation has been deemed acceptable to satisfy legal requirements, but its practicality when need arises is questionable.

Although it may be reasonable to provide a manual system for the secondary operations of a movable bridge, such as driving and withdrawing bascule or vertical lift bridge span locks, or even operating the end lifts of a swing bridge, the use of manual power to open and close any but the smallest movable bridges under emergency conditions, or in the event of failure of the main power source, should be avoided. It may be impossible, in an emergency, to locate sufficient numbers of people to operate a bridge manually. In the event of an operational failure of the main drive, it may take several hours just to assemble a satisfactory crew at the site, to say nothing of performing the operation of the bridge. In many cases, it has proven impossible to coerce operating or maintenance personnel to use a manual drive to operate a bridge, even in an emergency.

All hand drive auxiliary operating systems should have interlocks that prevent the application of normal power when the hand drive is engaged, or even partially engaged. This includes main bridge drives as well as span lock drives and any other devices on the bridge that are normally power operated but are equipped with auxiliary hand drives. The hand drive interlock should prevent access to the hand drive shaft unless power to the machinery has been cut off. A protective cover over the hand drive shaft, with a limit switch that trips whenever the cover is disturbed, is sufficient, provided the limit switch when activated trips a main breaker or in another appropriate way prevents power from being applied to the prime mover.

Emergency Drives

In some locations, emergency equipment such as fireboats or rescue craft must reach the scene of an emergency by water. In recent years, the U.S. Coast Guard has taken a more stringent approach to enforcement of regulations pertaining to the operation of a movable bridge when the primary power source is inoperable. Many movable bridge owners have been asked by the Coast Guard to demonstrate that their bridges can be opened in an emergency with auxiliary power. For many bridges, particularly those that operate very seldom, they have been unable to do so. There should be a realistic approach in regard to auxiliary drives. If manpower will not be available to operate a manual drive, it should be discarded and replaced with another type of drive. If the internal combustion engine power source for the auxiliary drive is unreliable, another power source should be obtained.

A portable power source can be provided, such as a relatively small air or hydraulic motor, that couples onto a hand drive or other shaft to operate the bridge. The power unit can be truck mounted, with air or hydraulic hoses connecting it to the motor. Caution must be used in providing such a unit to make sure that it has the torque capacity to operate the bridge under conditions that may be expected. The unit should be capable of producing the maximum bridge starting torque (MBST) at the bridge rack pinion or final drive. In the case of a bridge that seldom opens, extra torque should be available to overcome high friction in bearings that may not have been lubricated in some time. The unit should have sufficient power to operate the bridge in the opening time required by the Coast Guard. This time frame may include mobilization time and time for transport of a crew to the bridge, so the bridge may have to operate fairly rapidly. A bridge that normally operates with 100 hp at the main drive for normal operation may require 25 or 30 hp at the emergency drive to open in the required time. This requirement precludes the use of small electric drills or torque wrenches as the emergency drive power source. Large portable air-powered tools are available that can deliver this kind of power, with sufficient output torque to operate the bridge under adverse conditions.

OPERATORS' HOUSES

The operator's house is the control center for a movable bridge (see Figure 18-11). Thus, all information regarding operation of the bridge should be available in the op-

Figure 18-11 Vertical Lift Bridge Operator's House. This house provides a view of the roadway as well as the waterway for the bridge operator while he remains at his controls.

erator's house, and provided instantaneously as operation is occurring. The position of the bridge itself, the condition of various limit switches and position indication switches, and the operating condition of drive components such as motors and brakes, should be readily apparent by means of indicators in the form of lights or gauges. The external conditions should also be apparent. Wind speed and temperature readouts should be available, and the conditions at the ends of the movable bridge span, particularly at roadway joints, should be visible from the operator's house. Ideally, the bridge operator should have direct visual contact with the critical portions of the movable bridge, such as the ends of the moving span, and of the marine channel in the vicinity of the bridge. AASHTO requires that the bridge operator shall have "a clear view in all directions" (2.10.46). AREMA adds "where practicable" (15.6.2.9c). This is sometimes impossible; video camera connection is a substitute, but not an ideal one for viewing these areas. Generally, operator's houses are best placed on the moving span. This is practically impossible for bascule bridges, which invariably have the operator's house on the bascule pier. Many movable bridges have the operator's house located near the machinery, as this is usually the least costly location. It may not, however, be the best position for viewing the waterway or roadway.

Communication equipment should be located in the operator's house, not only for communication with boat operators, but also for communication with emergency and law enforcement personnel. Internal communication—to the machinery room or

Figure 18-12 Ornate Operator's House. This house controls the La Salle Street Bridge over the Chicago River.

rooms, auxiliary operating houses, and gatehouses–should also be available. The operator's house should be easy to reach, but also secure against intrusion. It should allow for installation and removal of the largest components mounted within, by the use of double doors or trap doors if necessary. There should be a means of lifting and moving heavy components, such as an overhead crane of sufficient capacity to lift the heaviest component in the house. At some locations, because of the impatience and irritability of motorists who must wait for bridge openings, the use of bulletproof glass in the windows of the operator's house is a necessity.

The operator's house should contain adequate sanitary facilities, and meet all other requirements of the Occupational Safety and Health Administration (OSHA) and any other labor regulatory authorities having jurisdiction at the bridge location. Facilities for recreation or entertainment should be provided at bridges that must be staffed continuously but do not open often, as prolonged exposure to this sometimes solitary environment may be damaging to mental health. The operator's house should have good access, including, preferably, a limited amount of reasonably secure parking immediately adjacent to the entrance to the house.

On many movable bridges, the operator's house is the only part of the structure that readily allows for architectural embellishment. Typically, it is desired to construct a movable bridge for the least cost. In times of general prosperity, such as in the 1920s or the last few decades of the twentieth century, other priorities may pre-

Figure 18-13 Modern Operator's House. This house controls the Madison Street Bridge over the South Branch of the Chicago River.

vail, resulting in "signature" bridges, intended to be architectural masterpieces. Many bascule bridges have fanciful operator's houses that look like lighthouses or airport control towers alongside the bridge. At locations such as the Chicago River in downtown Chicago, there is a splendid array of operator's houses for the many bascule bridges, with styles ranging from the baroque to the modernistic, changing at every bridge along the waterway (Figures 18-12 and 18-13).

19

REHABILITATION

SUBSTRUCTURE

As compared with the electrical, mechanical, or superstructure portion of a movable bridge, its substructure is generally the most resistant to damage and deterioration. It can also be the most difficult to repair. Deterioration can include loose masonry, because of joints that have been allowed to lose their mortar without replacement, poured concrete that has rotted or spalled, deteriorated pilings, and damage due to ship impact and/or shifting subsoil. All of these conditions can result in a substructure that is unable to support its intended loads. Once it has been determined that reworking the substructure of a bridge can be cost-effective, in relation to of the overall condition of the bridge, then the constructibility of the substructure repairs should be evaluated. The conditions under which the work will be done are important. Must traffic be maintained? Can the bridge be left in the opened position for prolonged periods of time? Many older bridges have substructures totally inadequate to withstand seismic loadings, as determined by the latest AASHTO or AREMA criteria. Extensive reconstruction of the substructure may place a bridge out of service for a prolonged period of time. It is difficult, and may be impossible, to determine accurately the extent of substructure deterioration before commencing repair work, so the final cost of repair may be much higher than estimated. It is more than likely that an extensive substructure reconstruction project will not be worth the cost. All the costs of such work should be tabulated, and combined with the cost of other work on the structure, to determine whether a replacement bridge would be more cost-effective. The cost to the public of detours or delays as a result of construction should not be ignored. AASHTO calls for "engineering judgment" to be used in the rehabilitation of movable bridges (Division V). AREMA, in Section 7 of Chapter 15, covers rehabilitation of existing structures but says very little specifically about movable bridges (7.4.5).

SUPERSTRUCTURE

The most common form of deterioration of the superstructure on a movable bridge, which is typically of steel construction, is corrosion. Highway bridges generally suffer much more than railroad bridges, because of the presence of deicing salts or sand and dust. These materials descend from the roadway through the typical open grating deck and accumulate on the horizontal surfaces of the bridge members, attracting and holding moisture, which begins to eat away at the steel. Repairs can be made to these members, but unlike the case with fixed bridges, simply adding reinforcing plates cannot be done to more than a very limited extent, because of the need to maintain the balance of the moving span. It is generally better to replace the deteriorated members with new, and to install closed decks to help prevent future corrosion.

Retrofitting of existing movable bridges to increase load capacity is occasionally done, but the addition of more steel in the form of larger floorbeams and stringers, or reinforcement of the main members, must be balanced, in the case of a bascule or vertical lift bridge, by the addition of weight to the counterweights. Truss type main members can usually be readily reinforced, but girder type bridge leaves may present a problem. On a double leaf bascule bridge in the Midwest, the contractor successfully offered to replace the entire old superstructure with a new one, at no increase from his bid price, rather than make the effort to stitch in all the new steel pieces to increase the load rating of the bridge girders from HS15 to HS20, as was called for in the bid documents.

In general, any part of a movable bridge superstructure can be replaced with new. It is usually possible to maintain traffic on the bridge, if necessary, while performing the replacement. The need to keep the bridge open may add substantially to the cost of the reconstruction or may add to the time required to do the job. In some cases, it may be impossible to keep a bridge open to traffic during repairs. At other times it may be necessary to repair a bridge while holding it in the open position, so that a temporary bridge or a detour may be necessary. AASHTO and AREMA provide about the same guidance to superstructure as to substructure rehabilitation.

MACHINERY

Bridge machinery can be rehabilitated in one of several modes. Existing components can be reworked, such as by building up worn surfaces by welding or plasma deposition and then remachining them, but these tend to work only in some cases as temporary or interim repairs because of the metallurgical differences from the original components. It is possible to replace only selected components that are severely worn or otherwise defective. It is possible to replace subsystems, such as a mating set of open gears and their associated shafts and bearings. The entire bridge drive may be replaced, from motor and brakes to racks and pinions. The entire bridge machinery system may be replaced, except for the main span support such as a swing bridge center pivot or bascule bridge trunnions. A 100 percent replacement of machinery may be performed, but this is extremely rare unless the moving bridge superstructure is being replaced as well. The difficulty of fitting new machinery parts to existing parts must always be kept in mind (see the following discussion).

Usually, in-kind replacement of more than just a few selected machinery components cannot be justified on the basis of deterioration. Advantages such as reduced maintenance requirements can be rationale enough to replace an old bridge drive consisting of many exposed gearsets, and babbitted bearings. Even this saving is not significant in situations where a maintenance person must visit the bridge regularly, which is usually the case. The chief advantage in reducing maintenance requirements is in reducing the skill level required of the maintenance personnel, as well-trained personnel are increasingly hard to find.

A single component, such as a rack pinion, can be replaced with a new one as long as the new component can be properly mated with the existing components or structure. This frequently involves careful field measurements, as the original manufacturing details of the component to be replaced may have been lost. Some of the dimensions of the new component may have to be revised from the original to fit a worn mating component. In the case of a new rack pinion, the teeth of the existing rack to remain are usually measured carefully, or a facsimile is made in the field, such as a plaster cast, which is measured in the shop to obtain the correct dimensions. As is often the case on bascule or swing bridges, the rack may be heavily worn but still have more than adequate strength for the loads it is expected to carry. The new pinion may then be made with thicker teeth than the original, so as to keep the backlash to a reasonable amount. The pinion also will be stronger and last longer. Care must be taken, however, as the wear on the existing rack teeth is usually not even, and some dressing of these teeth by careful grinding or filing may be required to provide proper tooth contact with the new mating pinion. Longevity of components may also be affected. For instance, if there is excessive wear or damage to an existing shaft journal, new friction bearings may not last as long as the original components. It may be necessary to rework existing components that are to remain in place but will be in contact with new components. In some cases, this will require removing components, such as shafting, to a shop to be built up, perhaps, and remachined to have the proper fit to a new part, such as a bearing bushing. As suggested above (page 300), this type of rework may not have great durability.

Most machinery components can be reworked or replaced while not under load, with very little effort. It may be necessary to block the movable span in the open or closed position while making repairs on or replacing drive machinery. Replacement or heavy repairs on certain components, such as center pivots or support rollers for swing bridges, bascule bridge trunnion bearings, or vertical lift bridge sheave bearings or shafts, will require substantial preparation, possibly including the installation of falsework or other temporary supports (see Figure 19-1).

Any partial replacement of machinery, particularly the main drive machinery, must be done carefully, as the interface between existing and new is an area of vulnerability because of the difficulty of making a fit that is as good as if all the components were new. The existing equipment that is not to be replaced may be somewhat worn, or these components may not have exactly the dimensions shown on existing, even if supposedly as-built, documents. The nature of bridge drive machinery, with equipment heavily loaded operating in one direction, then stopping and reversing direction, still being heavily loaded when running in the other direction, is unusual in industry, and the result can be failure at the interface between poorly fitted compo-

Figure 19-1 Vertical Lift Bridge under Repair. This bridge was found to have severely deteriorated counterweight sheave shafts. It was left in the raised position, supported on posts at each corner, while the counterweight ropes, sheaves, and shafts were replaced.

nents (see page 250). Gears on movable bridges are occasionally found to be loose on their shafts, and keys have actually come loose and worked their way out of their keyways and have been lost—all primarily due to this reversing loading. On some movable bridges, when trunnions or counterweight sheave shafts were replaced, the specified fit between the new shaft and the existing hole was not achieved, and a loose fit resulted. Work that is difficult to perform and inspect in the shop is even more difficult in the field. The practicality of performing major repairs in the field should be carefully considered before such work is undertaken. As mentioned previously, AASHTO, in its *Standard Specifications for Movable Highway Bridges,* calls for "engineering judgment" to be used in the rehabilitation of movable bridges (Division V); AREMA, in Section 7 of Chapter 15 of its manual, covers rehabilitation of existing structures, but says very little specifically about movable bridges. A rehabilitation that may be theoretically possible could, in the practical sense, be impossible to perform acceptably at a reasonable cost.

MACHINERY UPGRADES

On almost any older movable bridge, the gear reduction in the drive machinery can be replaced in groups. If this gearing is of the open type, it usually consists of one or more frames with gearsets mounted in them. Each of these groups of open gears can usually be replaced by a single enclosed gear reducer, with either parallel or right-angle shafts to match the existing arrangement. This increases the mechanical efficiency of that part of the bridge drive somewhat, and reduces maintenance requirements considerably. Enclosed drives also reduce the noise level of the machinery when operating, and make the area safer by reducing the amount of machinery that is exposed. In some cases, as with very large, low speed gearsets, it may be impossible to fit a standard enclosed gear reducer in the existing space and a special enclosed drive will have to be designed.

A typical bascule bridge leaf has one or three distinct groups of gear reductions that can be replaced with enclosed reducers in this way, with the separate groups connected by new floating shafts and couplings. The rack pinions can be mounted directly on the reducer output shafts, or mounted in their own frames and connected to the rest of the drive with flexible couplings, or arranged with a combination of the two. Waddell and other older span drive vertical lift bridges usually have a central gear reduction frame located inside the machinery house, on the moving span. This can be replaced by an enclosed parallel shaft gear reducer, with perhaps a horizontal right-angle input for connection to the direct drive engine. Connection is then made to the drum frames with floating shafts and couplings. The gear reduction on the drum frames, usually consisting of a single pinion at each mating to two drum drive gears, is usually left as it is or replaced in kind. Open gearing on a tower drive vertical lift bridge can usually be replaced in a manner similar to that used for bascule bridges. Older swing bridge turning machinery can usually be replaced with enclosed reducers and appropriate connecting shafting. One right angle reducer can be installed directly above each of the two rack pinions, with its output shaft connecting to an existing or new rack pinion shaft. A third, parallel shaft, reducer is then mounted between the two right angle reducers, and its output shafts connected to the input shafts of the two right angle reducers by means of floating shafts and couplings. Motors and brakes are connected directly to the input shaft of this reducer.

Specialized machinery for movable bridges sometimes requires more imaginative upgrades. Hopkins frame bascule drives can easily be replaced with hydraulic cylinder operation. Very old Scherzer rolling lift bascules operated by means of struts can also be retrofitted with hydraulic cylinder drives. These have sometimes worked well, and sometimes not. Low-speed, high-torque hydraulic motors have also been used to replace final drives for bridge machinery, particularly on bascule and swing bridges.

End lift and center wedge drives for swing bridges can be replaced with separate drives, hydraulic or electromechanical, at each component. This sometimes leads to difficulties resulting from a lack of absolute coordination between the operations of the components, so that the last end wedge to be driven home, for instance, may stall because of excessive load. End wedges have been known to gouge their seats when being driven in this manner. Usually, both end lifts at one end of a swing bridge are driven from a single source, which may be a parallel shaft or worm drive reducer replacing an old worm gear and spur gear drive, but this still allows one end of the swing bridge to be driven before the other. It is likely that the old drive had a single power source, at the center of the bridge, that was mechanically linked to the worm gear and spur gear reduction at each end by longitudinal shafting. This shafting performed well in coordinating the lift at each end, but was also a prime source of complaint because of the difficulty of maintenance. These shafts can be replaced "in kind," but by closed-gear type couplings and antifriction bearings, which greatly reduce maintenance requirements, as opposed to the old jaw type couplings and babbitt or cast iron bearings once used.

Other special machinery, such as span locks, end latches, and swing bridge centering devices, can have their drives replaced with various types of linear actuators, either specially built for the purpose or from standard industrial lines that can be modified for use in a movable bridge.

ELECTRICAL SYSTEMS

It is usually best, in considering replacement of the electrical system of a movable bridge, to provide a 100 percent replacement. Electrical components are more prone to deterioration with normal use, and to obsolescence, than any other part of a movable bridge, except possibly the deck. It is also desirable, even more than in other aspects of a movable bridge, to have a unified system for the controls. Interlocking is important, and by providing an entire new electrical system with a rehabilitated movable bridge, coordinated operation of the entire system can be more easily achieved. It is somewhat easier to troubleshoot the system during installation when all electrical components, including wiring and conduits, are new, than when a part of the system is older. Often, in the course of the life of a movable bridge, prior to rehabilitation, electricians may rewire parts of a system to maintain operability without documenting such work so that components are not connected the way they were originally intended to be. It is usually just as cost-effective to rip out the entire existing system and replace it with new as it is to trace all the wiring and reestablish the original circuits or determine the existing circuits.

Older electrical apparatus, particularly smaller components such as relays, switches, and the like, may appear to be perfectly functional, but are actually ready to fail instantly as a crack develops in old insulation or a metal pivot or spring fails in fatigue. The difficulty, expense, and uncertainty of confirming the suitability of existing equipment and projecting the amount of remaining life it may have can make replacement more economical. Engineering judgment is again necessary.

When contemplating bridge rehabilitation work, the bridge owner should consider the available options and attempt to maintain a flexible stance in contract administration. It may be practical to limit a given contract to a certain aspect of rehabilitation, such as for the electrical system, for superstructure rehabilitation, or for painting only. Rehabilitation contracts should have additional work built in so that a price is provided up front for additional work that may become necessary. This measure is particularly appropriate for superstructure and substructure work, which is most likely to turn out to be more extensive than anticipated.

COMPLETE REPLACEMENT

See Chapter 8 for a discussion of the complete replacement of a movable bridge.

20

OPERATING AND MAINTENANCE MANUALS AND BID DOCUMENTS

Operating and maintenance manuals are very important in ensuring the durability of movable bridges. Normally, these manuals are produced by the contractor supplying a new or rehabilitated movable bridge. Specific requirements for these manuals should be provided in the design of the bridge and spelled out in the bid documents. The manuals are expected to last the life of the bridge and should be put together accordingly. A heavy-duty permanent binding should be required. Although a three-ring binder is handy for making later revisions to a manual, it also can allow sheets to fall out, particularly after the holes in the paper have been damaged. Fold-out sheets are also prone to damage and should not be used. Separate bindings should be supplied for larger sheets, such as assembly drawings, when necessary, but preferably all information required to be included in the manuals should be presented on sheets of one size, which should be $8\frac{1}{2}$ in. $\times$ 11 in. or metric (SI) size A4.

OPERATING MANUALS

An operating manual should be written specifically for a particular project, describing exactly what is to be done to operate the bridge. All information should be supplied in a single volume. Use of the main drive and of the auxiliary or emergency drives should each be described separately, so that the reader is not required to go back and forth between sections of the manual. Troubleshooting sections should be included for each mode and step of operation, covering in detail the description of each type of failure that can occur, how to determine what the problem is, how to correct the problem, and what to do if it cannot be corrected.

Specific operating or user's instructions for separate components should be supplied in a separate appendix, for reference, but all information necessary for the operation of the bridge should be transcribed into the basic operating manual, rewritten as required to apply to the specific installation and purpose.

MAINTENANCE MANUALS

Maintenance manuals should be written specifically for a particular project, describing exactly what is to be done to maintain the bridge. Separate manuals should be provided for structural, machinery, and electrical maintenance, especially for a large and complex installation, for which a single maintenance manual would be extremely bulky. Compilation of this material can result in a considerable quantity of information, so the material included in the manuals should be as concise as possible. Extraneous material not relevant to the maintenance and repair of the components on the bridge being maintained should be avoided. Illustrations should be limited to those of value to maintenance personnel.

Maintenance information in maintenance manuals should be in verbal form to the greatest extent practical. These manuals should include a troubleshooting guide, in the most explicit terms possible.

All diagnostic testing that can be used to anticipate failures should be included in the maintenance manual, with recommended amounts of time between tests. The manual should include a list of all tools required for diagnosis, maintenance, and repair. The bridge owner should determine which of these tools are to be supplied with the new bridge and how and where they are to be stored. A list of spare parts should be provided, including the quantity of each to be kept on hand. The bridge owner should determine which of these parts are to be supplied with the new bridge, as well as how and where they are to be stored.

The maintenance manual should be tailored to the needs of the bridge owner. If the performance of preventive maintenance is desired, an extensive discussion on preventive maintenance procedures for all components of the bridge should be included. See further discussion in Section IV. In conjunction with maintenance manuals, maintenance charts should be called for, particularly for normal machinery maintenance. The charts should identify all locations, types, procedures for, and intervals for normal maintenance. They should be provided by the contractor in permanent form, mounted at locations convenient for use by maintenance personnel.

BID DOCUMENTS

Almost all new bridge construction, and most bridge rehabilitation, is accomplished by means of contracting. Bid documents are prepared by the owner's staff or by a consultant. These documents describe the work to be done and are released to potential bidders by advertising or other means. The documents contain plans that pictorialize the work to be done, specifications providing details of the work, and other documentation describing terms and conditions of the potential contract. The preparation work usually includes a cost estimate as well, which may or may not be released to bidders.

Plans

The plans provide a picture of the work to be done, showing all aspects of the work. They are usually prepared on drawings of approximately 24 in. × 36 in., but are almost always handled in "half-size" sets of 11 in. × 17 in. for convenience. It is usually the intent of the design to show all physical details of the work on the plans. Substructure and some other work may be performed directly from the plans, but steel and machinery fabrication is invariably required to be done via the mechanism of "shop drawings."

Many structural components are near duplicates of each other, particularly in a long structure, so the design plans show only the overall dimensions and critical details of superstructure components. The steel fabricator then prepares shop drawings that explicitly detail each piece that goes into the structure. Many shortcuts are taken in preparing these drawings, yet the components are completely detailed on them for fabrication. There may be 10, 20, or more shop drawings for each design drawing for a particular part of a structure.

Mechanical machinery components require a great deal of specific detail for fabrication. The design plans may provide sufficient detail to define each component, or may provide only performance requirements and leave the specific sizing and configuration of components to the contractor. Shop drawings are then prepared by the contractor or a subcontractor. A design drawing may show, at most, the overall dimensions of a component, such as a shaft, and provide information as to its fitup to mating components by means of a standard system of specifying fits, such as the ASME designation for fits for cylindrical parts. The shop drawings will then show each piece separately, with its overall dimensions, as well as the actual dimensions, with tolerances, of such critical areas as the diameter of a shaft at its end that is to fit into another part such as a bearing or a coupling hub. The toleranced interface dimensions are provided on the shop drawing, to four decimal places if the dimensions are in English units, indicating the appropriate maximum and minimum dimensions for the finished part. The designer or another competent party reviews the shop drawings for compatibility with the design and so indicates prior to fabrication of the components. Usually, the owner requires "approval" of the shop drawing by the entity reviewing them, but most consulting firms are reluctant to provide such an endorsement for liability reasons, as the amount of effort (money) allowed to review the drawings is usually severely limited by the owner. Typically, the shop drawings are stamped "Reviewed and found to be in general compliance" or with some such phrase, with a disclaimer as to responsibility for details. Checking of the details on the shop drawings should not be the responsibility of the reviewer, but rather of the contractor (who may delegate it to others, such as the shop preparing the drawings).

Mass-produced manufactured components that are to be used in the work are usually specified on the design plans by a generic description or by citing a particular manufacturer's item with model number and detailing the options to be selected. The contractor then procures a certified dimensional print from the manufacturer, which details the exact configuration of the component to be supplied, and submits copies of that print to the engineer in charge of the work for approval. This procedure is usually followed for most of the electrical components to be used on a project, as the electrical components are almost always standard products or variations of standard products. A few mechanical components, such as couplings, are also typically modifications of standard manufactured components. If only a sole source is acceptable for a manufactured component, this should be clearly stated in the special specifications (see below).

Specifications

Special specifications, or special provisions, are intended to supplement standard specifications. Generally, all critical technical details for the work are provided on the design plans—dimensions, materials to be used, and any limitations to be imposed on how the work is done, including the necessity for field measurements to determine the size or configuration of components. Any design details that cannot be conveniently included on the design plans are placed in the special specifications, a book of text that may run for hundreds of pages. Special material requirements may be included in the special specifications. The special specifications may also include details of suggested or required construction procedures. Typically, the details for payment of the work are spelled out in the special specifications. Because specifications tend to become lost after a project is completed, but drawings are usually retained, as much specific detailed information about components as possible should be shown on the drawings rather than in the special specifications.

Specifications also include the "boilerplate," the multitude of general requirements for executed contracts, including various governmental regulations requiring adherence to laws pertaining to such things as the use of domestic materials, adherence to labor standards, and affirmative action requirements. These generally apply only to projects that are funded entirely or in part with government money and are mostly applicable to highway projects. They can also include Amtrak and mass transit railroad work and bridge alterations under the Truman-Hobbs Act.

Tolerances

Somewhere in the bid documents it is necessary to describe the tolerances to be met in fabrication and erection of the entire structure. Most contractors and others in the bridge business are familiar with the tolerances used for fabrication and erection of fixed bridges, but these are quite crude in comparison to the standards used for movable bridges. A movable bridge is a large piece of precision machinery that happens to have a roadway surface on it. Some types of movable bridges, such as retractile pontoon types, may not require great precision in construction to be able to operate effectively. Other types, such as trunnion bascule bridges and rim bearing swing bridges, must be very accurately fabricated and erected to avoid difficulties in operation and perhaps premature failure. The superstructure of a bascule bridge, in the areas where the trunnions or rolling lift treads are mounted and where the rack and pinions are mounted, must be fitted to an accuracy of a few thousandths of an inch to avoid serious operating problems such as loosened or overstressed trunnions or unevenly worn rack and pinion teeth. These tolerances must be met at erection, because there is very little that can be done to correct errors after erection and large errors cannot be repaired without going to great expense.

Quality

As implied elsewhere, a movable bridge construction project can become quite contentious. It is imperative that a set of bid documents be as free of errors as possible. This is particularly true of the actual plans and special specifications, as a dispute over even a small, relatively simple, and inexpensive component can delay a project. There have been unquantifiable reams of material generated on the subjects of Quality Control and Quality Assurance in the past few decades, much of it of questionable value. To achieve the preparation of a suitable set of contract documents, it is important (1) to have the preparation of plans and specifications performed by people with adequate training who understand what they are supposed to be doing, (2) to have these people supervised by persons competent in the field of movable bridges, who direct and observe the work closely enough to have errors corrected before they cause cumulative problems, and (3) to have knowledgeable people review the work at various stages so that missing or incorrect information can be spotted.

CONSTRUCTION

The most important aspect in movable bridge construction, from the owner's point of view, is maintaining control of the work. Movable bridges are very unusual structures, and it is rare to find two movable bridges that are exactly alike. The contractor building the bridge, and possibly the engineer in charge of construction management and inspection, may be unfamiliar with the type of bridge being constructed or may never have been previously involved with a movable bridge at all. Very few owners have standard specifications or procedures that cover the details of construction of a movable bridge. Whether the construction involves building a completely new bridge or reconstructing an old one, it is extremely important to have a viable set of construction documents. This requirement is discussed in Chapter 20 of Part II, "Design." It is also, almost equally, important to have a competent resident engineer in charge of the construction, as discussed in Chapter 21.

21

CONSTRUCTION MANAGEMENT

Generally, construction of a new movable bridge or reconstruction of an old one is performed by a contractor who bids a lump sum price for the job based on contract documents, also called bid or tender documents. Some bridge owners may have the capability to perform work with their own personnel, but such work is usually limited to minor repairs. To ensure that the work is performed properly, a resident engineer is usually appointed to be in charge of the work at the site. The resident engineer has an office at the bridge site, unless the project is a small one not justifying the expense of a full-time resident. The resident engineer has a staff of inspectors and other personnel as required to fit the needs of the project. The resident and inspectors may be employees of the owner, but are usually provided under contract, generally by a consulting engineering firm, which in many cases is not the same consultant that designed the project.

There are several aspects of construction management for a movable bridge project, and several independent entities may be involved, so coordination of these different functions becomes a formidable management task in itself. For a typical movable bridge construction project in the United States, the following entities are major players:

The contractor, who actually builds the bridge.

The owner, who is usually represented by an engineer-in-charge (EIC). This person, or one of his superiors, is referred to as "the Engineer," the person having ultimate authority over the project.

The resident engineer, who usually employs the on-site inspectors, coordinates shop inspection of components, and verifies completion of the work by the contractor, on an incremental basis.

The designer, usually a consulting engineering firm, responsible for providing design drawings and other documents on which the construction of the bridge is based. The designer is frequently also responsible for verifying that the components of the bridge are being detailed according to the design intent. This verification may be made by a consulting firm other than the designer, as a check on the appropriateness of the design. In some cases a second consulting firm will perform what is called a *peer review* to check the adequacy of the design.

If a movable bridge is a highway bridge, the Department of Transportation of the state within which the project is being constructed. A Department of Transportation (DOT) has published standards for bridge construction, which may be mandatory for the work, and usually has material specialists who perform or supervise the more specialized aspects of shop inspection. The state may also, even when not the bridge owner, have a role in approving payment to the contractor for the work if state or federal funds are involved.

None of these entities necessarily have the same goals in commencing a bridge construction project. The contractor, the resident engineer, and the designer are all primarily interested in making a profit so that they can stay in business. The owner is interested in minimizing the disruption to traffic caused by the project and is usually interested in completing the project on time at the least cost. The state Department of Transportation, if involved, and the designer are interested in a properly constructed, successful project, so as to avoid damage to their reputations. Some of these goals can be mutually exclusive, and others can be shared to a greater or lesser degree by some or all the entities involved in the project. It can be greatly beneficial to have all, or at least most, of the entities on a project working together toward the same goals. A number of techniques have been used to accomplish this end, ranging from "partnering" to "design build" and other concepts, with varying degrees of success.

One of the ways in which the resident engineer controls the work is by comparing what the contractor produces with the design plans that are part of the bid documents. In addition to design plans, shop drawings are produced that show the fabricated components of the bridge in more detail than is present on the design plans. On occasion, design changes are made during the course of the work. When modifications are to be made to bridge components during construction someone, usually the designer, produces a drawing detailing the change. The contractor may produce revised shop drawings showing the changes; these are reviewed and approved, usually by the designer and the EIC. The effected changes are then noted in a set of design plans, called the "record set," or "as builts," which incorporates these and all other changes made in the course of construction.

It is not uncommon, during the course of a construction project, for a dispute to arise over what constitutes the contractor's scope of work. The resident engineer may request a change when it appears that a particular detail shown on the plans is not suitable. The contractor may decide not to perform a particular item of work because his or her interpretation of the intent of the plans, when the contractor bid the job, turned out to be in variance with the designer's, and the contractor believes the work per the designer's interpretation is impractical or impossible. In these cases, the bridge owner, or the entity actually contracting the work, must decide what is reasonable.

Most contracts contain a clause stating something to the effect that the contractor is responsible for reviewing the contract documents before submitting a bid and must make note or take exception to any part of or inclusion in the bid documents that he or she feels is unworkable. Unfortunately, many owners do not seriously restrict bidding on bridge projects, except perhaps for some vague determination of responsible bidders, and the lowest bidder is usually the preferred awardee. A bidder may misinterpret what work is required by the bid documents. Many contractors have learned that the most important part of the bidding process is developing the highest possible bid that will give them the job, that is, the highest possible bid that will be the lowest of the acceptable submitted bids. With this approach, any apparent discrepancies or undesirable items in the bid documents are typically ignored until the time comes to perform that work. Then the contractor attempts to make an argument that that part of the bid documents was misleading or otherwise attempts to show that he or she deserves redress. On many projects, the most competent, or at least the most energetic, staff is working for the contractor, so arguments are won often enough by the contractor that this method of gaining an increase in the contract price is often attempted.

To avoid, as much as possible, any litigation in the course of the work, the following guidelines are suggested:

Have a clear-cut scope of work in the contract documents, with competently produced details in the plans and specifications.

Establish a preapproval process for awarding contracts, so that the capability of a bidder to perform the work is ensured.

Have a schedule and sequence of the items of work to which the contractor's adherence is mandatory.

Require a contractor to perform substantially more than half the work with his or her own employees.

Include a clear statement in the contract bid documents, overruling any other provisions to the contrary, specifying that the contractor is to complete the project as intended in the contract documents and is to obtain clarification of any clauses that he or she finds or suspects of being misleading or questionable prior to submitting a bid.

Have a dispute settlement procedure in place, as part of the construction contract, to ensure that any differences of opinion as to the meaning of the contract documents are resolved quickly, so as not to affect the contract schedule.

Provide for changed conditions or unforeseen events so that changes can be made in the contract when necessary and can be implemented quickly and economically and avoid delays in completion of the work.

Before allowing the contractor to begin work, require a revision of price of all bid items that are substantially over or under the engineer's estimated cost. Care must be taken, however, to avoid recriminations from other bidders. The overall amount of the bid should not be changed.

Many of the aforementioned provisions are included in the contract documents, but they may not be rigorously enforced or may not be invoked until the project has fallen behind schedule. The chief duty of the resident engineer in charge of the project, after ensuring that the work is being done according to the contract documents, is to make sure that the project is proceeding on schedule. Often, neither of these functions is performed adequately, perhaps because of their being too difficult on a particular project, resulting in a project that is in trouble. The owner should make certain, by keeping up to date on the project's progress, that these difficulties do not arise. Frequent meetings, held on a regular basis, with all the principals of the project in attendance, can help to ensure that the project is completed on time and within budget.

22

STRUCTURE

SUBSTRUCTURE

Movable bridge piers are more heavily loaded than piers for fixed bridges of the same span length. The loading is more variable than for fixed bridges, because of the repositioning of the moving leaf itself, changes in support in different modes, and special situations such as high winds on open bascule spans. Most common types of movable bridges, except for symmetrical swing bridges, have heavy counterweights that concentrate a large mass in a small area. On vertical lift bridges, the counterweight remains in the same location, except for moving up and down, so that the effect on the foundations is minimal. On bascule bridges, the spans and counterweights also translate longitudinally to some extent, producing a variable loading on the foundations. This movement is particularly pronounced with Scherzer type rolling lift bridges, in which the entire mass of the moving leaf, counterweight and main span, can shift back and forth as much as several feet during opening and closing.

It is imperative, during the substructure construction process, that the design plans are not compromised. In many cases, with fixed bridges, shortcuts can be taken and the built-in conservatism of an AASHTO or AREMA design can cover for construction defects. Construction errors that are serious enough to be easily detectable may still be insignificant for a fixed bridge. This is not usually the case for a movable bridge, where substructure deficiencies have been known to produce bridges with very short lives.

Accurate placement of the substructure of a movable bridge is more important than it is in the case of a fixed bridge. Crude corrections in elevation, or in lateral or longitudinal position, can be accommodated in a fixed bridge substructure but can cause operational difficulties in a movable bridge, particularly a bascule or vertical lift span. It is very helpful in construction of a movable bridge substructure to be able to make final adjustments as erection is completed. Final pours at the tops of piers, and installation of anchor bolts, should wait for confirmation of correct dimensions, preferably after superstructure fabrication has progressed to the point that dimensions to the final roadway elevation and lateral position can be confirmed.

SUPERSTRUCTURE

Fabrication

The superstructures of movable bridges are almost invariably built of steel. For even a small movable bridge, the fabricated components are rather large, and a finished bascule leaf or swing or vertical lift span can be enormous. It is generally preferred to build the entire superstructure unit in the shop and deliver it to the site as a single piece. This is primarily to ensure the necessary high quality of fabrication of the movable bridge leaf or span, which is essentially a large component of a giant machine. Shop fabrication can also be a valid economy move, as on-site labor may be expensive. Moreover, site conditions may not be conducive to the performance of precision work or to high productivity. Generally, the only restriction on shop fabrication is the ability to ship the finished product to the site. Often, this restriction is avoided, as the entire superstructure may be fabricated in the shop and then match marked and knocked down for shipment to the site and reassembly. There are shops capable of fabricating large movable bridge superstructures in many parts of the country. Again, however, the difficulty is often in the capability of shipping the bridge to the site. The superstructure for the Columbus Drive Bridge, a very large double leaf bascule bridge over the Chicago River in downtown Chicago, was fabricated in large sections at Ambridge, Pennsylvania, near Pittsburgh, and loaded on barges and delivered to the site. Proper preparation for shipment is essential for such large assemblies so as to avoid possible overstresses during transit or in loading or unloading. For another movable bridge project, the contractor intended to build an entire swing bridge in Alabama and ship it complete to New York for erection, until it was discovered that there was a restriction in the waterway that prevented it.

Because of the large and frequent stress variations in many major structural components of typical movable bridges, care must be taken in fabrication to avoid stress risers in details, particularly in connections. After the fatigue failures of many bridge superstructure components in the latter part of the twentieth century, AASHTO and AREMA have been more particular and have limited certain details to very low stresses. Even when lower allowable stresses are used in design, failure to follow mandated fabrication practices can result in the rapid failure of components in service. This is particularly true of welded connection details, which can have fatal defects without exhibiting obvious external flaws. For instance, defective complete penetration welds in the hubs of the box counterweight girders for a heel trunnion bascule were detected when the machinery fabricator placed the trunnions inside the girder hubs. The trunnions had been cooled by immersion in liquid nitrogen in order to easily shrink them sufficiently to make the interference fit required in the assembly. The rapid cooling of the hubs, as they came in contact with the cold trunnions, caused large radial tensile stresses at the hub welds, and cracks developed and instantaneously propagated from large flaws in the welds.

A bascule bridge, because it rotates in a vertical plane and is quite susceptible to varying stresses due to gravity, is the type of movable bridge for which it is most necessary to be careful to avoid having built-up stresses in the fabricated superstructure. Swing bridges are susceptible to a much lesser degree, as they have only minor operating stress changes, resulting from actuation of end lift devices. Vertical lift and pontoon bridge spans, because they do not undergo significant stress changes during operation, can be fabricated and erected as if they were fixed bridges, with little concern for decreased durability. This qualification applies, of course, to the movable bridge span itself, but not to the operating machinery and supports of any movable bridge, or to the towers, sheaves, shafts, and bearings of a vertical lift bridge.

Erection

Bascule Bridges The erection of the superstructure is more critical in a bascule bridge than in any other common type of movable bridge, as each bascule leaf is supported at one end on two or more bearings and must regularly rotate upward and downward on them without binding. Whether the bridge is a trunnion or a rolling lift type, error in erection can result in serious operating difficulties from the start. AASHTO (4.1.3) and AREMA (15.6.9.2a) say "utmost accuracy" for trunnion and similar alignments. All components must be fabricated to very tight tolerances, and erection must be accomplished by highly skilled workers who are actively and competently supervised. The use of highly skilled labor, such as millwrights as opposed to ironworkers, on at least some of the field operations for the erection of the superstructure is warranted, in order to minimize construction errors at many points. Use of highly skilled labor can be particularly productive in the erection of the bearings supporting the superstructure and in other critical operations such as the mounting of racks and installation of rack pinions.

The placement of the trunnion bascule leaf on its trunnion bearings is also a critical operation. The successful operation of the bridge over its lifetime depends on the trunnions and their bearings being properly aligned with and attached to their connecting structural members. This operation is complicated by the heavy loads involved, which cause these components to deflect as they are loaded. To make matters worse, the stiffness of a bascule bridge leaf can vary considerably, depending on the orientation of the leaf in relation to gravity. A bascule leaf may be very stiffly resistant to gravitational loading in the open position, but can be quite flexible in response to the same loads when closed. Bearing support is particularly difficult with Chicago-type simple trunnion bascule bridge leaves, which are often supported on complicated structures consisting of beams and columns, some of which extend through openings in the bascule leaf itself and have considerable dead load deflection.

A heel trunnion bascule bridge requires the greatest precision of erection of any common movable bridge type. Two separate large structures, the bascule leaf and the counterweight frame, must be erected so that the pivot axes are exactly parallel and the correct distance apart. Usually, the main trunnion bearings for the bascule leaf are mounted on the lower end of the forward legs of the counterweight support A-frame, so that the distance from the centerline of the main trunnion bearing to the centerline of the counterweight trunnion bearing can be more readily controlled, but this is not always the case.

A heel trunnion bascule is simpler to erect than a simple trunnion, in one sense, because the overall structure is divided into two separate main components, the bridge leaf itself and the counterweight, but these two components must each be accurately erected. The connections between these structures, the links, must then be accurately and competently erected so that a perfect parallelogram is formed in each plane of the main bridge members, and the parallelograms must exactly match. Accuracy of fabrication and erection is essential to good operation of the bridge.

Regardless of the type of trunnion bascule, superstructure erection requires the same iterative process:

(1) The deflection of the bascule leaf, or counterweight superstructure, and of the trunnions and the trunnion supports must be determined beforehand.
(2) Compensation is made for these deflections, and the bearings are temporarily set.
(3) The superstructure is erected.
(4) The alignment is checked in the open and closed positions and at a few intermediate points.
(5) Adjustments are made to correct misalignment.
(6) The alignment is checked in the open and closed positions and at a few intermediate points. Steps 5 and 6 are repeated until the alignment is within specification in all positions, under dead and live load.
(7) Support components are fixed in their final position and connections are completed.
(8) Correct alignment is confirmed.

The Scherzer-type rolling lift bridge does not require the same degree of accuracy of field fitting as a trunnion bascule, but it must have properly fabricated components, particularly the segmental girders and tread plates supporting the bascule leaf. Poorly fitted segmental girders and tread plates can result in rapid deterioration of a bridge that operates with any degree of regularity. The tread plates must be accurately aligned in erection or the bridge will not maintain its alignment as it raises and lowers, resulting in difficulty in seating, misalignment when seated, and rapid wear of the tread plates, centering devices, and drive machinery.

Typically, the bascule girders are fabricated completely in the shop, with the curved track and the rack pinion bearing mounted on each girder. It is desirable, but not always convenient, to assemble the entire superstructure in the shop, making certain of the correct alignment of the components. If it cannot be shipped to the site as a unit, the superstructure is match marked and disassembled for shipment. When the components are reassembled in the field, even if the bascule superstructure is shipped to the site as a unit, all dimensions and finishes should be rechecked. It should not be assumed that the parts of the bridge can merely be reassembled and that smooth and true operation will result. Parts can be damaged in shipment, and misalignment can result—slight enough to be unnoticeable to the naked eye, yet detrimental to operation.

For the most part, the deck of a bascule bridge is no more difficult to erect than that of a fixed span, but care must be taken when operating a partially completed bascule bridge so that deck components are not dislodged. Extreme care must be taken, however, in erecting the deck joints. Particularly at the heel joints, the components must be erected exactly as designed so as to avoid interferences or large gaps. At the toe joints, it is best to allow for some adjustment at erection, to provide for errors that may occur during erection.

Vertical Lift Bridges A new vertical lift span, whether for a highway or a railroad, is in many cases a replacement for an existing bridge. The normal problems of bridge construction are thus further complicated by the need to maintain marine traffic, and perhaps also vehicular or railroad traffic, in the construction zone. The more restrictions that can be put on traffic, the faster and cheaper the construction project can be. If traffic must be maintained on the bridge during construction, the existing bridge must remain in an operable condition while the new bridge is erected unless marine traffic can be stopped or detoured. If the new bridge is to be on the same alignment as the old bridge, the complications become much more difficult to surmount, but the use of a vertical lift span for the new bridge, rather than another type of movable bridge, makes it easier. Several different schemes have been used such as the following:

Building the new movable span off-site and floating it into position at the required time

Building the new span in place in the raised position

Building the new span at a different channel location than the existing movable bridge.

The vertical lift span itself, because it is a closer approximation of a fixed span than any other common type of movable bridge, presents fewer difficulties in fabrication and erection. The bridge towers, however, must be carefully erected so that they are plumb, as misalignment may make it difficult or impossible for the bridge to raise and lower. This applies as well to the counterweights, although the counterweight guides can be adjusted to a small degree if necessary to eliminate binding.

The counterweight sheaves, shafts, and bearings for a vertical lift span are critical. These are extremely heavily loaded, so misalignment or poor fit will show up quickly in operational difficulties and rapid wear. The ropes supporting the vertical lift span and counterweights on the sheaves must be made to accurate lengths and installed to equal tensions to spread the load equally over all the ropes, eliminate overloading, and prevent excessive wear at a few of the ropes. Typically, these ropes are prestretched and accurately measured in the shop or factory.

It is important to properly align the ends of the bridge decks at erection. As the approaches and movable span of a vertical lift bridge are, in many cases, of different types of construction, it is sensible to allow for adjustment of surface elevation on the lifting span, approach deck, or both. The vertical lift span is fairly flexible in torsion along its longitudinal axis, so it will adjust itself to seat firmly on the live load shoes if the counterweights are not attached. After the counterweights are in action, proper balancing is required to ensure good seating. See Chapter 24 for further details on balancing vertical lift bridges.

Swing Bridges There are few new swing bridges built, and these can be of vastly different configurations—recently, there have been concrete deck girder, steel deck truss, and steel through-truss swing bridges built in the United States. It is almost useless to generalize about superstructure erection. For economy of construction, it is worthwhile to consider remote fabrication of the swing span, provided that it can be floated to the site in one or a few pieces. A swing span is somewhat more difficult to position than a bascule or vertical lift span because of its double-cantilever form and the presence of a center pier on which the span is set. The difficulty can be overcome by using a large floating crane, but this is very expensive. A much less expensive alternative is to design the bridge to act as a simple span, for most of its full length, under dead load and environmental loading conditions. The completed superstructure is placed on two barges, one under each normally cantilevered arm, and floated into position straddling the center pier. With allowable overstresses for construction, there may be very little or no need to increase the strength of members that will be more heavily loaded during the float-in operation.

A swing bridge is easier to erect in place without disturbing navigation than other movable bridges, because it can be erected on its permanent support, in place in the open position, without much difficulty. The fender system, which is usually needed to protect the completed bridge in the open position, can be used during construction as a base for falsework to support the span under erection. Of course, a new swing bridge cannot be erected in place on the same alignment to replace an existing bridge unless traffic, both road and marine, can be disrupted for a considerable length of time.

Parts of a swing span superstructure go through stress reversal, from tension to compression and vice versa, each time the bridge is opened and closed, and these components must be carefully fabricated and erected to avoid deterioration at the connections due to these load reversals. This is especially true at the lower chords of steel trusses, in the length between the end bearings of the swing span and about a third of the distance to the center pier. If the main bridge members are of bolted or riveted construction, they must be accurately fitted to avoid deterioration at the joints due to reversal of loads.

The dead load camber of a swing bridge is seldom exactly what it was intended to be to a small fraction of an inch. For this reason, it is best to allow for placement of the ends of the deck of the swing span after the main bridge members are in place, so that they can be set to the correct elevation. Alternately, if the actual end of a span elevation is not critical, the swing span can be erected and the approaches built with elevations adjusted to match. It is important to make sure that the amount of lift to be provided at each end, so that the roadway elevation is correct, is adequate to prevent lifting of the end of the swing span when live load is on the other arm, under any load and temperature conditions. This is usually not difficult in a long bridge, but a short swing span may have such stiffness that the amount of end lift required places too great a strain on the end lift devices. In these cases, it is important to erect the bridge accurately so that no error in elevation is to be corrected by adjustment of the end lifts.

Superstructure—General

Movable bridges are designed and intended to be moved about, as opposed to a fixed bridge, which is usually erected in place and never changes from that position throughout its life, except for small deviations due to temperature changes, high winds, or heavy live loads. When a movable bridge moves, or is prepared to move, it can experience significant stress changes in its main members and some secondary members. These changes may actually take the form of stress reversals. Under such reversing load, the connections between structural members can become loose if the members are not very carefully assembled. Bolted or riveted parts must have the holes reamed to close tolerances, and the bolts and rivets must be inserted without damaging the bolts or holes. High-strength bolts must be fully tensioned, with the parts being connected in intimate contact. Welded connections must be carefully made to avoid internal stresses, which, when added to stress changes caused by repositioning of the bridge during operation, may result in overstress of components. Stress relieving may be essential for certain components.

Movable bridges, regardless of type, are almost always designed to be balanced structures. The balance state of the finished bridge is more critical for some types of movable bridges than for others, but a significant difference between design and erected weight is a problem for any type. Thickness tolerances for rolled steel plate and shapes are fairly liberal, and it is to be expected that some variation in finished weight will result. Movable bridges almost invariably end up being heavier than intended, however, which would not seem to be the expected outcome if only random variations from nominal metal thicknesses occurred. Shop drawings and billed weight should be carefully checked in the course of the construction process so that weight deviations can be detected as soon as possible and counterweight changes or other corrections can be made economically.

The finished details of the counterweights, the exact size and weight and particularly the density of the concrete, if used, should be left to be decided in the field. The constructing entity should be responsible for maintaining an exact log of the weight of the bridge components and their position. From this log, a spreadsheet can be developed that sums the moments of all components so that the total mass and desired center of gravity of the counterweight can be determined (see Figure 11-17). This is a valuable tool for erection of all types of movable bridges that require counterweights. It is preferable that the final desired counterweight mass is determined only after all other material has been erected. The volume of the counterweight, if of the steel box type, should then be known exactly, and it is then a matter of determining the exact density that is required of the material that will fill it. Pours of different densities can be made at different levels or locations within the counterweight to adjust the location of the center of gravity. Test blocks should be made, if the counterweight fill is concrete, and their densities determined after the material has cured. When the desired density has been accomplished, then the counterweight can be poured. Use of other counterweight materials, such as metal blocks, will cause a slight shift in the location of the center of gravity of the counterweight as they are added or removed to accomplish the correct balance. This factor must be taken into consideration in finalizing the balance.

Regardless of how well the counterweight is placed, provision must be made for adjusting of the balance of the bridge. Pockets are usually designed into the counterweight for placement of small portable weights. These are usually adequate for the final fine-tuning of the balance at the completion of the construction project. Normally, the pockets are designed so that each is left half full when the project is completed and the weights of all constructed components of the moving span are precisely as intended in the design. The blocks in the counterweight should be carefully placed, neatly stacked, so that they will not shift during bridge operation. The external openings to the pockets should be closed and sealed so that water or other objectionable material will not enter them and cause deterioration or a change in the balance of the bridge.

23

FABRICATION AND INSTALLATION OF MACHINERY AND ELECTRICAL COMPONENTS

MACHINERY FABRICATION

Some machinery components for movable bridges are large enough in themselves to require careful planning of fabrication and shipment to the site. Swing bridge rack and track segments are usually manufactured and fitted together in the shop to ensure proper dimensions and fit, then match marked and disassembled for shipment to the site. Rolling lift bascule treads are usually machined and fitted to the segmental girders and support girders in the shop, then shipped to the site fully assembled on them, in spite of the difficulties that may be involved with transport of such large assemblies. Vertical lift bridge sheaves are almost always fitted to their shafts in the shop. This is done whether the shaft and sheave are fixed to each other or the sheave rotates on bearings on the shaft, in which case the bearings are naturally mounted as well. Sometimes, when the shaft is fixed to the sheave, the bearings are mounted on the shaft in the shop.

It is generally considered desirable to assemble all the machinery in the shop, mounting it on the bridge superstructure where possible. Match marking and disassembling for shipment can be done when necessary. In many cases, the machinery can be shipped assembled on the bridge superstructure, but care must be taken that damage in shipment does not occur.

Movable bridge machinery, because of its special nature, is generally fabricated by a firm specializing in the business. The machinery is usually one-of-a-kind, designed and built for a particular application. This applies even to enclosed drive gearboxes, which usually require special gear ratios, special shaft extensions, special lubrication provisions, and sometimes special gear construction or specific nonstandard distances between the input and output shafts. There are a few standard manufactured machinery products that have application to movable bridges; these include brakes, some forms of lock bar operators for bascule bridges, and not much else in the electromechanical gear drive category. This is one reason for the popularity of hydraulic systems in movable bridge drives, as a wide range of standard, if not quite off-the-shelf, hydraulic equipment can be applied to movable bridges. While gears might be considered standard items, each gear is usually detailed and fabricated to suit its particular application. Most modern gears are hobbed, hence the diametrical pitch designation. Some crude gears, such as movable bridge racks and pinions, can be made per circular pitch practice, so that a gear tooth pattern, for cast teeth, can be used for gears with a range of several teeth. Similarly a cutter, for cut teeth, can be used for gears with a range of several teeth. For instance, a cutter for 2″ circular pitch might be good for cutting gears with 18, 19, 20 or 21 teeth. Using this cutter for a 17 tooth or 22 tooth gear would produce too-large errors in tooth shape.

Movable bridge machinery can, to some extent, be built to less exacting standards than the criteria required for other machinery applications. Movable bridge machinery moves more slowly than most other types of machinery of similar configurations in manufacturing, mining, and comparable applications. The duty cycle for the machinery driving a movable bridge is very low, as compared with the machinery in a power plant or industrial application. A typical movable bridge may be in operation only a few minutes of a day or may sit for weeks without operating. Lower levels of manufacturing quality may not be noticed as quickly in a movable bridge drive as in the machinery for another application. There are some aspects of the use of movable bridge machinery, however, that contradict this conclusion and have resulted, in some cases, in the failure of relatively new machinery. Movable bridge drives are reversing applications. The bridge is driven open with the machinery loaded in one direction, and then it is driven closed with the machinery loaded in the other direction. This is very difficult service for couplings, gears, and other shaft connections, and it can cause them to come loose if they are not properly constructed or fitted, or not designed for this type of application.

The long periods of idleness experienced by machinery on movable bridges that seldom open can also cause difficulties. Condensation can occur inside machinery housings, resulting in corrosion of finished machinery surfaces and contamination of lubricants. Proper detailing of baffles, provision for the right type of breather caps, and good distribution of lubricant can prevent such deterioration.

ELECTRICAL COMPONENT FABRICATION

Proper integration of the electrical apparatus on a movable bridge is critical to providing reliable operation with a durable electrical system. AASHTO and AREMA state that when "practical" or "practicable" all major electrical components be furnished by the same manufacturer. This suggestion is not included in 2000 LRFD AASHTO, perhaps at least partly because it is more difficult to obtain competitive bid prices for this equipment with such a requirement, given the limited number of manufacturers that could provide all the required components. At this time (2003), it appears that no single manufacturer can produce, within its own facilities, all the major apparatus for any movable bridge electrical system, so the suggestion becomes inoperative by default. Nevertheless, the extreme complexity and sophistication of the electrical portion of a movable bridge forces the greatest practical care to be used in coordinating the supplying, installing, and making operative the electrical system.

One way to help minimize the difficulty in the electrical part of the work is to have a single qualified supplier provide all the apparatus, whether it manufactures each component itself or not, and require this supplier to guarantee the compatibility of the components it supplies, as well as meeting all other requirements of the specifications such as durability, environmental compatibility, and so forth. This should not result in a difficulty with competitiveness, as several suppliers should be available for any movable bridge electrical project, and there are enough manufacturers of each of the various components that single sourcing should not be, theoretically, a necessity. There may be a few electrical components for movable bridges that only have one preferred manufacturer, but this still does not exclude competitive bidding for the installation.

Some of the simpler electrical components can cause serious operational problems with movable bridges. The wiring terminals inside electrical cabinets, if made of poor materials that corrode, come loose, or break easily, or if not properly connected, either at the wire or at the terminal, can quickly fail and prevent proper operation of the bridge. It is important during erection of the bridge that the materials approved for the project are actually used and that no unauthorized substitutions are made.

Proper testing of electrical apparatus at the fabrication shop is an effective means of ensuring an acceptable finished product. It is much easier to correct defects in the shop than in the field. It is best to test the entire electrical system as a unit, as so many components must interact properly for the desired bridge operation. Such testing may require shipping control and power components from several manufacturing locations to one testing site. Care must be taken to avoid damage in shipping. The expediency of performance testing is another good reason to have a single supplier responsible for all electrical components. It has often happened that an electrical package for a movable bridge has been fabricated and shop tested and found acceptable, but has then failed in the field. Much time is lost in argument, and sometimes in litigation, over who is to make a field installation operable when it worked in the shop and then did not work in the field. The point at which responsibility for various components changes from the fabricator to the erector should be clearly spelled out if these entities are not to be one and the same.

MACHINERY INSTALLATION

Most new movable bridge machinery is required to be tested at the fabrication shop, and again in the field, before the bridge is accepted. The requirements of this testing are usually detailed in the special provisions for the construction contract. Some testing may be quite rigorous to make certain, for instance, that enclosed gear drives are capable of performing as required. AASHTO (4.1.2) and AREMA (15.6.9.1) both give special attention to the installation of machinery.

Bascule Bridges

Before commencing erection of the machinery of a bascule bridge, it is essential that the superstructure and main support bearing alignment have been checked and confirmed to be within specifications. With the bridge superstructure properly erected, it is a relatively simple matter to erect the machinery. Each type of bascule bridge that is currently likely to be used in new bridge construction has its own particular configuration of drive machinery, plus special variations of ancillary machinery such as span locks.

Simple Trunnion The drive machinery for a simple trunnion bascule is almost always mounted on the pier. If the bridge is a conventional rack pinion drive, electromechanical or hydraulic, the racks are usually mounted on the main girders and driven by rack pinions mounted on the pier with the other drive machinery. Proper installation and alignment of the drive machinery are expedited if the bascule leaf is mounted and aligned on its trunnion bearings first and the machinery is then installed. Precision and rigidity are essential to provide long-term trouble-free operation. The racks must be mounted on the bascule girders so that the pitch radii are correctly located to the center of rotation of the bascule leaf. Racks are often mounted in a structural shop where the necessary precision involved may be little understood. There is less chance of mounting errors if the racks, trunnions, and other machinery components can be mounted to the superstructure in the machinery shop. The faces of the rack teeth should be parallel to the axis of rotation of the bridge. All of these alignments should be exact, to the limits of accuracy of the precision measuring equipment used. It is frequently impossible on a trunnion bascule bridge to make direct measurements to check the alignment of major components, because columns supporting trunnions are in the way or the trunnion bearings themselves are outboard of the bascule girders and several feet distant from the flanges of the girders. Generally, before machinery installation, the bridge leaves are fully erected (with trunnions and bearings) and aligned, with deck and counterweight installed. The installation and alignment process for the drive machinery of a typical simple trunnion bascule bridge follows a certain sequence (*Exception:* Racks and trunnions may have been installed on the bascule girders in the shop):

1. Mount racks onto bascule girders or rack girders, taking care to align properly.
2. Mount rack pinions, in bearings on bearing frames, onto the bascule pier, with rack teeth properly meshed with pinion teeth and the pinion shafts on the same axis.
3. Mount gear frames or enclosed gear reducer drives with output shafts in proper alignment with rack pinion shafts.

4. Connect secondary reducer or gear frame output shafts to pinion shafts with full-flex gear-type couplings or with floating shafts.
5. Mount primary gear frames or reducers, if used, with output shafts aligned with input shafts of reducers or gear frames mounted in step 3.
6. Connect primary reducer output shafts to secondary reducer input shafts, usually with floating shafts.

Rolling Lift The first rolling lift bridges designed by William and Albert Scherzer were operated by means of struts connected via pins to the bridge superstructure. The drive machinery propelling the strut was mounted on the pier. The bulk of the operating machinery for the modern Scherzer type rolling lift bridge is almost always mounted on the bascule span itself. Typically, horizontal racks are mounted on frames rigidly attached to the pier, one rack on each side of the bascule bridge superstructure. The drive machinery, including motors, brakes, gear reduction, and main drive pinions, is mounted on the moving leaf, with the centerlines of the main drive pinion shafts mounted exactly on the center of roll of the leaf, at the center of radius of the curved track segments. One pinion extends out each side of the superstructure to mate with one of the racks. It is critical for the long-term, reliable operation of the bridge that the rolling lift tracks and the racks and pinions be precisely fitted, by being accurately machined and properly assembled for smooth and true operation. Delicate parts can be damaged during erection, such as by using a motor shaft as a lifting eye in installing a machinery frame. Reassembly of components, even if turned bolts are used, is not guaranteed to be precise without checking.

Heel Trunnion The erection of machinery for a heel trunnion bridge does not require great precision, as long as the carrier (see page 391) is correctly aligned with the operating strut. The two rack pinion shafts should be coaxial, as they are for simple trunnion machinery.

Hopkins Frame The Hopkins frame bascule can make the field erection of a bridge easier by mitigating the need for precision work in the alignment field. The Hopkins frame is connected by links to the axis of span rotation, so that it is less difficult to align the pinions to the racks. The alignment of the racks, when mounted to the bascule leaf, is still required to be precise, as deviation in the radius of the rack to the pivot axis is not corrected by the use of the Hopkins frame. The link connection points at the racks must be aligned with the bridge trunnions. The Hopkins frame must be accurately aligned when connected to the bascule pier, or severe bending stresses may develop in the frame during operation.

Almost all of the machinery is mounted on the Hopkins frame itself in the shop, where precise alignment is not as difficult. The Hopkins frame, with the machinery mounted on it, is then taken to the field and erected on the bridge. As long as no damage occurs in transit or erection, the machinery alignment is then adequate. A field check of the machinery should be made before it is put in operation, to make certain that no damage, such as bending of the shafts, has occurred during transportation or erection.

Vertical Lift Bridges

Machinery for vertical lift bridges is very difficult to install, because so much of it must be installed and properly aligned on top of the towers. Moreover, the counterweight rope sheaves are very heavy—among the heaviest of movable bridge machinery components—but they must be mounted in proper alignment to avoid operating difficulties and premature wear of the ropes. As the towers are very tall, they will deflect significantly as the dead load of the vertical lift span and counterweights, as well as the counterweight sheaves, shafts, ropes, and bearings, is added to them. If the main members of the supporting tower are symmetrically placed, equal deflection should occur and no compensation should be required in the placement of the sheaves and operating machinery. If the main members of the towers are in any way not symmetrical, the probable relative deflection of the members should be checked as they are loaded, and compensation made during the setting of the counterweight sheave bearings and the drive machinery so that all components will be properly level after installation is complete. This should be an iterative process similar to that described on page 320. Some older vertical lift bridges had the counterweight sheaves supported on what amounted to single columns at the fronts of the towers. Adjustment was provided in the connections to the rear legs, which acted only as bracing and did not directly support the load. It is unusual to find this type of adjustment capability in a new vertical lift bridge, but the single forward support columns are also unusual in new designs.

Once the counterweight sheaves have been mounted and the counterweight ropes connected, the rest of the machinery installation for a vertical lift span is relatively simple. If the bridge is a tower drive vertical lift, the drive machinery bears a resemblance to simple trunnion bascule machinery. For a tower drive vertical lift bridge, the racks are actually mounted on the counterweight sheaves and form continuous circles around their perimeters, near the outer flanges of the rims. The rest of the drive machinery is installed in the same manner as for a simple trunnion bascule bridge.

On a span drive vertical lift bridge, particularly the Waddell type, it is important that the operating ropes do not interfere with any structural steel. On many of these types of bridges, after construction is complete, it is found that one or more operating ropes rubs on a stationary piece of structural steel during operation. If the rope interferes while it is under load, the error must be repaired immediately, as the operating rope will soon break. If the interference occurs only when the rope is slack, rope failure will not occur immediately, but the rope life will be shortened. The offending piece of structural steel should be removed, or an intermediate deflector sheave should be relocated, or an additional one installed, to eliminate the interference.

Machinery for a Waddell-type vertical lift bridge usually consists of three major pieces that have been shop assembled. A central drive assembly consists of a gear reduction set, either open or closed, and two drum frames contain two geared drums each, plus a pinion shaft that drives the drums (see Figure 5-10). Great precision is not required in placing the entire machinery installation, as small errors in alignment do not affect the play of the operating ropes on and off the drums. The output shafts of the central drive must be precisely aligned with the inputs of the drum frames, or wear of the connecting couplings will result. The placement of deflector sheaves and intermediate sheaves or rollers to support the operating ropes must be done carefully to avoid chafing of the operating ropes on the sheaves or on the steelwork.

Machinery for span drive bridges with the drums mounted on the corners of the lift span must be installed with care, so as to avoid excessive misalignment at the couplings. The placement of the drum frames must also be done with care, to avoid chafing of the ropes coming off the drums.

Most vertical lift bridge ropes are socketed by the traditional method, using molten zinc as a matrix material to hold the end of a rope in its socket. The socketing process is a science in itself, as many high-tech applications, such as in cable stay bridges, have lent their technology to wire rope socketing in general, for better or worse. Some success has been achieved with resin material as the socket filler, and this has become a standard for wire rope socketing in industry. Resin has the advantage of filling the socket completely, which is almost impossible to do with zinc filling. Very little heat is generated in pouring a resin socket, so the lubricants and preservative materials within the rope are less likely to be destroyed or driven off and there is no possibility of damage being done to the rope wires by heat. The socket areas of the ropes have not been the critical areas in regard to durability, but poor socketing practice can cause premature failure, so it is important to see that the wire rope assemblies for the bridge are properly manufactured.

Swing Bridges

Swing bridges have considerably more machinery to install than other common types of movable bridges. However, the machinery consists of several separate sets of equipment that are independent, to a greater or lesser extent, from the rest of the machinery, so that installation can be handled on a system-by-system basis. Some of the separate systems may be linked by having a common drive source, which was a typical situation on older swing bridges, or each system may be mechanically independent, the usual form on modern swing bridges. The separate systems include the following:

1. Main support, consisting of
    a. Center pivot and balance wheels

        or

    b. Rim bearing and rollers
2. End lifts and/or supports
3. Center wedges (only for system 1a)
4. End latches or centering devices (or both) (or neither one)
5. Drive or turning machinery
6. Rail end separators (for railroad bridges only)
7. Catenary lifts or separators (for electrified rail lines only)

Much of this machinery can be installed on the swing span in the shop or erection yard and transported to the site with the superstructure. A typical swing bridge is floated into position and allowed to come to rest on the main support. Another feasible approach, which is sometimes used, is to place the superstructure on the main support, with the bridge in the open position, protected by the center pier fender system. All the finishing work is then done in place, with the fender also acting as a work platform.

The swing bridge machinery components most difficult to install are the main support and the drive or turning machinery. These are frequently combined to some extent, compounding the problem. It is important with rim-type support bearings to be particularly careful with installation. Swing bridges, whether center pivot or rim bearing, usually have the drive rack combined with the track mounted on the pier. If the rack and track have been properly prefitted in the shop, then proper mounting of the track in the field will make it much easier to mount the rack properly.

Unlike a bascule or vertical lift bridge, which can be placed in position without most of its machinery and be allowed to carry traffic, a swing bridge must be stabilized in the closed position to carry traffic. If the end and center support machinery have not been placed, but it is desired to carry traffic on the bridge, then alternative means must be provided to support the bridge under traffic. This can be accomplished by providing simple blocking that is adequately secured, or by providing mock wedges or other supports that can be bolted to the pier or superstructure. Of course, the bridge will be a serious obstruction to navigation when such supports are in use.

Machinery Installation—General

Movable bridge machinery is not as delicate as some precision equipment of a similar configuration, but intended for industrial or production use. In spite of its robustness, however, movable bridge machinery cannot tolerate excessive abuse during the erection process. There are three phases of machinery installation, each of which has its own set of special concerns: premounting, mounting, and postmounting.

The premounting phase includes delivery, unloading, preparation for storage, storage, and retrieval for installation. Many components of machinery for modern movable bridges are quite rugged, but others can be very sensitive to moisture, heat, cold, and other environmental conditions. Failure to properly protect and store such components can destroy them before they are ever installed on the bridge. For most standard machinery components that are more delicate, there are procedures specified by the manufacturer for protection during prolonged storage. These procedures should be followed, and only as the machinery is mounted, provided with the proper lubricants and preservative materials, and put into operation should these measures be completely undone.

The mounting phase includes positioning, fastening, aligning, connecting, and securing. A skilled millwright can and should perform all these tasks to avoid damaging the machinery. Meeting the required tolerances for the assembly of components is extremely important in erecting machinery. If these are ignored until the installation is complete, the whole installation may have to be done over again. The placement of geared racks should not be done until it is clear as to what the tolerance of the required finished fit between the rack and pinions is to be, because it is usually impossible to achieve the proper fit between pinion and rack by adjusting the pinion placement only. The placement of main bearings must be carefully performed to ensure that the bridge's superstructure is in the proper position and correctly oriented to operate properly. Errors in placement resulting from inattentive work may not become obvious until the bridge is ready to operate. Corrective measures at that time will be expensive and time-consuming. Backlash in gearsets is usually produced by expanding the center distance of the gears when mounting, unless backlash was removed from the gear when it was cut.

The postmounting phase includes component testing, lubrication, system testing, and putting the machinery into operation. Any rework that must be done is included in this phase and may be an iterative process.

ELECTRICAL INSTALLATION

One of the most common sources of difficulty in movable bridge construction is the installation of the basic wiring for the bridge's power and controls. Conduits to carry this wiring are in some cases very large, but must fit within tight spaces. A large conduit is difficult to bend to exact shape, and minimum allowed bend radii are rather large, so in some cases the conduit can be installed where it is not supposed to be. This may block the installation of another component, or result in interference to traffic or in operation. Such difficulties should be detected early and corrective measures taken. Preparations of accurate working drawings for the conduits is essential.

The electrical conduits should be installed as late as possible during construction so that interferences during construction such as with structural, electrical, and mechanical components, not previously coordinated in the project, can be avoided. It is generally easier to relocate electrical components than structural members or mechanical equipment.

As conduits are installed, they should be kept capped or otherwise sealed to prevent the entry of water or foreign substances that may impede the pulling of wires through the conduits and damage the insulation on the wiring. Even with clean conduits, wires must be pulled with care to avoid chafing or stripping the insulation. Unfortunately, the insulation on wiring may be damaged severely by improper pulling, without showing a fault. The ground or short may not show up until days, weeks, or years after the work is done, resulting in operational problems that may be very difficult to trace. Electrical wiring should be installed as late as possible before the bridge is completed to minimize the likelihood of possible loss due to vandalism or environmental damage.

The biggest problem in control system installation can be getting all the right wires connected to the right components. A considerable amount of time is usually required to "debug" an installation, because tracking down a mislabeled wire and getting it properly connected can be difficult to do except by trial and error. There are also many other difficulties that become apparent only after an installation is partially complete, such as conduits that cannot be run as intended because of structural interferences or insufficient space to make bends. Ample time should always be allowed to troubleshoot an electrical installation after it has been installed and before it is finally put in operation. Any signs of damage to electrical components should be looked into immediately. If internal damage is suspected, the component should be tested to confirm that it is not impaired.

A problem that occasionally arises is the inability to get a control panel or equipment enclosure into the location intended. The operator's house or machinery room door may not be large enough, or there may be a structural interference that does not allow the equipment to be placed as intended. In any of these cases, the straightforward solution is usually best. It is always preferable to leave the electrical cabinet, module, control desk, or other assembled component intact and make the access passage large enough to allow the electrical component to pass. It may be easier to repair a doorway, to replace a wall, or sometimes to remove and then reinstall a structural component, than to tear down an electrical cabinet or similar device in the field and then reassemble it. The only exception may be when a critical, nonredundant, major structural member is in the way, or if a very expensive architectural finish has been installed that will be destroyed if the member is disassembled. Proper coordination of all the work on a project should avoid such problems.

If possible, the control system should be installed only after the other parts of the bridge are substantially complete. The control system components are more susceptible to damage from the weather or vandalism than the other parts of a movable bridge. The control system should be left out on the bridge in an uncompleted state for as little time as possible. If the control system is installed before the bridge is operable, its components should be protected from the elements and vandalism by covering the windows of the control house and machinery room. These areas should be locked when work is not being done in them, and consideration should be given to patrolling the areas during off-hours when construction is in progress, whenever a bridge operator is not on duty. Responsibility for the protection of control system equipment should be clearly spelled out in the construction contract so that there are no after-the-fact recriminations and any damage done during the construction period is quickly repaired.

24

BALANCING

BASCULE BRIDGES

Proper balancing, as part of the erection process, is more important for bascule spans than for any other type of movable bridge, because of the greater difficulty in making corrections after erection. In spite of the most careful design computations, the erected weights of steel and concrete seldom agree precisely with the weights as designed. For a bascule bridge leaf, deviations in weights far from the center of rotation, such as at the toe of the leaf, have a very large effect on balance. Small changes in the overall density of the counterweight material, which can easily occur with concrete, can make a big difference in balance.

It is imperative that during erection, a careful log of the actual weights of components going into the bridge be kept, with precise notation as to the distance of the components from the center of rotation of the span. The distance should be measured or calculated along all three primary axes: vertical, or y, longitudinal, or x, and lateral, or z. The total moments of the bridge structure, less the counterweight, should be summed about the center of rotation at the midpoint between the rotating supports, trunnions, or tread plates (see Figure 11-17). If the counterweight is a steel box or frame filled with concrete, the structural steel portions of the counterweight can be added to the sum. The density of counterweight concrete required to produce the required moments to balance the remaining weight can then be computed, given the correct position of the center of gravity of the counterweight.

Once the desired density of the cured concrete in the counterweight is finally established, test samples of concrete should be poured to determine the exact mix required to produce the required density. It may be necessary to vary the density of the concrete at certain locations within the counterweight to achieve the desired longitudinal, lateral, and vertical moment arms of the counterweight as a whole. The density of the counterweight concrete can be varied to some degree, from about 80 lb per cubic foot to about 175 lb per cubic foot, without greatly affecting the strength of the concrete, so that it can stand as an exposed block. If denser concrete is required, steel punchings or iron ore can be added as aggregate, but the strength will suffer. Serious consideration should be given to enclosing the counterweight in a steel box with thick walls if high density is required. Alternatives such as using solid cast iron blocks for the counterweight, or adding lead to the counterweight, produce a more durable counterweight than concrete, particularly extremely dense concrete. These options are very expensive as compared with using concrete alone. A permanent steel box enclosure has the advantages of eliminating the need for temporary formwork when placing the concrete, and adding its dense mass to the counterweight structure.

After the moving span and its counterweight have been constructed, the balance should be tested to make certain it is as intended. The design documents should specify the balance state for the finished bridge. In some cases, the specifications require only that there be some designated amount of downward force at the live load shoes with the bridge seated. This can be tested by the use of load cells. Other specifications require that the center of gravity of the gross moving span be at a certain position relative to the center of rotation of the leaf. This will require more elaborate testing, such as by the use of strain gauges or other means to record the net torque required to open and close the bridge. This type of procedure is discussed in Chapter 34.

Bridge rehabilitation projects usually call for rebalancing the bridge after other work has been completed. If the specifications merely require that the balance state of a bascule bridge be maintained at or returned to the state it was in when the contract started, there is usually not much difficulty unless the bridge has been radically changed in configuration as a result of the rehabilitation. Some older bascules were designed to carry trolley cars or interurban trains, powered by overhead electric lines. If a rehabilitation calls for removal of trolley wires and their supports, there may be a serious problem with achieving proper balance. Some bridges have had to have heavy weights placed above the deck, near the toe of the span, to achieve proper balance. Many older bascule bridges were designed to be counterweight heavy when closed, and span heavy when open, to minimize power requirements. It is often desired to change these bridges during reconstruction projects so that they are span heavy when closed, and counterweight heavy when open, to improve stability and decrease the likelihood of accidents. This is a major change to the bridge, and it may require a great deal of ingenuity to obtain the desired balance without having to remove the entire counterweight and replace it with a new one.

VERTICAL LIFT BRIDGES

The primary balancing is somewhat easier for vertical lift spans than for bascules, as the position of the weight in the counterweight is not critical. All that must be accomplished is for the dead load component of each counterweight, measured at the supports, that is at the counterweight ropes, be equal to the dead load component of the movable span at the counterweight ropes. If the vertical lift span is erected in the lowered position, it should be a simple matter to weigh it at each corner and provide for matching weight at each counterweight connection.

The typical vertical lift span is fairly flexible in torsion about its longitudinal axis. If the span does not seat firmly at all four corners when the live load shoes have been properly adjusted and neither the drive machinery nor the span guides are restraining the span, it is likely that the mass of the counterweight is not properly distributed. Removing some weight from the end of a counterweight adjacent to the location of the part of the span that will not seat should correct the problem. This correction should be confirmed by rechecking the span balance at each corner of the vertical lift span.

To predict the amount of weight required in the counterweight before completing construction of the span, the weight of the material that goes into the vertical lift span can be tabulated as the span is erected. From this computation, the corresponding amount of force required at each corner of the span, to lift it, can be determined. A flexible model can be used to determine the reactions. For earlier results, the details on the shop drawings can be used to estimate the span weight and balance. It may be possible to use these estimates to determine the required counterweight mass, and then rely on adjusting the number of portable weights in the pockets in the counterweights to obtain the correct span balance, but this can be risky. Most vertical lift bridges are symmetrical in plan, so that the weight at each end of each counterweight should be the same. Unfortunately, concrete deck densities can vary, rolled steel thicknesses can vary, and other small details can vary enough so that the cumulative effect, particularly on a large vertical lift bridge, can be significant.

Many vertical lift spans have a form of auxiliary counterweight that is intended only to counteract the change in position of the counterweight ropes as they pass over the counterweight sheaves during bridge operation. These devices usually require very little adjustment during installation, as the counterweight ropes have a fairly well known weight per foot, so that the auxiliary counterweights can be accurately detailed prior to erection of the bridge and do not require trial-and-error adjustment or extensive field fitting to perform satisfactorily. An auxiliary counterweight should not be used to provide an adjustment in overall span balance. In fact, it should not be installed until the other construction is complete and the overall span balance has been achieved.

The balance of the bridge should be tested after the vertical lift span, counterweights, and all other components have been installed. Load cells may be sufficient to accomplish this, or balance can be tested by measuring the drive force required at each corner of the vertical lift span to move the bridge up and down. Chapter 34 discusses balance testing in more detail.

25

REPLACEMENT AND REPAIR OF EXISTING BRIDGES

REPLACEMENT

The replacement of an existing bridge requires accommodation of the traffic over that bridge while it is being replaced. For the purpose of expediting construction, a detour is usually the best choice. In many cases, a detour may be extremely inconvenient for users of the bridge. The extra distance may be too great, or the amount of traffic may be too large to be handled by the existing roads. In such cases, it may be necessary to maintain traffic during construction of the new bridge. If the new bridge is to be built alongside the existing one, this may not pose a great problem. A new bascule or vertical lift bridge can be built alongside an existing one without causing difficulties in operation, whether the new bridge will span the same navigation channel as the old or will be placed at a new location over a new channel. A concern with adjacent construction is to avoid undermining the foundations of the existing bridge when building foundations for the new. If the new foundations are to be immediately adjacent to the old, great care will be necessary. It is a relatively simple matter to redirect the roadways at either end of the bridge from the old to the new structure and then demolish the old bridge if necessary. If possible, the old foundations should be left in place to avoid disturbing the new.

If the new bridge is to be built on the existing alignment while maintaining traffic, maintaining traffic can be the most difficult part of the project. As indicated in Chapter 8, the selection of the new bridge type can influence the practicability of the replacement project. The easiest type of new movable bridge to construct in such cases is generally the vertical lift span. The foundations for the new towers can usually be constructed away from the existing piers and abutments, and if they are constructed under existing approach spans, traffic need not be disturbed.

A new highway bridge is usually wider than the bridge it replaces. Typically, part of the new bridge is built alongside the old, as mentioned earlier. Then the old bridge is removed and the rest of the new constructed. For the best fit of the first stage of the new bridge to the second stage, a bascule bridge type for the new structure is desirable, as the roadways can be continuous laterally across the bridge. The foundation difficulties are most severe, however, as the existing bridge remaining temporarily in use is in jeopardy during the first-phase of the new bridge construction, then the first phase new bridge is in jeopardy while the old bridge is demolished and the second phase of the new bridge is constructed. This procedure also results in shifting the roadway centerline at the bridge.

In the replacement of a railway bridge, additional traffic capacity is usually not required, so the new bridge is the same width as the old. It is also usually desirable to maintain the same alignment. Usually, as much of the old bridge as possible is reused, such as the piers, minimizing the difficulty in construction as well as reducing cost. Occasionally, difficulties are found with existing components that were to remain, and decisions must be made on the spot as to whether to repair or replace a defective component. The construction contract should, to the greatest extent possible, call out the procedures and prices that will be in effect in such a situation.

Difficulty should be anticipated in demolition of an old movable bridge. Unanticipated hazardous materials may turn up. What appeared to be deteriorated concrete in counterweights or piers may prove to be extremely difficult to demolish.

REPAIR

Whenever possible, repairs are done on movable bridges while maintaining them in operable condition, so that both highway or rail and marine traffic can continue. This is, in some cases, impossible. When the counterweight sheaves or sheave shafts of a vertical lift bridge must be replaced, usually the only practical procedure is to raise the bridge to the fully opened position, temporarily support it there (see Figure 19-1), temporarily support the counterweight in the lowered position, and remove the counterweight ropes. The sheaves, shafts, and/or bearings can then be replaced as required.

In some cases retrofit can prevent, or at least postpone, major repairs. Vertical lift bridge sheave shafts that are prone to fatigue failure have been precompressed by installing large tension rods in axial holes in the shafts. The holes must not be enlarged too much to accommodate these tension rods, as adequate shear capacity of the shaft must be maintained unless the tension rods can be made to carry some of the shear. By compressing the shaft axially, the maximum bending tensile stress in the shaft is reduced, reducing somewhat the likelihood of cracks originating and propagating. The cyclical stress range in the shaft remains nearly unchanged, however.

IV

MAINTENANCE

For a discussion of the standards applicable to the maintenance of movable bridges, see the *AASHTO Movable Bridge Inspection, Evaluation, and Maintenance Manual, Guide for Bridge Maintenance Movement,* the *AASHTO Maintenance Manual,* and the *Manual for Maintenance Inspection of Bridges,* published by AASHTO.

The proper performance of maintenance for a movable bridge includes inspection, testing, cleaning, adjustment, lubrication, and minor repairs and replacements. There are three types of maintenance in general use, and maintenance of most movable bridges is based on one of these types, or a combination thereof:

Repair Maintenance Maintenance staffs for bridges spend most of their time answering calls for emergency repairs. The rest of their time is spent waiting for these calls. Their immediate supervisors have found that being unavailable to respond immediately to emergency calls is the last position they want to be in, so maintainers are not sent to do normal maintenance where they may be unavailable to respond to emergency calls. This type of maintenance is referred to as *component failure maintenance* in *AASHTO Movable Bridge Inspection, Evaluation, and Maintenance Manual.*

Normal Maintenance Maintenance staffs proceed to each bridge periodically, on a fixed schedule, and perform what is sometimes referred to as preventive maintenance. The maintenance crew lubricates all components according to the maintenance charts prepared for the bridge. Normal maintenance plus inspection maintenance (see the following paragraph) is considered *preventive maintenance* in *AASHTO Movable Bridge Inspection, Evaluation, and Maintenance Manual.*

Inspection Maintenance Inspectors proceed to each bridge periodically, on a fixed schedule, and go over the entire bridge, checking each component on an inspection list, using inspection and testing methods included in a maintenance manual. They immediately make repairs and adjustments as needed, as called for in the manual. The inspection, repair, and adjustment functions may be performed by the same personnel, or one person may perform the inspection and another person or persons do the repair and adjustment work, or some other combination may be in effect. In the second approach, an inspection report is prepared by an inspector, who forwards the report to the repair and adjustment personnel for action. The report may be modified by a third entity, usually a supervisory person, or may be converted into a whole new document before being turned over to the people making repairs and adjustments.

For many bridges, there is a limited form of normal maintenance performed by a special group, referred to as "oilers" or a similar term, and the people referred to as "maintainers" perform the repair and adjustment work called for in inspection maintenance. These people may also perform the repair maintenance function.

Preventive maintenance is conspicuously missing from the above categories of types of maintenance. As noted earlier, AASHTO, in its *AASHTO Movable Bridge Inspection, Evaluation, and Maintenance Manual,* defines preventive maintenance as normal maintenance and repairs. There are almost as many definitions of preventive maintenance be found as there are publications discussing maintenance. Some entities consider any maintenance work that includes record keeping to be preventive maintenance.

Strictly speaking, preventive maintenance means keeping the system in sufficiently good condition that breakdowns do not occur, ever, except those due to external forces. This is a simple two-phase process. Normal maintenance is one phase, in which components are lubricated and other work is carried out to prevent or reduce wear or other deterioration of components that can deteriorate. The second phase is the replacement of those parts that will continue to age, wear, or otherwise deteriorate in spite of being supplied with adequate normal maintenance, before they fail. This approach involves record keeping and fairly continuous inspection to keep track of the degree of deterioration of the components that can fail. The deteriorated components are then replaced before they fail and after they have used up enough of their life to make replacement economical. Almost no movable bridge is maintained to this degree, and it may make little or no economic sense, for any movable bridge, to expend the effort required to do so. Movable bridges operate at irregular intervals, unlike much factory machinery that operates continuously and thereby makes life predictions fairly simple. It is difficult to predict accurately when a movable bridge component will wear out. Thus, although preventive maintenance is mentioned often in these pages, it does not necessarily constitute a practical theory of maintenance for movable bridges.

If there will be an attempt to provide preventive maintenance, it requires a carefully designed program to be fully effective. The program must be based on a thorough understanding of all the components of the movable bridge, including a good estimate of the rate of deterioration or the expected life of the bridge components in its particular environment. Preventive maintenance also requires that skilled personnel be available to perform the work and that they have all the necessary tools and diagnostic equipment on hand. A supply of spare parts must also be on hand to perform the necessary component replacements. All of these ingredients to a preventive maintenance program are expensive, and an attempt to provide them should not be made unless the cost can be justified.

The costs of lack of maintenance are very real and can be quantified fairly easily. The biggest cost of lack of maintenance is delays to traffic. For highway bridges, this cost is paid by the public and is largely ignored by the bureaucracies responsible for maintenance of bridges. Thus, highway bridges tend to be severely undermaintained. Railroads, on the other hand, bear the cost of bridge malfunctions themselves, either in lost revenue and wasted operations, the cost of delayed trains, or the cost of using another railroad's line on which to run their trains while a railroad's own bridge is being repaired. As a result, railroads tend to be diligent in performing normal maintenance on their bridges; they also tend to have adequate staff to perform inspection maintenance with a full division of labor, as indicated earlier. In some cases, movable bridges have been capitalized with the intent of having preventive maintenance performed and have sufficient stocks of replacement parts supplied with the bridge to begin a preventive maintenance program, but the annual budgets for maintenance have not been forthcoming to the degree required for preventive maintenance to be performed. The stocks of spare parts are gradually used up in making repairs after failures have occurred or have deteriorated because of poor storage and have become useless.

The second largest cost of lack of maintenance is incurred in premature replacement of bridges or in costly rehabilitation of bridges that were allowed to decay. Highway maintenance departments have tended to ignore this cost in the past, because the bureaucracy responsible for maintenance was usually totally independent from the bureaucracy responsible for new bridge construction and major rehabilitations of existing bridges. The situation has changed recently because of the action of the federal government, which is the prime source of funding for highway bridge construction and rehabilitation. The federal government has found that many of the new bridges it has funded in the past 30 years have failed inspections as a result of deterioration that is not due to traffic loading, but caused by degradation of the bridge structures due to corrosion or other material degradation not directly attributable to traffic. The federal government originally refused to fund maintenance assuming it would be performed by the bridge owners. When this did not happen, federal money mandates for maintenance were initiated.

Many bridge owners and others confuse repairs with maintenance (see the preceding discussion). Most repairs, except to correct for external damage, are actually an indication of a lack of maintenance. The less normal and inspection maintenance that is done, the more repair maintenance will be required. Some entities confuse normal maintenance with preventive maintenance. Normal maintenance is the carrying out of procedures such as lubrication and replacement of burned-out light bulbs. Preventive maintenance includes the programmed replacement of parts that wear, or otherwise deteriorate, before they fail (see page 342). Preventive maintenance may or may not be cost-effective, depending on the cost incurred due to the failure of parts not replaced in time and the cost of performing the preventive maintenance. If the major costs of lack of maintenance as discussed above are not included in the evaluation of maintenance costs, then it is very likely that preventive maintenance will not appear to be cost-effective.

Concern about these issues is within the realm of maintenance management, which is, for bridges, part of the bridge management field. (See Chapter 37 for a discussion of bridge management.) Some management of maintenance is required, such as accurate and complete record keeping, so that it is known what maintenance has been performed and when. This information is also useful for bridge management purposes, to help managers understand the cost of maintenance and what is received for that cost. A good maintenance manual should provide maintenance record sheets specifically written for the particular bridge and the maintenance program that was developed for that bridge. Maintenance personnel should be instructed in how to fill out these sheets properly, and instructed to do so promptly when performing the maintenance. The processing of such data sheets and how the data are to be used are beyond the scope of most maintenance manuals.

Repair maintenance, in order to be effective and economical, must include a diagnostic process that identifies the cause of a malfunction. Simply making a repair to a broken component cannot guarantee that it will not fail again soon. If the cause of a failure is found and that cause is eliminated, bridge operation will be more reliable and there will be a longer period of time before the next repair is necessary. This is the difference between a repair and a "patch." Making a proper diagnosis of the cause of a problem requires an understanding of the bridge operation, as well as the functions of the various components. To perform an effective diagnosis, maintenance personnel must be properly trained and have the needed tools available. A good maintenance manual is necessary as the basis for adequate diagnosis of malfunctions. To minimize a bridge's down time when repair maintenance is being done, a full stock of replacement parts should be readily available. The ability to make the required repair in a reasonable amount of time is a necessity. If all repair parts are not readily available on the bridge or in the shop, then there must be ready access to a supply house or the manufacturer. The necessary skill and tools to make the repairs must also be available. If the bridge maintenance or repair crew cannot fulfill these requirements, then a capable supplier, such as the manufacturer or the company that installed the equipment, should be ready to respond to a call for field service.

In an ideal sense, the best maintenance program for movable bridges is no maintenance. This can be achieved by designing and building a movable bridge that can meet its life expectations without requiring normal maintenance. Such a goal can be accomplished by minimizing the number of moving parts on the bridge and making the moving parts that do exist as robust and trouble free as possible. To do so, one has to start with a single leaf simple trunnion bascule bridge, as that is the simplest type of movable structure attainable unless a removable span-type movable bridge is to be used. In Part I of this book, application of movable bridge types is discussed, and it becomes apparent that a removable span or a single leaf simple trunnion bascule is not practical for many movable bridge applications. Yet there are many examples of these movable bridge types in existence and perhaps many more locations at which one of these types would be as serviceable as the movable bridge types in use there.

The type of drive used to power a movable bridge is also a factor in the amount of maintenance it requires. Electrical and hydraulic equipment and internal combustion engines are high-maintenance items, so they should be eliminated if possible. Hydraulics can easily be replaced with gear drives or other mechanisms. These can be and are built to be sufficiently robust and can be provided with internal lubrication systems so that they can be expected to last 20 or 30 years without maintenance. Many such devices have lasted longer in movable bridge applications with no affective maintenance and little or no deterioration. Roller bearings supporting vertical lift bridge counterweight sheave shafts are an example. The only type of power that can be considered to require zero maintenance is manual power, which can be supplied as needed and then removed from the site. If sufficient manpower is not always available to provide operation of the bridge in a timely manner, portable power units can be brought to the site when needed, then returned to a central facility for storage and more convenient maintenance. This is not always practical, so some other form of power must be supplied for movable bridges that must open fairly often, and do so rather quickly. These bridges must also be supplied with various interlocks and other safety devices. An electrical prime mover is about the only choice for such a bridge, so for many movable bridges, electrical equipment and its maintenance are necessities.

26

SUBSTRUCTURE AND SUPERSTRUCTURE

SUBSTRUCTURE

Some types of movable bridges produce highly variable loading on the piers. This is particularly true of Scherzer type rolling lift bridges, but also applies to the Rall type. As the leaf rolls open and closed, all elements of the bridge support take a beating, from the footings to the superstructure. The piers on such a bridge should be kept clean so that water will not collect and seep into small cracks formed at areas of localized high stresses.

Horizontal support girders for rolling lift bridges, on which the horizontal rolling lift tracks are mounted, should be watched; if signs of fatigue develop in a girder, it should be reinforced before it is destroyed or damage occurs to the pier. Any loose bolts or rivets that develop in the girder should be repaired as soon as discovered, to prevent further deterioration.

Masonry in piers is prone to deterioration, and if degradation is unchecked it can lead to pier failure. Stone piers should be kept repointed to prevent the loosening of stones, and cracks in concrete piers should be patched to limit the access of moisture to the reinforcement. Reinforcing steel that has been exposed because of spalling of surface concrete should be protected from corrosion. This can be accomplished by painting the reinforcing steel, or by grouting or patching the concrete. If there is extensive deterioration of the concrete, a decision should be made as to the extent of permanent repairs that will be made and when they will be made. Sufficient protection should be given to the remaining materials so that further deterioration is minimized while waiting for permanent repairs to be made.

In recent years, there has been a great deal more concern about the susceptibility of a bridge to seismic events. Various devices have been developed that can, perhaps, assist a bridge in surviving an earthquake. Many of these are mechanical devices and, as such, may require maintenance to ensure that they will perform their functions when needed.

Many older movable bridges are affected by scour more than fixed bridges, because the movable span may be shorter, causing a greater restriction in the waterway channel, or nonskewed so that the piers are more obstructive of water flow, as opposed to a fixed span at the same location would be. The uncorrected undermining of foundations can lead to settlement, which can cause difficulties with the operation of movable bridges or necessitate their replacement. Scour can occur very quickly, so biennial inspections should not be relied on at suspect locations. The underwater condition should be monitored regularly in these locations and checked after heavy flows such as those caused by cloudbursts. Remedial measures should be taken immediately when a problem develops. Filling in locations that have been exposed by scour is essential and should be done as soon as possible to avoid further damage. To prevent future scour, heavy material should be used; if possible, measures should be taken, such as rechannelization, to minimize the likelihood of future scour action at the piers.

Fender systems should be kept from deteriorating to the point that they become a menace to navigation. The faces of fender systems should be kept smooth to prevent snagging by passing vessels. Loose planks should be reattached, and missing planks should be replaced.

SUPERSTRUCTURE

Many movable bridges experience stress reversal during operation. Some bascule bridges that open nearly 90° have components that go from tension to compression, and vice versa, during operation. These components may require retensioning of bolts, or possibly repainting, more often than other components. Both of these procedures should be considered inspection maintenance items. High strength bolt tensions should be checked at regular intervals at critical connections, particularly those that are subject to load reversal and possible fatigue. Older riveted structures may experience loosening of rivets in stress reversal areas. This can be quite common at the segmental girders and support girders to which Scherzer rolling lift bridge tread plates are attached. In any such cases, loose rivets should be removed and replaced with high strength bolts. At segmental and support girders to which tread plates are connected, there may be several thicknesses of steel plates and shapes that are connected by the fasteners, which may be several inches long. If distortion has occurred, extraction of the old fastener may be difficult and installation of the new fastener may require enlargement of the hole.

Swing bridges can experience fatigue at the truss chord members or main girders, as the driving of the end lifts partly converts the bridge from a cantilever to a continuous span. A part of each main bridge member goes from pure tension to pure compression, or from compression to tension, as the bridge is opened to clear the navigation channel, and the reverse occurs as the bridge is closed again. The range of stress may be quite high, so that low-cycle fatigue can occur. There are few, if any, swing bridges that open and close millions of times in their lives, so that the more typical fatigue failures are seldom found at these locations.

Almost all movable bridges are balanced structures, many with counterweights that combine with the superstructure, placing the dead load of the moving bridge onto the bearings that support the bridge as it opens and closes. Most of the live load on one of these movable bridges is supported by separate bearings, which make contact only when the bridge is seated. A typical nonmovable bridge superstructure is supported on bearings, which also support the live load and never lose contact. With little or no dead load on the live load bearings of a movable bridge, contact at the live load bearings is intermittent. Subtle shifts in the weight of the movable bridge can result in lack of contact at one or more live load bearings except when heavy vehicles pass over the bridge. This results in large impact forces on the live load bearings, and these forces are translated to the adjacent superstructure. Many older movable bridges that carry heavy traffic have damage at the structural steel of the superstructure adjacent to the live load bearings. To prevent such damage, careful attention must be paid to the contact at the live load bearings and adjustments should be made when contact is not solid. In addition to balance changes that can cause the poor contact, wear occurring at the live load shoes can be irregular, so that one shoe may lose contact when seated while the others have firm seating. Care must be taken to adjust the balance when that is the cause of poor seating, and to adjust the live load shoes when they have worn. Frequently, the wrong adjustment is made, and the poor seating recurs after a short period of time. Corrections made by shimming should be above the sole plate or below the masonry plate, not at the interface between these items.

In addition to causing damage to the superstructure, poor seating can cause deterioration of the pier caps, as impact causes the masonry plate and the material under it to deteriorate. The most obvious sign of poor live load shoe contact, other than movement of the bridge under traffic, may be deterioration of the machinery, especially at the racks and pinions of bascule bridges. If movements under live load are excessive, the rack and pinion teeth in contact when the bridge is closed can wear rapidly and even fail.

Corrosion of the steel superstructure is frequently a problem in movable bridges, as it is in other types. Corrosion usually does not occur generally, unless a bridge is exposed to an extremely harsh environment. Corrosion of the type that causes damage serious enough to require repairs most often occurs locally. Corrosion of this type is usually accelerated by the presence of foreign material that traps moisture and allows corrosion cells to exist. Such cells attack the metal, sometimes causing quite rapid corrosion. These are often found at particular locations susceptible to such conditions, such as at horizontal gusset plates or at lacing bars at the undersides of truss members. Maintenance efforts that prevent the accumulation of foreign material can be quite effective in reducing corrosion damage. Dirt and airborne debris should not be allowed to accumulate on the horizontal surfaces of a bridge. A regular program to clean such surfaces can be very beneficial, particularly on highway bridges where deicing salts or sand are used. Other debris, such as materials left over from repairs, can also accelerate corrosion. After a repair is made, a maintenance crew should clean up the area. Excess lubricants spilled on the steelwork can trap moisture and cause corrosion to start. Spilled lubricants and old lubricants removed from gears and bearings, particularly greases, should be disposed of properly and not allowed to remain on the bridge.

Cleaning and painting a steel bridge superstructure can be regarded as a type of preventive maintenance or normal maintenance. The paint protects the steel beneath from corroding, and cleaning minimizes the likelihood that corrosion-producing conditions are present. The painting, with this process of maintenance, consists of constant repainting and touching up deteriorated areas as they are found. Painting can also be regarded as a rehabilitation process, whereby the bridge is repainted completely, on a regular basis, at intervals that prevent deterioration from advancing at the superstructure. The first process, painting as maintenance, generally develops from experience and requires regular inspection to be sure that deterioration is not allowed to begin. The second process, painting as rehabilitation, is based on research and testing and requires a statistical approach to the life of various real-world components so that the painting intervals are economic. If the research is properly done and valid conclusions are reached, the second process can be just as effective, at less cost than the first. Either process can prove cost effective when the cost of delays to traffic due to breakdowns or extensive repairs are figured into the overall cost of the bridge.

27

MACHINERY, CONTROL, AND POWER SYSTEMS

MACHINERY

Preventive maintenance is basically intended to prevent breakdowns. There are two basic types of preventive maintenance work. The first is performed to keep components from deteriorating. The most obvious and ordinary means of doing this is by lubricating the machinery—putting oil in the crankcase of an engine, for instance. Bridge drive machinery is an area where normal maintenance—that can be considered preventive maintenance—is usually performed. Most movable bridges have regular visits by persons, usually called oilers, who apply lubricant to open gears and friction type bearings. Oilers should follow lubrication charts, showing the type of lubricant to be used for a particular application, the frequency of lubrication, and the means of applying the lubricant. *AASHTO Movable Bridge Inspection, Evaluation, and Maintenance Manual* provides details on lubrication in Chapter 4.3.

The other aspect of preventive maintenance is the programmed replacement of a component before it breaks. In some areas there is overlap or ambiguity between the two types. There are few, if any, mechanical components in the properly designed main bridge drive that are susceptible to rapid wear as long as the components are kept properly lubricated. Machinery inspections once a year, or once every two years, are usually sufficient to identify components that are approaching the end of their useful life. These inspections are usually outside the realm of the maintenance personnel on a bridge. Typically, a mechanical inspection is performed by a contractor (usually a consulting engineer), and this inspector's report includes recommendations for replacement of certain parts before failure occurs. After a period of time, or perhaps after several such inspections, a rehabilitation project is developed for the bridge, including most of the repairs or replacements mentioned by the consultant.

Constant development of new products results in new options continually becoming available to bridge owners for use in their struggle against high maintenance expense. By the time the life cycle of a product such as a lubricant or paint has been established in a particular bridge application, another product comes along that promises a longer life.

The so-called conventional system of movable bridge machinery, consisting of open gears and shafting supported in pillow block type bearings, requires a fairly active maintenance program. These machinery components require frequent attention and relubrication, not only to reduce wear, but also to prevent corrosion and keep the components clean. Providing covers for these components helps to extend the interval between maintenance sessions, but does not eliminate the need for periodic maintenance. As part of the lubrication process for open gears, the old lubricant should be stripped off before the new lubricant is applied. Open gear lubricant can trap contaminants, such as airborne abrasive particles, that can greatly diminish the effectiveness of the lubricant or even turn it into an abrasive compound. In addition, contaminants that trap or absorb moisture can accelerate corrosion of the metal parts. After the old lubricant has been removed, and before the new is applied, is an excellent time to inspect the gears for wear and other damage. Friction type bearings supporting open gearing should be lubricated carefully. Old bearing lubricant ejected when applying new lubricant should be cleaned off the machinery. Excess bearing lubricant should not be allowed to mix with the gear lubricant, as they are usually of quite different formations.

On some movable bridges, the open gearing is actually enclosed in oil pans and covers. This allows oil to be used instead of grease and eliminates the need for constant maintenance.

Fully enclosed gear drives, or reducers, require very little maintenance. The lubricant level should be checked regularly and lubricant added as necessary to maintain the proper level. Breather caps should be checked regularly and should be cleaned when necessary to keep them from becoming plugged. In damp environments, the enclosed drive should be checked regularly to make certain that it is not collecting water. Whenever water is found inside a reducer, the unit should be drained, cleaned and dried, and refilled with lubricant after it is ascertained that there is no rust or foreign material inside the housing. Shaft seals should be checked and replaced if they are damaged. Packing glands, if present, should be checked to see that they are not leaking excessively, and tightened as necessary. If a gland will not stop leaking unless it is tightened to excess, it should be repacked.

All flexible couplings connecting machinery parts should receive regular lubrication, except those small jaw-type couplings having nonmetallic inserts and other nonmetallic couplings. In applying lubricant, care should be taken not to damage the coupling seals. Couplings requiring lubricant should have two fittings, one for entry of the lubricant and a second, consisting of a simple pipe plug, that can be removed while applying the lubricant to allow the old lubricant to escape, be captured, and disposed of properly.

If possible, maintenance personnel should accompany machinery inspectors who are performing in-depth or special inspections of machinery, so that repairs or extraordinary maintenance can be performed immediately in cases where the inspectors uncover serious defects. Some inspection work can most readily be performed in the process of maintenance (see the earlier discussion). Well-qualified and experienced inspectors can also point out maintenance shortcomings to maintenance personnel as they go through the machinery, which can help to improve the utility of the maintenance effort.

Brakes are usually considered electrical components, but most aspects of brake maintenance may be more properly included in the mechanical area. Brake shoes should be replaced before they become excessively worn and threaten damage to brake drums or discs. Brake linkages should be lubricated regularly. The fluid in thruster brakes should be checked regularly and topped off as needed. Thruster fluid should be changed according to the manufacturer's recommendations, as it can deteriorate with age. Brake shoe pressure should be checked occasionally and adjusted when required to maintain the desired braking torque.

Buffer cylinders, if present on a bridge, should be lubricated regularly. One type of lubricant may be applied internally to the cylinder, and another to the rod bearing. The check valve and needle valve should be cleaned, readjusted, and oiled.

Auxiliary drive systems require maintenance similar to that performed on other machinery. Engine generator and direct engine drives should have all maintenance performed according to the manufacturer's recommendations. Auxiliary engine generator drives should also be used on a regular basis to make certain they and the transfer switch gear are functional. Auxiliary direct engine drives should be used to operate the bridge on a regular basis. Engine generators should be used similarly, unless another exercising load is available to allow it to be operated under load without operating the bridge. Mere idling or operating at speed with no load even at full operating temperature is not sufficient, as carbon build-up can occur, which can rob the engine of power that is needed when it is used to operate the bridge. Fuel tanks should be kept free of condensation. Fuel lines, radiator hoses, and exhaust pipes should be checked regularly, and leaks repaired when found. Fuel subject to degradation over time should be discarded periodically and replaced with a fresh supply. Operation of the bridge by means of auxiliary power at least once a month, and preferably more often, can help to ensure that the auxiliary power is available when needed.

Hydraulic machinery is a high-maintenance item. No bridge should be supplied with hydraulic machinery unless the owner is prepared to fund an intensive maintenance program, with highly skilled, well-trained, and thus highly paid maintenance personnel. Maintenance for a hydraulic system must be diligently performed, or problems will result. *AASHTO Movable Bridge Inspection, Evaluation, and Maintenance Manual* (4.4.1) states: "Hydraulic systems which are not properly maintained are prone to leak, may blow a seal or fail to operate at any time, particularly during adverse weather conditions and when operating loads from wind or ice increase the loads in the system." When a hydraulic system is provided, a complete, detailed manual should be supplied, covering all manufacturer's suggested maintenance for all parts of the system. This manual should be followed diligently by maintenance personnel working on the system. The hydraulic maintenance manual should include a very detailed and complete troubleshooting section. It should also include an extensive preventive maintenance section, with listings of all components that should be replaced on a regular basis, including replacement intervals in hours of operation or calendar days. *AASHTO Movable Bridge Inspection, Evaluation, and Maintenance Manual* contains a detailed discussion of hydraulic system maintenance (Chapter 4.4).

CONTROL AND POWER SYSTEMS

Control system components are among the most "disposable" elements of a movable bridge. Items such as limit switches may wear out and be replaced several times before any drive gears in the bridge must be replaced. A movable bridge superstructure, if reasonably well maintained, may be in the same condition 50 years after being put into service as it was when new. Conversely, it is the rare older movable bridge that has not had almost all of its electrical equipment replaced at least once. A key element of movable bridge preventive maintenance is the weeding out of defective electrical components before they cause a bridge operating failure. A solid preventive maintenance program can be an extremely useful tool in maintaining the reliable operating capability of a movable bridge's electrical system, particularly the control system. This is true of to any system, whether an older manual type or any of the newer solid-state types with any degree of automation.

Programmed replacement of components that are prone to failure, such as the aforementioned limit switches, can be economical and help to ensure trouble-free operation. Most electrical components are standard items intended for industrial use, so fairly reliable figures on expected life for given conditions are available. These figures can usually be applied to movable bridge components with only a little engineering judgment required. Detailed inspection of these smaller components is not of appreciably great value in an electrical maintenance program, as many components, beginning to show signs of even what appears to be minor wear, can be ready to fail at any time. The cost of one visit to correct a breakdown can easily pay for the routine replacement of many electrical components. Programmed rehabilitation of major pieces of apparatus, such as span drive motors, is also usually cost-effective in the long run. A motor burning out in service can be much more expensive to repair and may cause damage to other pieces of equipment as well as being the source of a debilitating service failure. If a bridge is designed with two drive motors installed but is capable of operating without overload on only one motor, the other motor can be removed for servicing and rebuilding without affecting operation. See page 355 for movable bridges designed for two-motor operation. If the bridge has only one drive motor installed, a spare motor should be provided for flexibility in servicing as well as for emergencies, unless the bridge can be taken out of service completely or an auxiliary drive system can be used without great inconvenience to the public. The same applies to such items as brakes and major switchgear components.

Preventive maintenance is particularly useful in the upkeep of programmable-controller-type bridge control systems. Built-in diagnostic facilities can allow various tests to be performed quickly and easily to help determine whether a component is nearing the end of its useful life. The system itself can also inform the maintainer of faults instantaneously, even remotely, with no input from the maintainer.

Most pieces of electrical apparatus have well-defined requirements for maintenance. Electric motor manufacturers provide detailed maintenance instructions for their equipment, from large span drive motors of 100 hp or more, down to smaller units used for driving span locks or other secondary devices. The manufacturers of gearmotors provide maintenance instructions for the gearheads, as well as for the motors driving them. Manufacturer's maintenance instructions are also provided for brakes, whether they are drum type, solenoid or thruster operated, disc type, or other types of brakes that may be used on a movable bridge. Large and small switching apparatus also comes with manufacturers' maintenance instructions, although for many smaller devices, such as track-type limit switches, it is probably more economical to replace them than to perform any more than the most elementary maintenance. Specially built control units should have maintenance manuals prepared by the manufacturer, specially written for the particular units, but such manuals should also contain the manufacturer's particular maintenance instructions for standard components built into the control unit. All these instructions should be followed by a bridge's maintenance personnel if reliable, trouble-free operation of the bridge is to be expected. Reputable manufacturers want their equipment to last as long as the original purchaser expects it to, so they put considerable effort into developing and disseminating maintenance instructions that provide the optimum service life for their equipment.

Certain pieces of electrical equipment are more sensitive to lack of maintenance or poor maintenance. Brakes, because they have wearing parts and are usually placed in machinery rooms, or even out-of-doors, where it is difficult or impossible to keep them clean, require more maintenance than other electrical components. The wearing parts of brakes are mechanical machinery and are often maintained, at least in part, by the machinery maintainers. DC motors and generators also have wearing parts, the brushes and commutators. These usually do not wear rapidly on movable bridges, as their duty cycles are very low as compared with such components in other industries and applications, such as steel mills and locomotives. The amount of active service seen by a DC motor on a movable bridge over a 20-year period may only be 1000 hours or so. Generators see more active time, as they usually operate much longer during a bridge opening—up to an hour, as opposed to only a few minutes for a drive motor. Generators are usually housed in a much cleaner environment than the bridge drive motors, so their deterioration over time is usually less than that of motors. Wound rotor-type AC motors also experience wear at the slip rings and brushes, but usually at a much slower rate than a DC motor with its segmented commutator.

On all such apparatus, DC motors and generators and AC wound rotor motors, the slip rings, commutators, and brushes should be inspected often and kept clean. Brushes should be changed early, before they become excessively worn. Commutators should be kept in good condition by polishing or remachining as necessary. This work is probably best left to a good motor rebuilding shop, unless the bridge maintenance crew has the requisite experience and facilities. Such work can be done in place, but it is better to do it in the shop if a replacement motor is available or the motor can be taken out of service without hampering bridge operation. It is important not to make the mistake of removing one motor of a two-motor installation if both motors are required for bridge operation. The single motor expected to do the work of two will perform, but will heat up considerably more, especially if operating conditions are severe. This motor will soon require rebuilding, possibly including rewinding, so that the other motor, having been repaired, will become overworked while the latest motor is in the shop, resulting in a constant removal of alternate motors as each is overworked trying to operate the bridge while the other motor is "out."

Brakes should never be removed and taken to a shop for rebuilding unless a spare brake of the correct type and capacity is available for temporary or permanent replacement. If an entire spare brake is not available, spare thruster or solenoid units should be on hand so that a malfunctioning brake can be easily put back in service and the offending thruster or solenoid can be rebuilt at leisure. Because many—in fact, most—pieces of such electrical equipment are common to many bridges, an owner of several movable bridges can have a central shop for repair of electrical components and keep a stock of rebuilt units for quick installation in place of a malfunctioning unit. Manufacturers may also have rebuilding programs, allowing defective units to be exchanged for new or rebuilt ones.

Special maintenance attention must be given to warning and barrier gates, as many of these devices are built to very low standards of durability and quality. The operating mechanisms for these devices should be checked fairly frequently, at least on a monthly basis, and loose components retightened and wearing parts relubricated where possible.

28

MAINTENANCE PARTICULARS

BASCULE BRIDGES

A key element in the successful operation of a bascule bridge is the maintenance of the main supports. Trunnion bascules should have their friction type trunnion bearings lubricated regularly. It should be ascertained that the lubrication passages are not clogged and that lubricant is passing freely into the loaded areas of the bearings. In many trunnion bascules with bronze trunnion bearings, there is grease all over the external surfaces of the bearings. The requisite amount of lubricant is being applied, but because of blocked channels, the new lube is passing through the path of least resistance to the outside of the bearing and the old, dirty grease remains in the bearing.

Most trunnion bearing assemblies have several ports for lubrication. These are usually arranged so that new lubricant can be injected at one end of a passage, and the other end opened by removing a plug so that the old lube will be pushed out by the new. This process should be repeated for each passage, until all the passages have been purged of old lube and filled with new. Unfortunately, in practice, this seldom is done, because of ignorance on the part of the maintenance staff or unwillingness to deal with the used lubricant. The common result is clogged passages, the external accumulation of grease noted earlier, and unlubricated sliding surfaces in the bearing.

Lubrication of gears and bearings is generally a requirement of good maintenance, but in some situations the application of grease to racks and pinions is not beneficial, and may be harmful. In a dusty environment, lubricant can hold particles from the air that act as abrasives, causing excessive wear on the gears being lubricated. The slow-moving racks and pinions on a typical bascule or swing bridge, if properly designed and not overloaded during operation, may last longer if not lubricated. Steel normally does not corrode rapidly in dry air, but grease can trap moisture against the steel surfaces of racks and pinions, allowing corrosion to develop. Unless the surfaces can be kept clean and dry, or are cleaned and relubricated often, applying lubricant to the racks and pinions does little to improve their longevity and is probably not worth the effort. Dry-type lubricants can be used, but their effectiveness is reduced if they are not reapplied often. As most modern bascules have all their other span drive machinery in enclosed drives, which do not need regular relubrication, the cost of maintenance can be noticeably reduced by eliminating the lubrication of racks and pinions.

On a double leaf bascule bridge, center span locks or shear locks are very fragile mechanical components. Most of these bridges develop excessive clearance at the center locks almost as soon as they are put in service. Maintenance forces should check these locks regularly, and repair or readjust them as needed, to eliminate excessive clearances.

Live load shoes and anchors on double leaf bascules are also subject to deterioration. The flexing of a leaf during application of live load tends to produce a reaction causing the leaf to bounce after the load is removed. This results, in extreme cases such as with heavy truck loading, in impact at the live load supports as well as at the center locks. This impact contributes to wear, which eventually results in poor seating, increasing the impact and increasing the rate of wear. Inspection and maintenance forces can keep this wear from developing by monitoring the condition of these components and correcting poor alignment by reshimming (see page 349) when necessary. All live load shoes on movable bridges are at a low position on the pier, near the sea wall, where they are exposed to water spray. The shoes do not have a protective coating on the contact surface, and so tend to corrode. The loss of material may not be an immediate threat to the live load carrying capacity, but it makes it difficult to maintain the proper contact between the upper shoe, or sole plate, and the masonry plate. Live load shoes on single leaf bascules are not as prone to deterioration as those on double leaf bascules, but should be checked and maintained in the same fashion.

Many simple trunnion bascule bridges, and other types of bascule bridges, have counterweight pits. If the floors of these pits are below water level, water can accumulate in the pits. If the water level is high enough that the counterweight dips into the water when the bridge opens, it can interfere with bridge operation. The constant presence of water in a semi-enclosed area can promote corrosion of the steel components. For these reasons, counterweight pits should be kept dry. If water is leaking in, steps should be taken to stop it. If water continues to accumulate, permanent sump pumps should be installed and kept operational, with prescribed maintenance procedures followed, to minimize the presence of water.

VERTICAL LIFT BRIDGES

The counterweight ropes for a vertical lift bridge are extremely vulnerable to deterioration and are difficult and expensive to replace. Great care is usually taken during fabrication and erection so that the ropes are in the best possible condition when the bridge is put in service. Without some amount of maintenance, it is virtually impossible for the counterweight ropes to last more than 10 to 15 years on a bridge that is operated often.

Unlike other wire ropes used for structural purposes on bridges, which can be painted or coated and expected to survive for a prolonged period of time, counterweight ropes operate over sheaves. The outer wires of the ropes are subject to harsh, abrasive wear that eventually causes them to deteriorate in spots (see Figure 12-2), until the wires get so thin that they eventually break at these stress concentration points as the rope flexes, passing over its sheave. One of the most cost-effective maintenance procedures on a movable bridge is to minimize this wear by keeping a coat of lubricant on the surface of counterweight sheave rope grooves. Unfortunately, many types of automatic lubricators have been tried at these points, but have failed to perform satisfactorily. Manual lubrication, either at the ropes or on the sheave grooves directly, appears to be the only effective means of minimizing abrasive wear between the ropes and sheave grooves. Rope lubrication should also include cleaning of the ropes, as old, dried rope lubricant or preservative material can trap moisture, accelerating corrosion. This material should be removed and disposed of whenever lubrication is performed. Hard to reach areas, such as at rope gatherers and separators, are especially important.

Manual application of lubricant to counterweight sheave grooves is a difficult, dirty job, and on many bridges it is made no easier by the lack of suitable access to the tops of the towers. Very few vertical lift bridges have powered elevators for access to the tower tops. Most highway vertical lift spans have only ladders, which are almost useless to provide access for maintenance at the tower tops. The usual expedient is to use the vertical lift span itself as an elevator, but this accomplishes only part of the job, as the best that can usually be done by this means is to get to the top of the counterweights. Many highway vertical lift spans have had platforms retrofitted informally so that there is a walkway from the end of the lift span deck to the counterweight top, accessible at the halfway open position. Railroad vertical lift spans usually have stairways, and sometimes material hoists, for tower top access, so at least access for maintenance is not impossible. The more often a vertical lift bridge operates, the more helpful good access to the tower tops for maintenance can be.

Counterweight ropes and other ropes that work under load over sheaves are subject to internal abrasion, between the outer wires of strands in contact and between the wires within each strand. Both AASHTO and AREMA specify that counterweight ropes should be made with fiber cores, saturated with appropriate rope preservative material or lubricant, to reduce internal wear on the ropes. The lubricant within the rope core is also expected to retard internal corrosion of the rope wires. It has been argued that the fiber core is not very helpful in this regard as the lubricant cannot be replenished (see page 174), but the failure of a vertical lift bridge counterweight rope because of internal wear or internal corrosion is extremely rare. This is most likely due to the infrequent operation of bridges, the large ratio between rope and sheave diameter specified by AASHTO and AREMA, and to the large safety factor they specify. Some credit should also be given to good specification writing for the ropes installed with the bridge, as some fiber cores are not as good as others.

Manual cleaning and application of lubricant/preservative material appears to be the only practical way of doing a good job, as there are areas at the ropes that are not amenable to reaching by means of an automated mechanical piece of equipment. Automatic devices that have been developed apply rope lubricant/preservative to the ropes on the sheaves, or to the sheaves for transfer to the ropes as the bridge operates. Devices could be developed to clean the ropes as the bridge operates, and have been developed to clean and lubricate wire ropes in other fields. These devices, unfortunately, require maintenance themselves and on movable bridges are usually found unused or in a state of uselessness. Manual cleaning and reapplication of lubricant/preservative material to the ropes, even if not done as often as might be ideal, if it is done well will add considerably to the life of counterweight ropes. It will reduce external corrosion and abrasion. If it does not add materially to the internal lubrication/preservation of the ropes, it will slow down, to some extent, the deterioration and loss of the internal rope lubricant/preservative material that was put in the ropes when they were manufactured.

Vertical lift bridges have the same difficulties with live load shoes as bascule bridges, but they are typically compounded by the need to provide for considerable longitudinal expansion over the length of the bridge. The shoes carry the live load and only a small portion of the dead load of the bridge, but contact pressures can be substantial. Abrasive wear can be significant and can occur at both the "fixed" and "expansion" ends of the vertical lift span. This wear is typically uneven and must be compensated by reshimming the shoes, a difficult process that requires closing the bridge to traffic. "Shortcut" shimming is sometimes done by placing metal plates at the contact surfaces. These usually do not last long, as the thin plates have point or line contact and are highly stressed under live load; thus they deteriorate rapidly (see page 349). Tower drive and span tower drives often must have the drives reset to bring the vertical lift span to full seating in the lowered position. This can be accomplished readily with clutch reducers, disengaging clutch couplings or adjustable couplings. Without these devices it may be necessary to dismantle part of the bridge drive to get all four corners of the span to seat.

SWING BRIDGES

End wedges for swing bridges, when not maintained in a well-lubricated condition, tend to stick when driven so that a poorly designed or poorly maintained drive system cannot pull the wedges out. To solve this problem, maintenance forces sometimes reduce the lift provided at the wedges, reducing the vertical force and the likelihood of sticking. This usually leads to damage, and sometimes failure, of the ends of the main bridge members and to deterioration of the masonry at the pivot piers in the immediate end wedge area. Proper lubrication of the wedges' sliding surfaces will prevent their sticking, but lubricants able to withstand the harsh conditions of the application must be used, and they must be reapplied often. This is a difficult, dirty job that has not yet been successfully automated.

The balance wheels for center pivot type swing bridges sometimes come out of adjustment. If a bridge is unbalanced, it tends to operate with part of the dead load supported on a few of the balance wheels. These wheels are not intended to be constantly supporting heavy loads, so they and their bearings tend to wear significantly under such conditions. This results in larger than intended clearances at the balance wheels and more tilt as the bridge swings. Most balance wheels are fitted with adjusting mechanisms, sometimes consisting of large-diameter threaded rods that form part of the connection between the balance wheel and the bridge superstructure (Figure 28-1). Whenever the balance wheel clearances become excessive, adjustments should be made to bring the total clearance at the wheel, with the bridge supported on center wedges and end supports, to $\frac{1}{16}$ in. Dead load on balance wheels should be eliminated by balancing the swing span.

The support rollers of rim bearing swing bridges require very little maintenance. The roller load surfaces and tread plates do not wear, as long as the rollers are well aligned and do not slide. Nor do they tend to corrode heavily, as long as they are kept clean and free of debris, dirt, and road salt. The thrust bearings at the outboard end of the rollers tend to wear and should be kept lubricated. Most rollers were originally supplied with oil holes for application of lubricant. In such cases, the oil has to be replenished every few days, as it tends to run out. Some bridges have had the oil holes retrofitted with grease fittings. The grease tends to stay in place longer than the oil, but it will still work its way out of the loaded bearing area after a few operations, so frequent reapplication of grease is necessary.

Racks and pinions of swing bridges, like bascule bridges, can suffer in dusty environments from too much lubrication, particularly if old lubricant is not removed regularly and new, clean lubricant applied (see page 357).

Figure 28-1 Swing Bridge Balance Wheels. The center bearing swing bridge superstructure requires stabilization when not on its live load supports. Balance wheels, spaced around a circular track, are used.

Centering latches for swing bridges are usually gravity operated. As such, only small forces are applied to activate them. The components must be kept free to operate without binding and sticking. Lubrication is helpful, but more to prevent corrosion that can cause binding than to reduce wear, which is slight on the activating mechanism due to the small forces involved. The latch pockets must be kept clean of debris, and of ice and snow in cold weather if the bridge operates then, to prevent failure of the latch bar to engage. The counterweight on the actuating mechanism must be kept in the proper position, so that the bar drops lightly under gravity. Insufficient counterweighting will allow the bar to engage when the bridge is operating too fast, and result in damage to the bar, the pocket, or the bar guides and supports. Excessive counterweighting will prevent the bar from dropping into the pocket, except when the span is traveling extremely slowly, or keep it from dropping altogether. If there are two latch bars, one on each end of the swing span, they must be kept aligned so that both drop into their pockets when the bridge is centered.

V

INSPECTION

Ideally, all inspection of movable bridges should be coordinated with maintenance. To some extent, this is accomplished by maintenance inspections (see Part IV), but a great deal of inspection work, particularly when performed as part of biennial inspections with outside consultants doing the work, is almost totally separated from maintenance efforts. These inspections may accomplish the goal of establishing the condition of a bridge, but do little directly in improving its condition or longevity. By coordinating maintenance with inspection to a greater degree, the efficiency of inspection will improve. The degradation of bridges as a result of inspection activity will be less likely, and their condition can be greatly improved at little increase in cost. Unfortunately, the Federal Highway Administration (FHWA)-mandated biennial inspection process makes it very difficult to combine inspection with maintenance. The biennial inspection process is determined by the *National Bridge Inspection Standards.*[1] All other inspection efforts should combine maintenance with inspection to the greatest extent possible. The inspection effort will be more productive, the finished inspection product will be more useful, and the condition of the bridge will, in the end, be better.

[1]United States Government, *National Bridge Inspection Standards,* Code of Federal Regulations, Title 23, Part 650, subpart C, Washington, D.C., 1988.

29

INSPECTION BASICS

The most important aspect of inspection work is to *do no harm.* This means no injuries to the inspector, no damage to the bridge or its components, and do not put the public at risk when inspecting or testing movable bridge components. Make certain that warning and safety devices are functioning properly before doing any operational tests on the bridge. If there is a defect in a component, such as a barrier gate, make certain that the bridge operation staff has adequately compensated, such as by providing a flasher vehicle to protect the draw opening, before continuing with tests. Contact the operation or maintenance supervisor directly if necessary. If it is necessary to curtail the inspection for safety reasons, prepare a written explanation of the problem, with dates and locations as well as the nature of the malfunction. Indicate the effort made to correct the problem, or the means by which the inspection was allowed to continue, with names and titles of bridge supervisory personnel who are involved personally.

It should be recognized by the seasoned inspector that movable bridges, being rugged products that are usually somewhat overbuilt, can sustain a certain amount of deterioration before a truly dangerous situation arises. It is nevertheless important to recognize that anecdotal evidence of a substandard or deteriorated condition having been in existence for a long period of time, perhaps several years, does not preclude failure in the near future.

Most of the details of bridge inspection are covered in the FHWA's *Bridge Inspector's Training Manual* (1993, revised 1995), which is readily available at a moderate price. Movable bridge inspection details, including specific information on electrical equipment and machinery, are provided in AASHTO's *Movable Bridge Inspection, Evaluation, and Maintenance Manual* (1998), which is also readily available but is not inexpensive. In the literature on inspection of movable bridges, the emphasis has been on machinery and electrical installations, and this made up almost all of the content of the earlier FHWA *Bridge Inspector's Manual for Movable Bridges* (1977), which is no longer available. It has since been recognized that that the entire movable bridge is a special case, with issues concerning superstructure, substructure, and traffic maintenance that do not apply to fixed bridges. These areas have been discussed elsewhere in this book, and this part (V) discusses them as they apply to bridge inspection. Full details of inspection, as a "how-to" guide, are not presented, as this information is provided elsewhere. Exceptions are made for situations that require additional emphasis.

There has been much research done on high-tech nondestructive inspection techniques and equipment in the bridge industry. This effort has produced little of consequence in the movable bridge area, but some breakthroughs have been achieved. These are discussed in the chapters covering the specific areas to which they pertain. It must be realized that the entire movable bridge inspection process is inherently nondestructive, as a major goal of any inspection, with very few exceptions, is to do no damage to the bridge or any of its equipment. Most of the exceptions are not in the traditional areas of movable bridge inspection. They include core sampling of concrete roadways, piers, and other structures; sampling of metallic members for laboratory tests; and other more specialized sampling and testing. The structural aspects are not discussed here. Destructive testing as applied to areas particularly relevant to movable bridges, such as counterweight ropes (rope testing), are discussed within the particular areas or categories of inspection.

As movable bridges are a very specialized area of structural engineering, there are particular areas of movable bridge inspection that require specialists in order to get the job done properly. The inspection of movable bridge machinery, including hydraulic systems, requires a mechanical engineer. Inspection of movable bridge electrical systems, of all types, requires an electrical engineer. Movable bridge superstructure and substructure require a structural engineer, or perhaps more than one structural engineer, with a concrete specialist and a steel specialist, and perhaps a deck specialist, on an inspection team. It may also be necessary to have an architect on the inspection team. Many times bridge maintenance people will be aware of an upcoming inspection, and rush out to perform maintenance work to put on a "good show." This should be avoided, if possible, so that the inspector can observe the results of typical maintenance.

INSPECTION SAFETY

Movable bridges are dangerous, heavy pieces of machinery. There are a great many codes and legal regulations providing for safety of the public and the workplace. The codes applicable to movable bridge inspection vary from location to location. Any crew preparing to perform inspection work on a bridge should be aware of all the applicable regulations and be prepared to follow them. Bridge inspectors should always be aware of and concerned for their own safety. There are many dangerous aspects of movable bridges. Not unlike a fixed bridge, a movable bridge can be quite high above the water or above normal ground level along the bank of a waterway, so a fall can be harmful or even fatal. The traffic on a bridge, whether railroad or highway, can pass by very rapidly. Automobile drivers and train engineers may be unaware that an inspection is taking place and may not see, or may ignore, warning signs or cones. Some drivers may actually object to the presence of inspectors or other personnel on the bridge and make attempts to run them over, or at least scare them off. A field inspector should not assume that he or she is safe because a flagman has been assigned to him or her or because a roadway or railroad track has supposedly been closed for inspection.

Safety practices during the inspection of a movable bridge should also take navigation vessels into account. No activity other than observation of the bridge operation should occur when a bridge is opening for a ship or boat. The bridge operator should concentrate on performing a safe operation and not worry about what inspectors are doing. The bridge inspector should in no way be responsible for any activity that can be a danger to the vessel or to the bridge while a vessel is passing.

Movable bridges have heavy machinery that can move quietly and unexpectedly. An inspector should not assume that he or she is safe just because the bridge operator has been informed of the inspector's presence, or because there is a warning placard placed on the bridge control console. Inspection operations requiring a person to place an arm, a hand, a leg, or body into a space in which machinery can operate should not be done without guarantees that no accident will happen. Preferably, in such situations, the power should be disconnected or a colleague should be stationed at the controls to prevent inadvertent operation. No hand drives should be tested until it is absolutely confirmed that electric or other main power source has been cut off from the machinery. More information, including details on providing a safety plan for a specific movable bridge inspection project, can be found in *AASHTO Movable Bridge Inspection, Evaluation, and Maintenance Manual,* Part 1, Chapter 1.3.

CLEANING OF BRIDGE AND COMPONENTS

Many movable bridge components require thorough cleaning and removal of undesirable materials before the inspection work can be completed. There have been cases of movable bridges, that operate extremely infrequently, having the machinery rooms completely filled with bird droppings. Counterweight pits of bascule bridges have been found partially filled with stagnant, polluted water. The electrical installations of some older movable bridges have asbestos insulation that is in a state of decrepitude, or transformers or other devices containing PCBs. In most cases, the condition of the bridge or component prior to cleaning for inspection is important, so the inspector must make at least some observations of the bridge or component before any cleaning work is done. Cleaning may involve the removal of materials considered harmful to the environment, such as petroleum lubricants and asbestos insulation. The cleaning and removal work must comply with all federal, state, and local regulations and should be supervised and performed by persons familiar with those regulations. The procedures and materials used, such as solvents, should be in compliance with all regulations.

It is generally desirable for bridge maintenance forces to perform the cleaning of bridge components to be inspected, but the need to coordinate cleaning and removal with the inspection work may make it inconvenient or undesirable for the bridge maintenance forces to do this work. As part of the inspection effort, properly qualified cleaning and hazardous material removal forces can be employed by and supervised by the inspection personnel, and their work coordinated with the inspection. It may be necessary to perform an initial inspection, then perform the cleaning and hazardous material removal, and finally perform the rest of the inspection work. If removal of hazardous materials is required, proper disposal is the responsibility of the owner of the bridge.

TYPES OF INSPECTION

Many bridge owners, such as state departments of transportation, have different levels or degrees of inspection specified for different inspection projects within a chronological cycle. Federal requirements call for each highway bridge to be inspected once every two years. Such inspections are called "biennial" inspections. Because they form the bulk of bridge inspection efforts, biennials can be considered the norm, or standard level of inspection. Biennial inspections are intended to prevent life-threatening situations, those that can pose a danger to the traveling public, from developing on bridges. Biennial inspection regulations have very little to say about movable bridges, except in structural terms.

The *AASHTO Movable Bridge Inspection, Evaluation and Maintenance Manual* describes three different "methods" of inspection:

Routine condition examination
Performance examination
Special examination

The manual also describes five types of inspection, as taken from AASHTO's *Manual for Condition Evaluation of Bridges.*[1] According to the *Movable Bridge Inspection, Evaluation, and Maintenance Manual,* these are as follows:

Initial Inspection—a routine condition examination and a performance examination

Routine Inspection—the basic condition inspection along the lines of a biennial structural bridge examination, with or without disassembly of movable bridge drive components

Damage Inspection—routine condition examination, performance inspection, and special examination as necessary

[1]American Association of State Highway and Transportation Officials, *Manual for Condition Evaluation of Bridges* (Washington, D.C.: AASHTO, 1994).

Special Inspection—routine condition examination, performance inspection, and special examination as necessary

In-Depth Inspection—a routine inspection plus disassembly and cleaning of most machinery components

More detailed information as to the content of these inspection types is contained in AASHTO's *Manual for Condition Evaluation of Bridges,* written primarily with fixed bridges in mind, but the descriptions of the inspection types apply equally well to movable bridges, at least to the common types. This manual contains a short section on movable bridges that provides some guidance in applying its principles to movable bridge inspection. The FHWA's *Bridge Inspector's Training Manual*[2] contains a fairly extensive section on movable bridges.

The various bridge owning entities define different levels of inspection differently and may call them by different names; there may be more or fewer types of inspections required by the owner of a particular bridge. The inspection process is generally more formalized for highway bridges, as federal regulations are involved, than for railroad bridges. Most railroad bridges are inspected on a regular basis by in-house personnel; in some cases the same people who perform maintenance also perform inspections. Regardless of the level of the inspection, specialists in the particular field, as noted earlier, are necessary to perform an inspection properly and obtain worthwhile data from it.

The inspection criteria for movable bridges may be simplified somewhat from the broad AASHTO categories. Most movable bridge components that cause a bridge to operate are not part of the direct bridge supporting system and are generally not required to undergo intensive inspections every two years. This standard applies particularly to the electrical and machinery components of a movable bridge. A slight modification can be made, for the purpose of movable bridge inspection, of the names and categories of the inspection types. There may be an inspection every two years, the first being a "walk-through"; the second, after two additional years, an "in-depth"; and the third, two years later, a "special."

The Walk-Through Inspection A walk-through inspection is intended to quickly establish the overall condition of a movable bridge. It is often performed as a preliminary to performing an in-depth or a special inspection. In some cases a walk-through inspection is used to determine the future of a bridge without going to the expense of a full-blown in-depth inspection effort. It is important that the walk-through inspection be performed by competent personnel, as faulty conclusions may result in an excess of money being spent, rather than a savings. There are cases of bridges having been condemned on the basis of walk-through inspections, although in fact, they were not in imminent danger and could have been rehabilitated economically.

[2]Federal Highway Administration, *Bridge Inspector's Training Manual* (Washington, D.C.: FHWA, 1995).

A walk-through inspection, as its name implies, is characterized as a "hands-off" visit to the site. All accessible components are observed—and perhaps some are exposed, preferably by the bridge owner's staff, such as by removing certain protective panels—but no detailed inspection of any components is made. If possible, the bridge is operated one or more times for observation. The inspection may last one day or several, depending on the complexity of the bridge and the amount of information desired. The report on the bridge may consist of a brief statement of findings, a simple letter highlighting the findings and conclusions, or an oral presentation to the owner or the owner's representatives. A walk-through inspection of movable bridge mechanical and electrical systems can satisfy biennial bridge inspection requirements, when combined with a proper structural inspection.

The In-Depth Inspection In-depth inspections are performed "by the book." Inspection work is performed according to the procedures described in the FHWA bridge inspection manuals. For special components of a movable bridge, the checklists in *AASHTO Movable Bridge Inspection, Evaluation, and Maintenance Manual* are followed. Inspection and testing equipment is used as required for the machinery and other components present on the bridge, based on what is called for in the manual for the particular component being inspected. To perform the inspection efficiently requires proper preparation. Preparation includes a review of all available drawings for the bridge, a tabulation of components to be inspected, with diagrams showing the components, calculation of measurement parameters for certain components, and gathering the necessary tools to perform the inspection.

The Special Inspection In many cases, because of concerns arising from damage that has occurred, or because of a lack of complete drawings showing the details of the bridge, or simply as a matter of policy, an intensive inspection of a movable bridge, or part of it, is required. This inspection includes a complete evaluation of certain components, perhaps necessitating disassembly of those components. The inspection may also require taking the bridge out of service to road traffic, waterway traffic, or both, for a specified period of time. On the other hand, in the case of damage, a bridge may be taken out of service to waterway or roadway traffic until it is deemed safe. This type of inspection necessitates careful planning, arranging for closing dates, and ensuring that the required personnel and equipment will be on hand when the inspection commences. The inspection may call for removal and destructive testing of components of the bridge, which may require procurement of replacement parts to be ready at the site when the inspection commences.

If an inspection contract is being negotiated, the intensity of the inspection effort, as indicated in the preceding descriptions or in another general description, should be clearly spelled out. Any special features of the movable bridge, such as swing bridge center pivots, vertical lift bridge counterweight sheave shafts, or bascule bridge operating struts, should be discussed, and the type and amount of inspection required should be clearly detailed. The amount time allowed for the bridge to be closed to navigation or bridge traffic should be clearly defined. The source of labor and equipment, such as boom trucks, to assist in performing of this type of inspection work should also be defined. Also to be defined is responsibility for bridge parts, such as bearing cap or base bolts, that may be found to be broken, or may be damaged during the inspection. A large turned bolt may cost several hundred dollars to replace. Worn and damaged bearing bushings may be functional when in place, but may be impossible to return to service after disassembly for inspection. It is best, if possible, in conducting inspections that require disassembly of major components, to have the bridge owner's maintenance crews work with the inspectors so that such damage uncovered may be repaired as quickly as possible. The maintenance crews can readily perform certain maintenance tasks, such as cleaning or reestablishing lubrication passages in bearings, during the inspection work. It is generally inadvisable to have consultants responsible for performing dismantling or other heavy work as part of the inspection, so as to avoid putting fragile or perhaps irreplaceable components at risk by having unskilled hands working on them.

PLANNING

Depending on the degree of intensity of an inspection, more or less planning may be necessary to ensure that the inspection is successful. If assistance is required, either from the bridge owner's repair, inspection, or maintenance personnel or from outside providers such as rigging contractors, testing companies, or subconsultants, an inspection plan and schedule is necessary. It is best if a written plan is prepared in some detail. Even if it is only necessary to coordinate the efforts of more than one discipline within the inspection team, it is a good idea to plan ahead. It may be useful for inspection of some components to have more than one specialty available at the same time to perform a joint inspection. Rigging or falsework may be available for only a short time, so all activities requiring access to certain less accessible places should be scheduled together or in close succession. For other inspection work it may be necessary to avoid having too many people in one area at the same time, so inspections by different disciplines in those areas are best scheduled at different times. As far as the individual inspector is concerned, if time is critical, all components in a given area should be inspected at the same time, to avoid the inspector's having to retrace steps. If time permits, however, it is usually helpful to look at all the related components on a bridge at the same time so that problems of interaction are more readily picked up. The inspection specifics in the following chapters are arranged in this manner.

It may be necessary to arrange ahead of time for special bridge openings for inspection purposes, particularly if an auxiliary or emergency drive is to be used. If closure of the navigation channel is required for inspection purposes, even for a short period of time, it may be necessary to arrange for notification to mariners or other public announcements of the closure. On some bridges with heavy highway or railroad traffic, special notification or scheduling may be necessary before lane closures or test operations.

PREPARATION FOR INSPECTION

Regardless of the intensity of the inspection effort, prior information about the condition of the bridge should be reviewed by the inspectors before commencing field work. This information can include previous inspection reports, documentation regarding repairs, operating logs, or any other information about the condition of the bridge. Bridge owners should be asked specifically about any such information and should be requested to provide the information to the inspectors as early as possible, preferably before preparation for the field work starts. Some movable bridge owners are reluctant to provide this information, preferring to have the information verified independently or fearing that the inspectors will just repeat the same information provided on the previous inspection. In the case of a movable bridge, however, the condition of the bridge, its machinery, and its equipment seldom remains the same between inspections, so this fear is groundless. It is actually a valuable addition to an inspection and evaluation to have all the prior data, as the rate of deterioration of components can be more accurately calculated (see Chapter 39).

It is always helpful, regardless of the intensity of the inspection, to have diagrams of the areas to be inspected prepared beforehand. These can be simple schematics of the superstructure, electrical circuitry, or machinery of the bridge. They can be combined with tabulated listings of bridge components, including spaces for the inclusion of inspection findings. For in-depth or special inspections, detailed diagrams and lists of components should be prepared so that items can be checked off or entries made for them as they are inspected, minimizing the likelihood of missing a component during the inspection. Each item of the diagram should have a unique identification code using numbers, letters, or a combination. The identification code should also appear in the data table (see Figure 34-5) and be used for identification of the component in any discussion.

As indicated earlier, many owners have their own standardized form or forms of inspection. A particular inspection required for a bridge may resemble one of the types of inspection described earlier, or it may be a combination of types, depending on the requirements and policies of the owner.

Most owners have become quality conscious over the past few years and require discrete quality control and quality assurance efforts by an inspection team, particularly if the inspection is not performed by in-house personnel. The effort required to comply with these requirements may be extensive and is usually additive to the effort required to actually perform the inspection work. The *AASHTO Movable Bridge Inspection, Evaluation, and Maintenance Manual* contains an extensive discussion on the subject of quality control and quality assurance as applied to movable bridge inspection.

For a list of desirable equipment for an inspection, see the specialized categories discussed in the following pages.

30

STRUCTURE

SUBSTRUCTURE

Many movable bridges, of all types, develop difficulties with movement of the substructure. Several structures, bascule, vertical lift, and swing bridge types, have had interference problems in which the movable span no longer seems to fit between its approach structures or abutments. When this phenomenon is suspected or has been reported, the inspector should make careful note of various aspects of the substructure and superstructure in order to determine, if possible, what has been occurring and report it so that productive further action can be taken. Contrary to a fixed bridge, which acts largely as a continuous structure even though it may have several spans with expansion joints between them, a movable bridge actively separates and produces a gap in the superstructure every time it opens. There seems to be a tendency, as in the adage "nature abhors a vacuum," for movable bridge approaches to creep toward the draw span.

On a multiple-span bridge with a movable bridge as one of the spans, the adjacent spans usually have fixed bearings at the pier where the movable span meets the fixed span (see Figure 30-1). This allows better control over the joint between the fixed span and the movable span. The first thing an inspector should check, when structural interference is suspected, is whether general movement of the structure toward the navigation channel is, in fact, the case. The movable span, such as, particularly, a swing bridge mounted on a center pier, may have moved to one side. If both ends of the movable span show signs of interference with the approach superstructures or abutments, or if clearances are normal at one end and tight at the other, then creep of some sort is a real possibility. If one end of the movable span has tight longitudinal clearances while the other has larger-than-expected clearances, allowing for possible prior selective removal of superstructure to eliminate an interference, then the problem may be that the movable bridge pier has shifted, possibly as a result of impact from

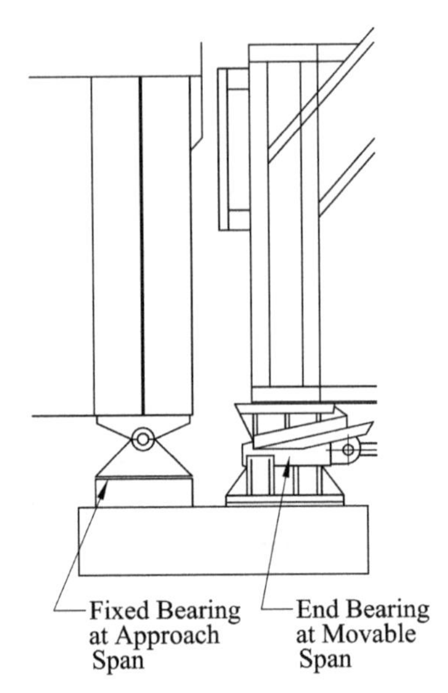

Figure 30-1 Approach Span Bearings. The approach span bearings adjacent to the movable span are fixed type bearings to minimize the change in the gap between the spans.

a ship or other vessel. This possibility is especially likely in the case of a swing span with its pivot pier in the middle of the navigation channel, as this is one of the movable bridge configurations more vulnerable to ship impact. It is possible that the pier is unstable because of undermining or scour. No noticeable movement should be detectable at the pier when the bridge is operated or when heavy traffic passes over the bridge. If such movement does occur, further investigation is warranted.

If center pier movement is the cause of the difficulty, the next step is to determine the amount of movement, whether it is ongoing, and whether the pier is sufficiently unstable to present a near-term danger to the public. Careful measurements of the pier's position can be taken. These measurements should be repeated over intervals of time, taking into account temperature changes and any other phenomena that may cause a difference in the pier's measured position. A series of measurements should be taken, so that any change in the rate of movement can be detected. If possible, measurements should be taken in summer and winter to get the full effect of seasonal temperature changes. The overall time frame of measurements should be a few weeks at least, but should be as long as is practical in the situation, until solid, meaningful evidence is gathered. Some movements may not be evident until a lapse of years has occurred, so measurements should be carefully recorded for future use. Internal inspection of the piers may be warranted, such as by taking corings, and underwater inspection should be performed, if it has not been already, to determine whether scour is a factor or whether there is visible deterioration in the lower portions of the pier.

After the main movable bridge pier or piers have been examined, the next step is to determine whether there has been movement at the other piers or the abutments.

Plumbness of a pier can be checked directly, or the spacing between pier tops can be compared with dimensions shown on design or as-built drawings. It is sometimes found that all the pier tops at a given approach are tilted in the same direction, proportionately the same amount. "Frozen" expansion joints of highway bridges usually occur because the joints become packed with dirt, so they can expand, but it is difficult for them to contract. Railroad bridges do not generally experience this problem, but the spaces between the end of the bridge and the abutment often become solidly packed with ballast. Frozen joints can exert tremendous longitudinal force on the pier tops. This can cause the tops of the piers to move, as the frozen joints allow sections of superstructure to move away from each other, but not to come together, over the length of an approach span.

As mentioned earlier, movable bridges can become unstable because of deterioration of the footings, to a greater extent than fixed bridge piers. Occasional high wind loadings on a movable bridge when in the open position can cause displacement at the footings, which results in imperfect support when the bridge is in position to carry traffic. Such displacement is more likely to occur with bascule bridges, but can also affect vertical lift and even swing bridges. The condition can sometimes be detected by observing movements of the bridge under traffic. Any unusual lateral or vertical movements can be cause for concern. If foundation deterioration is suspected, further investigation is warranted. There are numerous texts and manuals on foundation investigation available for reference in such situations.

Lateral movement at the piers of movable bridges can also occur. Vessels can impact piers directly, causing them to move out of alignment. Rivers with strong spring floods can wreak havoc with bridge piers, as large ice floes, groups of large trees, or other debris can be swept downstream and easily cause displacement of movable bridge piers. Fixed bridges, because of their continuity, have redundancy in resisting these forces that is not always considered in design. Movable bridge piers can be undermined by scour so that they fall or tilt of their own weight or become displaced when the first live load reaches them.

Bascule bridge counterweight pits should be checked for leaks and for proper drainage. If leaks are evident, their source should be determined. If a leak is coming from below the top of the pit, it may originate at a crack in the pit wall. Open areas allowing rain or other water to enter should be noted.

Pier protection, such as fenders and dolphins, should be checked for adequacy. Any loose or missing boards or planks on fenders, where a vessel could be snagged, should be noted. Any rotten, broken, or cracked piles or timbers, or fire damage, should be noted. Core samples should be taken where internal deterioration is suspected. Bolts and lag screws should be checked for corrosion.

SUPERSTRUCTURE

Some older movable bridges, particularly highway bridges, were designed for the cheapest possible construction consistent with the desired load capacity. This type of design/detailing process was often used for the sidewalks of a bridge. Many older movable bridges were built with a through-truss configuration, with the highway lanes occupying the entire width between the trusses. Concrete sidewalks were added, with the minimum of additional structure width, by pouring them right around the truss members—verticals, diagonals, and portals—that projected through the sidewalk. In many cases these partial encasements allowed saline moisture to remain in contact with the structural steel members for prolonged periods of time. This has, in many cases, resulted in significant corrosion loss, which may not be easily detectable. The section loss may occur at the steel just above the sidewalk surface, or it may occur within the area encased by the concrete. To determine whether such deterioration is occurring or has occurred, inspectors should carefully probe such details. Any signs of significant corrosion at the steel near the concrete surface should be explored further. Signs of swelling or spalling at the concrete in the area of the structural steel are strong indications that the metal is corroding. If corrosion is confirmed or suspected, a program should be initiated to remove the partial encasement and fully inspect the members affected.

In addition to fatigue cracks in the base metal, these bridges can develop looseness in fasteners at joints connecting component structural members that experience alternating compressive and tensile loads during bridge operation or live loading. These loose fasteners can sometimes be identified by fretting rust emanating from them. If loose fasteners are suspected, the inspector should not be afraid of striking the fasteners sharply with a heavy hammer, weighing 2 or 3 lb, as a large force may be required to produce visible movement at the "loose" fastener. Care should be taken with bolts, even high-strength bolts, so as not to damage the nuts, bolt heads, or threads when hitting them.

Bascule Bridges

There are many areas of the superstructure of a bascule bridge that call for special diligence by the inspector. Highway bascule bridges in particular have potential superstructure problem areas unique among bridges. The most obvious is the deck itself. A bascule bridge requires special heel deck joints to allow it to rotate about a horizontal axis when opening and closing. Deterioration or contamination at these joints can result in difficulty with movement of the bridge. Minor damage can cause interference. Swelling at connections due to corrosion, which is common on highway bridges, can cause interference that makes a bridge difficult or impossible to operate. Double leaf bascules may make contact at the toe joints between the leaves, indicating possible span or pier movement. Some bascule bridges have jointed deck plates at sidewalk areas that can deteriorate and interfere with operation. Inspectors should carefully observe the joint components and make note of any signs of abrasion between moving and non-moving parts. If possible, these joints should be observed during bridge operations to confirm such interferences.

As mentioned elsewhere (Chapter 10), bascule bridges can be subject to severe stress reversal within the superstructure when opening and closing. This is more marked when the bridge opens to a large angle, approaching 90°, as is in many cases the situation, particularly with older heel trunnion type bascule bridges. Many of these bridges open to a large angle to save on movable span length while still providing the desired navigation channel width. Stress reversal can occur not only on the bridge leaf, but also on the counterweight support frame, which operates through the same angle but may have even more pronounced stress reversal, depending on the orientation of the counterweight.

All superstructure areas on a bascule bridge that opens to angles greater than about 45° should be checked carefully for signs of fatigue distress. This distress will most likely take the form of loose or broken fasteners, but cracks in the base metal can also show up. Theoretically, the trunnions of bascule bridges can fatigue as the result of stress changes during opening and closing cycles, but this is almost impossible unless (1) the trunnions were underdesigned in the first place and (2) the bridge opens to a very large angle, and does so often. The trunnions must have a substantial stress change, and must experience it often, for fatigue cracks to develop. A fraction of a revolution per opening is miniscule as compared with the several complete revolutions and two complete stress reversals (tension to compression, and then compression to tension) per revolution of the counterweight sheave shafts on a vertical lift bridge. Heel trunnion bascule bridges that open to large angles can, over a long period of time, develop fatigue cracks. Particularly vulnerable are those heel trunnion bascule bridges that were built at an extremely low elevation and thus have to open for every vessel that passes, which results in more operating cycles per year. In addition, if these bridges were built with the bascule span designed to open to a very large angle, the stress change in some components per operating cycle can be very large. The traditional heel trunnion bascule has a counterweight suspended from a counterweight frame, or sometimes two counterweights, one on either side of the roadway, one suspended from each counterweight truss. Particularly vulnerable is the truss member between the counterweight and the counterweight trunnion, which is in compression when the bridge is lowered and in tension when the bridge is raised.

Tread plates and connecting angles at segmental and support girders for Scherzer-type rolling lift bridges are particularly susceptible to deterioration resulting from fatigue, regardless of the angle to which the bridge opens. The earliest deterioration of such bridges usually occurs at the area of the tread plates in contact when the bridge is closed, as a result of live load impact. The tread plates themselves may have cracks or complete breaks, particularly on older bridges that were built with thinner tread plates than are usually specified today. Cracks are likely to develop at the fillets of the connecting angles, particularly at the segmental girders rather than the supporting girders, because of a greater likelihood of corrosion as well as marginally higher localized stresses.

Counterweights should be checked for deterioration. Steel boxes should be checked for signs of swelling or heavy corrosion. Concrete counterweights should be examined for spalling. Some older concrete counterweights contain steel slugs or scrap iron to increase density. These inserts can corrode, causing the concrete around them to deteriorate. The structural steel supporting concrete counterweights should be checked for signs of corrosion where it emerges from the concrete. Counterweight pockets should be examined to see if they are free of debris. Counterweight pocket covers, if present, should be secure, watertight, and free of corrosion.

Vertical Lift Bridges

The vertical lift span is the closest to a fixed span of any type of ordinary movable bridge. This is the case whether the bridge is a truss or girder span. The only design features that distinguish tower drive vertical lift bridge spans from fixed bridges are the connections for the counterweight ropes and, possibly, some modification to the live load shoes. Vertical lift bridges are, however, vulnerable to deterioration modes that do not affect fixed bridges. If the vertical lift span does not seat firmly on its live load supports, impacts can develop under traffic that can result in fatigue damage to the superstructure where the live load shoes attach to it. This deterioration is especially likely to occur on railroad bridges. The inspector should carefully check the condition of the ends of the main girder or lower chord above the live load shoes to detect any damage or developing fatigue cracks. Dye penetrant or magnetic particle testing may be of assistance in locating these cracks. Fasteners, particularly rivets, can also be loosened or broken as a result of impact damage. Bolts and rivets can be tested by striking the exposed ends with a hammer (see page 376).

The inspector should be very careful when inspecting the towers to note any potentially damaging deterioration. Tilting or out-of-plumbness of the towers can be checked fairly readily, as both the vertical lift span and the counterweights are suspended from ropes and are guided by rails mounted vertically on the tower legs. These rails mate with shoes or slots on the lift span and counterweights, which prevent complete freedom of horizontal movement of the moving superstructure. Tilting of the tower can be detected in interference of these span or counterweight guides. The mating parts on the lift span or counterweights rub noticeably on the guide rails over the greater part of the operating distance of the span or counterweight if the tower has gotten far enough out of plumb. If the bridge operates fairly frequently, polished surfaces can be seen at the guides and at the rails wherever there is interference due to tilt. It is important to determine whether the interference causing the polishing is due to the tilting of the towers rather than a bent or mislocated rail or guide.

Counterweights of vertical lift bridges should be checked for deterioration in a manner similar to that for bascule bridges (see page 377).

Swing Bridges

Swing bridges, among all common types of movable bridges, are the most vulnerable to being struck by vessels passing through the draw opening. The single most important aspect of an inspection of the superstructure on a swing bridge is to look for ship impact damage. This may be relatively recent damage, noticeable by paint having been scraped off the lower parts of the superstructure. The damage may have occurred some time ago, and the metal surfaces may have visibly corroded or been painted over. The damage may be extensive, with lower chord members severely distorted or lateral bracing destroyed. The damage may be minor, with only a few rivet heads partially sheared off. Any such damage, even if it appears minor, should be noted, as it could affect the rating of the superstructure (see Chapter 38). Lower chord members of swing bridge trusses are primarily compression members, so any damage that causes a lower chord member to deviate from a straight line, or reduces the effectiveness of its bracing, can significantly reduce its load capacity.

Swing bridges, particularly railroad swing bridges, must be properly supported at the ends to avoid impact damage from live loads. The area of the lower chord at the intersection with the end floor beam should be closely inspected for signs of damage. The lower chord of the truss or the bottom flange of the main girder may have cracks and other deterioration where the upper casting of the end lift assembly connects to it. See Chapter 33 for a discussion of end lift adjustment.

The loading girders of center pivot swing bridges and the rim girders of rim bearing swing bridges should be closely inspected for corrosion, other deterioration, or damage of any sort. These members are very heavily loaded and can buckle or collapse if severely weakened at any point. Rim girders in particular should be checked for any concentration of load, such as rust packing at any of the points where the bridge superstructure rests on the rim girder, or signs that several rollers in sequence are not carrying significant loading. The bracing inside the rim girder should all be sound, with no loose or broken elements and no heavily corroded areas.

SUBSTRUCTURE AND SUPERSTRUCTURE—GENERAL

Equipment

For a walk-through inspection, the structural inspector needs few tools other than a notebook and perhaps safety equipment such as a hard hat and climbing equipment (i.e., a harness). For an in-depth inspection, the same equipment is needed, plus a good chipping hammer, such as an Estwing rock hammer. Hard corrosion is very difficult to remove, and it is often much thicker than it appears at the surface. When the loss is significant, direct measurement of the remaining material thickness, with the use of a caliper if necessary, and a steel rule is usually much more reliable than high-tech measurement, particularly if corrosion is present. Lumber crayons are about the best for marking the structure when necessary. The less extraneous equipment carried along the better. A small, cheap pocket camera takes pictures good enough for most purposes. Flashlights, ladders, electronic measuring devices, and other such material should be brought along when necessary. Bucket trucks, UB-50-type vehicles, and other pieces of heavy-duty access equipment that work from the roadway waste man-hours in preparation time and waiting for roadway closures. They should be avoided whenever possible and used only when direct access is impossible. Where areas cannot be easily reached by other means, climbing on the steel is best when it can be done. Inspection personnel who can climb the steel to do their work can be three or four times more productive than those who cannot.

Special Details

Movable railroad bridges have special details at the railroad rails at the ends of the movable span. The railroad tracks at these joints should provide a continuous, straight gauge line at these locations when the bridge is seated, providing solid support under the rails.

31

MACHINERY—BASCULE BRIDGES

SAFETY

Most bascule bridge operating machinery is slow moving, but operates with a great deal of force. Some of the machinery may operate fast, such as some cross shafts connecting gear reductions in the main bridge drive. All of the machinery is dangerous and must be treated with respect. It is sometimes desirable, for the purpose of inspection, to observe the machinery in operation. During such observations, it is essential to avoid the possibility of injury. When observing a bascule bridge's operating machinery from a vantage point on the bridge, it is necessary to be located in a safe area where falling into or onto moving machinery is impossible. A secure handhold should be readily available in case of sudden movements. Bascule bridges can operate very quietly and smoothly. It is important, when standing on the moving span, or under it on a pier, or beneath a counterweight, to avoid being struck or crushed by a moving or stationary part of the bridge. This is especially true with a Scherzer-type rolling lift bascule, as the segmental girder rolls very slowly and quietly and will crush anything in its path. Inspectors and maintenance personnel have been killed on bascule bridges, particularly when testing hand drives without taking proper precautions to avoid the engagement of the main power bridge drive while using the hand drive. Bascule bridges have the additional hazard of tilting as they operate, requiring extra caution when finding a place to stand during bridge operations. The machinery of most rolling lift bascule bridges is mounted on the moving leaf and thus also tilts during operation.

BASCULE MACHINERY—GENERAL

There is very little machinery of a general type that is unique to bascule type movable bridges. Bascule bridge trunnions are similar in size and configuration to the counterweight sheave shafts of vertical lift bridges. Bascule bridge drive machinery is very similar to the drive for a tower drive vertical lift bridge, except for having, in most cases, a mechanical differential mechanism.

SUPPORT AND STABILIZATION SYSTEMS

Bascule bridges, particularly double leaf bascule bridges, tend to bounce off their live load supports when heavy traffic passes over them. All parts of a bascule bridge should be observed while heavy traffic passes over them, and all movements should be noted. It is particularly important to observe vertical movements at the center locks on double leaf bascule bridges under traffic and, if possible, to measure the relative vertical movement between the tips of the opposing leaves and check the movement at live load shoes while traffic passes over the bridge.

Trunnions

Bascule bridge trunnions are, in most cases, force fitted into the bascule girders by means of very tight interference fits and supported on bearings. In rare cases the trunnions come loose in the bascule girders, even though heavy collars are used to help secure the trunnions to the girders. The motion of a trunnion relative to a girder may not be directly apparent, but rust may bleed from the joint or the paint on the adjacent surfaces may crack at the joint. A dial indicator properly placed may tell the magnitude of the relative movement during bridge operation.

In texts on inspection of movable bridges, much is made of the alignment of trunnions on a bascule bridge. Good alignment manifests itself in a smoothly running bridge that does not bind as it opens or closes. Poorly aligned trunnions can be the cause of damaged trunnion bearings, loose collars on trunnions, or loose trunnions themselves, and possibly structural damage near the trunnions. The trunnion alignment can be checked, in most cases, by stretching a wire through axial holes in the trunnions and comparing the alignment of the wire at the trunnions. This operation is made impossible at many trunnion bascules by the placement of machinery or structural obstructions between the trunnions. Indirect methods can be used when necessary for the alignment check, but result in a loss of accuracy. Accuracy with these methods of checking is not good to begin with, because of the small measurements being made, inaccuracy in wire placement, and sag in the wire. Inaccuracy or roughness in a supposedly axial hole in a trunnion results in inaccurate measurements even if a more precise method, such as by means of theodolite or laser equipment, is used to make the measurement. Originally, the "axial" holes in the trunnions were intended only as a sort of nondestructive test of the trunnion forgings, to help locate internal flaws, which were once prominent.[1] Especially on older bridges, the concentricity of the "axial" hole should be suspect. On some bascule bridges, the lack of structural rigidity causes angular deflections at the trunnions, as the bridge opens and

[1]John Fritz, *Autobiography* (New York: John Wiley & Sons, 1911), pp. 178–180.

closes. These deflections may be exceeded in magnitude by the inaccuracy of the alignment of the axial holes in the trunnions, so that an apparent misalignment in one direction may actually be in the opposite direction.

When damage at or near the trunnions suggests that trunnion misalignment may be a problem, it may be necessary to perform an accurate check of the trunnion alignment, which can be done with the proper equipment and procedures. The alignment should be checked from the trunnion journals, not the axial holes, if at all possible. This requires the removal of bearing caps so that the journals are exposed. Proper surveying techniques, performed to the required degree of accuracy, can then allow the trunnion alignment to be checked. Measurements should be made at two locations on the journals, 90° apart. The bascule bridge must NEVER be operated, even for small movements, with any trunnion bearing caps removed or loose.

On a trunnion bascule bridge with friction bearings at the trunnions, the trunnion journals should be exposed when called for in the scope of work, or whenever possible in an in-depth inspection. This should be done both with the bridge open and with the bridge closed, but the bridge should NOT be operated with the bearing cap off or the bolts loosened. Exposing the journals thus can allow the effectiveness of the lubrication to be determined. There should be no rust on the journal surface and no dry spots or burn spots from lack of lubrication. There may be grooves scored in the journal in the direction of movement. These are almost inevitable over time and are not of great concern unless they become very large. If the bearing is misaligned, contact between the journal and bearing will be incomplete and wear occurs in the bronze at the area the trunnion contacts. The non-contacting areas at both the bearing and journal may appear rough, dark, or dirty. The misalignment can cause excessive bending stress in the trunnion and result in a fatigue failure, unless wear occurs rapidly and provides for full seating and thus reduced bending stress in the trunnion (Figure 31-1). The reduced contact area may result in significantly higher bearing pressure, so that the lubricant fails, resulting in higher friction, more rapid wear, and possibly scoring of the surfaces. Otherwise, this is of little concern, unless the misalignment is so great that grease grooves in the bronze are worn away before enough wear occurs to provide a low enough bearing pressure that wear is arrested. If the trunnion journal is not perfectly aligned, incomplete contact and excessive bending stress in the shaft at the journal and excessive bearing stress at the bronze bushing can also occur (Figure 31-1). In addition, the bearing housing may be twisted or torn loose as the bridge opens and closes, with deterioration at the bearing foundation bolts. If self-aligning bearings are used, compensation is made for poor bearing mounting but misalignment of the trunnions will still be evident, as the bushing will oscillate in its spherical support as long as lubrication allows it and the spherical surfaces do not seize. Such oscillation can cause wear at the mating spherical surfaces.

Many heel trunnion bascule bridges are equipped with very large specialized sleeve bearings (see Figure 31-2). The bolts for these sleeves are, in some cases, found broken. Broken bolts should be noted, and bolts that appear to be broken should also be checked to determine whether they are sound. If necessary, ultrasonic testing can be performed on the bolts. The trunnion sleeve sometimes rotates or moves slightly after a few bolts have broken, placing side pressure on the bolts in

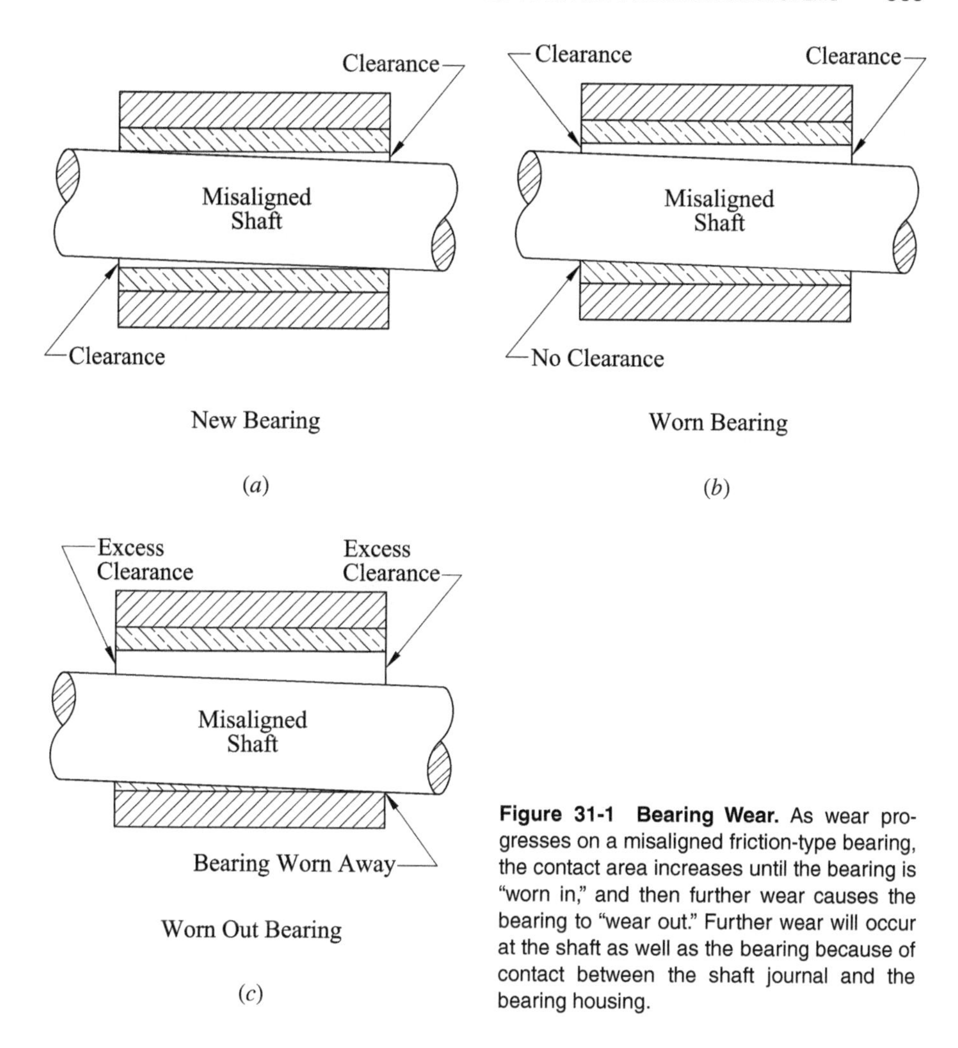

Figure 31-1 Bearing Wear. As wear progresses on a misaligned friction-type bearing, the contact area increases until the bearing is "worn in," and then further wear causes the bearing to "wear out." Further wear will occur at the shaft as well as the bearing because of contact between the shaft journal and the bearing housing.

place so that they may seem tight when, in fact, they are not and may even be broken. Close observation of the bearing as the bridge opens and closes may reveal movement and allow additional loose or broken bolts to be detected. Movement of the sleeve during operation may result in enlargement of the holes and wearing of the sides of the bolts, which may be detectable, and may also result in stepping of the sides of the bolts, which will make it difficult or impossible to remove them without cutting.

Thorough checking of the lubrication of friction type trunnion bearings should be performed, as they produce very large friction torques because of their large size. If possible, lubricant should be pumped into the bearing, and drain plugs removed one at a time to see if fresh grease can be made to flow out of each of the drains. Large quantities of old or contaminated grease, as well as plugged grease lines, should be noted. Do not use a wire to check a lube line, as it may break off and block the line.

Figure 31-2 Bascule Bridge Trunnion Bearing. This bearing supports one side of the counterweight of a large heel trunnion bascule bridge. The size of the bearing can be gauged by the handrail below it.

If it appears that there are serious problems, a suggestion should be made to dismantle the bearing for further inspection. Such dismantling is a very serious effort and should not be undertaken without adequate preparation. Main trunnion bearings may actually be experiencing uplift when the bridge is in certain positions and under certain loads, so it may be necessary to remove the caps of these bearings with the bridge in a raised or partially raised position. Falsework may also be necessary to hold the span or counterweight in position while the bearing is dismantled. NO attempt should be made to operate a movable bridge while a bearing cap is removed or any bolts are loosened.

Rolling Lift Treads

Rolling lift bascule treads should be carefully examined for cracks. Such cracks develop as a result of metal fatigue, as the material undergoes large stress changes as the bridge rolls open and closed. Cracks can also develop in the segmental girders and support girders to which the tread plates are attached, particularly in members in intimate contact with the tread plates. Flange angles are particularly susceptible. The cracks are, in some cases, hidden by excessive layers of paint, old attempted weld repairs, corrosion, or a combination of these. Older rolling lift bridges with thin rolled steel tread plates or cast tread plates can have severe cracks, even to the point at which pieces of tread break out. An attempt should be made to determine the cause of the cracks. A corroded or worn tread plate can produce uneven contact, resulting in extra local bending stress on the tread plate. Deterioration at the supporting steel can result in the tread plates bridging a gap in the support, also adding bending stress. Weld repairs can warp the tread plate or the connecting steel, adding built-in stresses.

The lugs and pockets on Scherzer rolling lift bridge tread plates should be checked for interference. There may be structural interference elsewhere, or other misalignment, that causes the bridge to operate out of line so that the lugs contact the sides or ends of the pockets. If this appears to be the case, the situation should be investigated to find the root of the problem. It may be that the treads have worn sufficiently that the operating radius of the curved treads is no longer correct. If this is the case, the lugs tend to contact the far side of the pockets at the ends of travel, barring other realigning forces. Some tread plates have been built up by welding or by attaching "shim" plates to their operating surfaces. These should be readily apparent from an inspection of the tread operating surfaces. Smoothness and flatness of the tread surfaces will be apparent from the contact patterns at the racks and pinions of Scherzer rolling lift bridges. At worn areas in the track, the pinion will tend to bottom in the rack. If the treads are excessively worn, the pinion shaft may be plastically deformed so that the pinion takes an oscillating path along the rack, bottoming at every full turn and having minimal contact at points halfway between the areas of bottoming. Wear will occur intermittently at the top lands of the pinion and rack teeth, making accurate gear tooth caliper measurements impossible.

Typically, the greatest amount of wear at Scherzer rolling lift tracks occurs at the closed position, as the bridge tends to slide on its treads under heavy live loads and may not seat firmly, adding to the live load impact forces. If wear progresses sufficiently at this location, the pinion, shaft, and bearing will participate in carrying live load and dead load. The result can be a broken or excessively worn bearing, a bent pinion shaft, damage to the pinion and rack teeth, or a combination of these (see previous paragraph).

Live Load Shoes

Double leaf bascule bridges require stabilization devices to hold each leaf in the closed position. These devices usually take the form of live load shoes, mounted on the bascule main girders or trusses so that they contact masonry plates on the pier ahead of the trunnions or rolling lift tracks. These are generally considered machinery components, although they may be more appropriately classified as structural parts. Both components may have shims or other adjustment devices present. These shims may be made of plain carbon steel and can be corroded, particularly on older bridges. If corrosion is present, it should be determined whether the corrosion is merely at the exposed surfaces or rust packing has occurred. Rust packing is especially likely if a large number of very thin shims have been used to make adjustments in the seating at the live load shoe. Rust packing can result in the build up of very high pressures within the assembly, so caution should be used when checking bolt tensions. Striking a severely overtensioned bolt hard with a hammer may result in breaking of the bolt (see page 376).

Single leaf bascule bridges employ live load shoes or sole plates and masonry plates at the toe ends of the main girders or trusses. These serve a similar purpose to the live load shoes on double leaf bascules, but the positioning is not as critical because maladjustment of the live load shoes on a single leaf bascule has a much lesser effect on the position of the bridge than on a double leaf bascule. The single leaf bascule can more easily flex to allow seating of the live load shoes, as the shoes are farther away from the rigid counterweight and/or trunnion area. The double leaf bascule superstructure is typically very rigid in the area of the live load shoes.

The contact surfaces of the live load shoes carry the entire live load of most double leaf bascule bridges. This load is multiplied because of leverage, as the distance from the pivot point of the bridge to the deck at the toe is usually considerably greater than the distance from the pivot point to the live load shoes, in some cases by a factor of 4 or 5 to 1. The contact area between the shoe and the masonry plate should be adequate to carry this load without overstress. In many cases, because of poor construction practices, wear, corrosion, or plastic deformation of some part of the bridge, the contact area is smaller than it should be. The size of the contact area should be noted. The contact surfaces should be in firm contact with the bridge seated. If they are not, impact can occur that magnifies the live load stresses on the bridge components. Note should be made of the firmness of contact and the amount of gap, if any, when there is no live load on the seated bridge. If possible, arrange for very gentle seating of the leaf, stopping the span as one live load shoe makes contact, then measure the gap at the other live load shoe.

Span Locks

All double leaf bascule bridges employ some kind of span locks to form a structural connection between the two leaves. This connection usually carries only vertical shear forces between the main girders of the opposing leaves, and for this reason these locks are sometimes called "shear locks." The most common type of shear lock consists of a steel bar that is mechanically extended from one leaf into a socket on the other leaf. The bar may be actuated by a self-contained lock bar mechanism. On many double leaf bascule bridges, the lock bars are activated by a drive assembly that consists of a motor and gear reduction that rotates a cross shaft, with cranks on the cross shaft that are connected to the lock bars. Some older bascules have manually operated span locks, driven and pulled by means of a lever in the operator's house. A few older trunnion bascule bridges have linkages connected to the pier that cause the span locks to automatically withdraw as the bridge opens, and automatically drive the span locks as the bridge closes. This type of drive necessitates liberal clearances between the lock bar and its guides and sockets to avoid binding during operation. These clearances are so large that there is little point in inspecting them closely or making precise measurements. The operation of these devices should be observed to make certain that they operate smoothly, and they should be checked for loose components. On other double leaf bascules, particularly Scherzer type rolling lift bridges, the span locks are rigid structural members that interlock as the bridge closes. These locks require coordination of the movement of the two leaves of the Scherzer bascule when beginning to open and when nearly closed, so that the lock components engage properly. On some of these bridges, position-sensing limit switches have been installed to control the bridge movement, so the interlocking takes place automatically, rather than relying on the skill of the bridge operator. Unfortunately, these switches can and do become misaligned so that their effectiveness is limited. All center locks have a common shortcoming: Contaminants get into the lubricant at the lock mating surfaces, or there is no lubricant, which, combined with movement under live load, temperature changes, or the process of actuation in engagement or disengagement, causes wear. Typically, a new bascule bridge may be in service for only three or four years before wear results in excessive clearances at the locks. Any clearance should be considered excessive that allows sufficient misalignment of the tips of the leaves so that noticeable impacts occur as vehicles cross the center joint.

The clearances at mechanically actuated span lock scan be approximated by measuring the depth of the male lockbar and the corresponding female socket and noting the difference. Similar procedures can be used on the rigid structural-type center locks by comparing the depth of the male to the depth of the slot in the corresponding female member. The difference between the thinnest part of the full-depth male member and the greatest depth at the female member may be the actual clearance. It is better if several measurements are taken, with corresponding readings at different mating areas of the male and female parts. A more accurate clearance measurement can be taken directly by gaining access to the lock bar and socket with the bridge closed and measuring the clearances with feeler gauges, a taper gauge, or a ruler. Clearances should be taken at top and bottom simultaneously, unless it is known that the lock bar is firmly contacting either the top or bottom of the socket. It is also possible to measure the clearance by placing a dial indicator on the tip of one leaf, and placing the pointer vertically against the tip of the other leaf, with the pointer at about the midpoint of the pointer's travel. Traffic will usually cause the lock-bar and socket to bottom out against each other, first in one direction and then the other. The full range of motion can be obtained by summing the dial movement in both directions. For mechanically actuated lock bar type center locks, this will include the effect of wear in the lock bar guides.

Self-contained lock bar actuators are very difficult to inspect, as the actual parts that produce the longitudinal motion driving the lock bar are buried deep inside the unit. They can be fully inspected only by disassembling the units, which is a time-consuming operation, and there is a risk of losing a critical part by dropping it into the waterway from the somewhat vulnerable position of the actuator, usually underneath the bridge deck at midspan. These units usually do not wear much, because they are enclosed and there is usually very little force required to drive or withdraw the lock bar. If it appears that the lock bar is sticking in engagement with the socket or is hard to engage because of misalignment, such a condition should be noted. Unusual noises, such as grinding or snapping, may indicate a worn mechanism, which can result from repeated driving and retraction of misaligned lock bars. The amount of play in the drive mechanism should also be checked, by observing any delay between the time the lock bar actuator motor starts turning and the beginning of the movement of the external thruster bar. Clearance may also be measured between the lock bar, the thruster bar, and the pin connecting them. Wear of the pin should be checked as well, if possible.

Crank-type lock bar actuating assemblies are usually built up of components similar to the main drives of bascule bridges, except that they are smaller and do not contain differential units. Most components of these assemblies can be inspected by methods similar to those used for machinery in general (see Chapter 34). The machinery, including the cranks, usually does not wear excessively, as there is normally little load required to drive and pull the lock bars. This type of machinery is vulnerable to wear due to poor lubrication or due to contamination of the lubricant. The condition of the lubricant on these installations should be noted, particularly if it appears to be contaminated. Sand or roadway grit should be obvious if present in the lubricant, but a laboratory analysis may be necessary to confirm the presence of contaminants. This is especially of concern if the lock drive machinery is worn. Crank and linkage parts are usually substantially oversized for the stresses they will experience in operation and are slow moving, so extra clearances at bearings and such will not cause dynamic problems. The cranks and linkages should be checked carefully to make certain that they are sound and operate correctly; it should be ascertained that they provide the required amount of movement to the lock bar when driving and pulling. Clearances at all pin connections should be checked, and worn pins measured so that they can be checked for overstress. Enclosures covering any parts of the machinery should be checked for effectiveness against water, dirt, dust, or any other contaminant that may be present.

Centering Devices

Railroad bascule bridges are almost always of the single leaf type and can be quite long. They require centering devices at the toe end so that the track rails are properly aligned with the bridge seated. The centering devices should have minimal clearance with the bridge seated. This clearance should be constant for a distance up from the fully seated position equal to the height of the rails. This will prevent interference as the bottom of the moving rail on the toe of the bascule span passes the top of the fixed rail on the abutment or the approach span. The upper portion of the fixed part of the centering device, mounted on the pier, should be well lubricated and show no signs of heavy interference. If there are signs of substantial wear at the operating surfaces of the centering device, check the misalignment as the bridge seats. The bridge may be permanently misaligned. This can be a result of wear or damage of the main bascule supports, particularly with rolling lift bridges. The misalignment should be reported, and Scherzer rolling lift treads or Rall linkages checked for damage or wear.

Highway bascule bridges have little need for centering devices. A bridge that has been properly constructed, even a double leaf bascule, should align itself adequately at seating. Centering devices may do more harm than good, causing interference that inhibits proper seating of the bridge. Double leaf bascule center locks usually have sufficient lateral clearance to avoid interference under the worst cases of misalignment that can occur because of temperature differences or other normal conditions. If excessive lateral misalignment occurs, its cause should be determined and reported.

Many double leaf bascules have been constructed with the leaves misaligned, and resultant interference occurs at centering devices or span locks until the components wear sufficiently to remove the interference.

Live Load Anchors

Many double leaf bascule bridges are equipped with live load anchors, located behind the center of rotation. These anchors may be installed in conjunction with forward-placed live load shoes, or they may be the only live load stabilizers at the pier. If they are the sole stabilizers, they may be considered a variation of live load shoes and should be inspected in the same way. If they work with forward-mounted live load shoes, their function is somewhat different. Normally, there should be a small gap at the anchors when the bascule leaf is seated firmly at the live load shoes. The function of the live load anchor in these installations is to prevent uplift at the trunnion bearings, or raising of the curved rolling lift treads off the horizontal treads, when large live loads are present on the bridge. A small clearance should exist at the live load anchors, and should be the same at each live load anchor on a seated bridge leaf. The anchors should make firm, full contact when heavy live loads cross the bridge. Corrosion and maladjustment affect this contact, and their presence should be noted. The clearance at live load anchors should be sufficient to allow firm seating at the forward live load shoes, when present. The clearances at the live load anchors used in conjunction with forward live load shoes should be small enough to prevent upward movement at the trunnions of simple trunnion bascules in excess of the bearing clearances at the trunnions. At rolling lift spans, the curved tread should not lose contact completely with the horizontal tread. Rear live load anchors are not normally used with heel trunnion or overhead articulated counterweight bascule bridges.

BASCULE DRIVE MACHINERY

Most electromechanical drive bascule bridges have similar operating machinery. On a simple trunnion bascule, there is one curved rack mounted on each of two main girders. In some rare cases, there are multiple main girders, with a rack on each. In other rare cases, usually only on smaller bascule spans, there is only one rack and pinion, at the longitudinal centerline of the bridge. Some simple trunnion bascule bridges, particularly Hopkins types, have the racks mounted on special rack girders, inboard from the main girders. Typical simple trunnion bascule drive machinery consists of main drive motors, brakes, and primary reduction, including a differential unit, mounted on the center of the pier, with secondary reduction and rack pinions mounted outboard from the primary reduction but inboard of and in line with the racks. A very few simple trunnion bascule bridges have the operating machinery mounted on the moving leaf, with internal racks mounted rigidly in the counterweight pit. The manufacturing drawings of these installations should be checked carefully before attempting wear measurements. Articulated counterweight bascules usually have the racks mounted on the main girders of the bascule leaf, but some of these bridges, particularly the overhead counterweight type, have the racks mounted on the supports of the links to the counterweight frame (see Figure 18-10).

The racks and pinions of simple trunnion bascules tend to wear most heavily at the teeth that are in contact with the bridge closed. This is primarily due to deflection and movement of the leaf under live load causing impact between the rack and pinion teeth. Movement under live load is particularly prevalent with double leaf bascules, as the span locks and live load shoes tend to be worn and poorly adjusted. In rare cases, rack or pinion teeth on these bridges have been known to break. When performing an inspection, the condition of the gear teeth in contact with the bridge closed should always be determined. If it is impossible to measure the teeth with a gear tooth caliper, because the bridge cannot be held open long enough, approximations should be made by checking the backlash at the rack and pinion interface. The backlash should be checked at both sides and both ends of the teeth in full mesh. A direct approximation can sometimes be made of the tooth thickness at the closed contact point by measuring the tooth thickness with a ruler, if the wear extends to the end of the tooth . Observations can also be made of movements and impacts at these teeth as live loads pass over the bridge.

Most Scherzer rolling lift bridges have flat racks mounted on frames on the piers, with drive machinery otherwise similar to that for simple trunnion bascules, but mounted on the bascule leaf itself. A few very old bridges of this type are opened and closed by means of operating struts similar to those on heel trunnion bascules, as discussed below. The racks and pinions of Scherzer rolling lift bridges are subject to the same wear as those of simple trunnion bascules; they can also experience additional wear and damage from worn rolling lift tracks, as discussed in the earlier section "Rolling Lift Treads." Bent rack pinion shafts can result in extremely high loads on the rack and pinion elsewhere (see below and page 385) than at the bridge closed position, with associated heavy wear and damage. A bent pinion shaft can also produce cross bearing between the rack and pinion, when the pinion shaft is 90° away from the fully closed position. This can be discerned by observing the contact patterns on the rack and pinion teeth. Half a revolution back from the fully closed position, the pinion will "bottom" in the rack. A full revolution back from fully closed, the pinion will make reduced contact with the rack. Between these points, cross-bearing will occur. These patterns will repeat themselves throughout the travel of the pinion on the rack, at one-revolution intervals.

Heel trunnion bascules have the same drive machinery as other bascule bridges, but it may be mounted on the bascule leaf, on the counterweight, or on the A-frame or tower supporting the counterweight. On almost all heel trunnion bascules, the racks are mounted on operating struts, which are pin connected, at the ends opposite the pinions, to another part of the superstructure. The pin connection for the operating strut is a critical member in the bridge drive, but it can be difficult to inspect. The pin is usually not directly accessible, so special arrangements must be made to gain safe access to it. The wearing surfaces of the pin may be hidden, so direct measurements of wear may be impossible without disassembling the connection. This may require special arrangements, as some operating struts weigh several tons. Disassembly may also put the bridge out of operation for several days. If an inspection contract is being negotiated, items such as this should be clearly identified so that the scope of the inspection is fully understood. Rall bascules are usually opened and closed by means of operating struts similar to those used for heel trunnion bascules.

The racks on the operating struts are held in engagement with the rack pinions by means of devices called carriers (see Figure 31-3). A carrier may seem complicated, but it is merely an assembly, riding on bearings on the rack pinion shaft, that has three sets of wheels holding the operating strut and rack in engagement. All the bearings on these assemblies are usually friction type, with the journals riding in bronze bushings. Unfortunately, the journals and bearings holding all these components in place are not all readily inspected. Clearances at the bearings holding the rack pinion shafts in position usually can be easily checked for clearances and state of lubrication. The bearings for the axles supporting the carrier wheels are usually accessible as well. The bearings in the carrier frame that hold it in position on the rack pinion shaft are almost always hidden by rack pinion shaft bearings. Clearances can be approximated by observing movement as the bridge is started in motion and stopped, but this can be unsatisfactory. If a large enough jack is available, and a dial indicator gauge can be placed in the proper position, and the brakes can be released on the machinery, it may be possible to obtain a better approximation of the amount of clearance in the carrier bearings. If the operating strut is resting on the carrier, it can be lifted up to the amount of clearance, which the dial indicator will measure. If it is possible to take the bridge out of service for a prolonged period of time, the carrier assembly can be disassembled and a visual inspection can be made of all the bearings. In the case of large bridges, this procedure requires considerable effort and some very heavy equipment.

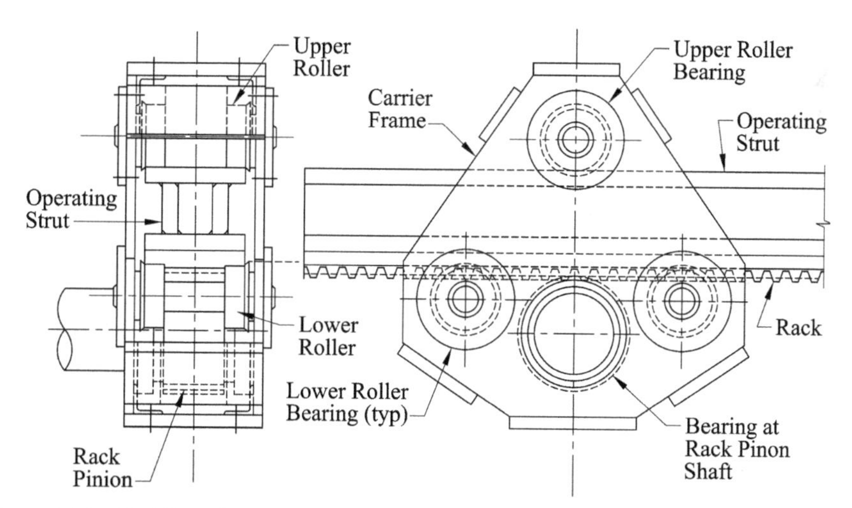

Figure 31-3 Heel Trunnion Bascule Bridge Carrier Assembly. The triangle-shaped assembly is supported on the rack pinion shaft, which is supported on the superstructure on bearings on the near and far sides of the carrier, at lower center. The operating strut is supported on the carrier on two pairs of wheels, one forward (to the right) and one rearward (to the left) of the rack pinion. A third set of wheels is above the strut to keep the rack from separating from the pinion.

Some older heel trunnion bascules have large emergency brakes installed at the carriers to act directly on the operating struts. These brakes were usually found to be unnecessary for normal operations, fell into disuse, and became inoperable because of neglect. If the brake is functional on both operating struts, it should be tested to see if it provides adequate braking power. Initial tests should be made near the midpoint of travel of the bridge, so as to minimize the likelihood of damage if the brakes do not work properly. If the brakes seem adequate, a test can be made near the end of travel of the bridge, where imbalance forces are likely to be higher.

Hopkins drive frames tend to deteriorate at the pin connections between the frames and the pier. Twisting of the frames during operation and impacts due to starting and stopping the bridge can produce cracks in the steelwork at these connections or elsewhere on the frame. The cracks may cease propagating if they relieve the overstress that caused them. Corrosion also tends to occur at the frame-to-pier pin connections, particularly if water is often present on the pier floor.

BUFFER CYLINDERS

Many bascule bridges are equipped with buffer cylinders. Their purpose is to prevent excessive impacts on the pier when closing the span. If the buffer cylinders do not work properly, and in unison, they can do more harm than good, perhaps even causing warping of the span. The buffer cylinders should be checked, first, to see if all of them on a leaf are all working reliably in unison, and second, to see if they are building up to the same pressure. It is especially important that the two buffers on one leaf work together, as, otherwise, uneven forces can be exerted on the leaf. The buffers should be observed during several operations of the bridge to make certain that they extend fully as the bridge lifts. Any sticking of the piston and rod in the retracted or partly retracted position should be noted, even if the rod quickly extends when lightly tapped or jarred. The buffer cylinders, when working correctly, should build up to the highest possible pressure while the bridge seats, without preventing seating of the bridge. The bridge should be able to seat at normal seating speed without having the pressurized buffer cylinders cause the bridge to raise up again, even slightly. Pressure gauges can be used to determine the maximum pressure developed during seating. All buffer cylinders on one leaf of the span should reach the same maximum pressure and are preferably checked during the same seating operation.

HYDRAULIC DRIVE MACHINERY

Many bascule bridges, particularly the simple trunnion type, are fitted with hydraulic drive machinery. This machinery is almost invariably one of two types: hydraulic motor or hydraulic cylinder. This type of machinery on these bridges frequently suffers from lack of proper maintenance. The bridge machinery inspector should make a serious effort at determining the level of maintenance that is performed regularly on the bridge. The inspector should also make note of the results of any lack of maintenance, which may include the following: components worn excessively because of lack of lubrication, damaged seals and improperly acting valves due to failure to keep filters clean, general coating of external surfaces with hydraulic fluid resulting from damaged and worn seals, low fluid levels due to leakage, and improper operation because of valves or other components being out of adjustment. Hydraulic cylinders generally suffer more than hydraulic motors from the adverse environment of the typical movable bridge. Standard industrial-type hydraulic cylinders, without heavy-duty or special deterioration-resistant fittings, are frequently specified for movable bridge applications. These cylinders should be checked especially carefully for damaged seals, scored rods, worn bushings at clevis connections, and loose or damaged hose connections. This type of damage can occur just as readily at the hydraulic cylinders that activate secondary and stabilizing devices such as span locks, as it can at main bridge operating cylinders.

32

MACHINERY—VERTICAL LIFT BRIDGES

SAFETY

Most of the operating machinery of a vertical lift bridge is slow moving, but operates with a great deal of force. Some of the machinery may operate fast, such as some cross shafts connecting gear reductions in the main bridge drive. All of the machinery is dangerous and must be treated with respect. It is sometimes desirable, for the purpose of inspection, to observe the machinery in operation. During such observations, it is essential to avoid the possibility of injury. When observing vertical lift bridge machinery from a vantage point on the bridge, it is necessary to be located in a safe area where falling into or onto moving machinery is impossible. If possible, a secure handhold should be nearby. Most vertical lift bridges operate very quietly and smoothly. It is important, when standing on the moving span, under it on a pier, or beneath a counterweight, to avoid being struck or crushed by a moving or stationary part of the bridge.

SUPPORT AND STABILIZATION SYSTEMS

The dead load support for a vertical lift bridge consists of the towers, counterweight sheaves, shafts, bearings, and counterweight ropes. In contrast to the simplicity, stability, and durability, of the vertical lift span itself, the counterweight sheaves, shafts, bearings, and counterweight ropes are vulnerable to deterioration. The live load supports for a vertical lift bridge are, on the other hand, relatively durable and trouble-free components. The other stabilizing components of the vertical lift span, including span guides, counterweight guides, and centering devices, are relatively trouble free.

Counterweight Sheaves

The counterweight sheaves of a vertical lift bridge can normally be expected to serve for many decades without difficulty. Some cast steel counterweight sheaves have been known to develop cracks as a result of faulty fabrication or excessive stress cycles. These cracks are difficult to detect because of the usually dirty and greasy condition of the sheaves. The surfaces of the sheaves are, in most cases, covered with old, dirty, and contaminated rope lubricant. There is also usually difficulty in gaining access to all the surfaces of the sheaves, and more than half the outer perimeter of the sheaves is covered at any time by the counterweight ropes. Cracks due to faulty manufacture can show up anywhere on the sheaves, but fatigue cracks usually appear first in the outer cylindrical section that is in contact with the ropes. Thorough cleaning can expose many larger cracks to the naked eye, but if fatigue is suspected, there may be internal cracks that have not yet shown themselves at the surface of the sheave. Magnetic particle inspection is usually effective in detecting these cracks. Poorly lubricated sheaves may have their rope grooves coated with dry powdered rust. This may be cleaned off with a wire brush to expose cracks.

Welded sheaves frequently have flaws within the welds that are the originating points for cracks that may take several years to appear at the surface, if they ever do. These cracks may be too deep below the surface to be detected by magnetic particle testing. Ultrasonic testing may be employed, but the geometry of the sheave may make this type of testing impossible or the results doubtful. If the level of concern warrants it, x-ray testing may be employed. The presence of large internal cracks, or what may more correctly be termed manufacturing flaws, can result in unintended flexibility of the sheave, so that internal movement occurs during bridge operation. This movement sometimes results in sharp snapping noises emanating from the sheaves, but the ability of the steel in the sheave to transmit sounds makes it very difficult to locate precisely the source of the noise. The sound may not be emanating from a crack in the sheave, but may be caused by movement between the sheave and the shaft. This movement can be the result of insufficient force or shrink fit between the shaft and the hub of the sheave. This can be particularly likely if the sheave or sheave shaft has been replaced in the field. Such movement should be detectable visually, by direct observation of movement as the sheave turns, or by finding powder rust emanating from the area of the mating surfaces. Snapping sounds can also come from abrupt movement of the counterweight ropes in the rope grooves as the bridge operates. Cleaning and liberal lubrication of the rope grooves should silence noises coming from the ropes, at least temporarily. If the noise disappears and then returns, the grooves on the sheave should be checked for deviations, including unequal overall diameters.

The rope grooves in counterweight sheaves should be smooth and free of corrosion. The groove radii should be very slightly larger than the outside radius of the rope. Any deterioration of the grooves should be noted, as it will accelerate deterioration of the ropes.

Counterweight Sheave Shafts

Vertical lift bridge sheave shafts have had a difficult history over the last 35 years or so. One of the counterweight sheave shafts failed on a vertical lift bridge in 1978 as a result of cyclical fatigue.[1] This failure occurred only a few months after the bridge machinery had been inspected by a respected engineering firm and had been given a reasonably clean bill of health. It was recognized fairly quickly after this incident that inadequate attention had been given to metal fatigue in the design of some vertical lift bridges. It also became clear that the understanding of this phenomenon was not sufficiently advanced at the time this bridge and others were designed, in the early 1930s or earlier, to reasonably hold the designers liable for the failure. Inspectors in the late twentieth century, however, had little other than precedent to defend them against charges of incompetence. There has now been plenty of work done to understand the problem, prevent it, and diagnose it. Since 1978 several vertical lift bridge counterweight sheave shafts have been inspected in which cracks were detected. No catastrophic failures have resulted, but the bridges were of necessity taken out of service for some length of time while the shafts were replaced.

Ultrasonic testing has been used to test for cracks in counterweight sheave shafts and appears to have been successful. No known failures have occurred where the shaft has gotten a clean bill of health following ultrasonic inspection. Ultrasonic inspection is relatively common in industry, and specialists are available who will come to the bridge with their own equipment, and test the counterweight sheave shafts in place for a modest fee. This testing will result in little or no disruption to bridge operations. Direct visual inspection can be used where an ultrasonic test provides inconclusive results or where it is impossible to do an ultrasonic test. These inspections involve some disruption of bridge operations, as well as some expense, as the bearing caps or covers have to be removed from the shafts and the bridge has to be raised and held to various partially opened positions during the inspection. The bridge should NOT be operated or moved while a bearing cap is removed or loose. Dye penetrant testing can confirm the presence of cracks in sheave shafts during such inspections, but care must be taken, as the cracks will close tight when the observable part of the shaft is in compression. This mode of testing has had a good success rate, as several cracks have been found by this means that were not previously detected, and many shafts in which this method has not found cracks have not had cracks appear later. It appears that cracks in sheave shafts take some time, after initiation, to propagate sufficiently to result in failure. Thus, inspection at fairly regular intervals should significantly reduce the likelihood of catastrophic failure. Once such a crack is found, the bridge should be taken out of service immediately until repairs are made.

Unfortunately, conclusive results in the inspection of vertical lift bridge counterweight sheave shafts are very difficult to achieve unless the damage is gross and obvious. If the condition of a shaft still causes concern after any or all of the aforementioned procedures have been performed, it may be warranted to unload the sheave and

[1]Floyd K. Jacobsen, Illinois Department of Transportation, *Investigation of Trunnion Failures Involving Movable Vertical Lift Bridges* (Springfield, Virginia: National Technical Information Service, 1980).

lift it free of the shaft bearings and perform further inspection on the shaft. Removal of the shaft to a laboratory for inspection and testing is almost impossible, but most procedures can be performed adequately in the field with proper preparation and at less cost than simply replacing the suspect shafts. Very few vertical lift bridges have spare counterweight shaft assemblies available, so that removal of a shaft to a laboratory would take the bridge out of operation for a considerable time.

Counterweight Sheave Shaft Bearings

Two configurations are used for counterweight sheave shaft bearings—rotating shaft and stationary shaft. The vast majority of installations have rotating shafts. The stationary shaft bearings are located in the hub of the sheave and are difficult to inspect. Two types of bearings are used for counterweight sheave shafts—friction and antifriction. The antifriction type has proven to be extremely reliable and durable over many decades of service. It has the important advantage that it requires little maintenance, which is difficult to achieve at the tops of the bridge towers. Use of the antifriction bearing results in no wear to the counterweight sheave shaft, an important factor in determining the fatigue life of the shaft (see page 180). Inspection of antifriction bearings by the typical inspector should be limited to external inspection. The inspector should note the condition of fasteners, the presence of any corrosion, and any signs of seal damage, leakage, lack of lubricant, or water penetration of the bearing. The inspector should listen closely to the antifriction bearing as the bridge operates. Any detectable noise can be a sign of internal deterioration. The antifriction counterweight sheave shaft bearing can be partially disassembled for inspection when warranted, but this should be performed by competent technical personnel, such as a representative of the bearing manufacturer.

For standard friction-type counterweight sheave shaft bearings, the clearances should be checked between the journal and the bushing or cap, at the top, sides, and bottom. This should be done at both the inboard and outboard sides of the bearing if possible. It may be necessary to remove an end cover to gain access to the outboard end of the bearing. Access to the inboard end may be impossible. The thickness of the bushing at the bottom of the bearing should be measured and recorded and compared with the original thickness if available. This can give an indication as to whether significant wear has occurred at the bushing. These bearings are vulnerable to poor lubrication, as access to the tower tops on most vertical lift bridges is poor at best and even with good maintenance it is difficult to get adequate lubricant penetration to all loaded parts of the bearing. The quality of the lubrication of these bearings is best checked by exposing the journals, at least partially, to visual inspection. This can be done only with rotating sheave shaft bearings and must usually be done by removing the bearing caps. The bridge must not be inoperable while any of the caps are removed, so adequate notice must be given to mariners. The bearing caps are very heavy, and the work space at the tower tops is limited and, in some cases, precarious. If the owner's maintenance personnel cannot be made available to perform the work of removing and reinstalling the caps, competent, experienced labor should be obtained elsewhere. The fastening bolts for these bearing caps may be in poor condition, so it is a good idea to have spare bolts available in case breakage occurs during disassembly or reassembly. The bridge must NOT be operated with any caps removed, as damage to a bearing may result. The journal may become dry and weld itself to the bushing, or the bushing may turn in the bearing base as the journal rotates.

The cap may be solid bronze or cast steel with or without a bronze half bushing inside. After the cap has been removed, the lubrication grooves, if they are accessible, should be checked to see if they are plugged and if fresh lubricant has been reaching all areas of the grooves. The surfaces of the journals should be closely inspected to see if there are any dry spots or discolored areas suggesting burns due to previous lack of lubricant. A new journal should have a mirror finish, but after several years of use the surface usually develops small annular grooves. Poorly lubricated journals may be rusty. Any such deterioration increases friction at the bearing and increases wear. Clearances between the journal and the lower half bushing can be checked along the parting line to determine whether the bushing is square to the journal in plan. The inner end of the journal, through the fillet, should be closely observed for corrosion and signs of cracks (Figure 32-1). An inspection using dye penetrant or other enhanced surface inspection can be done at this time. To observe the condition of the rest of the journal, replace the bearing cap and have the bridge raised so that the sheave shaft has made one half turn, then remove the cap and continue the inspection. The bridge is closed to traffic while held in the partially opened position, so arrangements must be made to provide for adequate time for cap removal, inspection, and cap replacement, as well as delays in bridge and waterway traffic. Allowances should also be made for the possibility that repairs may be necessary.

Counterweight Ropes

Vertical lift bridge counterweight ropes are vulnerable to corrosion and wear. Wear can be reduced, and corrosion eliminated, by proper lubrication of the ropes. The inspector should first check the quality of lubrication of the counterweight ropes before proceeding with further inspection of the ropes. The ropes should be covered to the point of saturation with reasonably fresh lubricant, and there should be a minimum of old lubricant clinging to the ropes or built up in adjacent areas such as the counterweight sheaves. Rope gatherers and separators should also be saturated with fresh lubricant, and there should be no old lubricant, especially dirty, contaminated old lubricant, in these areas. If it appears that the gatherer and separator areas have an excessive buildup of old lubricant, and especially if there are signs of corrosion at or bleeding out from these areas, the portions of the ropes contacting the separators and gatherers should be visually inspected. It will probably be necessary to disconnect the ropes and loosen them to expose these areas. This can usually be done without taking the bridge out of operation if only one rope is allowed to be in a detensioned state at any given time. The ability to operate the bridge with one heel removed should be confirmed by calculation before any such work is done.

The most severe corrosion of counterweight ropes usually occurs at the span end connections, as these are the most exposed locations of the ropes. It is usually fairly easy to gain access to the ropes at these connections by climbing to the truss at the

Figure 32-1 Inspection of Vertical Lift Bridge Counterweight Sheave Shaft. The shaft journal has been cleaned, and dye penetrant has been applied to test for cracks.

end of the vertical lift span. It may be necessary to do some digging, as the ropes can be covered thickly with old, dried lubricant at these locations. The old lubricant should be cleaned off, exposing the rope wires. Old, dried lubricant is not a good rust preventative and may accelerate corrosion by trapping moisture. Some of the ropes may be easier to reach for cleaning and inspection than others, but all should be closely inspected, as the less accessible ropes may have had poorer lubricant application in the past, making them more susceptible to corrosion. Immediately after the inspection of the ropes is completed, they should be coated with fresh lubricant to prevent corrosion. This procedure can be repeated at the counterweight end connections if necessary; however, the ropes are usually cleaner at the counterweight ends, so a simple visual inspection may be adequate.

The counterweight ropes wear because of sliding contact with the grooves in the counterweight sheaves. Usually, the most heavily worn areas of the ropes is just off the sheaves at the span side with the bridge closed, as the ropes tend to oscillate in the wind, causing small movements at the tangent points where they depart the sheaves. The areas of the ropes just off the span side of the sheaves are usually accessible for inspection, with the bridge in the lowered position, from the front floor areas at the tops of the bridge towers. To inspect the lengths of the ropes that are normally between the span connection and the counterweight sheave, a bosun's chair or painter's pick is sometimes used to gain access to these areas while the bridge is closed. As these areas of the ropes are unlikely to be the most heavily deteriorated, it is usually unnecessary to go to these lengths for adequate inspection. The ropes can be observed as the bridge opens and closes, with the inspector stationed at a convenient fixed point, watching out for bad spots in the ropes or unusual conditions. It is normally not necessary to stop the bridge in the partly opened position for a closer look at the ropes. The ropes can usually also be observed on the counterweight side, from a location at the tower tops, as the bridge opens and closes, but, again, it is usually not necessary to stop the bridge in the partly raised position for a closer look.

External wear of counterweight ropes takes the form of flat spots that develop on the outer wires of the ropes where they contact the grooves of the counterweight sheaves. These flat spots are usually most pronounced along the centerline of the rope where they contact the bottom of the sheave grooves. The lengths of these wear spots, along the axis of the rope, are called length of wear, or *wear lengths.* Many rope manufacturers provide tables relating wear lengths on different types and sizes of ropes to the loss in strength of the rope. These tables tend to be conservative, so they should be used with caution.

Because the sheaves supporting the counterweight ropes are usually quite large in comparison with the rope diameter, internal wear of the counterweight ropes is usually not a problem. Yet heavily worn ropes may experience internal wear. A counterweight rope may have a deteriorated core that allows the strands of the rope to contact each other inside the rope (Figure 32-2). This results in wear of the inner rope wires, which is usually indicated by bright red rust present in the valleys between the strands of the rope, or perhaps an even greater prevalence of this rust. If the ropes can be made sufficiently slack, they should be expanded and checked internally for wire and core damage. Inter-strand "nicking" indicates the core has deteriorated. Intra-strand nicking indicates the strands have deteriorated. The lengths of rope just off the span side of the counterweight sheaves with the bridge closed are the most likely locations of internal wear as well as external wear, as these lengths of rope experience the most flexing. They go on and off the sheaves for any bridge opening, no matter how short a distance the bridge lifts. External wear is likely to be much more severe than internal, but they go together. External wear begins immediately, whereas internal wear does not become severe until the deterioration of the rope's core has progressed to the degree that abrasion between strands in the rope can occur. Counterweight ropes elongate as they wear. A rapid increase in elongation can be a sign that the rope core has been destroyed. This elongation can be measured by noting changes in the vertical position of the counterweights over time. The rope may also reduce in diameter, but this reduction is difficult to measure accurately.

If the ropes are severely corroded or worn, or have kinks or severe damage such as cuts, it may be wise to condemn the ropes without performing disassembly for further inspection, particularly if bright red rust is bleeding from between the strands of the rope (see the preceding paragraph) or if even a few of the rope wires have broken at flat wear spots. Counterweight ropes have a very high safety factor, so it should be made certain that the ropes are overstressed before condemning them. In some cases, a single rope has been removed from a bridge and destructively tested by breaking it in tension and measuring the load required. If all the ropes are equally deteriorated, this procedure can determine the safety factor of the remaining ropes. Testing the

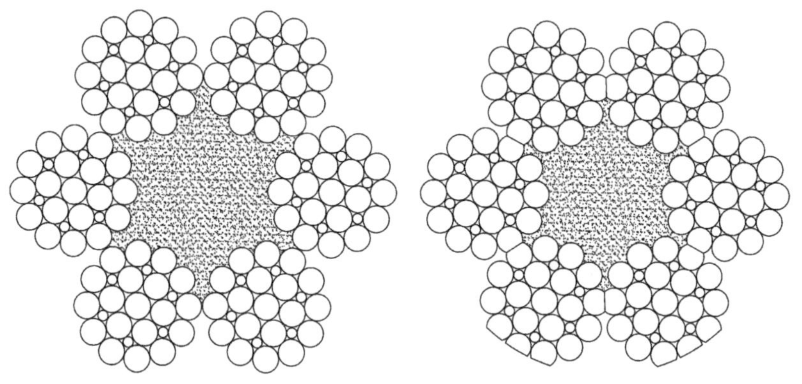

Figure 32-2 New Rope versus Worn. The cross section on the left shows a typical new vertical lift bridge counterweight rope. The section on the right shows a rope badly worn after years of use with poor lubrication.

worst worn rope, determined by inspection on the bridge, can give only a conservative estimate of the safety factor. Removing a rope for destructive testing is not recommended, as it is very difficult to install a new rope and get it to fully share the load with the other ropes. A new rope takes a considerable amount of time to stretch so that it has the same elastic properties as the older ropes and requires occasional retensionings as it becomes slack. Part VI covers the evaluation of vertical lift bridge counterweight ropes in detail.

The end connections of the counterweight ropes should also be closely inspected, particularly if signs of corrosion or other damage suggest a loss of strength in the connections. Turnbuckles and other adjusting devices are particularly vulnerable to corrosion, as they are not always painted properly when the bridge superstructure is painted under contract and are not likely to receive any other form of protection. Many such fittings are specified to be galvanized, but galvanizing is not a permanent corrosion preventative, as the zinc corrodes fairly rapidly and eventually disappears, particularly where exposed to the atmosphere. Zinc is also known to promote hydrogen embrittlement of high strength steels, so it should be used around rope wires with caution.

Live Load Shoes

Vertical lift bridges generally have expansion-type live load shoes at one end and fixed-type live load shoes at the other end. Some vertical lift bridges have only a simple flat shoe at each corner. This is adequate for very large spans that carry only light traffic, but heavy traffic, particularly railroad traffic, requires that longitudinal traction forces be restrained, and fixed end live load shoes are convenient devices to accomplish this. Wear and corrosion are not particularly detrimental to vertical lift bridge live load shoes, but they should be checked, particularly at hinged connections at the expansion end of the bridge, to make certain that they are not in danger of falling off the bridge. Expansion end live load shoes should also be checked to see that they are free to move and that they are not prone to sticking in one position. All live load shoes should be checked to make certain that they seat firmly when the bridge is down. Expansion live load shoes, or rockers, are very heavy and can rest on the masonry plate so that it appears that a solid seating is achieved, but the span can still drop a large fraction of an inch before bearing firmly on the rocker at a worn hinge pin. Observations should be made with the bridge under heavy traffic to make certain that there is no vertical movement resulting from live load at the corners of the lift span. Sometimes shims are placed on the masonry plates to accommodate wear at the live load shoes. These shims are, at best, a temporary expedient, even if welded or otherwise fixed in place, and their presence should be noted. They usually do not last long. The inspector should ask maintenance personnel how long such shims have been there and how often they have been replaced.

Span and Counterweight Guides

There are two types of span guides used on vertical lift bridges. The purpose of these guides is to prevent excessive lateral and longitudinal motion of the lift span while lifting and lowering. The lateral span guides are placed on each side at each end of the lift span, at top and bottom if the lift span is a truss type. The longitudinal guides are located only at the "fixed" end of the bridge, and only at the bottom chord of the lift span if it is a truss type. Roller-type span guides are vulnerable to seizing if they are not kept lubricated. Friction-type span guides can wear if the span tends to run against them. The guides usually mate with rails that are connected directly to the forward tower legs or are actually part of the legs. These rails can become bent or worn. The guides should be observed while the bridge opens and closes, with the inspector looking for binding or excessive clearance.

Roller-type span guides should be checked for freedom of motion. They should spin freely when not in firm contact with the guide rails. Roller guides should be checked for flat spots on the roller contact surfaces. Friction-type span guides should be checked for wear grooves in the friction pads. Clearances at the span guides are not particularly critical. They should not have such tight clearances that they bind at the slightest tilt or skew of the lift span, but clearances should not be so excessive that considerable lateral or longitudinal motion can occur. Most span guides are designed to have smaller clearances at the nearly closed position than they do when the span is lifted higher. When the span is seated, lower span guides should not have more than $\frac{3}{8}$ in. of clearance, more or less, and upper span guides should have no more than about $\frac{3}{4}$ in. of clearance. If as-constructed drawings of the bridge are available, that show clearances at the span guides, those figures should be taken as minimums.

Counterweight guides are similar in function to span guides but are usually simpler and lighter in construction. Rather than operate against the bridge tower legs, counterweight guides usually mate with separate vertical rails attached to the towers. These rails can become bent or displaced. The fit of the counterweight guides to the rails should be observed for the full length of the operation, with the inspector checking for binding or excessive free play. A pronounced kink in a counterweight guide rail, or a bent counterweight guide, is suggestive of prior seismic activity.

CENTERING DEVICES

Centering devices on vertical lift bridges are usually located below the deck, on the centerline of the roadway, at each end of the lift span. Separate centering devices are not particularly necessary on vertical lift bridges that carry highway traffic, but they are essential on railroad vertical lift bridges. The centering function is adequately handled on highway vertical lift spans by properly designed and functioning span guides. If centering devices are present on a highway vertical lift span, they should be checked to see that they are not the source of structural interference when seating the bridge. On one older vertical lift bridge, the live load shoes were heavily worn so that the centering devices were acting as live load supports at the centers of the end floorbeams. The bridge did not seat at its live load shoes unless sufficient live load was on the span to deflect the end floorbeam to the extent necessary to achieve contact at both live load shoes. Under lighter live loading, the span would rock back and forth on the centering device, first making contact at the live load shoe on one side of the pier, then on the other. This resulted in overstress at the end floorbeams and at their connections to the main trusses, because of these components not having been designed to carry the entire live load of the bridge. On another highway vertical lift span, the bridge lowered with just enough lateral misalignment to hang up on the edge of the centering device, which possibly had been improperly placed during construction. As it was a tower drive bridge, it did not seat fully even after live load vibrations jostled it off the hang-up point. The machinery and counterweight ropes carried the live load until a sufficiently heavy live load came along and pulled the bridge down.

SPAN LOCKS

Most vertical lift bridges were constructed with some form of span locks, although on many highway bridges they have been removed or are inoperable. The span lock usually does not assist in stabilizing the bridge but provides an interlocking function only. Vertical lift bridges never have significant negative live load reactions and are almost always span heavy when seated, so there is usually no actual need for a span lock to hold the bridge down. If a span lock is present, the inspector should make certain that it is operable, and it should be checked to see that it actually forms a restraint to movement of the bridge. The lock should form a secure bar to movement, with no loose components or slack in linkages that will allow the span lock to disengage unintentionally. The lock itself should be substantial enough to prevent bridge operation when engaged and should be fastened securely so that no parts can come loose during normal vibration and impact. As with centering devices, the span lock should be checked to see if it is interfering with seating of the span on the live load shoes.

SPAN OPERATING MACHINERY

The three common types of vertical lift bridge—span drive, tower drive and tower-span drive, have different forms of operating machinery. The span drive pulls up and down directly from the vertical lift span and has the bulk of the operating machinery located directly on the vertical lift span, with power from a single central point. The tower drive and tower-span drive operate by driving the counterweight sheaves, raising and lowering the vertical lift span by means of traction on the counterweight ropes. The span-tower drive is powered from a single point, but the tower drive has half the power at one tower and half at the other.

All three common types of vertical lift bridge drive incorporate some sort of electromechanical hoisting machinery, which is of roughly the same configuration for each type of bridge. The motors, brakes, and gear reduction units are similar to those used for a typical bascule or swing bridge drive, except for the lack of differential units, other than as noted in the following discussion. The drive machinery is usually smaller and less powerful for a vertical lift bridge than for a similarly sized bascule or swing span, because a vertical lift bridge does not move into the strongest sustained winds and the drive machinery is not expected to have to overcome the same wind pressures. The inertial forces to accelerate a vertical lift bridge are also usually smaller than those of similarly sized bascule or swing bridges.

Most vertical lift bridge drive machinery is rather durable, because it is usually well protected in a fully enclosed machinery house, avoiding environmental damage. Vertical lift bridge machinery does not tend to wear as rapidly as machinery for swing or bascule bridges, because the acceleration and deceleration forces are less. A vertical lift bridge does not have as much moving mass as a bascule bridge of similar size, and it does not have the large rotational inertia to overcome that a swing bridge of the same span length would have. The drive machinery for a vertical lift bridge does not usually experience the same shock loading, due to live load impacts, that a bascule bridge frequently suffers, because of the cushioning effects of the counterweight or operating ropes. As a result of all these factors, the final drive rack and pinion teeth of vertical lift bridges seldom display the tooth damage and wear often evident in a bascule or swing bridge that has seen similar service. The same rack and pinion teeth are not always in contact with a vertical lift bridge in the closed position, as a result of rope creep on a tower drive or tower-span drive bridge, or as a result of operating rope stretch and adjustment on a span drive bridge. Thus, on most vertical lift bridges, there is not a pronounced difference in the amount of wear on the gear teeth that are in contact with the bridge closed and those that are not. This is not the case with rack-drive vertical lift bridges, such as the Strauss type, but even on these types the wear differences are not usually as pronounced as on a bascule or swing bridge.

Many tower drive and span-tower drive vertical lift bridges incorporate gear reduction units with differential mechanisms, called clutch reducers, in their drive systems. On a tower drive bridge, the clutch reducer is normally locked, so it acts like a simple gear reduction unit. The differential is made active as the bridge is being seated so that the forces pushing down on the bridge can be equal on both sides of the end of the bridge, and each side is free to move down without being restrained by the other via the drive machinery. This allows the span to seat fully at all live load shoes. A span-tower drive bridge may have similar clutch reducers in the end machinery at the towers, plus a third clutch reducer, acting in the same fashion, at the center machinery. This allows equal forces and equal seating at each end of the bridge as well as at the sides. These reducers should be checked to see that the clutches engage and disengage as intended. Tower drive vertical lift bridges with eight sheaves usually drive at all eight sheaves. Usually a differential or differential reducer is employed to assure equal driving force at the two sheaves it is driving. These differentials (4 per bridge) do not have differentials.

On some tower drive or span-tower drive vertical lift bridges, adjustable couplings are used to adjust the drive for proper seating of the bridge. These couplings have many holes at the connection between the coupling halves. Usually only three bolts are used to make the connection. No less than two of these bolts should be present, and they should be checked to see if they are tight. Whether or not the bridge is equipped with clutch reducers or adjustable couplings, observations should be made to determine whether the bridge seats firmly on all four live load shoes. Commonly, full seating is not experienced, as one or more live load shoes is not contacting its masonry plate. If this is the case, an attempt should be made to determine the cause of the failure to seat properly. It may be a worn live load shoe, malfunction in one or more clutch reducers, a failure in the control system, allowing one end of the bridge to seat but not the other, or it may be a balance problem, with one end or side of the bridge span heavy and the other counterweight heavy. No attempt should be made to correct for poor seating unless the bridge maintenance or repair crew is assisting, with the proper responsible supervision.

A big problem with span drive vertical lift bridges, and one of the main reasons they are looked upon unfavorably by many bridge engineers, is the operating ropes. These ropes are frequently out of adjustment, and they are prone to wear and breakage. The inspector should examine the ropes closely their full length to see if they are worn, corroded, or have breaks in the wires or other damage such as kinks. An operating rope may appear sound externally, but may have a deteriorated core that allows the strands of the rope to contact each other inside the rope. This will result in rapid wear of the inner rope wires, and can usually be spotted because of the bright red rust present in the valleys between the strands of the rope or perhaps even more prevalent rust. If the operating ropes can be made sufficiently slack, they should be expanded and checked internally for wire and core damage. The rope fittings, including sockets and adjusting devices, should be examined for damage, corrosion, and wear.

The inspector should check the operating ropes to see whether they are working equally and at equal tensions. This can be accomplished on Waddell type vertical lift spans by observing the ropes as the span begins to lift, to see if all ropes start pulling at the same time, and checking the sag in the ropes under load to see if all ropes have the same sag over the same length. On span drive vertical lifts with a separate drum at each corner of the lift span, strain gauges can be used to record the drive torque at the pinion for each corner, and the results compared.

Many vertical lift bridges are equipped with buffer cylinders. Their purpose is to prevent excessive impacts on the pier when the span is closing. If the buffer cylinders do not work properly, and in unison, they can do more harm than good, perhaps even causing structural damage or warping of the span. The buffer cylinders should be checked, first, to see if they all are all working reliably in unison, and second, to see if they are building up to the same pressure. It is especially important that the two buffers on one end work together, as otherwise uneven forces can be exerted on the end of the vertical lift span. The buffers should be observed during several operations of the bridge to make certain that they extend fully as the bridge lifts. Any sticking of the pistons and rods in the retracted or partly retracted position should be noted. The buffer cylinders, when working correctly, should build up to the highest possible pressure while the bridge seats, without preventing smooth, full seating of the bridge. All cylinders should reach the same maximum pressure, which is preferably checked during the same seating operation. The bridge should be able to seat at normal seating speed without having the pressurized buffer cylinders cause the bridge to lift up again off the live load seats. Some vertical lift bridges are also equipped with buffer cylinders mounted at the tops of the towers, to cushion the impact of the span if an overtravel occurs while the bridge is being raised. These buffer cylinders are frequently inoperative, as they tend to stick in the retracted position and are not easily accessible for maintenance.

Most vertical lift bridges, except for those making very short lifts, are equipped with auxiliary counterweights. The auxiliary counterweights ease the load on the bridge machinery by partly or completely eliminating the imbalance caused by the movement of the counterweight ropes over their sheaves as the bridge raises and lowers. Thus, more wear is likely to be found on the drive machinery without auxiliary counterweights than on the drive machinery of vertical lift bridges that have them. The extra wear will be on both sides of the gear teeth. Wear on one side of the gear teeth is caused by raising the bridge from the lowered position and decelerating it on return. Wear is caused on the other side of the gear teeth by lowering the bridge from the raised position and decelerating it as it reaches the raised position while opening. There are many forms of auxiliary counterweights. The most common, perhaps, is a set of ropes that reach from one or both towers to a point on the span. The attachment point of the ropes to the towers is so located that these ropes pull down on the span in the raised position, and pull up on it when it is lowered. A functionally similar arrangement uses a pivoted pylon extending up from the center of the lift span, with ropes extending to the towers from the top of the pylon.

This type of auxiliary counterweight is almost solely used on span drive vertical lift bridges and usually carries electrical power and control cables for connection between the moving span and the fixed towers. These devices usually last many years with little deterioration. The ropes may become corroded, as they are virtually inaccessible for maintenance and, not incidentally, also for inspection. The ropes pass over sheaves at one or both ends and are attached to counterweights at the towers. Any wear experienced should show up here, conveniently, where the best access to the ropes is usually available. The ropes can be inspected at the sheaves for abrasive wear and checked for corrosion and the condition of the ropes' preservative material. If it is necessary to check the entire length of the ropes, a bosun's chair can be rigged up to provide access. If access at the span end is difficult, the chair can be used to ride down from the tower to the closed span, and then down from the open span to the tower. This may not be possible or advisable if the ropes carry electrical cables. If necessary, a temporary rope can be rigged to support the bosun's chair. Ropes not connected to a pylon may be accessible for inspection from the top of the vertical lift span in the half open position.

Span-tower drive vertical lift bridges usually have shorter lifts and do not require auxiliary counterweights. Tower drive bridges may incorporate auxiliary counterweights similar to those discussed earlier, but usually are installed wholly within the towers. These are sometimes used on span drive bridges as well. One variety is similar to those above, but has the ropes connected at the ends of the span instead of at the center. Another type, once fairly prevalent on all types of vertical lift bridges but now less common, uses heavy chains made of cast iron (see Figure 32-3). One end of each chain hangs from the bottom of a counterweight, and the other is attached to a fixed point on the tower. As the span lifts, the counterweight lowers and the weight of the chain gradually passes from the counterweight to the tower. As the span lowers, chain weight is transferred back to the counterweight. These chains have the advantage of being able to exactly balance the weight of the counterweight ropes as the span raises and lowers, but they are difficult to maintain and inspect. As the chains age and weather, they are prone to seizing up and becoming inflexible. In observing the chains as the span raises and lowers, any stiffness in the joints should be apparent. A close view from a vantage point on the tower is best.

A few vertical lift bridges, particularly some vertical lift bridges making short lifts, are fitted with hydraulic drive machinery. This machinery is almost invariably the hydraulic cylinder type. This type of machinery on these bridges can suffer from lack of proper maintenance. The bridge machinery inspector should make a serious effort at determining the level of maintenance that is performed regularly on the

Figure 32-3 Vertical Lift Bridge Balance Chains. The chains are usually made of a self-lubricating type of cast iron. They still have a tendency to rust or stick so that they do not flex freely.

bridge. The inspector should also make note of the results of any lack of maintenance, which may include the following: components worn excessively because of lack of lubrication, damaged seals and improperly acting valves due to failure to keep filters clean, general coating of external surfaces with hydraulic fluid resulting from damaged and worn seals, low fluid levels due to leakage, and improper operation because of valves or other components being out of adjustment. Hydraulic cylinders generally suffer more from the adverse environment of the typical movable bridge, and standard industrial-type cylinders, without heavy-duty or specially deterioration-resistant fittings are frequently specified for movable bridge applications, so cylinders should be checked especially carefully for damaged seals, scored rods, worn-out bushings at clevis connections, and loose or damaged hose connections. This type of damage can occur just as readily at hydraulic cylinders activating secondary and stabilizing devices such as span locks, as at main bridge operating cylinders. Synchronization of the separate hydraulic cylinders lifting a vertical lift span is critical to avoid tilting of the span, and this feature should be checked carefully, by special test operations if necessary.

33

MACHINERY—SWING BRIDGES

The swing bridge is unique among the common types of movable bridges in that many complicated mechanical components of the machinery are in the direct line of support for live loads. Other special mechanical machinery components are required on a swing bridge to properly align it horizontally for carrying live loads. For these reasons, it is especially important for swing bridge machinery to be properly inspected on a regular basis, for any defects found to be evaluated promptly, and corrective action taken immediately when required.

SAFETY

Some of the swing bridge machinery may operate fast, such as some cross shafts or longitudinal shafting connecting parts of the main bridge drive or wedge drive. These must be treated with caution when the bridge is operating or about to operate.

It is important, when standing on the moving span, or under it on a pier, to avoid being struck or crushed between a moving and stationary part of the bridge. Rim bearing swing bridges are particularly hazardous, as the rim bearing roller stabilizing and aligning mechanism rotates about the center pier at only half the rate of the swing span itself. An inspector stationed under a rim bearing swing bridge, standing on the center pier, inside the circular track, must nimbly avoid being caught between the swing bridge rim girder structure and the roller mechanism while the bridge is operating. If it is necessary to observe the spider or center post during bridge operation, it is less dangerous to find a place to sit or stand on the swing bridge superstructure, inside the drum girder, so that the inspector can ride the superstructure as the bridge rotates. Caution must be exercised, as some rim bearing swing bridges have spiders, or roller alignment frames, that can be mistaken for the superstructure of the swing bridge itself.

MAIN SUPPORTS

Rim Bearings

On a rim bearing swing bridge, the inspector should look for wear, binding, looseness, or heavy corrosion at the center post area. The inspector should be careful to note whether the swing span or rollers are pulling to one side as the bridge operates. All anchor bolts should be sound and tight, and there should be no signs of movement of or at the bolts. The lower tread plates should be firmly secured to the base castings, and the upper tread plates should be firmly secured to the bottom of the drum girder. The mating ends of all tread plate segments should be perfectly aligned with each other, with no gaps or steps at the working surface between segments. The inspector should determine whether the rollers are properly aligned with the track and follow in a circle. The rollers should not wander from side to side on the track, nor should they skid sideways. The rollers should bear only at their hubs on the thrust bearings on the outboard side of the rollers. The thrust bearings, roller shafts or axles, and center posts should be well lubricated. There should be no visible wear either at the roller contact surfaces or at the tread plates.

Rim bearing rollers may not all contact the upper and lower tread plates. This is not necessarily a particularly serious problem unless several adjacent rollers are not making contact. The rollers not making contact should be noted, as well as whether they lose contact only at certain locations on the tread plates, or never make contact.

Center Bearings

On a center bearing swing bridge, the balance wheels (Figures 33-1 and 33-2) should not be transferring any of the bridge's dead load or live load to the center pier. Only wind loads should cause there to be any reaction at a balance wheel and only while the bridge is not supported by its end lifts and center wedges. When the bridge is closed, the end lifts are driven, and the center wedges are driven, each balance wheel should be completely unloaded. Unless the bearings at the balance wheels are heavily worn, there should be a gap between the bottom of a balance wheel and the track. Sometimes excessive wear at a balance wheel bearing causes it to rest on the track. To see if it is carrying load, the wheel can be tested by attempting to move or rotate it with a pinch bar. If the wheel can be turned or moved by hand by these means, then it is satisfactorily unloaded. If the wheel cannot be turned, it is probably bearing some of the weight of the bridge that is supposed to be on the center pivot, or live load that is supposed to be on the center wedges. Unless the balance wheels have been wrongly adjusted, this is indicative of wear at the center pivot bearing or settlement of the bearing base on the pier, causing the span to drop until part of its weight is supported on the balance wheels. This may be sufficient evidence to warrant an in-depth inspection of the center pivot. There have been cases in which the balance wheels have been adjusted so that they bear some of the dead load of the bridge. This can sometimes be confirmed by investigating maintenance or repair records. When the balance wheels have been carrying dead load, they have also been carrying live load. In such cases, distress can occur at the balance wheels or their bearings or brackets and at

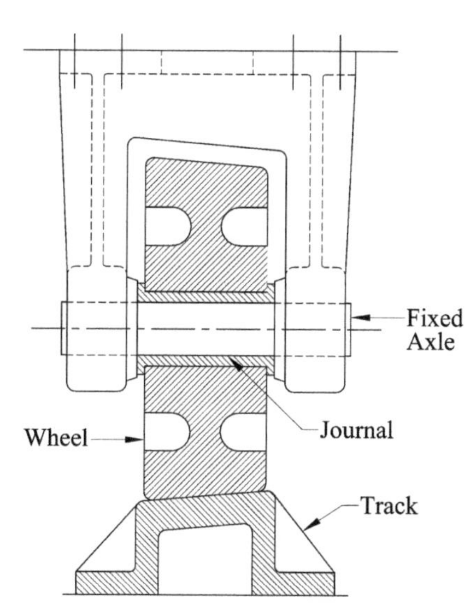

Figure 33-1 Swing Bridge Balance Wheel—Dead Axle. The cross section shows a typical balance wheel mounted to its axle on bearings, so only the wheel rotates when contacting the track as the bridge turns.

steelwork to which these supports are mounted. This distress should be readily apparent on close inspection of the balance wheel brackets and adjacent steel. The total clearance between the wheel and the track, and at the load side of the wheel bearing, should be approximately $\frac{1}{16}$ in. The clearance at the balance wheels should not be so great as to allow excessive tilting of the span, particularly at the ends of the span,

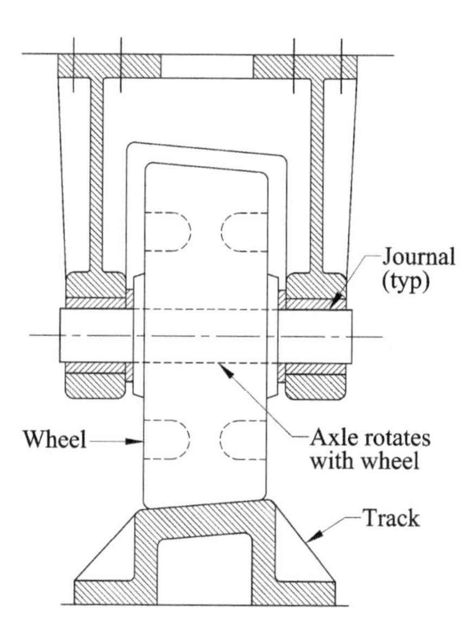

Figure 33-2 Swing Bridge Balance Wheel–Live Axle. The cross section shows a typical balance wheel fixed to its axle, so both rotate when contracting the track as the bridge turns.

where interference between swinging components on the span and stationary components on the rest piers can occur. The track segments should all be securely fastened to the pier, should be in good alignment, and securely fastened to each other at the joints. There should be no gaps under the track sections except at drains, and the masonry should all be sound under the track.

Center pivot bearings should show evidence of being well lubricated (see Figure 33-3). If possible, the lubricant in the bearing should be checked for contamination. There should be clearance between the oil ring and the adjacent part of the upper bearing housing. If the oil ring can be removed, it should be possible to measure the clearance between the edges of the upper and lower bearing discs, for a two-disc bearing, and between the edges of all three discs for a three-disc bearing. This edge clearance may be only about 0.01 in. for a 36-in. diameter bearing. If the bearing detail drawings are available, the unworn clearance can be calculated–consideration should be given to dead load deflection of the bearing.

It should be explicitly stated in the inspection contract or scope whether the center pivot bearing itself is to be dismantled for inspection. If this is to be done, it may require closing the bridge to traffic. It is impossible to rotate a swing bridge while the center pivot is dismantled, so the bridge cannot be opened to navigation unless it is opened and left open for the duration of the center pivot inspection. If the bridge is to be left in the closed position, blocking navigation, while the center pivot is dismantled for inspection, permission should be obtained from the U. S. Coast Guard to temporarily close the navigation channel before beginning work.

If the center pivot is to be dismantled for inspection, it should first be determined that the center wedges are capable of supporting the full dead and live load of the bridge. If jacks are to be used to raise the bridge so that the center pivot can be removed, it should be confirmed that there are jacking points on the bridge superstructure and on the pier that are capable of carrying the load. The top of the pier may appear to be solid, but it may be only a hollow, rubble-filled shell other than under the

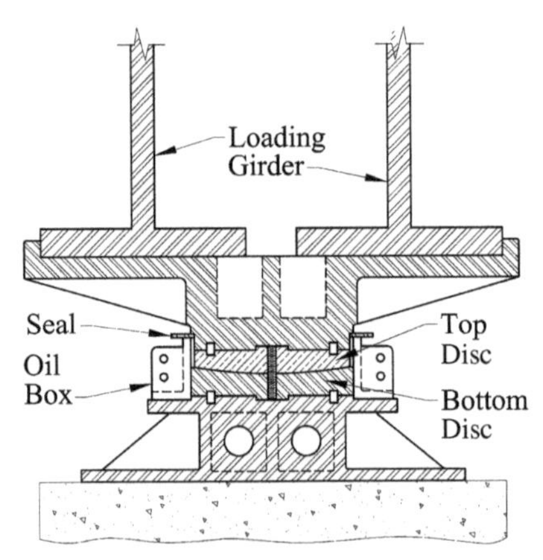

Figure 33-3 Swing Bridge Center Pivot Cross Section Showing Oil Ring Assembly. The cross section shows a typical center pivot bearing with two disks.

center pivot, the center wedges, and the rack and track. If it is determined that the center pivot must be dismantled and inspected, and it is feasible to do so, one of the following two methods should be used:

1) *For swing bridges with suspended support at the center pivot:* If there is sufficient vertical clearance to the center pivot, lock the bridge in the closed position after notifying navigation interests that the bridge will not be operable for a specified period of time; a minimum of two days is recommended. Confirm that the center wedges and linkages can support the full dead load and live load of the swing span except that carried at the end lifts. Confirm that the center wedges are fully driven and are in firm contact with their guides at the top and with the masonry seat castings at the bottom. Confirm that the end lifts and center wedges are locked in position and cannot move. Then loosen the tension rods supporting the weight of the swing span on the center pivot by gradually turning the nuts on the rods. These should all be turned simultaneously, or sequentially in small increments, until the center pivot assembly is unloaded and all dead load and live load at the center pier are supported on the center wedges. Then remove the rods.

   Lift the upper center pivot casting and upper grillage so that the pivot bearing surfaces are exposed to view. The upper bearing disc may stay with the upper casting, or it may fall onto the lower or center disc. This type of bearing usually has three discs. Be careful to avoid injury if the upper disc drops after being raised. If necessary, disassemble the upper components and temporarily remove them. View the bearing discs and determine whether they are cracked or broken and whether the lubrication grooves are worn away.

   Make note of the condition of the lubricant in the bearing. The bearing area of a center bearing swing bridge is very small; there may be dry spots at the maximum pressure points in the bearing.

   If the bearing is a three-piece type with a center bronze pumpkinseed disc, remove the center piece so that its bottom is exposed for inspection, also revealing the bottom forged steel bearing disc. Inspect these for wear, damage, and condition of lubrication. Check the fit of the top and bottom discs to the adjacent steelwork, noting any signs of looseness and whether the forged steel discs have spun on the connecting steel.

   Clean and inspect all of the disassembled components, including tension rods and nuts, the grillage, and the loading girder connection points. Look for cracks and corrosion, and repair or replace components as necessary.

   Reassemble the bearing, and reinstall the rods, making certain that all components are properly reassembled. If some components are excessively damaged or worn, install replacement parts, if available. Relubricate the bearing. Retension the rods. When reinstallation is complete, if the bridge is operable, it can be opened as needed for navigation.

2) *For swing bridges with all other types of center pivot bearings:* The bridge must be jacked up to remove the center pivot bearing for inspection. This can be done with the bridge in the open or closed position, but it is recommended that the

work be done with the bridge closed, if possible, for extra stability. It may be possible to maintain traffic on the bridge during the inspection if the bridge can be adequately supported and made sufficiently stable and the roadway vertical alignment is adequate or can be corrected. This can usually be done only for highway bridges, and it may be necessary to halt traffic for short periods during preparation. As indicated earlier, it is necessary to ensure proper placement of the jacks. If necessary, jack pads and supports will have to be added or modified. Check the center wedges to see if they can be extended beyond their normal fully driven position and if they can support the full dead and live load of the swing span in that condition. If such is the case, the bridge can be supported in this manner after jacking. If it is not the case, locking-type jacks should be used to avoid the need for falsework to support the bridge in the lifted position. If live load is to be carried, check whether the end lifts can be extended beyond their normal lift, to avoid negative live load reactions.

If possible, the oil ring should be removed before jacking the bridge. If the oil ring is not a split type, a new oil ring should be fabricated and the old ring cut off rather than lifting the bridge high enough to clear it. This type of bearing is likely to have only two discs. If the bearing discs are keyed into place, it should be necessary to lift the bridge only high enough to unload the bearing before the discs can be removed. If the discs are doweled, it will probably be necessary to lift the bridge high enough that the discs can be pulled from the dowels before sliding them out. If access to the dowels is available before removal of the discs, it may be possible to pull or drill the dowels out, which will make the disc removal operation simpler and safer. Check the center pivot details; perhaps parts can be unbolted, and the upper and lower discs, two castings, and the oil ring can be pulled out together. Check the discs for wear, cracks, and damage. Also make certain that the discs were properly secured to the mating steel. Check for proper lubrication. The bearing area of a center bearing swing bridge is very small; there may be a dry spot at the maximum pressure point in the bearing. If any bearing components are severely damaged and replacement parts are available, the damaged parts can be replaced. The bearing can then be cleaned and reassembled, with new components as required, and lubricated, and the jacks released. If the bearing is operable, the bridge can be opened as required for navigation.

END LIFTS

Swing bridge support components, such as end lifts and center wedges, reduce impact forces by providing solid support for live loads. Many swing bridges have poorly adjusted end lift drives that produce insufficient or no dead load reaction at the end supports. On lightly traveled highway bridges this may not have any ill effects, but impacts under even infrequent heavy bus or truck traffic, or any heavy passenger or freight train traffic on a railroad bridge, can cause severe deterioration of the superstructure in the area of the end supports. All these supports should be very carefully and closely inspected while the bridge is under live load of significant magnitude. A highway bridge should be observed under heavy truck traffic—ready-mix concrete trucks are usually best, but almost any large truck will do if it is loaded. Rail traffic is more consistent; almost any locomotive on a given railroad is likely to have the same axle loading and can produce close to the maximum vertical deflection at a railroad swing bridge. Any vertical or horizontal movement in the area of supports, due to live load, should be noted and reported. Temperature can have an effect on the amount of vertical movement, and so should be noted when making measurements.

Some swing bridge end wedges are intentionally adjusted to reduce the vertical dead load reaction to zero or nearly so. Such adjustment is usually done in response to complaints by the bridge operator that the end wedges stick, under some conditions, and cannot be pulled. This is almost always due to inadequate lubrication at the end wedge load-bearing surfaces. The condition of lubricant at these surfaces should be checked and any signs of inadequate lubrication noted. The wrong lubricant is frequently used, with inadequate pressure-withstanding capability, so that it is squeezed or scraped out, or a water-soluble grease is used that washes away quickly in the weather.

If no heavy traffic is available, test openings can be used to actually measure the dead load deflection produced when the bridge is seated. Because the deflection should be measured at each end lift, preferably during the same operation, at least four persons are needed to perform the inspection properly by measuring vertical movement at each of the four corners of the swing span. It is important to observe the measurement, rather than record it, as some drive mechanisms overtravel so that the end lifts, then drops back at the end of operation. A recording device may pick up the maximum travel without clearly noting the actual net amount of lift. Some swing bridges of the center pivot type are unbalanced, so that one end may actually drop while the end lifts are engaging, as tilt toward the heavy end of the bridge is removed by the lifts operating at that end. As the lift mechanism operates, the other end will drop until its end lifts make contact; then it will lift to its final position. Coordinated observation should be made of end wedge operation where the drives are not mechanically synchronized. One end may drive first, putting a heavy load on or stalling the drive at the other end (Figure 33-4).

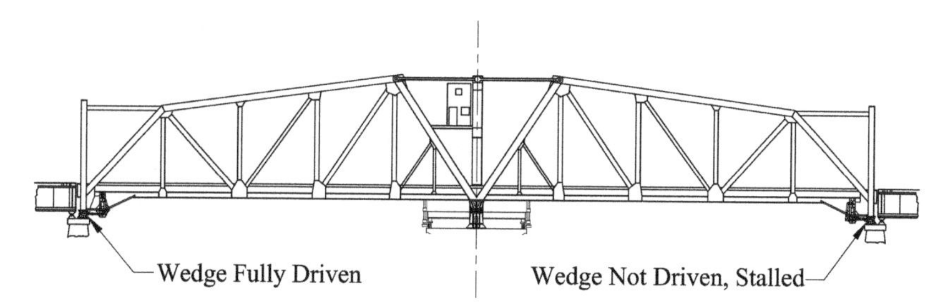

Figure 33-4 Swing Bridge End Lifts Operating Out of Synchronization. This can occur if the lifts at the two ends of the bridge are not mechanically linked by longitudinal shafting, or fully linked by other means, and the bridge is unbalanced longitudinally. This can cause the delayed lifts to fail to engage fully.

CENTER WEDGES

Center wedges are used only on center bearing type swing bridges. The same deterioration can occur, usually to a lesser extent, at the center wedges for a center bearing swing bridge as can occur at end wedges. Center wedges are not intended to lift the bridge superstructure to any degree when engaged. These wedges should make firm contact only with the seats and upper bearing surfaces. They should not be loose when driven, nor should they make such hard contact that they stick in place and are difficult to withdraw. The center wedge drive mechanism should not be heavily loaded during operation, so wear in the driveline components should be minimal. Any large amounts of wear may be due to excessive loading of the center wedge drive. The reasons for this loading should be determined. There is little or no deflection capability at the loading girder or girders, so overdriving the center wedges will result in their jamming or the drive mechanism's stalling out. The wedges may be driven too far, placing very large loads on the drive; the wedges may be sticking, requiring a large withdrawal force; or there may be damage to the mechanism, causing it to bind.

Some center wedge drives have spring or gravity forces designed into them. These drives should be checked to see if the center wedges consistently drive to full contact with the wedge seats and bearing surfaces of the guides, whether they overtravel, if they stick in the engaged position, and if they are pushed out under live load.

DRIVE MACHINERY

A swing bridge has its weight, or mass, distributed quite widely about its center of rotation, so it has a large moment of inertia requiring high starting torque. Many swing bridges were built with inadequate supports for their rack pinions, these supports tend to be overloaded and deteriorate rapidly. The pinion bearing brackets may be broken, and repairs to them may be broken. The racks and pinions usually have considerably more wear at the points where they contact when the bridge is beginning to open or to close. In some cases, broken rack teeth can be found at or near these locations. These broken teeth can be difficult to spot; they may still be in place, but loose, as there is little or no tendency for a broken tooth to be dislodged. As long as the contact ratio remains more than one, so that more than one pair of teeth in the mating pinion and gear are in loaded contact at any given time, the pinion tends to skip over these broken teeth when operating, but the load on adjacent teeth will be greater, increasing the likelihood that they will break in turn. The rack segments should all be securely mounted to the pier or to the track sections. The adjacent ends of the rack segments should be correctly aligned with each other. Check the pitch between adjacent rack teeth at the joints; it should be exactly correct. If bolts connect the segments at the joint, they should be tight and the segments should be contacting each other firmly, with no movement during bridge operation. Internal racks on swing bridges are especially susceptible to spreading, due to the separating forces developed between the racks and pinions during bridge operation. An older swing bridge with a pivot pier constructed of stone or unreinforced concrete can be particularly affected. Gaps between adjacent rack and track sections should be checked, and the condition of the connecting bolts and anchor bolts carefully examined.

A few older swing bridges have a single drive motor that operates the turning drive or the other machinery selectively, by means of clutches. These clutches should be closely inspected and observed during bridge operation, as they are likely to be heavily worn and out of adjustment.

CENTERING LATCHES

Centering latches are used to stop a swing bridge, when closing, near or at the fully closed position. These devices are usually automatic, operating by means of ramps at the centering pocket and levers and pawls at the drive mechanism, which is usually powered off the end lift drive. These devices can receive a great deal of wear, due to impacts that occur when the swing span is brought to a stop. Clearances should be checked at the latch bar guides, at the roller mounted on the latch bar, and at the pocket mounted on the pier. Heavily worn components can result in several inches of lateral free play at the ends of the swing bridge. Check to see if the free play exceeds the amount allowable at the end wedge, or other centering device. The operating mechanism can also wear heavily, particularly if it receives less than adequate maintenance. All links and pins in the operating mechanism should be checked for wear. Such wear does not increase the lateral free play of the bridge when the latch is engaged, but it may contribute to poor operation of the latch. Some centering latches have sprung pockets, so that the sides of the pockets give at impact, cushioning the blow and reducing wear at the latch bar, guides, and pocket. The sprung sides should return to their normal positions after arresting the motion of the span. If they do not, the bridge operating procedure should be investigated to see if the brakes are being applied prematurely when closing or if some other factor prevents the bridge from moving to the center position. A sprung pocket can be fouled with debris or have broken parts or missing springs.

CENTERING DEVICES

Some swing bridges are equipped with separate power-operated mechanical centering devices, which may be roughly similar in appearance to centering latches but are separately controlled and powered, operating only after the bridge has been stopped, on closing, in the nearly closed position. These devices usually appear only on railroad bridges, where precise alignment of the closed bridge is necessary in order to engage the end rail connections. Any wear in these devices negates their purpose, as precision in centering the bridge is lost. As long as the bridge is stopped very close to the properly aligned position before the device is actuated, loads on the centering device will be low and wear will be minimal. If the centering device must pull the end of the bridge over several inches every time it is actuated, then wear will be more rapid. Mechanical centering device clearances should be checked at the bar guides, at the roller or rollers, and at the pocket. The drive mechanism should also be checked for wear, particularly if it appears to be pulling the end of the swing span over a substantial distance in a significant percentage of closure operations. If the brakes are set when the centering device operates, wear is more likely.

The generally preferred type of centering device, according to AREMA and AASHTO (see page 204), utilizes the end wedges and, by the placement of ears on the sides of the end wedge seats, aligns the bridge laterally as the end wedges are driven. If this type of centering device is in use, the clearance between the end wedges and the ears should be checked when the wedges are in the driven position. Note that both wedges at one end of the bridge act to center the bridge at that end, so four different combinations of clearance at an end of the bridge can produce the minimum clearance (Figure 33-5).

END RAIL OPERATORS

Almost every railroad swing bridge has some sort of mechanical device to displace and reposition the railroad track rails at the ends of the swing bridge. The most common type of end rail is the miter rail. Before swinging the bridge open, the miter rails on the end of the bridge are lifted to clear their mating components on the approach span. Usually, a simple crank linkage performs this function by pushing up on a small platform located under the miter rails, near the end of the swing span. The platform is not fixed to the rails, but the rails, when lifted, rest on it. There should be some vertical clearance between the rails and the platform when the rails are in the lowered position. These linkages do not usually wear very rapidly, and as long as they lift the miter rails sufficiently to clear when the bridge swings and all components are secure, there is usually no problem with them. Other miter rails are linked directly to the actuators, so that they are pulled down when closing as well as pushed up when opening. These devices are more complicated, as they usually have some sort of cushioning device to ease the load on the actuators when the rails are pulled down tight into the lowered position. High forces can be developed in this arrangement, so detrimental wear of the components is more likely to occur. All components of either of these types of devices should be inspected as closely as possible for wear and other damage. Disassembly should not be performed unless rail traffic is stopped and the bridge will not be operating.

The other type of end rail separation consists of a sliding bar that is moved longitudinally, acting as a bridge across the ends of the rails at the joints at the ends of the swing span. Longitudinal forces developed by the railroad equipment rolling across the joint are resisted by the mechanism actuating the bar and usually result in short life for the mechanism components. These components should be checked closely for damage and wear. The wheels of locomotives powering a train tend to kick the sliding bar back, toward the rear of the train. The wheels of unpowered cars tend to drag

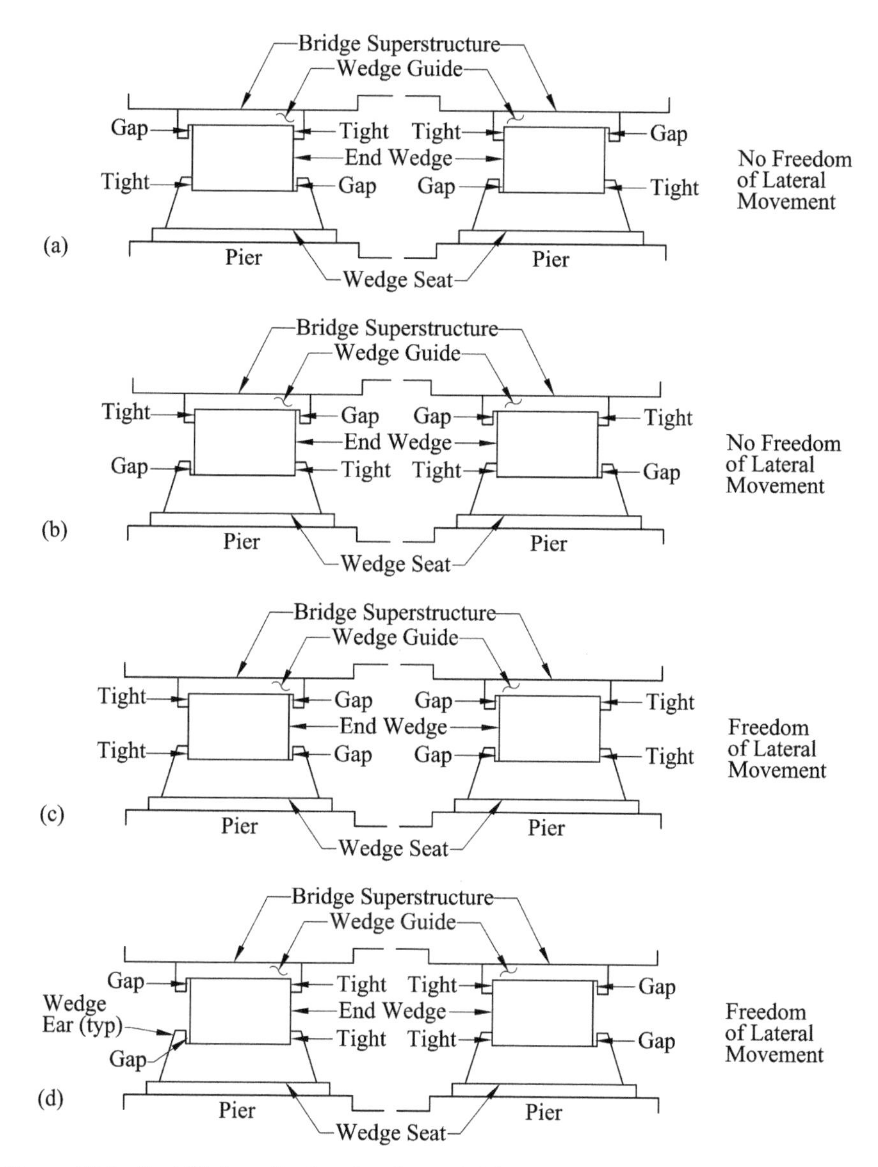

Figure 33-5 End Wedges Performing Lateral Alignment. The end wedges, guides, and seats can adequately perform the lateral alignment function on a swing bridge, provided these components are properly aligned and have satisfactory lateral clearances. The end wedges in this application typically have tapered sides at their leading faces.

the sliding bar along with them toward the front of the train. Standard instructions are usually given to train crews to minimize power and braking when crossing a bridge with sliding-bar-type end rail separators. The inspector should observe trains crossing the bridge and make a notation as to whether this is being done, and the amount of movement of the sliding bars under traffic. Usually it is not necessary to dismantle the components to determine wear amounts. If the components are dismantled for inspection, rail traffic must not be allowed across the bridge.

34

MACHINERY—GENERAL

SAFETY

Most movable bridge operating machinery is slow moving, but operates with a great deal of force. Some of the machinery may operate fast, such as some cross shafts or longitudinal shafting connecting parts of the main bridge drive or wedge drive. All of the machinery is dangerous and must be treated with respect. It is sometimes desirable, for the purpose of inspection, to observe the machinery in operation. When observing machinery in operation, it is essential to avoid the possibility of injury. When observing movable bridge machinery from a vantage point on the bridge, it is necessary to be located in a safe area where falling into or onto moving machinery is impossible. If possible, a secure handhold should be nearby. Most movable bridges operate quietly and smoothly. It cannot be expected that there will be an audible warning when a bridge is about to operate.

EQUIPMENT

For walk-through inspections, a pencil and pad of paper may be all that is necessary. If any close observation is to be done, a stiff-bladed putty knife is about the handiest tool to remove grease and dirt. A nonprecision taper gauge, such as the Starrett #267 (Figure 34-1), can be used to check approximate gear backlashes, clearances, and large bearing clearances. For in-depth inspections, a 1*P* gear-tooth caliper is usually sufficient (see discussion beginning on page 430 under "Gears"). A 24-in. vernier caliper, such as the Fowler 52-065-024 (Figure 34-2), can be used for span measurements on rack pinions and large gears, with a smaller caliper of approximately 12 in. capacity for smaller gears. This smaller caliper should still have large, square jaws,

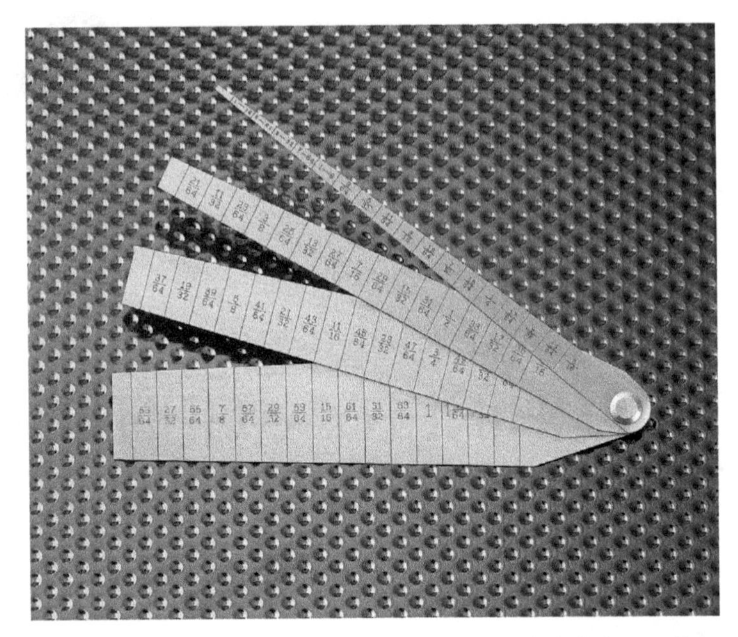

Figure 34-1 Starrett #267 Taper Gauge. (Photo courtesy the L.S. Starrett Company.)

as on the Fowler 52-060-100 Vernier calipers are generally best, as they are more rugged than dial-type calipers, and do not rely on batteries that can discharge such as with digital readout calipers. A precision taper gauge such as the Starrett #270 (Figure 34-3) can be added for more accurate backlash measurements. For friction bearing clearances, a long, thin set of feeler gauges such as the Starrett #245 (Figure 34-4) can be used, as it is sometimes difficult to get shorter, wider blades into the space. To avoid scratching or damaging bearings, thin blades only should be used, as they will at least tend to take the curvature of the bearing. It is better to use plastic gauges, which, like metal feeler gauges, are available individually or in sets. The handiest wrench for removing inspection covers of gear reducers and brake covers is an adjustable open-end wrench. At least a 12-in. length should be available, for large, tight

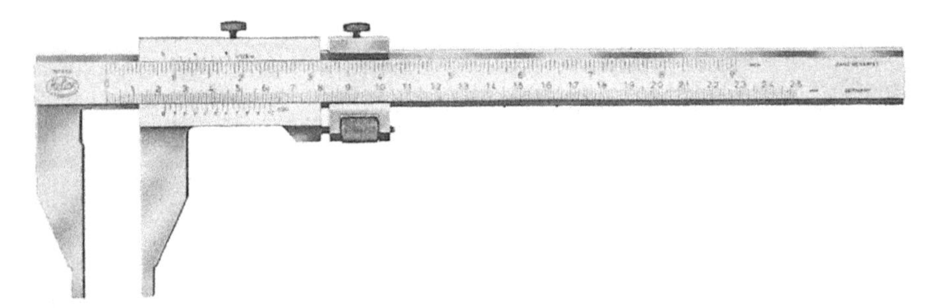

Figure 34-2 Fowler #52-065-024 Vernier Caliper. (Photo courtesy of Fred V. Fowler Company.)

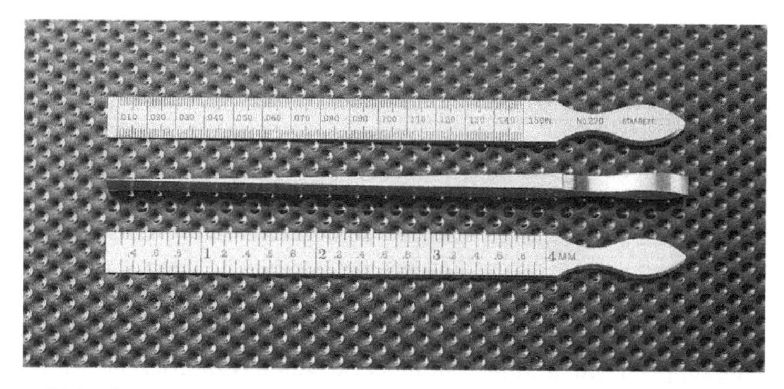

Figure 34-3 Starrett #270 Taper Gauge. (Photo courtesy the L.S. Starrett Company.)

bolts, but a smaller wrench can be handy to fit into tight spots. In using adjustable wrenches, it is important to avoid rounding the corners on hex nuts and bolt heads. A set of 6-point box wrenches or a socket wrench set should be available in case of need, but these are heavy to carry around and get lost easily. If buffer cylinder pressure readings are going to be taken, air pressure gauges reading 0 to 250 psi should be used, and pipe wrenches and fittings may be necessary to attach the gauges to the buffer cylinders. If possible, it is usually more expedient to purchase at or near the bridge site the ordinary tools and equipment needed. They can then be left at the bridge for future use, thus minimizing difficulties at airport security checkpoints. The other tools can be a problem, as they are not readily available near the typical movable bridge site. Express shipment may be the best option.

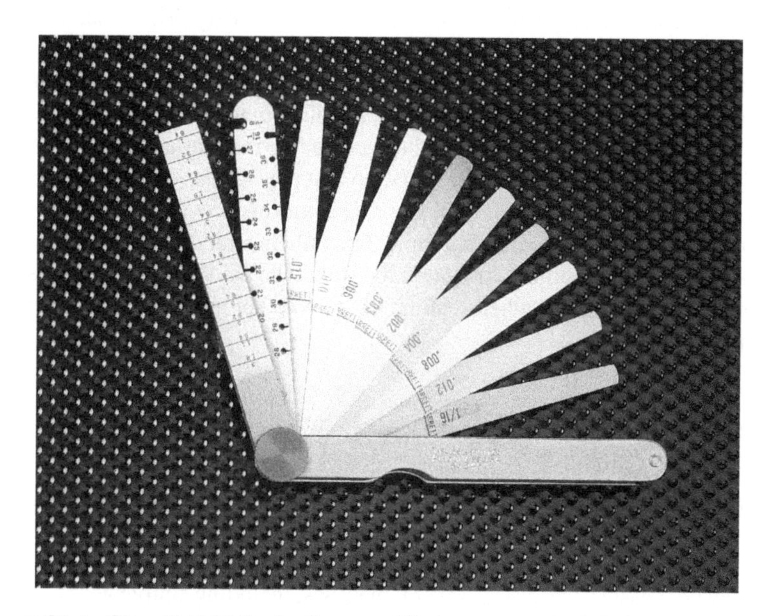

Figure 34-4 Starrett #245 Feeler Gauges. (Photo courtesy the L.S. Starrett Company.)

Mechanical machinery on a movable bridge is usually specially manufactured for the particular bridge. All components, including gears, shafts, and bearings, are made to specific detail drawings prepared for the bridge construction project. Only two exceptions are normally found. The first exception is gearboxes or speed reducers. These enclosed, self-contained gear reduction units are usually made by a manufacturer of standard gear reduction units, and may be identical to or only slightly vary, such as with external shaft lengths, from standard cataloged items as made by that manufacturer. In many, if not most, cases, however, the gearbox found on a movable bridge is a very special one-of-a-kind unit but its internal components are standard. There may be detail drawings available for a specially built unit, as some bridge owners insist on these being supplied with the unit. The other exception is antifriction bearings. Almost all antifriction bearings used on movable bridges are standard units, as catalogued by bearing manufacturers. These bearings are usually housed in special assemblies when used on movable bridges and are not mounted in standard pillow blocks.

During an in-depth machinery inspection, the first step in the field should be to confirm that the machinery on the bridge is as represented on the official drawings of the bridge. The shop drawings that were approved for manufacture of the machinery should be used to check the equipment. The data can be tabulated for easy field reference, with space provided on sheets for data obtained from field measurements, particularly to establish how much the components have worn. The FHWA's *Bridge Inspector's Manual for Movable Bridges* shows a typical sheet for use in field inspection of gears on a movable bridge. A variation is shown in the *AASHTO Movable Bridge Inspection, Education, and Maintenance Manual.* A suggested format is used in the table in Figure 34-5.

Regardless of the type of inspection being performed, the inspector should make an initial overall check of the condition of the machinery. The machinery should be examined to see if it is clean, whether there is corrosion, and if it is being well lubricated. Preferably, the bridge maintenance forces should not be forewarned of an impending inspection, so that the results of typical maintenance activity can be determined. At least one test operation should be performed, or a normal operation observed, prior to doing any intensive work, so that any serious defects can be detected and appropriate action taken if necessary. During this test operation, the inspector should listen for abnormal sounds coming from the machinery and make a visual observation of all components. Close attention should be paid when the bridge starts and stops moving, particularly when seating, to see whether any movement occurs at the machinery supports.

GEARS

Gear inspection on a movable bridge can take many different forms, depending on the level of the inspection. More than one type, or all types, of gear inspection may be done for a particular bridge as necessary to provide the required information. Gear wear may be measured by any of several methods, including simple gear tooth caliper measurements, span measurements, measurements over pins or balls, or the gear may actually be removed from the bridge and taken to a gear shop for machine measure-

GEARS

GEAR	NO. OF TEETH	PITCH	PRES. ANGLE	SPAN MEASUREMENT					CHORDAL MEASUREMENT					WEAR		BACKLASH	
						CURRENT			ORIGINAL		CURRENT THK.						
				NO. OF TEETH MEAS. OVER	ORIGINAL	SIDE	CTR.	SIDE	ADDEND.	THICK.	SIDE	CTR.	SIDE	IN	%	ORIGINAL	CURRENT

Figure 34-5 Typical Field Data Sheet for Movable Bridge Gear Inspection. The table includes both chordal and span measurement data for a group of gears.

ment. These procedures are generally performed only for an in-depth inspection or for a special case, such as when unexpected wear is found on a critical drive component. Gear condition can be determined sufficiently accurately for most bridge inspection purposes, without using the sophisticated tools and procedures required for the inspection processes described earlier, through one or more of the following procedures:

Noise Open gearsets make some noise during operation, but the noise becomes more pronounced as the gears wear, particularly if some of the gears have high peripheral speeds. A loud whine, or even a roar, suggests that closer inspection is warranted, as some obvious forms of gear wear may be present (see the following discussion).

Backlash Most gearsets have a specified range of allowable backlash when new. Some have specified maxim values, which, when reached after an amount of wear has occurred, require the support bearings to be reshimmed or that some other corrective action to be taken. If the backlash measured on a typical open gearset on a movable bridge is much larger than expected, it is a good clue that some deterioration has occurred somewhere in the immediate vicinity.

Visible Wear If one is sufficiently familiar with gearing to know what the teeth of a standard gear of the approximate size being inspected should look like, or if the actual manufactured detail of the gear is known, it is possible to make a visual estimate of the amount of wear. There are also many forms of gear wear indicated by standardized names and conditions. The FHWA's *Bridge Inspector's Manual for Movable Bridges* and AASHTO's *Movable Bridge Inspection, Evaluation and Maintenance Manual* illustrate many of these types of wear.

Tooth Condition The wearing faces, or flanks, of the gear teeth may show signs of distress, but it may be difficult or impossible to measure the wear at the teeth. Various standardized terms have been developed to represent various specific types of tooth deterioration. These are described in the American Gear Manufacturers Association's AGMA 1010-E95, *Appearance of Gear Teeth—Terminology of Wear and Failure.* Each of the different types of deterioration has its own cause, which, when determined, can lead to correction of the problem and continued use of the gears, or their replacement with a longer-lasting set.

Plastic Flow Some permanent deformation of the gear tooth material occurs in gearsets that do not have sufficient hardness to provide substantial durability. This often takes the form of "fins" growing at the tips of the gear teeth, as material is pushed up on the working faces and extends beyond the top land. This material can also be pushed down and develop into overlaps of metal at the roots of the teeth. Another special case of plastic flow occurs at gears that are adequately sized for bending strength, but have insufficient hardness to withstand local stresses on the tooth flanks at the line of contact. On well-lubricated teeth, this deficiency appears as a line across the teeth at almost exactly the location of the intersection of the pitch circle with the flanks of the teeth, the tooth working surfaces. On a pinion, or driving gear, the line is formed by a depression, and at the driven gear, the line is actually a ridge. On a movable bridge these gears can, in some cases, operate for years without failure, although they will be noisier than an unworn set of gears.

Destructive Wear When the gears are heavily loaded and operate at high speed, destructive failure can develop at the tooth surface. This lack of durability strength can be particularly damaging inside an enclosed reducer, as the entire assembly may be destroyed because of the action of pieces of gear tooth entering other components. It is especially important, when inspecting such machinery, to get the best possible view of the tooth surfaces so that signs of advancing deterioration can be detected.

Mesh Open gears on movable bridges are required by AASHTO and AREMA to have the pitch circles inscribed on their sides. These pitch circles should be tangent at the midpoint of the gear mesh (but the scribed circles may not be accurate). Gearsets can also be checked for axial misalignment, angular misalignment, and other forms of misalignment. FHWA's *Bridge Inspector's Manual for Movable Bridges* and AASHTO's *Bridge Inspection, Evaluation and Maintenance Manual* illustrate these conditions.

Most of the techniques described here, as well as gear tooth measurements, require the gear teeth to be cleaned. The gear teeth should be cleaned thoroughly, removing all traces of lubricant, as well as rust or any foreign material, from the surfaces of the teeth. The solvents used should be nonflammable and nontoxic. For best results, the following process should be used, particularly with open gearing lubricated with a form of grease:

1. Scrape as much material as possible off the tooth faces, flanks, sides, tips, and roots, using a flat tool such as a stiff putty knife.
2. Use a cloth rag dipped in solvent to remove smudges and material trapped in cracks or crevices in the surface of the teeth.
3. Use a wire brush to remove any remaining particles from the cleaned surfaces.
4. Wipe the surfaces to be inspected with a dry, clean cloth.
5. After inspection is complete, and before operation of the bridge, relubricate the gears.

This procedure will minimize the mess involved in cleaning the gear teeth and reduce the likelihood of contaminating any adjacent bearings with solvent. The same technique can be used to prepare for inspection any machinery surfaces suspected of having cracks. Care should be taken to avoid scratching a softer metal surface with a scraper.

If it is necessary to provide accurate measurements of gear tooth wear as part of an inspection, it is important to obtain accurate information as to the original dimensions of the components. The original shop drawings supplied with the bridge are usually sufficient, particularly if they have been confirmed by being certified as-built drawings. On older bridges, the gears may have been replaced at some time in the past. There is no guarantee that such replacement gears were manufactured to the same dimensions as those that were replaced. There have been cases of replacement gears being made to different standards than the originals, even to having different tooth forms. Although it is difficult to get straight spur gears of different pressure angles to mesh together, tooth types made according to different systems have been made to mesh (see Figure 17-8). If this situation is discovered in the inspection of a bridge, particularly in the train of gears forming the main drive of the movable span, it should be carefully documented and reported. If possible, the actual manufactured dimensions of the incorrect gear or gears should be ascertained. This may be difficult, as for many owners there are several different entities that could initiate such a change, and little or no documentation may have been made for the permanent record.

To expedite the field measurements of gear wear on a movable bridge, the drawings most likely to represent the actual gears in the field should be closely examined. The drawings for open spur gearing or racks and pinions on a movable bridge should be available in the bridge owner's files. These drawings should tell

1. The number of teeth in the gears
2. The pitch of the gears
3. The pressure angle of the gears
4. Whether the gears are standard, stub tooth, or long or short addendum (see pages 252, 254)

With this information, data can be assembled to use in making accurate measurements of tooth thickness in the field. If the gears have standard addenda, the chordal addendum and chordal tooth thickness for each gear can be found in a table like the one in the *AASHTO Movable Bridge Inspection, Evaluation, and Maintenance Manual* or FHWA's *Bridge Inspector's Manual for Movable Bridges.* The chordal tooth thickness is a straight-line measurement across the thickness of the tooth, rather than a curved measurement along the pitch circle of the tooth. The chordal addendum is the radial distance from the tip of the tooth, along the centerline of the tooth, to the chordal thickness line, rather than to the pitch circle (Figure 34-6). The tables give values for chordal addendum and chordal tooth thickness for 1 P gears, of size 1 diametral pitch. Dividing the values in the table by the diametral pitch of the gears being measured gives the chordal addendum and thickness for those gears. For gears sized in circular pitch, the values in the table are divided by π and multiplied by the circular pitch of the gears. The chordal addendum and tooth thickness are not dependent on the pressure angle of the gear. The pressure angle need not be known to make these measurements.

If necessary, the chordal addendum and chordal tooth thickness can be calculated directly without the use of the table. Unless the detail drawing for the gear states otherwise, there is assumed to be no backlash removed from the gear. For standard gears, the chordal tooth thickness is then

$$t_c = \text{PD} \times \sin\left(\frac{360}{4N}\right)$$

and the chordal addendum is

$$a_c = \frac{\text{OD}}{2} - \cos\left(\frac{360}{4N}\right) \times \frac{\text{PD}}{2}$$

where

the angle is measured in degrees
N = the number of teeth in the gear
OD = the outside diameter of the gear
PD = the pitch diameter of the gear

The preceding equations apply to standard gears with full-depth teeth and to standard gears modified with stub teeth. If the outside diameter of the gear is not given, it can be computed by adding $2a$ to the pitch diameter of the gear.

Other means must be used to compute the chordal tooth thickness and chordal addendum for modified addendum gearing and for other special tooth shapes. On some movable bridges, the racks and pinions are modified by having extremely large teeth on the pinions, and smaller than usual teeth on the tracks. The drawings for these gears usually show the true addendum and true tooth thicknesses for the rack and pin-

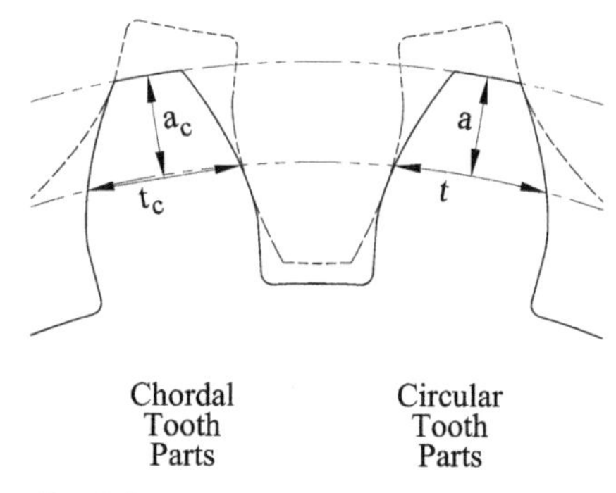

Figure 34-6 Spur Gear Tooth Parts. The chordal tooth thickness is measured on a straight line. The tooth thickness is measured along the pitch line.

ion, allowing the chordal values to be calculated. As on page 428, the half-tooth angle is determined, but the calculation is different than that for a standard gear. In this case, using radians rather than degrees is handy, as the half-tooth angle is simply

$$\frac{t/2}{\mathrm{PR}}$$

where

t is the true tooth thickness as shown on the drawing, measured along the pitch circle,

and

PR is the pitch radius of the gear.

Thus, the chordal tooth thickness for this gear is

$$t_c = 2 \times \mathrm{PR} \times \sin\left(\frac{t/2}{\mathrm{PR}}\right)$$

and the chordal addendum is

$$a_c = \mathrm{OR} - \mathrm{PR} \times \cos\left(\frac{t/2}{\mathrm{PR}}\right)$$

where OR is the outside radius of the gear. These values can also be calculated in degrees, the half angle of such an odd gear being

$$\frac{360t}{2(\mathrm{PD} \times \pi)}$$

The chordal tooth thickness and chordal addendum for a long or short addendum gear is much more difficult to compute, but, this type of gearing is extremely rare in the open gearing used on movable bridges (see page 252). The geometric details of such gears are required in order to make a computation, and these details as shown on a drawing are likely to include the values of the chordal addendum and tooth thickness. If it is necessary to compute these values, the FHWA *Bridge Inspectors Manual for Movable Bridges* provides a formula. When drawings are not available, there are two measurements that can be made in the field that usually allow the original gear tooth size to be determined. These are (1) number of teeth and (2) outside diameter of the gear. By combining these two measured values with an assumed addendum for the gear teeth, a pitch size for the gear can be calculated. When the computation uses a standard assumed addendum and results in a reasonable tooth size, or pitch, it can be assumed to be correct. Variations that may require additional trial computations can include stub teeth, either Nuttal or AGMA, and long and short addenda. Long and short addenda are rare in open gearing on movable bridges and are difficult to identify and accurately measure in the field. Stub teeth are usually obvious to the eye. If the stub tooth gear is an old, cast tooth gear, it is probably a Nuttal stub. If it is a machined gear with cut teeth, it is more likely to be an AGMA stub, especially if it is less than 75 years old. If necessary, gear tooth height measurements can be made and compared with the standard tooth height of a gear with the estimated pitch of the gear in question. To assist in approximations, it is helpful to keep in mind that AGMA stub teeth are usually sized in the diametral pitch system, whereas Nuttal stub teeth are normally measured in circular pitch. Unfortunately, many circular pitch tooth sizes are physically very close to sizes for diametral pitch tooth sizes, so determination of the exact pitch may be difficult, $1\frac{1}{2}$ in. *p*, for instance, is very close to 2*P*. There is only 0.0354 in. difference in tooth thicknesses between these sizes.

Once the chordal addendum and chordal tooth thickness are determined, direct measurement can be made with a gear tooth caliper (Figure 34-7). This is a device similar to a vernier caliper, with a second vernier scale at a right angle to the first. The second vernier scale is set to the chordal addendum of the gear being measured, and the first scale is adjusted to read the chordal thickness of the gear tooth. There are three sizes of gear tooth calipers in common use:

$\frac{1}{2}P$ for very large gears—overall caliper size about 10 in. × 10 in.
1*P* for most typical open gears—about 5 in. × 5 in.
2*P* for smaller gears–about 3 in. × 3 in.

The $\frac{1}{2}P$ caliper is very expensive and is rarely included in a set of bridge inspection equipment. It can come in handy in special inspections, such as if there is a problem with the rack and pinions on a large bridge and accurate chordal tooth thickness readings are required. In cases where extremely accurate measurements are not necessary, when measurements are made on gear teeth too large for a 1 P gear tooth caliper, a ruler can be used, or a standard vernier caliper can be used with a block to allow the caliper to be placed so that the jaws will be at approximately the correct depth. These measurements will not be precise, but they can be accurate enough for

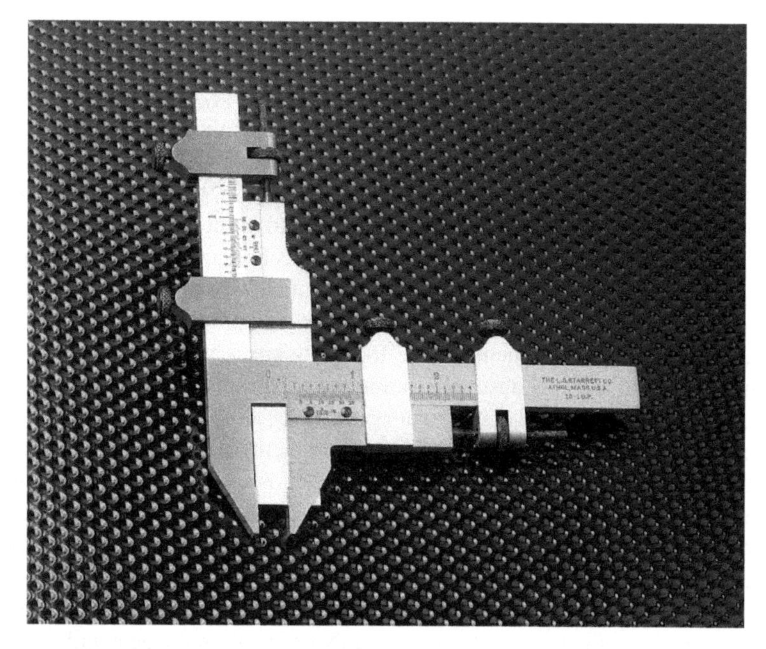

Figure 34-7 Starrett # 456BZ Gear Tooth Caliper. (Photo courtesy the L.S. Starrett Company.)

condition inspection ("biennial" inspection) purposes. The $\frac{1}{2}P$ caliper is very cumbersome for use on gears smaller than about 1*P*, so it should not be obtained for use as an all-purpose gear tooth caliper.

The 1*P* caliper is relatively inexpensive and fits most gears on a typical movable bridge. It can usually be stretched to fit over a $3\frac{1}{2}$-in. circular pitch gear tooth. It can fit readily on gears as small as 3 or 4*P*, equivalent to about 1 in. *p*.

The 2*P* caliper is about the same price as a 1*P* caliper. Most gears on a movable bridge are too large for a 2*P* caliper, but it can come in handy on smaller gears that are somewhat inaccessible and difficult to reach with a larger caliper. The 2*P* caliper can be particularly useful for measuring gears inside a reducer or other gearbox where space is limited. It also can be used for measuring position indicator drive gears and other very small gears, but these are normally lightly loaded and usually do not wear much, so wear is not a critical factor for them.

Gear tooth calipers have been available with digital readouts instead of vernier scales, for easier reading, but see page 422. Metric gear tooth calipers are available but have little value, as most bridge machinery gears made in the United States are diametral or circular pitch gears. It is unlikely that much bridge machinery in the future will be made to the metric system, until most U.S. industry converts to metric.

Care must be taken in placing a gear tooth caliper on a gear, as it must be straight and square to measure accurately. The gear must also be properly prepared before using the gear tooth caliper. The gear must be absolutely clean, as grease, dirt, and rust will all affect the readings, both on the faces of the teeth and on the tip of the gear tooth where the caliper is placed. See the discussion beginning on page 426 on the proper procedure for gear tooth cleaning. The gear must also be free of burrs along the tip of the tooth, as these will not allow the caliper to seat fully on the tooth. Burrs can be the result of the original machining of the gear, or they can develop as the gear wears, particularly if the gear is made of a soft, malleable material such as mild steel. The burrs can be removed with an ordinary large, flat file, but care must be taken not to damage the gear tooth flank surface or the top land of the tooth. If the top land (tip) of the gear tooth is inaccurately made, worn, severely damaged, or corroded, an accurate gear tooth caliper measurement cannot be made (see the following discussion). The gear tooth being measured should be cleaned and prepared thoroughly along its full length before measurement, as the tooth thickness is likely to vary along the length of the tooth and it is desirable to find the minimum thickness.

In some cases, it is inconvenient to measure the chordal tooth thickness of the gear because of wear or damage that obliterates the tip surface or top land of the gear or for other reasons. In such cases, it may be possible to make span measurements of the gear teeth. Span measurements are made with the use of a simple vernier or dial indicator caliper, preferably one with long, narrow jaws (see Figure 34-2). The span measurement is taken over two or more teeth of the gear, measuring essentially from the left side of one tooth to the right side of another tooth. The number of teeth included in the span measurement depends on the number of teeth in the gear and the pressure angle. The dimension of the span measurement is dependent on several factors:

1. The number of teeth in the gear
2. The pitch size of the gear
3. The pressure angle of the gear teeth
4. The number of teeth over which the span measurement is taken
5. The amount of wear on the gear teeth
6. The position of the caliper on the gear, particularly if the gear teeth are worn

The number of teeth over which span measurements are to be taken, and the span measurements for unworn gears, are tabulated in the *AASHTO Movable Bridge Inspection, Evaluation and Maintenance Manual* and the FHWA *Bridge Inspectors Manual for Movable Bridges.* For gears with more teeth or those with odd tooth profiles, the span can be computed if necessary. Caution must be used in making span measurements of odd gears, as there may be only one correct position of the caliper (Figure 34- 8) for taking the span measurement, but wear of the gear may result in a minimum measurement in a different position. Thus, it may be that the difference between the theoretical span and the actual measurement does not equal the amount of wear. The idea behind the span measurement is that a line tangent to the working surface on one side of a tooth on a gear and a parallel line tangent to the working surface on the opposite side of a nearby tooth, if the measurement is taken over a certain number of involute teeth, will be a constant distance apart. Many, if not most, gears on movable bridges particularly pinions, only have an approximate involute profile, due to the undercutting problem with $14\frac{1}{2}^\circ$ or 20° pressure angle gears with few teeth. The distance between the tangent points, and thus between the parallel lines formed

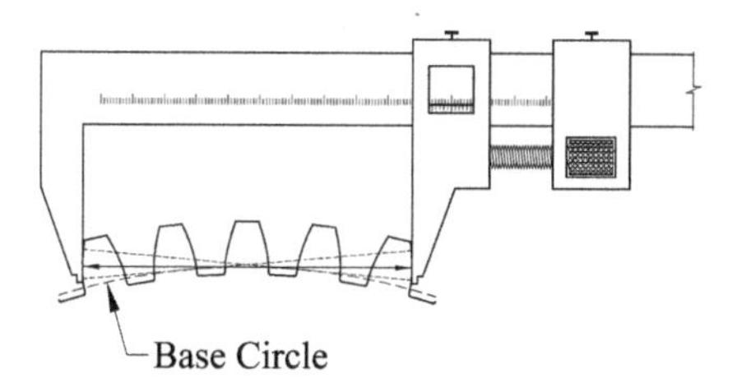

Figure 34-8 Span Measurement on a Spur Gear.

by the vernier caliper, will be a known value for an unworn gear. This makes it a useful and easy method of checking the accuracy of new gear teeth. The point of minimum measurement at worn gear teeth may or may not coincide with the pitch diameter of the gear at either or both of the teeth at which the measurement is taken. This, plus the fact that a span measurement is taken at two different gear teeth, usually results in apparent wear values that are not identical to those obtained by gear tooth caliper measurements.

To obtain the optimum number of teeth k over which the span is to be taken, find the number of teeth that produce an included angle, less twice the pressure angle, that is closest to zero. For example, for a $14\frac{1}{2}°$-pressure angle gear with 157 teeth, each pitch, or tooth plus space, represents

$$\frac{360°}{157} = 2.293°$$

Thus, 13 teeth is the correct number k over which to measure the span, as this provides $12\frac{1}{2}$ pitches, so that

$$12.5 \text{ pitches} \times 2.29° \text{ per pitch} - 2 \times 14.5° = -0.34°$$

This is the closest to zero possible. The span should be measured over 13 teeth. For a fairly large gear, this will require a very large caliper for measurement. To determine what the span measurement would be, it is necessary to know the curvature of the gear teeth. This curvature is projected out from the base circle, in opposite directions, to a straight line tangent to the base circle.

The span measurement M equals the base pitch times the number of teeth in the span measurement minus 1, plus the base thickness of one tooth:

$$\text{M} = P_B \times (k-1) + P_{Bt}$$

For the gear described here, with a tooth size of 2 in. circular pitch,

$$\begin{aligned} P_B &= 2 \times \cos(14\tfrac{1}{2}°) \\ &= 1.9363 \text{ in.} \\ k &= 13 \\ P_{Bt} &= 2 \times [R_P \times \sin(14\tfrac{1}{2}°) - R_B(\beta - \alpha/2)] \end{aligned}$$

R_P = pitch radius = $157 \times \frac{2}{2\pi} = 49.9747$ in.
R_B = base radius = $R_P \times \cos 14\frac{1}{2}^\circ = 48.3828$ in.
β = the pressure angle in radians = $14\frac{1}{2}^\circ \times \pi/180^\circ$
$= 0.2531$
α = the angle of one tooth in radians = $2 \times \pi/(157 \times 2)$
$= 0.02001$

This will result in a span measurement M of

$$M = 24.7402 \text{ in.}$$

for a 157 tooth, 2 in. c.p. $14\frac{1}{2}^\circ$ pressure angle gear.

This will require a vernier caliper with larger than 24 in. capacity, which makes a very unwieldy and expensive tool. Thus, the inspector should be satisfied with gear tooth caliper measurements for a gear of this size.

In addition to measuring gears and checking alignment, the inspector should also check the gears for other damage or deterioration. The gear may be loose on its shaft or have cracks. The gear hub should be checked for cracks, especially at the keyways. If the gear has spokes, these too can be cracked. Such cracks may be difficult to detect, as the spokes may have rough as-cast surfaces. Many older open gear drives have large spur gears. These were frequently made with very thin rims to minimize the weight of material used and cut the cost and to minimize the rotational inertia of the gears, particularly the higher-speed gears in a gear train. These thin rims can develop cracks, which usually are radial, parallel to the axis of the gear, and go straight across the rim of the gear, usually at the root of the gear teeth. In very rare cases, a gear tooth breaks out of the rim or hub of the gear. Such a crack will usually propagates rather quickly, so that the tooth breaks out before the crack can be detected. This is more likely to happen with idler gears, which are loaded in both directions alternately while operating, resulting in greater likelihood of fatigue failure. Such failure in other gears implies faulty design or manufacture.

Many movable bridge drives, particularly on older swing bridges, contain bevel gears. Straight bevel gear teeth have the same shape in cross section as straight spur gear teeth, but are cut on a taper. To find the chordal addendum and tooth thickness of such a gear, the *formative* number of teeth must be found. This number can most easily be determined by computing the pitch radius of the bevel gear in a plane at a right angle to the axis of the gear tooth (Figure 34-9). The plane is usually taken at the large end of the tooth, where the size of the tooth is at the nominal pitch, but it can be computed at any location along the tooth, with the pitch reduced proportionately. This formative pitch radius being developed, the number of teeth in a gear of that size, with the prescribed pitch, is computed, and this number of teeth used to determine the chordal addendum and chordal thickness. AASHTO and AREMA provide a formula for determining the formative number for a bevel pinion.

Differential gears on bascule and swing bridges require special consideration in inspection. If the bridge drive machinery is operating normally, the bevel pinions in

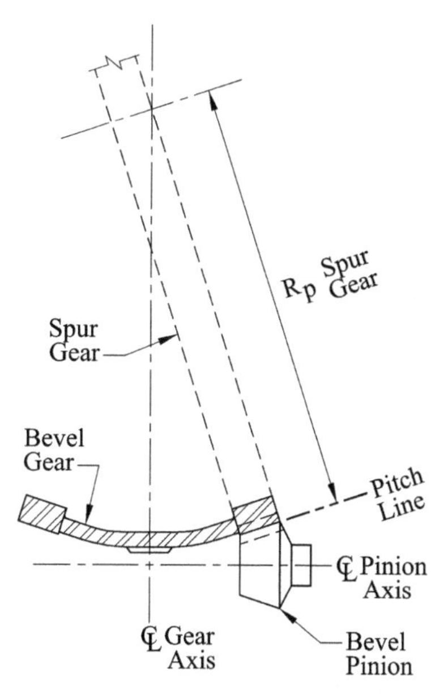

Figure 34-9 Bevel Gear Geometry. The correct tooth shape of a straight bevel gear can be determined by laying out a straight spur gear that mates with the mating bevel gear, that rotates about the same axis as the bevel gear in question. The number of teeth this straight spur gear would have is the *formative* number for the bevel gear being analyzed.

the differential make only small fractions of a revolution about their axes. The bevel gear and bevel pinion teeth that are in contact when the unit is being inspected are likely to be the only teeth that are ever in contact. They should thus be the only teeth on these gears that are worn. If other teeth, that are in the clear so that a tooth caliper can be placed on them, have noticeable wear, it means that the differential gears have been turning. Unless the mechanism has been dismantled, this implies a strong likelihood that there is excessive play somewhere in the drive train. This condition should be investigated. If the differential is acting normally, there will be no wear on the accessible bevel teeth. Backlash measurements should be used to obtain an approximation of the gear wear. The mesh of the bevel gears and pinions should be checked, as any increase of center distance will increase the backlash, even with no wear. Open differential units frequently have excessive center distance because of wear at the thrust bearings holding the bevel gears in mesh with the bevel pinions. The recorded backlash value can be adjusted for increased center distance by measuring how far out of mesh the bevel gears and pinions are, and computing or estimating the effect of this displacement on backlash. It is quite unusual for the bevel gears in enclosed differential units to have excessive center distances, particularly if these gears are supported on antifriction bearings. Clutch differential gears on vertical lift bridges can have several worn teeth, because the differential bevel pinions can rotate when the bridge is being seated, and may make considerable rotation if the bridge tilts or skews and this tilt or skew is corrected in seating the bridge.

ENCLOSED GEAR DRIVES

Enclosed gearboxes should be checked externally for damage or excessive axial or radial play at the external shafts and for leaks, damaged seals or oil level gauges, clogged breather caps, corrosion, cracks, broken mounting feet, and loose fasteners. The unit should be observed during bridge operation, and any movement, unusual noises, overheating, or any other abnormalities noted. The oil level in the reducer should be noted. If the inspection scope requires internal inspection of gearboxes, the power to the drive motors should be shut off via the local disconnect switches before beginning to provide access to the interior of the box, with appropriate notices given that the bridge will be inoperable. It may be necessary to be able to quickly put the gearbox back into operating condition, so covers, gaskets, and fasteners should be kept readily available when they are removed. Extreme care should be taken when removing gearbox inspection covers. Bolts or nuts should be removed carefully to avoid damaging threads or twisting off bolts or studs. The cover should also be removed very carefully to avoid damage to its gasket, unless a new gasket is to be installed. Be careful to avoid scratching the metal surface when removing an adhered gasket. Clean all bolt, nut, and screw threads carefully before reinstalling them. It is often helpful to put screws or bolts back in the same holes they came from, for a better fit. The same applies to nuts. Make sure the cover and gasket are properly aligned before tightening the fasteners.

After the cover has been removed, check the inside of the gearbox for signs of corrosion. Check for water in the box. There may be beads of condensation on the metal surfaces, or there may be water in the bottom of the box. The drain plug can be removed to check for water in the box, but the bottom of the box should also be checked for pieces of metal that may have come from a gear or bearing. All internal surfaces of the gearbox should be covered with oil, particularly if the unit has just operated prior to inspection; if they are not, it may be that the lubricant is not being distributed properly. If there is a lube pump, try to determine whether it is operating and circulating oil as it should. The oil can be checked by rubbing it in the hand or on a piece of metal to see if it has retained its lubricating qualities. A sample of the oil can be taken for laboratory analysis. A laboratory can test the oil for metal content, water content, and viscosity. If accessible, the gears and bearings inside the box can be checked for visible wear and deterioration. Fiber-optic devices or miniature video cameras can be used to see areas out of direct sight. Caution must be used with these devices, however, as the images can be deceiving. A streak of grease may appear in a video image as a crack or wear mark. Sleeve bearings can be checked for clearance with feeler gauges, but care must be taken not to damage the bearings, particularly at small-diameter shafts. Backlashes can be measured at gearsets. The measurement of gear teeth inside the box may be impossible because of lack of space. If space is available, measurement should not be attempted unless the manufacturing drawings showing chordal addendum and thickness, or other measurement parameters, are available. Disassembly of gearboxes beyond removal of inspection covers should not be undertaken (see under "Miscellaneous" later in this chapter).

SHAFTS AND COUPLINGS

Most movable bridge machinery shafts are somewhat oversized for the loads they are expected to carry. Nevertheless, these shafts can be susceptible to cracks. The most likely place for cracks to appear is in keyways. Rounded-end keyways are least likely to be cracked. Straight-ended keyways, even if the depth at the end of the keyway is gradually reduced to zero, are more likely to develop cracks, but all should be checked, particularly at the end of the keyway away from the end of the shaft. Stepped shafts can develop cracks at the base of the smaller diameter, particularly if bending is present. Unfortunately, this area is usually inaccessible for inspection because of a gear or collar having been mounted on the shaft at the smaller diameter. Any signs of abnormal movement or bleeding rust should be taken as a clue that the shaft may be damaged in this area. Further inspection by non destructive testing (NDT) methods, or by dismantling the shaft, may be necessary to confirm the presence or absence of a crack. Longer, thinner shafts can also become bent or overstressed in torsion. Such damage can most easily be detected in observing the shaft while the bridge is in operation. If a shaft is painted, torsional overstress may manifest itself in cracking of the paint.

Movable bridge machinery shafts are connected by many different varieties of couplings. All coupling hubs should be checked to see if they are tight on their shafts and the keys are tight. Flexible couplings should be checked to see if they are adequately lubricated and their seals are in place and in good condition, and whether the couplings have excess play. Most of this can be determined without dismantling the coupling, but it may be necessary in some cases to dismantle a coupling to confirm a defect. If a coupling is dismantled, the power should be turned off and proper notices given to mariners that the bridge will be inoperable for a period of time. Before the coupling is removed, the brakes should be released and the machinery rotated by hand until it is certain that the coupling is unloaded. Temporary support should be given to floating shafts supported by the coupling. On some bridges, it may be impossible to unload the coupling without providing temporary support for the bridge itself. If the bridge will not be stable with the shaft disengaged, it must be closed to traffic until the shaft is reengaged and the bridge is secure.

If the coupling is a grid type, pull the grids off and check for breaks. Check for wear at the slots in the coupling grooves. A new groove in a quality grid coupling has curved surfaces so that the grid can bend as the coupling flexes. Check the narrowest part of the slot for wear. Check the internal and external teeth of gear-type couplings for wear, cracks, or breaks. Chain-type couplings are sometimes used at more lightly loaded machinery connections. These couplings can quickly deteriorate if they are poorly lubricated. Slack in the chain components can be checked when the coupling is unloaded. Jaw-type couplings usually need not be dismantled for inspection. Wear in the jaws can be determined by comparing clearances between jaws with the original clearance. Check to see if the jaws are coaxial, as wear may have occurred inside where one shaft is extended into the other hub. Jaw-type couplings are usually unlubricated, so fretting corrosion and wear can be substantial. Rigid couplings are found on many older movable bridges. They should be checked carefully for signs of movement between any of their components. Loose connecting bolts, loose hubs or keys, and gaps between the coupling halves are indications that the shafts are not perfectly aligned so that flexure is occurring at the coupling, or that the coupling is overloaded.

BEARINGS

There are two categories of bearings on movable bridges, friction and antifriction. Friction bearings can be inspected with relative ease and can usually be disassembled for detailed inspection without risk of damage. Antifriction bearings are difficult to inspect and can be damaged if disassembled. An external inspection of the bearings should be made to look for signs of lubrication, for broken, missing, or loose bolts, cracked, misaligned, or otherwise damaged housings and seals, corrosion, and alignment of the bearing with the shaft. Check for missing lubrication fittings—if there are any missing, it is likely that lubrication is poor. Oil cups and other automatic lubricators are notoriously poor performers. If these are present, they should be checked to see whether they have any lubricant in them and whether the lubricant is getting from the reservoir to the bearing. Other lube fittings and lubrication lines should also be checked to see if they are clogged or loose. If possible, obtain a grease gun and test the fitting to see whether lubricant enters the bearing surface when applied. Be sure to remove drain plugs, if present, before applying lubricant so as to avoid displacing any seals that may be present. Note what comes out of the drain as new lubricant is applied. If there is more than one drain plug, remove and replace each in turn while applying fresh lubricant. Note whether any of the drain passages are clogged.

Friction Bearings

Friction bearings tend to wear because of the sliding of the shaft journal on the bearing material as the shaft rotates. This wear is more pronounced if the bearing is inadequately lubricated, runs dry of lubricant, or the lubricant is contaminated. The contact areas of most friction bearings are accessible, so the clearance, at one end at least, in the bearing can be measured with a feeler gauge. Care should be taken to ensure that the maximum clearance is measured. Some shafts are held up off the bottom of the bearing by static loading on the gears mounted on the shafts. If it does not interfere with traffic on the bridge, or if traffic can be stopped during the inspection, the machinery should be unloaded by releasing the brakes before making bearing clearance measurements. Clearance should be measured at the top, bottom, and each side of the journal. Use flexible feeler gauges (see page 422). While making the clearance measurement, check for lubrication of the bearing; check the condition of the lubricant by examining the feeler gauge after removing it from the bearing. If possible, clearance should be measured at both ends of the bearing. This is almost always impossible at heavily loaded bearings because the gear, sheave, or trunnion collar mounted on the shaft is immediately adjacent to one end of the bearing. Sometimes collars are mounted on the shaft adjacent to the bearing, so that the clearance cannot be measured without removing or displacing the collar. A collar can usually be removed or relocated temporarily to make a clearance measurement, as long as the machinery is not operated while the collar is not in place adjacent to the bearing. At other installations, the shaft is stepped or other obstructions prevent direct measurement of the bearing clearance. It is sometimes possible to displace a shaft in its bearing and measure the movement of the shaft by means of a dial indicator gauge, but this method is unreliable, as it does not necessarily indicate the full bearing clearance.

As little cleaning as necessary should be done on the bearing, unless it is to be disassembled, to avoid getting solvent in the lubrication channels. If the bearing is to be disassembled, cleaned, relubricated, and reassembled, then a complete cleaning, as described for gears on page 426, is permissible. The external surfaces of an assembled bearing can be cleaned with caution so that inspection can be made for cracks.

Disassembly of friction bearings can be done as part of an inspection if the bridge can be taken out of service and the machinery can be properly supported with the bearing cap removed. Friction bearings with bronze linings rarely wear significantly and rarely deteriorate in other ways, so little can be learned from disassembling them that would not be detectable from a close external inspection. Friction bearings with babbitt metal linings can have cracked or broken up linings or can be worn sufficiently that the lubrication grooves are eliminated. These defects are sometimes detected by means of external inspections, but are readily found if the bearing is disassembled. Unfortunately, a badly deteriorated babbitt bearing that is still serviceable may fall apart and crumble, or otherwise be destroyed or made unusable, when it is disassembled. If disassembly is performed as part of the inspection, contingency plans should be made in case a bearing cannot be put back in service after disassembly.

Some friction bearings that support the stub ends of shafts have covers over the stub ends of the bearings. This is particularly likely on vertical lift bridge sheave shaft bearings. Measurement of bearing clearances and direct visual inspection of the bearings necessitate removal of the covers. Caution should be used when removing these covers, as a mass of semifluid lubricant may be present inside them. Proper containment should be in place to avoid a spill onto the roadway or into the waterway.

Antifriction Bearings

Every effort should be made to determine the condition of antifriction bearings mounted on movable bridges without disassembling them. If it is necessary to disassemble the bearings, and the bridge owner's staff is not available to perform the work, a factory technician should perform the assembly and disassembly.

Many things can be determined from an external inspection of an antifriction bearing. Radial play can be determined by displacing the shaft radially. The movement should be negligible. Any greater movement indicates either a worn bearing or a bad or loose housing or shaft connection. Axially fixed and floating bearings should be checked to see whether they are properly restrained. If the bearing makes noise during operation, it is a sign of internal deterioration. A trained ear can estimate the type of deterioration, and its severity, from the sound produced by the antifriction bearing during operation. Abnormal conditions, such as overheating, can be detected by placing the hand or a thermocouple on the bearing while it is operating. If the bearing is oil lubricated, a small sample of the oil can be removed and examined for metallic particles. Most antifriction bearings use different materials for their races, rollers or balls, and cages. The size of the free particles, their composition, and their number are clues to the degree of deterioration of the bearing.

CRACKS

Although the machinery on a movable bridge is not high-precision equipment, as modern machinery goes, most of the surfaces of typical gears, shafts, and bearings have been machined and come from the manufacturing plant with smooth surface finishes. This allows the machine parts to be tested rather easily for surface cracks by means of dye penetrant. Dye penetrant works best on smooth, well-polished surfaces. It may be necessary to spend a great deal of effort to prepare a rusted surface so that a crack is visible when dye penetrant testing is done. Other methods, such as magnetic particle and ultrasonic testing, can be used to detect cracks (see page 396), but these methods are best left to technicians who specialize in this type of work. Ultrasonic testing frequently requires access to a component from more than one point to achieve complete results. These methods also require clean, smooth surfaces for best results. See page 426 for a description of an acceptable cleaning method. Once a machinery surface is adequately cleaned, any larger cracks present can usually be detected by the trained naked eye. Dye penetrant and magnetic particle techniques can be used for documentation and to discover microscopic cracks. Magnetic particle testing may be able to detect internal cracks present in castings or weldments.

ELECTRIC MOTORS AND BRAKES

Although electric motors and brakes are considered electrical equipment, certain aspects of their inspection are usually covered by machinery inspectors. These aspects include the mechanical functions of the units. The mountings should be checked to make sure that the units are secure and that all bolts are present and tight and properly fitted to the holes. Motors and brakes are usually specified to be attached to their mounts with turned bolts in fitted holes. In many cases these bolts have disappeared over the years, becoming lost when the motors or brakes are changed out or removed for repairs. Standard machine bolts are often substituted, fitting more loosely in the holes and allowing misalignment. The connections to the rest of the machinery should be checked. The brake wheels should be secure on their shafts, tight, and properly keyed. The motor coupling hubs should be tight on the motor shafts, with keys tightly in place. Brake shoe clearances should be checked with the brakes electrically released. A hand release may not produce the same shoe movement as the normal actuator. The clearances to the brake wheel should be equal at both shoes. The shoes should have substantial friction material remaining, with no rivet or shoe base material close to the friction surface. The brake shoe contact to the drum should be checked with the brake set. Both shoes should be firmly and squarely gripping the drum. The brake torque can be checked by using a large torque wrench. The normal brake rating is for dynamic action, with the wheel turning. Static brake torque is higher. The brake drum or wheel friction surface should be smooth and polished. This is usually the case only with motor brakes, as machinery brakes are usually not used to stop the moving bridge. Rust should be minimal on the machinery brake drum friction surface, and nearly nonexistent on the motor brake contact surface, and there should be no pits, gouges, grooves, or other surface discontinuities on either. There should be no oil or grease or other contaminants on the brake wheels or on the brake shoes. See pages 455–456 for more information.

HYDRAULIC EQUIPMENT

Hydraulic power transmission equipment has been used extensively on movable bridges in the last 25 years or so. Hydraulic machinery operates by making use of a fluid under pressure, so the first thing to look for in hydraulic bridge machinery is leaks. The inspector should be very careful in entering a hydraulic machinery area to avoid slipping on oily surfaces. Leaks are most likely to develop where the fluid acts on moving parts, so hydraulic cylinder and motor seals should be carefully checked when the components are working hard. The presence of fairly fresh hydraulic fluid on the floor, or coating the surfaces of machinery and supports, is a good clue that leaks are present but does not necessarily help in locating them. Fluid under pressure can spurt out and travel long distances, and leaking fluid can run along tubing or pipe lines and hoses and other surfaces for considerable distances before accumulating in noticeable amounts. Leaks can occur in flexible hoses, as well as in rigid pipe or tubing, as a result of damage. Fittings can work loose, which can result in leaks. All these areas can be checked closely with the system under pressure, but it may be more effective to wipe the surfaces clean and dry while the system is shut down, then note where hydraulic fluid first appears when pressure is restored.

Hydraulic prime movers such as cylinders and motors can be subject to very high forces. Cylinders and motors should be closely observed during operation to see whether abnormal movement is occurring or excessive clearances have developed in wearing areas such as clevis pins. Rods should be checked to see that they are clean, with only a light coating of oil on the rod surfaces. The rod surfaces should be smooth and highly polished, with no signs of dents, scratches, or score marks, and the rods should not be bent. Cylinders should be checked for proper alignment. Clevises and pins should be checked for wear, corrosion, and misalignment. Clevis and rod end bushings should be checked for wear. Hydraulic cylinders on some movable bridges have industry standard bushings installed, which may have very thin walls, or may consist of steel sleeves with thick (or thin) coatings of bronze or other nonferrous material on them instead of solid bronze bushings. Some have merely steel-on-steel contact. Thin bushings can quickly wear out, corrode, or seize, resulting in broken connections. Steel-on-steel pin connections can wear rapidly, or seize.

Hydraulic control valves of various types can wear, resulting in lack of adequate control of the machinery. Any erratic behavior of a hydraulic device can usually be traced to a malfunctioning or worn valve. If the bridge operator has difficulty controlling the speed or seating of the bridge, deteriorated valves can be the problem. The inspector should not attempt to dismantle hydraulic valves to determine the source of a problem, as damage or loss of components can result. Moreover, some valves are very complicated and can be reassembled incorrectly. Other valves have adjustments that must be made precisely in order for them to operate correctly, and these adjustments can be difficult and time-consuming to make. The possible or probable trouble source should be identified, with a recommendation made, if necessary, that it be further investigated by a trained technician.

Hydraulic power units are very susceptible to poor maintenance. Dirty filters and low fluid levels can cause damage to the hydraulic system, so the inspector should check for these conditions. If filter bypasses are present and the fluid is flowing through them, this should be noted. The hydraulic fluid should be examined, or laboratory tested, to see if it is contaminated or deteriorated, and if it is the proper type of fluid for the system. Hydraulic pumps should be observed during bridge operation and listened to for sounds of cavitation. Fluid reservoirs should be checked to see that they do not contain contaminants, that breather assemblies are working properly, that pump inlets are not clogged, that they are properly filled, and that covers are properly in place.

Hydraulic oil can be subjected to a great deal of heat during bridge operation, particularly if the system is not well maintained. A sample or samples of the hydraulic oil in the system should be taken and tested to determine the quality of the oil as well as the presence of contaminants. The *AASHTO Movable Bridge Inspection, Evaluation, and Maintenance Manual* (2.10.11.1) discusses the requirements of this testing. This manual also covers analysis of filter elements, pressure testing, leak testing, case drain flow analysis, and temperature analysis (in sections 2.10.11.2 through 2.10.11.6).

MACHINERY SUPPORTS

Machinery supports, including Hopkins frames, are often thickly coated with old grease or paint and sometimes obscured by debris. The machinery supports should be thoroughly cleaned and checked for corrosion, cracks, and loose fasteners. Old grease is not a corrosion-preventative material, as it can trap moisture and allow corrosion to occur unseen. After the machinery supports have been cleaned, they can be observed during bridge operation to see whether any undue deflection or other movement is occurring.

EMERGENCY OR AUXILIARY DRIVES

If an emergency or auxiliary drive is present on a movable bridge, it should be determined whether the unit is functional. If it is an engine drive unit, the bridge operator should be asked to start it. In many cases, the unit will not start. It should be inspected in an attempt to determine the reason for not starting and to determine the overall condition of the engine. If the scope of the inspection calls for it, and the engine is operable, an attempt should be made to operate the bridge with the engine drive. Under normal conditions, the engine drive should open and close the bridge with no difficulty. It may take considerable time, as the typical movable bridge operates much more slowly in emergency mode than normal mode because of a smaller prime mover with greater reduction gearing being employed in the emergency drive. It may be necessary to perform the operating test during off-hours to avoid disruption to traffic.

Manual emergency or auxiliary drives should be checked to see if they are functional. They are quite often seized because of rust so that they cannot be engaged or turned, as they are usually not the favorite pieces of machinery of maintenance personnel. Under no conditions should a manual or emergency drive be engaged unless the power has been disconnected from the main drive. Inspection and maintenance personnel have been killed or injured as a result of miscommunication between them and the bridge operator, or an inattentive bridge operator forgetting that the drive was being tested. If the inspection scope requires it, and the drive is operable, it should be tested in a full opening and closing cycle. The test should be timed, to see if the duration falls within the time it should take, provided the required number of personnel to perform the operation are available. A capstan drive may require four or more persons to move the bridge. If manually operated brakes are part of the emergency or auxiliary drive, they should be checked to see whether they are functional and capable of stopping and holding the bridge in any position, with adequate reserve capacity for snow, ice, or wind loads. Some bridges use air or hydraulic motors to power the emergency or auxiliary drive. These may be portable. If this is the case, or if electric drills or impact-type wrenches are normally used to operate the emergency drive, these should be tested; the assistance of bridge maintenance personnel may be required. Most manual emergency drives for movable bridges provide a great deal of torque multiplication so that the minimum number of persons can operate the bridge manually in the most adverse weather conditions, such as high winds. During test operations, the environmental conditions are most likely to be much less severe, so the effort at the input may be minimal.

MISCELLANEOUS

Many inspection contracts for movable bridge machinery call for the dismantling of certain components for the purpose of inspection or have a general requirement that machinery components must be dismantled as required for inspection. This is usually meant to include not only covers of open type gears, but also caps of friction and antifriction type bearings, inspection or access covers of enclosed gear drives, and guards on such components as couplings and chain drives. The inspector should never remove these covers or guards or dismantle machinery components, but should have all such work performed by the bridge owner's staff. There are a number of reasons that this should be done, but many contracts state that the inspector or inspecting firm must perform all this work with its own forces or hire a rigging contractor for assistance. The reasons for having this work done by the bridge owner's forces include the following:

Damaged equipment—Some components may already be damaged, broken, or missing before the inspection starts. Work on these components by the owner's forces absolves the inspector of blame or liability.

Difficult disassembly and reassembly—Most components, particularly bearing caps, are large, specially fabricated pieces of machinery and require special tools to remove large bolts and to lift heavy components free. These tools are frequently supplied with the machinery for the bridge when new, but can be lost or damaged in the course of time. The maintenance staff for the bridge should have a complete stock of tools and equipment suitable for use on the particular bridge and should be familiar with using these tools. Even a competent rigging contractor is likely to be unfamiliar with some of the machinery components on a movable bridge, may not have the special tools required, and may have trouble, at the least, in properly reassembling these components. Persons unfamiliar with the work may also injure themselves or others. A rigging contractor's forces may also inadvertently take some of the special tools for the bridge when they leave the site. Some of these special tools may be very difficult to replace.

Relubrication—Most moving parts should be properly relubricated immediately after disassembly and inspection. The bridge's maintenance staff are familiar with the lubricants in use on the bridge, know how to use them, and should have them in stock. The proper lubricant may not be available to the inspector. The inspector or his or her assistants may perform the lubrication inadequately, or with the wrong components, because of missing or damaged lubrication charts. This may result in unnecessary damage or wear to the equipment.

If it is absolutely necessary for the contract inspector to perform the dismantling of machinery components, explicit exclusion should be made in the inspection agreement against any liability on the part of the inspector for damage to components or for any failure of the bridge to operate. The condition of the equipment prior to dismantling should be carefully checked and of any previously damaged components documented. The inspector should carefully supervise all riggers or laborers employed as assistants, so as to minimize the likelihood of damage to any bridge component.

There are many special pieces of machinery on movable bridges that are sometimes overlooked during inspections. These include various indicators that are purely mechanical in nature. Many movable bridges are equipped with mechanical clearance indicators, which are often found in inoperable or poor condition. Some are equipped with a rope and pulley connector between the float or buoy and the indicator. Such pulleys often corrode and "freeze up" so that they do not turn. The ropes are usually made of a length of wire rope. The wire rope, although it may be galvanized, can become corroded so that it is stiff and does not allow the indicator to move freely. The rope can even break as a result of excessive corrosion or restricted movement. The inspector should make full notes of the condition of this equipment.

Many movable bridges have mechanical position indicators. These are usually made up of lengths of small shafting, with gears, bearings, and couplings. The components usually do not wear significantly, as they are relatively unloaded, but they sometimes become loose or misaligned so that false indications are given. In some cases, the drives for these devices become grossly misaligned, so that large forces develop on the couplings connecting them. Some small couplings driving electrical or mechanical indicating devices have nonmetallic inserts, which can be completely destroyed when operating under conditions of severe misalignment. These components may be critical for bridge operation, so the inspector should make careful note of their condition.

BALANCE TESTING

Bascule and vertical lift bridges are usually counterbalanced to minimize the amount of power required to raise and lower them. They almost always have a small amount of imbalance designed into them to allow them to seat more stably. An approximation of the span balance state can sometimes be obtained by observing the wear on gear teeth. Heavy wear on one face of each gear tooth only, or wear much heavier on one side than another, strongly suggests that a substantial imbalance of the span has been present for a substantial part of the bridge's life. A drift test, described later under "Imbalance Measurement," can be used to confirm whether this condition still exists. More precise balance testing allows the condition of balance to be determined. Balance testing can be performed by many different methods. The best way to obtain the balance condition accurately is to measure the force or torque required to raise and lower the bridge, under conditions when ice and wind are not a factor, and then compute the balance state of the bridge.

Bascule Bridges

All typical bascule bridges, whether simple trunnion, heel trunnion, or rolling lift, have a trigonometric balance relationship. The center of gravity of the bascule span is located some distance from the center of rotation of the bridge and rotates around it as the bridge opens and closes. For a simple trunnion or Hopkins frame bascule, the center of gravity rotates about the fixed center of rotation as the bridge rotates. A heel trunnion bascule has the bascule span and the counterweight frame connected by a parallelogram arrangement, so that the center of gravity of each rotates about its respective fixed center of rotation as the span opens and closes; the links can be considered fixed point masses attached to the other rotating components at their connecting pins. The operating struts of a heel trunnion bridge complicate this relationship somewhat, as the weight of the struts is partially removed and added to the weight of the bascule span or counterweight frame as the bridge opens and closes. Articulated counterweight bascules apply the mass of the counterweight vertically at the counterweight trunnion. The counterweight trunnion bearings and the link bearings add friction to the system. A bascule bridge of any type operating by means of hydraulic cylinders will have the balance affected by the change in geometry and weight of the hydraulic cylinders as the span opens and closes. A typical rolling lift bridge rotates not unlike a simple trunnion bascule, but the center of rotation translates backward as the bridge opens, and forward as the bridge closes, taking the center of gravity with it. Operating struts, when used on rolling lift bridges, can have the same effect on the balance of a rolling lift bridge as they do on a heel trunnion bascule. Some operating struts on rolling lift bridges are pinned to the bascule span at the center of rotation, so that they have no effect on the balance. The balance of a rolling counterweight bascule is more complicated and requires individual analysis.

Some bascule bridges are located so that the counterweight dips into the water as the bridge is opened. It is impossible to perform a useful balance test under these conditions. If the problem is a flooded counterweight pit, the pit should be pumped out, at least temporarily, for the duration of testing. If the counterweight is exposed, and the surrounding water level has risen since the bridge was built so that the counterweight dips into the water whenever the bridge opens, there is little that can be done except to replace the bridge or construct a watertight pit. If the counterweight touches the water only at high tide, the test should be made at low tide. It should be noted whether the counterweight, particularly if it is concrete, is saturated with water. Counterweight pockets should be checked to see if they contain water or debris.

For a bascule bridge, the torque required to operate the span is

$$\text{Torque to open} = T_o = T_f + T_i + T_w + T_s + T_a$$

The torque required to close the bridge is

$$\text{Torque to close} = T_c = T_f - T_i - T_w - T_s + T_a$$

T_f = friction torque, assumed constant when the bridge is moving, always acting against movement. A preliminary investigation should be made to confirm that this torque is constant, there are no stray resistances, such a binding in the machinery or the span itself that increases or decreases as the span opens or closes, or any condition that may cause a difference in the friction load developed to open the bridge versus that required to close it.

T_i = imbalance torque, assumed to be a trigonometric function of the opening angle.

T_w, T_s = torque due to wind, ice, and snow, made to be zero by conditions of test (if wind is unavoidably present, its sign would depend on wind direction).

T_a = acceleration torque, which can be of any magnitude and can be positive (accelerating) or negative (braking). This torque is eliminated by driving the bridge at a constant speed during the recording of test data.

The friction torque always acts against motion, but the other torques may be positive or negative, acting either with or against the movement of the bridge. It is thus important to observe a sign convention with these forces unless they are eliminated.

After eliminating T_w, T_s, and T_a, the result is two equations:

$$T_o = T_f + T_i$$
$$T_c = T_f - T_i$$

By subtracting the second equation from the first, and dividing by two, the imbalance torque is found:

$$T_o - T_c = T_f + T_i - (T_f - T_i) = T_f + T_i - T_f + T_i = 2\ T_i,$$
$$/2 = T_i$$

Therefore,

$$T_i = \frac{T_o - T_c}{2}$$

If this computation is made for several different angles of opening of the bridge, a curve of imbalance torque versus opening angle can be plotted. For almost any bascule bridge, it will be found that this is basically a cosine curve. For typical simple trunnion, Scherzer rolling lift, and articulated counterweight bascules, it should be an exact cosine curve. The cosine curve will be displaced left or right on the graph, depending on the location of the center of gravity of the bascule leaf. For heel trunnion or Rall bascules, or any other bascule with an operating strut or some other fairly substantial moving component that does not rotate exactly with the bridge, the effect of the movement of that component will be superimposed on the cosine curve of the basic bridge structure. The curve as plotted from field data may contain small variations due to roughness of the drive train or other factors. These should be smoothed out to produce a continuous curve (Figure 34-10). Unless heavy operating struts or some other known factor exists, the closest cosine curve should be used for the actual

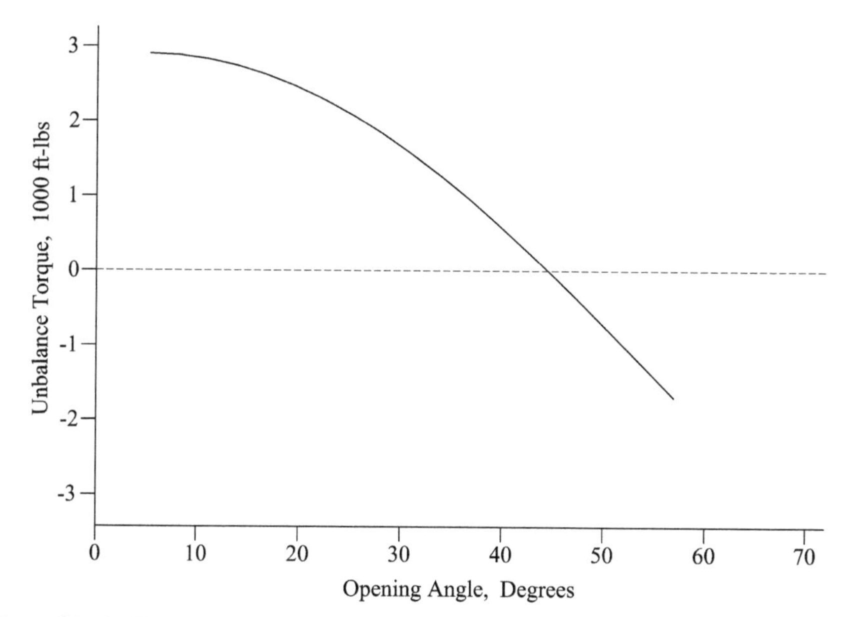

Figure 34-10 Plot of Points, Imbalance Torque versus Bascule Span Opening Angle. This is an idealized curve, smoothed out to eliminate irregularities.

imbalance curve. By extending the curve, amplitude and phase shift can be determined. The amplitude is the product of the total weight of the moving leaf times the distance from the center of gravity of the leaf to the center of rotation of the leaf.

The displacement of the point of maximum amplitude from the axis of the ordinate is the angle below the horizontal axis to the line from the center of rotation to the center of gravity with the bridge closed.

The angle α and the magnitude Wr of the imbalance can also be computed.

$$T_i = \mathbf{Wr} \cos \alpha$$

If the imbalance torque at two different angles alpha are known, solving the two equations simultaneously will give the angle of imbalance alpha and the magnitude, Wr. If it is desired to know the radius from the center of rotation to the center of gravity, Wr is simply divided by the weight of the moving span.

The value Wr, when associated with its angular position, becomes a vector, because it has both magnitude and direction. Once the imbalance vector **Wr** is known, a computation can be made by vector analysis to determine the amount of weight to be shifted, added, or subtracted at the counterweight to change the imbalance to a more desirable amount. The existing imbalance vector **Wr** is subtracted from the desired imbalance vector $\mathbf{Wr_f}$, to give the imbalance change vector, $\mathbf{Wr_c}$. Any change of mass amount and position that equals $\mathbf{Wr_c}$ will change the existing imbalance **Wr** to the desired imbalance $\mathbf{Wr_f}$ (see Figure 34-11).

Vertical Lift Bridges

On a vertical lift bridge without auxiliary counterbalancing devices, the imbalance is a linear function of the initial imbalance and the amount the bridge is lifted off its seated position. Balance chains do not affect this relationship except to reduce the

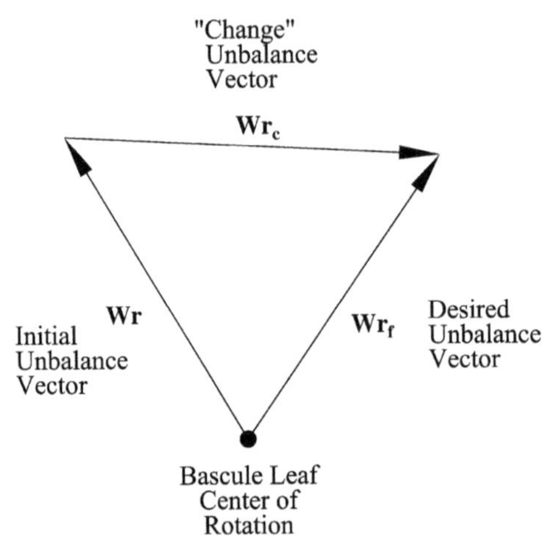

Figure 34-11 Vector Balance Process for a Bascule Bridge. The vector $\mathbf{Wr_c}$ shows the counterweight change required.

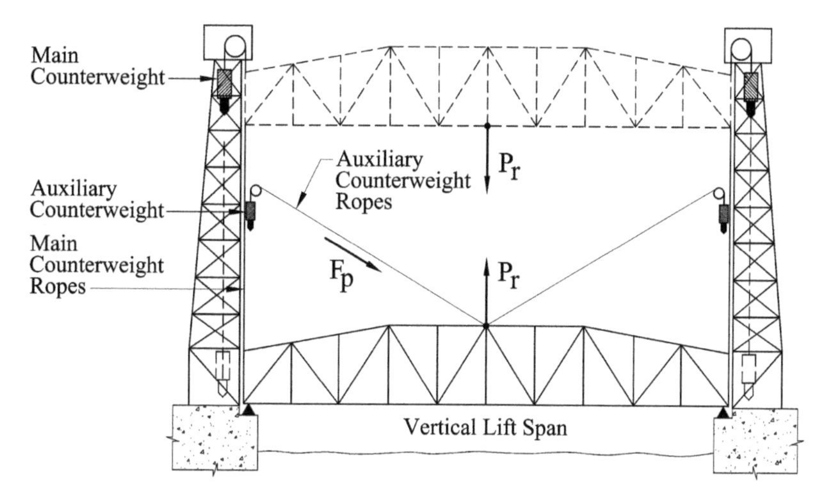

Figure 34-12 Vertical Lift Bridge Balancing with an Auxiliary Counterweight. The auxiliary counterweight ropes produce a constant tension on the span, but the vertical component varies sinusoidally as the bridge raises and lowers. The balance change due to the main counterweight ropes passing over their sheaves is linear. The auxiliary counterweight does not exactly balance the main counterweight ropes, but can be made to nearly do so.

change in imbalance as the bridge lifts. Rope-and-counterweight-type auxiliary counterbalancing devices have a trigonometric relationship to the balance of the bridge, as the angle of the connection of the ropes to the bridge changes as the bridge raises and lowers, while the rope tension remains constant (Figure 34-12).

Typical span drive vertical lift bridges can be tested by measuring the force at the operating ropes as the bridge opens and closes. Tower and span-tower drive bridges can be tested by measuring the torque required to drive the counterweight sheaves.

IMBALANCE MEASUREMENT

There are many different ways to measure span imbalance. One of the simplest is to measure the current at the drive motors. This method is usually not very accurate, as it is difficult to avoid fluctuations due to acceleration as the bridge is opened and closed and as the crude rack and pinion teeth make impacts during operation. The component of current due to imbalance may be quite small, so it can be difficult to read the differences as the bridge opens and closes. Worn machinery in the bridge drive train may produce short-term variations in the motor torque that cause a variation in the current equal to that resulting from change in balance as the bridge moves, so that the current change due to imbalance is imperceptible. If the bridge has low friction or high imbalance, it may be possible to determine the approximate imbalance by performing a drift test, observing the bridge when stopped at various opening angles and allowing the bridge to drift. A vertical lift bridge without auxiliary counterbalancing devices should remain stationary when stopped at the midway point, and should drift down when nearly closed and drift up when nearly open. A bascule span with a large imbalance will also tend to drift, except when the center of gravity is either directly above or directly below the center of rotation. By observing the drift at various opening angles, the location of the balance point or center of gravity and an approximate idea of the amount of imbalance can be obtained.

For bascule bridges and tower drive or span-tower drive vertical lift bridges, the most accurate and reliable method of imbalance measurement is to measure the torque required to operate the span. This can be accomplished by several methods, but generally the favorite is considered to be the use of strain gauges. This procedure was developed in the late 1970s in Chicago,[1] Florida,[2] and elsewhere. Several firms specialize in performing these tests, but anyone can obtain the equipment and, with a little training, develop the skill to do the testing. The gauges are placed on the lowest-speed drive shaft available on the bridge, mounted in such a way that any bending of the shaft is cancelled out and only torsion is measured. The gauges are installed with the shaft unloaded, so that zero initial strain equals zero torque. The shaft at the strain gauges should be free of corrosion losses and have a smooth surface with a constant diameter. The rack pinion shaft is ideal for this purpose, if it is accessible. Both rack pinion shafts should be gauged, and the gauges from both shafts interconnected to form a single Wheatstone bridge. In conjunction with the strain gauge readings, the span position is measured and recorded. This measurement can be done automatically with a device that reads the span opening angle or height as the bridge operates, or an attendant can mark the bridge position on the printout as the strains are being recorded. If the bridge and chart are operating at constant speed, it is easy to interpolate between marks to obtain the strain and resultant torque at any opening position.

The strain in the rack pinion shaft must be converted to torque on the bascule leaf or force in the vertical lift operating ropes. The torsional stiffness of the pinion shaft is determined as a function of the shaft cross section and material properties, and a direct conversion of strain to shaft torque is made. The ratio of pitch radius of pinion to rack or ring gear then gives the torque on the bascule leaf, tower drive vertical lift sheave, or span drive operating drum. The pitch radius of the rope on the sheave or drum is used to convert from torque to vertical lift bridge operating force.

For Waddell type vertical lift bridges and many other rope operated span drive vertical lift bridges, an arrangement similar to that for bascule bridges can be used. Alternatively, strain gauges can be installed on the operating rope connections at the towers to read tension forces directly. The connections should be unloaded when the gauges are installed, and the gauges should be placed so that any bending in the connections is cancelled out and only tension is sensed. Separate sets of strain gauge equipment must be set up at each set of operating ropes, as it cannot be assumed that the operating rope loads at each corner of the span are equal. Placing the strain gauges on the rope connections eliminates errors in reading the strain that result from sensor

[1]Henry Ecale et al., "Chicago Type Bascule Balancing: A New Technique," *Journal of the Structural Division,* ASCE Vol. 103 No. ST11. New York, 1977, pp. 2269–2272.

[2]L.E. Malvern et al., *Balancing of Trunnion-Type Bascule Bridges,* First Biennial Symposium on Movable Bridge Design and Technology, Tallahassee, Florida, 1985.

cables wrapping around shafts, but the bridge position must be sensed and recorded elsewhere. If the strain gauge equipment is located on the piers, a person can read the bridge position from the operator's house, which is usually located on the moving span, or from the machinery room, and communicate the bridge position by radio, as the bridge is lifting and lowering, to the personnel at the strain gauge equipment.

On a bascule bridge leaf with a differential unit, all the gauges reading drive torque should be wired together into a single Wheatstone bridge, as any inequality of driving effort between rack pinions may be due to internal differences in the drive, rather than differences in imbalance between the two sides of the leaf. The same can be true of bascule and vertical lift bridge drives without an active differential mechanism, but there may be value in recording the effort being applied at each final drive point separately. Additional investigation may be able to detect meaningful differences in final drive output force, and the reason for such differences may be discovered. For balance testing, the bridge should be treated as if it had an active differential, but it should not be assumed that the torques in both pinions are the same. If it can be clearly shown that there are no external forces, such as friction at span guides, acting on one side of the bridge, it may be possible to read separate drive torques for each pinion. Converting these to differences in imbalance between the sides of the bridge is problematic, as there may be an undiscernible transfer of drive torque from one side of the span to the other, either through the machinery or through the bridge superstructure. Roughness in the bridge drive train will cause torque variations that add "noise" to the strain gauge readings, similar to the effect experienced in trying to measure imbalance by reading motor torque. The strain gauge reading system can be adjusted to filter out much of this variation.[3]

Load cells can be used to determine whether seating forces are equal, but it must be realized that improper shimming of live load shoes may be the cause of unequal forces. If the live load shoes are shimmed so that they seat at the same time, then unequal seating forces are probably due to an uneven balance between the two sides of the bascule leaf, or two sides of one end of the vertical lift span.

[3]Edward J. Roland, *Balancing of Movable Bridges: The Strain Gage Technique* (paper presented at Heavy Movable Structures, Inc. 4th Biennial Symposium, Ft. Lauderdale, Florida, 1992).

35

ELECTRICAL

SAFETY

Safety is a special concern with electrical inspection of movable bridges. The inspector must be concerned for his or her own safety. In addition to the possibility of electrical shock, a further hazard is that many large pieces of electrical apparatus, such as transformers, contain PCB insulating oil. The sampling and testing of this fluid should be left to technicians specializing in such work. Older bridges may contain asbestos materials in their electrical insulation, which is also a hazardous material. The safety of the public must be maintained during all testing of bridge power and controls, particularly of traffic control systems. The safety of the bridge components must also be considered. Especially with older electrical components, a test procedure can unwittingly devolve into a destructive test procedure. When there is any possibility, during any part of an inspection, of causing a traffic control device to operate, or of a bridge drive or ancillary device to operate, traffic should be halted on the bridge prior to commencing that part of the inspection. A thorough, although somewhat archaic, discussion on electrical inspector safety is provided in Chapter VIII of the FHWA *Bridge Inspector's Manual for Movable Bridges.*

EQUIPMENT

For walk-through inspections, a pencil and pad of paper may be all that is necessary. In-depth inspections are usually intended to gather information to allow a comprehensive evaluation of the remaining life of the installation or to pinpoint areas requiring remedial action. To perform these functions, in-depth inspections require more equipment, up to and including a full battery of testing devices when quantitative evaluation of all components is desired.

The level of inspection required determines the equipment necessary to perform the inspection. For instance, if insulation resistance is to be measured, a megohm meter is needed. For simple continuity checks, a handheld multimeter that can read resistance and AC and DC voltages may be all that is required. Amperage readings can be made with a clamp-on ammeter. If large circuit breakers are to be fully tested, elaborate testing and recording devices are needed. This type of testing is best left to those who specialize in performing this work.

GENERAL

A bridge's electrical components may be several decades old and may have deteriorated insulation. The electrical inspector should be very careful to avoid injury to him- or herself and damage to these components when performing an inspection. The examination of some components will require the removal of covers, which may or may not be made of insulating material, and these covers are typically held in place by small screws or bolts that may have corroded or worn over the years. The act of inspection may cause sufficient trauma to these components to push them over the line from marginally functional to dysfunctional or unsafe. The act of inspecting, or measuring, an electrical component may not only change the quality being measured, it may permanently change the component in an undesirable way. Exclusion from liability should be given to electrical inspectors for such work. For most electrical installations on movable bridges, at least on older movable bridges, a visual inspection by a highly trained electrical engineer is at least as valuable as a full program of testing and measurement by a squad of poorly trained technicians and is less likely to cause permanent damage to the system. If at all possible, the bridge electrician should accompany the inspector during the inspection, preferably doing some or all of the dismantling, and repairing or replacing faulty components immediately. A deteriorated component often survives an inspection, only to fail later upon the first subsequent operation of the bridge. In many cases, even though an in-depth inspection is called for, the inspector may sensibly refuse to physically manipulate the devices but condemn them on sight.

Regardless of the type of inspection being performed, the inspector should make an initial overall check of the condition of the electrical equipment. The electrical equipment should be checked to see if it is clean, whether there is corrosion, and if all wiring is properly protected. Preferably, the bridge maintenance forces should not be forewarned of an impending inspection, so that the results of typical maintenance activity can be determined. Observation of at least one test operation or normal operation should be performed before any intensive work is done, so that any serious defects can be detected and appropriate action taken if necessary. During this operation, the inspector should listen for abnormal sounds coming from the electrical equipment, particularly solenoids, contactors, motors, and brakes, as well as make visual observation of all components. Close attention should be paid when the bridge starts and stops moving, particularly when seating, to see if any movement occurs at motor or brake supports. Motor currents should be checked when starting and during acceleration, as well as when operating at constant speed.

Many older bridges still retain their original control equipment in its original mounting. Some of this equipment is exposed, mounted on a large insulating board with little or no protection from or for personnel. In many cases, the room in which this equipment is placed is used for other purposes, with unauthorized personnel occasionally passing through. Conditions may be cramped, so that the likelihood of someone receiving an electrical shock from the exposed equipment is rather high. There is also a possibility of damage to unprotected equipment. The inspector should take note of such situations and describe them in his or her report. If a serious hazard exists, the responsible authorities should be informed immediately.

Control systems not quite as old as those described earlier have the control apparatus mounted in cabinets. These cabinets should be checked to see if they have corroded in the typically damp environment of a movable bridge and whether they provide solid and secure mounting for the components housed within them.

A bridge control system that has been in service for a substantial period of time is likely to have had component failures in the past. A failed component may have been replaced with a different piece of apparatus in a different location. The installation should be checked to see if old components no longer in use have been removed and if the wiring to and from them has been taken out. Notation should be made as to whether replacement components are securely mounted, and their wiring is neatly done and readily identifiable. Control systems in which repairs have been made in a haphazard fashion are difficult to maintain and inspect; they are also likely to be the source of future failures and should be reported as such.

Limit switches that perform interlocking functions should be inspected with special care. These devices, particularly plunger and lever arm limit switches, are subject to a great deal of wear and are sometimes abused during the maintenance or mechanical inspection of the devices that the switches monitor. In addition to being checked electrically, the actuating arms should be inspected for wear and damage, and the switches should be checked to see that they are securely mounted and that their actuating arms are correctly aligned with the component that trips the switch. In some installations, proximity switches take the place of limit switches. These are not subject to the same mechanical wear as ordinary limit switches, but can still become misaligned or damaged. Some control and interlocking functions, such as those indicating nearly closed and nearly open positions on bascule and swing bridges, are sensed by rotating cam limit switches. These devices are usually driven off the span itself and may have internal or external gearing so that the rotation of the cams is something less than 360° while the bridge superstructure goes through its full range of motion. These switches and their associated gearing and other parts are generally quite durable, but should be checked for proper lubrication, lack of corrosion, and security of housings and fasteners. Many rotating cam limit switches, selsyn-type transmitters, tachometers, and other rotating switches or sensors are connected to machinery parts or to the bridge superstructure by means of small shafts, couplings, and, perhaps, bevel or other gears. The connecting components can often become misaligned, causing overstress of these parts, particularly the couplings and bearings. These parts should be carefully examined for such distress (this may be performed by a machinery inspector, see page 444).

Power and control cables on movable bridges can be subject to a great deal of strain. All runs of conduits should be checked for corrosion or other damage, and all joints should be checked to see that they are secure. Flexible cables should be checked at their terminations to see that damage has not occurred. Exposed wiring should be checked for damage to the insulation and reported immediately. Cables connecting movable to fixed parts of the bridge, such as power and control cables leading from the control house on the pier to the span-mounted drive motors of Scherzer rolling lift bridges, are especially vulnerable to damage due to flexure and possible interference during bridge operation. Swing bridges may use flexible cable power and control connections, a collector ring arrangement, or a combination of these devices. Flexible cables can become abraded by dragging on the pier. They can also snag on stationary or moving parts. Collector rings can become fouled or deformed. In either case, current flow can become erratic. Exposed collector rings are dangerous for personnel to be near and can become short-circuited by stray pieces of metal. Span drive vertical lift bridges use pendant cables or collector rails, devices similar to those used on swing bridges, or aerial cables, as described in Chapter 12 under "Auxiliary Counterweights." Pendant cables can become snagged or abraded, and their end connections are prone to damage. Collector rails on vertical lift bridges, as on swing bridges, can become fouled or damaged, which can result in loss of contact. All these components should be carefully observed and their condition noted.

Many movable bridges have two span drive motors per leaf. It should be determined, by checking the design or shop drawings, whether one, or both, or alternating motors were intended by design to operate the bridge. Then the motors should be checked out to see if they are operating as intended. In some instances, a bridge operates with only one motor under power, when it should be operating with two. If such is the case, maintenance records should be reviewed to see if the motors have experienced an inordinate number of rewindings. The motor current should be checked to see if the motor is operating at a high percentage of its full load or is exceeding it under more than momentary or short-term conditions. A typical movable bridge operates, under normal conditions, without exceeding 40 or 50 percent of the motor full load torque, because of the motors being sized to handle the bridge under adverse conditions of high wind and snow or ice loadings. If a bridge is intended to operate under power from both motors, and is operating with only one motor powered and the other motor idle or missing, this should be noted. If both motors are supplying power, they should be checked to see that each motor is carrying its full share of the load. Motors should run without excessive heat or vibration. The motor operating temperature should be compared with the manufacturer's specifications. As motors in movable bridge service do not operate for long periods of time and usually do not operate at full output for the greater part of the duty cycle, any amount of heating that approaches the maximum for the motor should be suspect.

Motors should be checked to make certain that they are firmly mounted to the machinery frame or floor, that they are properly aligned with the machine they are driving, and that the mechanical component mounted on the motor output shaft is secure. Motor frames and mounting feet should all be sound, and mounting bolts should be tight and fit properly into mounting holes. Keys should be checked to see if they are loose. Gears or coupling hubs should be tight on motor shafts. See the chapters on machinery inspection for more information on these points. See page 440 for more information on inspection of motors and brakes. Inspection findings should be confirmed by the machinery inspector. Brushes, commutators, and slip rings should be inspected on wound rotor and DC motors, which usually requires removal of access covers. Some motors have transparent covers over the brush areas, but brushes may have to be removed for adequate inspection. Brushes should have plenty of carbon remaining, should move freely in their holders, should be firmly pushed against commutators or slip rings by their springs, and should be wearing evenly and making full square contact with the commutator or slip ring. Commutators and slip rings should be smooth and shiny, with no pits, rough spots, or damaged or loose sections. Arcing during operation should be minimal or nonexistent. DC commutators must be closely inspected to determine whether the effective gaps between bars have decreased due to movement of copper or deterioration of the insulation between bars.

Motor brakes and machinery brakes, if of the thruster type, should be checked for set times. Release times should be short, but should not be instantaneous. Thrusters and solenoids should be checked to see that they make full travel when activated, and do not overheat while actuated. Thruster oil levels should be checked to see if they are within the range allowed by the manufacturer.

The external resistors used in conjunction with wound rotor motors can be corroded, broken, or damaged from overheating. Older iron resistors are particularly susceptible to corrosion when they are heated in the damp atmosphere frequently found in the control room or electrical apparatus area of a movable bridge. These resistors should be given a thorough visual inspection. Any damage or deterioration should be fairly obvious, but resistance measurements can be taken and compared with original values.

Most movable bridges have electrical interlocking systems. These systems should be checked carefully to see whether they perform their intended function. On many older bridges, interlocking components deteriorate and become nonfunctional. Rather than repair or replace these components, bridge maintainers sometimes wire around them or utilize bypass switches included in the control panel to cut out the defective interlocking function. Any such alterations should be noted and reported so that they can be repaired immediately, as damage or injury to the bridge, to the bridge operating, maintenance, or inspection personnel, or to the public may result from the bridge's operating without the specified interlocks.

A bridge's navigation lighting is an important part of its control and protection system. Navigation lights are mounted in exposed positions, sometimes on the fenders and dolphins that physically protect the bridge from damage due to impact by vessels. This lighting should all be checked to see whether each light is functional and in good condition. The navigation lights themselves, the power cables supplying them, and the light supports should all be checked to see if they are secure and sound, without excessive corrosion. Walkways used for access to the lights should be safe and solid to encourage maintainers and inspectors to perform their functions. Navigation lights such as those on bascule bridge leaves, which change from red to green as the bridge opens to clear the channel for a vessel, should be checked to see that they operate properly and change color when intended to do so. Other communication devices that may be present should be inventoried, such as the horn and marine radio, to see that they are in place and in good working order.

If thorough electrical testing of the bridge's power and control system is to be done, a considerable array of testing equipment may be required. The inspector must know what the range of test values for a particular component should be. This type of testing must be done very carefully to avoid electrical shock. A megohm meter, or "megger," is used to check insulation resistance. This test usually requires disconnecting and reconnecting the wiring. A value of 1000 ohms is usually considered minimal. Typically, the insulation resistance value of the equipment—motor, submarine cable, controllers, or other cables—is checked and compared with previous values. A gradual deterioration in resistance is expected over time. Motor currents can be checked by observing the readout at the ammeter at the operator's console. It is difficult to compare these to nameplate values, as the motor may be operating at no load, maximum load, or anywhere in between, and it is difficult to determine the motor load by simple observation. AC currents can be measured at other points by using a handheld clamp-on ammeter, or in special cases a standard ammeter can be temporarily wired into the circuit. Heavy-duty circuit breaker testing requires specialized equipment. All testing involving high voltages and currents can be dangerous and should be done only by skilled personnel with the proper training. Other electrical testing procedures are discussed in the *AASHTO Movable Bridge Inspection, Evaluation, and Maintenance Manual* (2.10.8–2.10.10).

OPERATOR'S HOUSE, MACHINERY ENCLOSURES, AND ACCESS APPLIANCES

The operator's house and machinery enclosures should be checked to see that they are weather resistant and provide adequate protection against rain, snow, and dust. They should also provide reasonable protection against the intrusion of vermin, birds, and other animals. The entry of unauthorized humans should also be restricted. Any signs of the entry of rainwater should be noted, and the roof and walls checked for leaks. The presence of bird droppings or other such debris should be noted and the source determined if possible. Any signs of deterioration, such as corroding metal supports or walls, rotting timbers, or the like, should be noted, including an estimate of the extent of damage. Any broken or ill-fitting windows should be noted. Particular attention should be paid to the protection of the machinery, control panels, and electrical cabinets. If called for in the scope of the inspection, all operator's amenities and facilities should be tested to see that they are functional. If necessary, architectural; heating, ventilating, and air-conditioning (HVAC); or other building specialists should be involved in this work.

Separate machinery enclosures and housings should be checked to see that they are secure. Enclosures should adequately cover the machinery and should not interfere with operation. All machinery guards should be in place, securely mounted, and provide proper protection.

Access ladders, walkways, and handrails should be sound and solidly secured. All walkways and floors should be free of tripping and slipping hazards, such as spare parts, tools, and spilled grease or other lubricants. Corrosion and other damage should be minimal. Report any hazards, heavy corrosion or other substantial conditions. Pay particular attention to pipe handrails to see if there is any internal corrosion not apparent at the outer surfaces. Check closely the bases of pipe handrail stanchions, as these are the locations most apt to have internal corrosion. Corrosion is more likely to occur in a saline environment, so be especially vigilant in coastal areas. Note whether ladders, handrails, stairways, elevators, and walkways meet the applicable safety codes.

TRAFFIC CONTROL SYSTEMS

The operation of all gates, lights, bells, and other devices used in a traffic control system should be witnessed, preferably during a normal opening for a vessel. All equipment should operate in sequence, with no discretion available to the bridge operator, other than as to when to lower the warning or barrier gates after traffic, if any, has stopped or cleared the bridge. Wiring for all devices should be checked to determine whether any bypasses have been short-circuited, any other jerry-rigging is present, and any wiring is exposed to the elements. It should be determined whether the lights on all gates are operating properly—one light on each gate arm is usually wired to burn steady when the gate is activated, while the other lights are flashing. If possible, gate operation should be observed on a windy day, particularly at gates with longer arms. Any difficulty in raising or lowering a gate in such conditions should be noted.

Each gate and light should be checked internally and externally for loose, broken, or missing components. Housings, poles or pedestals, and bases should all be solid and firmly connected. Gate arms should be firmly fixed to the operating units, and the operating units' housings should be watertight. The internal components of all gate operating units should be closely inspected for worn or loose parts, and the condition of lubrication of all mechanical power transmission components should be determined. Any guy wires or other stabilizing devices should be secure and functional. Note any damaged or loose hardware.

It is practically impossible to witness a test impact on a barrier gate, particularly if the gate installation is of the attenuating type that is partially destroyed in a collision. On bridges with barrier gates installed and in use, the bridge operator should be interviewed, and any anecdotes regarding vehicle impact on the barrier gates should be recorded. Barrier gates should be carefully inspected in the lowered position; the engagement of all moving components in the active barrier position should be noted and compared with the design intent. The attenuating devices of impact-attenuating barrier gates should be carefully inspected to determine whether they are in good working order. Any corrosion, damage, or loose fittings at the attenuating device or mechanism should be noted. If the attenuating device incorporates a net, the net should be closely inspected for any corrosion, kinks, or loose fittings. If the attenuating device is a retractable, reusable type, a test should be arranged with a vehicle at low speed operating against the net, causing the attenuating system to operate. The force to actuate the attenuator should be measured if possible. The retracting mechanism can then be tested to see if it is in working order. If the attenuating device uses a stretching cable, the cable itself should be free of corrosion, kinks, nicks, or other damage, and the attachments of the cable to its anchors should be secure when the gate is in operating position. If the attenuating device uses a metal bender device, the metal bender functional components should be free of corrosion, kinks, nicks, or other damage, and the attachments of the metal bender to its anchors should be secure when the gate is in its operating position. If a test impact is not witnessed, any other diagnostic operations prescribed by the gate manufacturer should be performed. Fixed nonattenuating barrier gates should be closely inspected for corrosion and impact damage and checked to see that they are securely engaged when in position, blocking the roadway. All reflectors and other warning indicators on barrier and warning gates should be in good condition and clearly visible.

36

INSPECTION REPORTING

The inspection of a movable bridge may be performed by the owner's forces or by a consulting firm hired for the purpose. There are several levels of inspection that may be performed on a particular bridge, and the reporting is generally commensurate with the level of inspection. Many movable bridge owners developed standardized forms for inspections of their bridges, and some have produced such forms for movable bridges. These forms are generally distracting, and somewhat inefficient, because they are made up to cover all movable bridges under the jurisdiction of the particular owner. Many of the blanks in the forms must have "not applicable" entries, and there are usually components that are not covered on the forms. Only in areas where most of the bridges are of a similar type—such as in Florida, where most of the movable bridges are simple trunnion bascules—does a standard form offer a significant increase in the inspector's productivity. Some owners provide a semistandardized scope of work in the inspection requirements, tailoring this scope as needed for each movable bridge, and may leave the form of the inspection to the writer of the report. Generally, however, each movable bridge is unique, requiring a special list of items to be inspected and a report that will cover that work. Each item on the list should have its own identification code, using numbers or letters, or a combination of these, in a systematic, consistent fashion, so that the inspection, and the future use of the results of the inspection, can be most efficient.

As indicated on pages 368–370, there are generally three levels of bridge inspection that can be performed. The walk-through inspection is most readily adaptable to being reported on a standard form. The in-depth inspection may have a special form prepared for it by the owner, or the job requirements may include filling out a standard form, writing a supplemental narrative report, and filling out special forms prepared by the inspector. The special inspection is almost always required to be reported in a special narrative, but will include special forms if, for instance, measurements at the machinery are included in the inspection. Supplemental materials such as photographs and data tables may be required for an inspection at any level and are usually necessary and required for in-depth and special inspections. Regardless of the type of inspection, any maintenance or repairs performed during the inspection should be reported.

For all inspection reporting, it is imperative that only facts be presented in the main body of the report, whether it is a form or a narrative. Any opinions that are presented should be entered separately in a section labeled "Conclusions" or headed with an equivalent term. Recommendations and cost estimates, if required, should be provided in a separate section or sections.

WALK-THROUGH INSPECTION REPORTING

A standard form may be supplied by the owner for reporting the results of a walk-through inspection. This level of inspection may or may not require measurements. The inspection report is usually made up of a series of subjective statements, with each component rated "poor," "fair," "good," or "excellent" or with some other designations, or rated on a numerical scale from 0 to 10 or within some other range. When such a "point system" rating is required by the bridge owner, the inspection requirements usually spell out these rating levels and define each level in an attempt to eliminate subjectivity from the ratings. It is difficult to reduce subjectivity, however, so that over the years it is not unusual for protocols to be developed whereby the persons reviewing the reports learn what the inspector means within each level of rating and inspectors learn to be consistent in their ratings so that the correct information is communicated. For a change in the rating number of a particular component to be of value, the component must have been inspected two or more times with the same subjective standard in use. For this result to be fully valid, the same person will have had to do all the inspections. As this is rarely the case, it is best to be cautious about placing great weight on these subjective numerical ratings. The best rating system, or at least the most nearly foolproof, is the "digital" system. A component is given a rating of either zero or one. "Zero" means it does not work. "One" means the component is functional. The AASHTO numerical rating system uses numbers from 1 to 5, as follows:

1	Excellent
2	Good
3	Fair
4	Poor
5	Critical

The FHWA numerical rating system uses numbers from 0 to 9, as follows:

0, 1, or 2	Critical
3 or 4	Poor
5 or 6	Fair
7 or 8	Good
9	Excellent

Of course, one must be certain as to which system is being used so that an AASHTO rating of 5 will not be mistaken for an FHWA rating of 5, or an FHWA rating of 1 will not be mistaken for an AASHTO rating of 1, which can result in putting off repair or replacement of a "critical" component. The unfortunate reversal of order in the AASHTO versus FHWA systems can lead to confusion, especially if the same inspectors working on different bridges are required to continually change from one system to another. The *AASHTO Movable Bridge Inspection, Evaluation, and Maintenance Manual* provides guidance for applying ratings to machinery components (2.5.11, 2.6.5, 2.7.15). For best results, the use of numerical rating systems to communicate the condition of bridge components should be relied upon as little as possible. For more on bridge ratings, see Part VI.

If photographs are required as part of the report, they are usually attached with brief captions. A caption should define the exact component pictured. Photographs and captions should not be expected to tell the whole story of the inspection, but should provide basic information on the condition pictured. Photographs showing the general condition of the bridge should be limited to very few, unless expressly desired by the recipient of the report.

IN-DEPTH INSPECTION REPORTING

In-depth inspections invariably require detailed measurements of all key components, quantifying corrosion losses at structural components, tooth thicknesses for main drive gears, bearing clearances, and insulation resistances for electrical cables. These measurements are reported in tabular form, usually in a standardized format, but with the specific form prepared to suit the particular bridge being inspected. In many cases, the form is used in the field to record the data, then "cleaned up" for inclusion in the report (see Figure 34-5). A detailed narrative may be required, especially if the report is to include interpretation of the data and if recommendations for corrective action are required. Photographs are invariably required as part of the record of the inspection and are usually included in an appendix attached to the report. Photographs should be captioned with the exact name of the component shown, indicating which it is if there are several similar components on the bridge. A brief description of the condition pictured should be given so that the photograph and caption will stand alone and not require a reading of the text of the report for the reader to obtain useful information. Describing a component in a caption as *typical* is of little value in the report. Describing a condition as *typical* may be warranted.

SPECIAL INSPECTION REPORTING

Special inspections usually require a detailed narrative report and a written explanation of the purpose of the inspection. A special inspection may include all or part of the work included in an in-depth inspection as described earlier. The data may be presented in tabular form if a large number of measurements are made as part of the inspection.

REPORTING OF CRITICAL ITEMS

Inspection findings may result in a component's being designated as being in "critical" condition. If that component is essential for the safety of the public and its failure appears imminent, it must be reported immediately by telephone, in person, or by other appropriate means to the authorities in charge of the bridge. If the situation is deemed serious enough, the inspector should consider suggesting the immediate closure of the bridge or cessation of operations. Naturally, the inspector should be competent and sufficiently experienced to be able to make such judgments in the field. If in doubt, the inspector should contact his or her superiors for advice before taking any precipitate action. Any such communications to the bridge authorities should be followed up in writing with appropriate documentation. In all cases, reporting and follow-up of any such critical conditions should follow established procedures, if they exist. The *AASHTO Movable Bridge Inspection, Evaluation, and Maintenance Manual* discusses reporting of these critical items under "Deficiency Reports" (2.11.2.3).

REPORT FORMAT

An inspection report may include any amount of material and may be of any shape or size as is desired by the recipient, but it should clearly present the following three things:

Findings: All significant facts uncovered by the inspection should be discussed in narrative form.

Conclusions: The meaning of the findings should be explained.

Recommendations: What must be done to correct deficiencies should be spelled out.

There may be reports submitted that include only findings, such as measurements of material loss performed in a structural inspection. Such a report will be taken by other persons involved in the process, who will develop load ratings from these measurements, and the ratings will constitute the conclusions of the report. From the ratings will be developed a list of deficient members, which can be considered the recommendations section, as these members must have work done on them to correct the deficiencies. In most cases, the separate parts will be combined into a single document, for which each bridge owner has its own name. This document, which may include other information as well, will constitute the final product of the inspection process.

FURTHER EFFORTS

Any conditions on a bridge that appear suspect, but are not within the scope of work of the inspector, should be brought to the attention of the bridge authorities. If these conditions are, in the opinion of the bridge inspector, critical, as discussed earlier, they should be reported as such. If not, recommendations, such as for further inspection, should be included in the inspection report. The contract conditions of the work, however, may allow an increase of the scope of work in the field, in which case a suggestion for modification of the scope of work can be presented to the responsible authorities.

VI

EVALUATION OF EXISTING BRIDGES

Structural and substructural aspects are discussed in this part of the book only in those areas where they pertain to movable bridges. Machinery, electrical, and other areas of particular relevance to movable bridges are discussed in more detail.

The evaluation of an existing movable bridge must be based on a certain standard. Both AASHTO and AREMA have rather stringent requirements for new movable bridges, based on allowable stresses, durability of components, and safety. An existing bridge may be able to meet these standards in many respects, but fail in others. Allowances can be made in some areas, such as for slight overstresses, whereas other areas, such as safety, may require remedial action. Legal considerations of liability, likelihood of failure, and desired remaining life of the bridge are important parameters in determining a response to a substandard condition on a movable bridge. In many cases, either for a complete existing bridge or for part or parts of it, the following options can be weighed:

1. Do nothing.
2. Make repairs.
3. Replace the bridge.
4. Replace part(s) of the bridge.
5. Take the bridge out of service.
6. Change the loading on the bridge.

Costs can be developed for each of these options. The cost of doing nothing should include a probability and cost of a catastrophic failure, and the resultant cost of traffic delays and detours. The cost of taking the bridge out of service or changing the loading (restricting it to pedestrians and bicycles, for example) should include the cost of detours and delays to traffic. These costs should also be included, as appropriate, for the other alternatives. The future condition of the bridge can be estimated to help in making a decision concerning what to do about a structure of questionable status.

37

BRIDGE MANAGEMENT

Bridge management is an attempt to provide a rational basis for optimal allocation of limited resources for maintenance and repair of existing bridges and construction of new replacement bridges. Bridge management consists essentially of a database of information on a group of bridges and a model that determines optimum allocation of financial resources to maintain the group of bridges in optimum condition. The computerized combination is referred to a bridge management system (BMS). This, of course, is an idealized picture.

In fact, the data available on bridges is not consistent, and some of it is poor or downright false. The analytical methods available for evaluating bridges are limited and can, in some ways, be considered fanciful. To most economically allocate resources for the upkeep of a system of bridges, the condition of the bridges must be known exactly, the present and future rates of deterioration of the bridges must be known exactly, the present and future traffic on the bridges must be known exactly, and the present and future cost of maintenance, repairs, and construction work must be known exactly. Thus, the cost of money must be known for the expected life of the bridge. Many variables are associated with each of these values, and present values of most of these variables are barely understood, to say nothing of their being known to any precise degree. Projected future values of these variables are almost total speculation. Information from the field is not likely to be totally accurate, and the processing means will contain errors.

Nevertheless, output from a bridge management computer model can be helpful in the decision-making process, because there is little else in the way of a rational basis to go on. It is generally accepted, from statistical evidence, that the nation's highway bridges are deteriorating faster than repair and replacement efforts can presently accommodate. The complications due to non-structural difficulties in rehabilitation and replacement, such as environmental issues, make it essential to allocate manpower as well as financial efforts more efficiently in the effort to keep up with demand for maintaining the integrity of structures.

It is even more difficult to apply bridge management principles to movable bridges than to fixed bridges, because almost every movable bridge is a unique entity, whereas there are great similarities among most fixed bridges that allow them to be categorized and evaluated in a statistical fashion. Nevertheless, some parts of movable bridges, such as decks and piers, can, in some cases, be treated as part of the general population of bridges for bridge management purposes. If it is desired, the bridge management method can be applied to movable bridges in general, even if the movable bridge population is small. It must be realized, however, that to arrive at useful output information for specialized structures such as movable bridges, it is even more important to have correct input of condition data and accurate models of deterioration. General Rule #1 of computers, "Garbage in = garbage out," is eminently true in the evaluation of movable bridges. General Rule #2 of computers, "What is state of the art this month is junk the next," is also true for movable bridge management systems, as it is for the entire BMS field. Many software packages are available today, but they have their limitations even for simple structures such as highway overpasses and are not available ready-made to tackle movable bridges. These packages require expensive doctoring by experts in order to suit the ordinary BMS needs of entities that own many bridges, and must be radically customized to provide complete services for movable bridges. Progress is continuing, however, and perhaps in the near future user-friendly software that performs the whole job adequately may be available.

With increasing concern for proper bridge maintenance and increasingly tight budgets, it has become more difficult in recent years for those responsible for maintaining the integrity of bridge structures to keep in touch with the condition of their bridges. This situation has been compounded, ironically enough, by the requirements at the federal and, in some cases, the state or local level, for more detailed documentation of the condition of structures. At one time this situation held only for highway bridges. Now, with federal jurisdiction over railroad rights-of-way becoming more prevalent as a result of accidents, government funding of right-of-way improvements, and many states now becoming the owners of railroad rights-of-way, documentation of this type is often required for railroad bridges as well as highway bridges.

The goal of bridge management is to have a systematic, thorough means of maintaining awareness of the condition of a bridge and to provide a system for the efficient allocation of resources to maintain a bridge in an acceptable condition. Bridge management incorporates information gathered from biennial-type inspections with other sources, such as automated real-time monitoring, to maintain a database that fully describes the condition of a bridge. This information is then evaluated to program repair and special maintenance.

The key to manageability of a bridge database is the efficient processing of information. At one time, bridge inspectors spent long hours in the off-season, or during bad weather, compiling the data accumulated during field inspections to arrive at condition ratings and repair estimates. In some cases, particularly with larger bridges in deteriorated condition, a separate team of calculators in a field office receives the data from the field inspectors and determines the need for repairs, and another group of engineers may actually produce repair details while the inspectors are out in the field. Today, more effort is given to computerizing the data, with varying degrees of success. Some field inspectors have handheld computers through which they enter the data, which is, in some cases, transmitted instantaneously to a central facility.

Real-time monitoring of the condition of a bridge takes many forms. A more common device is the resistance strain gauge; many such gauges have been installed permanently on bridges. These gauges can be used to continuously monitor strain on bridge members, to measure rate of deterioration, or to count fatigue cycles. Strain gauges are also used to measure the balance condition of bascule bridges and, sometimes, vertical lift spans, but this measurement is generally not continuous. Rather, a team may use strain gauges that may have been previously applied to perform a one-time balance test. The testing may be repeated only every five or six years. Other devices include acoustic instruments to count bolt breaks or wire breaks in ropes or cables and position-monitoring equipment to measure gradual movement of piers or superstructures.

There are many other values on a movable bridge that can be continuously, automatically monitored, with the data collected and processed for bridge management purposes. Among these values are traffic counts, channel depth, and wind speed. Many aspects of the movable bridge operating machinery can be easily monitored, such as operation hours, motor amperes, hydraulic system pressures, and temperatures of hydraulic or lubricating oil.

There are many areas in which bridge management can be distinctly helpful in the upkeep of movable bridges, such as in monitoring of the condition of counterweight ropes on vertical lift bridges. These ropes deteriorate for one of two reasons: Either they wear out or they corrode away. Corrosion is impossible to predict accurately, but under constant conditions of maintenance and operation, it is possible to determine the rate of wear of counterweight ropes. This rate will not be constant, but will vary with time. The exact rate of wear is a function of the following:

Frequency of operation of bridge
Tension in the ropes
Size of ropes
Construction of ropes
Size and hardness of wires in ropes
Type of core in the rope
Condition of rope core
Size and shape of counterweight sheave
Hardness of counterweight sheave groove surfaces
Whether the bridge is span drive or tower-drive
Condition of lubrication of ropes
Condition of rope preservative material
Length of time the ropes have been in service
The amount of present deterioration

This is a long list of independent variables for a single aspect of a movable bridge. An equation must be developed for each of them, and these equations combined to produce an algorithm that will give a reasonable approximation of remaining rope life. The research required to produce the equations is daunting, especially considering that many attempts have already been made in this and related fields, with generally rather poor results. The other means of developing these equations, actual experience, also has its shortcomings, primarily the difficulty in maintaining constant conditions over a long enough time span to produce useful results. The FHWA Pontis system assumes the bridge is new when the process is initiated and builds its rates of deterioration on subsequent history. Another approach to determining rate of deterioration is to measure past deterioration and project it into the future. This may be valid if the rate of change over time of deterioration is known and the effect of proposed changes in maintenance is quantifiable. This approach also requires a known past history. Nevertheless, there is no question that bridge management is the wave of the future, or at least of the present. Certainly, our ability to analyze is growing at a faster rate than our ability to repair or replace. Most states are developing formal bridge management systems, and a few have developed their programs beyond the experimental stage.

Only one caution must be raised in discussing the application of bridge management systems. These are *management* systems, and they do not take the place of actual field work. Any funds allocated to BMS should be new funds, and not merely money taken from the existing budget for upkeep of the system. Taking funds from maintenance to apply to BMS will result in a reduction of maintenance for at least the short term, until the BMS becomes productive, which may never occur. Bridge management is a tool for using the data at hand, not a replacement for work on the bridge. It is a tool for management and must be used by management if it is to be of value. If the management is taken out of BMS, very little of value is left. Bridge owners should look carefully at the real expected benefits of BMS versus the cost. Other changes have been mandated in design, inspection, bridge construction, and maintenance, such as metrication, which have not returned any benefit.

Many people think that the bridge management function, as defined here, poses too difficult a task; that even if the financial resources are made available, there are limitations to the capabilities of data processing and analysis that make such micromanagement too daunting to contemplate. In some areas of industry, where the resources for maintenance are difficult to come by, a decision is made simply to eliminate maintenance from the budget and perform repairs and component replacements as necessary to keep the system operating. This is exactly what is done on many bridges. Within some jurisdictions, maintenance is nonexistent, as all so-called maintenance resources are tied up in making repairs to bridges after failures of greater or lesser moment have occurred. When bridges can become functionally obsolete before they are worn out, there is some value to this approach, as resources are not wasted on maintaining a bridge that is eventually going to be replaced anyway.

Yet where bridges can be expected to remain functional into the indefinite future, such an approach is not likely to be cost-effective. The FHWA, which funds most highway bridge replacement projects, takes a dim view of the zero maintenance approach. The FHWA, it should be noted, has also shifted in recent years from a strictly 20-year life criterion for new bridges to longer expected life spans. Railroad bridges, including movable ones, have traditionally been built to last forever. For several decades in the recent past, many railroad bridges were undermaintained because the future use of the bridges, and the rail lines on which they were built, has been in doubt. As many of those undermaintained railroad bridges and rail lines have been abandoned, or redundancy in the national rail system has otherwise been reduced, the future of the railroad bridges remaining in service is more definite. As a result, the rail industry has also taken another look at maintenance and has begun allocating more resources to the maintenance of movable bridges and other rights-of-way. This change has had an effect on railroad organizations—for instance, the American Railway Engineering Association (AREA) is now known as the American Railway Engineering and Maintenance-of-Way Association (AREMA).

38

STRUCTURE

SUBSTRUCTURE

Many older movable bridges have serious substructure difficulties. These problems have been exposed and magnified by the increased interest, during the last decade of the twentieth century, in seismic event survivability. In many instances, a movable bridge is a solitary crossing of a waterway, so that closure of the bridge in an emergency may impose great hardship. Some movable bridges, such as over the Harlem River in New York City, carry extremely heavy traffic, so closure of a critical span because of an earthquake can result in massive traffic disruptions. There is ongoing debate about the reasonability of seismic retrofit of such older structures, but there is no question that many older movable bridges have seriously deficient substructures.

Many older movable bridges have piers constructed of unreinforced concrete, or of granite or limestone blocks, and are inadequately strengthened for their normal loadings, to say nothing of earthquake forces. It is difficult to retrofit these substructures, because of inaccessibility plus the need to continue to support the superstructure, as well as the need, often, to maintain traffic while the work is done. Most bascule and vertical lift spans, because they must resist high wind moments, have piers with adequate capacity, unless they have been subject to deterioration of piles or concrete. Swing bridges, on the other hand, have in many instances been designed with little regard for lateral forces and some are found to have extremely slim chances of surviving a seismic event without retrofitting.

Foundations that have been continuously protected are very unlikely to deteriorate. There are wooden pilings that have survived for hundreds of years. Foundations that have been exposed as a result of scour or other damage, however, should be regarded with suspicion. Movable bridges, particularly bascule and swing bridges, exert greater forces on foundations than similar-sized fixed bridges. Careful thought should be given to the substructure before committing to a general rehabilitation of an existing movable bridge.

SUPERSTRUCTURE

Many highway bridge owners use a numerical rating system to place a quantitative value on the condition of the superstructure of a bridge and its components. These rating systems vary from entity to entity in numerical system, description of conditions, and broadness of application. See Chapter 36 for additional discussion of the use of numerical ratings. One typical state had the following system (note the implied definition of "maintenance"):

Adjectival Ratings

9 The item is in new condition with no maintenance necessary.
8 The item is in good condition with no maintenance necessary.
7 The item is in fair condition but requires maintenance.
6 The item is performing the function for which it is intended, but requires maintenance.
5 The item is still performing the function for which it is intended at minimum level, but requires maintenance.
4 The item is still performing the function for which it is intended at minimum level, but requires major rehabilitation.
3 The item is still performing the function for which it is intended at minimum level, but requires replacement.
2 The item is not performing the function for which it is intended, and requires maintenance.
1 The item is not performing the function for which it is intended, and requires major rehabilitation.
0 The item is not performing the function for which it is intended, and requires replacement.

Condition Ratings

9 New Condition—No maintenance recommended.
8 Good Condition—No maintenance recommended.
7 Fair Condition—Recommended maintenance for minor items.
6 Fair Condition—Recommended maintenance for major items.
5 Poor Condition—Recommended minor rehabilitation effort.
4 Poor Condition—Recommended major rehabilitation effort.
3 Poor Condition—Inadequate to tolerate unrestricted, legal loads—recommend immediate repair or major rehabilitation effort.

2 Critical Condition—Inadequate to tolerate unrestricted, legal loads—recommend posting bridge for light loads, immediate major rehabilitation necessary.
1 Critical Condition—Inadequate to tolerate any loads—recommend closing bridge until rehabilitated.
0 Critical Condition—Inadequate to tolerate any loads—recommend closing bridge until replaced.

As discussed in Chapter 36, these types of ratings are very subjective and are prone to misuse by inexperienced inspectors. They do provide a quick picture of the condition of a bridge, however, and can be adapted to almost any conditions, particularly for the various types of movable bridges and their special components, including machinery and electrical items. These ratings should be used with caution and should not be used in evaluation of structural and machinery components when quantitative inspection data are available.

AASHTO provides a complete set of references for the evaluation of existing highway bridges. Its *Manual for Condition Evaluation of Bridges* gives estimated values for the allowable stress of steels in old bridge members, based on the age of the bridge, if the actual material properties are not known. If it is necessary to be more precise or if the member is determined to be overstressed in its existing condition when using the allowable stress value from the manual, then the expense of obtaining and testing coupons from the bridge may be in order. This testing may result in an allowable stress value several ksi higher than provided by the manual. This procedure and its purpose should be discussed thoroughly with the bridge owner, and all parties should be in agreement on what constitutes acceptability before proceeding. If at all possible, sample coupons should be taken from areas that are not overstressed. It will probably be necessary to repair the areas from which the coupons are taken, by replacing the lost metal. Bolted-on cover plates are preferable in making repairs, rather than attempts at welding in repair patches. Improperly done field welding can cause more damage to the existing structure, and even if the welding is done correctly, it may be difficult to reestablish load in the added pieces.

If the manufacturing history of the steel members on the bridge can be established, it may be possible to take coupons without having to do any repairs at all. If the members of concern are made up of rolled shapes or plates that came from the same mill and in the same heat as other secondary members, it may be possible to take coupons from these secondary members without making repairs. These coupons will, in all probability, have the same mechanical properties as the steel of concern.

In evaluating an existing movable bridge, it is important that all relevant data have been properly obtained from the field. Many structural inspectors who are familiar with fixed bridges have little or no knowledge of movable bridges. Many movable bridges, such as vertical lift spans, are quite similar to fixed bridges, and there is little need for concern in evaluating the basic traffic-carrying components of such bridges. Other movable bridges, such as bascules and swing spans, which make up the vast majority of older movable bridges in use, have load paths and configurations that are nothing like those of fixed structures. Some movable bridges have unique superstructures that defy categorization and must be carefully analyzed to determine the loading of individual members.

Joseph Strauss, mentioned many times in these pages, was a prolific bridge designer who came up with many variations on the bascule bridge type to suit specific needs. At the time he did his work, many aspects of structural engineering and engineering mechanics were unknown territory for most engineers, so he perhaps cannot be faulted for producing some bridge designs that developed problems in their later years. It must also be understood that many of his designs were adaptations of earlier developments by others. Strauss sold many designs for heel trunnion bascules, and the bridges built from those designs provided many years of reliable service. The chief advantage of the heel trunnion bascule, that it allows a movable bridge to be built at a very low elevation over the water, without a counterweight pit and without exposing the bulk of the machinery to water damage, cannot be faulted. Many, if not most, of Strauss's heel trunnion bascule bridges were designed to open to a very large angle, approaching 90 degrees in some instances, and this is where they later got into trouble, due to metal fatigue in some main load-carrying members. Of course, if such a bridge had been built at a higher elevation, it may not have experienced as many openings over its life and metal fatigue may never have developed.

There are large variations in dead load stresses in both the main span and the counterweight frame of these heel trunnion bascule bridges as they open to large angles, and if they open often enough, they will start to fatigue and cracks will appear. Because these bridges go through only one stress cycle per opening and closing cycle, it takes many years of operation for the fatigue to develop. In one instance, a Strauss bascule built in 1908 did not start to experience metal fatigue in its main member until it had been in operation for more than 70 years. By that time, a great part of the machinery and a few other components of the bridge had been repaired or replaced, some more than once. If those inspecting and evaluating the bridge had been aware of the propensity of the bridge structural members to fatigue, some replacements, reinforcements, retrofits or repairs could have been made before the cracks appeared.

Any bascule bridge should be evaluated for dead load stress range during operation, and if the stress cycle is high, the possibility of fatigue should be considered. The lower flange or chord of a bascule bridge's main member is under compression at all times except for live load, unless it is acting completely as a cantilever. Tension may develop due to possible wind load effects. The upper flange is under tension when the bridge is down, but this tension reduces if the bridge supports live load as a simple span and may change to compression gradually as the bridge is raised. The higher the opening angle, the more the stress in the upper flange or chord becomes compressive. A typical simple trunnion bascule may open to only 50° or so, but some open to 70° or more. Those opening to larger angles, if they have operated frequently over a long period of time, can be considered candidates for metal fatigue and should be evaluated. Many Scherzer-type rolling lift bridges also open to large angles, and fatigue should also be considered in evaluating these bridges.

The remaining fatigue life of a movable bridge, or parts of a movable bridge, can be predicted if certain facts are known:

Details subject to fatigue
Past loading and operations history
Future loading and operations

SPAN BALANCE

The span balance of bascule and vertical lift bridges is frequently the subject of evaluation. As stated previously, the ideal balance state of such a movable bridge is open to debate. As a part of the evaluation of an existing bascule or vertical lift bridge, the cost of adjusting the span balance after an extensive renovation or reconstruction must not be ignored. The problem is especially severe with bascule bridges in which location, as well as amount, of counterweight is important. Some older bascule bridges were originally designed and constructed to carry trolley cars that received their power from an overhead wire. The removal of the trolley cars from the bridge meant that the trolley wire was no longer needed. Removal of the trolley wire and its support structure from the bascule leaf resulted in a gross change in the location of the center of gravity of the bascule span, which was not easily corrected. Some older bascule bridges were originally designed with the center of gravity of the moving leaf positioned to minimize the power requirements for operation of the bridge (see Figure 11-15), so that correction of the location of the center of gravity is a difficult job that can be made even more difficult if this correction is combined with compensation for permanently removed parts of the structure.

If a bascule or vertical lift span has a balance condition that is very far from ideal, any plan for rehabilitation of the bridge should include reconfiguration of the counterweight to allow for correction of the imbalance. The bridge should definitely be span heavy when seated, and this span heaviness should decrease as the bridge is raised. The amount of span heaviness should be sufficient to allow the bridge to seat firmly, but not much more. The bridge should be slightly counterweight heavy when in the fully open position, but if this cannot be achieved, it is not critical. The bridge should not be so span heavy when fully opened that it will overcome friction and come crashing down if the brakes are inadvertently released.

39

MACHINERY

Rating systems like those discussed in Chapter 36 can be applied to bridge machinery, but are very subjective and prone to misuse by inexperienced inspectors. They do provide a quick picture of the condition of the bridge machinery, however, and can be adapted to almost any conditions. Older bridges that were not designed to meet current AASHTO or AREMA requirements must be evaluated to determine whether these design shortcomings present a hazard to the bridge, the public, bridge users, or bridge personnel. If such is the case, these flaws should be rectified as soon as possible. Caution is necessary, however, as machinery components that appear to be overstressed or dangerously worn may have sufficient residual strength to last many years before failure. Unless there are other overwhelming reasons to proceed with machinery replacement, a careful analysis of the offending components should be performed before condemning them.

On movable bridges that have had a complete replacement of the electrical system without replacement of the mechanical machinery, the output of the electrical system should be determined to see whether it can overload the machinery. Many older bridge control systems were replaced with solid-state controls, and the drive motors controlled to maintain certain speeds during operation without regard to the output torque required to maintain those speeds. Such drive systems have severely overloaded the machinery, which has, in some cases, resulted in machinery failures.

Most movable bridge mechanical components can be evaluated by measuring the amount of wear they have sustained. This wear can be converted, in some cases, to remaining strength. For gear teeth, wear increases impact during operation, reducing the transmitted load capacity. When the remaining strength is reduced so that the gear cannot be relied on to carry its intended loads, it is time to replace the gear. For other components, such as friction-type bearings, wear produces increased running clearances that make the proper alignment of mating parts impossible. When misalignment due to wear becomes too great, it is time to replace these parts. Other components, such as antifriction bearings, are difficult to measure for deterioration. When such a component has noticeably excessive clearance, it may already be about to fail catastrophically. Such components can be evaluated on the basis of remaining useful life. When the time in service or number of operations has approached the design life of the component, it can be replaced before it fails.

GEARS

Most older existing movable bridges are operated by electromechanical drive systems. The components of these systems that usually show signs of excessive wear first are the teeth of the gears. In many high-speed machinery drive systems, any visible sign of wear is considered grounds for replacement. In the slow-moving world of open gear drive bridge machinery, there can be many years of reliable operation left for what appears to the eye to be a very heavily worn set of gears. By recalling the Lewis formula (Chapter 17) and recognizing that a gear tooth is considered to be a cantilevered beam,

$$s = My/I \tag{1}$$

$$= \text{bending stress on gear tooth being evaluated}$$

with

M = F, the force on the tooth, times the distance d from the point being evaluated on the tooth to the point of application of the load.

For a typical pinion, superimposing a constant-stress parabola on the tooth profile shows that the root is the weakest point. The load can be assumed to be at the pitch line, so that

$M = Fd$, where d = the dedendum

$y = \frac{1}{2}$ the tooth thickness at the point being evaluated. For a typical bridge pinion, the root thickness may be less than t, or t_c, the thickness at the pitch line.

I = The moment of inertia of a beam of rectangular cross section,

$= \frac{1}{12} bh^3$,

where

$h = t$, the thickness of the tooth

and

b = the width of the tooth face, f

so that

$$I = \tfrac{1}{12} f t^3$$

and

$$s = \frac{Fdt/2}{\tfrac{1}{12} f t^3}$$

If the load F equals the capacity W,

$$W = \frac{sft^2}{6d} \tag{2}$$

We can then develop the following equation:

$$W = kt^2, \tag{3}$$

where

W = the gear's strength
k = $sf/6d$, a constant for any particular gear in a set (s = allowable stress)
t = the thickness of the tooth

We can further say that the original thickness of the tooth is included with the constant k, and modify equation 3 to

$$W = K\left(\frac{t_{\text{present}}}{t_{\text{original}}}\right)^2 \tag{4}$$

where

t_{present} = the measured tooth thickness

and

t_{original} = the original tooth thickness

The constant K is then the original strength W as calculated in Chapter 17, and W of equations (3) and (4) is the present strength, having allowed for wear. By determining how the original strength compared to the minimum necessary, we can determine the maximum reduction in strength that can be tolerated. Assuming that the allowable stress used in the original strength determination was somewhat conservative, a higher allowable stress can be used to determine the value of W for a gear in the worn state. For instance:

Assume that for a particular highway movable bridge, the power required according to AASHTO is 55 hp. The original motor is rated at 35 hp, and the machinery was sized, per AASHTO, based on that motor horsepower. A particular spur gear in the bridge drive has 9 percent loss in its tooth thickness because of wear. The particulars of the gear are:

$n = 17$
$p = 4$ in.
$s = 22{,}500$ psi allowable stress per specification
$f = 10$ in. actual face width, not effective
$V = 35$ fpm

The original gear strength was then [*AASHTO Standard Specifications for Movable Highway Bridges* (1988), p. 24]

$$W = psf\left(0.154 - \frac{0.912}{n}\right)\frac{600}{600+V}$$

$$= 85{,}340 \text{ lb}$$

If the load on the tooth was, at 35 hp, at 180% of full load motor torque, with some drive train losses,

$$F = 57{,}890 \text{ lb},$$

with 55 hp it would be

$$\text{F} = 90{,}970 \text{ lb, clearly an overloaded condition.}$$

Factoring in the gear wear, 9%, the allowable load at the design stress of 22,500 psi, is

$$W = 70{,}670 \text{ lb}$$

so that the stress fully loaded with the properly sized motor operating at 180 percent of its rating would be

$$s = 22{,}500\left(\frac{90{,}970}{70{,}670}\right) = 28{,}963 \text{ psi}$$

If the proportional limit of the original gear material was 30,000 psi, and this was accepted as an allowable stress level, the gear would not be overstressed at present, during starting and accelerating the bridge during highest load conditions, assuming that the AASHTO motor overload limit of 180% was kept.

Conversely, without wear, the gear would be stressed only at 27,000 psi at 180% of full-load motor torque if it were sized at 150% according to AASHTO for a motor properly selected for the bridge load.

The foregoing procedure may be somewhat conservative in the analysis of a worn gear, because the highest bending stress in a gear during operation is usually at the root, whereas the most convenient wear measurement point is at the pitch line. Gears tend to wear mostly at the pitch line and very little at the root. Pinions thus tend not to have an increase in maximum bending stress as they wear, as compared with racks or gears with a large number of teeth. Straight involute rack teeth have a trapezoidal shape that maximizes efficiency in resisting bending stress and are thus vulnerable to overstress if even a small amount of wear distorts that shape (see Figures 39-1 and 39-2). In fact, gear teeth have been known to experience bending stress failure in service although not as often as pinions. Although the gear with more teeth is inherently stronger, the pinion is usually made of a stronger material to compensate for poorer geometric qualities. For a critical gear that may be very difficult or expensive to replace, it may be warranted to measure the tooth thicknesses at various heights, particularly at locations between the root circle and the pitch circle, and evaluate the worn gear's stress at these points when fully loaded. A parabola inscribed in the tooth cross section, with its apex at the pitch line or circle, tangent to the tooth flanks at the root will help indicate locations of vulnerability.

Although gears are frequently worn somewhat evenly throughout their faces, it may be found that areas with great amounts of wear do not have particularly high bending stress. A decision may be made to maintain a worn gear in service without there being an unacceptable increase in the likelihood of fracture, or at least it may be possible to leave the gear in service for the short term until a replacement can be provided. On the other hand, analysis may show that the gear is in immediate danger of failure, so that special operating procedures may be warranted and gear replacement should be expedited. In addition to basic static stress levels, there are dynamic considerations concerning gear tooth stresses. The contact ratio of a new gearset is usually substantially more than 1.0, so that more than one pair of gear teeth are in contact at any time as the gearset operates. This reduces stress and allows fairly smooth transfer of power from the driving gear in the set to the driven gear. As the gearset wears, this contact ratio decreases, so that impacts on mating gear teeth, as they come into mesh, increase. In severe wear the contact ratio may drop below 1.0, resulting in high impacts that may lead to tooth breakage. If this is the case for a gearset in question on a bridge, with high calculated stresses in the gears in their existing state, prudence dictates replacement as expeditiously as practical.

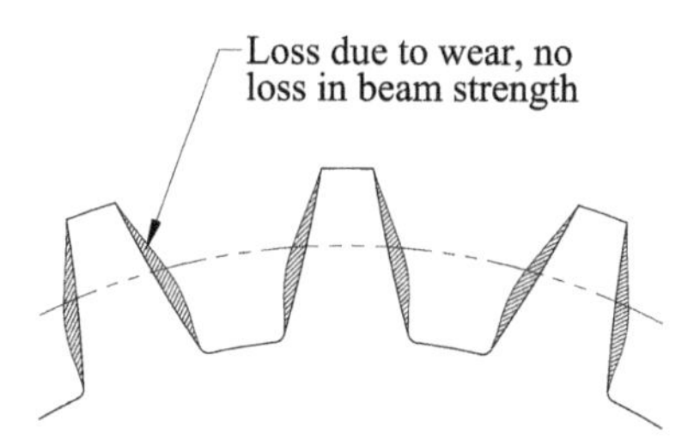

Figure 39-1 Pinion Tooth Wear. The weakest point is at the root; no wear has occurred there.

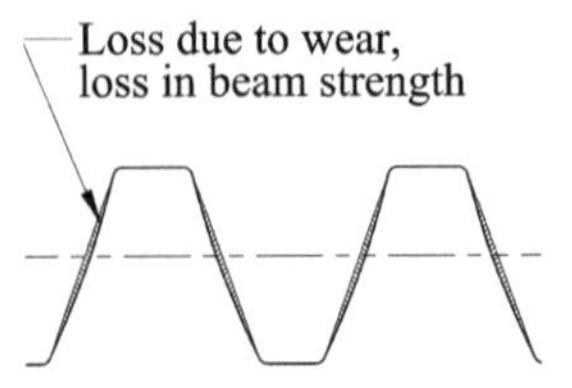

Figure 39-2 Rack Tooth Wear. As wear occurs on the flanks, the tooth is weakened in its bending strength.

Once the preceding calculations have been performed, a prediction of the remaining life of the gear can be made. If the maintenance and use of a machinery installation is constant over time, a plot of the gear strength as a function of time can be developed. If diligent inspections are carried out at regular intervals, with measurements of gear tooth thicknesses made and recorded, these can be converted to values of remaining gear strength as shown on page 479. The rate of wear of a gear tends to increase over time, as a gear loses efficiency as its involute curve deteriorates and poor contact at the worn teeth causes impact loading. Once the shape of a plot has been developed so that the rate of change of the gear tooth shape has been determined, prediction of the approximate time it will take to reach an unsafe strength level is simple. Graphical or mathematical methods can be used to develop the curve of tooth strength versus time, based on the plot.

The foregoing discussion assumes that no other deterioration is occurring in the machinery drive train that can have an effect on the longevity of the gearset. If wear occurs at the bearings supporting the shafting on which the gears are mounted, or the bearings or gears become misaligned, loads on the gears can increase. The distance d in equation 2 on page 479 will increase as the center distance increases because of worn bearings. The value b in the constant k in the same equation is the full face width of the gears in the set, but this should be reduced if the gears are badly misaligned axially. The effect of both these changes is to increase the stress on the gears and thus increase the likelihood of failure. The previous discussion did not address surface contact stress, as in open gearing this usually manifests itself in wear which continues until a bending stress failure occurs. Gear failure in an enclosed drive is another matter.

BEARINGS

There are three types of bearings used in bridge machinery—soft metal, hard metal, and antifriction. Soft metal bearings, usually made of babbitt metal, have a short life and are seldom used any longer on new bridge machinery installations. They can wear substantially, which results in increased stress on the gears being supported by them, as discussed earlier. On older bridges, the replacement of soft metal bearings was once considered a maintenance item, as it was relatively easy to pour in a new babbitt bearing when the old one had worn excessively. Hard bearings, usually of bronze, rarely wear out unless lubrication is grossly inadequate for the usage. As friction bearings wear, gear alignment gradually deteriorates. Antifriction bearings have a probable life, in terms of hours of operation, that is readily calculated with catalog equations that require the input of bearing load and speed. These calculated life spans are contingent on proper lubrication of the bearing and proper sizing of the bearing for its load and operating conditions, according to manufacturer's recommendations. AASHTO and AREMA specify long bearing lives, which theoretically exceed the likely life of the rest of the machinery, particularly the gears. As required maintenance of these bearings is usually minimal, they can be expected to survive to their calculated catalog life span unless environmental or other deterioration occurs. The most likely causes of premature failure of a properly designed antifriction bearing are lubrication failure and contamination. A failure of an antifriction bearing allows substantial displacement of the shaft being supported, approximately equal to the diameter of the rolling element in the bearing. This can cause increased gear stress, as discussed earlier, or immediate catastrophic failure of the drive train.

ROLLING LIFT TREAD PLATES

Many rolling lift bridges are found to have cracks in the tread plates or in the connections of the tread plates to the segmental or supporting girders. These cracks can develop because of fatigue of the material making up the components, or because of high stresses due to misalignment, or a combination of both. In the case of metal fatigue, it is likely that the cracks will continue to increase in number and grow in length. An estimation of the time it will take for the pieces to begin to come loose can be made if an estimate of the rate of growth in crack size and number can be made. If reliable previous inspection information is available, the crack sizes and numbers can be compared. An accurate inventory of cracks should be made for reference for a future inspection and reevaluation. This inventory is time-consuming and requires special effort. The cracks should be identified and their lengths recorded on the bridge as well as in portable form. Photographs should not be relied upon, but length measurements and locations should be recorded. Cracks developing because of overstress due to misalignment or defective fabrication can actually stop propagating after time if the crack helps ease the misalignment. If a crack stops before seriously weakening a component, a failure of the component may never occur. Fatigue cracking, however, is likely to continue until complete failure occurs. The primary loading of tread plates and connecting girders is in compression, so load can be transferred through a crack as long as excessive material is not lost through corrosion or abrasion.

COUNTERWEIGHT ROPES

Many vertical lift bridges have had their counterweight ropes replaced in recent years. The usual reason for deciding to replace these ropes is excessive wear. Many rope manufacturing companies have published charts indicating rope strength as a function of wear lengths. Sometimes the decision to replace vertical lift bridge counterweight ropes is based on data from these charts. The charts may be accurate for ropes used in industrial or mining operations, where the size of the sheaves over which the ropes operate are smaller, and the loads carried by the ropes are larger, than recommended by AASHTO or AREMA for counterweight ropes. In cases of counterweight ropes with extensive visible wear, it can usually be found that there is ample remaining strength to support the rope loading. If a detailed analysis of the condition of the counterweight ropes is performed, it is usually found that the rate of deterioration will allow the ropes to remain in service for many more years. Because of the helical twisting of the rope strands, and the same kind of twisting of the wires within the strands, only a small fraction of the wires in a rope are affected by external wear at a given cross section. Because of the twisted arrangement of the rope wires, load can be transferred between wires so that local deterioration is not fully additive along the length of the rope. If the wear at the rope wires has progressed to the point at which more than half of the wire cross-sectional area has worn away, the wires may begin to break in fatigue. Careful inspection of heavily worn ropes may uncover a few such broken wires. When they are found in quantity, it may be concluded that the rope is worn out.

Internal wear in a vertical lift bridge counterweight rope can be debilitating. This type of wear usually begins only after the rope has been in service for a long period of time on a very active bridge. The rate of wear is a function of the condition of the rope core, among other things. A deteriorated rope core allows the wire rope strands to contact each other within the rope. This contact causes a chafing action at the wires in contact, particularly as the ropes operate over the sheaves. This type of wear may produce red rust at the valleys of the rope strands, cause a reduction in the overall diameter of the rope, and cause an increase in the overall length of the loaded rope with a reduction in its diameter.

To establish the remaining strength of a counterweight rope, it is necessary to determine the total loss of rope wire cross-sectional area along a given length of rope, called the recovery length. The recovery length is the distance along a rope that a given wire, from a location of damage, is restored to its full carrying capacity. Depending on the construction of a rope, this distance may be only 1 ft, more or less. Within this length, the total of the area lost at the worst spot of damage on all the wires must be determined to obtain the total area section loss of the rope. For a typical 6×19 counterweight rope, there are 114 wires that may have experienced deterioration. Of these, 72 wires are external ones. These will experience wear as they contact the counterweight sheave. If these wires have all worn through 50 percent of their cross-sectional area at the same location along the rope length, then the rope has lost about one-third of its strength. As the rope has an initial safety factor of almost 8, it still has a safety factor of about 5, which is not cause for immediate replacement of the ropes. It may take six or more lay lengths for all the outside wires of the rope to make sufficient contact with the sheave groove to experience maximum wear (see Figure 39-3). This may be 4 ft or more of rope length. It is unlikely that all of the wear will be additive along this length. If there is additional deterioration within the areas of external wear, such that internal or external corrosion of the wires is apparent, or

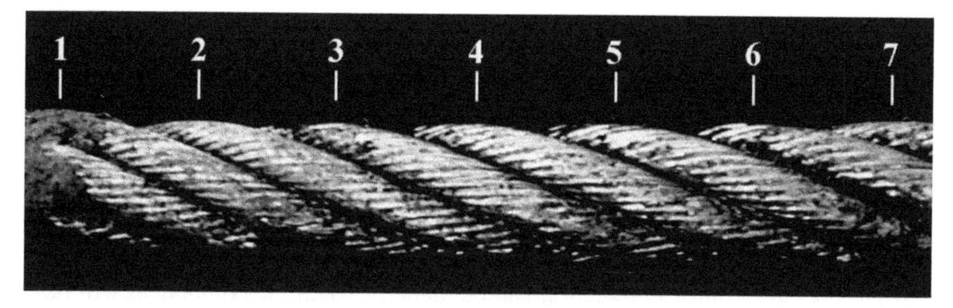

Figure 39-3 Rope Lay Length. The distance along the length of rope that it takes for one strand to make a complete rotation is the rope *lay length.* The strand at (1) reappears at (7). The distance along the length of a strand that it takes for one wire to make a complete rotation is the strand *lay length.*

some of the outer wires have broken, then the state of the ropes is cause for immediate concern.

COUNTERWEIGHT SHEAVES

Vertical lift bridge counterweight sheaves make several rotations as the bridge raises and lowers, so metal fatigue can occur in the sheaves if stress concentration exists, such as at flaws, producing cracks. If the bridge operates often, these cracks can grow, possibly until complete breakage occurs. Attempts have been made to repair such cracks by weld repairing the sheaves in place, but success has been practically nil. Repairs have been made by bolting on splices, which appear to have been successful.

Deteriorated sheave grooves can accelerate the deterioration of counterweight ropes. The rate of deterioration of the ropes is partly a function of the degree of the deterioration of the grooves and the frequency of operation of the bridge. It is possible to remachine the grooves in place, but it must be done carefully so that the diameters of all grooves on a sheave are the same, so as to avoid causing excessive sliding of the ropes. The remachining must not be made too deep, or it will weaken the sheave. The grooves must be made with precisely the correct cross section in order to properly support the ropes.

COUNTERWEIGHT SHEAVE SHAFTS

The typical counterweight sheave shaft rotates with the sheave. These shafts may make six or more complete rotations within a bridge operation cycle, each rotation putting the shaft under full bending stress reversal. The fatigue life of such shafts can be predicted, if the geometry, surface finish, material properties, and loading are known. If the total number of cycles that have been experienced can be counted, the remaining life can be approximated.

HYDRAULIC SYSTEMS

Hydraulic components such as valves, motors, and cylinders have an estimable life based on the number of hours of operation. This estimated life should be available from the manufacturer. It will be conditional on degree of maintenance, severity of service, and other factors. As movable bridge machinery is operated only intermittently, the basic life of a component will be quite long. Factoring in realistic values for maintenance and operating conditions can reduce this life considerably. By monitoring operating time, a prediction can be made of when the component will be liable to failure.

In evaluating an existing hydraulic system that has deteriorated and attempting to determine whether it should be replaced in kind, with another hydraulic system, or with an electromechanical drive, consideration should be given as to whether the system in place has received the maintenance it should have, for optimum life. If it has not been adequately maintained, then a realistic appraisal should be made of the likely quality of future maintenance before deciding on the type of replacement drive.

AUXILIARY/EMERGENCY DRIVE SYSTEMS

Many movable bridges have auxiliary drive systems that are inoperable. Hand-drive systems are corroded and rusted so that they are immovable. Direct engine drives or auxiliary generators have engines that will not start. Secondary utility power sources have been scavenged for spare parts or have been vandalized. The United States Coast Guard, which is responsible for enforcing most federal navigation regulations, has taken a more stringent approach in recent years to enforcement of regulations pertaining to the opening of movable bridges when the primary power source is inoperable. Many movable bridge owners have been asked by the Coast Guard to demonstrate that their bridges can be opened in an emergency, with auxiliary power, and have been unable to do so. A realistic approach should be taken in regard to auxiliary drives. If manpower will not be available to operate a manual drive, it should be discarded and replaced with another type of auxiliary drive. If the internal combustion engine power source for the auxiliary drive is unreliable, another power source should be obtained. See Chapter 18 for such possible alternatives.

40

ELECTRICAL SYSTEMS, TRAFFIC CONTROL SYSTEMS, AND OPERATORS' HOUSES

Rating systems such as discussed in Chapter 36 can be applied to the electrical equipment of movable bridges, but they are very subjective and are prone to misuse by inexperienced inspectors or evaluators. Ratings do provide a quick picture of the condition of the bridge equipment, however, and can be adapted to almost any conditions. As much of the evaluation of electrical equipment is, to a great extent, subjective, numerical ratings can be of substantial value in reporting the condition of this equipment if the criteria are well defined and the system is used by competent personnel. Older bridges that were not designed to meet current AASHTO or AREMA requirements or electrical codes must be evaluated to determine whether these design shortcomings present a hazard to the bridge, the public, bridge users, or bridge personnel. If such is the case, these flaws should be rectified as soon as possible.

Many pieces of electrical apparatus can be evaluated by simply looking at them. The condition of the housings of switches and other devices can give an indication of the likely amount of reliable life left in these components. In fact, to some extent, certain pieces of electrical apparatus can be evaluated by simply going over the records of the bridge and determining how long some components have been in service. These components can "fail inspection" because they are no longer within code and are considered hazardous for operating and maintenance personnel. Many components are composed, in part, of insulating materials that have limited life expectancies. With a little additional research into the components installed on the bridge, they can be identified and flagged for replacement. Many older movable bridges that maintain a good part of their original electrical equipment fail to meet present-day codes for personnel protection because this equipment is exposed. This condition alone should be sufficient reason for condemnation of such equipment and its speedy replacement with a satisfactory modern installation. Chances are that many components of such an installation are defective in other ways, which helps to justify complete replacement of the installation.

Other components, such as contactor points, switch components, motor brushes or slip rings, brake actuators, and resistor grids also deteriorate with age. These components may be ready to fail even if they have not operated for their designed number of cycles. Any of these components can be subjectively determined to be ready for retirement even if they do not fail more quantitative tests. A formal delineation is made between "Engineering Evaluation" and "Predictive Life" methods of evaluating electrical components in the *AASHTO Movable Bridge Inspection, Evaluation, and Maintenance Manual,* but on any given bridge, or even on a particular assembly, some components may be evaluated in one way and some by the other. A numerical rating system such as that of AASHTO or FHWA, or on a similar basis, coupled with rating level descriptions such as those presented in Chapter 36 can be used to delineate various degrees of deterioration.

Electrical equipment, including wiring and apparatus, is more easily exposed to severe overload conditions, resulting in deterioration, than other movable bridge components. Sometimes a simple walk-through inspection can reveal to the trained eye severely deteriorated equipment that should, without question, be replaced.

Electrical cables, such as submarine cables, can be effectively evaluated by determining the rate of deterioration of the insulation on the wires. The insulation resistance may be lower than optimum, but if does not appear to be decreasing rapidly with time, the cable may have several years of useful life remaining. To make a sound judgment on this basis, several tests must be made over a period of time. The testing process has to be consistent so that results are compared under similar conditions of temperature, humidity, test voltage, and even time of day and time of year. The time frame must be fairly extensive for positive results. A miniscule measurable change over six months means little. A 20 or 30 percent deterioration over four or five years suggests that planning for replacement should begin. Any confirmed major drop in insulation resistance, regardless of the time frame, is cause for concern.

Amperage draws by drive motors can increase over time and can suggest deterioration of the motors. The motor current can also be affected by deterioration in the mechanical drive machinery or by structural changes in the bridge. Test results should be compared with the findings of mechanical and structural inspections before definitive conclusions are drawn.

Surprisingly, many movable bridges have electrical equipment installed that does not match their needs. The main drive motors should be compared with the power requirements according to AASHTO or AREMA. In many cases, the motor is found to be too small. In other situations, a bridge with two motors, but using only one at a time, actually requires both motors to be operating together. There have been cases of motor replacement in which an old motor of, say, 100 hp at 600 rpm, was replaced with a new motor of 100 hp at, say, 1200 rpm. As the person specifying the replacement motor was unaware of torque requirements and assumed that horsepower was horsepower, no additional gear reduction was provided and the motor was inadequate to power the bridge, but the bridge ran fast when it was able to get moving. Once the condition of the motors is determined, then the size of the power source itself can be evaluated. It should be more than large enough to meet the maximum needs of the bridge during operation. If any of these components are undersized, then replacement with a larger system is warranted unless it can be shown that the bridge need not operate in the most adverse conditions as prescribed by AASHTO or AREMA. In rare cases, the motor or motors are too large, or the machinery is inadequately designed, so that the motor(s) provide excessive power to the machinery. This condition should be corrected by reducing the size of the motor(s) or replacing the machinery, depending on the actual power requirements of the movable bridge.

TRAFFIC CONTROL SYSTEMS

The traffic control system on a movable highway bridge should adequately keep traffic from interfering with operation of the bridge and reasonably protect the public from hitting the bridge or driving into the water when the bridge is opened. On many older movable bridges, minimal highway traffic protection is provided. This may be adequate if the public is aware of the presence of the movable bridge and is generally law-abiding, so that instances of people trying to cross the bridge as it is being prepared for an opening, or driving around the gates, are unknown. In most situations, this is not the case—certainly not in the typical urban area, and not in suburban or other areas where traffic may have increased greatly since a bridge was built. Observations of bridge openings during different periods can indicate whether additional protection is desirable. When in doubt, the addition of barrier gates and improvement of warning gates may be warranted. If the bridge is not equipped with red-yellow-green traffic lights, the addition of such lights can prove quite helpful in easing a traffic control problem, as people tend to pay more attention to them than to the typical flashing red lights. It may be necessary to perform a traffic study to see if the enhancement of existing traffic protection devices is warranted. At movable bridges carrying high-speed automobile or heavy truck traffic, it should be made certain that motorists are given adequate advance warning that there is a drawbridge ahead.

OPERATORS' HOUSES, MACHINERY ENCLOSURES, AND ACCESS APPLIANCES

Many older bridges have operator's houses, machinery enclosures, ladders, walkways, or stairways that do not meet present-day codes and standards. Keeping up to date with the safety requirements of the Occupational Safety and Health Administration (OSHA), AASHTO, and the many other authorities that have jurisdiction over a particular bridge can be a full-time job in itself. Safety equipment, particularly ladders and walkways, should be reconstructed or replaced if deficient. Machinery enclosures should include guards to protect personnel from machinery. Operator's houses are required to provide minimal levels of sanitary facilities and other comforts. The cost of providing all these features must be included in any program for rehabilitation of a movable bridge. If the bridge is to remain in service indefinitely, most or all of these improvements should be provided as soon as practical.

41

FAILURE AND BRIDGE FUNCTIONALITY ANALYSIS

Existing movable bridges, whether they appear to have any glaring defects or not, should be subjected to failure analysis to determine whether there are any dangerous or potentially dangerous conditions existing that could lead to failure, or other shortcomings that should be eliminated. Before proceeding with a rehabilitation program, all the shortcomings of a movable bridge should be determined and assessed, as it may turn out that, all things considered, replacement of the bridge is a wiser choice.

On many inland waterways, movable bridges are in locations or are of such a size that they are vulnerable to damage resulting from impacts by large commercial vessels or barge tows. As these vessels have become larger, older bridges designed to accommodate smaller vessels provide insufficient clearances for safe and easy passage. A movable span may also have been placed at a location in the channel that was adequate for vessels at the time of the bridge's construction, but now requires intricate maneuvering by the typical vessel to navigate the draw opening. A new bridge with a wider navigation opening, placed in the optimum location in the waterway, can significantly reduce the likelihood of impact damage. With heavier traffic on railroad and highway bridges, it may also be possible to significantly reduce the number of bridge openings required for navigation if the existing bridge is replaced by one at a higher elevation. The U.S. Coast Guard has ordered many such bridge replacements, referring to them as "alterations" in keeping to the letter of the law, on the Mississippi and other inland waterways, providing wider, better-placed navigation channels. In some cases, where feasible, the alterations provide higher clearances with the bridge closed than previously. For a bridge vulnerable to ship impact damage, the cost of providing adequate protection for the existing bridge, and its approaches as necessary, should be compared with the cost of replacement with a new structure with lower vulnerability.

Many movable bridges have structural or substructural deficiencies that are not readily apparent, such as weak foundations or susceptibility to scour. Before an electrical or mechanical rehabilitation program is implemented, all such deficiencies should be determined, and the cost of their correction, or the risk associated with leaving them as is, should be reckoned. If earthquake resistance is a concern, a seismic assessment of the bridge should be made before any major reconstruction of the bridge is begun. This is especially important for bridges categorized as critical, but the seismic vulnerability of other bridges should be considered as well.

If a bridge has insufficient capacity for the traffic it carries, or should carry, the cost of rebuilding the structure should be very closely estimated before commencing a reconstruction program. In some cases, reconstruction of a bridge had already begun, when it was found that the overall cost of replacement was cheaper.

Most movable bridges do not require their machinery to support traffic loads, so there is little concern for the traffic on a bridge in evaluating bridge machinery for operational failure. Swing bridges are an exception, as machinery drive components hold the end lifts in place and the dead load of rim-bearing-type bridges is held up by large roller-type thrust bearings, which can develop inadequacies over time. Bascule bridge trunnions are also mechanical parts that support dead and live loads. These are subject to wear, fatigue, and other forms of failure. All of these mechanical machinery components are under the greatest stress when operating, so that a failure of the catastrophic type is more likely to occur when the bridge is opening or closing than when it is at rest, carrying traffic. There have been failures in which a bridge has fallen on a vessel passing through the draw, for instance, but they are rare. Even more rare is a movable bridge collapsing while carrying traffic (see Figure 2-1).

Most bridge machinery tends to be fail-safe, in that a failure will cause the bridge to become inoperable and stuck in the position it was in when the failure occurred. This is what happened when one of the counterweight sheave shafts failed on the Shippensport Bridge in Illinois in 1978. The vertical lift span ended up immobile in the partially opened position. Similar mechanical failures, with similar results, have happened on other bridges. Most movable bridge electrical equipment is also usually fail-safe, in that failure of an electrical component will result in power being lost to or through that component, and the bridge stops operating. Distressing failures of another kind occurred, however, when programmable logic controllers (PLCs) were first used on movable bridges. There were several instances of equipment being energized when it was not supposed to be. These faults were traced to shorts, grounding malfunctions, lightning strikes, and other failures. Such failures would have been unlikely to cause such problems with older types of movable bridge control systems, and many people have shied away from PLCs because of this sensitivity. PLC-type systems have since been provided with ground fault protection, lightning protection, and other safeguards in an effort to prevent such failures. Any existing PLC system on a movable bridge should be evaluated to see if it is secure, and what steps should be taken to make it so if it is not.

Most movable bridges are basically balanced structures, but bascule and vertical lift bridges are often designed to have slight imbalances to promote stability in the seated position. A few newer movable bridges of the bascule or vertical lift type have been built or proposed without counterweights, so that opening them involves a dead lift against gravity. The control and drive systems of these bridges should be evaluated to see if they have adequate safeguards to minimize the likelihood of a dead drop should a failure occur.

Operator error has caused many bridge accidents that have led to loss of life. The skill and training of the bridge operator should always be included in an evaluation of the likelihood of a movable bridge to be involved in an accident. Even with largely automated controls, the bridge operator decides when a movable span should be opened and when it should be closed. For this reason, there can be, and there have been, collisions with vessels when operators have erred. Bridge operators have also been known to open a bridge with traffic on it or to allow traffic on a bridge before it has been fully closed. Interlockings may be in place to prevent such incidents, but on many bridges the interlockings have been bypassed, because of previous failure, and the operator is relied upon to perform correctly. The evaluator of a movable bridge should take careful note of the use of bypasses on the bridge in relation to the skill levels of the operators of that bridge.

REFERENCES

American Association of State Highway and Transportation Officials (AASHTO). *Manual for Maintenance Inspection of Bridges.* AASHTO: Washington, D.C., 1983.

American Association of State Highway and Transportation Officials (AASHTO). *Standard Specifications for Movable Highway Bridges.* AASHTO: Washington, D.C., 1988. Updated and revised periodically with minor addenda to 1993m.

American Association of State Highway and Transportation Officials (AASHTO). *Guide Specification and Commentary for Vessel Collision Design of Highway Bridges.* AASHTO: Washington, D.C., 1991.

American Association of State Highway and Transportation Officials (AASHTO). *Standard Specifications for Highway Bridges.* AASHTO: Washington, D.C., 1996.

American Association of State Highway and Transportation Officials (AASHTO). *Movable Bridge Inspection, Evaluation, and Maintenance Manual.* AASHTO: Washington, D.C., 1998.

American Association of State Highway and Transportation Officials (AASHTO). *Manual for Condition Evaluation of Bridges*, Second edition. AASHTO: Washington, D.C., 2000.

American Association of State Highway and Transportation Officials (AASHTO). *AASHTO LRFD Movable Highway Bridge Design Specifications.* AASHTO: Washington, D.C., 2000.

American Gear Manufacturers Association (AGMA). *American Standard Nomenclature of Gear-Tooth Wear and Failure.* AGMA 110.03; ASA B6.12-1964. AGMA: Washington, D.C., 1964.

American Gear Manufacturers Association (AGMA). *USA Standard System Tooth Proportions for Coarse-Pitch Involute Spur Gears.* AGMA 201.02; USAS B6.1-1968. AGMA: Arlington, Virginia, 1974 (reaffirmed).

American Gear Manufacturers Association (AGMA). *AGMA Gear Handbook, Volume 1 Gear Classification, Materials and Measuring Methods for Unassembled Gears.* AGMA 390.03. AGMA: Arlington, Virginia, 1980.

American Gear Manufacturers Association (AGMA). *Fundamental Rating Factors and Calculation Methods for Involute Spur and Helical Gear Teeth.* ANSI/AGMA 2001-B88. AGMA: Alexandria, Virginia, 1988.

American Gear Manufacturers Association (AGMA). *Gear Classification and Inspection Handbook.* AGMA 2000-A88. AGMA: Arlington, Virginia, 1988.

American Gear Manufacturers Association (AGMA). *Standard for Spur, Helical, Herringbone, and Bevel Enclosed Drives.* AGMA 6010-E88. AGMA: Alexandria, Virginia, 1988.

American Gear Manufacturers Association (AGMA). *Geometry Factors for Determining the Pitting Resistance and Bending Strength of Spur, Helical and Herringbone Gear Teeth.* AGMA 908-B89. AGMA: Alexandria, Virginia, 1989.

American Gear Manufacturers Association (AGMA). *Metric Usage.* AGMA 904-B89. AGMA: Alexandria, Virginia, 1989.

American Gear Manufacturers Association (AGMA). *Appearance of Gear Teeth—Terminology of Wear and Failure.* AGMA 1010 E95. AGMA: Alexandria, Virginia, 1995.

American Railway Engineering and Maintenance-of-Way Association (AREMA). "Steel Structures," Chapter 15 in *Manual for Railway Engineering.* AREMA: Landover, Maryland, published approximately annually.

American Society of Mechanical Engineers (ASME). American National Standard Drafting Practices. *Gear Drawing Standards—Part 1 for Spur, Helical, Double Helical and Rack.* ANSI Y14.7.1. ASME: New York, 1971.

Cook, Richard J. *The Beauty of Railroad Bridges.* Golden West Books: San Marino California, 1987.

Federal Highway Administration. *Bridge Inspector's Manual for Movable Bridges.* U.S. Department of Transportation, Federal Highway Administration: Washington, D.C., 1977 (out of print).

Mobil Oil Corporation. *Mobil Technical Bulletin: Gear Tooth Failures.* Mobil: Fairfax, Virginia, 1973.

New York State. *Steel Construction Manual.* New York State Department of Transportation Office of Engineering Structures, Design and Construction Division: Albany, New York, 1984.

United States Code, Title 33. *Navigation and Navigable Waters,* Chapter 11, *Bridges over Navigable Waters.* U.S. Government Printing Office: Washington, D.C., 1989.

U.S. Coast Guard (USCG). *Bridges over Navigable Waters of the United States: Atlantic Coast.* U.S. Department of Transportation, USCG: Washington, D.C., 1984 (last year published).

U.S. Coast Guard (USCG). *Bridges over Navigable Water of the United States: Great Lakes.* U.S. Department of Transportation, USCG: Washington, D.C., 1984 (last year published).

U.S. Coast Guard (USCG). *Bridges over Navigable Waters of the United States: Gulf Coast & Mississippi River System.* U.S. Department of Transportation, USCG: Washington, D.C., 1984 (last year published).

U.S. Coast Guard (USCG). *Bridges over Navigable Waters of the United States: Pacific Coast.* U.S. Department of Transportation, USCG: Washington, D.C., 1984 (last year published).

Waddell, J. A. L. *Bridge Engineering.* 2 volumes. New York: John Wiley & Sons, 1916.

War Department, Corps of Engineers. *List of Bridges over the Navigable Waters of the United States.* United States Government Printing Office: Washington, D.C., 1948 (data current as of 1941).

Whitney, Charles S. *Bridges.* 1929.

BIBLIOGRAPHY

ARTICLES ON MOVABLE BRIDGE SUBJECTS

Title	Author	Periodical*	Date	Description
After Fire, 3rd Avenue Bridge Is Closed for the Morning Rush	John Sullivan	NYT	11/9/99	Damaged swing bridge
All Welded Vertical Lift Span Replaces Barge Bridge	V. R. Gorham	CE	February 1948	VL 107′ repl pontoon, White City FL
A New Bridge for Long Island	—	NYT	1/26/86	used bascule bridge sent from Florida to New York
Army to Launch New Bridges	—	ENR	6/12/52	Pontoon military bridges
Arresting Abutment Shifting on a Bascule Bridge	Orrin H. Pilkey	ENR	5/19/32	Toledo dl basc foundation repair
Atlantic Beach Bridge Supported on 90 ft Piles weighing 30 Tons	Lloyd I. Monroe	CE	March 1952	Dbl l basc Reynolds Channel NY LI
Barge Takes Bridge for a Ride	—	CE	July 1988	Transport of bascule for LIRR Reynolds Channel
Bascule Bridge Machinery	Ken Avery	FDC	1994	Bascule machinery design and testing
Bascule Bridge Redux	—	ACE	August–September 1996	Noncounterweighted bascule @ Sheboygan, WI
Bascule Configuration	Lynn W. Biwer	FDC	1994	Bascule bridge alternatives
Bascule Counterweight Falls on Winnipeg Bridge	—	ENR	4/29/37	Strauss dbl lf hwy art cwt overhad type
Bascule Vets Declare Armistice	Rob McManamy	ENR	11/23/92	Michigan Avenue Bridge, Chicago
Battle for America's Bridges Gains Momentum	Walt Moore	CON	September 1983	Bridge repairs
Beefing up a Battered Bascule Bridge	—	PW	April 1980	Open grid deck replacement on a bascule bridge
Bids Sought on 3,750-Ft Highway Bridge in Virginia	—	CE	August 1949	
Biggest Bascule	Russ Swan	BD&E	January 1996	Erasmus Bridge, Rotterdam

*See key to periodical abbreviations at end of list.

Title	Author	Periodical*	Date	Description
Biloxi Bay Bottleneck Removed	—	ENR	4/5/62	Dl basc US 90 172′
Bridge Balancing	Lynn W. Biwer	FDC	1994	Trunnion bascule balancing
Bridge by Barge	—	LIRR	March 1988	Transport of Reynolds Channel bascule
Bridge Flips, Killing Workers	David Kohn	ENR	5/24/93	Daytona FL basc constr accident
Bridge Permit Application Guide	—	USCG	April 1995	Guide for new construction or modifications
Bridge Plan Enrages NJ Residents	Kerry McKean	BCCT	12/7/88	Opposition to repl of Burl Bris vl br
Bridge Pulls Tidewater Virginia Together	—	ENR	5/8/52	Dbl sw Coleman completed Yorktown VA
Bridges	Janice Rosenberg	CH	Fall–Winter 1990–91	Hist of Chi mov br
Bridge Vaults over Floods	Josef Sorkin	ENR	6/22/50	VL 3span cont rr br Blue Riv KC Mo
Bridge Work May Be Stadium Hit	Carl Campanile	NYP	3/05/99	Macombs Dam swing bridge work scheduled around NY Yankees season
British Open Kent Lift Bridge	—	ENR	7/14/60	Kent-Sheppey vlrrt 21 hwy 90′
Call It Bridgelock	Keith Moore	DN	June 7 1989	tower drive vertical lift span skewed, jammed
Catastrophic Gear Train Failure	—	ME	December 1992	Failure of Mich Ave Bascule
Cause of Bridge Failure at Soo Not Determined	—	ENR	10/30/41	RR dbl leaf basc failure
Chicago's Newest and Heaviest Bascule	—	CE	May 1947	New State Street bridge
Chicoutimi Swing Span Requires Huge Protection Pier	—	ENR	7/5/34	Sw br in Quebec-Saguenay River
City Bridge Was Falling Down as Pols Argued	Dan Janison	NYP	9/5/94	Damaged NYC swing bridge
Coleman Bridge	—	MSC	September 1998	Steel bridge prize winner—movable span
Columbus Drive Bascule Bridge	—	ENR	11/22/84	Steel bridge award winner, Chicago
Computerized Training of Bridge Operators	Douglas Kamm	FDC	1994	Attempted solution to training of operators of movable bridges
Design Chosen for New, 12 Lane Wilson Bridge	Alice Reid	WP	11/19/98	Selection of PTG for new dbl lf basc Potomac R Alexandria VA
The Design & Construction of Breydon Bridge–Great Yarmouth Western Bypass	B. Simpson	SE	1/19/88	Des & Constr of new ohd cwt sl basc

*See key to periodical abbreviations at end of list.

Title	Author	Periodical*	Date	Description
Design and Construction of a Bascule Bridge	Peter Sluszka	IBSE	1991	Lock St Repl
The Design and Construction of the Lock Street Bridge Replacement	Martin Kendall	IBC	6/12/89	Lock St Repl
Designing of the Draw Bridge	Georgy Stepanov	IBSE		Construction of a draw span
Design of Movable Bridges	G. M. Stepanov	SEI	1/1/91	Movable bridge practice in Russia
Deteriorating Bridges: Some Common Problems	Ralph Hudson	PW	May 1987	Discussion of Broadway bascule bridge, Portland, Oregon
Discussion and Rebuttal on the Hackensack Bascule Failure	Hardy Cross	ENR	11/14/29	Commentary, tech disc on Hack Collapse
Discussion and Rebuttal on the Hackensack Bascule Failure	P. W. Ott	ENR	11/14/29	Commentary, tech disc on Hack Collapse
Discussion and Rebuttal on the Hackensack Bascule Failure	O. E. Hovey	ENR	11/14/29	Commentary, tech disc on Hack Collapse
DOT: Homework Was Done	Mike Shaw	DCA	6/16/98	Options for bascule bridge in Sturgeon Bay, WI
Double Draw	—	ENR	7/1/48	GP Coleman Br prop @ Yorktown VA
Double Swing Bridge is First of Type	—	ENR	10/13/49	GP Coleman Yorktown VA
Draw Span of Arlington Memorial Bridge	John L. Nagle	TME	November–December 1930	Dl basc fake mas arch DC
Dutch Bridge Is Coast Route Link	—	ENR	7/12/62	Dl hwy basc 98′
Early Movable Bridges of Chicago	Donald N. Becker	CE	9/1/43	History of Chicago movable bridges
Een Knap Staaltje van Samenwerking	—	BMS	September–October 1996	New Movable bridges in the Netherlands
Eine Bewegliche Brucke uber den Lotsekanal in Hamburg-Harburg	J. Carl	SB	4/1/86	New single leaf heel trunnion bascule in Hamburg, Germany
Electrical Controls and Standards	Alfred M. Trotta	FSSC	1993	Florida standards
Electrical Equipment for Movable Bridges List 796	—	GE	1/3/55	Electrical power and control systems for movable bridges
Engineering Contracts and Specifications	Robert W. Abbett	JWS	1963	Contains section on movable bridges
Erecting 3800-Ton Bascules in the Arlington Bridge	R. S. Foulds	ENR	3/3/32	Arl mem dl basc Potomac

*See key to periodical abbreviations at end of list.

Title	Author	Periodical*	Date	Description
Fatal Accident on Drawbridge	—	ND	3/29/89	Driver killed ramming open bascule heel
Fifty-year History of Movable Bridge Constr I	Egbert R. Hardesty	AJC	September 1975	Hist of mvbl bridges I
Fifty-year History of Movable Bridge Constr II	Egbert R. Hardesty	AJC	September 1975	Hist of mvbl bridges II
Fifty-year History of Movable Bridge Constr III	Reece H. Wengenroth	AJC	September 1975	Hist of mvbl bridges III
Floating Bridge over Hood Canal, Washington	Michael J. Abrahams	ASCE	4/18/85	Repairs to retractile pontoon bridge
Future Bridge Location Stays Put	Joe Knaapen	DCA	6/19/98	Public input in bridge replacement, bascule at sturgeon Bay, WI
Gears Mesh	Dominic Sama	PI	4/25/78	Story of Earle Gear & Machine Co
George P. Coleman Bridge	—	CV	5/4/96	Commemorative brochure on swing bridge widening project
Getting a Landmark in Shape for Its 100th Birthday	Christopher Gray	NYT	5/21/89	Carroll St retractile bridge, Brooklyn
Good Looks Count in This Bridge	—	ENR	3/30/61	Woodrow Wilson Bac Br @ DC
Great Bridge Bridge Chesapeake, VA	—	TT	Spring 2002	New bascule bridge project
High Wide and Handsome	—	AB		S Heim Long Beach CA VL
Hydraulics Operate Bascule Span	—	ENR	5/2/63	Sl basc plymouth co mass—hyd
It's Driving History	William Bunch	ND	2/14/87	Long Island, New York road and bascules declared historic landmarks
Jack-Knife Draws Originated Nearly 100 Years Ago	J. P. Snow	ENR	1/21/32	New England type swing for Eatern RR
Japan to Bridge Vietnam's Mekong	—	ENR	8/5/65	VL near saigon 401′ span 2L hwy
Laboratory Report	W. M. Schwenke	BSC	4/7/82	Test report on vertical lift bridge counterweight ropes
Lack of Input in Span Plan Angers Bucks Officials	Patricia Wandling	BCCT	12/4/88	Opposition to repl of Burl Bris vl br
Lafayette Street Bridge: Reconstructing a Rolling Lift Bascule	Frederick Parkinson	IRB	10/12/88	In-kind replacement of 50 year old Scherzer rolling lift bascule
Lift Bridge Looks Like Sculpture	Rita Robinson	CE	April 1985	New Kaukauna WI vl span
Lift Bridge Provides 175 ft Clearance	—	ENR	4/15/48	Long Beach CA VL S Heim Bridge lifts 125 ft

*See key to periodical abbreviations at end of list.

Title	Author	Periodical*	Date	Description
Lift Span Erected by New Procedure	—	ENR	5/7/36	Erection of Triboro VL bridge
Lift Span to Link Island with British Mainland	—	ENR	4/17/58	VL span 120′ 21 Kent-Shepsey
Linking Brooklyn & Queens	—	MK	October 1985	Repl of Greenpoint Ave bascule
London Visitors to Get New View from Old Bridge	—	NYT	2/21/82	London's Tower Bridge rehabiitation
Long Bascule Replaces Swing Span at Lorain, Ohio	—	ENR	10/3/40	
Lorain Bascule Sets New Span Record	—	ENR	12/5/40	Hwy basc 333′ repl swing
Lowestoft Double Leaf Trunion Bascule Bridge	J. M. Erde	CEP	May 1993	Constr of repl for sw br
Machinery Drive Systems	Ken Avery	FSSC	1993	Machinery concerns for new and rehabilitated movable bridges
Mare Island Vertical-Lift Bridge Earthquake Resistant, NRC Rated	Richard W. Christie	HHC	Spring 1987	New VL span at Vallejo, CA
Millions Hinge on Draw-bridge Decision	Carl Campanile	NYP	11/1/98	New York City movable bridges
Minnesota Slip Pedestrian Bridge	David W. Krech	WIQ	1st Quarter 1993	Double leaf Dutch draw in Duluth
Movable Bridge Controls for Florida	Richard Newcomb	FDC	1994	Standardized control system
Myton Bridge, Hull, UK	D. C. C. Davis	IBSE		Cable stayed swing bridge
Navigational Clearance Requirements	—	DC	1955 (reprinted 1960)	Clearance requirements for highway and railroad bridges
Neubau der Reihersteig-klappbrucke im Ham-burger Hafen	R. Kindmann	BI	1985	New single leaf heel trunnion bascule in Hamburg, Germany
New Bascule Bridge Breaks Bottleneck	—	ENR	5/15/52	LI NY Long Beach dbl basc hwy opened $6m 153′
New Bascule Span Joins Rehabilitated Mate in Seattle	Stephen H. Daniels	ENR	10/12/98	First Avenue South
New Bedford Bridge Rehabilitation	—	—	1995	Mass swing bridge 93 yrs old
New Bridges Remove Bottle-necks on Boston Shore Highways	—	ENR	7/18/35	Mass bascules etc.
New Burlington Northern Bridge at Prescott, WI	D. V. Sartore	ARB	3/26/85	RR VL replaces swing
New Lift Span Will Rise over Canal and under the Tracks	Aileen Cho	ENR	6/30/97	Replacement for 9th St. bascule in Brooklyn

*See key to periodical abbreviations at end of list.

Title	Author	Periodical*	Date	Description
New York Central to Build New Harlem River Span	—	CE	June 1951	4 track VL bridge NYC
Non-Destructive Bridge Pin Testing	Frank L. Carroll	PW	January 1989	Ultrasonic testing—from IBC June 13 1988
North Jersey's Movable Bridges	Alan Tillotson	TR	July 1988	Railroad movable bridges of northern New Jersey
Operating Movable Bridges	C. P. Hamilton	ENR	11/18/37	GE electrical mov br dev—Buzzards Bary Br
Particularites et Evolution des Ponts Mobiles	Pierre Mehue	IBSE		Movable bridge developments over last 30 years
Plans for Harlem River Span of the Triborough Bridge	—	NYT	2/15/35	Triboro VL bridge
Prize Bridge Award: Movable Span	—	MSC	October 1996	8th Street Bridge over the Sheboygan River
Rebound of the Bascule Bridge	Patrick A. Cassity	CE	August 1996	Noncounterweighted bascule @ Sheboygan, WI
Recent Bascule Bridges of Canadian Design and Construction	R. S. Eadie	EJC	January 1935	Hwy bascules details—Canada
Record Double Leaf Bascule	—	CE	May 1947	Loraine OH 333′ US2 4 L completed
Reexamining the Roles of Engineers, Fabricators, and Inspectors in Welded Construction	Duane K. Miller	FDC	1994	Responsibilities of entities re: code requirements
Renovate or Replace	Joe Knaapen	DCA	6/16/98	Options for bascule bridge in Sturgeon Bay, WI
Renovate or Replace	Mike Shaw	DCA	6/16/98	Options for bascule bridge in Sturgeon Bay, WI
Replace Electrical Torque Control Drive System with an Automatic Hydraulic Speed control System	John A. Schultz	IRB	10/12/88	Hydraulic drive retrofit of existing movable bridges
Revitalizing 100 Year Old Movable Railroad Bridges	Charles M. Minervino	IBSE		Cos Cob, Walk & Devon bridges
Rise and Fall of Greenpoint Avenue Bridge	—	MEI	5/13/87	Bascule bridge replacement
Safety Color Code for Marking Physical Hazards	—	ANSI	1979	ANSI Z53.1-1979
Seattle Slews Long, Heavy Span	Peter Green	ENR	8/5/91	Concrete swing pivot
Seattle Swings Again	Rita Robison	CE	July 1992	Concrete double swing br @ Seattle WA
Seattle Swings Up and Wide	—	ENR	7/14/88	Concrete double swing br @ Seattle WA
Seven-lane Road Rises to Dodge Marine Traffic	Gerald Burns	ENR	4/16/87	Danziger VL bridge, New Orleans
Sheboygan Gets New Single-Leaf Bascule Span	Charles B. Schroeder	R&B	April 1995	Des & Constr of sheb hyd no cwt basc

*See key to periodical abbreviations at end of list.

Title	Author	Periodical*	Date	Description
Short Span Takes Long to Build	—	ENR	11/28/63	Broadway VL bridge, NYC
Simplicity Distinguishes Bascule Span	W. J. Deady	ENR	5/11/50	Ca sngl leaf basc—cheap—Mossdale, Joaquin River
Skew Bascule a Strange and Unique Structure	Wm. Guy Williams	ENR	6/2/32	Over D&R canal between Trenton & Princeton
Skew Bascule Design Puts Bridges Inside Budget Limit	—	ENR	3/13/52	Miami Canal, FL single leaf 45 degre skews
Skew Bascule Erection a Complicated Job	Joseph H. Townsend	ENR	6/2/32	Over D&R canal between Trenton & Princeton
Span of Time	William Gordon	SL	6/24/93	Reconstruction of Strauss heel trunnion bascule bridge
Span Plan Is Flooded by Doubts	Kerry McKean	BCCT	12/4/88	Public opposition to vertical lift bridge replacement
Split Span Aids Rehab	—	ENR	6/3/91	Reconstruction of Michigan Avenue bascule, Chicago
State Street Bridge in Chicago	—	ENR	12/2/48	210′ × 108′ wide 4200 T leaves nears compl
Structural and Mechanical Rehabilitation of Old Bascule Bridges	Leif Jonsen	IBSE		Case histories of bascule rehab
Suez Bridge Spans Wider Canal	—	ENR	3/11/65	ElFerdan swing twin 520′ ril
Swing Bridge's Fast Fix Saves Water, Road Jam	—	ENR	2/19/87	Ctr pivot pier repair NC swing br
Swing Down, Sweet Bascule	Rob McManamy	ENR	11/9/92	Michigan Avenue Bridge, Chicago
Swing Span Slews on Skew, Opens on Cue	—	BD&E	November 1997	New Danish swing span
Technical Memorandum 505-4	T. J. Hirsch	BPR	2/28/69	Testing of dragnet vehicle arresting system
Terminal Island Bridge Nears Completion	—	ENR	10/30/47	Long Beach CA VL S Heim Bridge
The ASB Bridge	John Hartman			History of KC vl bridge
The Bridges of New York	Sharon Reier	QP	1977	Discussion of Harlem River and other movable bridges in New York City
The Mississippi River Bridges at Clinton Iowa	Ronald D. Sims	NWL	Spring 1993	CNW Rwy swing br @ Clinton IA
The Movable Bridges of Chicago	—	COC	1983	History of Chicago movable bridges
The New Galata Bascule Bridge at Istanbul	Reiner Saul	IBSE		4 leaf bascule 80 m span

*See key to periodical abbreviations at end of list.

Title	Author	Periodical*	Date	Description
The New Manchester Road Bridge in the Port of London	R. E. West	ICE-P	February 1971	New British single leaf heel trunnion bascule
Trunnion Bearings—Alignment	Ken Avery	FDC	1994	Difficulties in aligning trunnion bearings on bascule bridges
Twin Double Leaf Bascules in Potomac River Bridge	John M. Duggan	CE	March 1962	Woodrow Wilson Br DC HNTB
Twin Swing Spans at Suez	S. Akerib	ENR	7/31/52	Twin rr sw bridge blt 1941 being repl
Two or Four? Two or Three?	Joe Knaapen	DCA	6/16/98	Options for bascule bridge in Sturgeon Bay, WI
Unexpected Skyward Tilt	—	ENR	9/28/92	Chi Mich Ave basc accident
Vertical Lift Bridges with Synchronous Motors	W. L. Six	ENR	10/5/33	MKT @ Mo Riv Boonville MO
Veterans Memorial Bridge	—	ENR	11/22/84	Steel bridge award winner, Kaukauna, WI
What Goes Up Does Not Always Come Down	John A. Schultz	IRB	10/12/88	Frozen bearing on vertical lift bridge
Wind Loads on Movable Bridges	Arie W. F. Reij	IBSE		Maximum operational wind loads
Winds Wreak Havoc on Pontoon Span with Troubled History	Stephen H. Daniels	ENR	3/15/99	
Woman Dies, Brother Hurt in Bridge Crash	Alvin E. Bessent	ND		Car strikes open bascule leaf
Woodrow Wilson Bridge Washington, D.C. Metro Area	—	TT	Spring 2002	Construction of new bascule bridge
Workhorse Bridge Overhauled	—	BR	Summer 1988	Redecking Lorain OH bascule
World's Longest Girder Lift Span	—	AB		Ward's Island VL NYC
Wounded Lift Bridge Gets Long-Distance Diagnoses	—	CE	August 1997	65-year old rolling bascule with deteriorated segmental and track girders
"You can highball right along"	—	TR	September 1975	Description of railroad vertical lift bridge collision
Zwei Neue Waagebalken Klappbrucken in Mannheim	G. Freudenberg	SB	February 1989	New bascule bridges in Manheim, Germany
Zugbruke bei Grossfehn in Ostfriesland	—			Single leaf dutch draw in Germany

*See key to periodical abbreviations at end of list.

Key to Periodical Abbreviations

Abbreviation	Publication/Publisher
AB	*American Bridge* (Ad)
ACE	*American Consulting Engineer*
AJC	American Society of Civil Engineers, *Journal of the Construction Division*
ANSI	American National Standards Institute
ARB	*AREA Bulletin*
ASCE	*American Society of Civil Engineers Metropolitan Section*
BCCT	*Bucks County Courier Times*
BD&E	*Bridge Design & Engineering*
BI	*Bauingenieur*
BMS	*Bouwen Met Staal*
BPR	U.S. Department of Transportation, Federal Highway Administration, Bureau of Public Roads
BR	*Bridges*
BSC	Bethlehem Steel Corporation
CE	*Civil Engineering* (US)
CEP	*Civil Engineering and Public Works Review*
CH	*Chicago History*
COC	City of Chicago
CON	*Construction Equipment*
CV	Commonwealth of Virginia
DC	United States Department of Commerce
DCA	Door County Advocate
DN	*Daily News* (New York)
EJC	*The Engineering Journal* (Canada)
ENR	*Engineering News Record*
FDC	Florida Design Conference
FSSC	4th Annual State Structures Conference, Orlando, Florida
GE	General Electric Company
HHC	*Hardesty & Hanover Centennial*
IBC	International Bridge Conference (Pittsburgh)
IBSE	International Association for Bridge & Structural Engineering
ICE-P	Institution of Civil Engineers (London), *Proceedings*
IRB	1988 International Road and Bridge Maintenance/Rehabilitation Conference and Exposition, Atlanta, Georgia
JWS	John Wiley & Sons
LIRR	*Long Island Rail Road News Briefs*
ME	*Mechanical Engineering*
MEI	Municipal Engineers Inspection
MK	*The eM-Kayan*
MSC	*Modern Steel Construction*
ND	*Newsday*
NWL	North Western Lines
NYP	*New York Post*
NYT	*New York Times*
PI	*Philadelphia Inquirer*
PW	*Public Works*
QP	Quadrant Press
R&B	*Roads & Bridges*
SB	*Stahlbau*
SE	*Structural Engineer*
SEI	Structural Engineering International
SL	*Star-Ledger* (Newark, NJ)
TBE	TAEBSE
TME	*The Military Engineer*
TR	*Trains Magazine*
TT	*Tidewater Times*
USCG	United States Coast Guard
WIQ	*Welding Innovation Quarterly*
WP	*Washington Post*

ARTICLES FROM BIENNIAL SYMPOSIA, HEAVY MOVABLE STRUCTURES

Author	Title
	First
Robert A. Richardson	Movable Bridge Electrification
Timothy H. Nance	Why Use Adjustable Frequency Control for Movable Bridge Drives
Lance V. Borden	Torque Characteristics of Wound Rotor Motors Used in Movable Bridge Drive Systems
Jim Berdine	Direct Current Bridge Drives
T. J. Haas	Microprocessor Controlled DC Adjustable Speed Drives for Movable Highway Bridges
Allen E. Boldt	Application Considerations for Adjustable Violtage AC SCR Movable Bridge Drives
Irwin C. Smiley	Automated Movable Bridge Drives
Joe M. Warren	A Perspective from the Manufacturing, Assembly and Testing of Movable Bridge Components
Thomas C. Glew	Comparison of AGMA and AASHTO Gear Ratings
Joseph J. Hosanna	The Columbus Drive Bascule Bridge and Recommended Modification to AASHTO Movable Bridge Specifications Related to Stress Concentrations in Keyways
Richard W. Christie	Air Motors for Emergency Span Operation
Bradford R. Hollingsworth	AGMA Requirements for Movable Bridge Speed Reducers
James M. Phillips	Bascule Bridge Locking Devices
Richard R. Ramsey	Protection of Steel Members in Existing Movable Bridge Structures
Larry Oline	Computer Aided Structural Analysis of the Hopkins Frame
William H. Mahoney	Single Leaf Bascule on SR40 at Astor, FL
Ulo S. Pessa	Repair of Counterweight Trunnions on Strauss Bascule Bridge
Ben G. Christopher	The New River Bridge Ft. Lauderdale, Florida
Michael A. Abrahams	Hood Canal Bridge A Floating Lift Draw Bridge
Thomas F. Mahoney	Proposed Concrete Swing Bridge
Gerd Schmitt	Advantages of Hydraulics for Movable Bridges
Michael A. Hanley	Hydraulic Systems for Movable Bridges
John A. Schultz	Development of Hydraulic Systems Designed from 1965 to the State of the Art Hydraulic Systems of Today
B. P. Yaskin	Miami Avenue over the Miami River
William H. Hamilton	Design Features of a Unique Swing Bridge Hydraulic and Control System
John K. Liu	Application of a Hydroviscous Drives to Movable Bridges
Lynn R. Causey	Traffic Gates and Barriers for Movable Bridges
Captain Al Cattalini	Coast Guard Regulations and Recommendations for Communication with Vessels
James R. Loepp	Land and Marine Traffic Management Systems for Movable Bridges
Robert J. Dodge	Solar Power Bridge Navigation Lights
B. P. Yaskin	AASHTO Design Criteria Applied to Machinery Bascule Bridges
B. P. Yaskin	AASHTO Design Criteria Applied to Hydraulic Bascule Bridges
Abba G. Lichtenstein	Rehabilitation of a Scherzer Bascule Bridge
Thomas J. Clancy	Rehabilitation of a Scherzer Bascule Bridge Mechanical–Electrical Systems
Lynn W. Biwer	Maintainability of a Movable Bridge
Jim Burleson	Record Keeping and Work Order Control
Edward M. Beck	Non-Destructive Testing in a Bridge Maintenance Program
William Bloom	Maintenance Procedures for Electrical Controls on Movable Bridges
K.C. Roberts	Paper Trail from Inspection to Repair
Harold Greasy	Microprocessor Controlled DC Adjustable Speed Drives for Movable Highway Bridges
Alan Butler	Application of Programmable Controllers for Automated Bridge Control
William J. Bender	Programmable Controller Communications to Remote I/O An Alternative to Communications Conductors in Submarine Cable?

Author	Title
Stephen Hewson	Overview of Bridge Control Systems Past, Present and Future
William Bowden	Design Considerations for Improved Operational Performance
C.Ward Yelverton	Drawbridge Control as Total System Design
William J. Wright	In-situ Repairs to Machinery of Movable Bridges
H. Everett Druge	Old Movable Bridge Recycled
David Shiner	Epoxy Resins Engineered Materials
Maurice J. Davis	Characteristics of Anchor Load Transfer to Structural Concrete
L.E. Malvern	Balancing of Trunnion-Type Bascule Bridges
Glenn M. Poche	The Application of Current Transducer Technology on Bascule Bridges as the Basis of a Computerized Predictive Maintenance Program
	Second
Gerd Schmitt	Moving Bridges
Erich Wirzberger	Comparison of Mechanical and Hydraulic Drive Systems
Ernst Herz	Shuaiba Maqal Road Bridge
Erich Wirzberger	Designing and Executing Hydraulic Powerpacks
Ton Huinink	The Hydraulic Cylinder
Dieter Lagging	Oil Hydraulic Drive System on the Herrenbrücke, Lubeck
L. Cypers	Bascule Bridges at the Berendrecht Lock in the Port of Antwerp, Belgium
Erich Wirzberger	Bascule Bridge Stade
L. Cypers	Movable Bridges in Belgium
Ernst Herz	Shatt-Al-Basrah Bridge a Swing Bridge with a Hydraulic Drive
Dieter Lagging	The Drives on the Moving Bridges at the Kruckau, Pinnau and Stor Barriers
Nelson Russell	Rehabilitation of the Control System for the Twin Bascule Bridges at Lake Ponchartrain Causeway
Randal Mayeaux	Low Tech Control System for Swing Span Bridges
John A. Schultz	Data Acquisition System for Movable Bridges
Robert P. Bloomquist	A Computerized Predictive Maintenance Program on a Trunnion Type Bascule Bridge
Frank C. Breeze	Lightning Surge Supression Fundamentals and Application
Dieter Wuensche	Properties and Applications of Electric Linear Actuators
Bill Arnold	Considerations for Maintenance for Movable Bridges. . . . Why Should I Protect My Investment?
William Bowden	Machinery and Motor Brakes: Limits and Applications
H. G. Murphy	Using VFD Technology to Improve Drawbridge Performance
Robert L. Cragg	Trunnion Bascule Bridge Static Stabilizing Systems
W. C. Adams	State of the Art Hydraulic Systems for Movable Bridges
Joseph J. Hosanna	Recommended Revisions to Unit Stresses in Machinery Parts
Michael Hanley	Inspection and Approval of Hydraulic Drive Machinery
James K. Merrill	Field Welding for Bridge Rehabilitation
John A. Schultz	Low Speed–High Torque Fluid Motor Drive for Movable Bridges
Nick E. Mpras	Recommendations for Temporary and Permanent Installations of Bridge Navigation Lights and Reflective Materials
David Shiner	Machinery Bases — Chemical Dependency
William J. Doherty	Failure Analysis of High Strength Bolt's Loss of Tension on Two Movable Bridges
Norman J. Feuer	Abnormal Movements of Bascule Bridges
Harland F. Ellis	Navigation Locks on the Mississippi
James J. Powers	The Columbus Drive Bascule Bridge
James M. Phillips	Hydraulic Cylinder Drives for Existing Bascule Bridges
Howard W. Lichus	Renovation of the Duluth Aerial Bridge
George V. Aseff	Corrosion Mechanisms in the Deterioration of Exposed Bridge Steel
Michael E. Looram	Use and Misuse of Galvanized Structural Bolts

Author	Title
Gene Gilmore	Lightweight Bridge & Sidewalk Decking
George Walter	An Advanced Concept in the Design of Large Unorthodox Structures
Neal H. Bettigole	Case Study; Substitution of Exodermic Deck for Concrete Filled Grid on a Vertical Lift Highway Bridge
Thomas J. Clancy	Rehabilitation of a Railroad Swing Bridge Under Traffic
James Anderson	Claims: Image From an Owner's Point of View
Frank McDonough	Bascule Bridge Delay Claims—Problems and Recommendations
Robert Shearer	Left Handed Bidding
Michael W. Doherty	Construction Control: A Method of Determining Traffic Alternatives
O.R. (Lee) Evans	Planning and Scheduling for Project Control (and Avoidance of Claims)
Jack Miller	U.S.Army and Marine Corps Development Programs for Assault and Tactical Bridges
Kenneth N. Smith	Inspection Manuals for Movable Bridges
W. J. (Buddy) Provost	Subject: Contract Estimating System
David W. Jacobs	Connecticut Moveable Railroad Bridge Rehabilitation Program
Herman S. Clark	Liability and Loss Prevention
	Third
John A. Schultz	Self Destruction of a Strauss Bascule Bridge
	Fourth
Bill Carson	Completing a Successful Automation Project
William Bowden	A 10 Year Retrospect: Relay vs PLC's
Hassan Eldessoky	Broadway Bridge Failure of Automatic Skew Correction System
Ron Hughes	AC Adjustable Voltage vs Adjustable Frequency Control for Movable Bridge Applications
Ron Hansen	Flux Vector AC Drives: Movable Bridges Applications
Roger k. Wiebusch	Remote Operation of Drawbridges; the Regulatory Viewpoint
Charles W. Hiatt	A Control System for a Unique Double Leaf All Concrete Swing Bridge The West Seattle Swing Bridge Control System
J. A. C. Wels	Key Technical Features of Hydraulic Cylinders in Movable Bridges and Gates
Andrzej Studenny	Hydraulic Lifting and Rotating System for West Seattle Swing Bridge
Thomas L. Huber	The Benefits of Using High Pressure Combined with High Volume In Hydraulic Systems
E. Allanson	Breydon Bridge, An Electro-Hydraulic Application in Operation
John A. Schultz	PLC's Provide Complete Automatic Control of Movable Spans
Parveen Gupta	Bridging the Niantic
Timothy C. Gresham	Trunnion Alignment on Bascule Bridges
Nicolae Dini	Importance of Lubricating/Flushing...Bascule Trunnions
Terry L. Koglin	Bridge Trunnions
Paul M. Bandlow	Thames River Bridge Counterweight Trunnion Repair
R. S. Pelczar	Evolution of AGMA Rating Methods
Edward J. Roland	Balancing of Movable Bridges: The Strain Gage Technique
Steve Allsopp	Shop Drawing, Review & Necessary Interaction for Successful Projects
William E. Nickoley	Movable Bridges vs Murphy's Law: Avoid Headlines by Establishing "Accountability Checkpoints"
Barton C. Bonney	Proactive Approach to PM
Win Welsch	What Amount of Stress Reduction Benefits Fatigue Life?
William E. Nyman	South Slough Bridge Erection of a New Bascule Adjacent to an Operational Swing Span
Thomas A. Skinner	Rehabilitation of the ASB Bridge in Kansas City, Missouri

Author	Title
R. J. Slattery	Rehabilitation of the Hanover Street Bridge
Ralph J. Eppeheimer	Replacement of Swing Span with Vertical Lift Span to Modify Railroad Bridge over the Red River Waterway
	Fifth
Michael J. Abrahams	Temporary Movable Bridges
James F. Allison	Bridge Drive Systems Incorporating Hydraulic Motors and Gear Reducers
Carl Angeloff	Overcoating Structural Steel Containing Lead Based Paints—An Economical Alternative to Total Removal and Containment
Karl L. Aschenbach	Narrow Entry Waterway Protection
Paul M. Blair	Emergency Replacement of Deformed, Crystallized Scherzer Tread Plates
James S. Borbas	Spherical Roller Bearing Design as Used on Movable Bridges
Robert L. Cragg	Observations and Comments: AASHTO Design Practices for Movable Bridge Operating Machinery
Phillip J. Dibb	The Application of Triple Modular Redundant (TMR) Programmable Logic Controls (PLC) to a Heavy Movable Structure
Nicolae Dini	Strain Gage Method for Clutch Adjustment of Vertical Lift Bridges
Bill Doherty	Speed Reducer: Failure Investigation and Analysis
Parveen Gupta	New Concepts of Hydraulically Operated Steel Structures
Michael S. Hershey	145th Street Bridge Repair
Ron Hughes	Mechanical and Electrical Redundancy Considerations for Movable Bridge Systems
Bruce A. Kaiser	Protection System Design for Equipment Reliability and Longevity
Jeffrey D. Keyt	Rolling Lift Repairs
Daniel H. Copeland	Grid Reinforced Concrete Bridge Deck Overlay Choices for Movable Bridges
Roger Hickman	The Design and Construction of Mutford Bridge and Associated Works
Peter Wusthof	The Bascule Bridge for Seville, a New Concept of Hydraulic Control for Bridges
Terry L. Koglin	Stabilization of Double Leaf Bascule Bridges
	Sixth
William H. Detweiler	Guidelines for Selecting Spherical Plain Bearings and Spherical Roller Bearings
David A. Winkel	Guidelines for Selection of Large sized Spherical Plain Bearings for Heavy Movable Structures
Lance V. Borden	Torque Characteristics of Wound-Rotor Motors, Revisited
Richard C. Kreppel	Rehabilitation of the Dorset Avenue Bridge over Inside Thorofare, Ventnor City, Atlantic County, New Jersey
Wayne Swafford	Case History 8th Street Bascule Bridge
David Nyarko	E D Sawyer's Legacy: Barrier Gate Rehabilitation
Eric T. Kelly	Span Lock Failure—A Case Study
Bill Killeen	Use of Temporary Acrow Panel Lift Bridge at the Tomlinson Bridge Project in New Haven, Connecticut
Tom Mahoney	First Avenue South Bascule Bridge, Seattle Washington
Paul M. Bandlow	Aerial Bridge Emergency Counterweight Sheave Assembly Repairs
Jeffrey S. Mancuso	Emergency Repair of a 7 Million Pound Swing Span (I Street Bridge Center Bearing Repairs)
Hugh V. Jamieson	Controlling Proportional Valve Amplifier Cards—Methods and Practice
Scott Pastorius	Understanding Thruster Technology and Performance Benefits
Douglas Godfrey	New Anti-Fretting Wear Test for Greases
Robert A. Bettigole	Exodermic Bridge Deck on Bascule Bridges
R. David Swift	Libby Street Bridge, Wallaceburg, Ontario, Canada
George C. Patton	The Temporary Bridge at 17th Street Causeway

Author	Title
Siamak Pourhamidi	Bascule Bridges—Functionality, Elegance and Durability
Richard J. Beaupre	Aesthetic Engineering—A Unique Solution for a Signature Bascule Bridge
Richard J. Slattery	The Replacement of the Ocean Avenue Bascule Bridge
Larry Welsh	Economical and Functional Steel Details
Paul M. Blair	Rehabilitation of a Bascule Bridge Without Interrruption of Marine or Vehicular Traffic
Mohammed N. Nasim	Rehabilitation of the Stratford Avenue Bridge over the Yellowmill Channel
Jeffrey S. Flanders	Drive System Replacement of the Tarpon Docks Bridge
Erich Wirzberger	Hydraulic Drive System for Bascule Bridges
Lou Wendel	Hydraulic Slewing Drives for the Coleman Swing Span Bridge
Parveen Gupta	Design Considerations for Bridge Hydraulics
Charles A. Simons	New Design Features and Guidelines for Hydraulic Cylinders
Andrzej Z. Studenny	Hydraulic Lifting System for SR 520 Bridge Seattle, Washington
E. Allanson	Unique Moving Bridge on the A29 Motorway LeHavre, France
Jim Phillips	Hydraulic Systems for Movable Bridges—Seminar
Raymond G. Hentschel	Flux Vector Drive Applications on a Movable Bridge
Rob Tannor	A Comparison of the 1988 AASHTO Electrical Requirements and the National Electric Code
John A. Schultz	A Century of Progress with Scherzer Rolling Lift Bascule Bridges Historical Development of Movable Bridges, Part II
John A. Schultz	The Selection and Evolution of the Chicago Type Trunnion Bascule Bridge Historical Development of Movable Bridges, Part III
Michael J. Abrahams	Ford Island Bridge Pearl Harbor, Hawaii
Michael J. Abrahams	The George P. Coleman Bridge Reconstruction Yorktown, Virginia
Charles M. Minervino	Development of a "Manual for Inspection, Evaluation and Maintenance of Movable Bridges"
Charles Birnstiel	Construction and Collapse of An Early Cable-Stayed Bridge With a Movable Span
B. Robert Tessiatore	106th Street Basule Emergency Repairs
	Seventh
Michael J. Abrahams	Suez Canal Swing Bridge
Michael J. Abrahams	Seismic Performance of Movable Bridges
Rob Tannor	Better Electrical Documents Allow Quicker Construction and Fewer Claims
Frank M. Marzella	Rehabilitation of the 1st Ave. South Bascule Bridge Seattle, Washington
Barbara Thayer	Technical Aspects of Control Houses on Movable Bridges
Robert J. Tosolt	Wire Rope Replacement at Broadway Bridge
Darryl E. Cain	Conversion of a Railroad Center Pivot Swing Span to Rim Bearing Under Traffic
Michael C. Harrison	Rehabilitation of a Railroad Swing Span under $800,000
Jeffrey W. Newman	The Use of Load Sharing Devices for Tower Drive Vertical Lift Bridges
Carl Angeloff	Coating Systems for Marine and Immersion Applications
Jeffrey S. Flanders	Hydraulic Drive Solution for the Popps Ferry Bridge
Janine Krempa	Hydraulic System for a Large Movable Bridge
Erich Wirzberger	Hydraulic Drives for Bridges Considering the New DIN Standard 19704
Charles A. Simons	Design of Optimized Seals for Leak Free Hydraulic Cylinders
A. C. Johnson	A Quickly Deployable Emergency Bridge Lift System
Dongming White	Development of the Movable Bridge Balance Program
Raymond E. Erbe	AC Motors Replace Hydraulics, DC Motors for Heavy Structures
George C. Patton	A Structurally Efficient and Aesthetic Bascule Leaf Design for the 17th Street Causeway Bridge
James M. Phillips	Single Leaf—Integral Trunnion Bridge
Thomas A. Cherukara	Mechanical Codes and Standards for Movable Bridge Design and Construction

Author	Title
William Bowden	The Use and Abuse of Bypass Switches
G. Alan Klevens	Movable Bridge Rehabilitation in Florida: Four Case Studies
R. Wayne McLennon	Replacement of the Ocean Avenue Bascule Bridge/Boynton Beach Bridge
Donald L. Miller	An Overview of NCHRP Project 12-44—Recommended Specifications for Movable Highway Bridges
John W. McConnell	Rehabilitation of the Arthur Kill Lift Bridge
Nicholas Altebrando	Rejuvenating Movable Bridges
Anthony J. Saeli	Applications of Bearings in Heavy Movable Strutures
Bernd Kottwitz	Spherical Plain Bearings for the World's Largest Bascule
Gary L. Peters	Replacement of a Combined Pneumatic/Electric Control System to an All-Electric Control System with PLC on a Swing Span
Patrick S. McCurdy	Transient Lightning Protection of Outdoor Measurement and Control Systems
Alan D. Fisher	Applications of Hydraulic Jacking Systems or How to Move Heavy Structures including Heavy Movable Structures
Stephen A. Mikucki	Repairs to a Scherzer Rolling Lift Bridge
Richard J. Beaupre	The Evolution, Analysis, Design and Construction of the Bascule Pier for the 17th Street Causeway Bridge
Adil H. Khan	Inspection/Testing of Structural Steel and Machinery Components
Matthew Bodziony	Aerial Cables/Droop Cables
Nick Mpras	USCG Bridge Administration Program
	Ninth
William E. Nickoley	Milwaukee's New Knapp Street Towerless Vertical lift Bridge

The symposia titles are

First Biennial Symposium and Exhibition on Movable Bridge Design and Technology, Talahassee, Florida, November 4, 1985.

Second Biennial Movable Bridge Symposium, St. Petersburg Beach, Florida, November 10, 1987.

Heavy Movable Structures, Third Biennial Symposium, St. Petersburg, Florida, November 12, 1990.

Heavy Movable Structures, Fourth Biennial Symposium, Ft. Lauderdale, Florida, November 10, 1992.

Heavy Movable Structures, Fifth Biennial Symposium, Clearwater Beach, Florida, November 2, 1994.

Heavy Movable Structures, Sixth Biennial Symposium, Clearwater Beach, Florida, October 30, 1996.

Heavy Movable Structures, Seventh Biennial Symposium, Lake Buena Vista, Florida, November 4, 1998.

Heavy Movable Structures, Ninth Biennial Symposium, Daytona Beach, Florida, October 22, 2002.

GLOSSARY

AASHTO American Association of State Highway and Transportation Officials. The official coordinating agency for the highway/transportation departments of the 50 states, Puerto Rico, and the District of Columbia; publisher of standard manuals and specifications on the design, maintenance, and inspection of fixed and movable bridges.

Adiabatic process A thermodynamic term indicating an absence of exchange of heat. An adiabatic compression would reduce the volume of a quantity of a gas, such as air, without allowing the heat developed to escape. The temperature of the air would thus increase, further raising its pressure from the change that would occur with a decrease in volume without a change in temperature.

Antifriction Bearing A support for a rotating shaft that uses rotating elements to provide free rotation of the shaft without the occurrence of sliding between the shaft and its basic support.

AREMA American Railway Engineering and Maintenance-of-Way Association (formerly AREA, American Railway Engineering Association), the coordinating body for U.S. and Canadian railroads, mostly private, for engineering functions regarding track and structures.

Bascule A movable bridge that pivots about a horizontal axis to open.

Center locks Connections at the joint between the two leaves of a double leaf movable bridge, generally used on a bascule bridge but sometimes on a swing bridge. See also *shear locks*.

Center of gravity The point in space, in reference to a movable bridge span and its counterweight or counterweights, or the combination of these plus any other moving parts of the bridge, at which the sum of the products of the points of mass times their distance to the center of gravity, is zero. The center of gravity can be computed in any Cartesian axis or combination of axes. Usually the center of gravity in a plane normal to the rotational axis is significant for bascule or swing bridges.

Dead load The moving weight of a bridge, consisting of the superstructure, deck, and all other parts that are supported on the moving span. Also the self-weight of any bridge structure.

Drive train The system of gears and shafts or other machinery between the prime mover and the final drive point, such as a rack pinion.

Environmental load The force of wind, ice, or other natural phenomena on a bridge structure.

FHWA Federal Highway Administration, the U.S. government agency, within the Department of Transportation, responsible for disbursing federal tax money for highway construction purposes and for seeing to it that such moneys expended by state and local authorities are used in accordance with U.S. law regarding roadway standards and usage.

Floating shaft A part of a power train consisting of a component designed to carry no external load except torsion, such as a shaft, that is supported, not on bearings, but by means of flexible couplings at each end.

Guide A device to control the movement of the body to which it is attached. Span guides and counterweight guides are mounted on the spans and counterweights of vertical lift bridges, respectively, to keep these bodies on a linear vertical path when traveling up and down as the bridge opens and closes.

Guide rail A linear device to assist a bridge component in traveling on its intended path. A guide rail usually mates with a guide, allowing one degree of freedom of movement.

Journal The surface of a shaft that is supported in a bearing; taken to refer only to those types of bearings that work by sliding action, called "friction" bearings or "journal" bearings.

Live load The traffic loading on a bridge.

Miter rail A piece of railroad rail cut so that it can easily be separated and rejoined, such as for providing end connections at movable bridges.

Overhaul (1) To rebuild, such as a piece of mechanical equipment, or an electric motor; (2) to backdrive. A bascule bridge is said to be overhauling when it moves without the aid of its drive motor, such as when lowering with the wind pushing down on the deck, or when its imbalance is such that gravity is forcing the bridge down.

Prime mover Engine or motor that provides the input force or torque to drive the machinery of a movable bridge or other device.

Races The inner and outer housings of an antifriction bearing that bear against the rotating elements of the bearing.

Rall bascule Bascule bridge supported on rollers supported on tracks in such a way that the bascule leaf withdraws away from the navigation channel as the leaf swings open.

Retractile Movable bridge that is removed from the navigation channel by withdrawing in a straight line to the side of the channel. The bridge may have a single leaf, withdrawing to one side, or a double leaf, with two sections withdrawing away from each other to opposite sides of the navigation channel.

Rim bearing A support for a swing bridge consisting of a large antifriction thrust bearing, with the lower race mounted on the pivot pier and the upper race consisting of a cylindrical drum supporting the swing bridge superstructure, the lower surface of the drum consisting of a rim resting on the upper surface of the rollers.

Rolling lift Bascule bridge that rotates to open position by rocking back on tread plates supporting curved surfaces on the bottom of the bascule girders, so that as the bascule leaf rotates, it also translates along the longitudinal axis of the bridge. Also a Rall type bascule.

Shear locks Removable connection keys or pins that join two sections of a movable bridge, preventing relative vertical motion between the spans normal to the axis of the key or pin by transferring forces in shear from one span to the other.

Span drive Vertical lift bridge with the drive machinery located on the moving span.

Span-tower drive (or tower-span drive) Vertical lift bridge with drive machinery located on a platform suspended between the tower tops.

Swing Movable bridge that pivots on a vertical axis to clear the navigation channel for traffic.

Tower drive Vertical lift bridge with the drive machinery located at the tops of the support towers.

Trunnion Horizontal shaft about the axis of which a member pivots. It may be fixed to the member or fixed to its supports.

Vertical lift Movable bridge that lifts without tilting to provide sufficient clearance over the navigation channel for marine traffic.

APPENDIX I

UNITED STATES PATENTS ON MOVABLE BRIDGES

Number	Patentee	U.S.P.O. Class.*	Bridge Type	Description of Invention	Date
1,004	Ring, A.	14/34	Swing	Bobtail	1838
4,926	Winkler, G.R.	14/34	Swing	Reissue, see 86,117	1872
5,372	Schneider, L., & Montgomery, J.A.	14/34	Swing	Reissue, see 29,917	1873
5,997	Ross, J.	14/32	Swing	Bobtail, shear pole	1849
10,250	Champion, S.	14/43	Retractile	Erection device—launcher	1853
10,527	Woodruff, J.D. & Butterworth, J.H.	14/42	Vertical lift	Unlift, submersible span	1854
12,570	Rall, T.	14/40	Bascule	Reissue, see 669,348	
12,952	King, J.N.	14/34	Swing	Spring operated 2-arm	1855
13,258	Gamble, J.K.	14/47	Rail end	Safety derail	1855
14,928	Proctor, N.B.	14/47	Pontoon	Retractile	1856
15,873	Anderson, B.G.	14/32	Swing	Double bobtail	1856
18,196	Earle, C.H.	14/36	Swing	Center pivot	1857
26,156	Berry, D.	14/44	Retractile	Boat operated	1859
28,148	Bovey, G.C.	14/44	Retractile	Boat operated	1860
29,917	Schneider, L., & Montgomery, J.A.	14/34	Swing	Bobtail, rope operated, self-acting	1860

*See key at end.

Number	Patentee	U.S.P.O. Class.*	Bridge Type	Description of Invention	Date
31,222	Bayley, G.W.R., & Nelson, T.W.	14/32	Swing	One arm over land, eccentric pivot	1861
33,606	Selser, J.	14/34	Swing	Bobtail, boat operated	1861
35,765	Koch, A.	14/34	Swing	Self acting	1862
45,051	King, Z.	14/32	Swing	Cable-stayed	1864
56,777	Mitchell, D.A.	14/43	Retractile	Suspended	1866
58,492	Smith, A.F.	14/49		Signal	1866
63,507	Grotz, R.	14/56	Gate	Automatic, counterweighted	1867
76,542	Speakman, T.S.	14/54	Retractile	Double acting with automatic gates	1868
76,560	Valliquette, J.B.	14/54	Swing	Automatic gate	1868
79,768	Marshall, C.K.	14/43	Retractile	Suspended	1868
81,383	Lehmann, J.	14/66	Swing	Automatic gate	1868
86,117	Winkler, G.R.	14/34	Swing	Bobtail, rope operated, boat operated	1869
89,201	Cogsil, O.	14/31	Ferry	Underwater track	1869
94,529	Trowbridge, W.P.	14/43	Retractile	Cable stay	1869
94,534	Weide, A.	14/52	Swing	Automatic gate	1869
95,821	May, G.A.	14/54	Swing	Gate	1869
97,044	Cass, H.W.	14/32	Swing	Truss	1869
97,855	Anderson, L.	14/59	Swing	Automatic gate	1869
100,065	Pratt, T.W.	14/31	Retractile	Diagonal on tracks	1870
100,480	Wilcke, J., & Ellenbogen, M.	14/62	Swing	Automatic gate	1870
100,910	MacNeale, N.	14/42	Vertical lift	Hydraulically operated	1870
104,086	Wermerskirchen, H.	14/65	Swing	Automatic gate	1870
107,119	Sturges, J.D.	14/60	Swing	Automatic gate	1870
109,520	Kirsch, M.	14/65	Swing	Automatic gate	1870
113,916	Pairan, F.	14/69	Swing	Automatic gate	1871
114,414	Coyne, E.R.	14/62	Swing	Automatic gate	1871
121,892	Northway, A.D.	14/61	Swing	Automatic gate	1871
124,072	Ladewig, R., & Rosenberg, A.	14/58	Swing	Automatic gate	1872
128,115	Clarke., T.C.	14/35	Swing	End-rail lift & centering device	1872
129,374	Thomas, W.E.	14/42	Vertical lift	Water operated	1872
129,638	Whitney, H.	14/36	Bascule	Boat puller	1872
132,038	Winkler, G.R.	14/34	Swing	Roller pivot	1872
134,338	Whipple, S.	14/42	Vertical lift	Lifting deck	1872
135,776	Clarke, T.C.	14/35	Swing	End lift pivoted arm roller	1873

*See key at end.

Number	Patentee	U.S.P.O. Class.*	Bridge Type	Description of Invention	Date
136,278	Swartz, S.	14/38	Bascule	Heel trunnion	1873
146,843	Sicklesteel, G.	14/43	Retractile	Locomotive powered	1874
150,949	Gasser, L., & Severin, C.	14/54	Swing	Automatic gate	1874
151,741	Baier, F.	14/60	Swing	Automatic gate	1874
151,985	Kaufmann, C.H.	14/56	Swing	Automatic gate	1874
153,729	Soulerin, L.	14/42	Vertical lift	Vertical lowering—under water	1874
154,028	Ellingson, J.	14/59	Swing	Automatic gate	1874
154,684	Ladwig, J.	14/67	Swing	Automatic gate	1874
155,550	Smith, R.W.	14/35	Swing	Turntable drive	1874
162,576	Post, A.J.	14/42	Vertical lift	Rope drive from one tower	1875
164,708	Atkins, C.J.	14/35	Swing	End latch	1875
171,003	Esche, C.	14/52	Swing	Gate	1875
172,204	Swartz, S.	14/38	Bascule		1876
173,253	Adams, M. B., & Krause, F. L.	14/36	Bascule		1876
173,507	Simon, L.	14/60		Gate	1876
173,508	Simon, L.	14/57		Gate	1876
174,307	Schwennesen, J.	14/64	Swing	Automatic gate	1876
175,150	Pfautz, D.M.	14/45	Retractile	Scissors extension "lazy tongs"	1876
180,491	Moody, G.	14/39	Bascule	Drop leaf non-counterweighted	1876
181,409	Clark, J.L., & Standfield, J.	14/42	Vertical lift	Hydraulic lifting system	1876
183,111	Betser, B.F.	14/58	Swing	Automatic gate	1876
185,206	Babbitt, D.	14/36		Gate—farm	1876
186,971	Arndt, D.	14/60	Swing	Automatic gate	1877
189,320	Mounsdon, R.C.	14/64	Swing	Automatic gate	1877
193,825	McGrath, M.	14/57	Swing	Automatic gate	1877
195,160	Perry, O.H., & Watson, E.P.	14/49	Swing	Lock and signal	1877
197,816	Behrens, N.	14/66	Swing	Automatic gate	1877
198,449	Buchanan, W.	14/49	Swing	Operating device w signal	1877
206,046	Sherman, A.R.	14/52	Swing	Barrier gate	1878
202,526	Douglas, W.O.	14/12	Swing	Truss suitable for	1878
202,828	Johnston, E.H.	14/47	Swing	Sliding rail lock	1878
203,296	Schaefer, P.	14/50		Gate	1878
203,370	Pringle, W.A.	14/49	Swing	Automatic gate	1878
214,530	Snyder, H.F.	14/34	Swing	Boat operated—center support, latch	1879

*See key at end.

Number	Patentee	U.S.P.O. Class.*	Bridge Type	Description of Invention	Date
225,535	Roehr, F., & Jeyte, E.A.	14/66	Swing	Automatic gate	1880
225,775	Whitmore, E.B.	14/42	Vertical lift	Rope operated	1880
227,125	Petersen, A.F.	14/52	Swing	Hydraulically lifted gate	1880
230,203	Sherman, A.R.	14/52	Swing	Barrier gate	1880
250,027	Wible, E.A.	14/33	Swing	Split swing cable stay	1881
254,627	Edwards, J., & Kelly, J.R.F.	14/33	Swing	Drive, clutch, end lifts and locks	1882
255,831	Zurcher, M.A.	14/35	Swing	Lifts ends from center-lift truss	1882
298,647	Williams, J.N.	14/49	Swing	Signal and gate	1884
300,326	Ostertag, J.	14/58	Swing	End lift and automatic gate	1884
301,751	Philip, J.S.	14/66	Swing	Automatic gate	1884
304,690	Barnard, T.H.	14/49	Swing	Automatic gate	1884
305,795	Cooper, T.	14/33	Swing	Rim bearing	1884
316,597	Barnett, F.	14/32	Swing	Double series swing	1885
322,708	Groves, J.	14/50	Bascule	Automatic gate	1885
322,787	Bullock, W.D.	14/52		Gate	1885
323,711	Mershon, G.W.	14/35	Swing	Rail joint	1885
325,472	Anthony, M.O.	14/52	Swing	Gate	1885
326,965	Fisk, D.B.	14/49	Swing	Automatic gate and signal	1885
328,186	Carpenter, J.H.	14/36		Gate	1885
332,058	Daigle, M.	14/43	Retractile	Cable stay	1885
332,899	Hawks, G.E.	14/53	Swing	Automatic gate	1885
333,865	Klopp, H.F.	14/54	Swing	Automatic gate	1886
334,084	Morton, W.S.	14/49	Swing	Automatic gate	1886
337,089	Sensiba, F.W.	14/50		Gate	1886
341,979	Winkler, P.	14/53	Swing	Gate	1886
343,322	Johnson, J.G.	14/49	Swing	Automatic gate and signal	1886
343,377	Kandeler, C.F.T.	14/35	Swing	End lift	1886
344,817	Dammeyer, H.	14/49	Swing	Signal	1886
344,855	Petru, E., & Zidek, J.	14/68	Swing	Automatic gate	1886
345,150	Kourian, M.	14/49	Swing	Automatic gate	1886
346,724	Chandler, T.P.	14/31	Vertical lift	Cable—slack underwater	1886
352,970	Class, J.F.	14/42	Swing	Turntable	1886
354,105	Deveraux, W.	14/51	Swing	Gate	1886
357,638	Smith, R.W.	14/34	Swing	Rim bearing	1887
358,201	Johnston, L.J.	14/49	Swing	Signal	1887
360,347	Haynes, W.H.	14/31	Bascule	Cable stay in truss	1887

*See key at end.

Number	Patentee	U.S.P.O. Class.*	Bridge Type	Description of Invention	Date
361,420	Johnson, H.	14/49	Swing	Signal connection	1887
365,818	Johnston, L.J.	14/49	Swing	Automatic signal	1887
370,669	Wheeler, M.	14/51	Swing	Gate	1887
373,130	Cordrey, C.K.	14/53	Swing	Automatic gate	1887
374,435	Von der Muhlen, C.	14/60	Swing	Automatic gate	1887
376,225	Balston, O.F.	14/43	Retractile	Truss	1888
379,861	Cornell, J.M.	14/38	Bascule	Dock bridge	1888
381,584	Sawyier, R.A.	14/33	Swing	Rim bearing	1888
383,880	Harmon, W.	14/37	Bascule	Heel trunnion double leaf cross conn.	1888
386,154	Myre, M. & Christofferson, C.	14/54	Swing	Automatic gate	1888
389,116	Stapleton, P.W. & Carmody, M.L.	14/35	Swing	Screw end lift	1888
389,980	Sawyier, R.A.	14/33	Swing	Rim bearing	1888
390,618	Riding, W.	14/53		Railway crossing gate	1888
393,557	Hambay, J.T.	14/49	Swing	Signal coupling	1888
400,672	Haney, W.	14/53	Swing	Automatic gate	1889
403,549	Orford, J.M.	14/33	Swing	Electrical connection	1889
405,042	Long, M.H.	14/49	Swing	Signal	1889
407,967	Schlesser, N.	14/49	Swing	Automatic gate	1889
408,370	Brothwell, C.R.	14/35	Swing	Rim bearing	1889
409,700	Pegram, G.H.	14/33	Swing	Double	1889
417,293	Turner, G.	14/54	Swing	Automatic gate	1889
417,481	Friar, L.M.	14/42	Vertical lift	Connecting stair-ways when open	1889
418,027	Hoyt, B.R., & Fracher, L.	14/52	Swing	Gate	1889
428,079	Day, P.H.	14/49	Swing	Gate	1890
431,101	Alden, J.F.	14/42	Vertical lift		1890
431,838	Brothwell, C.R.	14/51	Swing	Gate	1890
432,016	Buel, A.W.	14/51	Swing	Gate	1890
433,206	Madsen, C.	14/52	Swing	Gate	1890
433,328	Crafts, G.H.	14/43	Retractile	Rope operatede	1890
436,570	Theis, W.	14/53	Swing	Automatic gate	1890
437,727	Smith, T.D.	14/36		Gate	1890
437,937	Reynolds, H.H.	14/35	Swing	Automatic gate	1890
439,063	Hambay, J.T.	14/35	Swing	Rail lock & detector	1890
440,368	Ratto, G.R.	14/53	Swing	Automatic gate	1890
442,755	Mills, B.	14/53		Railway crossing gate	1890
442,847	Weidenmayer, G.A.	14/36	Bascule		1890
447,636	Dewey, H.E.	14/53	Swing	Automatic gate	1891

*See key at end.

Number	Patentee	U.S.P.O. Class.*	Bridge Type	Description of Invention	Date
451,963	Wallich, J.C.	14/51		Gate	1891
452,630	Andrews, D.H.	14/43	Retractile	Lifting approach	1891
453,244	Madsen, C.	14/52	Swing	Gate	1891
460,797	Himmes, M.	14/60	Swing	Automatic gate	1891
468,813	Metzner, W.C., & Buschick, G.E.	14/33	Swing	Rim bearing for cable railway	1892
469,314	Clark, G.B.	14/42	Vertical lift	Adjusting coupling	1892
469,321	Scheubeck, J.	14/58	Swing	Automatic gate	1892
474,125	Haynes, W.	14/54	Swing	Automatic gate	1892
479,858	Thillman, N.W. & Firne, C.W.	14/41	Swing	Turntable	1892
482,017	Loomis, H.A.	14/32	Swing	Cable stay	1892
484,470	Smith, A.J.	14/51	Swing	Gate	1892
485,662	Scace, J.B.	14/32	Swing	Wire support	1892
486,552	Zimmerman, C.A.	14/43	Retractile	Ramp approach	1892
486,738	Rohlf, H.J.	14/52	Swing	Automatic gate	1892
486,793	Philbrick, F.A.	14/51	Swing	Gate	1892
486,843	Anderson, A.G.	14/53	Swing	Automatic gate	1892
494,796	Roessing, H.	14/50	Swing	Automatic gate	1893
496,074	Thompson, G.H.	14/36	Bascule		1893
500,290	Napion, L.	14/53	Swing	Gate	1893
500,633	Turner, C.A.P., & Warner, P.A.	14/35	Swing	End lifts	1893
501,516	Lagerquist, A.J.	14/54	Swing	Automatic gate	1893
502,779	Simon, S. & Pletka, J.	14/51	Swing	Gate	1893
503,273	Turner, C.A.P., & Warner, P.A.	14/35	Swing	Truss hinge end lift	1893
503,377	Lamont, R.P.	14/37	Bascule	Counterbalance	1893
503,378	Lamont, R.P.	14/37	Bascule	Driving mechanism	1893
506,571	Waddell, J.A.L.	14/42	Vertical lift	S. Halsted Street	1893
508,929	Fontaine, E.	14/42	Rsp	Winter bridge	1893
511,713	Scherzer, W.	14/39	Bascule	Rolling lift	1893
514,754	Jessup, N.C.	14/43	Retractile	Ramp approach	1894
516,121	Ryan, G.F.	14/51	Swing	Automatic gate	1894
516,179	Barndt, H.F.	14/52	Swing	Gate	1894
517,809	Schinke, M.G.	14/39	Bascule		1894
518,111	Moore, B.	14/66	Swing	Automatic gate	1894
530,234	Mershon, W.R.	14/47	Swing	Rail joints	1894
530,348	O'Brien & Wilson	14/53	Swing	Gate	1894
534,704	Worden, B.L.	14/37	Bascule	Hinged-leaf non cwted dbl leaf	1895
535,831	Jessup, N.C.	14/43	Retractile	Rack operated	1895
536,129	Clausen, W.F.	14/53	Swing	Automatic gate	1895
536,313	Worden, B.L.	14/39	Bascule		1895

*See key at end.

Number	Patentee	U.S.P.O. Class.*	Bridge Type	Description of Invention	Date
541,860	Lesna, A.	14/60	Swing	Automatic gate	1895
544,733	Lamont, R.P.	14/36	Bascule		1895
551,004	Schinke, M.G.	14/39	Bascule		1895
554,390	Jennings, E.B.	14/41	Bascule	Dbl leaf arch pin conn lwr chord at center	1896
554,767	Brooke, E.F., & Trott, J.S.	14/52	Swing	Air operated gate and signal	1896
557,106	Clark, R.P.	14/53	Swing	Automatic gate	1896
558,192	Mase, F.W.	14/53		Gate	1896
558,202	McNicol, J.A.	14/35	Swing	Lifts ends from center-lift truss	1896
561,375	Coup, J.	14/49	Swing	Signal	1896
561,671	Kroehnke, P.J.	14/48	Swing	Bridge guard	1896
562,805	Coup, J.	14/48	Swing	Automatic gate	1896
563,019	Condon, W.F.	14/52	Swing	End rail	1896
564,164	Shaw, E.S.	14/38	Bascule	Simple trunnion	1896
566,960	Andrew, T.H.	14/48	Swing	Automatic gate	1896
567,875	Aston, W.E.	14/43	Retractile	Ramp approach	1896
570,029	Lavey, C.J.	14/52	Swing	Gate	1896
572,318	Smith, F.M.	14/51	Swing	Gate	1896
572,971	Dowd, F.	14/51	Swing	Gate & derail	1896
577,443	Morse, C.M., & Sylvester, F.M.	14/43	Retractile	End wedged	1897
587,685	Filiatreault, W.	14/60	Swing	Guard gate	1897
587,926	Breithaupt, W.H.	14/36	Bascule		1897
590,787	Brown, T.E.	14/36	Bascule	Rope CWT conn.	1897
591,076	Gerdau, B.	14/32		Fluid bearing	1897
597,221	Pflanz, J.	14/43	Retractile	Removable approach	1898
598,012	Sampson, W.L.	14/47	Retractile	Tilting approach	1898
598,167	Waddell, M.	14/39	Bascule		1898
598,168	Waddell, M.	14/39	Bascule		1898
602,460	Mason, E.W.	14/48	Swing	Automatic gate	1898
615,824	Edwards, J.	14/47	Swing	End lift	1898
617,201	Strobel, C.L.	14/43	Retractile	Lifted span	1899
617,606	Sampson, W.L.	14/46	Retractile	With center lock	1899
620,116	Fusch, W.A.	14/51	Swing	Automatic gate	1899
620,500	Rohrbach, G.	14/47	Swing	Rail lock	1899
621,466	Waddell, M.	14/38	Bascule		1899
622,685	Hoch, F.	14/31	Swing	Deck slots for cables	1899
625,097	Edwards, J.	14/35	Swing	Apron	1899
626,970	Cowing, J.P.	14/51	Swing	Automatic gate	1899
630,879	Franson, C.F., & Wilmann, E.	14/43	Retractile	Tilting super-structure	1899
632,985	Brayton, W.L.	14/38	Bascule	Dbl leaf ctr lock	1899

*See key at end.

Number	Patentee	U.S.P.O. Class.*	Bridge Type	Description of Invention	Date
633,811	Cowing, J.P.	14/38	Bascule		1899
634,717	Chesman, H.	14/52	Swing	Gate	1899
635,394	Ruthenberg, A.	14/42	Vertical lift	Rack driven	1899
637,050	Waddell, M.	14/40	Bascule		1899
638,633	Grover, G.E.	14/52	Swing	Automatic gate	1899
644,405	Cowing, J.P.	14/38	Bascule		1900
648,447	Vent, F.G.	14/40	Bascule		1900
652,019	Edwards, J.	14/35	Swing	Toggle end lift	1900
652,020	Edwards, J.	14/33	Swing	Drive clutch	1900
652,201	Vent, F.G.	14/40	Bascule	Rocking bascule bridge	1900
653,729	Cowing, J.P.	14/51	Swing	Automatic gate	1900
657,122	La Pointe, F.	14/39	Bascule		1900
660,827	Waddell, M.	14/36	Bascule		1900
661,113	Waddell, M.	14/38	Bascule		1900
663,484	Bevans, T.R.	14/43	Retractile	Double leaf	1900
665,405	Cowing, J.P.	14/38	Bascule		1901
668,232	Strauss, J.B.	14/38	Bascule	Hanging CWT	1901
669,348	Rall, T.	14/40	Bascule	Rolling lift (reissued)	1901
672,848	Cowing, J.P.	14/38	Bascule	Chi type w tail lock	1901
673,923	Page, J.W.	14/36	Bascule	Basc w rack on top chord	1901
678,085	Wolfe, A.C.	14/33	Swing	Screw pivot-lift	1901
683,261	Einert, E.	14/66	Swing	Automatic gate	1901
683,627	Vent, F.G.	14/41	Bascule	DL basc rear rack w ctr lock	1901
683,811	Cowing, J.P.	14/38	Bascule		
685,767	Keller, C.L.	14/41	Bascule		1901
685,768	Keller, C.L.	14/41	Bascule	Ctr lock	1901
689,856	Cummings, E.D.	14/38	Bascule	Precurso ABT type dbl lf w ctr lock	1901
691,035	Wilkins, J.D.	14/36	Bascule	Counterbalance, chain type	1902
692,386	Ten Broeke, H.W.	14/52	Swing	Gate	1902
693,467	Waddell, M.	14/38	Bascule	With rope drive	1902
694,744	Hall, C.F.	14/40	Bascule		1902
704,611	Brown, T.E.	14/42	Vertical lift	Leveling device	1902
707,044	Worthington, C.	14/35	Swing	Lifts ends from center-lift truss	1902
708,348	Hall, C.F.	14/40	Bascule		1902
713,561	Doyen, D.	14/45	Retractile	Scissors bridge	1902
719,153	Smetters, S.T.	14/40	Bascule		1903

*See key at end.

Number	Patentee	U.S.P.O. Class.*	Bridge Type	Description of Invention	Date
721,918	Scherzer, A.H., & Kandeler, C.F.	14/39	Bascule		1903
731,321	Page, J.W.	14/38	Bascule		1903
731,322	Page, J.W.	14/36	Bascule		1903
735,414	Scherzer, A.H., & Kandeler, C.F.	14/39	Bascule		1903
738,954	Strauss, J.B.	14/38	Bascule	Dropping rear leaf	1903
752,563	Keller, C.L.	14/40	Bascule		1904
756,651	Kelly, F.T.	14/47	Swing	Pivoting end rails	1904
771,237	Lera	Russian	Bascule	Moving cwt support	1980
772,671	Reid, R.H.	14/46	Swing	Rail lock	1904
780,193	Joyce J.A.	14/41	Bascule	Center lock	1905
789,398	Watson, W.J.	14/39	Bascule		1905
803,834	Larson, C.W.	14/36	Bascule	Electrical control	1905
817,516	Rall, T.	14/40	Bascule		1906
824,135	Newton, R.E.	14/36	Bascule		1906
828,873	Borg, F.G.	14/36	Bascule		1906
842,457	Harding, S.B.	14/43	Trans-porter	Aerial ferry	1907
843,167	Nikonow, J.P.	14/38	Bascule		1907
851,114	Biron, M.R.	14/32	Swing	Double bobtail	1907
859,489	Donovan, W.J.	14/53		Gate (ferry)	1907
865,839	Wieland, G.A.	14/33	Swing	Chain drive centering at ends	1907
874,327	Goodman, M.	14/50		Gate	1907
887,131	Shoemaker, L.H.	14/36	Bascule		1908
887,868	Strauss, J.B.	14/36	Bascule	Trolley wire support	1908
890,947	Waddell, M.	14/38	Bascule		1908
894,199	Gaston, R.M.	14/41		Motor control	1908
894,239	Strauss, J.B.	14/39	Bascule		1908
895,673	D'Or, F.	14/33	Swing	Gyratory lift	1908
908,713	Waddell, M.	14/40	Bascule		1909
911,628	Swenson, E.	14/38	Bascule		1909
932,359	Waddell, J.A.L.	14/42	Vertical lift		1909
952,485	Harrington, J.L., & Waddell, M.	14/36	Bascule		1910
952,486	Waddell, J.A.L., & Harrington, J.L.	14/42	Vertical lift	Multiple counterweight arrangement	1910
953,307	Waddell, J.A.L.	14/42	Vertical lift		1910
963,399	Scherzer, A.H.	14/40	Bascule	Rolling lift double leaf	1910
968,987	Scherzer, A.H.	14/41	Bascule		1910

*See key at end.

Number	Patentee	U.S.P.O. Class.*	Bridge Type	Description of Invention	Date
968,988	Scherzer, A.H.	14/41	Bascule	Rllng lft dbl leaf jaw ctr lock	1910
974,538	Strauss, J.B.	14/38	Bascule		1910
978,493	Scherzer, A.H.	14/41	Bascule	Rllng lft rear-end lock	1910
979,458	Freak, G.R.	14/52		Electrically operated gate	1910
983,194	Agnew, R.M.	14/41	Bascule	Locking device	1911
995,813	Strauss, J.B.	14/38	Bascule	Chicago infringement cross girder	1911
1,001,800	Von Babo, A.F.L.	14/38	Bascule	Trunnion bascule, internal rack in truss	1911
1,003,901	Harrington, J.L.	14/42	Vertical lift	Fluid conduit	1911
1,021,488	Scherzer, A.H.	14/39	Bascule		1912
1,027,477	Harrington, J.L.	14/42	Vertical lift		1912
1,027,478	Harrington, J.L.	14/42	Vertical lift		1912
1,038,226	Strauss, J.B.	14/42	Vertical lift		1912
1,041,454	Frank, M.E.	14/53		Gate	1912
1,041,885	Scherzer, A.H.	14/39	Bascule		1912
1,042,238	Krase, H.C.	14/41	Bascule	Tail lock	1912
1,046,264	Casler, W.A.	14/46		Rail lock	1912
1,047,950	Keller, C.L.	14/38	Bascule	Means for operation	1912
1,048,440	Benni, F.J.	14/39	Bascule		1912
1,049,422	Waddell, J.A.L.	14/42	Vertical lift		1913
1,055,194	Lichtfuss, A.	14/50		Gate & signal	1913
1,078,293	Leslie, H.	14/37	Bascule		1913
1,087,233	Hedrick, I.G.	14/42	Vertical lift	Counterweight	1914
1,091,829	Gardner, R.D.	14/35	Swing	End lift	1914
1,093,202	Paine C.E.	14/42	Vertical lift		1914
1,094,473	Rall, T.	14/38	Bascule		1914
1,102,254	Faust, C.	14/51		Gate	1914
1,104,318	Scherzer, A.H.	14/39	Bascule		1914
1,106,259	Turner, C.A.P.	14/32	Swing	Lifts ends from center-lift truss	1914
1,109,792	Scherzer, A.H.	14/40	Bascule		1914
1,111,872	Strauss, J.B.	14/42	Vertical lift		1914
1,114,535	Scherzer, A.H.	14/39	Bascule		1914
1,124,356	Strauss, J.B.	14/36	Bascule		1915
1,124,922	Larsson, C.G.E.	14/36	Bascule		1915
1,128,478	McKibben, C.	14/40	Bascule	Rolling lift	1915
1,140,316	Rall, T.	14/42	Vertical lift		1915
1,140,984	Icke, H.A.	14/31	Bascule	Cable lift from arch	1915
1,145,166	Reehl, D.L.	14/46	Swing	Rail lock & wedge drive	1915
1,150,643	Strauss, J.B.	14/38	Bascule		1915

*See key at end.

Number	Patentee	U.S.P.O. Class.*	Bridge Type	Description of Invention	Date
1,150,975	Strauss, J.B.	14/38	Bascule		1915
1,151,657	Brown, T.E.	14/36	Bascule		1915
1,157,449	Strauss, J.B.	14/41	Bascule	Dbl leaf heel trunnion	1915
1,157,450	Strauss, J.B.	14/42	Bascule		1915
1,158,084	Strauss, J.B.	14/32	Swing	Traction drive	1915
1,158,546	Osborn, F.C.	14/42	Vertical lift	Chain connection to counterweight	1915
1,170,702	Strauss, J.B.	14/42	Vertical lift		1916
1,170,703	Strauss, J.B.	14/38	Bascule	Double deck	1916
1,171,553	Strauss, J.B.	14/38	Bascule		1916
1,200,077	Casler, W.A.	14/46		Rail lock	1916
1,203,695	Brown, T.E.	14/36	Bascule		1916
1,206,505	Blume, G.F.	14/36		Roller bearing	1916
1,210,410	Brown, T.E.	14/36	Bascule		1917
1,210,715	Shoemaker, L.H.	14/42	Vertical lift	Operating mechanism	1917 1917
1,211,639	Strauss, J.B.	14/36	Bascule		
1,211,640	Strauss, J.B.	14/38	Bascule		1917
1,224,629	Gardner, R.D.	14/36	Bascule	Counterbalance device	1917
1,241,052	Strauss, J.B.	14/38	Bascule		1917
1,241,237	Mercer, C.H., & Woethle, C.H.	14/36	Bascule		1917
1,251,634	Brown, T.E.	14/36	Bascule	Backwards rolling lift	1918
1,254,772	Brown, T.E.	14/36	Bascule		1918
1,254,773	Brown, T.E.	14/36	Bascule		1918
1,261,124	Harrington, J.L., Howard, E.E., & Ash, L.R.	14/42	Vertical lift	Guide for counter-weight ropes	1918
1,261,286	Pike, E.R.	14/42	Vertical lift	Rope load equalizer	1918
1,265,617	Cosham, D.H.	14/54	Swing	Automatic gate	1918
1,269,976	Strauss, J.B.	14/42	Vertical lift	Suspension lift bridge	1918
1,270,925	Brown, T.E.	14/36	Bascule	Heel trun w complex linkage	1918
1,285,696	Harrington, J.L.	14/42	Vertical lift		1918
1,302,302	Brown, T.E.	14/36	Bascule		1919
1,311,284	Morreell, S.	14/36	Bascule		1919
1,332,179	Howard, E.E.	14/46		Rail lock	1920
1,337,187	Anderson, H.	14/41	Bascule		1920
1,352,450	Henderson, A.A.	14/42	Vertical lift	Hinged approach connection, hydraulic lift alternate	1920

*See key at end.

Number	Patentee	U.S.P.O. Class.*	Bridge Type	Description of Invention	Date
1,370,999	Scherzer, A.H.	14/39	Bascule		1921
1,373,972	Richards, W.R.	14/42	Bascule	Trolley wire support	1921
1,394,519	Abt, H.A.F.	14/36	Bascule	Reverse opr cwt	1921
1,401,374	Swonger, J.M.	14/37	Bascule	Hinged fixed bridge	1921
1,406,259	Brown, T.E., Jr.	14/36	Bascule	Overhead articulated counterweight	1922
1,412,327	Abt, H.A.F.	14/38	Bascule		1922
1,412,328	Abt, H.A.F.	14/38	Bascule		1922
1,422,717	Joosting, P.	14/42	Vertical lift		1922
1,470,140	Brown, T.E.	14/36	Bascule	Overhead articulated counterweight	1923
1,471,282	Roake, S.A.	14/41	Bascule	Bridge construction—end lock	1923
1,510,043	Cunningham, A.O.	14/36	Bascule		1924
1,519,189	Brown, T.E.	14/36	Bascule	Heel trunnion w knuckle link	1924
1,534,944	Paris, J.	French	Bascule		1968
1,542,972	Strauss, J.B.	14/41	Bascule	Dbl lf artic cwt under, mech jaw ctr lks	1925
1,543,199	Winebrake, C.J.	14/50		Gate	1925
1,550,353	Evans, J.E.	14/42	Vertical lift	Rope strain equalizer	1925
1,552,243	Strauss, J.B.	14/36	Bascule	Simple trunnion w support posts	1925
1,561,671	Strauss, J.B.	14/38	Bascule	Trunnion support	1925
1,565,571	Howard, E.E.	14/42	Vertical lift	Span drive w drums at corners	1925
1,583,705	Strauss, J.B.	14/38	Bascule	Heel trunnion	1926
1,588,329	Murray, S.	14/38	Bascule	Cwt moving relative to leaf	1926
1,589,036	Young H.E.	14/38	Bascule	Simple trunnion	1926
1,593,681	Strauss, J.B.	14/36	Bascule	Underdeck articulated cwt	1926
1,597,298	Strauss, J.B.	14/42	Vertical lift	Rack driven	1926
1,613,504	Dayde, R.	14/40	Bascule	Rolling lift	1927
1,615,891	Strauss, J.B.	14/38	Bascule	Ohd artic cwt	1927
1,618,267	Brown, T.E.	14/38	Bascule	Brown bascule w links	1927
1,619,678	Prinz, H.L.	14/43	Retractile	W drop deck	1927
1,620,284	Pearson, E.A.	14/42	Vertical lift	Sheave-rope changeout	1927
1,624,325	Folino, A.	14/42	Pontoon	Bridge construction	1927
1,633,565	Ash, L.R.	14/36	Bascule	Simple trunnion w support posts	1927

*See key at end.

Number	Patentee	U.S.P.O. Class.*	Bridge Type	Description of Invention	Date
1,646,340	Ash, L.R.	14/41	Bascule	Mech jaw ctr lock	1927
1,646,403	Hardesty, S.	14/42	Vertical lift	Span drive	1927
1,648,574	Brown, T.E., Jr	14/38	Bascule	Brown bascule w links	1927
1,659,250	Erdal, I.	14/41	Bascule	Tail lock	1928
1,663,359	Strauss, J.B.	14/38	Bascule	Underdeck articulated cwt	1928
1,671,693	Ash, L.R.	14/42	Vertical lift	Rope anchor	1928
1,673,358	Hardesty, S.	14/42	Vertical lift	Counterbalance	1928
1,673,359	Hardesty, S.	14/42	Vertical lift	Counterweight apparatus	1928
1,673,383	Williams, W.G.	14/42	Vertical lift	Rope attachment	1928
1,676,994	Lamond, R.C.	14/42		Control mechanism	1928
1,680,821	Strauss, J.B.	14/36	Bascule	Geared counterweight	1928
1,681,424	Morton, A.A.	14/42	Vertical lift	Linked counterweight	1928
1,694,516	Prescott, A.T.	14/42	Bascule	Ferry bridge	1928
1,700,464	Brown, T.E., Jr	14/49	Bascule	Indicator	1929
1,714,699	Strauss, J.B.	14/36	Bascule	Overhead articulated cwt	1929
1,729,637	Williams, W.G.	14/42	Vertical lift	Conductor support	1929
1,730,595	Strauss, J.B.	14/41	Bascule	End lock	1929
1,731,697	Allen, G.H.	14/36	Bascule	Cwt blocks positioned by cables	1929
1,739,103	Strauss, J.B.	14/36	Bascule	Simple trunnion w support posts, cwt	1929
1,765,407	Dodge, R.E.	14/42	Vertical lift	Pivoted counterweights	1930
1,778,234	Strauss, J.B.	14/38	Bascule	Overhead counterweight articulated	1930
1,798,506	Strauss, J.B.	14/36	Bascule	Double deck	1931
1,807,780	Duwe, F.	14/35	Swing	End lift	1931
1,807,907	Foshay, A.	14/58		Gate	1931
1,810,034	Strauss, J.B.	14/35	Bascule	Simple trunnion w internal rack	1931
1,816,672	Dillman, P.B.	14/36	Swing	Portable	1931
1,827,916	Stevens, P.E.	14/31	Bascule	Balanced w no cwt	1931
1,832,650	Ortlip, W.H.	14/36	Bascule	Drive mechanism	1931
1,855,520	Howard, E.E.	14/42	Vertical lift	Span drive	1932
1,865,621	Duwe, F.	14/38	Bascule	Chain drive	1932
1,866,613	Auryansen, F.	14/32	Swing	Skewed	1932
1,893,014	Breitwieser, H.	14/35	Swing	End lift	1933
1,897,056	Keller, C.L.	14/36	Bascule	Arch	1933

*See key at end.

Number	Patentee	U.S.P.O. Class.*	Bridge Type	Description of Invention	Date
1,905,571	Smith, R.G.	14/42	Vertical lift	For toy trains	1933
1,909,431	Strauss, J.B.	14/42	Vertical lift	Rack drive	1933
1,974,114	Kuyk, E.	14/38	Bascule	Cam drive	1934
1,991,336	Six, W.L.	14/42	Vertical lift	Control	1935
2,011,467	Breitwieser, H.	14/40	Bascule	Chain-link drive	1935
2,035,193	Russell, W.A.	14/39	Bascule	Cylinder drive	1936
2,040,445	Sakamoto, T.	14/42	Vertical lift	Leveling rope connection	1936
2,047,601	Strauss, J.B.	14/42	Vertical lift	Rack driven	1936
2,057,028	Howard, E.E.	14/42	Vertical lift	Rope auxiliary counterweight	1936
2,062,635	Breitwieser, H.	14/42	Vertical lift	Chain aux counter-weight	1936
2,066,110	Hopkins, L.O.	14/36	Bascule	Drive	1936
2,085,613	Stiles, T.G.	14/35		Circuit controller	1937
2,093,612	Lovelace, C.	14/58		Gate	1937
2,103,661	Brown, T. E., Jr	14/42	Vertical lift	Counterbalance	1937
2,109,797	Lubin, I.	14/36	Bascule	Center lock, etc	1938
2,189,974	Buford, W.J.	14/52		Gate	1940
2,191,708	Dunford, J.A.	14/36	Bascule	Operating mechanism	1940
2,193,350	Silver, J.W.	14/52		Gate	1940
2,198,810	Finke, R.W.	14/42	Vertical lift	Rope connection	1940
2,212,504	Roberts, L.A.	14/59		Barrier gate	1940
2,242,615	Moore, C.H.	14/38	Bascule	Simple trunnion—stationary rack	1941
2,271,697	Leffler, B.R.	14/42	Vertical lift	Fixed sheave shaft	1942
2,337,994	Hanover, C.D.	14/36	Bascule	Skew	1943
2,344,772	Heidenreich, A.	14/45	Folding	Not for navigable waters	1944
2,366,848	Ferri, H.J.	14/36	Bascule	For toy trains	1945
2,411,480	Temple, A.E.S.	14/36	Bascule	Float operated	1946
2,443,657	King, G.E.	14/42	Vertical lift	Tower drive control	1948
2,445,953	Lancaster, F.A.	14/49	Bascule	Navigation light	1948
2,474,128	Tonnelier, J.E.	14/42	Vertical lift	Counterweight rope lubricator	1949
2,482,562	Shoemaker, L.H.	14/42	Vertical lift	Rope-chain equalizing	1949
2,556,175	Frost, L.P.	14/37	Bascule	Hinged leaf portable	1951
2,593,215	Summerhays, W.A.	14/46	Swing	Turntable rail joint	1952
2,610,341	Gilbert, H.H.	14/41	Bascule	Ctr lock	1952
2,675,569	Foshay, A.	14/58	Swing	With gate	1954
2,724,135	Hopkins, L.O.	14/36	Bascule	Drive	1955
2,740,145	Loser, H.A.	14/38	Bascule	Center girder	1956

*See key at end.

Number	Patentee	U.S.P.O. Class.*	Bridge Type	Description of Invention	Date
2,755,493	Miller, M.	14/42	Vertical lift	Sectional, linkage connections	1956
2,759,430	Kelsey, C.P.	14/46	Swing	End lock	1956
2,861,285	Sawyer, E.D.	14/52		Yielding barrier gate	1958
2,875,325	Koenig, E.L.	14/46	Swing	Pivoting cast miter rail indicator	1959
2,887,191	Lovell, J.	14/36	Bascule	Hydraulic mechanism	1959
2,889,565	Harty, R.V.	14/42	Vertical lift	In-plant 2 level	1959
3,004,272	Clarke, J.E.	14/42	Vertical lift	In-plant transfer table	1961
3,276,059	Allen, E.F.	14/36	Bascule	Hydraulic cylinder operating device	1966
3,308,496	Mooney, G.G.	14/36	Bascule	Hydraulic cylinder operating device	1967
3,376,795	Allen, E.F.	14/36	Bascule	Hydraulic cylinder operating device	1968
3,394,420	Popov, V.	14/42	Vertical lift	Rope equalized	1968
3,466,686	Allen, E.F.	14/42	Vertical lift	Hydraulic operating system	1969
3,548,434	Overson, R.	14/36	Bascule	Reversing cwt	1970
3,570,032	Allen, E.F.	14/42	Vertical lift	Stepped hydraulic cylinder operated	1971
3,593,481	Mikulin, T.T.	14/45	Retractile	Scissors extending bridge	1971
3,668,729	Mori, K.	14/43	Retractile	Drive	1972
3,740,782	Newman, O.G.	14/42	Vertical lift	Servicing device	1973
3,744,075	Zaremba, D.	14/41	Portable	Span lock	1973
3,848,288	Richards, L.I.	14/41	Bascule	Safety lock	1974
3,864,777	Perkons, G.	14/38	Bascule	Basculux Drive—Earle	1975
3,908,217	Lapeyre, J.M.	14/42	Vertical lift	Unlift	1975
4,751,758	Nedelcu, L.I.	14/36	Bascule	Rope drive	1988
5,148,751	Alten, K.	104/31†	Bascule	Bridge over railroad tracks	1992
5,186,421	Caccomo, P.G.	14/49		Miter rail check	1993
5,263,217	Miglietti, G.	14/32	Swing	Asymetrical, apron end	1993
5,327,605	Cragg, R.L.	14/41	Bascule	Locking device	1994
5,421,051	Patten, R.W.	14/37	Bascule	Folding-hinged	1995
5,454,127	Dvorak, I.	14/42	Bascule	Selective torsional rigidity, hydraulic opr, non cwt	1995
5,474,266	Koglin, T.L.	246/111‡	Bascule	Bridge over railroad	1995

*See key at end.

†Railways/shifting.

‡Railway switches and signals/grade crossing—track protection.

Number	Patentee	U.S.P.O. Class.*	Bridge Type	Description of Invention	Date
5,984,569	Chou, P.	14/53		Gate, barrier	1999
6,363,564	Weaver, B.F.	14/56	Swing	Pivoting cast miter rail joint, 3 pc	2002
6,453,494	Koglin, T.L.	14/41	Bascule	Center lock	2002

*KEY TO U.S. PATENT OFFICE SUBCLASSIFICATIONS: ALL CATEGORY "14"—BRIDGES

31 Bridges, Draw
32 Bridges, Draw, Swing
33 Bridges, Draw, Swing, Hand/Motor Operated
34 Bridges, Draw, Swing, Boat Operated
35 Bridges, Draw, Swing, End Supports/ Locking Devices
36 Bridges, Draw, Bascule, Fixed Pivot
37 Bridges, Draw, Bascule, Hinged Sections
38 Bridges, Draw, Bascule, Fixed Pivot, Rack & Pinion
39 Bridges, Draw, Bascule, Non-Pivoted
40 Bridges, Draw, Bascule, Non-Pivoted, Rack & Pinion
41 Bridges, Draw, Bascule, Locking Devices
42 Bridges, Lift
43 Bridges, Draw, Horizontally Sliding
44 Bridges, Draw, Horizontally Sliding, Boat Operated
45 Bridges, Draw, Lazy Tongs (scissors extension)
46 Bridges, Draw, Locking Devices
47 Bridges, Draw, Aprons
48 Bridges, Draw, Buffers
49 Bridges, Draw, Signals
50 Bridges, Draw, Gates, Hand/Motor Operated
51 Bridges, Draw, Gates, Horizontal Swing
52 Bridges, Draw, Gates, Vertical Sliding
53 Bridges, Draw, Gates, Vertical Swinging Bridge Operated
54 Bridges, Draw, Gates, Horizontally Sliding
55 Bridges, Draw, Gates, Horizontally Sliding, Locking
56 Bridges, Draw, Gates, Horizontal Swing, Bridge Operated
57 Bridges, Draw, Gates, Horizontal Swing, Bridge Operated, Locking
58 Bridges, Draw, Gates, Vertical Sliding Bridge Operated
59 Bridges, Draw, Gates, Vertical Sliding Bridge Operated, Locking
60 Bridges, Draw, Gates, Vertical Swinging Bridge Operated
61 Bridges, Draw, Gates, Vertical Swinging Bridge Operated, Locking
62 Bridges, Draw, Gates, Horizontally Sliding, Bridge Operated, Shaft Rotation
63 Bridges, Draw, Gates, Horizontally Sliding, Bridge Operated, Shaft Rotation, Locking
64 Bridges, Draw, Gates, Horizontal Swing, Bridge Operated, Shaft Rotation
65 Bridges, Draw, Gates, Horizontal Swing, Bridge Operated, Shaft Rotation, Locking
66 Bridges, Draw, Gates, Vertical Sliding Bridge Operated, Shaft Rotation
67 Bridges, Draw, Gates, Vertical Sliding Bridge Operated, Shaft Rotation, Locking
68 Bridges, Draw, Gates, Vertical Swinging Bridge Operated, Shaft Rotation
69 Bridges, Draw, Gates, Vertical Swinging Bridge Operated, Shaft Rotation, Locking

APPENDIX II

NORTH AMERICAN MOVABLE BRIDGES

Waterway[a]	Road Name/ Route Carried	Bridge Type[b]	Location	Const. Date	Traffic Type[c]	Owner*	Miles from Mouth of Waterway	Channel Width/Lateral Clearance (feet)	Clearance Height (feet)
Acushnet River	US 6	SW	New Bedford, MA	1901	HWY	Massachusetts	0	95	6
Acushnet River	Coggleshell St.	SW	New Bedford, MA	1890	HWY	New Bedford	1	40	4
Ahnapee River	2nd St.	SW	Algoma, WI	1899	HWY	Algoma	0.2	73	6.1
Alabama River	—	SW	Coy, AL	1916	RR	BNSF	105.3	100	18
Alabama River	—	SW	Selma, AL	1903	RR	CSX	205.9	110	26
Alabama River	—	VL	Montgomery, AL	1981	RR	ICG	277.8	313	33
Alabama River	US 31	SW	Montgomery, AL	1926	HWY	Alabama	278.3	150	18
Alabama River	—	SW	Hails Station, AL	1897	RR	M&ORR	283	135	10
Alabama River	—	SW	Hails Station, AL	1926	HWY	Alabama	283	150	7
Alabama River	—	SW	Montgomery, AL	1948	RR	CSX	293.3	105	31
Alabama River	—	SW	Montgomery, AL	1904	HWY	Alabama	296	104	3
Alabama River	—	SW	Montgomery, AL	—	RR	L&N	299	86	superstructure removed
Alabama River	—	SW	Montgomery, AL	1922	RR	CSX	299	106	2.3
Alabama River	—	SW	Montgomery, AL	1934	HWY	Alabama	300	110	25
Alafia River	—	SW	Gibsontown, FL	1957	RR	LN	1.2	40	6
Alafia River	—	SW	Riverview, FL	1919	RR	TS	1.4	40	4
Alamitos Bay	—	RSP	Seal Beach, CA	1932	HWY	Los Angeles Co.	0	39	18.6
Alamitos Bay	—	RSP	Long Beach, CA	1929	HWY	Long Beach	0.8	17.1	3.6
Alamitos Bay	—	RSP	Long Beach, CA	1904	RR	PE	1	37	2
Alamitos Bay	—	RSP	Long Beach, CA	1929	HWY	Long Beach	1.5	22.2	3.6
Alamitos Bay	—	RSP	Long Beach, CA	1924	RR	PE	1.5	46.5	4.6
Alamitos Bay	—	RSP	Long Beach, CA	1904	RR	PE	2	37	1.9
Alamitos Bay	—	RSP	Long Beach, CA	1932	HWY	Long Beach	2.25	40	7
Alamitos Bay	—	RSP	Long Beach, CA	1931	HWY	California	2.5	26.8	7
Albemarle Sound	SR 32	SW	Edenton, NC	1938	HWY	North Carolina	47.7	140	14
Albemarle Sound	—	SW (closed)	Edenton, NC	1910	RR	NS	52.3	35	4
Albemarle Sound	—	B	Edenton, NC	1969	RR	NS	52.3	120	6

Albemarle Sound	—	B	Edenton, NC	1910	RR	NS	52.3	140	6
Alley Creek	Long Island RR	SW	Douglaston, NY	1923	RR	LIRR	0.2	40	3
Alligator River Northwest Fork	—	P	Gum Neck, NC	1935	HWY	North Carolina	3	50	—
Alloway Creek	—	SW	Hancocks Bridge, NJ	1954	HWY	Salem Co.	5.1	40	4
Alloway Creek	—	SW	Upper Hancocks Bridge, NJ	1906	HWY	Salem Co.	6.5	35	3
Alloway Creek	SR 49	SW, F	Quinton, NJ	1925	HWY	New Jersey	9.5	30	3
Altamaha River	—	SW	Everett City, GA	1950	RR	CSX	23.5	50	0
Altamaha River	—	B	Doctortown, GA	1914	RR	CSX	59.4	97	17
Altahama River	—	SW	Doctortown, GA	1930	HWY	Georgia	60.9	83	1.3
Altahama River	—	SW	Glenville, GA	1921	HWY	Georgia	88.7	90	19.1
Altahama River	—	SW	Baxley, GA	1928	HWY	Georgia	117	85	4.7
Altamaha River	—	B	Hazelhurst, GA	1909	RR	NS	139.9	85	1
Altahama River (Bridge over Darien River)	—	SW	Darien, GA	1914	HWY	Georgia	7	97	19.1
Altahama River, South	—	SW	Darien, GA	1914	HWY	Georgia	6.3	98	5.5
American River	Jiboom St.	SW, F	Sacramento, CA	1931	HWY	Sacramento Co.	0.1	155	35
Amite River	SR 22	SW	Clio, LA	1975	HWY	Louisiana	6	60	4
Amite River	SR 22	SW	French Settlement, LA	1959	HWY	Louisiana	21.4	60	15
Amite River	SR 42	SW	Port Vincent, LA	1964	HWY	Louisiana	32	60	7
Anacostia River	S. Capitol St. (Douglas Br.)	SW	Washington, DC	1978	HWY	Dist. Of Columbia	1.2	142	42
Anacostia River	11th St. S.E.	B	Washington, DC	1908	HWY	Dist. Of Columbia	2.1	100	21.6
Anacostia River	—	VL	Washington, DC	1892	RR	CR	3.4	33	28.9
Anaheim Bay	—	RSP	Seal Beach, CA	1904	RR	PE	0	37.3	4
Anaheim Bay	—	RSP	Seal Beach, CA	1915	HWY	Orange Co.	0	60	5
Anaheim Bay	—	RSP	Seal Beach, CA	1933	HWY	California	0.75	36	11

*See key of abbreviations at end.

Waterway[a]	Road Name/ Route Carried	Bridge Type[b]	Location	Const. Date	Traffic Type[c]	Owner*	Miles from Mouth of Waterway	Channel Width/Lateral Clearance (feet)	Clearance Height (feet)
Annisquam River (Blynman Canal)	SR 127	B	Gloucester, MA	—	HWY	Massachusetts	0	39	7
Annisquam River (Blynman Canal)	SR 127	B	Gloucester, MA	—	HWY	Massachuestts	0	40	7
Annisquam River (Blynman Canal)	—	B	Gloucester, MA	1910	RR	B&M	0.7	38	16
Apalachicola River	John Gorrie Br. US 98	SW	Apalachicola, FL	1936	HWY	Florida	0	120	28
Apalachicola River	—	SW	Apalachicola, FL	1906	RR	AN	4.3	119	11
Apalachicola River	—	SW	River Jct., FL	1945	RR	CSX	105.9	115	1
Apalachicola River	—	B	Chattahoochee, FL	1923	HWY	Florida	111	100	6
Appomatox River	—	SW	Hopewell, VA	1928	HWY	Virginia	1.5	80	8
Appomatox River	—	SW	Hopewell, VA	1930	RR	CSX	2.5	79	10
Appomatox River	—	SW	Petersburg, VA	1962	CB	Friend S&G	9.2	50	11
Apponagansett Bay	—	SW	Dartmouth, MA	1936	HWY	Dartmouth	1.0	31	9
Appoquinimink River	—	SW	Fennimore, DE	1922	HWY	Delaware	3.5	39.5	7.3
Appoquinimink River	—	SW, F	Odessa, DE	1929	HWY	Delaware	6.77	40	4.8
Aquia Creek	—	SW	Stafford, VA	1923	RR	RFP	3.2	30.7	14.9
Arkansas River	—	SW, VL per COE	Yancopin, AR	1927	RR	MP	23.1	165	1
Arkansas River	—	VL	Rob Roy, AR	1971	RR	UP	67.4	300	52
Arkansas River	—	VL	Pine Bluff, AR	1915	HWY	Arkansas	115.6	229	10.8
Arkansas River	—	VL	Little Rock, AR	1971	RR	RI	118.2	318	52
Arkansas River	—	VL	Little Rock, AR	1972	RR	UP	118.7	331	52
Arkansas River	—	VL	Little Rock, AR	1972	RR	UP	119.6	315	52
Arkansas River	—	SW	Morrilton, AR	1919	HWY	Arkansas	221	162	4.1
Arkansas River	—	SW	Dardanelle, AR	1929	HWY	Arkansas	255.8	160.5	5.4

Arkansas River	—	VL, F	Van Buren, AR	1913	HWY	Ft. Smith Van Buren Bridge Dist.	300.2	180	2.9
Arkansas River	—	VL	Van Buren, AR	1970	RR	BNSF	300.8	312	52
Arkansas River	—	SW	Van Buren, AR	1915	RR	SLSF	300.8	165	9.8
Arkansas River	—	SW	Ft. Smith, AR	1892	RR	MP	361.8	165	2.1
Arroyo Colorado River	SR 106	VL	Rio Hondo, TX	1950	HWY	Texas	22.5	125	27
Arthur Kill	—	VL	Elizabeth, NJ/NY	1952	RR	PANYNJ	11.6	500	135
Ash Creek	—	—	Bridgeport, CT	—	RR	CR	—	—	—
Ashepoo River	—	SW	Brickyard Ferry, SC	1936	HWY	South Carolina	15	60	7
Ashepoo River	—	SW	Fenwick, SC	1917	RR	SAL	20	60	4.7
Ashepoo River	—	B	Ashepoo, SC	1927	RR	CSX	32	43	2
Ashley River	US 17	B	Charleston, SC	1961	HWY	South Carolina	2.4	100	18
Ashley River	US 17	B	Charleston, SC	1926	HWY	South Carolina	2.4	100	14
Ashley River	—	SW	Charleston, SC	1917	RR	SAL	3.75	100	9.85
Ashley River	—	B	Drayton Hall, SC	1927	RR	CSX	12	60	3
Ashtabula River	5th St.	B	Ashtabla, OH	1924	HWY	Ashtabula Comm.	1.4	139	6
Ashtabula River	—	B	Ashtabla, OH	1911	RR	CR	2.2	112	7
Aspatuck Creek	—	SW	Quogue, NY	1914	HWY	Southampton	0.6	30	7
Atchafalaya River	SR 182	SW	Patterson, LA	1949	HWY	Louisiana	26.8	60	5
Atchafalaya River	—	SW	Krotz Springs, LA	1954	RR	UP	95.7	129	6
Atchafalaya River	—	VL	Melville, LA	1929	RR	UP	107.4	160	54
Atchafalaya River	—	SW	Simmesport, LA	1928	HWY, RR	LARR	133.1	132	6
Atchafalaya River–Berwick Bay	—	VL	Morgan City, LA	1971	RR	SP	17.5	321	73
Back Cove	—	SW	Portland, ME	—	RR	GT	0	88	5
Back River 1	St. Simons Island Cswy.	VL	Brunswick, GA	1950	HWY	Glynn Co.	1.5	100	85
Back River 2	River Rd.	SW	Barters Island–Hodgdon Island, ME	—	HWY	Boothbay	2	36	6
Back River 3	—	B	Baltimore, MD	1911	HWY	Baltimore Co.	9.5	24	4.4

*See key of abbreviations at end.

Waterway[a]	Road Name/ Route Carried	Bridge Type[b]	Location	Const. Date	Traffic Type[c]	Owner*	Miles from Mouth of Waterway	Channel Width/Lateral Clearance (feet)	Clearance Height (feet)
Back River 3	—	B	Baltimore, MD	1897	RR	Baltimore Transit	9.5	21	3.5
Baines Creek	—	SW	Portsmouth, VA	1912	RR	ACL	0.5	21.5	2
Banana River	—	SW	Eau Galle, FL	1928	HWY	L. Mathers	0.5	31.2	8
Banana River	SR 3	SW	Indian Harbor Beach, FL	1952	HWY	Florida	0.5	78	7
Banana River	SR 3	SW	Indian Harbor Beach, FL	1978	HWY	Florida	0.5	80	7
Banana River	SR 3 (temp.)	SW	Indian Harbor Beach, FL	1982	HWY	Florida	0.5	74	7
Banana River	—	SW	Cocoa, FL	1927	HWY	Florida	13	30	8
Banana River	—	B	Cape Kennedy, FL	1964	HWY	NASA	27.6	90	24
Barataria Bayou	SR 302	SW	Lafitte, LA	1948	HWY	Louisiana	35.7	70	7
Barataria Bayou	—	SW	Wagner Ferry, LA	1931	HWY	Louisiana	39.7	75	9.1
Barnard Bayou	—	SW	Handsboro, MS	1908	HWY	Harrison Co.	2	82.5	7.8
Bartholomew Bayou	—	SW	Wardsville, LA	1902	RR	MP	44.1	60	3
Bartholomew Bayou	—	SW	Wardsville, LA	1907	RR	A&LM	44.3	72	4.8
Bartholomew Bayou	—	SW	Wilmot, AR	1920	HWY	Ashley Co.	98.2	60	4
Bartholomew Bayou	—	SW	Parkdale, AR	1909	HWY	Ashley Co.	107.5	60	4
Bartholomew Bayou	—	SW	Portland, AR	1910	HWY	Ashley Co.	133.6	60	5
Bartholomew Bayou	—	SW	Montrose, AR	1906	RR	MP	141.1	44	2.5
Bass River 1	Bridge St.	SW	Beverly, MA	1930	HWY	Massachusetts	0.6	40	5
Bass River 2	US 9	B	Bass River, NJ	1926	HWY	New Jersey	2.6	30	9
Bastrop Bayou	—	RSP	Angleton, TX	1922	RR	Freeport Sulphur	12.6	45	—
Bastrop Bayou	—	SW	Angleton, TX	1893	HWY	Texas	20	55	—
Bath Creek	—	SW	Bath, NC	1924	HWY	North Carolina	3	39	4.5
Battery Creek	—	SW	Port Royal, SC	1981	HWY	South Carolina	3	104	12
Bayou Boeuf	SR 307	VL	Kraemer, LA	1975	HWY	Louisiana	1.3	60	4
Beach Thorofare	—	B	Atlantic City, NJ	1926	HWY	Atlantic Co.	0.6	75	11.6

Bear Creek	—	SW	Sparrows Pt., MD	1940	RR	PRR	0.5	81.5	7.1
Bear Creek	—	SW	Sparrows Pt., MD	1939	RR	Baltimore Transit	0.5	85	6.5
Bear Creek	—	B	Sparrows Pt., MD	1971	HWY	Baltimore Co.	1.5	129	15
Bear Creek	—	B	Sparrows Pt., MD	1960	HWY	Baltimore Co.	2.1	60	25
Bear Creek	—	SW	Sparrows Pt., MD	1967	RR	CSX	2.3	80	8
Bear Creek	Wise Ave.	SW	Baltimore, MD	1895	HWY	Baltimore Co.	3	68	5.9
Bear Creek	—	B	North Pt., MD	1948	HWY	Baltimore Co.	3.4	75	12
Beaver Dam Creek 1	—	SW	Pt. Pleasant, NJ	1933	HWY	Ocean Co.	0.5	40	14
Beaver Dam Creek 2	—	SW	Westhampton, NY	1932	HWY	Suffolk Co.	0.3	26	6
Beaver Dam Creek 3	—	SW	Taylor Island, MD	—	HWY	Dorchester Co.	0.5	25	2
Beaver Slough 1	—	RSP	Thornton, CA	1933	HWY	San Joaquin Co.	2.1	68	1.3
Beaver Slough 2	—	B	Clatskanie, OR	1944	HWY	Columbia Co.	2	50	8
Beaver Slough 2	—	B	Clatskanie, OR	1931	HWY	Columbia Co.	3	50	6
Bellamy River	US 4	B	Dover, NH	1935	HWY	New Hampshire	0.1	40	9
Belle River	Scott St.	SW (abandoned)	Marine City, MI	1899	RR	Rapid Rwy. System	0.2	50	2.5
Belle River	Bridge St.	SW	Marine City, MI	1924	HWY	Marine City	0.5	40	3
Bellmans Creek	NYS&W	SW	Fairview, NJ	—	RR	Del Otsego	0.8	19	1
Bellmans Creek	—	B	Fairview, NJ	1918	RR	NYC	0.84	20.6	11.1
Bellmans Creek	NYS&W	SW	Fairview, NJ	—	RR	Del Otsego	1	50	20
Bennet Creek	—	SW	Belleville, VA	1932	HWY	Virginia	1.5	35	8.5
Bernard Bayou	US 90	B	Handsboro, MS	1950	HWY	Harrison Co.	2	70	12
Berry's Creek	—	B	Rutherford, NJ	—	RR	NYLE&W	1.9	25.5	2.5
Berry's Creek	—	SW	Carlstadt, NJ	1910	HWY	Bergen Co.	3.4	47	2
Berry's Creek	Paterson Plank Rd.	SW	Paterson, NJ	1910	HWY	Bergen Co.	3.45	48.6	2
Bienvenue Bayou	—	SW	Chalmette, LA	1935	HWY	Louisiana	9.8	60	2.8
Big Black River	—	B	Allens Station, MS	1917	RR	Y&MVRR	13.4	64	2
Big Black River	—	SW	Hankinson Ferry, MS	1925	HWY	Warren Co.	25.5	73	13
Big Black River	—	SW	Fishers Ferry, MS	1902	HWY	Warren Co.	44	72	4
Big Carlos Pass	SR 865	B	Estero Island, FL	—	HWY	Florida	0	50	23
Big Cypress Creek	—	RSP	Shreveport, LA	1962	HWY	Texas	48.2	70	11

*See key of abbreviations at end.

Waterway[a]	Road Name/ Route Carried	Bridge Type[b]	Location	Const. Date	Traffic Type[c]	Owner*	Miles from Mouth of Waterway	Channel Width/Lateral Clearance (feet)	Clearance Height (feet)
Big Marco River	—	SW	Marco, FL	1927	RR	FTMSRR	3	50	3.4
Big Marco River	SR 92	SW	Collier City, FL	1938	HWY	Florida	6.5	50	7
Big Marco River	—	SW	Collier City, FL	1928	HWY	—	7.5	50	4
Big Sunflower River	—	SW	Holly Bluff, MS	1907	RR	Y&MVRR	18.7	120	4
Big Sunflower River	—	SW	Holly Bluff, MS	1928	HWY	Sharkey Co.	19.2	125	4
Big Sunflower River	—	VL	Anguilla, MS	1925	HWY	Sharkey Co.	40.1	125	5
Big Sunflower River	—	SW	Inverness, MS	1930	HWY	Sunflower Co.	76.8	100	6.6
Big Sunflower River	—	SW	Woodburn, MS	1917	HWY	Sunflower Co.	82.7	100	8
Big Sunflower River	—	SW	Baird, MS	1888	RR	C&GRR	96.1	85	3
Big Sunflower River	—	SW	Indianola, MS	1920	HWY	Sunflower Co.	103.2	100	2.2
Big Sunflower River	—	SW	Sunflower, MS	1916	HWY	Sunflower Co.	119.1	100	4
Big Sunflower River	—	SW	Blaine, MS	1914	HWY	Sunflower Co.	125.9	100	4
Big Sunflower River	—	SW	Doddsville, MS	1916	HWY	Sunflower Co.	132	100	6
Big Sunflower River	—	SW	Lehrton, MS	1918	HWY	Sunflower Co.	141.1	100	6
Big Sunflower River	—	SW	Smith Ferry, MS	1919	HWY	Bolivar Co.	152	105	6
Big Sunflower River	—	SW	Wilmot Ferry, MS	1919	HWY	Bolivar Co.	155.3	102	6
Big Sunflower River	—	SW	Lombardy, MS	1917	HWY	Sunflower Co.	163.5	100	4
Billy's Creek	SR 80	VL	Ft. Myers, FL	1942	HWY	Florida	0.1	24	11
Billy's Creek	—	SW	Ft. Myers, FL	1922	HWY	Lee Co.	0.1	38	6
Biloxi, Back Bay of	L&N RR Br.	SW	Biloxi, MS	1978	RR	Harrison Co.	0	132	14
Biloxi, Back Bay of	US 90	B	Biloxi, MS	1962	HWY	Mississippi	0.4	132	40
Biloxi, Back Bay of	SR 15	B	Biloxi, MS	1975	HWY	Mississippi	3	132	60
Biloxi, Back Bay of	Popps Ferry Rd.	B	Biloxi, MS	1979	HWY	Harrison Co.	8	180	25
Biloxi River	—	SW	Lorraine, MS	1912	HWY	Harrison Co.	4	71.2	4.2
Biscayne Bay	—	B	Rivo Alto, FL	1926	HWY	Dade Co.	—	58	9
Biscayne Bay	—	B	Treasure Island, FL	1926	HWY	Florida	—	60	25
Biscayne Bay	—	B	Miami, FL	1920	HWY	Dade Co.	0.1	60	8

Biscayne Bay	—	B	Miami, FL	1929	HWY	Dade Co.	—	60	18.5
Biscayne Bay	—	B	Miami Beach, FL	1929	HWY	Dade Co.	—	60	18.5
Biscayne Bay	—	B	Miami, FL	1926	HWY	Bay Biscayne Improvement	—	60	8.5
Bishop Cut	8 Mile Rd.	SW	San Joaquin, CA	1927	HWY	San Joaquin Co.	1	75	6
Bishop Cut	—	SW	San Joaquin, CA	1983	HWY	San Joaquin, CA	1	22	6
Black Bayou	US 90, SR 20	VL	Gibson, LA	1946	HWY	Louisiana	7	50	56
Black Bayou	US 80, SR 20	SW	Gibson, LA	1951	HWY	Terrebonne Pa.	7.5	36	4
Black Bayou	—	SW	Humphreys Greenwood Plantation, LA	—	HWY	Realty Operators	15	36.5	−0.3
Black Bayou	US 90, SR 24, Greenwood Plantation	SW	Humphreys, LA	1952	HWY	Terrebonne Pa.	15	36	4
Black Bayou	US 90, SR 24	P	Humphreys, LA	1965	HWY	Shell Oil	15.7	60	—
Black Bayou	—	SW	Humphreys, LA	—	HWY	Terrebonne Pa.	18.6	37	2.7
Black Bayou	US 90, SR 24	SW	Humphreys, LA	1953	HWY	Terrebonne Pa.	18.7	36	4
Black Bayou	US 90, SR 24	SW	Humphreys, LA	1923	HWY	Terrebonne Pa.	22.5	40	7
Black Bayou	US 90, SR 24	SW	Waterproof Plantation, LA	1912	HWY, RR	Realty Operators	23.8	40	1
Black Bayou	—	SW	Houma-Waterproof Plantation, LA	1913	HWY	Realty Operators	24.4	37.5	6.1
Black Bayou	US 90, SR 24	SW	Houma-Mandalay Plantation, LA	1947	HWY	Realty Operators	25.4	38	6
Black Bird Creek	—	SW (repl. w. fixed)	Taylors Bridge, DE	1893	HWY	Delaware	5	42	1.1
Black Bird Creek	—	SW	Blackbird Landing, DE	1925	HWY	Delaware	8	25	6.5
Black Creek									
North Fork Tributary of St. Johns River	—	B	Middleburg, FL	1922	HWY	Clay Co.	1.6	27.7	7
Black Creek Tributary of St. Johns River	—	SW	Green Cove Springs, FL	1928	HWY	Florida	0.1	63	9.25

*See key of abbreviations at end.

Waterway[a]	Road Name/ Route Carried	Bridge Type[b]	Location	Const. Date	Traffic Type[c]	Owner*	Miles from Mouth of Waterway	Channel Width/Lateral Clearance (feet)	Clearance Height (feet)
Black Creek Tributary of St. Johns River	—	SW	Green Cove Springs, FL	1898	HWY	Clay Co.	0.5	53	5.2
Black Creek Tributary of St. Johns River	—	SW	Doctors Inlet, FL	1893	RR	ACL	5	51	6
Black Lake	SR 27	B	Hackberry, LA	1952	HWY	Shell Oil	0.8	13	10
Black Lake	SR 27	B	Hackberry, LA	—	HWY	—	0.9	50	4
Black River 1	—	B	Port Huron, MI	1929	RR	CSX	0.1	100	1
Black River 1	Military St.	B	Port Huron, MI	1934	HWY	Port Huron	0.3	65	4
Black River 1	7th St.	B	Port Huron, MI	1931	HWY	Port Huron	0.5	83	7
Black River 1	10th St.	SW	Port Huron, MI	1958	HWY	Port Huron	0.9	90	13
Black River 1	—	B	Port Huron, MI	1929	RR	CN	1.6	80	10
Black River 1.5	—	B	Mt. Clemens Subdivision, MI	1929	RR	CN	—	100	—
Black River 2	—	SW	South Haven, MI	—	HWY	South Haven	0.7	57.9	5.7
Black River 2	Dyckman Ave.	B	South Haven, MI	1968	HWY	Michigan	0.9	60	6
Black River 3	Erie Ave.	B	Loraine, OH	1940	HWY	Loraine Co.	0.6	176	30
Black River 3	—	VL	Loraine, OH	1975	RR	NS	1.3	205	123
Black River 4	—	SW	Georgetown, SC	1925	HWY	South Carolina	—	71	4.95
Black River 4	US 701	SW	Georgetown, SC	1959	HWY	South Carolina	8	60	8
Black River 4	US 701	SW	Rhems, SC	1954	HWY	South Carolina	25.7	70	3
Black River 5	—	SW	La Crosse, WI	1900	RR	CP	1	127	4
Black River 5	—	B	La Crosse, WI	1921	HWY	Wisconsin	1.9	60	0
Black River 6	—	SW	Paroquet, AR	1911	RR	MP	3.4	116	−0.1
Black River 6	—	SW	Black Rock, AR	1883	RR	SLSF	68.4	80	0.8
Black River 6	—	SW	Pocahontas, AR	1934	HWY	Arkansas	90.1	95	2.7
Black River 6	—	SW	Pocahontas, AR	1912	RR	SLSF	90.4	202	2.9
Black River 6	—	B	Corning, AR	1924	RR	MP	144.4	32	6.4
Black River 6	—	SW	Corning, AR	1919	HWY	Arkansas	152.2	82	8.3

Black River 6	—	SW	Qulin, MO	1917	HWY	Butler Co.	197.1	40	2.8
Black River 7	—	R	Currie, NC	1925	HWY	North Carolina	26	30	0
Black River 7	—	R	Ivanhoe, NC	1902	HWY	North Carolina	38	29.3	0
Black River 7	—	SW	Ivanhoe, NC	1914	HWY	North Carolina	53	48.1	0
Black Rock Canal	Ferry St.	B	Buffalo, NY	1914	HWY	Buffalo	2.6	149	13
Black Rock Canal	International Br.	SW	Buffalo, NY	1911	RR, HWY	CN	3.8	162	13
Black Warrior River	—	VL	Demopolis, AL	1927	RR	BNSF	219	176	63
Black Warrior River	—	VL	Eutaw, AL	1919	RR	NS	267.8	155	72
Blackwater River 1	—	SW	Milton, FL	1912	RR	LN	9	85	4
Blackwater River 1	—	B	Milton, FL	1919	HWY	Florida	9	81	2.7
Blackwater River 2	—	SW	Shorters Wharf, MD	—	HWY	Maryland	9	21	—
Blackwater River 3	SR 189	SW	South Quay, VA	1940	HWY	Virginia	9.2	60	15
Blackwater River 3	SR 603	SW	Burdette, VA	—	HWY	Virginia	19	57	9
Blind Slough	—	SW	Knappa, OR	1899	RR	BNSF	1.1	45	11
Blind Slough	—	RSP, F	Brownsmead, OR	1973	HWY	Clatsop Co.	2	47	13
Blue Bayou	—	RSP	Larose, LA	1938	HWY	Louisiana	35	45	2.1
Blue River	—	VL	Kansas City, MO	1949	RR	Armco Steel	2.9	83	8
Blynman Canal	—	—	Gloucester, MA	1913	HWY	Essex Co.	0.03	40	7.5
Boca Raton Inlet	SR A1A	B	Boca Raton, FL	1963	HWY	Florida	0.3	45	23
Boeuf Bayou	—	SW	Morgan City, LA	1929	HWY	Louisiana	9.75	66.4	8.3
Boeuf Bayou (St. Mary Parish)	—	SW	Morgan City, LA	1910	RR	SP	10.2	56	6
Boeuf River	—	SW	Mason, LA	1935	HWY	Louisiana	32.3	60	−8
Boeuf River	—	SW	Buckner, LA	1932	HWY	Richland Pa.	83.4	70	12.7
Boeuf River	—	SW	Girard, LA	1883	RR	Y&MVRR	121.7	58	0
Boeuf River	—	SW	Rayville, LA	1894	RR	MP	132.7	60	3
Bohemia River	SR 213	B	Cayots, MD	1932	HWY	Maryland	4	40	12
Bonfoca Bayou	SR 433	SW	Slidell, LA	1976	HWY	Louisiana	7	125	8
Bonfouca Bayou (Vincent)	—	SW	Slidell, LA	1938	HWY	Louisiana	7	50	2.8

*See key of abbreviations at end.

Waterway[a]	Road Name/ Route Carried	Bridge Type[b]	Location	Const. Date	Traffic Type[c]	Owner*	Miles from Mouth of Waterway	Channel Width/Lateral Clearance (feet)	Clearance Height (feet)
Boothbay Harbor	—	SW	Boothbay Harbor, ME	1928	FTBR	Boothbay Harbor	2	32	4
Brandywine River	7th St.	SW	Wilmington, DE	1902	HWY	Delaware	0.1	48	10
Brandywine River	8th St.	SW	Wilmington, DE	1929	RR	PRR	1.12	48	18.6
Brandywine River	Church St.	B	Wilmington, DE	1933	HWY	Delaware	1.3	40	12
Brandywine River	16th St.	VL	Wilmington, DE	1925	HWY	Delaware	1.71	40.7	54
Brays Bayou	—	VL	Harrisburg, TX	1915	RR	GH&SA(SP)RY	0.1	40.4	17.1
Brays Bayou	—	VL	Harrisburg, TX	1917	HWY	Harris Co.	0.1	40	13
Brazos River	—	SW	Freeport, TX	1916	HWY, RR	Brazoria Co., MP	5.9	108	—
Brazos River	—	SW	Brazoria, TX	1906	RR	STLB&MRY(MP)	22.6	94.3	15
Brazos River diversion channel	SR 36	SW	Freeport, TX	1929	HWY	Texas	4.4	97	10
Brices Creek	—	SW	Newbern, NC	1908	HWY	North Carolina	1	62	5.9
Broad Canal	Cambridge Pkwy.	B	Cambridge, MA	1956	HWY	Metropolitan Dist. Comm.	0	40	4
Broad Canal	1st St.	B	Cambridge, MA	1925	HWY	Metropolitan Dist. Comm.	0	40	7
Broad Canal	3rd St.	B	Cambridge, MA	1832	HWY	Cambridge	0.25	36.1	7.74
Broad Canal	6th St.	B	Cambridge, MA	1926	HWY	Cambridge	0.44	32.5	6.44
Broad Canal	—	SW	Cambridge, MA	1854	RR	B&A	0.62	32.8	2.76
Broad Creek	—	SW	Norfolk, VA	—	RR	NS	0.25	28.5	2.6
Broad Creek	—	SW	Norfolk, VA	1921	HWY	Atlantic Holding Corp.	0.25	28	11
Broad Creek River	—	SW	Bethel, DE	—	HWY	Delaware	4	52	1.3
Broad Creek River	—	SW	Laurel, DE	1915	RR	CR	8	40	14
Broad Creek River	US 13	SW	Laurel, DE	1916	HWY	Delaware	8.15	40	2
Broad Creek River	—	B	Laurel, DE	1923	HWY	Delaware	8.2	37	5
Broad Creek River	Delaware Ave.	SW	Laurel, DE	—	HWY	Delaware	8.3	29.8	3.6

Broad River	SR 170	SW	Beaufort, SC	1958	HWY	South Carolina	14	120	12
Broad River	—	SW	Beaufort, SC	1917	RR	CSX	17	57	7
Broadkill River	Rte. 14	B	Milton, DE	1928	HWY	Delaware	6	43.1	3.5
Bronx River	Bruckner Blvd.	B	Bronx, NY	1953	HWY	New York City	1.1	69	27
Bronx River	Ludlow Ave.	B	Bronx, NY	1931	HWY	New York City	1.1	70	27.5
Bronx River	Westchester Ave.	B, F	Bronx, NY	1938	HWY	New York City	1.5	60	14
Bronx River	—	RL, F	Bronx, NY	1908	RR	ATK	1.6	69	8
Brooks Slough	—	SW	Skamokawa, WA	1913	HWY	Wahkiakum Co.	0.2	60	6
Broward River	SR 105	B	Eastport, FL	1949	HWY	Florida	0.2	40	13
Broward River	—	B	Eastport, FL	1927	HWY	North Shore Corp.	0.25	30	9
Broward River	—	SW	Eastport, FL	1915	RR	CSX	0.6	40	8
Broward River	—	SW	Eastport, FL	1926	RR	ACL	0.7	44	3.5
Brunswick River	Sidney Lanier Br.	VL	Brunswick, GA	1956	HWY	Georgia	4.9	250	139
Buffalo Bayou	—	SW	Houston, TX	1936	RR	UP	0.1	100	25
Buffalo Bayou	US 90 (69th St.)	B	Houston, TX	1927	HWY	Houston	0.8	96	29
Buffalo Bayou	—	SW	Houston, TX	1911	RR	HBT	1.2	84	27
Buffalo Bayou	Lockwood Dr.	VL	Houston, TX	1928	HWY	Houston	2.3	106	57
Buffalo Bayou	Velasco St.	SW	Houston, TX	1941	RR	UP	3.1	74	29
Buffalo Bayou	—	B	Houston, TX	1915	RR	IGN	4.0	91.4	−5
Buffalo Bayou	Mary St.	B	Houston, TX	1915	RR	HBT	4.3	91	21
Buffalo City Ship Canal	Michigan Ave.	B	Buffalo, NY	—	HWY	Buffalo	1	73.5	5.8
Buffalo River	Michigan St.	VL	Buffalo, NY	1931	HWY	Buffalo	1.3	183	1.1
Buffalo River	Ohio St.	VL	Buffalo, NY	1957	HWY	Buffalo	2.1	250	10.5
Buffalo River	—	B	Buffalo, NY	1912	RR	PC	4	100	14
Buffalo River	—	B	Buffalo, NY	1912	RR	NS	4.4	97	8
Buffalo River	—	B	Buffalo, NY	1912	RR	CSX	4.5	97	8
Buffalo River	—	B	Buffalo, NY	1925	RR	EL	5.1	110	29
Buffalo River	South Park Ave.	VL	Buffalo, NY	1951	HWY	Buffalo	5.3	100	9
Buffalo River	—	B	Buffalo, NY	1925	RR	EL	5.8	100	17
Buffalo River	Bailey Ave.	B	Buffalo, NY	1927	HWY	Buffalo	6.2	90	10
Buffalo River	—	VL	Buffalo, NY	1926	PL	Iriquois Gas Corp.	6.6	170	93

*See key of abbreviations at end.

Waterway[a]	Road Name/ Route Carried	Bridge Type[b]	Location	Const. Date	Traffic Type[c]	Owner*	Miles from Mouth of Waterway	Channel Width/Lateral Clearance (feet)	Clearance Height (feet)
Buffalo River	S. Ogden St.	VL	Buffalo, NY	1931	HWY	Buffalo	7	170	10
Bullocks Cove	—	SW	Crescent Park, RI	1906	FTBR	Hope Land	0.75	26.7	4
Burns Cutoff, Tributary of San Juaquin River	Dagget Rd.	SW	Stockton, CA	1903	HWY	U.S. Navy	3	61	12
Bush River	—	B	Perryman, MD	1913	RR	ATK	6.8	35	12
Cache Slough	—	B	Rio Vista, CA	1969	HWY	Soland	—	80	6
Calcasieu River	SR 378	VL	Lake Charles, LA	1968	HWY	Louisiana	5	60	50
Calcasieu River	—	SW	Lake Charles, LA	1966	RR	UP	36.4	93	1
Calcasieu River	—	B	Lake Charles, LA	1916	HWY	Louisiana	43.5	99.6	9.7
Calcasieu River	—	SW	Lake Charles, LA	1897	RR	KCS	45.75	96.9	3.6
Calcasieu River	—	SW	Lake Charles, LA	1907	RR	LC&N(SP)	53.5	95	7.2
Calcasieu River	—	VL	Lake Charles, LA	1923	HWY	Louisiana	57	94.6	45
Caloosahatchee River (see also Okeechobee Waterway)	—	B	Goodno Island, FL	1952	PRIV RD	Connie Mack Dev.	1	12	3
Caloosahatchee River	—	B	Fort Myers, FL	1931	HWY	Florida	17.7	80	4.5
Caloosahatchee River	—	SW	Fort Myers, FL	1927	RR	SAFRR	18.4	60	6
Caloosahatchee River	—	SW	Fort Myers, FL	1904	RR	ACL	22.8	50	3.4
Caloosahatchee River	—	SW	Olga, FL	1915	HWY	Florida	29.5	55.8	4.4
Caloosahatchee River	—	SW	Alva, FL	1927	HWY	Lee Co.	36.3	68	7
Caloosahatchee River	—	SW	Fort Denaud, FL	1908	HWY	Lee Co.	43.6	50	5.2
Caloosahatchee River	—	SW	La Belle, FL	1936	HWY	Florida	48.8	50	10.8
Caloosahatchee River	—	SW	Coffee Mill Hammock, FL	1919	RR	ACL	57.8	50	7
Caloosahatchee River	—	SW	Moore Haven, FL	1921	HWY	Glades Co.	73.1	50	5
Caloosahatchee River	—	SW	Moore Haven, FL	1921	RR	ACL	73.5	50	2
Calumet River	—	VL	Chicago, IL	1974	RR	EJE	0.63	200	125
Calumet River	92nd St.	B	Chicago, IL	1914	HWY	Chicago	0.78	180	14
Calumet River	95th St.	B	Chicago, IL	1958	HWY	Chicago	1.12	204	18
Calumet River	—	B (out)	Chicago, IL	1916	RR	BO	1.31	135	14

Calumet River	—	VL	Chicago, IL	1916	RR	PC	1.335	138	12
Calumet River	—	VL	Chicago, IL	1916	RR	PC	1.345	138	12
Calumet River	—	VL (out)	Chicago, IL	1916	RR	PC	1.355	138	12
Calumet River	—	VL	Chicago, IL	1916	RR	NS	1.365	138	12
Calumet River	100th St.	B	Chicago, IL	1927	HWY	Chicago	1.76	186	13
Calumet River	106th St.	B	Chicago, IL	1929	HWY	Chicago	2.6	195	13
Calumet River	—	SW	Chicago, IL	1901	RR	CW (PRR)	5.17	87.4	13.5
Calumet River	—	VL	Chicago, IL	1972	RR	CWI	5.31	200	125
Calumet River	Torrence Ave.	VL	Chicago, IL	1934	HWY	Chicago	5.34	200	126
Calumet River	—	SW	Chicago, IL	1903	RR	NKP	5.66	79.2	4.8
Calumet River	—	VL	Chicago, IL	1970	RR	NS	5.7	200	125
Calumet River	130th St.	P (T)	Chicago, IL	1931	HWY	Chicago	6.39	70	4.6
Calumet River	—	SW	Chicago, IL	1909	RR	CSSSB	6.45	92	9.4
Cambridge Creek	SR 342	B	Cambridge, MD	1940	HWY	Maryland	0.1	50	8
Canaveral Harbor Barge Canal	SR AIA	B	Merritt Island, FL	1961	HWY	Florida	1	90	26
Canaveral Harbor Barge Canal	SR 401	B	Port Canaveral, FL	1965	HWY	Florida	5.5	90	26
Canaveral Harbor Barge Canal	SR 401	B	Port Canaveral, FL	1972	HWY	Florida	5.501	90	25
Cane River	—	SW	Derry, LA	1904	HWY	Natchitoches Pa.	26.4	90	3.5
Cany Creek	—	SW	Sargent, TX	1918	HWY	Matagorda Co.	9.4	51	4
Cany Creek	—	SW	Sargent, TX	1891	HWY	Matagorda Co.	25	26.5	0
Cape Cod Canal	Cape Cod RR	VL	Bourne, MA	1935	RR	U.S. Army COE	0.7	500	135
Cape Fear River	US 17, 421, 74, 76	VL	Wilmington, NC	1969	HWY	North Carolina	26.8	350	135
Cape Fear River	US 421	B	Wilmington, NC	1929	HWY	North Carolina	30	120	26
Cape Fear River	—	B	Navassa, NC	1913	RR	CSX	34	102	6
Cape Island Creek	US 9	B, F	Cape May, NJ	1929	HWY	Cape May Co.	0.3	38	5
Cape Neddick River	—	SW	York, ME	1908	RR	ASLRY	0.3	30	—
Carabelle River	—	SW	Carabelle, FL	1927	HWY	Florida	1.5	40	4
Carlin Bayou	—	VL	Delcambre, LA	1939	RR	SP	6.4	40	46

*See key of abbreviations at end.

Waterway[a]	Road Name/ Route Carried	Bridge Type[b]	Location	Const. Date	Traffic Type[c]	Owner*	Miles from Mouth of Waterway	Channel Width/Lateral Clearance (feet)	Clearance Height (feet)
Carlin Bayou	SR 14	VL	Delcambre, LA	1935	HWY	Louisiana	6.4	40	44
Carquinez Strait	—	VL	Martinez, CA	1930	RR	UP	7	291	135
Carvers Harbor	—	B	Vinalhaven, ME	—	HWY	Vinalhaven	—	30	—
Cat Point Creek	—	R	Mount Airy, VA	1936	HWY	Richmond Co.	6.5	31.5	7
Cat Point Creek	SR 634	R	Naylors, VA	—	HWY	Virginia	0.3	31	7
Catching Slough	—	SW	Coos Bay, OR	1939	HWY	Oregon	1	50	13
Catskill Creek	—	B	Catskill, NY	1930	HWY	New York	0.7	70	14
Cedar Bayou	—	B	Baytown, TX	1931	HWY	Harris Co.	0.5	80	13
Cedar Bayou	—	VL	Baytown, TX	1968	RR	UP	7	146	82
Cedar Bayou	—	MB	Cedar Bayou, TX	1929	HWY	Harris Co.	10.5	70	0
Cedar Creek	—	SW	Cedar Beach, DE	1950	HWY	Delaware	0.5	23	2
Centerville River	—	B	Centerville, MA	—	FTBR	Ninkle et al.	0.3	26	9
Chambers Creek–Steilacoom Creek	—	VL	Steilacoom, WA	1941	RR	BNSF	—	85	10
Chambly Canal	Granby	SW	Granby Industrial Spur, Canada	1928	RR	CN	—	96	—
Champlain Lake	—	SW, F	South Bay, NY	1925	RR	DH	2.5	96	7
Champlain Lake	—	B	South Bay, NY	1930	HWY	New York	3.2	90	4.6
Champlain Lake	—	SW	South Hero, VT	1899	RR	Vermont	78.1	80	4
Champlain Lake	North Hero	SW	South Hero, VT	1899	RR	Vermont	90.1	80	5
Champlain Lake	—	B	Grand Island, VT	1953	HWY	Vermont	91.8	—	—
Champlain Lake	—	B (out)	Grand Island, VT	1944	HWY	Vermont	91.8	80	13
Champlain Lake	North Hero	SW	South Hero, VT	1922	HWY	Vermont	91.8	80.6	3.5
Champlain Lake	—	SW	Pelot Pt., VT	1899	RR	rut	95.7	77	5.3
Champlain Lake	—	SW	North Hero, VT	1896	HWY	Vermont	99.2	79	4.7
Champlain Lake	—	SW	Isle La Motte, VT	1912	HWY	Grand Isle Co.	99.4	30	3.5
Champlain Lake	—	SW	Missisquoi Bay, VT	1912	RR	CV	105.6	36	6
Champlain Lake	—	B	Missisquoi Bay, VT	1938	HWY	Vermont	105.9	45	13

Champlain Lake	—	SW	Rouses Pt., NY	1907	RR	Vermont	106.6	89	7
Champlain Lake	—	SW	Rouses Pt., NY	1936	HWY	Lake Champlain Bridge Comm.	106.8	126	17
Channel Street Mission Creek China Basin	3rd St.	B	San Francisco, CA	1933	HWY	San Francisco	0	103	1
Channel Street Mission Creek China Basin	4th St.	B	San Francisco, CA	1933	HWY	San Francisco	0.2	75	0
Charenton Drainage & Navigation Canal	—	SW	Baldwin, LA	1941	RR	SP	0.4	75	5
Charles River	City Square	SW	Boston, MA	1900	HWY	Boston	0.44	50	23
Charles River	—	R	Boston, MA	1828	HWY	Boston	0.56	36.5	6
Charles River	BM-1	B	Boston, MA	1931	RR	Metropolitan Boston Transit Auth.	0.8	65	3
Charles River	Green Line	B	Boston, MA	1912	RR	Metropolitan Boston Transit Auth.	1	50	33
Charles River	M. O'Brien Hwy.	B	Boston, MA	1912	HWY	Metropolitan Dist. Comm.	1	50	5
Chattahoochee River	—	SW	Alaga, AL	1931	HWY	Alabama and Georgia	35.5	100	6
Chattahoochee River	—	SW	Alaga, AL	1913	RR	ACL	36	98	7
Chattahoochee River	—	SW	Columbia, AL	1896	HWY	Columbia	50.5	100	6
Chattahoochee River	—	SW	Columbia, AL	1889	RR	CG	50.8	100	4.5
Chattahoochee River	—	VL	Omaha, GA	1969	RR	CSX	117.1	153	57
Chaumont River	—	B	Chaumont, NY	1849	HWY	Lyme	0.8	53.5	2.7
Chaumont River	—	SW	Chaumont, NY	1912	RR	NYC	0.9	50	16.2
Cheboygan River	State St.	B	Cheboygan, MI	1939	HWY	Michigan	0.9	60	5
Cheboygan River	State Rd.	SW	Cheboygan, MI	—	HWY	Michigan	0.92	46.5	4.5

*See key of abbreviations at end.

Waterway[a]	Road Name/ Route Carried	Bridge Type[b]	Location	Const. Date	Traffic Type[c]	Owner*	Miles from Mouth of Waterway	Channel Width/Lateral Clearance (feet)	Clearance Height (feet)
Cheboygan River	—	SW	Alanson, MI	—	HWY	Emmet Co.	29.35	22	4.6
Cheboygan River	—	SW	Alanson, MI	—	HWY	Emmet Co.	29.6	21.5	5.5
Cheesequake Creek	—	B	Morgan, NJ	1913	HWY	Middlesex Co.	0	51	5
Cheesequake Creek	SR 35	B	Morgan, NJ	1941	HWY	New Jersey	0	50	25
Cheesequake Creek	—	B	Morgan, NJ	1912	RR	New Jersey Transit	0.2	50	3
Chef Menteur Pass	—	SW	Chef Menteur, LA	1926	RR	CSX	2.5	104	10
Chef Menteur Pass	US 90	SW	Chef Menteur, LA	1929	HWY	Louisiana	2.8	97	11
Chefuncte River	SR 22	SW	Madisonville, LA	1954	HWY	Louisiana	2.5	79	5
Chefuncte River	SR 22	SW	Madisonville, LA	1964	HWY	Louisiana	2.5	125	6
Chefuncte River and Bogue Falls	—	SW	Madisonville, LA	1935	HWY	Louisiana	2.5	60	2.3
Chehalis River	—	SW	Aberdeen, WA	1911	RR	UP	0	125	11
Chehalis River	—	B	Aberdeen, WA	1952	HWY	Washington	0.1	150	35
Chehalis River	—	SW	Aberdeen, WA	1926	HWY	Aberdeen	1.5	130	8.5
Chehalis River	—	SW	South Aberdeen, WA	1892	RR	NP	3	125	9
Chehalis River	—	SW	Montesano, WA	1912	RR	UP	13.1	100	8
Chehalis River	—	SW	Montesano, WA	1910	HWY	Grays Harbor	13.35	100	8
Chelsea (River) Creek	Meridian St.	SW	East Boston–Chelsea, MA	1914	HWY	Boston	0.25	100	5
Chelsea River	P. J. McCardle Br.	B	Boston, MA	1946	HWY	Boston	0.3	175	21
Chelsea River	Chelsea St.	B	Boston–Chelsea, MA	1937	HWY	Boston	1.2	96	9
Chelsea (River) Creek	—	B	Boston–Chelsea, MA	1921	RR	B&A	1.19	70	3
Chesapeake & Delaware Canal	—	B	Delaware City, DE	1934	HWY	U.S. Govt.	0.62	60	6.75
Chesapeake & Delaware Canal	—	VL	Reedy Point, DE	1927	HWY	U.S. Govt.	0.95	168.2	140

Chesapeake & Delaware Canal	—	VL	Canal, DE	1927	RR	CR	7.65	500	133
Chesapeake & Delaware Canal	—	VL	Summit Bridge, DE	1926	HWY	U.S. Govt.	9	163	140
Chesapeake & Delaware Canal	—	VL	Chesapeake City, DE	1926	HWY	U.S. Govt.	13.8	221.3	140
Chesapeake Bay–Honga River Waterway	—	SW	Hoopersville, MD	1940	HWY	Maryland	0	25	6.4
Chester River 1	SR 213	B	Chestertown, MD	1930	HWY	Maryland	26.8	60	12
Chester River 1	—	SW	Crumpton, MD	1923	HWY	Queen Anne and Kent Cos.	36	37	4
Chester River 2	—	SW	Chester, PA	1908	RR	CR	0.1	51	1
Chevreuil Bayou	—	RSP	Vacherie, LA	1930	HWY	Louisiana	12	36	6
Chevron Oil Company Canal	SR 3093	SW	Leeville, LA	1972	HWY	Lafourcher Pa.	0.5	50	12
Chicago River	Lake Shore Dr.	B	Chicago, IL	1938	HWY	South Park Comm.	0.3	210	22
Chicago River	Columbus Dr.	B	Chicago, IL	1979	HWY	Chicago	0.7	180	20
Chicago River	Michigan Ave.	B (2)	Chicago, IL	1912	HWY	Chicago	0.9	195	17
Chicago River	Wabash Ave.	B	Chicago, IL	1929	HWY	Chicago	1	193	21
Chicago River	State St.	B	Chicago, IL	1950	HWY	Chicago	1.1	200	21
Chicago River	Dearborn St.	B	Chicago, IL	1963	HWY	Chicago	1.2	200	18
Chicago River	Clark St.	B	Chicago, IL	1929	HWY	Chicago	1.2	200	19
Chicago River	La Salle St.	B	Chicago, IL	1925	HWY	Chicago	1.3	200	17
Chicago River	Wells St.	B	Chicago, IL	1922	HWY, RR	Chicago	1.4	200	17
Chicago River	Orleans St.	B	Chicago, IL	1916	HWY	Chicago	1.5	210	17
Chicago River, North Branch	—	B	Chicago, IL	1909	RR	UP	1.77†	100	6
Chicago River, North Branch	Kinzie St.	B	Chicago, IL	1909	HWY	Chicago	1.82	105	14
Chicago River, North Branch	—	SW	Chicago, IL	1909	RR	MILW	1.85	100	4.2

*See key of abbreviations at end.

† Mileage from Lake Michigan.

Waterway[a]	Road Name/ Route Carried	Bridge Type[b]	Location	Const. Date	Traffic Type[c]	Owner*	Miles from Mouth of Waterway	Channel Width/Lateral Clearance (feet)	Clearance Height (feet)
Chicago River, North Branch	Grand Ave.	B	Chicago, IL	1914	HWY	Chicago	2	134	17
Chicago River, North Branch	Kennedy Expwy.– Ohio St.	B	Chicago, IL	1953	HWY	Chicago	2.1	138	27
Chicago River, North Branch	Erie St.	B (removed)	Chicago, IL	1910	HWY	Chicago	2.21	131.6	—
Chicago River, North Branch	Chicago Ave.	B	Chicago, IL	1914	HWY	Chicago	2.4	145	17
Chicago River, North Branch	N. Halstead St.	B	Chicago, IL	1959	HWY	Chicago	2.6	140	17
Chicago River, North Branch	Ogden Ave.	B	Chicago, IL	1933	HWY	Chicago	2.9	138	22
Chicago River, North Branch	Division St.	B	Chicago, IL	1904	HWY	Chicago	3.3	100	17
Chicago River, North Branch	North Ave.	B	Chicago, IL	1908	HWY	Chicago	3.9	137	16
Chicago River, North Branch	—	SW	Chicago, IL	1899	RR	CP	4.4	81	5
Chicago River, North Branch	Cortland St.	B	Chicago, IL	1902	HWY	Chicago	4.5	101	13
Chicago River, North Branch	Webster Ave.	B	Chicago, IL	1916	HWY	Chicago	4.9	128	11
Chicago River, North Branch	Ashland Ave.	B	Chicago, IL	1936	HWY	Chicago	4.96	140	17
Chicago River, North Branch	—	B	Chicago, IL	1916	RR	UP	5	145	11
Chicago River, North Branch	Damen Ave.	B	Chicago, IL	1929	HWY	Chicago	5.6	118	17

Chicago River, North Branch	Fullerton Ave.	SW	Chicago, IL	1899	HWY	Chicago	5.36	56	13.8
Chicago River, North Branch	Diversey Blvd.	SW	Chicago, IL	1895	HWY	Chicago	6.07	55.6	13.3
Chicago River, North Branch	N. Western Ave.	B	Chicago, IL	1904	HWY	Chicago	6.48	100	14.1
Chicago River, North Branch	Belmont Ave.	B	Chicago, IL	1917	HWY	Chicago	6.84	74	14.2
Chicago River, North Branch Canal	N. Halstead St.	B	Chicago, IL	1908	HWY	Chicago	2.8	56	15
Chicago River, North Branch Canal	Ogden Ave.	B	Chicago, IL	1933	HWY	Chicago	2.9	132	27
Chicago River, North Branch Canal	Division St.	B	Chicago, IL	1903	HWY	Chicago	3	74	17
Chicago River, North Branch Canal	—	SW	Chicago, IL	1903	RR	CP	3.6	111	7
Chicago River, South Branch	Lake St.	B	Chicago, IL	1916	HWY, RR	Chicago	1.6†	195	18
Chicago River, South Branch	Randolph St.	B	Chicago, IL	1980	HWY	Chicago	1.74	134	21
Chicago River, South Branch	Washington St.	B	Chicago, IL	1913	HWY	Chicago	1.82	160	17
Chicago River, South Branch	Madison St.	B	Chicago, IL	1922	HWY	Chicago	1.91	188	17
Chicago River, South Branch	Monroe St.	B	Chicago, IL	1919	HWY	Chicago	2	165	17
Chicago River, South Branch	Adams St.	B	Chicago, IL	1928	HWY	Chicago	2.09	151	17
Chicago River, South Branch	Jackson Blvd.	B	Chicago, IL	1916	HWY	Chicago	2.18	180	17

*See key of abbreviations at end.

†Mileage from Lake Michigan.

Waterway[a]	Road Name/ Route Carried	Bridge Type[b]	Location	Const. Date	Traffic Type[c]	Owner*	Miles from Mouth of Waterway	Channel Width/Lateral Clearance (feet)	Clearance Height (feet)
Chicago River, South Branch	—	B	Chicago, IL	1895	RR	Metropolitan Elevated Rwy	2.22	92	6.6
Chicago River, South Branch	Van Buren St.	B	Chicago, IL	1957	HWY	Chicago	2.27	170	17
Chicago River, South Branch	Eisenhower Expwy.	B (2)	Chicago, IL	1956	HWY	Chicago	2.4	169	17
Chicago River, South Branch	Harrison St.	B	Chicago, IL	1961	HWY	Chicago	2.45	170	17
Chicago River, South Branch	Polk St.	B	Chicago, IL	1910	HWY	Chicago	2.61	130	Leaves removed 1972
Chicago River, South Branch	Roosevelt Rd.	B	Chicago, IL	1927	HWY	Chicago	2.94	170	17
Chicago River, South Branch	—	B	Chicago, IL	1930	RR	CSX	3.36	171	11
Chicago River, South Branch	—	B	Chicago, IL	1919	RR	CN 50%, UP & BNSF REM, OPR BY CSX	3.37	170	20
Chicago River, South Branch	18th St.	B	Chicago, IL	1961	HWY	Chicago	3.59	125	22
Chicago River, South Branch	—	VL	Chicago, IL	1911	RR	ATK	3.77	156	10
Chicago River, South Branch	Canal St.	B	Chicago, IL	1941	HWY	Chicago	3.88	167	17
Chicago River, South Branch	Cermak Rd.	B	Chicago, IL	1907	HWY	Chicago	4.04	129	14

Chicago River, South Branch	S. Halsted St.	B	Chicago, IL	1934	HWY	Chicago	4.47	163	24
Chicago River, South Branch	Throop St.	B	Chicago, IL	1903	HWY	Chicago	5.08	134	16
Chicago River, South Branch	Loomis St.	B	Chicago, IL	1974	HWY	Chicago	5.3	144	22
West Fork of Chicago River, South Branch	Ashland Ave.	B	Chicago, IL	1938	HWY	Chicago	5.58	183	21
West Fork of Chicago River, South Branch	Damen Ave.	B	Chicago, IL	1930	HWY	Chicago	6.14	140	24
South Fork of Chicago River, South Branch	—	B, F	Chicago, IL	1907	RR	ICG	5.79	92.8	13
South Fork of Chicago River, South Branch	Archer Ave.	B	Chicago, IL	1906	HWY	Chicago	5.88	90	13.4
South Fork of Chicago River, South Branch	35th St.	B	Chicago, IL	1914	HWY	Chicago	6.55	118	10.4
South Fork of Chicago River, South Branch, West Arm	Iron St.	SW	Chicago, IL	1909	HWY, RR	CJRWY	6.98	69.2	5.4
Chicago Sanitary and Ship Canal	—	VL	Lockport, IL	1934	RR	EJE	290.1†	225	50
Chicago Sanitary and Ship Canal	16th St.	SW	Lockport, IL	1906	HWY	Illinois	292.1	160	0
Chicago Sanitary and Ship Canal	Romeo Rd.	SW	Romeo, IL	1899	HWY	Illinois	296.2	160	13
Chicago Sanitary and Ship Canal	Lemont Rd.	SW, F (out)	Lemont, IL	1899	HWY	Illinois	300.5	160	13
Chicago Sanitary and Ship Canal	—	SW, F	Lemont, IL	1899	RR	BNSF	300.6	160	14

*See key of abbreviations at end.
†Mileage from Mississippi River.

Waterway[a]	Road Name/ Route Carried	Bridge Type[b]	Location	Const. Date	Traffic Type[c]	Owner*	Miles from Mouth of Waterway	Channel Width/Lateral Clearance (feet)	Clearance Height (feet)
Chicago Sanitary and Ship Canal	—	SW, F (out)	Willow Springs, IL	1899	HWY	Willow Springs	307.9	160	14
Chicago Sanitary and Ship Canal	—	SW, (repl. w. fixed)	Argo, IL	1898	RR	CSX	312.4	113	14
Chicago Sanitary and Ship Canal	Harlem Ave.	B F	Stickney, IL	1932	HWY	Cook Co.	314	130	21
Chicago Sanitary and Ship Canal	Harlem Ave.	B, F	Stickney, IL	1968	HWY	Cook Co.	314	140	19
Chicago Sanitary and Ship Canal	—	SW, F	Lemoyne, IL	1898	RR	BNSF	314.8	107	14
Chicago Sanitary and Ship Canal	S. Cicero Ave.	B, F	Chicago, IL	1927	HWY	Chicago	317.3	140	14
Chicago Sanitary and Ship Canal	—	SW, F	Chicago, IL	1900	RR	BRC	317.6	97	13
Chicago Sanitary and Ship Canal	Pulaski Rd.	B, F (repl. w. fixed)	Chicago, IL	1930	HWY	Chicago	318.4	140	14
Chicago Sanitary and Ship Canal	—	SW, F	Chicago, IL	1899	RR	IN(ATSF)(BNSF)	318.9	95	14
Chicago Sanitary and Ship Canal	S. Kedzie Ave.	VL, F	Chicago, IL	1971	HWY	Chicago	319.5	130	18
Chicago Sanitary and Ship Canal	—	SW, F	Chicago, IL	1899	RR	CN	319.6	100	15
Chicago Sanitary and Ship Canal	California Ave.	B, F	Chicago, IL	1922	HWY	Chicago	320	140	13
Chicago Sanitary and Ship Canal	—	B, F (abandoned)	Chicago, IL	1909	RR	SDCHI	320.4	120	13

Chicago Sanitary and Ship Canal	—	B, F	Chicago, IL	1910	RR	CR	320.41	120	13
Chicago Sanitary and Ship Canal	—	B, F	Chicago, IL	1910	RR	CRI (CJ per COE 1941)	320.42	120	13
Chicago Sanitary and Ship Canal	—	B, F	Chicago, IL	1909	RR	CSX	320.43	120	13
Chicago Sanitary and Ship Canal	Western Ave.	VL, F	Chicago, IL	1940	HWY	Chicago	320.5	155	18
Chicago Sanitary and Ship Canal, North Arm	—	B	Chicago, IL	—	RR	CN IC	—	—	—
Chickahominy River	—	SW	Barrets Ferry, VA	1939	HWY	Virginia	1.5	80	12
Chickasaw Creek	—	SW	Prichard, AL	1927	RR	CSX	0	131	6
Chickasaw Creek	US 43	RSP	Chickasaw, AL	1958	HWY	Alabama	4.1	67	6
Chickasaw Creek	—	RSP	Mobile, AL	1928	HWY	Mobile Co.	4.1	60	7.5
Chico Bayou	—	B	Pensacola, FL	1949	HWY	Florida	0.3	90	13
Chico Bayou	—	B	Pensacola, FL	1920	HWY	Escambia Co.	0.6	80	6
Chico Bayou	—	B	Pensacola, FL	1918	RR	MSB&PRR(SLSF)	0.6	80	1.5
Chincoteague Bay	SR 175	SW	Chincoteague, VA	1940	HWY	Virginia	3.5	52	15
Chipola River	—	SW	Scotts Ferry, FL	1913	HWY	Calhoun Co.	6	60	15
Chipola River	—	B	Scotts Ferry, FL	—	RR	Chical Lumber	8	40	14
Chipola River	—	B	Scotts Ferry, FL	—	RR	MBT	26.5	36	14
Chippewa River	—	SW	Trevino, WI	1904	RR	CB&Q	1.5	140.5	10
Chippewa River	Walcott St.	SW	Durand, WI	1902	HWY	Durand	17.4	145	3.2
Chocolate Bayou	—	SW	Liverpool, TX	1892	HWY	Brazoria Co.	10.8	51	6.5
Chocolate Bayou	—	SW	Liverpool, TX	1911	HWY	Brazoria Co.	13.6	46.5	6.1
Chocolate Bayou	—	SW	Liverpool, TX	1905	RR	STLB&MRR(MP)	14	54	21
Choctaw Bayou	—	RSP	Morley, LA	—	RR	MP	8	41	25
Choctawhatchee River	SR 20	SW	Ebro, FL	1930	HWY	Florida	20	78	22
Choctawhatchee River	—	B	Caryville, FL	1923	HWY	Florida	66.9	80	5
Choctawhatchee River	—	SW	Caryville, FL	1913	RR	CSX	67	78	11
Choctawhatchee River	—	SW, F	Geneva, AL	1901	RR	LN	96	78	3.5

*See key of abbreviations at end.

Waterway[a]	Road Name/ Route Carried	Bridge Type[b]	Location	Const. Date	Traffic Type[c]	Owner*	Miles from Mouth of Waterway	Channel Width/Lateral Clearance (feet)	Clearance Height (feet)
Choctawhatchee River	—	SW	Geneva, AL	1903	HWY	Alabama	97	98	2.5
Choctawhatchee River	—	SW	Bellwood, AL	1903	HWY	Geneva Co.	106	60	20
Choctawhatchee River	—	SW	Bellwood, AL	1907	HWY	Geneva Co.	106.5	64	11
Choctawhatchee River	—	VL	Bellwood, AL	—	RR	COFG	114	27	5
Choctawhatchee River	—	SW	Newton, AL	—	RR	ACL	140	57	5
Choptank River	US 50	SW	Cambridge, MD	1935	HWY	Maryland	15.6	100	18
Choptank River	SR 331	SW	Dover, MD	1934	HWY	Maryland	35.3	80	10
Choptank River	—	B	Denton, MD	1913	HWY	Maryland	49	50	4.3
Choptank River	—	SW	Denton, MD	1948	RR	CR	50.9	50	6
Choupique Bayou	—	SW	Calcasieu, LA	1941	HWY	Louisiana	4.4	46	5.22
Choupique Bayou	—	SW	Calcasieu, LA	1915	HWY	Calcasieu Pa.	13.75	45	6.7
Chowan River	US 17	SW	Edenton, NC	1953	HWY	North Carolina	2	80	4
Chowan River	—	SW	Edenton, NC	1927	HWY	North Carolina	2	80	4.5
Chowan River	—	SW	Tunis, NC	—	RR	ACL	34	54	2
Chowan River	—	SW	Winton, NC	1925	HWY	North Carolina	36.5	60	2
Chowan River	—	SW	Riddicksville, NC	1930	RR	Camp Manuf.	47	49.5	5
Christina River	—	SW	Wilmington, DE	1888	RR	CR	1.4	84	6
Christina River	Christina Ave.	B	Wilmington, DE	1974	HWY	Delaware	2.3	145	14
Christina River	3rd St.	B	Wilmington, DE	1915	HWY	Delaware	2.3	145	15
Christina River	Walnut St.	B	Wilmington, DE	1957	HWY	Delaware	2.8	175	13
Christina River	Market St.	B	Wilmington, DE	1928	HWY	Delaware	3	175	8
Christina River	—	SW	Wilmington, DE	1873	RR	B&O	3.48	67.3	7.6
Christina River	—	SW	Wilmington, DE	1888	RR	CR	4.1	62	6
Christina River	—	SW	Wilmington, DE	1914	RR	CR	4.2	57	3
Christina River	—	SW	Wilmington, DE	1852	RR	CR	5.4	37	2
Christina River	—	B, F	Newport, DE	1929	HWY	Delaware	7.5	49	4
Christina River	—	F, convertible to bascule	Churchman, DE	1933	HWY	Delaware	12.5	61	8

Chuckatuck Creek	US 17	B	Crittenden, VA	1928	HWY	Virginia	1	80	21
Church Creek	—	SW	Johns Island, SC	1919	HWY	South Carolina	5.5	32.4	6.8
City Waterway	S. 11th St.	VL	Tacoma, WA	1913	HWY	Tacoma	0.6	200	64
City Waterway	S. 14th St.	SW	Tacoma, WA	1927	RR	NP	0.8	100	12.6
City Waterway	S. 15th St.	SW	Tacoma, WA	1915	RR, HWY	UP	0.87	100	6
Clapboard Creek	—	B	Mayport, FL	1927	HWY	North Shore Corp.	0.1	28.9	9
Clatskanie Creek	—	SW	Clatskanie, OR	1898	RR	BNSF	0.7	60	6
Clear Creek	SR 146	B	Seabrook, TX	1958	HWY	Texas	1	75	42
Clear Creek	—	B	Seabrook, TX	1928	HWY	Texas	1	40	6.1
Clear Creek	—	B	Seabrook, TX	1971	HWY	Texas	1	75	43
Clear Creek	—	SW	Seabrook, TX	1895	RR	SP	1	60	6
Clear Creek	SR 528	B	Friendswood, TX	1962	HWY	Galveston and Harris Cos.	17.8	50	24
Clearwater Harbor	—	SW	Anona, FL	1916	HWY	Florida	0	50	5.2
Clearwater Harbor	Central Ave.	B	St. Petersburg, FL	1939	HWY	City of Treasure Island	5.3	80	8.5
Clearwater Harbor	—	B	Clearwater, FL	1927	HWY	Pinellas Co.	6.4	50	6.2
Clearwater Harbor	—	B	St. Petersburg, FL	1926	HWY	Pinellas Co.	6.5	50	8
Clearwater Harbor	—	B	St. Petersburg, FL	1927	HWY	Pinellas Co.	7	60	8
Clearwater Pass	—	B	Sand Key, FL	1962	HWY	Florida	0	50	24
Clearwater River	—	SW	Lewiston, ID	1908	RR	UP	0.6	100	18
Clearwater River	—	VL	Lewiston, ID	1972	RR	UP	0.6	232	60
Clinton River	Market St.	SW	Mt. Clemens, MI	—	HWY	Mt. Clemens	7.2	32	9.7
Clinton River	Macomb St.	SW	Mt. Clemens, MI	1916 repaired	HWY	Mt. Clemens	7.33	47	13.6
Clubfoot Creek	Harlowe Bridge	SW	Beaufort, NC	1868	HWY	North Carolina	5	29.8	4.9
Coal Bank Slough	—	SW	Marshfield, OR	1893	RR	SP	0.1	42	1
Coal Bank Slough	—	B	Marshfield, OR	1938	HWY	Oregon	0.15	32	13.24
Coal Creek Slough	—	SW	Stella, WA	1930	HWY	Washington	0.06	54	1.75
Coasters Harbor	Gate 1	SW, F	Newport, RI	1933	HWY	U.S. Navy	0	31	3
Coffee Pot Bayou	—	B	St. Petersburg, FL	1932	HWY	St. Petersburg	0.4	34	6

*See key of abbreviations at end.

Waterway[a]	Road Name/ Route Carried	Bridge Type[b]	Location	Const. Date	Traffic Type[c]	Owner*	Miles from Mouth of Waterway	Channel Width/Lateral Clearance (feet)	Clearance Height (feet)
Cohansey River	Broad St.	B, F	Bridgetown, NJ	1936	HWY	New Jersey	18.2	40	6
Cohasset Narrows	—	B, F	Buzzards Bay–Wareham, MA	1913	RR	CR	0.75	25	6
Coldwater River	—	SW	Marks, MS	1908	HWY	Quitman Co.	12.7	60	4
Coldwater River	—	SW	Marks, MS	1903	RR	IC	16	54.8	0.6
Coldwater River	—	B	Darling, MS	1910	HWY	Quitman Co.	23.5	60	−0.7
Coldwater River	—	B	Ackland, MS	1916	HWY	Quitman Co.	25.4	60	−5.5
Coldwater River	—	SW	Parnell, MS	1905	HWY	Quitman Co.	29.8	—	—
Coldwater River	—	B	Jonestown, MS	1920	HWY	Coahoma Co.	34.3	60	−2.3
Coldwater River	—	B	Birdie, MS	1919	HWY	Quitman Co.	39	55	4
Colgate Creek	—	SW	Baltimore, MD	1927	HWY	Baltimore	0.1	40	4.25
Colgate Creek	—	SW	Baltimore, MD	1929	RR	CR	1	30	5
Colorado River 1	Fm 521	B	Wadsworth, TX	1959	HWY	Texas	10.7	100	28
Colorado River 2	—	RSP	Lower Cibola, CA	1970	HWY	U.S. Bureau of Reclamation	93.6	71	17
Colorado River 2	—	RSP	Cibola, CA	1966	HWY	U.S. Bureau of Reclamation	100.1	47	17
Colorado River 2	Farmers Br.	RSP	Cibola, CA	1981	HWY	Cibola Valley Irrigation and Drainage Dist.	104.15	50	17
Columbia River	—	SW	Vancouver, WA	1908	RR	BNSF	105.6	200	21
Columbia River	I 5 SB	VL	Vancouver, WA	1958	HWY	Oregon	106.5	263	159
Columbia River	I 5 NB	VL	Vancouver, WA	1958	HWY	Oregon	106.5	263	159
Columbia River	—	VL	Hood River, OR	1951	HWY	Port of Hood River	169.8	246	149
Columbia River	—	VL	Celilo, OR	1957	RR	SI	201.2	290	78
Columbia River	—	VL	Kennewick, WA	1975	RR	UP	323.4	379	73
Columbia River	—	VL	Kennewick, WA	1952	RR	BNSF	328	290	70

Columbia Slough	—	RSP	Portland, OR	1929	HWY	Multnomah Co.	6.5	66	13
Colyell Bayou	—	RSP	Port Vincent, LA	1950	HWY	Louisiana	1	37	4
Combahee–Salkchatchie River	—	SW	Wiggins, SC	1926	RR	SAL	14.2	44	6
Combahee–Salkchatchie River	—	SW	Sheldon, SC	1926	HWY	South Carolina	23.2	100	8.4
Compamy Canal	—	P	Lockport, LA	1921	HWY	Louisiana	0.2	90	—
Compamy Canal	—	VL	Bourg, LA	1951	HWY	Louisiana	8.1	50	5
Compton Creek	—	SW	Belford, NJ	1928	HWY	Monmouth Co.	0.2	40	4
Coney Island Creek	Cropsey Ave.	B, F	Brooklyn, NY	1932	HWY	New York City	0.4	75	11
Coney Island Creek	Stilwell Ave.	SW, F	Brooklyn, NY	1916	HWY	New York City	0.6	40	5
Coney Island Creek	—	B	Brooklyn, NY	1908	RR	BMT	0.74	34	3.6
Coney Island Creek	—	B	Brooklyn, NY	1907	PL, FTBR	BBGAS	0.94	—	3.5
Coney Island Creek	—	B	Brooklyn, NY	—	RR	Nassau Electric Rwy.	0.98	—	2.5
Congaree River	—	SW	Fort Motte, SC	1917	RR	NS	4.3	90	10
Conneaut River	—	SW	Conneaut, OH	—	RR	BLE	1.2	—	—
Connecticut River	—	B	Old Lyme, CT	1907	RR	ATK	3.4	120	19
Connecticut River	—	B	Lyme, CT	1911	HWY	Connecticut	4.1	200	29.6
Connecticut River	—	SW	East Haddam, CT	1913	HWY	Connecticut	16.8	200	22
Connecticut River	—	SW	Middletown, CT	1910	RR	Connecticut DOT	32	100	25
Connection Slough	R 2027	SW	Mandeville, CA	1966	HWY	Delta Farms	2.5	95	7
Contentnea Creek	—	SW	Grifton, NC	1894	RR	ACL	4.5	43.6	—
Contentnea Creek	—	B, F	Grifton, NC	1924	HWY	North Carolina	5	57	2
Contentnea Creek	Edwards Br.	SW	Grifton, NC	1924	HWY	North Carolina	13	48	3
Contraband Bayou	—	B	Lake Charles, LA	1936	HW	Calcasieu Pa.	3	35.1	4
Cooper River 1	State St.	SW	Camden, NJ	1898	HWY	Camden Co.	0.3	46	7
Cooper River 1	River Ave.	SW	Camden, NJ	1930	RR	CR	0.9	34	3
Cooper River 1	Federal St.	B	Camden, NJ	1908	HWY	Camden Co.	1	50	6
Cooper River 1	Adm. Wilson Blvd.	B, F	Camden, NJ	1927	HWY	New Jersey	1.1	67	3
Cooper River 2	—	B	Cordesville, SC	1978	RR	CSX	42.8	100	50

*See key of abbreviations at end.

Waterway[a]	Road Name/ Route Carried	Bridge Type[b]	Location	Const. Date	Traffic Type[c]	Owner*	Miles from Mouth of Waterway	Channel Width/Lateral Clearance (feet)	Clearance Height (feet)
Cooper River 2	—	VL	Moncks Corner, SC	1943	RR	CSX	49.8	100	50
Coos Bay	—	SW	North Bend, OR	1916	RR	SP	9	197	13
Coos River	—	VL	Eastside, OR	1953	HWY	Oregon	2.2	60	65
Coosa River	—	SW	Riverside, AL	1930	HWY	Alabama State Bridge Corp.	120.3	100	5
Coosa River	US 78	SW	Riverside, AL	1972	HWY	Alabama	122.3	100	9
Coosa River	—	SW	Riverside, AL	1964	RR	NS	123	99	13
Coosa River	—	SW	Riverside, AL	1896	RR	SOU	123	99	10
Coosa River	—	SW	Ohatchee, AL	1963	RR	SCL	148	83	11
Coosa River	—	SW	Lock, AL	1882	RR	SAL	148	83	0
Coosa River	SR 93	SW	Rainbow City, AL	1937	HWY	Etowah Co.	163.7	100	8.6
Coosa River	—	SW	Gadsden, AL	1910	RR	CSX	175	100	14
Coosa River	—	SW	Leesburg, AL	1927	HWY	Alabama	227	110	4.5
Coosa River	—	SW	Cedar Bluff, AL	1930	HWY	Alabama State Bridge Corp.	260.4	100	5
		—							
Coosaw River	—	SW	Seabrook, SC	1917	RR	CSX	5.3	51	5
Coosaw River	US 17, 21	SW	Beaufort, SC	1931	HWY	SC	7	34	5
Coosawhatchie River	—		Coosawhatchie, SC	—	RR	CSX	—	—	—
Copano Bay	—	B	Rockport, TX	1931	HWY	Texas	0	42	1
Coquille River	—	VL	Bandon, OR	1954	HWY	Oregon	3.5	75	75
Coquille River	US 101	SW	Coquille, OR	1923	HWY	Coos Co.	24	110	20
Cordelia Slough	—	SW	Suisun, CA	1900	RR	SP	1.5	40	4
Corey Creek	—	B	Great Hog Neck, NY	1969	FTBR	Laughing Water Property Owners Assoc.	0	20	5
Corpus Christi Aransas Pass Channel	—	B	Corpus Christi, TX	1926	HWY, RR	Nueces Co. Navigation Dist.	—	90	0

Corpus Christi Channel	Tule Lake Channel	VL	Corpus Christi, TX	—	HWY, RR	Corpus Christie	14	300	138
Corson Inlet	—	B	Strathmere, NJ	1948	HWY	Cape May Co.	0.9	50	15
Corte Madera Creek	—	B	Greenbrae, CA	1924	RR	NWP	0.5	40	8
Corte Madera Creek	—	B	Corte Madera, CA	1930	HWY	California	0.5	40	10.5
Courtableau Bayou	—	SW	Courtableau, LA	1910	RR	NOT&M(MP)	12.5	60	0
Courtableau Bayou	—	SW	Port Barre, LA	1910	HWY	St. Landry Pa.	17.8	30	4
Courtableau Bayou	—	SW	Port Barre, LA	1909	RR	OG&NE(T&P)	18.4	60	0
Courtableau Bayou	—	SW	Washington, LA	1927	RR	ML&TRR&SS	27.35	77.66	20
Cow Bayou	—	SW	West Orange, TX	1963	HWY	Orange Co.	2.9	61	8
Cow Bayou	SR 87	SW	Bay City, TX	1940	HWY	Texas	4.5	53	8
Cow Bayou	—	SW	Orange, TX	1923	HWY	Orange Co.	7	46	0
Cow Bayou	—	SW	Orange, TX	1940	HWY	Texas	9.5	50	4.7
Cow Bayou	—	SW	Orange, TX	1900	HWY	Orange Co.	17	47.7	0
Cow Neck	—	B	Southampton, NY	1918	HWY	Southampton	0	15	6
Cowlitz River	—	B	Kelso (near), WA	1935	RR	BN	1.5	136	7
Cowlitz River	—	VL	Kelso, WA	1927	HWY	Kelso	5.5	100	46
Coyote Creek	—	SW	Alviso, CA	1905	RR	SP	2.5	49	3.5
Coyote Hill Slough	—	B	Alvarado, CA	1926	PR	Leslie Salt	1.8	39	6
Crisfield Harbor	—	R	Crisfield, MD	1884	HWY	Crisfield	0.1	24	3.6
Crooked River	—	SW	Alanson, MI	—	HWY	Michigan	29.6	21	5
Crystal Cove	—	B	Winthrop, MA	1903	RR	BRB&L	0	29.9	1.35
Cumberland River	—	SW	Eureka, KY	1872	RR	IC	31.5	118	−7.7
Cumberland River	—	SW	Clarksville, TN	1890	RR	CSX	126.5	118	18
Cumberland River	Buchanan St.	SW	Nashville, TN	1904	RR	TCRR	185.2	128	7.6
Cumberland River	Locust St.	SW	Nashville, TN	1897	RR	L&N (reconstructed 1931)	190.4	116	9.2
Cumberland River Big South Fork	3rd St.	SW	Burnside, KY	1914	HWY	Burnside Bridge	0.2	108	−9.3
Current River	—	SW	Biggers, AR	1930	HWY	Arkansas	10.2	98.4	4.3
Current River	—	SW	Biggers, AR	1903	RR	SLSF	12.2	98	6.1

*See key of abbreviations at end.

Waterway[a]	Road Name/ Route Carried	Bridge Type[b]	Location	Const. Date	Traffic Type[c]	Owner*	Miles from Mouth of Waterway	Channel Width/Lateral Clearance (feet)	Clearance Height (feet)
Currituck Sound	—	SW	Point Harbor, NC	1930	HWY	North Carolina	1	40	4
Curry Creek	—	B	Nokomis, FL	—	RR	CSX	0.4	90	16
Curtis Creek	Pennington Ave.	B	Baltimore, MD	1977	HWY	Maryland	0.9	200	38
Curtis Creek	I 695	B	Baltimore, MD	1977	HWY	Maryland	1	200	58
Curtis Creek	—	B	Baltimore, MD	1931	HWY	Baltimore	1	150	17.8
Curtis Creek	—	SW	Baltimore, MD	1930	RR	CSX	1.3	150	13
Cut-Off Slough	—	RSP	Suisun, CA	1951	HWY	California	0.7	75	9
Cuyahoga River	—	VL	Cleveland, OH	1957	RR	CR	0.8	250	98
Cuyahoga River	Main Ave.	SW (removed)	Cleveland, OH	1885	HWY	Cleveland	0.98	97	10.1
Cuyahoga River	—	B (abandoned)	Cleveland, OH	1957	RR	CSX	1.3	229	8
Cuyahoga River	Center St.	SW	Cleveland, OH	1957	HWY	Cleveland	1.4	113	17
Cuyahoga River	Columbus Rd.	VL	Cleveland, OH	1940	HWY	Cleveland	1.9	220	9.8
Cuyahoga River	Columbus Rd.	VL	Cleveland, OH	1950	RR	CR	2.2	200	97
Cuyahoga River	Big 4	VL (abandoned)	Cleveland, OH	1952	RR	PC	2.4	200	98
Cuyahoga River	Carter Rd.	VL	Cleveland, OH	1986	HWY	Cleveland	2.4	201	97
Cuyahoga River	Eagle Ave.	VL	Cleveland, OH	1931	HWY	Cleveland	2.8	187	97
Cuyahoga River	—	B (out)	Cleveland, OH	1919	RR	CR	3.2	134	20
Cuyahoga River	—	VL	Cleveland, OH	1957	RR	NS	3.3	200	97
Cuyahoga River	W. 3rd St.	VL	Cleveland, OH	1940	HWY	Cleveland	3.7	200	97
Cuyahoga River	—	B (out)	Cleveland, OH	1911	RR	Erie	4.3	117	29
Cuyahoga River	Jefferson Ave. abutments only	B	Cleveland, OH	1907	HWY	Cleveland	4.48	100	11.8
Cuyahoga River	—	B	Cleveland, OH	1904	RR	NSS	4.7	110	11
Cuyahoga River	—	B	Cleveland, OH	1906	RR	BO	4.8	110	10

Cuyahoga River	—	B	Cleveland, OH	1914	RR	RT	5.4	129	15
Cuyahoga River	—	VL	Cleveland, OH	—	RR	WLE	5.5	200	97
Cuyahoga River, Old	—	B	Cleveland, OH	1907	RR	B&O	0.44	190	2.9
Cuyahoga River, Old	Willow Ave.	SW	Cleveland, OH	1897	HWY	Cleveland	0.57	137	11.5
Cuyahoga River, Old	Willow Ave.	VL	Cleveland, OH	—	HWY	Cleveland	0.57	—	—
Cypress Creek	SR 10	B	Smithfield, VA	1927	HWY	Virginia	0	40	3
D Inde Bayou	—	P	Lake Charles, LA	1927	HWY	Calcasieu Pa.	1	65	—
D Inde Bayou	—	RSP	Lake Charles, LA	1943	RR	SP	4.3	33	6
Damariscotta River	—	SW	South Bristol, ME	1930	HWY	Maine	1.5	31	3.9
Danvers River	Essex Br. (SR 1A)	SW	Salem, MA	1918	HWY	Massachusetts	0	40	10
Danvers River	—	SW	Salem, MA	1921	RR	MTA	0	40	3
Danvers River	Kernwood Ave.	SW	Salem, MA	1928	HWY	Essex Co.	1	50	8
Darby Creek	—	B	Essington, PA	1920	RR	CR PRR	0.3	50	3
Darby Creek	—	B	Essington, PA	1924	RR	CR RDG	0.3	50	3
Darby Creek	—	SW	Essington, PA	1901	RR	Philadelphia Transit Co.	0.35	24	3.7
Darby Creek	—	F, B (no machinery)	Prospect Park, PA	1932	HWY, TROL	Pennsylvania	1.33	50	8
De Glaise Bayou	—	SW	De Glaise, LA	1911	RR	Baton Rouge Electric	3.75	60	4
Dead River	—	SW	Tavares, FL	1928	RR	CSX	0.3	30	3
Dead River	—	B	Tavares, FL	1929	HWY	Florida	0.5	40	9.5
Dead River	—	SW	Tavares, FL	—	HWY	Lake Co.	1	29.5	—
Dean Creek	—	HNGL	Reedsport, OR	1933	HWY	Oregon	0	40	5
Dear Meadow Brook	—	SW	Newcastle, ME	—	RR	MEC	0.3	30	2
Debbies Creek	—	B	Manasquan, NJ	1951	HWY	Monmouth Co.	0.4	31	9
Deep Creek	—	B	Jarvisburg, NC	1948	FT BR	—	0.2	12	4
Deep River	US 830	SW	Deep River, WA	1933	HWY	Washington	3.5	60	13
Deep River	—	B	Deep River, WA	1900	HWY	Wahkiakum Co.	4.25	38.3	11
Delaware City Branch Canal	—	B	Delaware City, DE	1934	HWY	U.S. Govt.	0.6	60	6

*See key of abbreviations at end.

Waterway[a]	Road Name/ Route Carried	Bridge Type[b]	Location	Const. Date	Traffic Type[c]	Owner*	Miles from Mouth of Waterway	Channel Width/Lateral Clearance (feet)	Clearance Height (feet)
Delaware River	—	B	Camden to Pettys Island, NJ	1919	RR	CR	103.2	80	12
Delaware River	—	VL	Delair, NJ	1961	RR	CR	104.6	500	135
Delaware River	Tacony–Palmyra	B	Tacony Palmyra, NJ	1929	HWY	Burlington Co.	107.2	240	53
Delaware River	Burlington–Bristol	VL	Burlington Bristol, NJ	1931	HWY	Burlington Co.	117.8	500	134
Delaware and Raritan Canal	—	S, F	Princeton, NJ	1905	RR	New Jersey Transit	—	35	15
Depot Slough	—	B	Toledo, OR	1915	HWY	Lincoln Co.	0.5	39	7.5
Des Allemands Bayou	US 90	SW	Des Allemands, LA	1935	HWY	Louisiana	13.9	30	5
Des Allemands Bayou	—	SW	Des Allemands, LA	1973	RR	SP	14	32	3
Des Cannes Bayou	—	SW	Evangeline, LA	1916	HWY	Louisiana	8.5	52	4.8
Des Glaises Bayou	—	SW	Sarto, LA	1905	RR	CSX	21.1	60	5
Des Glaises Bayou	—	SW	Sarto, LA	1915	HWY	Avoyelles Pa.	21.8	56	12
Des Glaises Bayou	—	SW	Longbridge, LA	1895	RR	MP	44	61	10
Des Plaines River	Brandon Rd.	B (span repl 1986)	Joliet, IL	1932	HWY	Illinois	285.8†	110	8
Des Plaines River	Mcdonough St.	B	Joliet, IL	1933	HWY	Illinois	287.3	150	15
Des Plaines River	—	VL	Joliet, IL	1933	RR	RI	287.6	150	49
Des Plaines River	Jefferson St.	B	Joliet, IL	1933	HWY	Illinois	287.9	150	15
Des Plaines River	Cass St.	B	Joliet, IL	1933	HWY	Illinois	288.1	150	15
Des Plaines River	Jackson St.	B	Joliet, IL	1933	HWY	Illinois	288.4	200	15
Des Plaines River	Ruby St.	B	Joliet, IL	1935	HWY	Illinois	288.7	200	15
Des Plaines River (also Chicago Sanitary and Ship Canal)	—	VL	Joliet, IL	1934	RR	EJE	290.1	225	50
Deschutes Waterway	—	VL	Olympia, WA	1929	RR	NP	0.6	57	17
Detroit River	Van Horn Rd.	SW	Trenton, MI	1931	HWY	Wayne Co.	5.6	150	14

Detroit River	Gross Isle Toll	SW	Wyandotte, MI	1912	HWY	Gross Isle Bridge	8.8	125	6
Devils Slough	—	RSP	Vallejo, CA	1924	HWY	Russ Investment	1	15	4
Dickinson Bayou	—	B	San Leon, TX	1941	HWY	Texas	1.2	50	11.9
Dickinson Bayou	—	SW	San Leon, TX	1908	RR	SP	1.5	40	8
Dismal Swamp Canal	—	B	Deep Creek, VA	1934	HWY	US	0.42‡	60	3.56
Dismal Swamp Canal	—	B	South Mills, NC	1934	HWY	US	21.6	60	3.56
Dividing Creek	—	SW	Dividing Creek, NJ	—	HWY	Cumberland Co.	10	39	2.2
Doctor's Lake	—	SW	Orange Park, FL	1928	HWY	Florida	0	63.75	9.29
Dog River	SR 163	B	Mobile, AL	1929	HWY	Alabama	0	60	11
Dorchester Bay Basin	Morrisey Blvd.	B	Boston, MA	—	RR	Metropolitan Dist. Comm.	0	65	12
Dorseys Creek	—	SW	Annapolis, MD	1931	HWY	U.S. Navy (1914, academy per COE)	0.3	40	5
Dorseys Creek	—	B	Annapolis, MD	1914	HWY	Maryland (1931, navy per COE)	0.4	40	5
Dorseys Creek	—	B	Annapolis, MD	1931	RR	U.S. Navy	0.4	40	6
Dorseys Creek	—	B	Annapolis, MD	1912	RR	b&a(b&o)	0.5	40	11.2
Doullut Canal	—	SW	Empire, LA	1950	RR	Plaquemine Pa.	0.1	41	1
Doullut Canal	SR 31	B	Empire, LA	1950	HWY	Louisiana	0.1	50	3
Duck Creek	—	SW	Duck Creek, WI	1904	RR	CNW	1.6	42.4	5.5
Dulac Bayou	SR 57	SW	Dulac, LA	1971	HWY	Louisiana	0.6	60	7
Dulac Bayou	—	SW	Dulac, LA	1937	HWY	Louisiana	2	40	2.2
Dularge Bayou	—	SW	Theriot, LA	1936	HWY	Terrebonne Pa.	21.3	40	3
Dularge Bayou	—	SW	Theriot, LA	1961	HWY	Terrebonne Pa.	22.6	40	5
Dularge Bayou	—	B	Theriot, LA	1959	HWY	Terrebonne Pa.	23.2	20	4
Dularge Bayou	—	B	Theriot, LA	1977	HWY	Terrebonne Pa.	23.2	27	4
Dularge Bayou	—	SW	Theriot, LA	—	HWY	Terrebonne Pa.	23.4	34.5	1.3
Dularge Bayou	—	SW	Theriot, LA	—	HWY	Terrebonne Pa.	24	33	0
Dularge Bayou	—	SW	Theriot, LA	—	HWY	Dr. Deacres	28	34	3
Dularge Bayou	—	SW	Sunrise, LA	—	HWY	Milling & Marmande	29	34	3.4

*See key of abbreviations at end.
†Mileage from Mississippi River.
‡Mileage from lock at north end.

Waterway[a]	Road Name/ Route Carried	Bridge Type[b]	Location	Const. Date	Traffic Type[c]	Owner*	Miles from Mouth of Waterway	Channel Width/Lateral Clearance (feet)	Clearance Height (feet)
Dularge Bayou	—	SW	Sunrise, LA	—	HWY	Paul Vice	30.2	34	1
Dularge Bayou	—	SW	Sunrise, LA	—	HWY	Tom Darce	31	36	1.4
Dularge Bayou	—	SW	Sunrise, LA	—	HWY	South Coast	32.1	34	2.9
Dularge Bayou	—	SW	Sunrise, LA	—	HWY	Duplantis Estate	33.3	35	3.2
Duluth and Superior Harbor	Lake Ave.	VL	Duluth, MN	1931	HWY	Duluth	—	300	138
Dunns Creek	US 17	SW	San Mateo, FL	1931	HWY	Florida	0.9	60	11
Dupre Bayou	—	SW	Violet, LA	—	HWY	Louisiana	6.4	41.6	5
Dupre Bayou	—	SW	Violet, LA	—	RR	LA SOU RR	6.4	40.6	5.9
Durham Creek	—	RSP	Bushy Park, SC	1971	HWY	SCE&GLCOMM	1.7	45	5
Durham Creek	—	RSP	Bushy Park, SC	—	HWY	SCE&GLCOMM	2.4	46	5
Dutch Kills	—	SW	Long Island City, NY	1893	RR	LIRR	0.1	46	2
Dutch Kills	—	B	Long Island City, NY	1921	RR	LIRR	0.15	50	14
Dutch Kills	Borden Ave.	R	Long Island City, NY	1908	HWY	New York City	0.25	49	4
Dutch Kills	Hunters Point Ave.	B	Long Island City, NY	1984	HWY	New York City	0.4	60	8
Duwamish–West Waterway and River	Spokane St. (repl. w. double swing)	B	Seattle, WA	1924	HWY	Seattle	0.3	150	42
Duwamish–West Waterway and River	Spokane St. (repl. w. double swing)	B	Seattle, WA	1930	HWY	Seattle	0.3	150	42
Duwamish–West Waterway and River	—	B	Seattle, WA	1932	RR	BNSF	0.4	150	8
Duwamish–West Waterway and River	1st Ave. S.	B	Seattle, WA	1955	HWY	King Co.	2.5	150	24
Duwamish–West Waterway and River	14th Ave. S.	B	Seattle, WA	1931	HWY	King Co.	3.8	125	20
East Branch Newtown Creek	Grand Ave.	SW	Brooklyn, NY	1899	HWY	New York City	3.1	58	10

East River 1	Roosevelt Island Tramway	TRAM	Roosevelt Island, NY	1976	TRAM	RIDC	5.5	850	135
East River 1	Welfare Island	VL	Queens, NY	1955	HWY	New York City	6.4	403	99
East River 2	Monroe Ave.	B	Green Bay, WI	1932	HWY	Green Bay	0.25	60	9
East River 2	—	SW	Green Bay, WI	1897	RR	MILW	0.3	60	6
East River 2	Webster Ave.	SW	Green Bay, WI	1905	HWY	Green Bay	0.61	60	4.9
East River 2	—	SW	Green Bay, WI	1915	RR	MILW	0.9	60	3.2
East River 2	Main St.	SW	Green Bay, WI	1916	HWY	Green Bay	0.98	60	3.2
Ebey Slough	—	SW	Marysville, WA	1907	RR	BNSF	1.5	108	5
Ebey Slough	—	SW	Marysville, WA	1927	HWY	Washington	1.6	110	10
Ebey Slough	—	B	Everett, WA	1914	HWY	Snohomish Co.	11.9	100	8
Ebey Slough	—	SW	Lowell, WA	1922	RR	MILW	12.52	81	6.4
Edisto River	—	SW	Fenwick, SC	1930	RR	SAL	22.5	62.5	5.7
Eel Pond Channel	—	B	Woods Hole, MA	1914	HWY	Falmouth	—	31.5	5.25
Elizabeth River	Front St.	B	Elizabeth, NJ	1922	HWY	Union Co.	0	74	3
Elizabeth River	S. 1st St.	B	Elizabeth, NJ	1910	HWY	Union Co.	0.4	60	5
Elizabeth River	—	RL, F	Elizabeth, NJ	1952	RR	CR	0.7	59	14
Elizabeth River	Baltic St.	B, F	Elizabeth, NJ	1951	HWY	Union Co.	0.9	60	5
Elizabeth River	Summer St.	B, F	Elizabeth, NJ	1951	HWY	Union Co.	1.3	60	5
Elizabeth River	South St.	SW, F	Elizabeth, NJ	1951	HWY	Union Co.	1.8	47	3
Elizabeth River	Bridge St.	SW	Elizabeth, NJ	1912	HWY	Union Co.	2.06	31.6	3
Elizabeth River, Eastern Branch	US 460 (Berkeley Br.)	B	Norfolk, VA	1954	HWY	Elizabeth River Tunnel Comm.	0.4	150	48
Elizabeth River, Eastern Branch	Main St.	B	Norfolk, VA	1918	ERR, HWY	Norfolk–Berkeley Bridge Corp.	0.75	126.4	35
Elizabeth River, Eastern Branch	—	B	Norfolk, VA	1947	RR	NS	1.1	140	4
Elizabeth River, Eastern Branch	—	B	Norfolk, VA	1906	RR	NW	1.12	140	4.3
Elizabeth River, Eastern Branch	US 460 (Campostella Br.)	B	Norfolk, VA	1935	HWY	Norfolk	1.8	140	14

*See key of abbreviations at end.

Waterway[a]	Road Name/ Route Carried	Bridge Type[b]	Location	Const. Date	Traffic Type[c]	Owner*	Miles from Mouth of Waterway	Channel Width/Lateral Clearance (feet)	Clearance Height (feet)
Elizabeth River, Eastern Branch	—	SW	Norfolk, VA	1907	RR	NS	2.7	60	6
Elizabeth River, Southern Branch (see Intracoastal waterway)	—	SW	South Norfolk, VA	1898	RR	NPB	2	110	12.3
Elizabeth River, Southern Branch (see Intracoastal waterway)	—	VL	South Norfolk, VA	1928	HWY	Norfolk–Portsmouth Bridge Corp.	2.25	220	15
Elizabeth River, Southern Branch (see Intracoastal waterway)	—	SW	Money Pt., VA	1907	RR	VGN	3	110	9.3
Elizabeth River, Southern Branch (see Intracoastal waterway)	—	B	Gilmerton, VA	1938	HWY	Virginia	5.2	125	7.3
Elizabeth River, Southern Branch (see Intracoastal waterway)	—	B	Gilmerton, VA	1909	RR	NW	5.25	125	7.3
Elizabeth River, Southern Branch (see Intracoastal waterway)	—	SW	Deep Creek, VA	1907	RR	NPB	7.5	80	7.3
Elizabeth River, Southern Branch (see Intracoastal waterway)	—	SW	Deep Creek, VA	1903	HWY	Virginia	8	81	6.3
Elizabeth River, Western Branch	W. Norfolk Br.	SW	Portsmouth, VA	1979	HWY	Portsmouth	0.5	140	4
Elizabeth River, Western Branch	—	SW	Portsmouth, VA	1924	HWY	Virginia	0.5	140	4.7
Elizabeth River, Western Branch	—	SW	Portsmouth, VA	1934	HWY	Virginia	2	60	4.7

Elizabeth River, Western Branch	—	SW	Bruce, VA	1907	RR	ACL	3.25	50.7	4
Elizabeth River, Western Branch	SR 337	B	Hodges Ferry, VA	1929	HWY	Chesapeake	4.7 (4.25 per COE)	32	5
Elk Slough	—	RSP	Courtland, CA	1938	HWY	Yolo Co.	0.02	44.5	0.9
English Bayou	—	SW	Lake Charles, LA	1903	HWY	Calcasieu Pa.	1	45	2
English Kills	Metropolitan Ave.	B	Brooklyn, NY	1933	HWY	New York City	3.4	86	10
English Kills	Montrose Ave.	SW	Brooklyn, NY	1891	RR	LIRR	3.8	46	4
Escambia and Conecuh Rivers	—	SW	Pensacola, FL	1925	HWY	Florida	0	75	3
Escambia and Conecuh Rivers	US 29	SW	Molino, FL	—	HWY	Escambia Co.	25	55	8
Escambia and Conecuh Rivers	US 29	SW	Bogia, FL	1916	HWY	Santa Rosa Co.	30	70	14
Escambia Bay	—	SW	Mulat, FL	1911	RR	CSX	8.5	85	6
Escatawpa River	SR 63	SW	Moss Pt., MS	1964	HWY	Mississippi	1	115	9
Escatawpa River	—	SW	Moss Pt., MS	1905	RR	MSRR	3	73	5
Eureka Slough	—	VL	Eureka, CA	1975	RR	NWP	0.5	67	16
Eureka Slough	—	VL	Eureka, CA	1920	HWY	California	0.5	65	7.7
Eureka Slough	—	B	Eureka, CA	1921	RR	Pacific Lumber	2.3	9.5	3.4
Eureka Slough	—	B (inoperable)	Eureka, CA	—	PRIV RD	N. M. Deroy	2.9	18	11
Evans Slip	Water St.	SW	Buffalo, NY	—	HWY, RR	Buffalo	—	56	7
Far Rockaway Bay	—	B	Far Rockaway, NY	1927	HWY	Atlantic Beach Bridge Corp.	0.46	100	7.8
Feather River	SR 99	RSP	Nicolaus, CA	1958	HWY	Sutter Co.	9.3	100	4
Feather River	—	SW	Nicolaus, CA	1918	HWY	Sutter Co.	9.7	100	7
Fish River, East Prong	—	B	Magnolia Springs, AL	1921	HWY	Magnolia Springs Road and Bridge Assn.	4	20	9
Fishing Creek	SR 335	SW	Honga, MD	1942	HWY	Maryland	1	28	6

*See key of abbreviations at end.

Waterway[a]	Road Name/ Route Carried	Bridge Type[b]	Location	Const. Date	Traffic Type[c]	Owner*	Miles from Mouth of Waterway	Channel Width/Lateral Clearance (feet)	Clearance Height (feet)
Fishing Creek	—	R	Lawrence, NC	1916	HWY	North Carolina	4	31	5
Florida Keys Boot Key Harbor	—	B	Marathon, FL	1960	HWY	Florida	0.1	48	26
Florida Keys Indian Key Channel	—	B	Crevallo, FL	1937	HWY	Florida	—	40	5
Florida Keys Metecumbe Channel 5	US 1	B	Craig Key (Long Key per COE), FL	1936	HWY	Florida	0	46	8
Florida Keys Matecumbe Channel 5	Long Key	B	Long Key, FL	—	HWY	Overseas Road and Tollbridge Dist.	0.5	50	8.5
Florida Keys Snake Creek	—	B	Islamorada, FL	1981	HWY	Florida	0.1	60	27
Flint River	—	SW	Bainbridge, GA	1966	RR	CSX	28	91	12
Flint River	—	SW	Bainbridge, GA	1913	RR	CSX	28.7	100	19
Flint River	—	SW	Bainbridge, GA	1897	RR	SAL	29	100	2.5
Flint River	—	B	Bainbridge, GA	1926	HWY	Georgia & Decatur Co.	30	104	10
Flint River	—	VL	Newton, GA	1921	HWY	Georgia	73	77	26
Flint River	—	SW	Cordele, GA	—	RR	CSX	155	83	16
Flushing Creek	Whitestone Pkwy.	B, F	Flushing, NY	1939	HWY	New York City	0.2	140	34
Flushing Creek	Northern Blvd.	B	Flushing, NY	1939	HWY	New York City	0.4	80	25
Flushing Creek	Roosevelt Ave.	B, F	Flushing, NY	1927	HWY	New York City	0.8	70	24
Flushing Creek	—	B	Queens, NY	1936	RR	LIRR	1	—	—
Flushing Creek	—	SW	Flushing, NY	1890	RR	LIRR	1.1	30	3
Folse Bayou	—	RSP	Raceland, LA	1938	HWY	Louisiana	3.52	40	4
Fore River 1	SR 3A	B	Quincy, MA	1933	HWY	Massachusetts	3.5	175	33
Fore River 2	Million Dollar Br.	B	South Portland, ME	—	HWY	Maine	1.5	100	31

Forked Deer River, South Fork	—	SW (obsolete per COE)	Yellow Bluff, TN	1899	HWY	Dyer Co.	7.3	36	−1.7
Fort Point Channel	Northern Ave.	SW	Boston, MA	1911	HWY	Boston	0.12	76	7
Fort Point Channel	Congress Ave.	B	Boston, MA	1931	HWY	Boston	0.31	75	6
Fort Point Channel	Summer St.	R, F	Boston, MA	1897	HWY	Boston	0.37	51	4
Fort Point Channel	Dorchester Ave.	R	Boston, MA	1890	HWY	Boston	0.81	42	4.6
Fort Point Channel	—	B (3)	Boston, MA	1900	RR	NYNHH	0.87	44.3	8.4
Fort Point Channel	Broadway	SW	Boston, MA	1915	HWY	Boston	1	50	23.8
Fort Point Channel	Dover St.	SW	Boston, MA	1895	HWY	Boston	1.19	40.7	15.4
Fowl River	—	B	Mobile, AL	—	HWY	Mobile Co.	0.5	19.1	6.5
Fox River	—	SW	Green Bay, WI	1924	RR	WC	1	84	8
Fox River	Main St.	B	Green Bay, WI	1998	HWY	Green Bay	1.6	87	12
Fox River	Walnut St.	B	Green Bay, WI	1988	HWY	Green Bay	1.8	78	9
Fox River	Tilleman Memorial Br.	B	Green Bay, WI	1974	HWY	Green Bay	2.3	124	32
Fox River	—	SW	Green Bay, WI	1903	RR	MILW	2.6	75	8
Fox River	—	SW	Green Bay, WI	1907	RR	CNW	3.3	75	31
Fox River	—	SW	De Pere, WI	1930	RR	CNW	7.1	40	2.9
Fox River	George St.	B	De Pere, WI	1935	HWY	Wisconsin	7.2	75	24
Fox River	Bridge St.	B	Wrightstown, WI	1933	HWY	Wisconsin	17.4	70	16
Fox River	—	SW	Kaukauna, WI	1901	RR	CNW	23.4	40	14
Fox River	Wisconsin Ave.	VL	Kaukauna, WI	1979	HWY	Kaukauna	23.8	90	60
Fox River	Lawe Ave.	B	Kaukauna, WI	1932	HWY	Kaukauna	23.9	90	15
Fox River	Mill St.	B	Little Chute, WI	1929	HWY	Outagamie Co.	26.6	35.4	0.5
Fox River	Sidney St.	B	Kimberly, WI	1912	HWY	Kimberly and Outagamie Cos.	27.54	68.5	9.56
Fox River	John St.	SW	Appleton, WI	1907	HWY	Appleton	30.8	70	1
Fox River	—	SW	Appleton, WI	1928	RR	CNW	31.1	59	4
Fox River	Lawe Ave.	B	Appleton, WI	1952	HWY	Appleton	31.4	70	2
Fox River	S. Oneida St.	SW	Appleton, WI	1898	HWY	Appleton	31.7	59.5	5.16
Fox River	Oneida St.	B	Appleton, WI	1960	HWY	Appleton	31.8	30	8

*See key of abbreviations at end.

Waterway[a]	Road Name/ Route Carried	Bridge Type[b]	Location	Const. Date	Traffic Type[c]	Owner*	Miles from Mouth of Waterway	Channel Width/Lateral Clearance (feet)	Clearance Height (feet)
Fox River	State St.	SW	Appleton, WI	1904	RR	MILW	32.4	61	5.92
Fox River	Lush St.	SW	Menasha, WI	1906	RR	MILW (&MSTP&SSM per COE)	37.26	60	1.62
Fox River	Tayco St.	B	Menasha, WI	1928	HWY	Wisconsin	37.51	100	2
Fox River	Mill St.	SW	Menasha, WI	1901	HWY	Menasha	37.81	50	3.16
Fox River	Racine St.	B	Menasha, WI	1951	HWY	Wisconsin	38.5	100	3
Fox River	Broad St.	SW	Oshkosh, WI	1899	RR	WC	56.02	70	2
Fox River	Main St.	B	Oshkosh, WI	1973	HWY	Wisconsin	56.27	89	11
Fox River	Jackson St.	B	Oshkosh, WI	1955	HWY	Oshkosh	56.52	97	11
Fox River	Pine St.	SW	Oshkosh, WI	1902	RR	WC	56.88	68	5
Fox River	Wisconsin St.	B	Oshkosh, WI	1938	HWY	Wisconsin	57.03	75	12
Fox River	Congress Ave.	B	Oshkosh, WI	1982	HWY	Wisconsin	58.3	75	13
Fox River	W. Algoma St.	B	Oshkosh, WI	1914	HWY	Oshkosh	58.31	75	9.1
Fox River	W. Division St.	B	Omro, WI	1918	HWY	Winnebago Co.	72.6	70.3	3
Fox River	Park St.	SW	Omro, WI	1899	RR	MILW	72.9	62.4	2
Fox River	Ferry St.	B	Eureka, WI	1940	HWY	Wisconsin	79.3	70	3.4
Fox River	Wisconsin St.	SW	Berlin, WI	1879 (rebuilt)	HWY	Berlin, Aurora Twp.	85.7	51.5	3
Fox River	Huron St.	B	Berlin, WI	1931	HWY	Wisconsin	87.4	79	2.1
Fox River	Main St.	SW	Princeton, WI	1901	RR	CNW	104.9	60	3.53
Fox River	Main St.	B	Princeton, WI	1930	HWY	Wisconsin	105.3	70	2.8
Fox River	Main St.	SW	Montello, WI	1905	HWY	Montello	132.6	35	1.72
Fox River	Liberty St.	SW	Packwaukee, WI	1908	HWY	Packwaukee	139.6	70	2.4
Fox River	—	SW	Lake Buffalo, WI	1912	RR	UP	141	75	14.7
Fox River	—	SW	Moundville, WI	1912	HWY	Moundville	151.1	70	1.2
Fox River	—	SW	Ft. Hope, WI	1895	HWY	Ft. Winnebago Twp.	154.6	59.5	3.2

Fox River	—	B	Ft. Hope, WI	1910	RR	SOO	158.5	60	−3.6
Fox River†	Center Ave.	SW	Portage, WI	—	HWY	Portage	162	61	3.1
Fox River†	—	VL, F	Portage, WI	1938	RR	CP	162.03	70	6.65
Fox River†	Wisconsin Ave.	B	Portage, WI	1929	HWY	Wisconsin	162.9	70	6.95
Fox River–Pistakee Nippersink Lake	West Channel	SW	Fox Lake, IL	1900	RR	MILW	106.7	30	0.9
Franklin Canal	SR 317	SW	Franklin, LA	1950	HWY	St. Mary Pa.	4.8	55	7
Frazer River	—	SW	Yale, BC, Canada	1924	RR	CN	—	380	—
Frazer River, North Arm	—	SW	Lulu Island, BC, Canada	1931	RR	CN	—	280	—
Fraser River	—	SW	Mission, BC, Canada	1902	RR	CP	—	228	—
Fraser River	—	SW	Van Horne, BC, Canada	1968	RR	CP	—	279	—
Freeport Harbor– Old Brazos River	—	SW	Freeport, TX	1946	HWY, RR	MP and Brazoria Cos.	4.4	108	11
Freshwater Slough	—	B, F	Eureka, CA	1929	PR	Dutra Farms	0.5	36	10
Freshwater Slough	—	B (inoperable)	Eureka, CA	—	CC	M. Brazil	1.66	35	11.6
Galena River	—	SW	Galena Junction, IL	1886	RR	ICG	0.1	114	8
Galena River	—	SW	Galena Junction, IL	1886	RR	CBQ	0.1	114	8.4
Galena River	—	B	Galena, IL	1914	RR	IC	3.75	93	0.3
Gallants Channel	—	B	Beaufort, NC	1959	HWY	North Carolina	0.1	60	13
Gallants Channel	—	B	Beaufort, NC	1917	RR	BEAUF& MOREH	0.1	60	4
Gallants Channel	—	B	Beaufort, NC	1928	HWY	North Carolina	—	60	7.3
Gallinas Creek	—	B	San Rafel, CA	—	RR	NWP	—	30	3.3
Garrison Channel	—	B	Tampa, FL	1909	RR	CSX	0.2	110	5
Gasparilla Sound	—	SW	Gasparilla, FL	1907	RR	CH&NRWY	0.9	50	8
Gasparilla Sound	—	SW	Placida, FL	1909	RR	CSX	1.3	49	4
Genessee River	—	SW	Rochester, NY	1905	RR	PC	0.9	132	5

*See key of abbreviations at end.

†Portage Canal.

Waterway[a]	Road Name/ Route Carried	Bridge Type[b]	Location	Const. Date	Traffic Type[c]	Owner*	Miles from Mouth of Waterway	Channel Width/Lateral Clearance (feet)	Clearance Height (feet)
Genessee River	Stutson St.	B	Rochester, NY	1918	HWY	Monroe Co.	1.2	138	19
Georgiana Slough	—	SW	Isleton, CA	1941	HWY	Sacramento Co.	4.5	80	10
Georgiana Slough	—	B	Isleton, CA	1929	RR	SP	5.75	80	2.6
Georgiana Slough	—	SW	Walnut Grove, CA	1963	HWY	Sacramento Co.	12.4	80	14
Gilpatricks Cove	—	R	Mt. Desert, ME	1896	FTBR	W. C. Vaughan	0	25	—
Gloriana Canal	—	CB	Cape Coral, FL	1979	HWY	GAC Properties	2	10	10
Goleta Slough	—	B	Goleta, CA	1929	HWY	Santa Barbara Co.	0.5	19	8
Good River	—	B	Strawberry Pt., AK	1923	HWY	Alaska	0.7	18	2
Goodyear Slough	Ehmann Br.	RSP	Benicia, CA	1932	PR	J. F. Eggert	3.2	26	8
Goose Creek	Wantagh Cswy.	B	Jones Beach, NY	1929	HWY	New York	16.1	76	16
Gowanus Canal	Hamilton Ave.	B	Brooklyn, NY	1942	HWY	New York City	1.2	47	19
Gowanus Canal	9th St.	B (repl. w. VL)	Brooklyn, NY	1905	HWY	New York City	1.4	45	7
Gowanus Canal	3rd St.	B	Brooklyn, NY	1905	HWY	New York City	1.8	41	10
Gowanus Canal	Carroll St.	R	Brooklyn, NY	1889	HWY	New York City	2	36	3
Gowanus Canal	Union St.	B	Brooklyn, NY	1905	HWY	New York City	2.1	43	9
Grand Bayou	SR 70	P	Paincourtville, LA	1953	HWY	Louisiana	7.6	68	—
Grand Caillou Bayou	SR 56	SW	Dulac, LA	1958	HWY	Terrebonne Pa.	25.9	50	5
Grand Caillou Bayou	—	SW	Dulac, LA	1975	HWY	Terrebonne Pa.	25.9	60	10
Grand Caillou Bayou	—	SW	Houma, LA	—	HWY	Terrebonne Pa.	29	39	2.2
Grand Caillou Bayou	—	SW	Houma, LA	1947	HWY	Terrebonne Pa.	29	45	3
Grand Caillou Bayou	—	SW	Houma, LA	—	HWY	Joseph Jacuzzo	35	34	−0.1
Grand Calumet River	Ada St.	SW	Burnham, IL	—	RR	SC&SRRWY	8.85	68	6
Grand Calumet River	—	SW	Burnham, IL	1907	RR	B&OCT	9.97	63	6
Grand Calumet River	—	SW	Calumet City, IL	1899	RR	IHB	10.48	55	3
Grand Calumet River	—	B	Hammond, IN	1909	RR	BOCT	10.7	80	6
Grand Calumet River	—	B	Hammond, IN	1909	RR	CIL	10.71	80	6
Grand Calumet River	—	SW	Hammond, IL	1906	RR	NS	10.76	84	6
Grand Calumet River	Hohman St.	B	Hammond, IN	1911	HWY	Hammond	10.93	77	4

Grand Calumet River	—	SW	Hammond, IN	1904	RR	IHB	11.18	46	8
Grand Calumet River	Calumet Ave.	B	Hammond, IN	1922	HWY	Lake Co.	11.48	66	1
Grand Calumet River	Columbia Ave.	SW	Hammond, IN	1909	HWY	Lake Co.	11.96	40	5
Grand Calumet River	Columbia Ave.	B	Hammond, IN	1960	HWY	Hammond	12	—	—
Grand River 1	—	SW	Grand Haven, MI	1909	RR	CN	2.8	61	5
Grand River 1	US 31	B	Grand Haven, MI	1959	HWY	Michigan	2.9	155	21
Grand River 1	—	SW	Eastmanville, MI	1916	HWY	Michigan	19.4	70.6	19.5
Grand River 1	—	SW	Grand Rapids, MI	1916	RR	PC	37.2	54	3
Grand River 1	—	SW	Grand Rapids, MI	1915	RR	Michigan Railway (elect)	39.5	70	27
Grand River 1	Wealthy St.	SW	Grand Rapids, MI	1904	HWY	Grand Rapids	39.8	70	19.3
Grand River 1	—	SW	Grand Rapids, MI	1902	RR	PM	39.9	70	19.2
Grand River 2	—	SW	Fairport, OH	—	RR	BO	2.2	56	14
Grant Line Canal	—	B	Tracy, CA	1962	HWY	San Joaquin Co.	5.5	88	16
Grassy Sound Channel	Ocean Dr.	B	North Wildwood, NJ	1940	HWY	Cape May Co. Bridge Commission	1	50	15
Grays River	—	SW	Rosburg, WA	1930	HWY	Wahkiakum Co.	5	48	10
Great Canal	—	B	Satellite Beach, FL	1979	HWY	Tortoise Island Communities	2.6	24	10
Great Channel	—	B, F	Stone Harbor, NJ	1940	HWY	Cape May Co.	0.7	50	15
Great Egg Bay	—	B	Ocean City, NJ	1928	HWY	Ocean City–Longport Automobile Bridge Company	0.25	90	18
Great Egg Bay	—	SW	Somers Point, NJ	1908	RR	AC&SHORE RR	1	40	4
Great Egg Harbor Bay (Great Egg Bay)	SR 52	B	Somers Pt. Ship Channel, NJ	1932	HWY	New Jersey	1	50	14
Great Egg Harbor Bay (Great Egg Bay)	US 9	B	Beesleys Point, NJ	1928	HWY	Ocean City Automobile Bridge Comm.	3.5	60	14
Great Egg Harbor Inlet	—	B	Ocean City, NJ	1928	HWY	Cape May Co.	0.3	100	23
Great Wicomico River	SR 200	SW	Tipers, VA	1934	HWY	Virginia	8	101	9

*See key of abbreviations at end.

Waterway[a]	Road Name/ Route Carried	Bridge Type[b]	Location	Const. Date	Traffic Type[c]	Owner*	Miles from Mouth of Waterway	Channel Width/Lateral Clearance (feet)	Clearance Height (feet)
Green River	—	SW	Spotsville, KY	1972	RR	CSX	8.3	108	0
Green River	—	SW	Livermore, KY	1908	RR	CSX	71.2	103	0
Green River	—	SW	Smallhouse, KY	1909	RR	CSX	79.6	128	1
Green River	—	B	Rockport, KY	1932	RR	ICG	95.8	150	38
Greens Bayou	—	VL	Houston, TX	1931	RR	Port of Houston Authority	2.8	71	27
Grosse Tete Bayou	SR 75	SW	Indian Village, LA	1945	HWY	Louisiana	0.5	70	11
Grosse Tete Bayou	SR 77 411	RSP	Grosse Tete, LA	1966	HWY	Iberville Pa.	11	40	11
Grosse Tete Bayou	—	SW	Grosse Tete, LA	1937	RR	MP	14.7	45	9
Grosse Tete Bayou	SR 77	RSP	Grosse Tete, LA	1968	HWY	Iberville Pa.	15.3	42	12
Grosse Tete Bayou	I 10	RSP	Grosse Tete, LA	1972	HWY	Louisiana	15.9	40	10
Grosse Tete Bayou	—	B	Rosedale, LA	1913	HWY	Iberville Pa.	16.25	49.8	12.4
Grosse Tete Bayou	SR 76	RSP	Rosedale, LA	1973	HWY	Louisiana	17.8	40	12
Grosse Tete Bayou	SR 77	RSP	Maringouin, LA	1967	HWY	Iberville Pa.	21.2	40	13
Grosse Tete Bayou	—	SW	Maringouin, LA	1914	HWY	Iberville Pa.	24	60.9	20.7
Grosse Tete Bayou	—	B	Livonia, LA	1917	HWY	Central Sugar Company	28.5	50	21.2
Guadelupe River	—	VL	Tivoli, TX	1921	RR	UP	10.4	36	42
Hackensack River	—	SW	Kearney, NJ	1913	RR	CNJ	1.04	93	25
Hackensack River	Lincoln Hwy. (US 1, 9)	VL	Kearney, NJ	1954	HWY	New Jersey	1.8	201	135
Hackensack River	Path	VL	Jersey City, NJ	1930	RR	PANYNJ	3.04	168	135
Hackensack River	Hack Freight	VL	Jersey City, NJ	1930	RR	CR	3.07	158	135
Hackensack River	Wittpenn	VL	Jersey City, NJ	1930	HWY	New Jersey	3.08	158	135
Hackensack River	Lower Hack	VL	Jersey City, NJ	1928	RR	New Jersey Transit	3.39	150	135
Hackensack River	Portal	SW	Snake Hill, NJ	1910	RR	ATK	5.04	99	23

Hackensack River	DB (Erie)	SW	Snake Hill, NJ	1908	RR	New Jersey Transit	5.37	99	7
Hackensack River	—	SW	Secaucus, NJ	1901	RR	DLW	6.89	56.2	5.5
Hackensack River	Upper Hack	VL	Secaucus, NJ	1959	RR	New Jersey Transit	6.9	127	110
Hackensack River	HX Jacknife	B	Secaucus, NJ	1912	RR	New Jersey Transit	7.69	101	4
Hackensack River	Rte. 3	B	Secaucus, NJ	1934	HWY	New Jersey	8.84	150	35
Hackensack River	SR 46	B	Little Ferry, NJ	1934	HWY	New Jersey	14.04	150	35
Hackensack River	Court St. E.	SW	Hackensack, NJ	1907	HWY	Bergen Co.	16.24	56	3
Hackensack River	NYS&W	SW, F	Hackensack, NJ	1914	RR	Del Otsego	16.34	43	2
Hackensack River	Midtown Br.	SW	Hackensack, NJ	1952	HWY	Bergen Co.	16.5	53	7
Hackensack River	Anderson St.	B	Hackensack, NJ	1910	HWY	Bergen Co.	17.34	45	3.5
Hackensack River	—	B	New Bridge, NJ	1888	HWY	Bergen Co.	19.09	40	3
Hackensack River	Bridge St.	B	River Edge, NJ	1913	HWY	Bergen Co.	20.59	39	5.5
Haines Creek	—	B	Lisbon, FL	1929	HWY	Florida	3.5	40	1
Haines Creek	—	SW	Lisbon, FL	1928	RR	ACL	3.5	40	0.66
Hampton Harbor	SR 1A	B	Hampton–Seabrook, NH	1972	HWY	New Hampshire	0	40	18
Hampton River	—	B	Hampton, NH	1934	HWY	New Hampshire	0	40	10
Harbor River	US 21	SW	Hunting Island, SC	1939	HWY	South Carolina	0.5	60	15
Harbor View Creek	—	R	South Norwalk, CT	1966	HWY	Harbor Beach Co.	0	20	6
Harlem River	103rd St.	VL	New York, NY	1951	FTBR	New York City	0	300	136
Harlem River	125th St.	VL	New York, NY	1930	HWY	TBTA	1.3	204	136
Harlem River	Willis Ave.	SW	New York, NY	1901	HWY	New York City	1.5	109	24
Harlem River	2nd Ave.	SW	New York, NY	1915	RR	IRT	1.75	99.2	29.1
Harlem River	3rd Ave.	SW	New York, NY	1898	HWY	New York City	1.9	100	25
Harlem River	Park Ave.	VL	New York, NY	1947	RR	MNRR	2.1	225	135
Harlem River	Madison Ave.	SW	New York, NY	1910	HWY	New York City	2.3	104	25
Harlem River	145th St.	SW	New York, NY	1905	HWY	New York City	2.8	104	25
Harlem River	Macombs Dam	SW	New York, NY	1895	HWY	New York City	3.2	164	27
Harlem River	157th St.	SW	New York, NY	1880	FTBR, RR	NYC	3.5	123.2	29.5

*See key of abbreviations at end.

Waterway[a]	Road Name/ Route Carried	Bridge Type[b]	Location	Const. Date	Traffic Type[c]	Owner*	Miles from Mouth of Waterway	Channel Width/Lateral Clearance (feet)	Clearance Height (feet)
Harlem River	207th St.	SW	New York, NY	1908	HWY	New York City	6	101	26
Harlem River	Broadway	VL	New York, NY	1949	HWY	New York City	6.8	288	135
Harlem River	Spuyten Duyvil	SW	New York, NY	1900	RR	ATK	7.9	100	5
Harrison County Industrial Canal	—	B	Handsboro, MS	1965	HWY	Harrison Co.	11.3	96	29
Harrison River	—	SW	Cascade, BC, Canada	1913	RR	CPR	—	127	—
Hatchet Creek	—	B	Venice, FL	1929	HWY	Florida	0.2	40	4.6
Hatchie River	—	SW	Rialto, TN	1912	RR	IC	35	49	−1.1
Hatchie River	—	SW	Shepp, TN	1908	RR	LN	71	55	0.2
Hawtree Creek	Nolins Ave.	B	Howard Beach, NY	1927	HWY	New York City	0	40	10
Hempstead Bay	Long Creek	B	Jones Beach, NY	1933	HWY	LISPC	1.4	75	26
Hempstead Bay	Sloop Channel	B	Jones Beach, NY	1933	HWY	LISPC	1.7	75	27
Hermitage Bayou (Judge Leander Perez Bayou)	SR 23	RSP	Hermitage, LA	1970	HWY	Plaquemines Pa.	0.3	57	10
Hickeys Creek	—	SW	Alva, FL	1922	HWY	Lee Co.	0.4	38	0
Hill Slough	—	RSP	Suisun, CA	1920	HWY	Solano Co.	—	28	5.2
Hillebrant Bayou	—	SW	Portacres, TX	—	HWY	Jefferson Co.	6.2	38	0.6
Hillsboro Bay	—	B	Tampa, FL	1925	HWY	Hillsborough Co.	—	60	8
Hillsboro Canal	—	SW	Pompano, FL	1925	HWY	Broward Co.	0.35	49.6	4
Hillsboro Canal	—	B	Deerfield Beach, FL	1925	HWY	Florida	1	60	8
Hillsboro Inlet	SR A1A	B	Hillsboro Beach, FL	—	HWY	Florida	0.3	60	13
Hillsboro Inlet	—	B	Hollywood, FL	1978	HWY	Florida	—	90	21
Hillsborough River	Platt St.	B	Tampa, FL	1926	HWY	Tampa	0.04	80	15
Hillsborough River	Brorein St.	B	Tampa, FL	1959	HWY	Tampa	0.1	80	15
Hillsborough River	Kennedy Blvd. (was Lafayette St.)	B	Tampa, FL	1913	HWY	Tampa	0.4	75	11
Hillsborough River	—	B	Tampa, FL	1916	RR	CSX	0.75	75	6

Hillsborough River	Cass St.	B	Tampa, FL	1926	HWY	Tampa	0.76	75	13
Hillsborough River	Laurel St. (was Fortune St.)	B	Tampa, FL	1927	HWY	Tampa	1.05	75	12
Hillsboro River	Garcia Ave.	SW	Tampa, FL	1910	RR, ST	Tampa	1.6	50	6
Hillsborough River	W. Columbus Dr. (was Michigan Ave.)	SW	Tampa, FL	1927	HWY	Tampa	2.3	50	10
Hillsborough River	US 92 (Hills-borough Ave.)	VL	Tampa, FL	1939	HWY	Florida	4.85	60	53
Hillsborough River	US 92 (Hills-borough Ave.)	B	Tampa, FL	2000	HWY	Florida	4.852	—	—
Hillsboro River	Sligh Ave.	SW	Tampa, FL	1927	HWY	Tampa	6.52	56.5	12.2
Hiwassee River	—	SW	Charleston, TN	1912	HWY	Bridge and McMinn Cos. (McMinn and Bradley Cos. per COE)	18.9	112	20
Honker Cut	8 Mile Rd.	SW	Stockton, CA	1936	HWY	San Joaquin Co.	0.3	75	7
Honolulu Harbor	Ford Island	R, P	Honolulu, HI	1998	HWY	U.S. Govt.	—	—	—
Honolulu Harbor	Slattery Br.	B	Honolulu, HI	1962	HWY	Hawaii	1.8	250	15
Hood Canal, Puget Sound	—	P	Port Gamble, WA	1961	HWY	Washington	5	600	—
Hoquarten Slough	—	B	Tillamook, OR	1916	HWY	Tillamook	2.55	30	4.5
Hoquiam River	—	SW	Hoquiam, WA	1909	RR	BNSF	0.3	135	11
Hoquiam River	Simpson Ave.	B	Hoquiam, WA	1928	HWY	Washington	0.5	125	25
Hoquiam River	8th St.	SW	Hoquiam, WA	1915	HWY, ST, RWY	Hoquiam	0.75	125	8.8
Hoquiam River	US 101 6th Ave.	VL	Hoquiam, WA	1971	HWY	Washington	0.9	150	69
Hoquiam River	—	SW	Hoquiam, WA	1915	RR	BNSF	2.5	105	5
Hoquiam River, East Fork	—	VL	Hoquiam, WA	1930	HWY	Grays Harbor Co.	0.7	80	8
Horse Creek	—	—	Eau Gallie, FL	1885	RR	FEC	0.3	12	11
Houma Canal	—	SW	Houma, LA	1938	HWY	Louisiana	1.7	40	4

*See key of abbreviations at end.

Waterway[a]	Road Name/ Route Carried	Bridge Type[b]	Location	Const. Date	Traffic Type[c]	Owner*	Miles from Mouth of Waterway	Channel Width/Lateral Clearance (feet)	Clearance Height (feet)
Houma Navigation Channel	SR 7	P	Dulac, LA	1982	HWY	Terrebonne, PA	23.1	160	0
Houma Navigation Channel	SR 661	SW, H	Houma, LA	1962	HWY	Louisiana	36	127	1
Housatonic River	Washington Br. (Rte. 1)	B	Stratford, CT	1922	HWY	Connecticut	3.5	125	32
Housatonic River	—	B (2)	Devon, CT	1905	RR	Connecticut DOT	3.9	83	19
Houston River	—	SW	Lake Charles, LA	1900	RR	KCS	5.2	27	6
Howards Bay	Lamborn Ave.	SW	Superior, WI	1890	HWY	Superior	—	123	1.3
Hudson Bayou	—	B	Sarasota, FL	1926	HWY	Sarasota	0.2	30	7
Hudson (Frarey) Creek	—	RSP	Gardiner, OR	1935	HWY	Douglas Co.	0.1	35	8
Hudson River	—	VL	Albany, NY	1933	HWY	New York	145.2	309	41
Hudson River	—	SW	Albany, NY	1900	RR	NYC&HR RR	145.7	117	27.2
Hudson River	—	SW	Albany, NY	1902	RR	ATK	146.2	98	25
Hudson River	—	VL, F	Troy, NY	1931	HWY	New York	150.2	300	139
Hudson River	—	SW	Troy, NY	1917	HWY	New York	152.2	181	31.8
Hudson River	—	VL	Green Island, NY	1979	HWY	New York	152.7	184	60
Hudson River	—	B (repl. w. fixed)	Cohoes, NY	1924	HWY	New York	155.4	183	23
Humble Canal	—	SW	Terrebonne Parish, LA	1937	HWY	Humble Oil and Refining	0	40	2
Humble Canal	SR 3147	P	Pecan Island, LA	1982	HWY	Louisiana	0.5	80	0
Humboldt Bay	SR 255	F, convertible to VL	Eureka, CA	1971	HWY	California	6.6	200	45
Huron River	Van Renssalaer St.	Lift inoperable	Huron, OH	1911	HWY	Erie Co.	0.72	84	10.5
Huron River	—	SW	Huron, OH	1893	RR	NYC	0.79	57	14.5
Hutchinson River (Long Island Sound)	—	SW	City Island, NY	1901	HWY, TROL	New York City	0	54.6	12

Hutchinson River	Pelham Pkwy.	B	Bronx, NY	1918	HWY	New York City	0.4	59	13
Hutchinson River	Pelham	B	Bronx, NY	1908	RR	ATK	0.5	68	8
Hutchinson River	Hutchinson River Pkwy.	B	Bronx, NY	1941	HWY	New York City	0.9	130	30
Hutchinson River	I95	B repl w F	Bronx, NY	1944	HWY	New York	2.2	110	31
Hutchinson River	Boston Post Rd.	B	Bronx, NY	1922	HWY	New York City	2.5	104.4	5.7
Hutchinson River	S. Fulton Ave.	B	Mt. Vernon, NY	1975	HWY	Westchester Co.	2.9	80	6
Hylebos Waterway	E. 11th St.	B	Tacoma, WA	1939	HWY	Tacoma	1.1	150	21
Illinois and Mississippi Canal	—	P	Milan, IL	1894	HWY	U.S. Govt.	2.1	35	—
Illinois and Mississippi Canal	—	SW	Milan, IL	1895	RR	U.S. Govt.	3.07	59.26	6.5
Illinois and Mississippi Canal	—	SW	Milan, IL	1895	HWY	U.S. Govt.	3.1	46	4.8
Illinois and Mississippi Canal	—	VL	Green River, IL	1905	HWY	U.S. Govt.	20.5	35	24.8
Illinois and Mississippi Canal	—	VL	Mineral, IL	1905	HWY	U.S. Govt.	46.3	35	24.7
Illinois and Mississippi Canal	—	VL	Wyanet, IL	1905	HWY	U.S. Govt.	57.8	35	26.9
Illinois and Mississippi Canal	—	R	Bureau, IL	1929	HWY	U.S. Govt.	73.9	35	—
Illinois River	SR 16	VL	Hardin, IL	1931	HWY	Illinois	21.6	300	65
Illinois River	—	VL	Pearl, IL	1979	RR	GWW	43.2	315	37
Illinois River	US 36 54	VL	Florence, IL	1930	HWY	Illinois	56	202	60
Illinois River	Wabash Br.	VL	Valley City, IL	1952	RR	NS	61.4	300	52
Illinois River	Pearl St.	SW	Meredosia, IL	1866	RR	—	71.1	114	2.1
Illinois River	State St.	SW	Beardstown, IL	1890	HWY	Beardstown	88.6	127	−2.4
Illinois River	—	VL	Beardstown, IL	1977	RR	BNSF	88.8	300	55
Illinois River	—	VL	Pekin, IL	1939	RR	UP	151.2	153	50
Illinois River	SR 9	VL	Pekin, IL	1930	HWY	Illinois	152.9	210	63
Illinois River	—	SW	Pekin, IL	1903	RR	CRIP	153	141	2.5

*See key of abbreviations at end.

Waterway[a]	Road Name/ Route Carried	Bridge Type[b]	Location	Const. Date	Traffic Type[c]	Owner*	Miles from Mouth of Waterway	Channel Width/Lateral Clearance (feet)	Clearance Height (feet)
Illinois River	—	VL	Peoria, IL	1981	RR	PPU	160.7	300	4
Illinois River	—	B (out)	Peoria, IL	1914	RR	PPU	160.7	140	7
Illinois River	—	B	Peoria, IL	1907	RR	IT	162.2	125	16.3
Illinois River	—	SW	Peoria, IL	1906	RR	TPW	162.2	118	−0.7
Illinois River	US 34, SR 29 (Franklin St.)	B (repl. w. F)	Peoria, IL	1913	HWY	Illinois	162.3	121	1
Illinois River	Lorentz Ave.	SW	Peoria, IL	1888	HWY	Peoria	166.1	135	−1.1
Illinois River	—	SW (out)	Depue, IL	1900	RR	CR	213.9	162	2
Illinois River	Marion St.	SW	Peru, IL	1869	HWY	Lasalle Co.	222.4	136	0.6
Illinois River	—	VL	La Salle–Shippingsport, IL	1929	HWY	Illinois	224.7	249	64
Illinois River	—	VL	La Salle, IL	1933 (reconstruction)	RR	CB&Q	225.4	225	50.2
Illinois River	—	SW	Utica, IL	1909	HWY	Illinois	229.6	129	8.7
Illinois River	—	VL	Ottawa, IL	1933	RR	BO (CB&Q per COE)	239.4	167	38
Illinois River	—	VL	Seneca, IL	1934	RR	RI	254.1	141	35
Illinois River	—	VL	Divine, IL	1933	RR	EJE	270.6	113	38
Imperial River	—	SW	Bonita Springs, FL	1927	RR	NS&G	3.5	50	5.1
Imperial River	—	VL	Bonita Springs, FL	1925	RR	FMS	4.5	50	30
Imperial River	—	B	Bonita Springs, FL	1927	HWY	Lee Co.	5.2	22	7
Indian Creek	—	B	Miami Beach, FL	1934	—	Miami Beach	0.57	59.8	11
Indian Creek	63rd St.	B	Miami Beach, FL	1951	HWY	East Miami Beach	4	50	11.2
Indian River Inlet	—	SW	Bethany Beach, DE	1940	HWY	Delaware	0.3	60	14
Indiana Harbor Canal	—	B	East Chicago, IN	1909	RR	EJE	0.62	61	4
Indiana Harbor Canal	—	B	East Chicago, IN	1909	RR	CR (B&O per COE)	0.64	65	3
Indiana Harbor Canal	—	B (2)	East Chicago, IN	1909	RR	PC	0.65	65	3
Indiana Harbor Canal	—	B	East Chicago, IN	1909	RR	IHB	0.67	65	3

Indiana Harbor Canal	—	B	East Chicago, IN	1909	RR	PC	0.91	65	2
Indiana Harbor Canal	—	B	East Chicago, IN	1979	HWY	Indiana	1.1	118	16
Indiana Harbor Canal	Dickey Pl.	B	East Chicago, IN	1931	HWY	Lake Co.	1.14	118	14
Indiana Harbor Canal	Canal St.	B	East Chicago, IN	1928	HWY	Lake Co.	1.79	65	5.6
Indiana Harbor Canal	—	B	East Chicago, IN	1913	RR	EJE	1.81	65	1.4
Indiana Harbor Canal	Chicago Ave.	B	East Chicago, IN	1913	HWY	Lake Co.	3.19	50	3
Indiana Harbor Canal	151st St.	B	East Chicago, IN	1918	HWY	Lake Co.	3.76	66	7
Indiana Harbor Canal, Lake George Branch	—	B	East Chicago, IN	1979	HWY	Indiana	2.5	68	12
Indiana Harbor Canal, Lake George Branch	Indianapolis Blvd.	B	East Chicago, IN	1939	HWY	Indiana	2.5	64	8
Indiana Harbor Canal, Lake George Branch	—	B	East Chicago, IN	1915	RR	BOCT	2.94	67.8	6.3
Inland Waterway White Lake to Vermillion Bayou	SR 82	SW	Intracoastal City, LA	1965	HWY	Louisiana	4	125	6
Inner Harbor Navigation Canal	—	B	New Orleans, LA	1923	HWY, RR	Port of New Orleans	2.9	93	1
Inner Harbor Navigation Canal	US 90 (Danzinger Br.)	B (repl. w. VL)	New Orleans, LA	1932	HWY	Louisiana	3.1	100	9
Inner Harbor Navigation Canal	—	B	New Orleans, LA	1923	HWY, RR	Port of New Orleans	4.5	92	1
Inner Harbor Navigation Canal	Seabrook Br.	B	New Orleans, LA	1967	HWY, RR	Orleans Levee District	4.6	96	46
INTRACOASTAL WATERWAY (INTC, INTCW, INTCWW)[d]									
INTCW Manasquan River	—	B	Brielle, NJ	1917	RR	New Jersey Transit	0.9	48	3
INTCW Manasquan River	—	B	Brielle, NJ	1951	HWY	New Jersey	1.1	90	30
INTCINLWW Manasquan River	Rte 34	B	Brick Twp, NJ	1937	HWY	New Jersey	3.41	50	15
INTCINLWW Bayhead–Manasquan Canal	Loveland Rd.	SW	West Pt. Pleasant, NJ	1928	HWY	New Jersey	1.33	—	—
INTCW Point Pleasant Canal	SR 88	VL	West Pt. Pleasant, NJ	—	HWY	New Jersey	3	—	—

*See key of abbreviations at end.

Waterway[a]	Road Name/ Route Carried	Bridge Type[b]	Location	Const. Date	Traffic Type[c]	Owner*	Miles from Mouth of Waterway	Channel Width/Lateral Clearance (feet)	Clearance Height (feet)
INTCW Point Pleasant Canal	SR 88	B (repl. by above)	West Pt. Pleasant, NJ	1924	HWY	New Jersey	3	47	10
INTCW Point Pleasant Canal	SR 13	VL	West Pt. Pleasant, NJ	1972	HWY	New Jersey	3.9	80	65
INTCW Barnegat Bay	—	B	Montoloking, NJ	1939	HWY	Ocean Co.	6.3	51	14
INTCW Barnegat Bay	Thomas Mathis Br.	B	Island Heights, NJ	1950	HWY	New Jersey	14.1	80	30
INTCINLWW Barnegate Bay	—	SW	Seaside Park, NJ	1913	RR	PRR	16.5	68.9	1
INTCINLWW Manahawkin Bay	—	B	Ship Bottom, NJ	1929	HWY	New Jersey	37.4	50.5	6.7
INTCW Beach Thorofare1	US 30	B	Atlantic City, NJ	1946	HWY	New Jersey	67.2	60	20
INTCWINLWW Beach Thorofare	Virginia Ave.	SW	Atlantic City, NJ	1906	TROL	AC&SRR	68.7	37	4
INTCW Beach Thorofare1	—	SW	Atlantic City, NJ	1993	RR	New Jersey Transit	68.9	—	—
INTCINLWW Beach Thorofare	—	SW	Atlantic City, NJ	1920	RR	PRSL	68.9	34.2	4.1
INTCW Beach Thorofare1	—	SW	Atlantic City, NJ	1924	RR	CR	68.9	50	5
INTCINLWW Beach Thorofare	—	SW	Atlantic City, NJ	1921	RR	PRSL	69	38.4	2.1
INTCW Beach Thorofare1	—	F, convert-ible to VL	Atlantic City, NJ	1965	HWY	New Jersey	69	80	35
INTCW Inside Thorofare	Albany Ave. (SR 40)	B	Atlantic City, NJ	1929	HWY	New Jersey	70	50	10
INTCW Inside Thorofare	Dorset Ave.	B	Ventnor City, NJ	1930	HWY	Atlantic Co.	71.2	50	9
INTCW Beach Thorofare2	—	B	Margate, NJ	1930	HWY	Margate Bridge Comm.	74	60	14
INTCW Broad Thorofare	SR 152	B	Longport, NJ	1917	HWY	New Jersey	77.8	49	9

INTCINLWW Beach Thorofare3	—	SW	Ocean City, NJ	1908	TROL	AC&SRR	80.3	30.4	5.3
INTCW Beach Thorofare3	SR 52	B	Ocean City, NJ	1933	HWY	New Jersey	80.4	70	14
INTCINLWW Main Thorofare	34th St. (Roosevelt Blvd.)	SW	Ocean City, NJ	1916	HWY	Cape May Co.	84.3	49	6
INTCW Crookhorn Creek	—	SW	Ocean City, NJ	1910	RR	CR	86.6	59	2
INTCINLWW Ben Hands Thorofare	—	SW	Strathmere, NJ	—	RR	PRSL	89.3	54	—
INTCINLWW Ludlam Thorofare	—	B (repl. w. C)	Sea Isle City, NJ	1916	HWY	Cape May Co.	93.6	50	8.1
INTC Ingram Thorofare	—	SW	Avalon, NJ	1915	HWY	Cape May Co.	97.8	49	6.4
INTCW Great Channel	—	B	Stone Harbor, NJ	1931	HWY	Cape May Co.	102	50	11
INTCW Grassy Sound Channel	SR 147	B	North Wildwood, NJ	1924	HWY	New Jersey	105.2	50	8
INTCW Grassy Sound Channel	—	B	Wildwood, NJ	1913	RR	CR	107.5	50	6
INTCW Grassy Sound Channel	SR 47	B	Wildwood, NJ	1950	HWY	New Jersey	108.9	57	25
INTCW Middle Thorofare	—	B	Wildwood Crest, NJ	1940	HWY	Cape May Co.	112.2	50	23
INTCW Cape May Canal	—	SW	Cape May, NJ	1942	RR	CR	115.1	50	4
INTCWILWW Lewes–Rehoboth Canal	Savannah Rd.	B	Lewes, DE	1981	HWY	Delaware	1.7	70	15
INTCWILWW Lewes–Rehoboth Canal	SR 18	B	Lewes, DE	1929	HWY	Delaware	1.7	46	6
INTCWILWW Lewes–Rehoboth Canal	—	SW	Lewes, DE	1917	RR	CR	2.2	46	10
INTCWILWW Lewes–Rehoboth Canal	Rehoboth Ave. (SR 14A)	B	Rehoboth, DE	1925	HWY	—	6.7	49	16
INTCWILWW	—	B	Rehoboth, DE	1940	HWY	Delaware	8.6	50	14
INTCWILW Assawoman Bay	US 50	B	Ocean City, MD	1939	HWY	Maryland	36.8	70	13

*See key of abbreviations at end.

Waterway[a]	Road Name/ Route Carried	Bridge Type[b]	Location	Const. Date	Traffic Type[c]	Owner*	Miles from Mouth of Waterway	Channel Width/Lateral Clearance (feet)	Clearance Height (feet)
INTCWILWW	—	B	Ocean City, MD	1928	HWY	Maryland	37	40	4
INTCW Elizabeth River, Southern Branch	—	VL	Portsmouth, VA	1947	RR	NB	2.6	300	142
INTCW Elizabeth River, Southern Branch	Jordon Br.	VL	Portsmouth, VA	1928	HWY	South Norfolk Bridge Comm.	2.8	220	145
INTCW Elizabeth River, Southern Branch	—	VL	Portsmouth, VA	1974	RR	NS	3.6	220	8
INTCW Elizabeth River, Southern Branch	Gilmerton Br. (US 13 460)	B	Chesapeake, VA	1938	HWY	Chesapeake	5.8	125	7
INTCW Elizabeth River, Southern Branch	—	B	Chesapeake, VA	1909	RR	NS	5.8	125	7
INTCW Elizabeth River, Southern Branch	I 64	B	Chesapeake, VA	1969	HWY	Virginia	7.1	125	65
INTCW Elizabeth River, Southern Branch	—	SW	Chesapeake, VA	1907	RR	NPB	8.1	80	7
INTCW Elizabeth River, Southern Branch	Steel Br. (SR 166)	B	Chesapeake, VA	1959	HWY	Chesapeake	8.8	128	12
INTCW Albemarle & Chesapeake Canal	SR 168	B	Great Bridge, VA	1916	HWY	U.S. Govt.	12	80	5
INTCW Albemarle & Chesapeake Canal	—	B	Great Bridge, VA	1928	RR	NS	13.9	80	5
INTCW Albemarle & Chesapeake Canal	—	SW	Fentress, VA	1955	HWY	Chesapeake	15.2	80	4
INTCINLWW Virginia Cut	—	B	North Landing, VA	1916	HWY	U.S. Govt.	19.25	80	5
INTCW North Landing River	SR 165	SW	Chesapeake, VA	1916	HWY	U.S. Govt.	20.2	80	6
INTCW North Landing River	SR 604	SW	Virginia Beach, VA	1953	HWY	Virginia Beach	28.2	80	6
INTCWALT Dismal Swamp Canal	US 17	B	Deep Creek, VA	1934	HWY	U.S. Govt.	11.1	60	3

INTCWALT Dismal Swamp Canal	US 17	B	South Mills, NC	1934	HWY	U.S. Govt.	32.6	60	3
INTCWALT Pasquotank River	—	SW	Elizabeth City, NC	1903	RR	NS	47.7	42	2
INTCWALT Pasquotank River	US 158	B	Elizabeth City, NC	1931	HWY	North Carolina	50.7	90	3
INTCWALT Pasquotank River	US 158	B	Elizabeth City, NC	1973	HWY	North Carolina	50.7	90	2
INTCINLWW North Carolina Cut	—	B	Coinjock, NC	1916	HWY	U.S. Govt.	48.6	80	7
INTCW North Carolina Cut	US 158	SW	Coinjock, NC	1940	HWY	U.S. Army COE	49.9	80	4
INTCW North Carolina Cut	US 64	SW	Columbia, NC	1962	HWY	North Carolina	84.2	100	14
INTCINLWW Alligator/ Pungo River Cut	—	SW	Fairfield, NC	1935	HWY	U.S. Govt.	112.66	80	7.31
INTCW North Carolina Cut	SH 94	SW	Fairfield, NC	1935	HWY	U.S. Army COE	113.8	80	7
INTCINLWW Alligator/ Pungo River Cut	—	SW	Belhaven, NC	1934	HWY	U.S. Govt.	124.05	80	7.35
INTCW North Carolina Cut	SR 264	SW	Belhaven, NC	1934	HWY	U.S. Army COE	125.9	80	8
INTCINLWW Upperspring Creek, Jones Boy Land Cut	—	SW	Hobucken, NC	1929	HWY	U.S. Govt.	155.8	79.9	6
INTCW North Carolina Cut	SR 304	SW	Hobucken, NC	1930	HWY	U.S. Army COE	157.2	79	6
INTCINLWW Adams/ Core Creek Cut	—	SW	Beaufort, NC	1935	HWY	U.S. Govt.	194	80	15
INTCW North Carolina Cut	SR 101	SW	Beaufort, NC	1935	HWY	U.S. Army COE	195.8	80	16

*See key of abbreviations at end.

Waterway[a]	Road Name/ Route Carried	Bridge Type[b]	Location	Const. Date	Traffic Type[c]	Owner*	Miles from Mouth of Waterway	Channel Width/Lateral Clearance (feet)	Clearance Height (feet)
INTCINLWW Newport River	—	SW	Morehead City, NC	1907	RR	BMH	201.7	50	2.7
INTCINLWW Newport River	—	B	Morehead City, NC	1928	HWY	North Carolina	201.7	79	7.3
INTCW North Carolina Cut	—	B	Morehead City, NC	1947	RR	BMH	203.8	80	4
INTCW North Carolina Cut	SR 58	SW	Atlantic Beach, NC	1953	HWY	North Carolina	206.7	90	13
INTCINLWW Bogue Sound	—	SW	Morehead City, NC	1928	HWY	North Carolina	207.5	84	6.8
INTCW North Carolina Cut	—	SW	Camp Lejeune, NC	1954	HWY	U.S. Marine Corps	240.7	80	12
INTCW North Carolina Cut	SR 50	SW	Sears Landing, NC	1955	HWY	North Carolina	260.7	92	13
INTCW North Carolina Cut	—	SW	Scotts Hill, NC	1980 (relocated)	HWY	Fig. 8 Beach Homeowners Assn.	278.1	105	20
INTCW North Carolina Cut	US 74, 76	B	Wrightsville Beach, NC	1958	HWY	North Carolina	283.1	90	20
INTCINLWW Middle Sound	—	B	Wrightsville, NC	1931	HWY, RR	North Carolina	283.2	80	4.2
INTCINLWW Land Cut	—	SW	Carolina Beach, NC	1931	HWY	North Carolina	295.7	80	13.2
INTCINLWW Cape Fear River-Little River	—	P	Southport, NC	1933	HWY	North Carolina	302.6	80	0
INTCW North Carolina Cut	SR 130	SW	Holden Beach, NC	1954	HWY	North Carolina	323.7	87	13
INTCW North Carolina Cut	SR 904	SW	Ocean Isle, NC	1958	HWY	North Carolina	333.7	80	13

INTCW North Carolina Cut	—	P	Sunset Beach, NC	1961	HWY	North Carolina	337.9	90	0
INTCINLWW Little River/ Winyah Bay	—	SW	Little River, SC	1935	HWY	U.S. Govt.	337.9	80	8.2
INTCW Little River	SR 9	SW	—	1935	HWY	South Carolina	347.3	78	7
INTCINLWW Little River/ Winyah Bay	—	B	Myrtle Beach, SC	1936	HWY	U.S. Govt.	356	80	5.2
INTCW Little River	Skyway Golf Club	TRAM	Myrtle Beach, SC	1975	TRAM	Skyway Golf Club	356	90	67
INTCINLWW Cove Inlet	—	SW	Mt. Pleasant, SC	1927	HWY	South Carolina	360.6	61.7	7.93
INTCINLWW Little River/ Winyah Bay	—	B	Myrtle Beach, SC	1935	HWY	U.S. Govt.	361.6	80	2.5
INTCW Little River	SR 503	B	Myrtle Beach, SC	1936	HWY, RR	South Carolina	365.4	80	16
INTCW Little River	US 501	SW	Myrtle Beach, SC	1935	HWY	South Carolina	371	80	12
INTCW Little River	SR 703	SW	Sullivans Island, SC	1945	HWY	South Carolina	462.2	93	31
INTCW Wappoo Creek	SR 171	B	Charleston, SC	1956	HWY	South Carolina	470.8	100	33
INTCINLWW Wappo Creek	—	SW	Charleston, SC	1927	HWY	South Carolina	471.3	81	12.5
INTCINLWW Stono River	—	SW	Charleston, SC	1940	RR	SAL	477.8	60	5.8
INTCW Little River	John F. Limehouse Br. (SR 20)	SW	St. Johns Island, SC	1958	HWY	South Carolina	479.3	90	12
INTCINLWW Stono River	—	SW	Charleston, SC	1915	HWY	South Carolina	479.8	66.9	5.8
INTCINLWW Stono River	—	SW	Charleston, SC	1917	RR	SAL	483.1	66	6
INTCW Dawho River	SR 174	SW	Adams Run, SC	1953	HWY	South Carolina	501.3	90	8
INTCINLWW Dawho River	—	SW	Adams Run, SC	1920	HWY	South Carolina	506.1	60	5
INTCINLWW Beaufort River	—	SW	Beaufort, SC	1927	HWY	South Carolina	536	80.3	20
INTCW Beaufort	Ladies Island Br. (US 21)	SW	Beaufort, SC	1960	HWY	South Carolina	536	90	30
INTCINLWW Archers Creek	—	SW	Parris Island, SC	1928	HWY	U.S. Navy	543.8	78.6	4.3

*See key of abbreviations at end.

Waterway[a]	Road Name/ Route Carried	Bridge Type[b]	Location	Const. Date	Traffic Type[c]	Owner*	Miles from Mouth of Waterway	Channel Width/Lateral Clearance (feet)	Clearance Height (feet)
INTCW Skull Creek	—	SW	Hilton Head, SC	1956	HWY	South Carolina	557.6	97	30
INTCW Wilmington River	US 80	B	Savannah Beach, GA	1963	HWY	Chatham Co.	579.9	100	21
INTCW Wilmington River	—	B	Thunderbolt, GA	1955	HWY	Georgia	582.2	100	21
INTCINLWW Wilmington River	—	SW	Thunderbolt, GA	1921	HWY	Georgia	586.8	93	6
INTCW Skidmore Narrows	US 80	B	Skidaway Island, GA	1971	HWY	Chatham Co.	592.9	100	22
INTCW Frederica River	—	VL	St. Simon Island, GA	1950	HWY	Glynn Co.	675.5	100	85
INTCW Frederica River	—	SW	Brunswick, GA	1924	HWY	Glynn Co.	679	84.4	11
INTCINLWW Back River	—	SW	Brunswick, GA	1924	HWY	Glynn Co.	679.6	85	11
INTCW Jekyll Creek	—	VL	Jekyll Island, GA	1955	HWY	Georgia	684.3	100	85
INTCWF Kingsley Creek	—	SW	Fernandina, FL	1927	HWY	Florida	725.7	81	6
INTCWF Kingsley Creek	—	SW	Fernandina, FL	—	RR	SAL	725.7	49.6	3.3
INTCWF Sisters Creek	—	B	Jacksonville, FL	1927	HWY	North Shore Corp.	747	81	9
INTCWF Kingsley Creek	—	SW	Fernandina, FL	1953	RR	CSX	3.7	90	5
INTCWF Sisters Creek	SR 105	B	Mayport, FL	1953	HWY	Florida	22.2	90	24
INTCWF Pablo Creek	—	B	Neptune, FL	1928	HWY	Florida	28.4	60	7.1
INTCWF Pablo Creek	US 90	B	Jacksonville Beach, FL	1949	HWY	Florida	30.5	90	37
INTCWF Pablo Creek	SR 210	B	Palm Valley, FL	1937	HWY	U.S. Army COE	41.8	80	10
INTCWFL Coast Line Canal	—	B	Palm Valley, FL	1937	HWY	U.S. Govt.	42.6	80	10
INTCWF Tolomato River	SR A1A	VL	Vilano Beach, FL	1949	HWY	Florida	58.8	89	83
INTCWF Matanzas River	Bridge of Lions (SR A1A)	B	St Augustine, FL	1927	HWY	Florida	60.9	76	25
INTCWF Matanzas River	SR 206	B	Crescent Beach, FL	1976	HWY	Florida	71.6	90	20

INTCWF Matanzas River	—	B	Crescent Beach, FL	1928	HWY	Florida	72.3	79.5	6.4
INTCWF Matanzas River	SR 100	B	Flagler Beach, FL	1951	HWY	Florida	93.6	91	14
INTCWF Smith Creek	—	SW	Flagler Beach, FL	1931	HWY	Flagler Co.	94.6	53	4
INTCWF Smith Creek	SR A1A	B	Bulow, FL	1955	HWY	Florida	99	91	15
INTCWF Smith Creek	—	SW	Bulow, FL	1921	HWY	Volusia Co.	100	56	3.7
INTCWF Halifax River	—	B	Ormond Beach, FL	1954	HWY	Florida	107.9	89	21
INTCWF Halifax River	—	SW	Ormond, FL	1920	HWY	Volusia Co.	108.9	55.8	3.2
INTCWF Halifax River	SR A1A	B	Daytona Beach, FL	1951	HWY	Volusia Co.	112.1	90	20
INTCWF Halifax River	—	B	Daytona Beach, FL	1959	HWY	Volusia Co.	112.7	90	22
INTCWF Halifax River	—	SW	Seabreeze, FL	1904	HWY	Volusia Co.	113.1	45.7	6.4
INTCWF Halifax River	—	B	Daytona Beach, FL	1948	HWY	Florida	113.1	90	20
INTCWF Halifax River	—	B	Daytona Beach, FL	1955	HWY	Volusia Co.	113.6	89	21
INTCWF Halifax River	—	B	Daytona Beach, FL	1928	HWY	Volusia Co.	113.7	59.7	8.5
INTCWF Halifax River	—	SW	Daytona Beach, FL	1913	HWY	Volusia Co.	114.1	57.5	8.5
INTCWF Halifax River	—	B	Daytona Beach, FL	1928	HWY	Volusia Co.	114.7	60.9	9.6
INTCWF Halifax River	—	B	Port Orange, FL	1951	HWY	Volusia Co.	118.5	92	20
INTCWF Hillsboro River	—	B	New Smyrna Beach, FL	1953	HWY	Florida	128	91	14
INTCWF Hillsborough River	—	SW	Coronado Beach, FL	1928	HWY	Volusia Co.	133.1	60.8	6.8
INTCWF Hillsboro River	—	B	New Smyrna Beach, FL	1967	HWY	Florida	129.5	95	24
INTCWF Hillsborough River	—	SW	New Smyrna, FL	1928	HWY	Florida	142.1	58	4.8
INTCWF Haulover Canal	—	B	Allenhurst, FL	1965	HWY	NASA	152.2	90	27
INTCWF Haulover Canal	—	SW	Allenhurst, FL	—	HWY	Brevard Co.	153.6	55	7.4
INTCWF Indian River	—	B	Jay Jay, FL	1963	HWY	FEC	159.6	90	7
INTCWF Indian River	SR 402	SW	Titusville, FL	1949	HWY	Florida	161.9	81	9
INTCWF Indian River	—	SW	Titusville, FL	1922	HWY	Florida	163.2	55	9.5
INTCWF Indian River	—	SW	Titusville, FL	1939	HWY	Florida	163.2	80	6.5
INTCWF Indian River	—	B	Orsinio, FL	1964	HWY	NASA	168	90	27

*See key of abbreviations at end.

Waterway[a]	Road Name/ Route Carried	Bridge Type[b]	Location	Const. Date	Traffic Type[c]	Owner*	Miles from Mouth of Waterway	Channel Width/Lateral Clearance (feet)	Clearance Height (feet)
INTCWF Indian River	—	SW	Cocoa, FL	1939	HWY	Florida	182.3	80	3
INTCWF Indian River	—	SW	Eau Gallie, FL	1925	HWY	Florida	198.8	60.83	10
INTCWF Indian River	SR 518	SW	Eau Gallie, FL	1955	HWY	Florida	197.4	80	9
INTCWF Indian River	SR 516	SW	Melbourne, FL	1942	HWY	Florida	201.2	80	6
INTCWF Indian River	—	SW	Melbourne, FL	1921	HWY	Florida	202.6	55	8.5
INTCWF Indian River	—	SW	Wabasso, FL	1926	HWY	Florida	227.5	60	9.8
INTCWF Indian River	—	SW	Winter Beach, FL	1924	HWY	Indian River Co.	230.6	60	9.8
INTCWF Indian River	SR 502	B	Vero Beach, FL	1951	HWY	Florida	234.9	90	22
INTCWF Indian River	—	SW	Vero Beach, FL	1920	HWY	Florida	236.4	60	9.8
INTCWF Indian River	SR A1A	B	Ft. Pierce, FL	1964	HWY	Florida	247.8	90	26
INTCWF Indian River	—	SW	Ft. Pierce, FL	1939	HWY	Ft. Pierce	249.7	80	8.5
INTCWF Indian River	—	SW	Ft. Pierce, FL	1934	HWY	Florida	250.3	60	8.5
INTCWF Indian River	SR 707	B	Jensen Beach, FL	1965	HWY	Florida	264.4	90	24
INTCWF Indian River	—	SW	Jensen, FL	1924	HWY	Martin Co.	265.7	60	9
INTCWF Indian River	SR A1A	B	Stuart, FL	1957	HWY	Florida	267.9	90	28
INTCWF South Jupiter Narrows	SR 707	B	Hobe Sound, FL	1948	HWY	Florida	278.9	80	10
INTCWF South Jupiter Narrows	—	B	Hobe Sound, FL	1929	HWY	Florida	280.2	87	11
INTCWF Jupiter Sound	SR 707	B	Jupiter Island, FL	1969	HWY	Florida	287.1	90	24
INTCWF Loxahatchie River	US 1	B	Jupiter, FL	1958	HWY	Florida	287.8	91	26
INTCWF Jupiter Sound	—	SW	Jupiter, FL	1923	HWY	Florida	287.8	55	8.7
INTCWF Lake Worth Creek	SR 706	B	—	1965	HWY	Florida	289.2	91	15
INTCWF Jupiter River	—	B	Jupiter, FL	1927	HWY	Florida	289.3	58	6.6
INTCWF	Donald Ross Br.	B	Juno Beach, FL	1958	HWY	Florida	292.3	90	14

INTCWF Lake Worth Creek	SR 703	B	North Palm Beach, FL	1966	HWY	Florida	295.6	90	25
INTCWF Lake Worth Creek	US 1	B	Lake Park, FL	1956	HWY	Florida	296.7	94	25
INTCWFFCL Canal Jupiter River to Lake Worth	—	B	Kelsey City, FL	—	HWY	Palm Beach Co.	297.4	58	4.2
INTCWFFCL Canal Jupiter River to Lake Worth	—	B	Kelsey City, FL	—	HWY	Florida	298	58	13.3
INTCWF Lake Worth	—	B	Riviera, FL	1926	HWY	Florida	301.5	59	9.5
INTCWF Lake Worth Creek	Flagler Memorial	B	West Palm Beach, FL	1938	HWY	Palm Beach Co.	304.8	80	17
INTCWF Lake Worth	Royal Palm	B	West Palm Beach, FL	1959	HWY	Florida	305.6	91	14
INTCWF Lake Worth	—	SW	Palm Beach, FL	1924	HWY	Palm Beach Co.	306.9	59	10.5
INTCWF Lake Worth	Southern Blvd.	B	West Palm Beach, FL	1951	HWY	Florida	307.7	81	14
INTCWF Lake Worth	—	B	West Palm Beach, FL	1927	HWY	Palm Beach Co.	308.9	59	10.3
INTCWF Jupiter Inlet to Lake Worth	SR 802	B	—	1974	HWY	Florida	311.8	90	35
INTCWF Lake Worth	—	B	Lake Worth, FL	1933	HWY	Palm Beach Co.	313	80	15
INTCWF Lake Worth	US 1	B	Lantana, FL	1950	HWY	Florida	314	90	13
INTCWF Lake Worth	—	SW	Lantana, FL	1925	HWY	Palm Beach Co.	315.1	60	11
INTCWF Lake Worth	SR 804	B	Boynton Beach, FL	1937	HWY	Florida	318	80	10
INTCWF Lake Worth	—	B	Boynton Beach, FL	1967	HWY	Florida	318.8	87	25
INTCWF Lake Worth	SR 806	B	Delray Beach, FL	1950	HWY	Palm Beach Co.	321.7	80	9
INTCWF Lake Worth	—	B	Delray Beach, FL	1952	HWY	Florida	322.6	90	12
INTCWFFLA Coast Line Canal, Lake Worth to New River Sound	—	B	Delray, FL	—	HWY	Palm Beach Co.	323.7	60	5.5
INTCWF Lake Worth	S.E. 12th St.	B	Delray Beach, FL	1981	HWY	Florida	324.1	90	30
INTCWF Lake Worth Inlet	—	B	Boca Raton, FL	1971	HWY	Florida	327.9	90	21

*See key of abbreviations at end.

Waterway[a]	Road Name/ Route Carried	Bridge Type[b]	Location	Const. Date	Traffic Type[c]	Owner*	Miles from Mouth of Waterway	Channel Width/Lateral Clearance (feet)	Clearance Height (feet)
INTCWF Lake Boca Raton	North Br. (SR 808)	B	Boca Raton, FL	1930	HWY	Florida	330.5	80	6
INTCWF Lake Boca Raton	South Br. (SR 808)	B	Boca Raton, FL	1939	HWY	Palm Beach Co.	331.2	85	9
INTCWF Hillsboro River	—	B	Deerfield Park, FL	1957	HWY	Florida	333	91	21
INTCWFFLA Coast Line Canal, Lake Worth to New River Sound	—	SW	Deerfield, FL	1917	HWY	Broward Co.	334	53	5.5
INTCWF Hillsboro River	N.E. 14th St.	B	Pompano Beach, FL	1968	HWY	Florida	338	90	15
INTCWF Hillsboro River	SR 814	B	Pompano Beach, FL	1956	HWY	Florida	339	89	15
INTCWFFLA Coast Line Canal, Lake Worth to New River Sound	—	SW	Pompano, FL	—	HWY	Broward Co.	340.1	55	4.8
INTCWF New River Sound	—	B	Ft. Lauderdale, FL	1966	HWY	Florida	342	90	15
INTCWF New River Sound	—	B	Ft. Lauderdale, FL	1956	HWY	Florida	343.5	88	22
INTCWFFLA Coast Line Canal, Lake Worth to New River Sound	—	SW	Oakland Park, FL	—	HWY	Broward Co.	344.6	60	7.5
INTCWF New River Sound	Sunrise Blvd. N.	B	Ft. Lauderdale, FL	1940	HWY	Ft. Lauderdale	345.6	85	16
INTCWF New River Sound	Sunrise Blvd. S.	B	Ft. Lauderdale, FL	1960	HWY	Ft. Lauderdale	345.6	86	17
INTCWF New River Sound	—	B	Ft. Lauderdale, FL	1940	HWY	Ft. Lauderdale	346.3	85	17.5
INTCWF New River Sound	Las Olas Blvd.	B	Ft. Lauderdale, FL	1958	HWY	Florida	347	91	31

INTCWF New River Sound	—	SW	Ft. Lauderdale, FL	1925	HWY	Broward Co.	348	59	6.8
INTCWF Stranahan River	—	B	Ft. Lauderdale, FL	1956	HWY	Florida	348.9	99	25
INTCWF	Dania Beach Blvd.	B	Dania Beach, FL	1957	HWY	Florida	352.4	91	22
INTCWF	Sheridan St. (SR 822)	B	Hollywood Beach, FL	1962	HWY	Florida	353.5	90	22
INTCWFFLA Coast Line Canal, Lake Mabel to Biscayne Creek	—	SW	Dania, FL	—	HWY	Broward Co.	353.5	55	3.5
INTCWF	Hollywood Blvd.	B	Hollywood Beach, FL	1978	HWY	Florida	355.2	90	25
INTCWFFLA Coast Line Canal, Lake Mabel to Biscayne Creek	—	B	Hollywood, FL	—	HWY	Broward Co.	356.3	66	9.5
INTCWF	Hallandale Beach Blvd.	B	Hallandale, FL	1960	HWY	Florida	357	90	22
INTCWFFLA Coast Line Canal, Lake Mabel to Biscayne Creek	—	SW	Hallandale, FL	1917	HWY	Broward Co.	358.1	55.9	9.3
INTCWF Biscayne Creek	—	B	North Miami Beach, FL	1948	HWY	Florida	361	92	19
INTCWF Biscayne Creek	—	SW	Fulford, FL	—	HWY	Florida	362.1	53.5	6.5
INTCWF Biscayne Bay	Broad Cswy.	B	North Miami Beach, FL	1954	HWY	Bay Harbor Islands	364.4	80	16
INTCWF Biscayne Bay	79th St. Cswy.	B	Miami, FL	1972	HWY	Florida	367.6	90	19
INTCWF Biscayne Bay	Venetian Cswy.	B (2 Br)	Miami, FL	1926	HWY	Dade Co.	371.6	60	8
INTCWF Biscayne Bay	Macarthur Cswy.	B	Miami, FL	1961	HWY	Florida	371.8	91	35
INTCWF Biscayne Bay	Dodge Island	B	Miami, FL	1964	RR	Dade Co.	372.4	90	22
INTCWF Biscayne Bay	Dodge Island	B	Miami, FL	1964	HWY	Dade Co.	372.4	90	22
INTCWF Biscayne Bay	Venetian Cswy.	B	Miami, FL	1926	HWY	Bay Biscayne Impr. Company	372.62	—	—
INTCWF Biscayne Bay	13th St.	B	Miami, FL	1920	HWY	Florida	373	60	8
INTCWF Biscayne Bay	Rickenbacker Cswy.	B	—	1947	HWY	Dade Co.	374.6	80	23

*See key of abbreviations at end.

Waterway[a]	Road Name/ Route Carried	Bridge Type[b]	Location	Const. Date	Traffic Type[c]	Owner*	Miles from Mouth of Waterway	Channel Width/Lateral Clearance (feet)	Clearance Height (feet)
INTCWF Barnes Sound	—	SW	Florida City, FL	1924	HWY	Florida	412.4	40	9
INTCWF Jewfish Creek	—	B	Key Largo, FL	1944	HWY	Florida	417.1	80	11
INTCWF Jewfish Creek	—	B	Jewfish, FL	1939	HWY	Florida	419.7	80	10
INTCWF Moser Channel	—	SW	Marathon, FL	1939	HWY	Overseas Road & Tollbridge Dist.	482.9	107.6	23
INTCW Gulf Gasparilla Sound	—	B	Placida, FL	1973	RR	CSX	778[†]	90	3
INTCW Gulf Gasparilla Sound	SR 771	SW	Placida, FL	1958	HWY	Florida	777.7	81	9
INTCW Gulf Lemon Bay	SR 776	B	Englewood, FL	1965	HWY	Florida	770.5	86	26
INTCW Gulf Lemon Bay	—	B	Manasota, FL	1965	HWY	Florida	762.1	90	26
INTCW Gulf Venice Inlet	US 41	B	Venice, FL	1967	HWY	Florida	757	90	25
INTCW Gulf Venice Inlet	—	B	Venice, FL	1967	HWY	Florida	756.2	90	30
INTCW Gulf Hatchett Creek	—	B	Venice, FL	1966	HWY	Florida	755.2	90	16
INTCW Gulf Little Sarasota Bay	SR 789	B	Nokomis, FL	1963	HWY	Florida	752.7	90	14
INTCW Gulf Little Sarasota Bay	SR 72	SW	Blackburn Point, FL	1963	HWY	Florida	748.2	51	9
INTCW Gulf Little Sarasota Bay	SR 72	B	Stickney Pt., FL	1969	HWY	Florida	742.9	90	18
INTCW Gulf Roberts Bay	—	B	Siesta Key, FL	1973	HWY	Florida	739.1	90	21
INTCW Gulf Sarasota Bay	Ringling Cswy.	B	Sarasota Bay, FL	1954	HWY	Florida	737	90	22
INTCW Gulf Sarasota Pass	SR 64	B	Cortez, FL	1954	HWY	Florida	723.2	90	25
INTCW Gulf Sarasota Pass	SR 684	B	Sarasota, FL	1954	HWY	Florida	721.3	90	25
INTCW Gulf Tampa Bay	US 19 (Sunshine Skyway)	B	Tampa, FL	1969	HWY	Florida	701.5	90	21

INTCW Gulf Tampa Bay	US 19 (Sunshine Skyway)	B	Tampa, FL	1950	HWY	Florida	701.5	90	21
INTCW Gulf Pinellas Bayway Bocaciega Bay	SR 669	B	St. Petersburg, FL	1962	HWY	Florida	698.1	89	25
INTCW Gulf Pinellas Bayway Bocaciega Bay	SR 669	B	Boca Ciela, FL	1962	HWY	Florida	697.3	91	25
INTCW Gulf Corey Causeway	—	B	St. Petersburg, FL	1973	HWY	Florida	694.7	90	23
INTCW Gulf Corey Causeway	—	B	St. Petersburg, FL	1966	HWY	Florida	694.6	90	23
INTCW Gulf Treasure Island Causeway	SR 669	B	St. Petersburg, FL	1939	HWY	Florida	692.4	80	8
INTCW Gulf Boca Ciega Bay	SR 699 (Madeira Cswy.)	B	Bay Pines, FL	1962	HWY	Florida	688.6	89	25
INTCWF The Narrows	—	B	Seminole, FL	1982	HWY	Pinellas Co.	126	90	4
INTCW Gulf Boca Ciega Bay	SR 699 (Indian Rocks)	B	Bay Pines, FL	—	HWY	Florida	681	90	25
INTCW Gulf Clearwater Harbor	SR 699 (Belleair Cswy.)	B	—	1948	HWY	Florida	678.3	80	21
INTCW Gulf Clearwater Harbor	SR 60 (Clearwater Cswy.)	B	Clearwater, FL	1963	HWY	Florida	674.7	90	25
INTCW Gulf St. Joseph Sound	Honeymoon Island Cswy.	B	Dunedin, FL	1963	HWY	City of Dunedin	668.7	91	24
INTCWG Appal–West Bay	—	SW	Apalachicola, FL	1936	HWY	Florida	357.8	120	29.2
INTCW Gulf Apalachee Bay to Panama City	US 99, 319	SW	Apalachicola, FL	1934	HWY	Florida	351.4	120	28
INTCW Gulf Apalachicola River	—	SW	Apalachicola, FL	1905	RR	AN	347	119	11
INTCWG Appal–West Bay	—	P	White City, FL	1939	RR	St. Joe Lumber & Export	337.5	80	—
INTCWG Appal–West Bay	—	P	White City, FL	1937	HWY	Florida	336.1	50	—

*See key of abbreviations at end.

†Mileage from Harvey Lock, LA.

Waterway[a]	Road Name/ Route Carried	Bridge Type[b]	Location	Const. Date	Traffic Type[c]	Owner*	Miles from Mouth of Waterway	Channel Width/Lateral Clearance (feet)	Clearance Height (feet)
INTCW Gulf Apalachee Bay to Panama City	SR 71	VL	White City, FL	1947	HWY	Florida	329.3	86	80
INTCW Gulf Apalachee Bay to Panama City (connecting channel)	US 98	B	Port St. Joe, FL	1939	HWY	Florida	320	80	10
INTCW Gulf Wetappo Creek	—	P	Overstreet, FL	1947	HWY	Florida	315.5	86	—
INTCWG Appal–West Bay	—	SW	Long Pt., FL	1928	HWY	Florida	301.5	75	10.5
INTCWG Appal–West Bay	—	SW	St. Andrews, FL	1928	HWY	Florida	290.7	90	10.5
INTCWG St. Andrews to Choctawhatchee	—	P	West Bay, FL	1937	HWY	Florida	278.5	55	—
INTCWGULF West Bay Creek	SR 79	VL	West Bay, FL	1945	HWY	Florida	271.8	86	80
INTCWG St. Andrews to Choctawhatchee	—	B	La Grange, FL	1937	HWY	Walton Co.	257	100	15.5
INTCWGULF St. Andrew Bay to New Orleans	—	P	Point Washington, FL	1970	HWY	Florida	254.3	150	—
INTCWGULF Choctawhatchee Bay	US 331	B	—	1937	HWY	Walton Co.	250.3	100	10
INTCWG Choctawhatchee to Pensacola Bay	—	SW	Ft. Walton, FL	1934	HWY	Florida	229.3	90	16.5
INTCWG Choctawhatchee to Pensacola Bay	—	SW	Pensacola, FL	1931	HWY	Escambia and Santa Rosa Cos.	195.7	100	15
INTCWG Pensacola Bay to Mobile Bay	—	B	Gulf Beach, FL	1939	HWY	Escambia Co.	178.6	80	10.5
INTCWG Pensacola Bay to Mobile Bay	—	P	Foley, AL	1938	HWY	Florida(?)	161.5	90	—
INTCWGla Inner Harbor Navigation Canal	SR 39 (St. Claudia Ave.)	B	New Orleans, LA	1923	HWY	Port of New Orleans	0.5	75	16

INTCWGla Inner Harbor Navigation Canal	SR 39 (Claiborne Ave., Judge W. Seeber Br.)	VL	New Orleans, LA	1957	HWY	Louisiana	0.9	305	156
INTCWGla Inner Harbor Navigation Canal	Florida Ave.	B	New Orleans, LA	1923	HWY, RR	Port of New Orleans	1.7	91	1
INTCWGlbALT Algiers Canal	SR 407	VL	Algiers, LA	1958	HWY	Louisiana	1	125	100
INTCWGlbALT Algiers Canal	—	VL	Belle Chasse, LA	1957	RR	UP	3.7	125	100
INTCWGlbALT Algiers Canal	SR 23	VL	Belle Chasse, LA	1968	HWY	Louisiana	3.8	125	100
INTCWGlbALT Algiers Canal	—	VL	Belle Chasse, LA	1966	HWY	Hero Wall Comm.	7.6	125	100
INTCWGlc Harvey Canal 1	—	B	Harvey, LA	—	RR	UP	0.2	75	9
INTCWGlc Harvey Canal 1	—	B	Harvey, LA	1935	HWY	Louisiana	0.24	75	7.5
INTCWGlc Harvey Canal 1	—	B	Harvey, LA	1972	HWY	Jefferson Pa.	2.8	150	45
INTCWG Harvey Canal 2	—	P	Larose, LA	1933	HWY	Louisiana	35.4	78	—
INTCWGlc Harvey Canal 2	SR 1	VL	Larose, LA	1961	HWY	Louisiana	35.6	125	35
INTCWGlc Harvey Lock to Morgan City	SR 316	P	Bourg, LA	1972	HWY	Louisiana	49	125	—
INTCWG Company Canal	—	P	Lockport, LA	1939	HWY	Louisiana	55.1	75.6	—
INTCWGlc Harvey Lock to Morgan City	—	VL	Houma, LA	1967	HWY	Louisiana	57.2	125	73
INTCWG Company Canal	Park Ave.	B	Houma, LA	1933	HWY	Louisiana	63.25	75	9.2
INTCWG Company Canal	Main St.	B	Houma, LA	1937	HWY	Louisiana	63.3	78	9.2
INTCWG Company Canal	—	SW	Houma, LA	1930	RR	ML&T SPRR	64.6	91	3.9
INTCWG Company Canal	—	P	Houma, LA	1933	HWY	Louisiana	65.4	12.6	—

*See key of abbreviations at end.

Waterway[a]	Road Name/ Route Carried	Bridge Type[b]	Location	Const. Date	Traffic Type[c]	Owner*	Miles from Mouth of Waterway	Channel Width/Lateral Clearance (feet)	Clearance Height (feet)
INTCWG Company Canal	—	SW	Bayou Sale, LA	1932	RR	ML&T SPRR	118.6	75	2.1
INTCWG Company Canal	—	SW	Louisa, LA	1936	HWY	St. Mary Pa.	138.9	80	7.9
INTCWG Inland WW Company Canal (Mississippi River to Teche Bayou)	—	B	Bourg, LA	1926	HWY	Louisiana	8.08	42	4.5
INTCWG Inland WW Houma Company Canal (Mississippi River to Teche Bayou)	—	SW	Houma, LA	1938	HWY	Louisiana	1.75	40.5	4.4
INTCWGlc Harvey Lock to Morgan City	SR 659	VL	Houma, LA	1958	HWY	Louisiana	57.6	122	73
INTCWGlc Harvey Lock to Morgan City	SR 659 (E. Main St.)	B	Houma, LA	1937	HWY	Louisiana	57.7	78	9
INTCWGlc Harvey Lock to Morgan City	No 14.82	VL	Houma, LA	1982	RR	UP	58.9	224	73
INTCWGlc Harvey Lock to Morgan City	SR 315	B	Houma, LA	1968	HWY	Louisiana	59.9	125	40
INTCWGld Berwick Bay	—	VL	Morgan City, LA	—	RR	SP	0.4	322	73
INTCWGld Lower Grand River	SR 75	P	Bayou Sorrel, LA	1965	HWY	Louisiana	38.4	125	—
INTCWGld Port Allen Canal	SR 77	SW	Indian Village, LA	1961	HWY	Louisiana	47	125	2
INTCWGld Port Allen Canal	—	VL	Morley, LA	1961	RR	UP	56	125	73
INTCWGld Port Allen Canal	—	VL	Port Allen, LA	1961	RR	UP	64	84	73

INTCWGle Plaquemine to Morgan City Landside Rte., Lower Grand River	SR 75	P	Pigeon, LA	1957	HWY	Louisiana	25.9	126	—
INTCWGle Plaquemine to Morgan City Landside Rte., Belle River	SR 70	P	Pierre Pass, LA	1958	HWY	Louisiana	43.2	125	—
INTCWGle Plaquemine to Morgan City Landside Rte., Milhomme Bayou	—	P	Morgan City, LA	1960	HWY	St. Martin, PA	55.2	125	—
INTCWGlf Atchafalaya River to Lake Charles Deep Channel Bayou Sale	SR 317	VL	North Bend, LA	1982	HWY	—	113	125	80
INTCWGlf Atchafalaya River to Lake Charles Deep Channel	SR 319	SW	Cypremort (Louisa), LA	1968	HWY	Louisiana	134	126	4
INTCWGlf Atchafalaya River to Lake Charles Deep Channel	SR 27	P	Gibbstown, LA	1957	HWY	Louisiana	219.9	151	—
INTCWGlf Atchafalaya River to Lake Charles Deep Channel	—	P	Grand Lake, LA	1964	HWY	Louisiana	231.4	125	—
INTCWGlf Atchafalaya River to Lake Charles Deep Channel	SR 27	VL	Hackberry, LA	1975	HWY	Louisiana	243.8	240	135
INTCWG Sabine River to Corpus Christi	—	B	West Port Arthur, TX	1933	HWY	Texas	295.1	100	16
INTCWG Sabine River to Corpus Christi	—	SW	High Island, TX	1934	RR	U.S. Govt.	325.4	100	10
INTCWG Sabine River to Corpus Christi	—	SW	High Island, TX	1935	HWY	Texas	325.5	100	10

*See key of abbreviations at end.

Waterway[a]	Road Name/ Route Carried	Bridge Type[b]	Location	Const. Date	Traffic Type[c]	Owner*	Miles from Mouth of Waterway	Channel Width/Lateral Clearance (feet)	Clearance Height (feet)
INTCWGt Port Arthur to Brownsville	—	B	Galveston, TX	1955	HWY, RR	Galveston Co.	356.1	125	12
INTCWGt Port Arthur to Brownsville	—	B	Galveston, TX	1984	RR	GHH/ATSF	357.2	100	10
INTCWGt Port Arthur to Brownsville	—	B (repl. by above)	Galveston, TX	1932	RR	GHH/ATSF	357.2	100	10
INTCWG Sabine River to Corpus Christi	—	B	Galveston, TX	1938	HWY	Texas	366.4	105	15
INTCWGt Port Arthur to Brownsville	Fm 1495	P	Freeport, TX	1959	HWY	Texas	397.6	130	—
INTCWG Sabine River to Corpus Christi	—	P	Freeport, TX	1941	HWY	Texas	403.5	100	—
INTCWG Sabine River to Corpus Christi	—	P	Freeport, TX	1941	HWY	Texas	404.3	100	—
INTCWG Sabine River to Corpus Christi	—	P	Freeport, TX	1941	HWY	Texas	407.3	100	—
INTCWG Sabine River to Corpus Christi	—	P	Freeport, TX	1940	HWY	Texas	413.8	100	—
INTCWGt Port Arthur to Brownsville Caney Creek	Fm 457	P	Sargent, TX	1941	HWY	Texas	418	40	4
INTCWG Sabine River to Corpus Christi	—	P	Sargent, TX	1940	HWY	Texas	426.6	100	—
INTCWGt Port Arthur to Brownsville	Fm 2031, SR 60	P	Mattagorda, TX	1967	HWY	Texas	440.7	141	—
INTCWG Sabine River to Corpus Christi	—	P	Matagorda, TX	1940	HWY	Texas	449.7	100	—
INTCWG Sabine River to Corpus Christi	—	P	Collegeport, TX	1940	HWY	Texas	466.5	100	—

INTCWGt Port Arthur to Brownsville	—	P	Port Isabel, TX	1953	HWY	Cameron Co.	666	145	—
Islais Creek	3rd St.	B	San Francisco, CA	1948	HWY	California	0.4	97	4
Isthmus Slough	—	B	Coos Bay, OR	1932	HWY	Oregon	1	140	18
Isthmus Slough	—	B	Marshfield, OR	1938	HWY	Coos Co. Court	5	23	13.75
Isthmus Slough	—	B	Coos Bay, OR	1955	HWY	Coos Co.	6	45	18
Jamaica Bay	Gil Hodges Marine Pkwy.	VL	Queens, NY	1937	HWY	TBTA	3	503	152
Jamaica Bay	—	B	Beach Channel, NY	1939	HWY	New York City	6	100.8	21.5
Jamaica Bay	—	SW	Hammels, NY	1892	RR	LIRR	6.5	50	3
Jamaica Bay	Beach Channel	SW	Queens, NY	1956	RR	MTA	6.7	101	26
Jamaica Bay	—	SW	Broad Channel Station, NY	1892	RR	LIRR	7	50	3
Jamaica Bay North Channel	Cross Bay Blvd.	B, F	Queens, NY	1978	HWY	New York City	10	95	20
Jamaica Bay	—	B	North Channel, NY	1925	HWY	New York City	10	102	26.3
Jamaica Bay North Channel	Howard Beach	B, F	Queens, NY	1956	RR	New York City	10.6	100	26
James River	US 17, 258	VL	Newport News, VA	1928	HWY	Virginia	5	250	145
James River	US 17, 258	VL	Newport News, VA	1972	HWY	Virginia	5	350	145
James River	Benjamin Harrison Memorial Br. (SR 156)	VL	Hopewell, VA	1967	HWY	Virginia	65	300	145
Jefferson–Shreveport Waterway, 12 Mile Bayou	SR 3094	VL, F	Shreveport, LA	1968	HWY	Louisiana	0.7	56	5
Jefferson–Shreveport Waterway, Cross Bayou	—	SW	Shreveport, LA	1931	RR	UP	1.2	53	8

*See key of abbreviations at end.

Waterway[a]	Road Name/ Route Carried	Bridge Type[b]	Location	Const. Date	Traffic Type[c]	Owner*	Miles from Mouth of Waterway	Channel Width/Lateral Clearance (feet)	Clearance Height (feet)
Jefferson–Shreveport Waterway, 12 Mile Bayou	—	VL, F	Shreveport, LA	1956	HWY	Louisiana	3.3	60	10
Jefferson–Shreveport Waterway, 12 Mile Bayou	SR 1	VL, F	Shreveport, LA	1945	HWY	Louisiana	6.4	60	31
Jefferson–Shreveport Waterway, 12 Mile Bayou	SR 1	VL, F	Mooringsport, LA	1958	HWY	Louisiana	23.3	60	2
Jefferson–Shreveport Waterway	—	VL	Mooringsport, LA	1914	HWY	Police jury of Caddo Pa.	26.3	90	9.6
Jefferson–Shreveport Waterway, Caddo Lake	—	SW	Mooringsport, LA	1897	RR	KCS	26.4	57	13
Joe Gould Narrows	—	SW	Cohasset, MN	—	HWY	Itasca Co.	2.2	—	—
John Day River	—	SW	Astoria, OR	1896	RR	BNSF	0	60	4
John Day River	US 30	SW	Astoria, OR	1933	HWY	Oregon	1	40	18
Johns Pass	SR 699	B	St. Petersburg, FL	1972	HWY	Florida	0.1	60	20
Johns Pass	—	B	St. Petersburg, FL	1928	HWY	Pinellas Co.	0.3	60	6
Johnson River	I 95	SW	Pleasure Beach, CT	1924	HWY	Bridgeport	0	70	7.3
Jones Creek	—	SW	Sparrows Point, MD	1906	RR	Baltimore Transit	0	40	3.6
Jordan River	SR 603	SW	Kiln, MS	1962	HWY	Mississippi	7.6	38	16
Jourdan River	—	SW	McLeod, MS	1937	HWY	Hancock Co.	7.6	35	5.5
Jupiter River	—	B	Jupiter, FL	1934	HWY	Florida	1.05	45	5.3
Jupiter River	—	B	Jupiter, FL	1925	RR	FEC	1.06	40	4.5
Kalamazoo River	—	SW	New Richmond, MI	1907	HWY	CO	10.8	41	9

Kalamazoo River	—	SW, F	New Richmond, MI	1974	RR	Allegan Co.	10.9	31	9
Kalamazoo River	—	SW	New Richmond, MI	1879	HWY	"highway"	10.9	31	7.3
Kalamazoo River	—	SW	Allegan, MI	1908	RR	CSX	37.5	35	2
Kaministiquia River	—	SW	Kashabowie, ON, Canada	1908	RR	CN	—	258	—
Kaministiquia River	—	B	Kaministiquia, ON, Canada	1914	RR	CP	—	188	—
Kansas River	—	VL	Kansas City, KS	1912	RR	UP	1.5	174	23
Kansas River	—	VL	Kansas City, KS	1914	RR	RI	2	297	23
Kansas River	—	VL	Topeka, KS	1951	RR	RI	84.6	140	23
Kaskaskia River	—	SW	Roots, IL	1903	RR	MP	3.1	142	3.3
Kaskaskia River	—	SW	Roots, IL	1903	RR	M-IRR	4.5	187	3.7
Kaskaskia River	—	F (movable but no machinery)	Evansville, IL	1924	HWY	Illinois	12	157	1.4
Kaskaskia River	—	SW	Evansville, IL	—	HWY	Randolph Co.	12.2	154	−1.8
Kaskaskia River	SR 145	SW, F	Baldwin, IL	1934	HWY	Illinois	18.6	106	0
Kaskaskia River	—	F (movable but no machinery)	Baldwin, IL	1907	RR	M&O	22.3	102	−3.4
Kawkawlin River	—	SW, F	Bay City, MI	1896	RR	D&M	0.625	43	6
Kelso Bayou	SR 27	SW	Hackberry, LA	1971	HWY	Louisiana	0.7	50	3
Kelso Bayou	—	SW	Hackberry, LA	1937	HWY	Louisiana	1.36	55.7	4.5
Kendrick Creek	Main Channel	SW, F	Mackeys, NC	1910	RR	NS	0.5	32	3
Kendrick Creek	—	SW	Mackeys, NC	—	HWY	North Carolina	0.6	31.5	4
Kendrick Creek	—	RSP	Mackeys, NC	1948	HWY	North Carolina	0.9	34	8

*See key of abbreviations at end.

Waterway[a]	Road Name/ Route Carried	Bridge Type[b]	Location	Const. Date	Traffic Type[c]	Owner*	Miles from Mouth of Waterway	Channel Width/Lateral Clearance (feet)	Clearance Height (feet)
Kennebeck River Entrance to Sasanoa River	—	SW	Bath, ME	—	HWY	Woolwich and Arrowsic Twps.	13.5	40	—
Kennebeck River	US 1	VL	Bath, ME	1927	HWY, RR	Maine	14	200	135
Kennebeck River	SR 197	SW	Dresden, ME	1938	HWY	Maine	27.1	63	15
Kennebeck River	—	SW	Gardner, ME	1939	HWY	Maine	37.6	65.7	20.45
Kennebunk River	—	SW	Kennebunkport, ME	1933	HWY	Maine	1	39	5
Kent Avenue Basin	Wallabout Canal	B	Brooklyn, NY	1893	HWY, TROLL	New York	0.15	40	5.57
Kent Island Narrows	US 50	B	Narrows, MD	1951	HWY	Maryland	1	48	18
Kent Island Narrows	—	B repl by above	Narrows, MD	1932	HWY	Maryland	1	50	6.83
Kent Island Narrows	—	SW	Narrows, MD	1937	RR	B&E	1.05	38.6	3.2
Kentucky Slough	—	B	North Bend, OR	1928	HWY	Coos Co.	1.5	32	9
Kern Slough	—	RSP	St. Helens, OR	1958	RR	BNSF	0	24	10
Kewaunee River	Park St.	B	Kewaunee, WI	1931	HWY	Wisconsin	0.28	90	5
Kewaunee River	—	SW	Kewaunee, WI	1894	RR	KGB&W	1.5	30	1.7
King Island Cut	—	SW	King Island, CA	1927	HWY	San Joaquin Co.	1	75	2
Kingston Lake	—	SW	Conway, SC	1929	HWY	South Carolina	0.6	57	1.3
Kinnickinnic River	—	SW	Milwaukee, WI	1899	RR	UP	1	61	4
Kinnickinnic River	Kinnickinnic Ave.	B	Milwaukee, WI	1909	HWY	Milwaukee	1.45	100	6
Kinnickinnic River	Seimer St.	SW	Milwaukee, WI	1909	RR	CP	1.49	93	11
Kinnickinnic River	—	SW	Milwaukee, WI	1909	RR	UP	1.52	93	11
Kinnickinnic River	S. 1st St.	B	Milwaukee, WI	1958	HWY	Milwaukee	1.6	70	9
Kinnickinnic River	S. 1st St.	SW	Milwaukee, WI	1894	HWY	Milwaukee	1.61	62	5.8
Kinsale Creek	—	R	Kinsale, VA	1934	HWY	Virginia	4	29	8.8
Kissimee River	SR 78	RSP, F	Okeechobee, FL	1964	HWY	Florida	0.8	50	20
Kissimmee River	—	SW	Okeechobee, FL	1932	HWY	Florida	0.8	40	4.8
Kissimmee River	—	SW	Okeechobee, FL	1925	HWY	Florida	12	60	3.5
Kissimee River	SR 70	RSP, F	Okeechobee, FL	1966	HWY	Florida	19.5	50	17

Kissimmee River	—	SW	Okeechobee, FL	1925	RR	SAL	35	50	3.33
Kissimee River	—	SW, F	Ft. Bassinger, FL	1964	RR	CSX	37	48	12
Kissimmee River	Pearces Ferry	SW	Bassenger, FL	1917	HWY	Florida	38	52.2	8
Kissimee River	US 98	RSP, F	Ft. Bassinger, FL	1953	HWY	Florida	39	50	17
Kissimmee River	—	SW	Kicco, FL	1930	HWY	Florida	96	59.9	10
Kissimee River	—	RSP, F	Lake Kissimmee, FL	1965	HWY	W. C. Zipprer Est.	105	30	12
Kissimee River	—	RSP, F	Lake Tohopekalliga, FL	1960	HWY	Pope Berry Groves	122.5	33	12
Kissimmee River Borrow Pit Channel	—	B	Lake Okechobee, FL	1958	HWY	C. H. Carlton	2.8	29	10
Knapps Narrows	SR 33	B (repl. 1998)	Tilghman, MD	1935	HWY	Maryland	0.4	50	7
Knobbs Creek	—	SW	Elizabeth City, NC	1909	RR	N-S	0.1	35	4
Knobbs Creek	—	SW	Elizabeth City, NC	1929	HWY	North Carolina	0.12	40	4
Kootenay Lake	—	VL	Nelson, BC	1930	RR	CP	—	93	—
La Batre Bayou	SR 188	B	Bayou La Batre, AL	1928	HWY	Alabama	2.3	50	6
La Loutre Bayou	SR 46	VL	Yscloskey, LA	1957	HWY	Louisiana	22.9	45	55
Lacassine Bayou	SR 14	SW	Hayes, LA	1959	HWY	Louisiana	17	60	5
Lacassine Bayou	—	SW	Hayes, LA	1904	RR	T&NO	20.4	53.6	2
Lacassine Bayou	—	SW	Hayes, LA	1955	RR	SP	20.4	66	3
Lacassine Bayou	—	B	Hayes, LA	—	HWY	Calcasieu and Jefferson Davis Pas.	21.5	29	3
Lacombe Bayou	—	SW	Lacombe, LA	1908	RR	ICG	5.2	54	5
Lacombe Bayou	US 190	SW	Lacombe, LA	1938	HWY	Louisiana	6.8	45	5
La Croix Bayou	—	SW	Bay St Louis, MS	1940	HWY	Hancock Co.	1	40	2
Lafayette River	US 460 Granby St.	B, F	Norfolk, VA	1979	HWY	Virginia	3.3	40	22
Lafayette River, North Branch	—	RSP	Norfolk, VA	1930	HWY	Norfolk	0	30	10
Lafourche Bayou	SR 1	VL	Leeville, LA	1970	HWY	Louisiana	13.3	125	73
Lafourche Bayou	—	SW	Leeville, LA	1931	HWY	Louisiana	13.4	51.7	8.9
Lafourche Bayou	SR 1	VL	Golden Meadow, LA	1971	HWY	Louisiana	23.9	80	73

*See key of abbreviations at end.

Waterway[a]	Road Name/ Route Carried	Bridge Type[b]	Location	Const. Date	Traffic Type[c]	Owner*	Miles from Mouth of Waterway	Channel Width/Lateral Clearance (feet)	Clearance Height (feet)
Lafourche Bayou	—	P	Golden Meadow, LA	—	HWY	Lafourche Pa.	27.7	89	—
Lafourche Bayou	SR 1	P	Galliano, LA	1957	HWY	Lafourche Pa.	27.8	80	—
Lafourche Bayou	SR 1	VL	CT-Off, LA	1972	HWY	Lafourche Pa.	30.6	80	73
Lafourche Bayou	SR 1	P	CT-Off, LA	1953	HWY	Lafourche Pa.	34.1	92	—
Lafourche Bayou	SR 1	P	CT-Off, LA	1951	HWY	Lafourche Pa.	36.3	91	—
Lafourche Bayou	—	P	Larose, LA	1971	HWY	Lafourche Pa.	37	93	—
Lafourche Bayou	SR 1	P	Larose, LA	1953	HWY	Louisiana	39.1	92	—
Lafourche Bayou	—	P	Larose, LA	1937	HWY	Louisiana	39.2	80	—
Lafourche Bayou	SR 1	P	Valentine, LA	1951	HWY	Lafourche Pa.	44.7	90	—
Lafourche Bayou	SR 1	SW	Lockport, LA	1940	HWY	Louisiana	50.8	85	6
Lafourche Bayou	—	P	Mathews, LA	—	HWY	Lafourche Pa.	54.1	79.8	—
Lafourche Bayou	SR 1	P	Mathews, LA	1961	HWY	Louisiana	54.2	95	—
Lafourche Bayou	US 90, SR 1	VL	Raceland, LA	1936	HWY	Louisiana	58.2	61	59
Lafourche Bayou	US 90, SR 308	VL	Raceland, LA	1968	HWY	Louisiana	58.7	60	50
Lafourche Bayou	SR 1	SW	Lafourche, LA	1949	HWY	Louisiana	66.1	61	10
Lafourche Bayou	—	SW, F	Lafourche, LA	1900	RR	SP	69	70	19
Lafourche Bayou	SR 20	VL, F	Thibodaux, LA	1937	HWY	Louisiana	73.4	50	11
Lafourche Bayou	—	SW	Thibodaux, LA	1938	HWY	Louisiana	76.2	72	8.1
Lafourche Bayou	—	P	Thibodaux, LA	1922	HWY	Lafourche Pa.	76.9	75	—
Lafourche Bayou	—	SW, F	Thibodaux, LA	1926	RR	Caldwell Sugars	77.7	91	2
Lafourche Bayou	SR 348	RSP	Labadieville, LA	1953	HWY	Louisiana	81.9	59	12
Lafourche Bayou	—	SW	Labadieville, LA	1906	HWY	Louisiana	81.95	90	13.3
Lafourche Bayou	SR 1010	RSP	Napoleonville, LA	1951	HWY	Louisiana	86.3	54	11
Lafourche Bayou	SR 1008	RSP	Napoleonville, LA	1948	HWY	Louisiana	90	50	14
Lafourche Bayou	—	SW	Napoleonville, LA	1942	RR	MP	91.6	40	5
Lafourche Bayou	—	P	Napoleonville, LA	—	HWY	Emile Campo	94.9	66	—
Lafourche Bayou	—	P	Belle Rose, LA	—	HWY	Louisiana	101.3	65	—
Lafourche Bayou	—	SW	Donaldsonville, LA	—	HWY	Palo Alto Co.	104.3	91.5	10.1

Lagoon Pond	Beach Rd.	B	Tisbury, MA	1935	HWY	Oak Bluffs	0	30	15
Laguna Madre	—	MB	Corpus Christi, TX	1927	HWY	Sam A. Robertson	14	50	1.5
Lake Michigan Ship Canal†	WI 42–57	B	Sturgeon Bay, WI	1978	HWY	Wisconsin	—	—	—
Lake Michigan Ship Canal†	Michigan St.	B	Sturgeon Bay, WI	1931	HWY	Wisconsin	4.3	139	11
Lake Michigan Ship Canal†	—	SW	Sturgeon Bay, WI	1912	RR	A&W	4.4	77	11.1
Lake Washington	—	P	Evergreen Pt, WA	1963	HWY	Foster Island	—	200	0
Lake Washington	Day St.	P	Mercer Island, WA	1941	HWY	Seattle	—	200	0
Lake Washington	Day St.	P	Mercer Island, WA	1965	HWY	Seattle	—	0	0
Lake Washington Ship Canal	—	B	Seattle, WA	1914	RR	BNSF	0.1	150	43
Lake Washington Ship Canal	Ballard Br. (15th Ave. N.W.)	B	Seattle, WA	1941	HWY	Seattle	1.1	150	29
Lake Washington Ship Canal	—	B	Seattle, WA	1914	RR	BNSF	1.6	150	15
Lake Washington Ship Canal	Fremont Br.	B	Seattle, WA	1917	HWY	Seattle	2.6	150	16
Lake Washington Canal	—	B	Seattle, WA	1933	HWY	Seattle	4.45	175	29
Lake Washington Ship Canal	SR 13 (Montlake Br.)	B	Seattle, WA	1925	HWY	Seattle	5.2	150	30
Larson Slough Inlet	—	B	North Bend, OR	1921	HWY	Coos Co.	1	40	14
Lavaca Bay	—	B	Port Lavaca, TX	1931	HWY	Texas	—	38	10
Lavaca River	Fm 616	RSP	Vanderbilt, TX	1951	HWY	Texas	11.2	50	18
Lavaca River	—	SW	Vanderbilt, TX	1910	RR	UP	11.2	50	12
Lazaretto Creek	—	SW	Savannah Beach, GA	1922	HWY	Georgia	0.3	90	6
Le Carpe Bayou	—	SW	Houma, LA	—	HWY	Louisiana	7.5	35.5	6.1
Le Carpe Bayou	US 90	VL	Houma, LA	1965	HWY	Louisiana	7.5	60	73
Lechmere Canal	Commercial Ave.	B	Cambridge, MA	1906	HWY	Cambridge	0	40.2	7
Lechmere Canal	Memorial Dr.	B	Cambridge, MA	—	HWY	Massachusetts	0	40	7
Leech Lake–Kabekona River	—	SW	Walker, MN	1914	RR	GN	0.1	50	2
Leech Lake–Steamboat River	—	SW	Wilkinson, MN	1914	RR	GN	3.9	58	4
Leipsic River	—	SW	Leipsic, DE	1922	HWY	Delaware	9	35	3.5

*See key of abbreviations at end.

†See Sturgeon Bay.

Waterway[a]	Road Name/ Route Carried	Bridge Type[b]	Location	Const. Date	Traffic Type[c]	Owner*	Miles from Mouth of Waterway	Channel Width/Lateral Clearance (feet)	Clearance Height (feet)
Lemon Bay	—	B	Englewood, FL	1928	HWY	Charlotte Co.	6	26	8.5
Lemon Creek	Bayview Ave.	R	Staten Island, NY	1851	HWY	New York City	0.1	30	3.7
Lewis and Clark River	US 101	B	Astoria, OR	1925	HWY	Oregon	1	100	17
Lewis and Clark River	—	SW	Chadwell, OR	1896	HWY	Clatsop Co.	5	50	4
Lewis River	—	SW, F	Woodland, WA	1915	RR	BNSF	2	100	12
Liberty Bayou	—	SW	Bonfouca, LA	1936	HWY	Louisiana	2	40	1
Liberty Bayou	—	SW	Slidell, LA	1971	HWY	Louisiana	2	60	4
Lindsey Slough	—	RSP	Egbert, CA	1964	HWY	Hastings Farms	2	53	19
Little Bayou Black	—	SW	Houma, LA	1939	HWY	Terrebonne Pa.	29.4	40.9	6.4
Little Bayou Black	—	SW	Houma, LA	1912	HWY, RR	John D. Minor	29.6	58	5.7
Little Calumet River	142nd St.	SW	Riverdale, IL	1903	RR	MC	8.11	61	5.9
Little Calumet River	134th St.	SW	Riverdale, IL	1915	RR	CWI	10.75	78	13.2
Little Calumet River	Indiana Ave.	SW	Riverdale, IL	1915	HWY	Chicago	11.01	70	10.7
Little Calumet River	—	B	Riverdale, IL	1925	RR	IC	11.21	71.4	27.4
Little Calumet River	Stewart Ave.	SW	Riverdale, IL	1927	RR	PCCSTL	12.72	50	13.8
Little Harbor	Wentworth Br.	B	New Castle, NH	1943	HWY	New Hampshire	0.8	29	13
Little Harbor	—	B	New Castle, NH	—	HWY	—	1	—	—
Little Hoquiam River	—	B	Hoquiam, WA	1926	HWY	Hoquiam	3.25	40	9.5
Little Lake Harris	SR 19	SW	Astatula, FL	1927	HWY	Florida	2	30	1
Little Manatee River	—	SW	Ruskin, FL	1958	RR	CSX	2.4	35	4
Little Manatee River	—	SW	Ruskin, FL	1941	HWY	Florida	2.6	50	5.5
Little Nestucca River	—	B, F	Oretown, OR	1938	HWY	Oregon	2	34	9
Little Potato Slough	—	SW	Terminous, CA	1954	HWY	California	0.1	100	10
Little Red River	—	SW	Judsonia, AR	1912	RR	MP	25	108	1.7
Little Red River	—	SW	Judsonia, AR	1924	HWY	Arkansas	25.2	118	−1.1
Little Red River	—	SW	Searcy, AR	1911	HWY	Arkansas	30.5	115	4.2

Little River 1	—	SW	Fulton, AR	1904	RR	BNSF	7.1	58	21
Little River 2	—	SW	Little Creek, DE	1880	HWY	Delaware	3	24	4.5
Little River 3	—	VL	Archie, LA	1914	RR	L&A	12.1	100	4
Little Sarasota Bay	—	SW	Nokomis, FL	1930	HWY	Sarasota Co.	0.9	48	7
Little Sarasota Bay	—	SW	Osprey, FL	1927	HWY	Sarasota Co.	4.5	55	9.2
Little Sarasota Bay	—	SW	Hayden, FL	1926	HWY	Sarasota Co.	12	55	9.2
Little Sunflower River	—	SW	Kelso, MS	1906	RR	Y&MV IC	10	66.8	7
Lobina Lagoon	—	VL	Culebra, PR	1977	HWY	Culebra	0.1	32	39
Loggy Bayou Lake Bistineau and Dorcheat Bayou	—	SW	East Pt., LA	1901	RR	LARR	1.1	50	3
Loggy Bayou Lake Bisteneau and Dorcheat Bayou	—	SW	Sibley, LA	1884	RR	IC	37.7	59	5
Long Beach Harbor	Entrance to West Basin	B	Los Angeles, CA	1912	RR	SP	3	148.5	4.9
Long Beach Harbor	—	VL	Long Beach, CA	1998	RR	UP	4.4	—	—
Long Beach Harbor	Henry Ford Ave.	B (out)	Long Beach, CA	1924	HWY, RR	Los Angeles	4.4	180	8
Long Beach Harbor	Terminal Island Freeway (Schuyler Heim Bridge)	VL	Long Beach, CA	1948	HWY	Long Beach	4.5	180	163
Long Beach Harbor	Cerritos Channel	SW	Long Beach, CA	1941	HWY	U.S. Navy	5	80	5.3
Long Creek 1	Loop Pkwy.	B	Hempstead, NY	1933	HWY	Park Comm.	0.7	75	21
Long Creek 2	—	SW	Inlet, VA	1925	HWY	Virginia	1	19.66	5.3
Longboat Pass	—	B	Longboat and Anna Maria Keys, FL	1958	HWY	Florida	0	45	17
Loxahatchee River	—	B	Jupiter, FL	1925	RR	FEC	1.2	40	4
Loxahatchee River, N.W. Fork	—	RSP	Jupiter, FL	1965	HWY	Martin Co.	4.9	38	11
Lumber River	—	SW, F	Nichols, SC	1893	RR	CSX	6.8	—	6
Lynnhaven Inlet	—	B	Ocean Park, VA	1928	HWY	Virginia	0.2	32	4
Lynnhaven Inlet	—	SW	Ocean Park, VA	1939	RR	NS	0.25	31	1.9

*See key of abbreviations at end.

Waterway[a]	Road Name/ Route Carried	Bridge Type[b]	Location	Const. Date	Traffic Type[c]	Owner*	Miles from Mouth of Waterway	Channel Width/Lateral Clearance (feet)	Clearance Height (feet)
Machias River	—	SW	Machiasport, ME	1907	HWY	Trustees of the Machiasport Bridge	4.5	45	4
Macon Bayou	—	VL	Wisner, LA	1922	HWY	Franklin Pa.	13.8	60	4
Macon Bayou	—	SW	Winnsboro, LA	1932	HWY	Louisiana	36.1	60	5
Macon Bayou	—	SW	Delhi, LA	1883	RR	IC	87.5	58	2.3
Main Channel	—	SW	Corson Inlet Strathmere, NJ	1919	RR	PRR	0.9	31.2	2
Main Channel	—	SW	Corson Inlet Strathmere, NJ	1924	HWY	Cape May Co.	1	32	4.6
Malden River	Mystic Valley Pkwy.	B, F	Medford, MA	1954	HWY	Metropolitan Dist. Comm.	0.3	80	15
Malden River	Revere Beach Pkwy.	B	Medford, MA	1905	HWY	Metropolitan Dist. Comm.	0.31	50.3	7.7
Malden River	Medford St.	B	Malden, MA	1936	HWY	Massachusetts	1.3	52	3.1
Manantico Creek	—	SW, F	Millville, NJ	1924	HWY	New Jersey	0.5	30	3
Manasquan River (INLWW per COE)	SR 34	B	Brick Township, NJ	1937	HWY	New Jersey	3.4	50	15
Manatee River	US 41	B	Bradenton, FL	1927	HWY	—	4.3	75	8
Manatee River	—	B	Bradenton, FL	1957	RR	CSX	4.5	75	5
Manatee River	—	SW	Manatee, FL	1912	RR	SAL	5.1	57	5.7
Manchester Harbor	—	B	Manchester, MA	1845	RR	B&M	1	50	6
Manistee River	Maple St.	B	Manistee, MI	1964	HWY	Manistee	1.1	101	19
Manistee River	US 31	B	Manistee, MI	1933	HWY	Michigan	1.4	120	32
Manistee River	—	SW	Manistee, MI	1930	RR	CO	1.5	100	9
Manitowoc River	8th St.	B	Manitowoc, WI	1926	HWY	Manitowoc	0.3	97	8
Manitowoc River	10th St.	B	Manitowoc, WI	1921	HWY	Manitowoc	0.5	89	11
Manitowoc River	—	B	Manitowoc, WI	1926	RR	SOO	1	93	2

Manitowoc River	—	SW	Manitowoc, WI	1897	RR	SOO	1.8	61	2
Manitowoc River	—	SW	Manitowoc, WI	1910	RR	CNW	1.9	60	7
Manitowoc River	21st St.	SW	Manitowoc, WI	1920	HWY	Manitowoc	2.12	66.5	11.4
Mantua Creek	—	SW	Paulsboro, NJ	—	RR	CR	1.4	32	1
Mantua Creek	—	VL	Paulsboro, NJ	1936	HWY	New Jersey	1.7	75	64
Mare Island Strait	—	B (repl. w. VL)	Vallejo, CA	1935	HWY, RR	U.S. Navy	2.8	75	13
Marshyhope Creek	—	B F	Brookview, MD	1932	HWY	Maryland	5.8	60	11
Marshyhope Creek	—	SW	Harrisons Landing, MD	—	HWY	Dorchester Co.	9.5	52	1
Martin Pena Channel	—	B	San Juan, PR	1913	HWY	Puerto Rico	0	39	8.3
Massalina Bayou	—	B	Panama City, FL	1950	HWY	—	0	40	7
Massalina Bayou	—	B	Panama City, FL	1928	HWY	Panama City	0.3	30	9
Matanzas Inlet	—	B	St. Augustine, FL	1927	HWY	Florida	0	40	12.8
Matanzas Pass	SR 865	SW	Punta Rassa, FL	1929	HWY	Lee Co.	0	57	7
Matawan Creek	Front St.	B	Keyport, NJ	1922	HWY	New Jersey	0.56	49	6.2
Matawan Creek	Rte. 25	B	Matawan, NJ	1929	HWY, TROL	New Jersey	0.9	51	12.2
Matlacha Pass	—	SW	Fort Myers, FL	1928	HWY	Lee Co.	—	29	8
Mattaponi River	Lord Delaware Br. SR 33	SW	West Point, VA	1945	HWY	Virginia	0.8	81	12
Mattaponi River	—	B	West Point, VA	1915	HWY	Virginia	1.4	80	7.1
Mattaponi River	SR 629	SW	Walkerton, VA	1937	HWY	Virginia	28.5	53	6
Mattawoman Creek	—	B	Indianhead, MD	1917	FTBR	U.S. Navy	4.5	80	4
Mattituck Harbor	—	SW	Mattituck, NY	1910	HWY	Southold Twp.	1	57	6
Maumee River	—	SW	Toledo, OH	1932	RR	TT	1.1	145	18
Maumee River	—	SW	Toledo, OH	1906	RR	NW	1.9	145	16
Maumee River	Ash–Consaul Sts.	SW	Toledo, OH	1938	HWY	Toledo	3.12	101.32	24.3
Maumee River	Craig Memorial	B	Toledo, OH	1958	HWY	Toledo	3.4	200	38
Maumee River	—	B (out) (SW per COE)	Toledo, OH	1913	RR	PC	4	115	12

*See key of abbreviations at end.

Waterway[a]	Road Name/ Route Carried	Bridge Type[b]	Location	Const. Date	Traffic Type[c]	Owner*	Miles from Mouth of Waterway	Channel Width/Lateral Clearance (feet)	Clearance Height (feet)
Maumee River	Cherry St.	B	Toledo, OH	1915	HWY	Toledo	4.4	200	21
Maumee River	—	SW	Toledo, OH	1899	RR	NS	5.9	115	11
Maumee River	Fassett St.	SW	Toledo, OH	1935	HWY	Toledo	6	115.45	34.3
Maumee River	—	SW (abandoned)	Toledo, OH	1903	RR	TT	11.5	110	49
Maurice River	—	SW	Mauricetown, NJ	1933	HWY	Cumberland Co.	11.9	59.7	3.6
Maurice River	—	SW	Millville, NJ	1914	HWY	New Jersey	23.9	35.8	4.8
Maurice River	SR 49	SW, F	Millville, NJ	1972	HWY	New Jersey	23.9	60	4
McKellar River	—	B	Kaministiquia, ON, Canada	1914	RR	CP	—	155	—
Meherrin River	—	SW	Dunns Fishery, NC	1931	RR	Camp Mfg.	3.9	50	5
Menantico Creek	—	SW	Millville, NJ	1925	HWY	New Jersey	0.5	30	3.42
Menominee River	1st Ogden St.	B	Marinette, WI	1973	HWY	—	0.4	100	12
Menominee River, Burnham Canal	S. 6th St.	B	Milwaukee, WI	1908	HWY	Milwaukee	1.3	77	26
Menominee River, Burnham Canal	—	SW	Milwaukee, WI	1911	RR	CP	1.5	65	8
Menominee River, Burnham Canal	Muskego Ave.	B	Milwaukee, WI	1980	HWY	Milwaukee	1.7	75	12
Menominee River, Burnham Canal	S. 11th St.	SW	Milwaukee, WI	1886	HWY	Milwaukee	1.8	46	3
Menominee River Canal	—	SW	Milwaukee, WI	1904	RR	CP	0.8	75	4
Menominee River Canal	Plankinton Ave.	B	Milwaukee, WI	1978	HWY	Milwaukee	0.801	75	4
Menominee River Canal	N. 6th St.	B	Milwaukee, WI	1908	HWY	Milwaukee	1.1	71	26
Menominee River Canal	16th St.	B	Milwaukee, WI	1919	HWY	Milwaukee	1.9	120	30
Mermentau River	SR 82	SW	Grand Cheniere, LA	1959	HWY	—	7.1	70	13
Mermentau River	—	SW	Grand Chenier, LA	1939	HWY	Louisiana	8.75	54.7	2.2

Mermentau River	—	SW	Mermentau, LA	1953	RR	SP	68	52	10
Mermentau River	US 90	VL	Mermentau, LA	1973	HWY	—	68.1	125	50
Merrimac River	—	SW	Newburyport, MA	1904	HWY	Massachusetts	3.37	76	13.2
Merrimac River	US 1	B	Newburyport, MA	1969	HWY	Massachusetts	3.4	100	35
Merrimac River	—	SW	Newburyport, MA	1864	RR	B&M	3.4	69	13
Merrimac River	Main St.	SW	Newburyport, MA	—	HWY	Essex Co.	5.8	56	15
Merrimac River	—	SW	Amesbury, MA	—	HWY	Essex Co.	5.83	56	7.8
Merrimac River	Main St.	SW	Haverhill, MA	1915	HWY	Essex Co.	12.6	54	17
Merrimac River	SR 97 113	B	Haverhill, MA	1948	HWY	Massachusetts	16.5	70	13
Merrimac River	—	SW	Groveland, MA	1914	HWY	Essex Co.	16.53	64	14.5
Miakka River	—	SW	Southland, FL	1927	HWY	Florida	2.98	40	10
Miakka River	—	SW	McCall, FL	1908	RR	CH&N	3	30	6
Miami Canal	SR 25	SW	—	1922	HWY	Palm Beach Co.	—	50	8
Miami Canal	36th St. N.W.	B	Miami, FL	—	HWY	Dade Co.	—	70	7
Miami Canal	S.E. 11th Ave.	B	Hialeah, FL	—	RR	Dade Co.	—	50	8
Miami Canal	E. 1st Ave.	VL	Hialeah, FL	—	HWY	Dade Co.	—	60	5
Miami River	US 1	B	Miami, FL	1929	HWY	Florida	0.1	94	20
Miami River	S. Miami Ave.	B	Miami, FL	1918	HWY	Miami	0.18	75	8
Miami River	S.W. 1st Ave.	SW	Miami, FL	1932	RR	FEC	0.39	50.25	6.67
Miami River	2nd Ave. S.W.	B	Miami, FL	1924	HWY	Miami	0.5	75	11
Miami River	1st St S.W.	B	Miami, FL	1929	HWY	Miami	0.9	75	18
Miami River	W. Flagler St.	B	Miami, FL	1967	HWY	Florida	1	75	35
Miami River	5th St. N.W.	B	Miami, FL	1925	HWY	Miami	1.5	75	12
Miami River	12th Ave. N.W.	B	Miami, FL	1929	HWY	Miami	2.1	94	17
Miami River	17th Ave. N.W.	B	Miami, FL	1929	HWY	Miami	2.6	75	17
Miami River	22nd Ave. N.W.	B	Miami, FL	1966	HWY	Miami	3.3	81	25
Miami River	27th Ave. N.W.	B	Miami, FL	1939	HWY	Dade Co.	3.7	75	18
Miami River, North Fork	—	B	Hialeah, FL	—	RR	CSX	5.3	60	6
Mianus River	Cos Cob	B	Greenwich, CT	1904	RR	Connecticut DOT	1	67	20
Middle River 1	Bacon Island Rd.	SW	Middle River, CA	1951	HWY	San Joaquin Co.	8.6	129	9
Middle River 1	—	B	Middle River, CA	1930	RR	BNSF	9.8	98	11

*See key of abbreviations at end.

Waterway[a]	Road Name/ Route Carried	Bridge Type[b]	Location	Const. Date	Traffic Type[c]	Owner*	Miles from Mouth of Waterway	Channel Width/Lateral Clearance (feet)	Clearance Height (feet)
Middle River 1	SR 4 (Borden Hwy.)	SW	Victoria Island, CA	1916	HWY	California	15.1	105	11
Middle River 1	—	SW	French Camp, CA	1913	HWY	San Joaquin Co.	22.2	75	3.8
Middle River 2	N.E. 10th St.	B	Ft. Lauderdale, FL	1940	HWY	Ft. Lauderdale	0.8	28	2.5
Miles River	SR 370	B	Easton, MD	1941	HWY	Maryland	10	34	5
Milford Haven	SR 223	SW	Grimstead, VA	1940	HWY	Virginia	0.1	80	12
Mill Basin	Shore Pkwy.	B	Brooklyn, NY	1940	HWY	New York City	0.8	135	34
Mill Neck Creek	—	B	Bayville, NY	1939	HWY	Nassau Co.	0.1	76	9
Mill River 1	Chapel St.	SW	New Haven, CT	1899	HWY	New Haven	0.4	72	7
Mill River 2	—	B, F	Thomaston, ME	—	RR	MEC	0	28	25
Mill Tail Creek	—	B	East Lake, NC	—	RR	Dare Lumber Co.	—	—	—
Milwaukee River	—	SW	Milwaukee, WI	1914	RR	UP	0.3	87	3
Milwaukee River	Broadway St.	B	Milwaukee, WI	1982	HWY	Milwaukee	0.5	100	14
Milwaukee River	Water St.	B	Milwaukee, WI	1909	HWY	Milwaukee	0.7	130	9
Milwaukee River	Buffalo St.	B	Milwaukee, WI	1914	HWY	Milwaukee	0.88	130	4.9
Milwaukee River	St. Paul St.	VL	Milwaukee, WI	1964	HWY	Milwaukee	1	50	23
Milwaukee River	Clybourn St.	VL	Milwaukee, WI	1964	HWY	Milwaukee	1	52	23
Milwaukee River	Michigan St.	VL	Milwaukee, WI	1978	HWY	Milwaukee	1.1	50	28
Milwaukee River	Wisconsin Ave.	VL	Milwaukee, WI	1975	HWY	Milwaukee	1.2	50	27
Milwaukee River	Wells St.	VL	Milwaukee, WI	1992	HWY	Milwaukee	1.4	—	—
Milwaukee River	Kilbourn Ave.	B	Milwaukee, WI	1929	HWY	Milwaukee	1.5	100	10
Milwaukee River	State St.	B	Milwaukee, WI	1924	HWY	Milwaukee	1.6	80	10
Milwaukee River	Juneau Ave.	B	Milwaukee, WI	1954	HWY	Milwaukee	1.7	90	10
Milwaukee River	Cherry St.	B	Milwaukee, WI	1940	HWY	Milwaukee	1.9	80	10
Milwaukee River	Walnut St.	SW	Milwaukee, WI	1882	HWY	Milwaukee	2.24	58.3	5.1
Milwaukee River	Pleasant St.	VL	Milwaukee, WI	1973	HWY	Milwaukee	2.3	50	27

Milwaukee River	Holton St.	B	Milwaukee, WI	1926	HWY	Milwaukee	2.5	79	60
Miner Slough	—	SW	Rio Vista, CA	1933	HWY	California	5.5	72	3
Minnesota River	—	SW	Ft. Snelling, MN	1855	RR	CMSTP&P	1.6	103	8.9
Minnesota River	Cedar Ave.	SW	Minneapolis, MN	1890	HWY	Hennepin Co.	7.4	125	11.8
Minnesota River	Lyndale Ave.	B	Minneapolis, MN	1919	HWY	Minnesota	10.8	100	8.3
Minnesota River	—	SW	Savage, MN	1907	RR, HWY	MNS	14.2	103	2
Minnesota River	—	SW	Bloomington, MN	1890	HWY	Hennepin Co.	16.8	125	9.5
Minnesota River	—	SW	Shakopee, MN	—	HWY	Shakopee	25	85	3.6
Minnesota River	—	SW	Chaska, MN	1866	RR	Milwaukee	28.7	110	3
Minnesota River	—	SW	Belle Plaine, MN	1879	HWY	Minnesota	49.2	86	4
Minnesota River	—	SW	Le Sueur, MN	—	HWY	Minnesota	74.7	200	−4.8
Minnesota River	—	SW	St. Peter, MN	—	RR	CNW	92.7	—	25.9
Minnesota River	—	SW	New Ulm, MN	—	RR	CNW	140.7	85	19
Mispillion River	—	B	Milford, DE	1930	HWY	Delaware	11	45	5
Mission Bay	—	RSP	San Diego, CA	1915	HWY, RR	Bay Shore RR	0	40	7.1
Mission Bay	—	RSP	San Diego, CA	1931	HWY	San Diego	1.25	50	25
Mission Bay	—	RSP	San Diego, CA	1931	HWY	San Diego	1.75	50	30
Mission Creek	—	B	San Francisco, CA	1933	HWY, RR	San Francisco	0	103	1.5
Mission Creek	—	B	San Francisco, CA	1917	HWY	San Francisco	0.15	75	8
Mission Creek	—	SW	San Francisco, CA	1891	HWY	San Francisco	0.5	50	7
Mississippi River	—	SW	Alton, IL	1894	RR	BNSF	202.7†	200	62
Mississippi River	—	SW	Louisiana, MO	1945	RR	GWW	282.1	195	69
Mississippi River	—	SW	Hannibal, MO	1970	RR	NS	309.9	159	5
Mississippi River	—	SW	Quincy, IL	1868	RR	CBQ	327.9	153	5
Mississippi River	US 136, 218	SW	Keokuk, IA	1916	HWY, RR	Keokuk	364	158	7
Mississippi River	US 61, SR 96	SW	Ft. Madison, IA	1928	HWY, RR	BNSF	383.9	200	6
Mississippi River	—	SW	Burlington, IA	1975	RR	BNSF	403.1	200	9
Mississippi River	—	VL (out)	Keithsburg, IL	1910	RR	CNW	428	224	54
Mississippi River	Crescent Br.	SW	Rock Island, IL	1899	RR	DRINW	481.4	197	6
Mississippi River	Government Br.	SW	Rock Island, IL	1896	RR, HWY	U.S. Govt.	482.9	110	3
Mississippi River	—	SW	Clinton, IA	1909	RR	UP	518	177	3

*See key of abbreviations at end.

†Mileage from confluence of Ohio River.

Waterway[a]	Road Name/ Route Carried	Bridge Type[b]	Location	Const. Date	Traffic Type[c]	Owner*	Miles from Mouth of Waterway	Channel Width/Lateral Clearance (feet)	Clearance Height (feet)
Mississippi River	—	SW	Sabula, IA	1881	RR	MILW	534.9	154	6
Mississippi River	—	SW	Dubuque, IA	1892	RR	CN	579.9	146	4
Mississippi River	—	P	Marquette, IA	1915	RR	CMSTP&P	634.7	238	0
Mississippi River	—	P	Prairie Du Chien, WI	1912	RR	CMSTP&P	634.7	160	0
Mississippi River	—	SW	La Crosse, WI	1928	RR	CP	699.8	150	8
Mississippi River	—	SW (out)	Winona, MN	1891	RR	BNSF	723.8	200	6
Mississippi River	—	SW (out)	Winona, MN	1870	RR	CNW	725.8	154	9.8
Mississippi River	—	P	Reads Landing, MN	1882	RR	CMSTP&P	762.7	302.5	0
Mississippi River	—	VL	Hastings, MN	1980	RR	CP	813.7	—	—
Mississippi River	SR 38	SW	Inver Grove, MN	1895	HWY, RR	CRIP	830.3	195	2
Mississippi River	—	SW	South St. Paul, MN	1887	RR	UP	835.7	174	66
Mississippi River	—	VL	St. Paul, MN	1925	RR	UP	839.2	158	8
Mississippi River	Randolph St. (Omaha Br.)	SW	St. Paul, MN	1948	RR	UP	841.4	160	4
Mississippi River	—	SW	Aitkin, MN	1896	HWY	Aitkin Co.	1055.9	111	3.6
Mississippi River	—	SW	Cohasset, MN	1910	HWY	Bass Rock Twp.	1186.3	40	8
Mississippi River	—	B	Cohasset, MN	1923	HWY	Itasca Co.	1196.5	57	5.6
Mississippi River	—	SW	Ball Club, MN	1908	RR	GNRWY	1230.8	40	5.7
Missouri River	—	SW	Jefferson City, MO	1896	HWY	Missouri	143	204	34
Missouri River	—	VL	Boonville, MO	1932	RR	Missouri	197.1	400	57
Missouri River	Harry Truman Br.	VL	Kansas City, MO	1945	RR	MILW	359.4	403	55
Missouri River	US 169, 71, 69 (ASB Br.)	VL	Kansas City, MO	1911	HWY, RR	BNSF & Missouri	365.6	395	69
Missouri River	—	SW	Kansas City, MO	1917	HWY, RR	BNSF	366.1	200	26
Missouri River	—	SW	Leavenworth, KS	1894	RR	CNW	396.7	202	8
Missouri River	—	SW	Atchison, KS	1880	RR	Atchison & Eastern Bridge Co.	422.5	160	13
Missouri River	—	SW	St. Joseph, MO	1918	RR	UP	448.2	200	11

Missouri River	—	SW (Double)	Council Bluffs, IA	1893	RR	CN	618.3	520	12
Missouri River	—	SW	Sioux City, IA	1896	HWY, ERR	Dakota Co. Missouri River Bridge	764.4	215	18
Missouri River	US 8	VL	Yankton, SD	1924	HWY	South Dakota	805.7	230	39
Missouri River	—	SW	Chamberlain, SD	1925	RR	CMSTP&P	1067.4	200	13.8
Missouri River	—	SW	Pierre, SD	1907	RR	CNW	1174	210	17
Missouri River	SR 55	VL	Snowden, MT	1915	HWY, RR	BNSF	1589	196	56
Mitchell River	—	B	Chatham, MA	1979	HWY	Chatham	0	15	8
Mitchell River	Bridge St.	B	Chatham, MA	1926	HWY	Chatham	0.2	15	8
Mobile River	US 90 (Cohrane Br.)	VL	Mobile, AL	1925	HWY	Alabama	2.9	300	135
Mobile River	—	SW	Hurricane, AL	1926	RR	CSX	13.3	146	4
Mokelumne River	SR 12	SW	Isleton, CA	1942	HWY	California	3	100	8
Mokelumne River	—	SW	Walnut Grove, CA	1954	HWY	Sacramento and San Joaquin Cos.	12.1	85	12
Mokelumne River	—	RSP	Walnut Grove, CA	1955	HWY	San Joaquin Co.	12.2	56	11
Mokelumne River	—	SW	New Hope Landing, CA	1893	HWY	California	13.1	55.5	—
Mokelumne River	—	RSP	Thornton, CA	1958	HWY	San Joaquin Co.	18	58	13
Mokelumne River	—	SW	Benson Ferry–Thornton, CA	1910	HWY	San Joaquin Co.	18.1	65	2.5
Mokelumne River	—	SW	Thornton, CA	1909	HWY	Sacramento and San Joaquin Cos.	20.2	62	—
Mokelumne River	—	SW	Benson Ferry, CA	1950	HWY	Sacramento and San Joaquin Cos.	23.3	80	18
Mokelumne River	—	SW	Thornton, CA	1908	RR	WP	24	61	16

*See key of abbreviations at end.

Waterway[a]	Road Name/ Route Carried	Bridge Type[b]	Location	Const. Date	Traffic Type[c]	Owner*	Miles from Mouth of Waterway	Channel Width/Lateral Clearance (feet)	Clearance Height (feet)
Mona Lake	—	SW	Lake Harbor, MI	1941	HWY	Muskegon Co.	0.4	29	7
Montezuma Slough	—	SW	Collinsville, CA	1931	RR	SN	0.1	100	4
Montezuma Slough	—	RSP	Suisun City, CA	1962	HWY	California	7.1	51	21
Moriches Bay	—	B	Southampton, NY	—	HWY	—	—	36	10
Mormon Channel	Washington St.	B	Stockton, CA	1927	HWY	Stockton	1	100	3
Mormon Channel	Edison St.	SW	Stockton, CA	1896	RR	ATSF	1.1	72.5	2.7
Mormon Channel	Lincoln St.	SW	Stockton, CA	1894	HWY	Stockton	1.2	72	2.7
Morris and Cummings Cut	—	B	Aransas Pass, TX	1912	HWY, RR	Nueces Co.	2.3	24	6
Morrison Channel	Wayne St.	SW	St. Joseph, MI	1910	HWY	St. Joseph	0.2	99	22
Morrison Channel	—	SW	St. Joseph, MI	1906	RR	PC	1	26	4
Mount Desert Narrows	—	SW	Bar Harbor, ME	1918	HWY	Maine	—	40	7.6
Mud Slough	—	SW	Alviso, CA	1905	RR	UP	0.7	50	1
Mud Slough	—	RSP	Alviso, CA	1980	FB	U.S. Fish and Wildlife Services	0.8	44	8
Mullato Bayou	—	SW	Mulat, FL	1910	RR	L&N	—	85	4
Mullica River	US 9	B	New Gretna, NJ	1919	HWY	New Jersey	7.5	50	4
Mullica River	—	B	Lower Bank, NJ	1926	HWY	Atlantic and Burlington Cos.	15	30	6
Mullica River	—	B	Green Bank, NJ	1928	HWY	Atlantic and Burlington Cos.	18	30	5
Muskegon River	—	SW	Muskegon, MI	—	RR	PRR	5.8	30	1.1
Muskingum River	Butler St.	SW	Marietta, OH	1921	RR	BO	0.18	70	−15.4
Muskingum River	Butler St. Br.	SW	Marietta, OH	1921	HWY	Ohio	0.2	70	15
Muskingum River	Putnam St.	SW	Marietta, OH	1914	HWY	Washington Co.	0.3	88	7
Muskingum River	Bridge St.	SW	Lowell, OH	1914	HWY	Washington Co.	13.7	50	−15.4
Muskingum River	Upper Lowell	SW	Lowell, OH	1907	HWY	Washington Co.	14	50	−12.9
Muskingum River	—	SW	Beverly, OH	1915	HWY	Ohio	24.25	80	−10.6

Muskingum River	—	P	Beverly, OH	1939	FTBR	M. H. Gidlow	24.8	80	—
Muskingum River	—	SW	Stockport, OH	1914	HWY	Ohio	40.2	80	0
Muskingum River	Lock St.	SW	McConnelsville, OH	1898	HWY	U.S. Govt.	49.7	36	0
Muskingum River	Center St.	SW	McConnelsville, OH	1920	HWY	Ohio	50.2	80	11
Muskingum River	—	SW	Gaysport, OH	1914	HWY	Muskingum Co.	63.66	83	−13.6
Muskingum River	—	SW	Philo, OH	1954	HWY	Muskingum Co.	68.5	56	9
Muskingum River	—	SW	Brush Creek, OH	1914	HWY	Ohio	72.42	80	−15.4
Muskingum River	US 22	B	Zanesville, OH	—	HWY	Ohio	75.1	—	—
Muskingum River	6th St.	RL	Zanesville, OH	1915	HWY	Muskingum Co.	76.79	60	−15.5
Muskingum River	2nd St.	SW	Zanesville, OH	1913	RR	CSX	77.1	45	0
Muskingum River	Main St.	VL, F	Zanesville, OH	1903	HWY	Muskingum Co.	77.3	50	0
Muskingum River	Rte. 22	B	Zanesville, OH	—	HWY	Muskingum Co.	77.3	50	0
Muskingum River	—	B	Zanesville, OH	1913	RR	CSX	77.4	50	0
Myakka River	—	SW	McCall, FL	1908	RR	CSX	3.1	31	3
Mystic River 1	—	SW	Mystic, CT	1982	RR	ATK	2.4	65	4
Mystic River 1	US 1	B	Mystic, CT	1922	HWY	Connecticut	2.8	65	4
Mystic River 2	Chelsea St.	SW	Charlestown, MA	1913	HWY	Boston	0.06	125	14
Mystic River 2	Malden Br. (Alford St.)	B	Charlestown, MA	1964	HWY	Boston	1.4	75	7
Mystic River 2	—	B	Charlestown, MA	1918	RR	Boston Elevated Rwy.	1.44	75	28.8
Mystic River 2	—	SW	Somerville, MA	1848	RR	B&A	1.75	43	−3.5
Mystic River 2	Bm 7	SW	Somerville, MA	—	RR	B&M	1.8	42	0
Mystic River 2	—	SW	Somerville, MA	1920	RR	B&M	2	44	2.96
Mystic River 2	Wellington Br.	B	Somerville, MA	1936	HWY	Metropolitan Dist. Comm.	2.5	50	5
Mystic River 2	Gen. Lawrence Br. (SR 16)	B, F	Medford, MA	1935	HWY	Metropolitan Dist. Comm.	3.6	50	13

*See key of abbreviations at end.

Waterway[a]	Road Name/ Route Carried	Bridge Type[b]	Location	Const. Date	Traffic Type[c]	Owner*	Miles from Mouth of Waterway	Channel Width/Lateral Clearance (feet)	Clearance Height (feet)
Mystic River 2, South Channel	Chelsea St.	B	Charlestown, MA	1927	HWY	Boston	0.19	75	25
Nacote Creek	US 9	B	Port Republic, NJ	1926	HWY	New Jersey	1.5	30	5
Nacote Creek	—	SW	Port Republic, NJ	—	HWY	New Jersey	3.5	30	8
Nansemond River	US 17	B	Town Point, VA	1928	HWY	Virginia	2.5	96	21
Nansemond River	SR 125	SW	Holiday Point, VA	1928	HWY	Virginia	7.7	80	7
Nansemond River	US 460	B	Suffolk, VA	1935	HWY	Virginia	18.2	40	6
Nansemond River, Western Branch	—	SW	Reids Ferry, VA	1931	HWY	Virginia	2	35	9.3
Nansemond River, Western Branch	—	SW	Everetts, VA	—	HWY	Virginia	7	25	3
Nanticoke River	US 50	B	Vienna, MD	1965	HWY	Maryland	22.2	80	18
Nanticoke River	SR 313	SW	Sharptown, MD	1912	HWY	Maryland	30	75	7
Nanticoke River	—	SW	Seaford, DE	1929	RR	CR	39.4	47	0
Nanticoke River	—	B	Seaford, DE	1925	HWY	Delaware	39.6	40	3
Napa River	—	B	Vallejo, CA	1927	HWY	California	1	75	11
Napa River	Tennessee St.	VL	Vallejo, CA	1980	HWY, RR	U.S. Navy	2.8	140	6
Napa River	—	SW	Vallejo, CA	1904	RR	SP	8.25	79.8	5
Napa River	—	VL	Brazos Station, CA	1979	RR	SP	10.6	157	97
Napa River	Imola Br.	VL	Napa, CA	1949	HWY	California	17.6	90	60
Napa Slough	—	B	Sonoma, CA	1920	PRIV RD	S. A. Skaggs	6	16	7
Narraguagus River	US 1A	SW	Milbridge, ME	1937	HWY	Maine	1.8	25	5
Narrow Bay	—	B	South Beach, NY	1959	HWY	Suffolk Co.	6.1	55	18
Naselle River	—	SW	Naselle, WA	1938	HWY	Washington	2.5	107	9
Nassau Sound	—	SW	Fernandina, FL	1950	HWY	Fernandina Port Auth.	0.4	60	15
Navesink River	—	B	Oceanic, NJ	1938	HWY	Monmouth Co.	4.5	75	22
Neale Sound	—	B	Cobb Island, MD	1932	HWY	Maryland	0.6	22	8.5

Neches River	—	VL	Beaumont, TX	1941	RR	KCS	19.5	200	145
Neches River	—	B	Beaumont, TX	1940	RR	SP	19.8	200	2.6
Neches River	—	SW	Beaumont, TX	1925	HWY	Texas	21.4	90	8
Neches River	—	SW	Evadale, TX	—	RR	BNSF	53.9	63	8
Nehalem River	US 101	SW, F	Nehalem, OR	1928	HWY, PL	Tillamook Co.	6.5	77	21
Nehalem River, North Fork	—	B	Nehalem, OR	1933	HWY	Tillamook Co.	0.25	90	9
Nemadji River	—	SW	Superior, WI	1903	RR	CNW	0.6	59	10
Neponset River	—	B	Boston, MA	1928	HWY	Metropolitan Dist. Comm.	0	65	12.9
Neponset River	—	B	Boston, MA	1908	RR	NYNHH	1.37	50	5.9
Neponset River	—	B	Boston, MA	1925	HWY	Metropolitan Dist. Comm.	1.5	78.6	10.9
Neponset River	Granite Ave.	B	Milton, MA	1957	HWY	Massachusetts	2.5	50	6
Neuse River	US 17	SW	New Bern, NC	1951	HWY	North Carolina	33.7	60	7
Neuse River	—	SW	Newbern, NC	1931	HWY	North Carolina	34	60	2.4
Neuse River	—	SW	New Bern, NC	1907	RR	NS	34.2	58	0
Neuse River	—	SW	Fort Barnwell, NC	1906	HWY	North Carolina	60	55.5	0.6
Neuse River	US 17	SW	Kinston, NC	1907	RR	AEC	80	42	2
Neuse River	Main St.	SW	Kingston, NC	1931	HWY	North Carolina	82.4	40	2.5
Neuse River	—	SW	Kingston, NC	1907	RR	NS	83.1	47	7
Neuse River	Caswell St.	SW	Kingston, NC	1916	HWY	North Carolina	83.2	48	3.8
Neuse River	Hardys Br.	SW	La Grange, NC	1868	HWY	North Carolina	102	47	0
Neuse River	Rockford Br.	SW	Whitehall, NC	1924	HWY	North Carolina	106.3	48	1.5
Neuse River	Arrington Br.	B	Goldsboro, NC	1895	HWY	North Carolina	125.8	29	5.6
Neuse River	—	SW	Goldsboro, NC	1894	RR	ACL	128.2	61	9.5
New Begun Creek	SR 170	SW	New Weeksville, NC	—	HWY	North Carolina	3.5	30	3
New Begun Creek	SR 170	B	Old Weeksville, NC	—	HWY	North Carolina	4.5	18	3
New Mill Creek	—	SW	Deep Creek, VA	1907	RR	N&P Belt Line	—	26	5.3
New Pass	—	B	Sarasota, FL	1929	HWY	Florida	0.5	110	12
New River 1	3rd Ave. S.E.	B	Ft. Lauderdale, FL	1960	HWY	Ft. Lauderdale	1.4	60	21

*See key of abbreviations at end.

Waterway[a]	Road Name/ Route Carried	Bridge Type[b]	Location	Const. Date	Traffic Type[c]	Owner*	Miles from Mouth of Waterway	Channel Width/Lateral Clearance (feet)	Clearance Height (feet)
New River 1	—	B	Ft. Lauderdale, FL	1926	HWY	Ft. Lauderdale	1.8	60	5.5
New River 1	—	B	Ft. Lauderdale, FL	1916	HWY	Broward Co.	2.2	59	5.8
New River 1	Andrews Ave.	B	Ft. Lauderdale, FL	1981	HWY	Broward Co.	2.3	60	21
New River 1	—	B	Ft. Lauderdale, FL	1945	RR	FEC	2.5	51	3
New River 1	5th Ave. S.W.	B	Ft. Lauderdale, FL	1964	HWY	Ft. Lauderdale	2.7	60	20
New River 1	—	B	Ft. Lauderdale, FL	1927	HWY	Broward Co.	2.8	40	11
New River 2	SR 172	SW	Sneads Ferry, NC	1944	HWY	North Carolina	5.5	50	8
New River 2	—	SW	Jacksonville, NC	1907	RR	CSX	21.1	48	3
New River Canal, North Branch	—	SW	Ft. Lauderdale, FL	1926	HWY	Florida	1	50	—
New River Canal, South Branch	—	B	Ft. Lauderdale, FL	1916	HWY	Florida	3.5	50	10
New River, North Fork	11th Ave. S.W.	SW	Ft. Lauderdale, FL	1926	HWY	Ft. Lauderdale	0.5	40	4
New River, South Fork	12th St. S.W.	B	Ft. Lauderdale, FL	1960	HWY	Ft. Lauderdale	0.9	60	21
New River, South Fork	—	B	Ft. Lauderdale, FL	1928	RR	SAFRR	1.25	63.5	4.1
New River, South Fork	—	B	Ft. Lauderdale, FL	1978	RR	CSX	2.8	64	2
New River, South Fork	SR 84	B	Ft. Lauderdale, FL	1957	HWY	Florida	4.4	40	21
New Rochelle Harbor	Glen Island	B	New Rochelle, NY	1929	HWY	Westchester Co.	0.4	59	13
Newark Bay	CNJ E.	VL (out)	Elizabeth, NJ	1926	RR	CNJ	0.7	134	135
Newark Bay	CNJ W.	VL (out)	Elizabeth, NJ	1926	RR	CNJ	0.7	216	135
Newark Bay	—	VL	Newark, NJ	1930	RR	CR	4.3	300	135
Newark Slough	—	SW	Newark, CA	1909	RR	SP	0.5	49	6
Newark Slough	—	RSP	Newark, CA	1981	FTBR	U.S. Fish and Wildlife Services	4.1	40	7
Newark Slough	—	RSP	Newark, CA	1981	FTBR	U.S. Fish and Wildlife Services	4.3	40	7

Newport Bay	—	RSP	Newport Beach, CA	1932	HWY	California	2.3	40	12.9
Newport Bay	—	RSP	Newport Beach, CA	1936	HWY	Newport Beach	3.5	40	9.4
Newport River	—	SW	Newport, SC	1905	RR	A&ECRR	13	19	9
Newton Creek	Broadway	B	Camden, NJ	1916	HWY	New Jersey	0.25	50	5.4
Newtown Creek	Pulaski Br.	B	Long Island City, NY	1954	HWY	New York City	0.6	150	39
Newtown Creek	Greenpont Ave.	B	Long Island City, NY	1988	HWY	New York City	1.3	149	26
Newtown Creek, East Branch	Grand Ave.	SW	Brooklyn, NY	1903	HWY	New York City	3.1	58	10
Nezpique Bayou	I 10	SW	Hennings, LA	1960	HWY	Louisiana	6.5	55	28
Nezpique Bayou	—	SW	Jennings, LA	1937	HWY	Louisiana	7	40	4
Niagara River	International Br.	SW	Buffalo, NY	1873	RR	CN	33	156	18
Niantic River	—	B	Niantic, CT	1907	RR	ATK	0	45	11
Niantic River	—	SW	Niantic, CT	1921	HWY	Connecticut	0.1	65	9
Nolte Creek	—	B	Magnolia Springs, AL	1921	HWY	Magnolia Springs	5	20	6
Nomini Creek	SR 202	SW	Mt. Holly, VA	1931	HWY	Virginia	3.5	39	5
Nomini Creek	—	SW	Mt. Holly, VA	1908	HWY	Westmoreland Co.	5.3	17	5.1
Nooksack River	—	SW	Marietta, WA	1918	HWY	Whatcom Co.	0.25	88.5	11.5
Nooksack River	—	SW	Ferndale, WA	1918	HWY	Whatcom Co.	5.75	70.6	16
Nooksack River	—	SW	Ferndale, WA	1911	RR	GN	6	102	12
North Point Creek	—	SW	Sparrows Pt., MD	1907	RR	Baltimore Transit	1	40	3.4
North River	SR 3A	B	Scituate, MA	1934	HWY	Massachusetts	1.6	32	12
North River	Bridge St.	B, F	Norwell, MA	1958	HWY	Plymouth Co.	4	27	6
North Slough	—	B	North Bend, OR	1927	HWY	Oregon	1.5	30	5
North Wimbee Creek	—	SW	Dale, SC	1917	RR	SAL	1.6	40	5.6
Northeast River	SR 133	B	Wilmington, NC	1929	HWY	North Carolina	1	150	26
Northeast River	SR 133	B	Wilmington, NC	1976	HWY	North Carolina	1	200	40
Northeast River	—	B	Wilmington, NC	1973	RR	CSX	1.5	200	6
Northeast River	—	SW	Castle Hayne, NC	1913	HWY	North Carolina	26.25	82	2
Northeast River	—	SW	Castle Hayne, NC	1950	RR	CSX	27	52	5

*See key of abbreviations at end.

Waterway[a]	Road Name/ Route Carried	Bridge Type[b]	Location	Const. Date	Traffic Type[c]	Owner*	Miles from Mouth of Waterway	Channel Width/Lateral Clearance (feet)	Clearance Height (feet)
Northeast River	—	SW	Rocky Pt., NC	1925	HWY	North Carolina	35	59.5	0
Northeast River	—	R	Burgaw, NC	1924	HWY	North Carolina	52	29	0
Northeast River	—	R	Wallace, NC	1930	HWY	North Carolina	72	29.1	0
Northeast River	—	R	Sloan's, NC	1925	HWY	North Carolina	75	30	1.5
Northeast River	—	R	Chinquepin, NC	1925	HWY	North Carolina	79	30	0
Northeast River	—	SW	Hallsville, NC	1917	RR	Trumbull Lumber	84	38	0
Northwest River	—	SW	Northwest, VA	1881	RR	NS	14	26	3
Nortons Creek	—	SW	Edgemere, NY	1896	HWY	New York City	0.2	44	2
Norwalk River	SR 136	B	South Norwalk, CT	1968	HWY	Connecticut	0	100	9
Norwalk River	SR 136	B	South Norwalk, CT	1915	HWY	Connecticut	0	70	8
Norwalk River	—	SW	South Norwalk, CT	1896	RR	Connecticut DOT	0.1	58	16
Novato Creek	—	B	Ignacio, CA	1917	RR	NWP	3.5	—	0.5
Novato Creek	—	RSP	Ignacio, CA	1918	HWY	California	3.5	30	0.5
Novato Creek	—	RSP	Novato, CA	1888	RR	NWP	4.5	26	6
Noxubee River	—	SW	Gainesville, AL	—	HWY	Sumpter Co.	0.2	72	5.3
Nueces Bay	—	B	Corpus Christi, TX	1913	RR	SP	0	32	4.4
Nueces Bay	—	B	Corpus Christi, TX	1921	HWY	Texas	0	32	5.5
Nueces River	—	SW	Calallen, TX	1905	RR	UP	10.9	35	6
NYSBC Erie Canal	Union St.	VL	Spencerport, NY	—	HWY	Spencerport	273	—	15
NYSBC Erie Canal	Washington St.	VL	Adams Basin, NY	—	HWY	Adams Basin	276	—	15
NYSBC Erie Canal	Park Ave.	VL	Brockport, NY	—	HWY	Brockport	280.4	—	15
NYSBC Erie Canal	Main St.	VL	Brockport, NY	1915	HWY	Brockport	280.5	—	15
NYSBC Erie Canal	E. Holley Rd.	VL	Holley, NY	—	HWY	Holley	285.1	—	15
NYSBC Erie Canal	Hulberton Rd.	VL	Hulberton, NY	—	HWY	Hulberton	288.2	—	15
NYSBC Erie Canal	Ingersoll St.	VL	Albion, NY	—	HWY	Albion	294.7	—	15
NYSBC Erie Canal	Main St.	VL	Albion, NY	—	HWY	Albion	294.9	—	15
NYSBC Erie Canal	Eagle Harbor Rd.	VL	Eagle Harbor, NY	—	HWY	Eagle Harbor	298.2	—	15
NYSBC Erie Canal	Knowlesville Rd.	VL	Knowlesville, NY	—	HWY	Knowlesville	302.3	—	15

NYSBC Erie Canal	Prospect St.	VL	Medina, NY	—	HWY	Medina	305.9	—	15
NYSBC Erie Canal	Main St.	VL	Middleport, NY	—	HWY	Middleport	310.6	—	16
NYSBC Erie Canal	Gasport Rd.	VL	Gasport, NY	—	HWY	Gasport	315.9	—	16
NYSBC Erie Canal	Adams St.	VL	Lockport, NY	—	HWY	Lockport	321.8	—	16
NYSBC Erie Canal	Exchange St.	VL	Lockport, NY	—	HWY	Lockport	322	—	16
NYSBC Erie Canal	—	B	Tonawanda, NY	—	RR	CR	340.1	—	15
NYSBC Erie Canal	Union St.	SW	Tonawanda, NY	—	RR	CR	340.6	—	—
NYSBC Oswego Canal	Lock O2	—	Fulton, NY	—	HWY	New York	11.7	—	2
NYSBC Oswego Canal	Culvert St.	B	Phoenix, NY	1986	HWY	New York	20.8	45	2
NYSBC Oswego Canal	Bridge St.	B (span removed)	Phoenix, NY	—	HWY	Phoenix	20.9	45	2
NYSBC Oswego Canal	Lock St.	B	Phoenix, NY	—	HWY	Phoenix	21	135	2
Oak Creek	—	SW	Royal Oak, MD	1924	HWY	Maryland	0	26	3.5
Oak Creek	—	B	Royal Oak, MD	1924	RR	PRR (B&E RR)	0	24.5	2.7
Oakland Inner Harbor	Park St.	B	Alameda, CA	1935	HWY	Alameda	7.3	240	15
Oakland Inner Harbor	Fruitvale Ave.	VL	Alameda, CA	1951	RR	U.S. Army COE	7.7	200	135
Oakland Inner Harbor	Fruitvale Ave.	B	Alameda, CA	1974	HWY	California	7.7	95	15
Oakland Inner Harbor	High St.	B	Alameda, CA	1939	HWY	Alameda	8.1	240	16
Obion River	—	SW	Bradley Ferry, TN	1922	HWY	Dyer Co.	10	80	−0.7
Obion River	—	SW	Petty Ferry, TN	1904	HWY	Dyer Co.	19.1	36	−5.1
Obion River	—	SW	McClures Ferry, TN	1916	HWY	Dyer Co.	26	61	−1.8
Obion River	—	SW	Lenox, TN	1908	RR	ICRR	33.5	85	−0.9
Obion River	—	SW	Lanes Ferry, TN	1901	HWY	Dyer Co.	47.2	37	−6.9
Oceanport Creek	—	SW	Oceanport, NJ	1914	RR	NJ Transit	8.4	58	4
Ocmulgee River	—	SW	Lumber City, GA	1928	HWY	Georgia	11.7	86	2.1
Ocmulgee River	—	SW	Lumber City, GA	1930	RR	NS	11.8	90	5
Ocmulgee River	—	SW	Jacksonville, GA	1936	HWY	Georgia	51.2	80	5
Ocmulgee River	—	SW	Abbeville, GA	1952	RR	CSX	95.9	73	0
Ocmulgee River	—	SW	Abbeville, GA	1920	HWY	Georgia	98.5	78	−0.5
Ocmulgee River	—	SW	Hawkinsville, GA	1921	HWY	Georgia	135.1	70	2.3
Ocmulgee River	—	SW	Hawkinsville, GA	1926	RR	NS	135.4	67	—
Ocmulgee River	—	SW	Macon, GA	1930	RR	NS	194.9	75	2

*See key of abbreviations at end.

Waterway[a]	Road Name/ Route Carried	Bridge Type[b]	Location	Const. Date	Traffic Type[c]	Owner*	Miles from Mouth of Waterway	Channel Width/Lateral Clearance (feet)	Clearance Height (feet)
Ocmulgee River	—	SW	Macon, GA	1915	RR	CSX	203.4	100	3
Oconee River	—	SW	Mt. Vernon, GA	1921	HWY	Georgia	28.7	72.5	8
Oconee River	—	SW	Mt. Vernon, GA	1890	RR	SAL	29	64.5	3
Oconee River	—	SW	Soperton, GA	1940	HWY	Georgia	44.3	100	12
Oconee River	Marion St.	SW	Dublin, GA	1890	RR	W&T Rwy.	77.8	66	5.1
Oconee River	—	SW	Dublin, GA	1923	HWY	Georgia	77.9	61	−2.1
Oconee River	—	B	Oconee, GA	1930	RR	Cleveland-Oconee Lumber	106	34	−15
Oconee River	—	SW	Oconee, GA	1918	RR	C of GA Rwy.	106.6	62	5.5
Ogden Slip	Outer Dr.	B	Chicago, IL	1937	HWY	Chicago	0.5	70	19
Ogeechee River	—	SW	Ways, GA	1926	RR	SAL	30.7	40.5	0.3
Ogeechee River	—	VL	Richmond Hill, GA	1926	RR	CSX	30.7	60	40
Ogeechee River	—	SW	Ways, GA	1926	HWY	Georgia	32.8	47	2.7
Ohio River	Indiana Channel	SW	New Albany, IN	1913	HWY, RR	K&ITRR	373.6	172	22.8
Ohio River	I 64, 264 (27th St.)	SW	Louisville, KY	1927	HWY, RR	Louisville Gas and Electric Comm.	374.2	110	0
Ohio River	I 64, 264 (27th St.)	B	Louisville, KY	—	HWY, RR	Louisville Gas and Electric Comm.	374.2	110	0
Ohio River	Canal at 18th St.	VL	Louisville, KY	1916	HWY	USG	376.4	200	30
Ohio River	—	VL	Louisville, KY	1919	RR	CR	376.6	241	29
Ohio River	—	VL (not used per COE; fixed bridge 2003)	Cincinnati, OH	1922	RR	Cin Sou Rwy. Trustees	508.7	500	25.3
Okanogan River	—	SW	Malott, WA	1932	HWY	Okanogan Co.	16	—	—
Okanogan River	—	SW	Omak, WA	1924	HWY	Washington	30	—	—

Okeechobee Waterway, St. Lucie River	SR A1A	B	Stuart, FL	1957	HWY	Florida	3.3	89	21
Okeechobee Waterway, St. Lucie River	—	B	Stuart, FL	1938	RR	FEC	7.5	50	7
Okeechobee Waterway, St. Lucie River	US 1	B (repl)	Stuart, FL	1934	HWY	Florida	7.5	80	14
Okeechobee Waterway, St. Lucie River	US 1	B (repl)	Stuart, FL	1964	HWY	Florida	7.5	80	14
Okeechobee Waterway, St. Lucie Canal	—	SW	Indiantown, FL	1936	RR	CSX	28.5	51	6
Okeechobee Waterway, St. Lucie Canal	—	VL	Port Mayaca, FL	1925	RR	FEC	38	56	48
Okeechobee Waterway, St. Lucie Canal	US 441	SW	Port Mayaca, FL	—	HWY	Florida	38.8	53	10
Okeechobee Waterway, Lake Okeechobee	SR 717	SW	Torry Island, FL	1935	HWY	Florida	60.7	50	11
Okeechobee Waterway, Caloosahatchee Canal	—	SW	Moorehaven, FL	1921	RR	CSX	78.3	50	5
Okeechobee Waterway, Caloosahatchee Canal	US 27	B	Moorehaven, FL	1955	HWY	Florida	78.4	90	23
Okeechobee Waterway, Caloosahatchee Canal	—	SW	Ortona, FL	1919	RR	CSX	94	54	7
Okeechobee Waterway, Caloosahatchee Canal	SR 29	B	La Belle, FL	1959	HWY	Florida	103	90	28
Okeechobee Waterway, Caloosahatchee River	—	SW	Ft. Denaud, FL	1965	HWY	Florida	108.2	80	9
Okeechobee Waterway, Caloosahatchee River	SR 78	B	Alva, FL	1970	HWY	Florida	116	90	21
Okeechobee Waterway, Caloosahatchee River	Wilson Piggott	B	Olga, FL	1961	HWY	Florida	126.3	89	27

*See key of abbreviations at end.

Waterway[a]	Road Name/ Route Carried	Bridge Type[b]	Location	Const. Date	Traffic Type[c]	Owner*	Miles from Mouth of Waterway	Channel Width/Lateral Clearance (feet)	Clearance Height (feet)
Okeechobee Waterway, Caloosahatchee River	—	SW	Tice, FL	1921	RR	CSX	129.9	49	3
Okeechobee Waterway, Caloosahatchee River	US 41	B	Ft. Myers, FL	1931	HWY	Florida	134.5	78	10
Okeechobee Waterway, San Carlos Bay	—	B	Punta Rassa, FL	1963	HWY	Lee Co.	151	90	26
Oklawaha River	—	SW	Eureka, FL	1929	HWY	Marion Co.	33.7	40	3
Oklawaha River	—	SW	Delks Bluff, FL	1927	HWY	Marion Co.	52.3	40	3
Oklawaha River	—	SW	Ocala, FL	—	HWY	Marion Co.	55	34	0.7
Oklawaha River	—	SW	Ocala, FL	1913	RR	Ocala Northern Rwy.	55.1	38	2
Oklawaha River	SR 40	SW	Sharpes Ferry, FL	1971	HWY	Southwest Florida Water Mgmt. Dist.	55.1	40	6
Oklawaha River	—	SW	Mucian Farms, FL	1970	PRIV RD	Southwest Florida Water Mgmt. Dist.	63.9	28	8
Oklawaha River	SR 464	SW	Moss Bluff, FL	1927	HWY	Florida	66	40	4
Oklawaha River	SR 42	SW	Starkes Ferry, FL	1927	HWY	Florida	73	39	5
Old Brazos River	—	SW	Freeport, TX	1946	RR	UP	4.4	108	11
Old Fort Bayou	US 90	SW	Ocean Springs, MS	1928	HWY	Jackson Co.	1.6	50	6
Old River	—	B	Orwood, CA	1929	RR	BNSF	10.4	90	11
Old River	SR 4	SW	Victoria Island, CA	1932	HWY	California	14.8	98	13
Old River Navigation Canal	SR 15	VL	Torras, LA	—	HWY	Louisiana	—	75	53
Old Tampa Bay	US 92 (Gandy Br.)	B	Tampa, FL	1924	HWY	Florida	2.5	65	13
Old Tampa Bay	Courtney Campbell Pkwy.	B	Tampa, FL	1934	HWY	Florida	8.2	55	13

Oldmans Creek	US 130	VL, F	Nortonville, NJ	1937	HWY	New Jersey	3.1	74	64
Oldmans Creek	—	SW, F	Pedricktown, NJ	—	RR	CR	4	36	2
Oldmans Creek	—	RSP	Pedricktown, NJ	1979	RR	CR	4	14	3
Oldmans Creek	—	SW, F	Pedricktown, NJ	1906	HWY	Salem and Gloucester Cos.	5.1	36	7
Onancock River	Warrington Branch	B NO	Onancock, VA	1923	HWY	H. Powell and S. Buckle	0.05	17.2	5.8
Ontonagon River	SR 37	SW	Ontonagon, MI	1940	HWY	Michigan	0.2	31	7
Oostenaula River	—	SW	Rome, GA	1888	RR	C of G	0.1	87.3	1
Oostenaula River	—	SW	Rome, GA	—	HWY	Rome Land	0.4	96	5
Oostenaula River	—	VL	Rome, GA	1918	HWY	Floyd Co.	0.7	100	0
Oostenaula River	—	SW	Calhoun, GA	1918	HWY	Gordon Co.	35	90	0
Orange River	—	B	Ft. Myers, FL	1948	HWY	Florida	0.9	40	11
Orange River	—	B	Ft. Myers, FL	1927	RR	SAFRR	2.25	33	4.3
Orange River	—	SW	Buckingham, FL	1928	HWY	Florida	5.7	33.5	6.5
Oregon Slough	—	SW	Portland, OR	1908	RR	BNSF	3.2	125	19
Ortega River	SR 211	B	Jacksonville, FL	1929	HWY	Florida	0.27	53	9
Ortega River	—	B, UC	Jacksonville, FL	1940	HWY	Florida	1.04	40	4
Ortega River	—	B, UC	Jacksonville, FL	1940	HWY	Florida	1.04	40	4
Ortega River	—	B	Jacksonville, FL	1927	RR	CSX	1.05	40	2
Osage River	—	VL, F	Osage City, MO	1925	RR	UP	5.6	96	1
Osage River	—	VL	Osage City, MO	1922	HWY	Missouri	10	96	40.3
Otonabee River	—	SW	Havelock, ON, Canada	1925	RR	CP	—	127	—
Ouachita and Black Rivers	—	VL	Jonesville, LA	1913	RR	Trans Action	40.5	150	51
Ouachita and Black Rivers	US 84	SW	Jonesville, LA	1933	HWY	Louisiana	40.9	140	8
Ouachita and Black Rivers	SR 8	SW	Harrisonburg, LA	1933	HWY	Louisiana	57.5	140	10
Ouachita and Black Rivers	US 165	VL	Columbia, LA	1936	HWY	Louisiana	110.2	140	62
Ouachita and Black Rivers	—	VL	Riverton, LA	1942	RR	—	114.3	156	57
Ouachita and Black Rivers	—	SW	Monroe, LA	1926	HWY	Monroe	166.9	130	0
Ouachita and Black Rivers	—	SW	Monroe, LA	1908	RR	ICG	167	130	2
Ouachita and Black Rivers	—	VL	Monroe, LA	—	HWY	—	167.2	—	—
Ouachita and Black Rivers	US 80	B	Monroe, LA	1936	HWY	Louisiana	167.4	130	4

*See key of abbreviations at end.

Waterway[a]	Road Name/ Route Carried	Bridge Type[b]	Location	Const. Date	Traffic Type[c]	Owner*	Miles from Mouth of Waterway	Channel Width/Lateral Clearance (feet)	Clearance Height (feet)
Ouachita and Black Rivers	—	SW	Sterlington, LA	1946	RR	MP	191.8	137	1
Ouachita and Black Rivers	US 165	SW	Sterlington, LA	1932	HWY	Louisiana	192.3	130	8
Ouachita and Black Rivers	—	SW	Felsenthal, LA	1904	RR	MP	225.8	130	2.2
Ouachita and Black Rivers	—	SW	Calion, AR	1906	RR	RI	291.7	140	7
Ouachita and Black Rivers	—	SW	Camden, AR	1913	HWY	Ouachata Co.	336.6	130	0.4
Ouachita and Black Rivers	—	SW	Camden, AR	1930	RR	SLSW	337	130	4.1
Overpeck Creek	—	B, F	Ridgefield Park, NJ	1910	RR	Del Otsego	0	37	4
Overpeck Creek	—	SW, F	Ridgefield Park, NJ	1901	RR	CSX	0	30	3
Overpeck Creek	Bergen Tpk.	SW	Ridgefield Park, NJ	1902	HWY	Bergen Co.	0.8	40.6	4.9
Overpeck Creek	SR 6	SW (no operating machinery)	—	1930	HWY	New Jersey	1.4	70	9
Overpeck Creek	Ft. Lee Tpk.	B	Leonia, NJ	1913	HWY	Bergen Co.	3.04	40.2	4.7
Oyster Creek	—	RSP	Velasco, TX	1926	HWY	Texas	8.9	25	7
Oyster Creek	—	RSP	Velasco, TX	1922	RR	UP	12.9	24	11
Oyster Creek	—	RSP	Velasco, TX	—	HWY	Brazoria Co.	14.3	11.3	4
Pacheco Creek	—	SW	Martinez, CA	1916	HWY	Contra Costa Co.	0.5	41.25	7
Pacheco Creek	—	SW	Martinez, CA	1946	HWY	Contra Costa Co.	0.5	40	4
Pacheco Creek	Avon Hwy.	SW	Martinez, CA	1948	HWY	Contra Costa Co.	1	40	4
Pacheco Creek	—	SW	Martinez, CA	1901	RR	UP	1.1	41	7
Palix River	—	B	Bay Center, WA	1924	HWY	Washington	2.5	36	10.67
Pamlico and Tar Rivers	—	SW	Washington, NC	1911	RR	NS	37	69	7
Pamlico and Tar Rivers	—	SW	Washington, NC	1929	HWY	North Carolina	37	61	11.3
Pamlico and Tar Rivers	—	SW	Washington, NC	1907	RR	ACL (Washington and Vandemere RR)	37	58	7
Pamlico and Tar Rivers	US 17	SW	Washington, NC	1966	HWY	North Carolina	37.2	50	6
Pamlico and Tar Rivers	Boyds Ferry Br.	SW	Grimesland, NC	1954	HWY	North Carolina	44.8	55	5
Pamlico and Tar Rivers	—	SW	Greenville, NC	1913	RR	ACL	59	59.5	11.6

Pamlico and Tar Rivers	—	SW	Old Sparta, NC	1924	HWY	North Carolina	79	45	2
Pamlico and Tar Rivers	—	SW	Tarboro, NC	1921	RR	ACL	86	69.5	18.8
Pamlico and Tar Rivers	—	SW	Tarboro, NC	1914	RR	ACL	91	73.2	17
Pamlico and Tar Rivers	US 258 (Bells Br.)	SW	Kingsboro, NC	1915	HWY	North Carolina	97	41	9.4
Pamunkey River	SR 33	SW	West Point, VA	1957	HWY	Virginia	1	90	10
Pamunkey River	—	SW	White House, VA	1928	RR	NS	32.7	53	3
Pantego Creek	—	SW	Belhaven, NC	1910	HWY	North Carolina	0.5	40	4.2
Papys Bayou	—	B	St. Petersburg, FL	1928	HWY	Boulevard and Bay Land and Development	—	20	8.6
Paradise Creek	—	B	Portsmouth, VA	1898	RR	Virginia Electric & Power	0.75	25	2
Pascagoula River	—	SW	Pascagoula, MS	1904	RR	CSX	1.5	81	7
Pascagoula River	—	SW	Pascagoula, MS	1926	HWY	Jackson Co.	1.75	82	8.5
Pascagoula River	US 90	B	Pascagoula, MS	1955	HWY	Mississippi	1.8	140	31
Pascagoula River	—	SW	Merrill, MS	1901	RR	ICG	81.2	53	8
Pasquotank River	—	B	Elizabeth City, NC	1931	HWY	North Carolina	15	100	8.5
Pasquotank River	—	SW	Elizabeth City, NC	1903	RR	NS	19	42.5	2
Passagasawakeag River	—	SW	Belfast, ME	1921	HWY	Maine	1.1	48	6.5
Passagasawakeag River	—	B	Belfast, ME	—	HWY	Belfast	2	28	7
Passaic River	—	SW (out)	Newark, NJ	1913	RR	CR	1.1	100	25
Passaic River	Lincoln Hwy.	VL	Newark, NJ	1943	HWY	New Jersey	1.8	300	135
Passaic River	Point No Point	SW	Newark, NJ	1901	RR	CR	2.6	103	16
Passaic River	Jackson St.	SW	Newark, NJ	1897	HWY	Essex and Hudson Cos.	4.6	72	15
Passaic River	—	VL	Newark, NJ	1939	RR	ATK	5	200	137
Passaic River	—	VL	Newark, NJ	1939	RR	ATK	5.01	200	138.2
Passaic River	—	VL	Newark, NJ	1939	RR	ATK	5.02	200	136.9
Passaic River	Bridge St. (Center St. per COE)	SW (out)	Newark, NJ	1911	RR, HWY	CR	5.3	81	10
Passaic River	Bridge St.	SW	Newark, NJ	1913	HWY	Essex and Hudson Cos.	5.6	80	7

*See key of abbreviations at end.

Waterway[a]	Road Name/ Route Carried	Bridge Type[b]	Location	Const. Date	Traffic Type[c]	Owner*	Miles from Mouth of Waterway	Channel Width/Lateral Clearance (feet)	Clearance Height (feet)
Passaic River	Morristown Line	SW	Newark, NJ	1903	RR	New Jersey Transit	5.8	77	15
Passaic River	I 280 (Stickle Memorial Hwy.)	VL	Newark, NJ	1944	HWY	New Jersey	5.8	200	135
Passaic River	Clay St.	SW	Newark, NJ	1908	HWY	Essex and Hudson Cos.	6	75	8
Passaic River	4th Ave.	B	Newark, NJ	1922	RR	CR	6.3	126	7
Passaic River	Erie Greenwood Lake	SW	Newark, NJ	1937	RR	New Jersey Transit	8	48	35
Passaic River	Rutgers St.	B	Belleville, NJ	1936	HWY	New Jersey	8.9	98	8
Passaic River	—	SW	Avondale, NJ	1905	HWY	Essex and Hudson Cos.	10.7	65	7
Passaic River	Boonton Line	SW	Lyndhurst, NJ	1901	RR	New Jersey Transit	11.7	47	26
Passaic River	Rutherford Ave.	B	Rutherford, NJ	1949	HWY	New Jersey	11.8	125	35
Passaic River	Union Ave.	SW	Passaic, NJ	1897	HWY	Passaic and Bergen Cos.	13.2	60	13
Passaic River	Ayerigg Ave.	SW	Passaic Park, NJ	1892	RR	Erie (NYLE&WRR)	13.57	62.6	18.6
Passaic River	Gregory Ave.	SW	Passaic, NJ	1906	HWY	Passaic and Bergen Cos.	14	71	12
Passaic River	Market St. (2nd St. per COE)	B, F	Passaic, NJ	1930	HWY	Passaic and Bergen Cos.	14.7	100	5
Passaic River	W. 8th St.	B, F	Passaic, NJ	1914	HWY	Passaic and Bergen Cos.	15.3	70	5
Pass Manchac	—	B	Manchac, LA	1971	RR	CN	6.7	85	8
Pass Manchac	—	SW	Manchac, LA	1907	RR	IC	6.75	62	4.8
Pass Manchac	—	SW	Manchac, LA	1926	HWY	Louisiana	6.75	63.5	3.9

Patapsco River, Middle Branch	Hanover St.	B	Baltimore, MD	1916	HWY	Baltimore	12	150	21
Patapsco River, Middle Branch	—	SW	Baltimore, MD	1904	RR	WMD	12.5	85	9
Patapsco River, South Branch	—	C	Baltimore, MD	1916	HWY	Maryland	12	28	8.5
Patout Bayou–Gaspergou Bayou	SR 83	SW	Weeks Island, LA	1948	HWY	Louisiana	0.4	50	5
Patuxent River	SR 231	SW	Bennedict, MD	1951	HWY	Maryland	24.4	49	16
Patuxent River, Western Branch	—	SW (abandoned)	Marlboro, MD	1923	RR	CR	—	77	—
Paw Paw River	—	B	Benton Harbor, MI	1902	RR	B4	0.02	43	−0.5
Paw Paw River	—	B	Benton Harbor, MI	—	RR	PM	0.22	45	2.5
Peace River	—	B	Punta Gorda, FL	1931	HWY	Florida	2	75	13
Pearl River	—	SW	Bogalusa, LA	1922	HWY	Pearl River Co.	79.5	100	5.4
Pearl River	SR 26	SW	Bogalusa, LA	1951	HWY	Louisiana	79.5	80	3
Pearl River	US 98	RSP	Columbia, MS	1933	HWY	Mississippi	139.8	190	10
Pearl River	—	SW	Columbia, MS	1911	RR	ICG	141.8	87	11
Pearl River	—	SW	Monticello, MS	—	HWY	Lawrence Co.	194.7	96	11.3
Pearl River	—	SW	Wanilla, MS	1906	RR	ICG	205.5	117	13
Pearl River	—	SW	Carthage, MS	1911	HWY	Leake Co.	373.3	55.5	2.3
Pearl River (East)	—	SW	Dunbar, LA	1914	RR	L&N	1	85.8	5.4
Pearl River (East)	—	SW	Pearlington, MS	1934	HWY	Louisiana	10	90	10
Pearl River (West)	—	SW	Pearl River, LA	1906	RR	SOU	25.2	96.3	3.6
Pearl River (West)	—	SW	Pearl River, LA	1926	HWY	Louisiana	25.3	92	3.6
Pee Dee River	—	SW	Georgetown, SC	1935	HWY	South Carolina	1	80	12.7
Pee Dee River	—	SW	Poston, SC	1913	RR	SAL	72.9	60	10
Pee Dee River	—	SW	Pee Dee, SC	—	RR	ACL	107.2	60	13
Pee Dee River	—	SW	Bennetsville, SC	1919	RR	CSX	143.1	60	11
Peekskill Creek	—	RL, F	Peeksville, NY	1914	RR	CSX	0	50	3
Pend Oreille Lake	—	B	Sandpoint, ID	1934	HWY	Bonner Co.	—	79	11.64

*See key of abbreviations at end.

Waterway[a]	Road Name/ Route Carried	Bridge Type[b]	Location	Const. Date	Traffic Type[c]	Owner*	Miles from Mouth of Waterway	Channel Width/Lateral Clearance (feet)	Clearance Height (feet)
Pend Oreille Lake	—	SW, F	Sandpoint, ID	1904	RR	BNSF	2.7	77	15
Pend Oreille River	—	B	Usk, WA	1920	HWY	Pend Ooreille Co.	67	60	6
Pend Oreille River	—	SW	Sandpoint, ID	1906	RR	SI	105	78	24
Penobscot River	Kenduskeag Stream Br.	SW	Bangor, ME	1905	RR	MEC	26.5	40	6.25
Penobscot River	—	B	Bangor, ME	1930	HWY	Maine	26.6	30	8.07
Penobscot River	—	SW	Bangor, ME	—	RR	MEC	30.9	39	20
Pensacola Bay	—	B	Pensacola, FL	1931	HWY	Pensacola Bridge Corp.	11	100	17
Pentwater Channel	—	SW	Pentwater, MI	1926	HWY	Pentwater	0.38	40	7.1
Pequonnock River	SR 1 (Stratford Ave.)	VL	Bridgeport, CT	1976	HWY	Connecticut	0.1	103	68
Pequonnock River	—	B (2) (repl. by below)	Bridgeport, CT	1896	RR	Connecticut DOT	0.3	70	18
Pequonnock River	—	B (2)	Bridgeport, CT	1998	RR	Connecticut DOT	0.3	70	18
Pequonnock River	Congress St.	B	Bridgeport, CT	1922	HWY	Bridgeport	0.4	67	8
Pequonnock River	Washington Ave.	B	Bridgeport, CT	1925	HWY	Bridgeport	0.6	69	4
Pequonnock River	Grand St.	B	Bridgeport, CT	1918	HWY	Bridgeport	0.9	71	13
Perdidio River	—	SW	Seminole, AL	1922	HWY	Hiram Sage	4	40	6
Perkins Cove	—	B	Ogunquit, ME	—	FTBR	Ogunquit Village Corp.	0.2	20	16
Perquimans River	US 17	SW	Hertford, NC	1929	HWY	North Carolina	12	60	3
Perquimans River	—	SW	Hertford, NC	—	RR	NS	13	40	2
Perquimans River	—	B	Belvidere, NC	1908	HWY	North Carolina	20	18.5	2.5
Petaluma River	—	SW	Blackpoint, CA	1911	RR	NWP	0.8	110	7
Petaluma River (creek per COE)	—	B	Green Pt., CA	1917	HWY	California	1	120	9
Petaluma River	—	SW	Petaluma, CA	1904	RR	NWP	12.4	54	4
Petaluma River	D St.	B	Petaluma, CA	1933	HWY	Petaluma	13.7	65	5
Petaluma River (Creek per COE)	Washington St.	B	Petaluma, CA	1914	HWY	Petaluma	14.5	51	11.9

Petit Anse Bayou	—	B	Lee, LA	—	RR	SP	9	33.3	3.9
Petit Caillou, Bayou	—	SW	Chauvin, LA	—	HWY	South Coast Company	24	34.5	1.9
Petit Caillou, Bayou	—	SW	Chauvin, LA	1917	HWY	Terrebonne Pa.	25.75	41	4.4
Petit Caillou, Bayou	—	SW	Chauvin, LA	—	HWY	Community	28.6	32.4	1.1
Petit Caillou, Bayou	—	SW	Chauvin, LA	1937	HWY	Terrebonne Pa.	30.3	40.8	4.8
Petit Caillou, Bayou	—	SW	Bourg, LA	—	HWY	South Coast Company	32.6	32.5	0.7
Petit Caillou, Bayou	—	VL	Bourg, LA	1941	HWY	Louisiana	33.75	45	48.4
Petit–Little Calico Bayou	—	SW	Chauvin, LA	1968	HWY	Terrebonne Pa.	20.2	50	3
Petit–Little Calico Bayou	SR 56	SW	Chauvin, LA	1955	HWY	Terrebonne Pa.	21.5	51	5
Petit–Little Calico Bayou	SR 55	VL	Montegut, LA	1964	HWY	Louisiana	25.7	45	48
Petit–Little Calico Bayou	SR 56	SW	Chauvin, LA	1969	HWY	Terrebonne Pa.	26.6	51	3
Petit–Little Calico Bayou	SR 56	VL	Houma, LA	1974	HWY	Terrebonne Pa.	29.9	45	47
Petit–Little Calico Bayou	SR 24	VL	Presquille Plantation–Bourg, LA	1941	HWY	Louisiana	33.7	45	48
Pierre Pass Bayou	SR 70	SW	Pierre Pass, LA	1967	HWY	Louisiana	1	50	3
Pike Creek	6th Ave.	B	Kenosha, WI	1921	HWY	Kenosha	0.4	80	13
Pike Creek	50th St.	SW	Kenosha, WI	1922	HWY	Kenosha	0.4	75	13
Pike Creek	52nd St.	SW	Kenosha, WI	1899	HWY	Kenosha	0.6	53	7
Piles Creek	—	SW	Bayway, NJ	1896	RR	CR	0.3	29.5	4.9
Pine River 1	—	B	Charlevoix, MI	1949	HWY	Michigan	0.3	90	12
Pine River 1	—	SW	Charlevoix, MI	1892	RR	CSX	1.1	50	6
Pine River 2	Front St.	SW	St. Clair, MI	1899	HWY, RR	Rapid Railway Company	0.059	50	6.5
Pine River 2	—	B	St. Clair, MI	1978	HWY	Michigan	0.06	50	10
Pine River 2	—	SW	Riverside, MI	—	RR	PHD	2	50	6
Piscataqua River	US 1	VL	Portsmouth, NH	1923	HWY	Maine and New Hampshire	3.5	260	150
Piscataqua River	US 1 bypass	VL	Portsmouth, NH	1964	HWY, RR	Maine and New Hampshire	4	200	135

*See key of abbreviations at end.

Waterway[a]	Road Name/ Route Carried	Bridge Type[b]	Location	Const. Date	Traffic Type[c]	Owner*	Miles from Mouth of Waterway	Channel Width/Lateral Clearance (feet)	Clearance Height (feet)
Piscataqua River	—	R	Portsmouth, NH	1964	RR	B&M	4	70	5
Piscataqua River	—	B	Dover, NH	1935	HWY	New Hampshire Toll Bridge Comm.	9.4	40	9.6
Piscataqua River	—	SW	South Newmarket, NH	1926	HWY	New Hampshire	18	50	9.5
Pithiachascootee River	—	HNDL	Port Richie, FL	—	HWY	Pasco Co.	1.25	12	5
Pitt River	—	SW	Cascade, BC, Canada	1914	RR	CP	—	238	—
Plaquemine Bayou	SR 77, 3066	SW	Indian Village, LA	1972	HWY	Louisiana	6.5	65	2
Plaquemine Bayou	—	B	Plaquemine, LA	1933	RR	MP	10.5	110	15
Plaquemine Bayou	SR 1	VL	Plaquemine, LA	1947	HWY	Louisiana	10.5	85	52
Plaquemine Brule Bayou	—	SW	Midland, LA	1961	RR	SP	5.1	54	5
Plaquemine Brule Bayou	SR 3066, 77	SW	Plaquemine, LA	1972	HWY	Louisiana	6.5	65	2
Plaquemine Brule Bayou	—	SW	Midland, LA	1911	RR	SP	8	47	7.1
Plaquemine Brule Bayou	SR 91	P	Estherwood, LA	1975	HWY	Louisiana	8	90	—
Pleasant River	Main St.	SW	Addison, ME	1930	HWY	Maine	4.9	36	5
Plum Creek	Plum Island Tpk.	B	Newbury, MA	1974	HWY	Massachusetts	3.3	40	13
Plum Island River	—	SW	Newburyport, MA	1922	HWY	Essex Co.	1.5	30	7.5
Pocomoke River	—	SW	Pocomoke City, MD	1940	RR	CR	15.2	60	4
Pocomoke River	US 113	B	Pocomoke City, MD	1922	HWY	Maryland	15.6	65	3
Pocomoke River	SR 12	B	Snow Hill, MD	1932	HWY	Maryland	29.9	40	2
Ponchartrain Lake	SR 11, north opening	B	New Orleans, LA	1928	HWY	Louisiana	—	151	13
Ponchartrain Lake	SR 11, south opening	B	New Orleans, LA	1928	HWY	Louisiana	—	107	12
Ponchartrain Lake	North	SW	New Orleans, LA	1909	RR	NS	—	105	2
Ponchartrain Lake	South	SW	New Orleans, LA	1909	RR	NS	—	106	1
Ponchartrain Lake	North Channel Span	B	New Orleans, LA	—	—	Expressway Comm.	—	125	42

Ponchartrain Lake	North Channel Span	B	New Orleans, LA	—	—	Expressway Comm.	—	125	42
Poquonnock River	—	—	Bridgeport, CT	—	—	See Pequonnock River	—	—	—
Port Industrial Waterway–Blair	E. 11th St.	B	Tacoma, WA	1953	HWY	Washington	0.3	150	8
Portage Bayou	—	B	Pass Christian, MS	1948	HWY	Harrison Co.	2	70	11
Portage Lake	—	VL	Houghton–Hancock, MI	1941	HWY	Michigan	10	250	3
Portage Lake	—	SW	Houghton–Hancock, MI	1906	HWY, RR	Michigan	16	118	1
Portage River	Monroe St.	B	Port Huron, OH	1932	HWY	Ohio	1.1	75	3
Portage River	—	B	Port Clinton, OH	1910	RR	NS	2.1	109	9
Portage River	—	SW	Oak Harbor, OH	1903	RR	NS	12	70	8
Portland Harbor (See Fore River)	—	B	Portland, ME	1917	HWY, RR	Cumberland Co.	1.5	100	31
Portland Harbor	—	SW	Portland, ME	1908	HWY, RR	Portland Bridge Dist.	2.8	60	5.6
Portland Harbor, Entrance to Back Cove	—	SW	Portland, ME	1914	RR	GT	0	88	5.6
Portland Harbor, Entrance to Back Cove	Turkey's Br.	SW	Portland, ME	1898	HWY, RR	Portland	0.3	67	5.1
Potato Slough	—	SW	Terminous, CA	1937	HWY	California	0.1	100	5
Potomac River	W. Wilson Br. (I 95)	B	Alexandria, VA	1961	HWY	Federal Highway Administration	103.8	175	50
Potomac River	Long Br.	SW, F	Washington, DC	1904	RR	CR	109.8	104	18
Potomac River	I 95 US1 (Rochambeau Memorial Br.)	B, F	Washington, DC	1950	HWY	Dist. of Columbia	109.9	106	24
Potomac River	—	SW	Washington, DC	1906	HWY	Dist. of Columbia	110	100.7	18.3
Potomac River	US 50 (Arlington Mem Br.)	B, F	Washington, DC	1932	HWY	National Park Service	111	142	30

*See key of abbreviations at end.

Waterway[a]	Road Name/ Route Carried	Bridge Type[b]	Location	Const. Date	Traffic Type[c]	Owner*	Miles from Mouth of Waterway	Channel Width/Lateral Clearance (feet)	Clearance Height (feet)
Powell Bay	—	B	Lake Charles, LA	1936	HWY	Union Sulphur	0.25	36	0.7
Powell Creek	—	B	Oceanside, NY	1927	RR	LIRR	0.1	20	1
Powwow River	—	B	Amesbury, MA	1916	RR	Merrimac Power and Bldg. Comm.	1.5	11	4
Presumpscot River	US 1	B	Portland, ME	1941	HWY	Maine	0	75	12
Providence River	Point St.	SW	Providence, RI	1926	HWY	Providence	2.3	101	9
Pungo Creek	—	SW	Bellhaven, NC	—	HWY	North Carolina	3	39	3
Puyallup River	SR 509 (11th St.)	VL (not required to open)	Tacoma, WA	1914	HWY	Washington	0.8	150	29
Puyallup River	—	SW	Tacoma, WA	1909	RR	MILW	0.9	120	12
Puyallup River	—	SW	Tacoma, WA	1906	RR	NP	2.1	96	8
Quantuck Canal	West Bay Br.	B	Potunk Point, NY	1928	HWY	Suffolk Co.	0.1	50	11
Quantuck Canal	Beach Ln.	B	West Hampton, NY	1936	HWY	Suffolk Co.	1.1	50	14
Quincy Bay	—	SW	Quincy, IL	1897	RR	CBQ	0	160	4.6
Quincy Bay	—	SW	Quincy, IL	1868	RR	BNSF	0.7	80	11
Quinnipiac River	Tomlinson Br.	B (being replaced with VL)	New Haven, CT	1925	HWY	Connecticut	0	117	11
Quinnipiac River	Ferry St.	B	New Haven, CT	1940	HWY	New Haven	0.7	101	25
Qsuinnipiac River	Grand Ave.	SW	New Haven, CT	1898	HWY	New Haven	1.3	70	9
Quogue Canal	Post Ln.	B	Quogue, NY	1940	HWY	Suffolk Co.	1.1	50	15
Quogue Canal	—	B	Quantuck Bay, NY	1936	HWY	Suffolk Co.	2.71	50.4	10.5
Quogue Canal	—	B	Potunk Point, NY	1928	HWY	Suffolk Co.	3.67	50.1	11
Raccoon Creek 1	US 130	VL	Bridgeport, NJ	1940	HWY	New Jersey	1.8	65	64
Raccoon Creek 1	—	SW	Bridgeport, NJ	1924	RR	CR	2	38	7
Raccoon Creek 1	—	SW	Swedesboro, NJ	1913	HWY	Gloucester Co.	8.3	50	6
Raccoon Creek 2	—	SW	Hertford, NC	—	HWY	—	1.1	—	—
Rahway River	—	SW	Carteret, NJ	1897	RR	CNJ	0.18	50.7	4.27

Rahway River	—	B	Linden, NJ	1921	RR	CR	1.95	65	6
Rahway River	Lawrence St.	SW, F	Rahway, NJ	1911	HWY	Union Co.	4.43	60	6.2
Rahway River	Milton Ave.	SW	Rahway, NJ	1922	HWY	Union Co.	5.09	43.7	4.9
Rainy Lake	—	B	Ft. Frances, ON, Canada	1915	RR	CN	—	54.5	—
Rainy Lake	—	B	Ft. Frances, ON, Canada	1914	RR	CN	—	93	—
Rainy River	—	SW	Baudette, MN	1901	RR	MO	14	160	14
Rainy River	—	B	Rainy, ON, Canada	1908	RR	CN	—	134	—
Rainy River	—	B	Ranier, MN	1908	RR	Rainy River Bridge Auth.	85	125	6
Rancocas Creek	SR 543	SW	Riverside, NJ	1935	HWY	Burlington Co.	1.3	50	4
Rancocas Creek	—	SW	Delanco, NJ	1909	RR	CR	1.6	42	3
Rancocas Creek	US 130	B (repl. w. fixed)	Bridgeboro, NJ	1928	HWY	New Jersey	3.3	60	8
Rancocas Creek	—	SW	Centerton, NJ	—	HWY	Burlington Co.	7.8	48	6
Rancocas Creek, South Branch	SR 537	SW, F	Hainesport, NJ	1932	HWY	Burlington Co.	10.7	45	5
Rantowles Creek	—	VL	Rantowles, SC	1925	RR	ACL	1.1	32	2.8
Rappahannock River	—	SW	Tappahannock, VA	1927	HWY	Virginia	42.5	100.7	9.9
Rappahannock River	—	SW	Port Royal, VA	1934	HWY	James Madison Memorial Bridge	79.5	100	8.1
Raritan River	—	SW	Perth Amboy, NJ	1908	RR	New Jersey Transit	0.5	132	8
Raritan River	Victory Br.	SW	Perth Amboy, NJ	1926	HWY	New Jersey	1.6	140	28
Red Pass	—	B	Venice, LA	1949	HWY	Tidewater Associated Oil	5.6	14	2
Red River	—	SW	Moncla, LA	1934	HWY	Louisiana	63	130	6
Red River	SR 107, 115	VL	Moncla, LA	1950	HWY	Louisiana	66	330	50
Red River	—	SW	Alexandria, LA	1902	RR	L&A	102.8	133	2
Red River	SR 28	VL	Alexandria, LA	1962	HWY	Louisiana	102.9	275	50
Red River	US 165	SW	Alexandria, LA	1900	HWY	Louisiana	103.2	140	1
Red River	—	SW	Alexandria, LA	1890	RR	UP	104.9	160	4

*See key of abbreviations at end.

Waterway[a]	Road Name/ Route Carried	Bridge Type[b]	Location	Const. Date	Traffic Type[c]	Owner*	Miles from Mouth of Waterway	Channel Width/Lateral Clearance (feet)	Clearance Height (feet)
Red River	SR 8	SW	Boyce, LA	1948	HWY	Louisiana	123.9	142	5
Red River	—	SW	Grand Ecore, LA	1900	HWY	Louisiana	186.5	160.7	6
Red River	US 71, 84	SW	Coushatta, LA	1932	HWY	Louisiana	217.6	130	10
Red River	—	SW	Shreveport, LA	1969	RR	BNSF	275.9	110	15
Red River	—	SW	Shreveport, LA	1915	RR	ICG	277.1	120	5
Red River	—	SW	Shreveport, LA	1907	RR	SLSW	286.7	120	6.1
Red River	—	SW, F	Garland City, AR	1929	RR	BNSF	372.2	134	2
Red River	—	SW	Fulton, AR	1879	RR	MP	405.2	133	6.3
Red River	—	VL	Index, AR	1921	HWY	Arkansas	429.7	191	11.3
Red River	—	SW, F	Index, AR	1899	RR	KCS	429.9	134	20
Red River of the North	—	VL	Oslo, MN	1912	HWY	Oslo	265.8	150	37.3
Red River of the North	—	SW	Oslo, MN	1905	RR	CP	266	98	6
Red River of the North	—	SW	Grand Forks, ND	1915	RR	NP	292.5	112	1
Red River Old River	—	SW	Torras, LA	1902	RR	Louisiana	2.5	160	3.3
Reserved Channel	L St. (Summer St.)	R	Boston, MA	1930	HWY	Boston	0.9	39	6
Reynolds Channel	—	B	Atlantic Beach, NY	1952	HWY	Nassau Co.	0.4	125	25
Reynolds Channel	—	SW (repl. by below)	Island Park, NY	1913	RR	LIRR	4.4	40	3
Reynolds Channel	—	B	Island Park, NY	1987	RR	LIRR	4.4	100	7.8
Reynolds Channel (Long Beach Channel per COE)	—	B	Wreck Lead, NY	1922	HWY	Nassau Co.	4.6	—	—
Reynolds Channel	—	B	Long Beach, NY	1956	HWY	Nassau Co.	4.7	100	20
Ribault River	—	SW	Jacksonville, FL	1913	HWY	Florida	0.84	43	6
Ribault River	—	B	Jacksonville, FL	1927	HWY	Duval Co.	3.8	25	5.3
Ribault River	Moncrief Rd.	B	Jacksonville, FL	1977	HWY	Duval Co.	3.9	62	8
Rice Creek	—	SW	Palatka, FL	1930	RR	CSX	0.8	30	2
Rice Creek	—	SW	Palatka, FL	1930	HWY	Florida	1	40	1.4

Richardson Bay	—	VL	Sausalito, CA	1931	HWY	California	3.2	40	57
Richardsons Creek	—	SW	Thunderbolt, GA	1926	HWY	Chatham Co.	3.5	27	5.5
Richelieu River	—	SW	St. Hyacinthe, QC, Canada	1908	RR	CN	—	145.5	—
Richelieu River	—	SW	St. Hyacinthe, QC, Canada	1912	RR	CN	—	244.61	—
Richmond Creek	Richmond Ave.	B, F	Staten Island, NY	1931	HWY	New York City	2	60	9
Rigolets Pass	—	SW	Dunbar, LA	1925	RR	CSX	0	153	11
Rigolets Pass	Fort Pike	SW	New Orleans, LA	1930	HWY	Louisiana	6.2	152	14
Rio Grande	—	SW	Brownsville, TX	—	HWY, RR	MP Brownsville & Matamoras	61.5	96	4.9
Risleys Channel	—	B	Longport, NJ	1916	HWY	Atlantic Co.	1.1	50	9
Roanoke River	—	SW	Williamston, NC	1922	HWY	North Carolina	27.5	80	4.5
Roanoke River	US 17	SW	Williamston, NC	1947	HWY	North Carolina	37.5	83	4
Roanoke River	—	SW	Palmyra, NC	—	RR	CSX	94	85	8
Roanoke Sound	US 64	SW, F	Manteo, NC	1949	HWY	North Carolina	2.8	50	9
Root River	Main St.	B	Racine, WI	1927	HWY	Racine	0.3	90	7
Root River	State St.	B	Racine, WI	1922	HWY	Racine	0.5	80	16
Root River	4th St.	SW	Racine, WI	—	HWY	Racine	0.7	44	17.1
Root River	St. Clair St.	SW	Racine, WI	1872	RR	CMSTP&P	0.93	50	3.6
Root River	8th–9th Sts.	SW	Racine, WI	1909	RR	CNW	1.7	107	13
Rouge River	—	B	Detroit, MI	1913	RR	Solvay Process Co.	0.3	120	3.1
Rouge River, Old Channel	—	B	Detroit, MI	1913	HWY, RR	DC	0.3	120	3
Rouge River	—	B	Detroit, MI	1913	HWY, RR	Solvay Process Co.	0.36	120	3.1
Rouge River, Old Channel	—	B	Detroit, MI	—	RR	DC	0.36	120	3
Rouge River	—	B	Detroit, MI	1922	RR	DTI	0.4	125	2.5
Rouge River, Old Channel	—	SW	Detroit, MI	1913	RR	DTI	0.8	102	3
Rouge River, Short Cut Channel	—	B	Detroit, MI	1922	RR	Great Lakes Steel Corp.	0.4	125	2

*See key of abbreviations at end.

Waterway[a]	Road Name/ Route Carried	Bridge Type[b]	Location	Const. Date	Traffic Type[c]	Owner*	Miles from Mouth of Waterway	Channel Width/Lateral Clearance (feet)	Clearance Height (feet)
Rouge River, Short Cut Channel	W. Jefferson Ave.	B	Detroit, MI	1923	HWY	Wayne Co.	1.1	125	5
Rouge River, Short Cut Channel	—	B	Detroit, MI	1921	RR	CR	1.5	123	4
Rouge River, Short Cut Channel	—	B	Detroit, MI	1923	RR	NS	1.9	125	4
Rouge River, Short Cut Channel	Fort St.	B	Detroit, MI	1923	HWY	Michigan	2.2	118	5
Rouge River, Short Cut Channel	Dix Ave.	B	Detroit, MI	1924	HWY	Wayne Co.	2.8	125	5
Sabine Lake	—	SW	Port Arthur, TX	1957	HWY	Jefferson Co., TX and Cameron Pa., LA	10	80	9
Sabine–Neches Canal	—	B	Port Arthur, TX	1931	HWY	Port Arthur	3.5	200	11
Sabine River	—	P	Orange, TX	1970	HWY	Levingston Shipbuilding	9.5	350	—
Sabine River	—	SW	Orange, TX	1927	HWY	Texas and Louisiana	10.5	125	12.8
Sabine River	—	SW	Echo, TX	1955	RR	UP	19.3	92	6
Sabine River	—	SW	Ruliff, TX	1937	RR	KCS	36.2	60	4
Sabine River	Tx SR 235, LA SR 7	SW	Deweyville, TX	1938	HWY	Texas and Louisiana	40.8	60	6
Sabine River	—	SW	Merryville, LA	1906	RR	BNSF	96.2	—	27
Sacramento River	SR 12	VL	Rio Vista, CA	1960	HWY	California	12.8	270	146
Sacramento River	SR 160	B	Isleton, CA	1923	HWY	California	18.7	200	15
Sacramento River	SR D 13	B	Walnut Grove, CA	1952	HWY	Sacramento Co.	26.7	200	21
Sacramento River	SR 160	B	Paintersville, CA	1926	HWY	California	33.4	200	24

Sacramento River	SR E 9	B	Freeport, CA	1930	HWY	Sacramento and Yolo Cos.	46	200	29
Sacramento River	Tower Bridge	VL	Sacramento, CA	1935	HWY	California	59	170	131
Sacramento River	I Street	SW	Sacramento, CA	1912	RR, HWY	UP	59.4	148	34
Sacramento River	I 880	VL	Bryte, CA	1971	HWY	California	62.7	250	106
Sacramento River	—	SW	Knights Landing, CA	1904	RR	SP	89.6	105	−1.8
Sacramento River	SR 113	B	Knights Landing, CA	1933	HWY	California	90.1	200	3
Sacramento River	—	SW	Meridian, CA	1913	RR, HWY	SN	134	150	5
Sacramento River	SR 20	SW	Meridian, CA	1913	HWY, RR	California	135.5	147	5
Sacramento River	—	RSP	Colusa, CA	1980	HWY	Colusa Co.	143.5	100	1
Sacramento River	SR 162	SW	Butte City, CA	1949	HWY	California	169.7	105	8
Sacramento River	SR 32	SW	Hamilton City, CA	1910	HWY	California	203.5	100	7
Sacramento River	—	SW	Squaw Hill, CA	1922	HWY	Tehama Co.	218.7	115	6.5
Sacramento River	—	VL	Tehama, CA	1912	HWY	Tehama Co.	230.1	160	6
Sacramento River	—	SW, F	Tehama, CA	1898	RR	UP	232.5	100	1
Sacramento River Barge Canal	Jefferson Blvd.	B	West Sacramento, CA	1961	HWY, RR	U.S. Army COE	44.2	86	17
Saginaw River	—	SW	Bay City, MI	1896	RR	DM	3.1	96	7
Saginaw River	Belinda St.	B	Bay City, MI	1976	HWY	Bay City	3.9	150	30
Saginaw River	—	SW	Bay City, MI	—	RR	PC	5	101	4
Saginaw River	3rd St.	SW	Bay City, MI	—	HWY	Bay City	5.29	85	14
Saginaw River	Veteran Memorial	B	Bay City, MI	1953	HWY	Michigan	5.6	146	19
Saginaw River (West Channel)	13th St.	SW	Bay City, MI	1912	RR	Bay City Terminal (GT)	6.21	100	6
Saginaw River	Lafayette Ave.	B	Bay City, MI	1938	HWY	Bay City	6.8	150	16
Saginaw River	Cass Ave.	SW	Bay City, MI	—	HWY	Bay City	7.19	73	5
Saginaw River	Cass Ave.	P	Bay City, MI	1939	HWY	Bay City	7.8	104	—
Saginaw River	I 75	B	Zilwaukee, MI	1960	HWY	Michigan	14.7	150	26
Saginaw River	6th Ave.	SW	Saginaw, MI	1905	HWY	Michigan	17.1	72	8
Saginaw River	Bayou West	SW	Saginaw, MI	—	RR	PM	17.41	53	6.5
Saginaw River	Johnson St.	B	Saginaw, MI	1912	HWY	Saginaw	17.88	90	10.5

*See key of abbreviations at end.

Waterway[a]	Road Name/ Route Carried	Bridge Type[b]	Location	Const. Date	Traffic Type[c]	Owner*	Miles from Mouth of Waterway	Channel Width/Lateral Clearance (feet)	Clearance Height (feet)
Saginaw River	—	B	Saginaw, MI	1944	RR	CSX	18	150	9
Saginaw River	Thompson St.	SW	Saginaw, MI	1917	RR	GT	18.4	72	7.9
Saginaw River	Genessee Ave.	B, F	Saginaw, MI	1939	HWY	Michigan	18.7	88	21
Saginaw River	Bristol St.	SW	Saginaw, MI	1894	HWY	Saginaw	19.03	66	10.3
Saginaw River	Emerson St.	SW	Saginaw, MI	1913	RR	PC	19.2	70	8
Saginaw River	Mackinaw St.	SW	Saginaw, MI	—	HWY	Saginaw	19.97	64	6.8
Saginaw River	Center St.	SW	Saginaw, MI	1904	HWY	Saginaw	20.74	74	12.5
Saint Andrews Bay	—	B	Lynn Haven, FL	1925	HWY	Bay Co.	5	32	2.7
Saint Andrews Bay	—	SW	Panama City, FL	1928	HWY	Florida	—	90	8
Saint Andrews Bay	—	SW	Panama City, FL	1928	HWY	Florida	—	75	8
Saint Croix River	—	SW (repl. w. VL)	Prescott, WI	1887	RR	BNSF	0.2	150	8
Saint Croix River	US 61	VL	Prescott, WI	1923	HWY	Wisconsin and Minnesota	0.3	160	47
Saint Croix River	—	SW	Hudson, WI	1922	RR	UP	17.3	132	0
Saint Croix River	SR 36	VL	Stillwater, MN	1931	HWY	Minnesota and Wisconsin	23.4	135	43
Saint Croix River	—	SW	Otisville, MN	1887	RR	SOO	40.7	102	5
Saint Croix River	—	SW	Osceola, WI	1895	HWY	Osceola	45.4	116	6
Saint Francis River	—	VL	Cody, AR	1913	RR	MP	29.6	156	63.2
Saint Francis River	US 79	VL	Cody, AR	1929	HWY	Arkansas	29.7	159	63
Saint Francis River	—	SW, F	Madison, AR	1853	RR	RI	59.7	78	1.8
Saint Francis River	—	SW	Madison, AR	1909	HWY	St. Francis Co.	60.1	90	−0.3
Saint Francis River	—	SW	Madison, AR	1933	HWY	Arkansas	61.9	90	5
Saint Francis River	—	SW	Parkin, AR	1917	HWY	Cross County Improvement Dist.	100.8	110	1.2
Saint Francis River	—	SW	Marked Tree, AR	1928	HWY	Arkansas	148.4	68	−3.6
Saint Francis River	—	SW	Marked Tree, AR	—	RR	SLSF	151.4	60	3.9

Saint Francis River	—	SW	Lunsford, AR	1930	RR	SLSW	168.3	39	6.6
Saint Francis River	—	RSP	Lake City, AR	1931	RR	SLSF	173.5	24	1.3
Saint Francis River	—	VL	Lake City, AR	1934	HWY	Arkansas	173.6	24	1.4
Saint Francis River	—	RSP	Cardwell, MO	1927	HWY	Missouri & Arkansas	191.1	41.6	1
Saint Francis River	—	SW	Bertig, AR	—	RR	SLSW	193.1	34	−2.9
Saint Francis River	—	RSP	Kennett, MO	1930	HWY	Missouri	209.6	37	7.3
Saint Francis River	—	B	West Kennett, AR	1910	RR	SLSF	211.2	28.5	−2.4
Saint Francis River	—	VL	—	1930	RR	SLSW	229	66	3.1
Saint George River	Wadsworth St.	B, F	Thomaston, ME	1928	HWY	Maine	12	42	5
Saint George River	—	SW	Thomaston, ME	1899	RR	MEC	12	28	25.5
Saint Joe River	—	SW	Chatcolet, ID	1917	RR	UP	0.7	94	10
Saint Johns River	Main St.	VL	Jacksonville, FL	1967	HWY	Florida	24.7	350	135
Saint Johns River	SR 13 (Acosta Br.)	VL (repl. w. fixed)	Jacksonville, FL	1921	HWY	Florida	24.9	174	164
Saint Johns River	—	B	Jacksonville, FL	1926	RR	FEC	24.9	195	5
Saint Johns River	I 95 Gilmorest	B	Jacksonville, FL	1954	HWY	Florida	25.4	173	44
Saint Johns River	SR 16 (Shands Br.)	RSP, F	Green Cove Springs, FL	1961	HWY	Florida	50.2	91	45
Saint Johns River	—	B	Green Cove Springs, FL	1939	HWY	Florida	50.5	100	11
Saint Johns River	—	SW	Palatka, FL	1931	RR	FEC	83	101	3.2
Saint Johns River	SR 15 (Reid St.)	B	Palatka, FL	1928	HWY	Florida	83.1	100	21
Saint Johns River	—	B	Buffalo Bluff, FL	1962	RR	CSX	94.5	90	7
Saint Johns River	SR 40	SW	Astor, FL	1926	HWY	Florida	126	90	4
Saint Johns River	SR 40	B	Astor, FL	1981	HWY	Florida	126	114	21
Saint Johns River	SR 44	B	Crows Bluff, FL	1955	HWY	Florida	146	90	15
Saint Johns River	—	SW	Crows Bluff, FL	1917	HWY	Florida	146	92.3	7
Saint Johns River	—	B	Sanford, FL	1961	RR	CSX	160.9	91	8
Saint Johns River	—	SW	Lake Monroe, FL	1893	RR	ACL	160.95	91	7
Saint Johns River	US 17	SW	Sanford, FL (Lake Monroe)	1934	HWY	Florida	161	90	14
Saint Johns River	—	SW	Sanford, FL	1923	HWY	Florida	178.2	90	4.5
Saint Johns River	—	SW	Geneva Ferry, FL	1912	HWY	Florida	181.5	47.8	2

*See key of abbreviations at end.

Waterway[a]	Road Name/ Route Carried	Bridge Type[b]	Location	Const. Date	Traffic Type[c]	Owner*	Miles from Mouth of Waterway	Channel Width/Lateral Clearance (feet)	Clearance Height (feet)
Saint Johns River	—	B	Cooks Ferry, FL	1912	RR	FEC	194.1	50	1.2
Saint Jones River	—	B	Barkers Landing, DE	1934	HWY	Delaware	4.5	50	5
Saint Jones River	—	SW	Lebanon, DE	—	HWY	Kent Co.	8	30	6
Saint Joseph River	—	SW	St. Joseph, MI	1893	RR	CSX	0.7	100	8
Saint Joseph River	State St.	SW	St. Joseph, MI	1909	HWY	St. Joseph	0.8	100	11.2
Saint Joseph River	US 33	B	St. Joseph, MI	1949	HWY	Michigan	0.9	150	33
Saint Joseph River	—	SW	St. Joseph, MI	1910	HWY	St. Joseph	1	60	22
Saint Joseph River	Main St.	SW	Benton Harbor, MI	1909	HWY	St. Joseph	1.3	74	7
Saint Joseph River	—	SW	Benton Harbor, MI	1901	RR	MC	1.6	58	4.7
Saint Lawrence River	Victoria Br.	VL	St. Hyacinthe, QC, Canada	1957	RR	CN	—	107	—
Saint Lawrence River	Victoria Br.	VL	St. Hyacinthe, QC, Canada	1957	RR	CN	—	96	—
Saint Lawrence Seaway	—	VL	Adirondack, QC Canada	1958	RR, HWY	Saint Lawrence Seaway Auth.	—	322	—
Saint Louis Bay	—	SW	Bay St. Louis, MS	1967	RR	CSX	0.5	100	13
Saint Louis Bay	—	SW	Bay St. Louis, MS	1926	HWY	Harrison Co. & Hancock Co.	0.75	86	13.3
Saint Louis Bay	US 90	B	Bay St. Louis, MS	1953	HWY	Mississippi	1	100	17
Saint Louis Canal	—	RSP	Bourg, LA	1975	HWY	Louisiana	9	35	1
Saint Louis River	—	SW (out)	Duluth, MN	1908	RR, HWY	BN	5.3	200	—
Saint Louis River	—	SW	Duluth, MN	1909	RR	BNSF	5.7	175	13
Saint Louis River	—	SW	Duluth, MN	1909	RR	NP	5.72	175	11.1
Saint Louis River	—	SW	Duluth, MN	1912	RR	BNSF	8.8	175	9
Saint Louis River	US 2	B	Duluth, MN	1927	HWY	Arrowhead Bridge Comm.	8.9	211	21
Saint Louis River	SR 39	SW	Duluth, MN	1911	RR, HWY	INT-MN	16.3	125	19
Saint Lucie River	—	B	Stuart, FL	1934	HWY	Florida	6.5	80	9.8
Saint Lucie River	—	B	Stuart, FL	1925	RR	FEC	6.5	50	6.6

Saint Lucie River, South Fork	—	B	Palm City, FL	1929	HWY	Martin Co.	8.5	59	12.2
Saint Marks River	US 98	B	Newport, FL	1954	HWY	Florida	9	40	9
Saint Marys Falls Canal	S. Channel	VL	Sault Ste. Marie, MI	1959	RR	Soo Bridge Comm.	1	316	12
Saint Marys Falls Canal	—	B	Sault Ste. Marie, MI	1913	RR	WC	1	282	9
Saint Marys River	US 17	SW	Kingsland, GA	1927	HWY	Florida	23	75	5
Saint Marys River	—	SW	Kingsland, GA	1982	RR	CSX	23.1	57	5
Sakonnet River	—	B	Tiverton, RI	1906	HWY	Rhode Island	11.5	100	8.9
Sakonnet River	—	SW	Tiverton, RI	1900	RR	NYNH&H	12.37	99	12
Salem Canal	—	B	Deep Water Pt., NJ	1915	HWY	Salem Co.	1	39.3	8
Salem Canal	—	F, convertible to VL	Deep Water Pt., NJ	1930	HWY	Salem Co.	1.4	60	6
Salem River	SR 49	B	Salem, NJ	1928	HWY	Salem Co.	3.5	60	5
Saline River	—	SW	Godfrey, AR	1905	RR	CRI&P	15.3	110	0
Saline River	—	SW	Suttons Ferry, AR	1919	HWY	Bradley Co.	58	83	6
Saline River	—	SW	Warren, AR	1939	RR	MP	64.8	59	4
Salt Bayou	—	B	Slidell, LA	1929	HWY	Louisiana	1.5	36.9	2
Sammamish River, South Channel	—	P	Lake Washington, WA	—	FTBR	Barron Properties	—	—	—
Sammamish River, South Channel	—	P	Lake Washington, WA	—	FTBR	Barron Properties	—	—	—
Sampit River	—	SW	Georgetown, SC	1926	HWY	South Carolina	2	60	8.14
San Bernard River	—	SW	Church-Hill, TX	1876	HWY	Brazoria Co.	10.4	55	1.5
San Bernard River	—	SW	Hinkles Ferry, TX	1912	HWY	Brazoria Co.	13.8	55	0.5
San Bernard River	—	SW	Brazoria, TX	1941	HWY	Brazoria Co.	17.9	62	0.4
San Bernardo River	—	SW	Brazoria, TX	1910	RR	UP	20.7	50	2
San Francisco Bay	—	VL	San Mateo, CA	1929	HWY	San Francisco Bay Toll Bridge Co.	22.7	270	135
San Francisco Bay	SR 84 (Ravenswood Pt.)	VL	Palo Alto, CA	1927	HWY	California	32	200	13.5

*See key of abbreviations at end.

Waterway[a]	Road Name/ Route Carried	Bridge Type[b]	Location	Const. Date	Traffic Type[c]	Owner*	Miles from Mouth of Waterway	Channel Width/Lateral Clearance (feet)	Clearance Height (feet)
San Francisco Bay	Dumbarton Pt.	SW	Palo Alto, CA	1909	RR	SP	32.5	125	13
San Joaquin River	—	VL	Antioch, CA	1926	HWY	California	7.6	265	135
San Joaquin River	Jacobs Road No. 1	SW	Stockton, CA	1910	HWY	San Joaquin Co.	39.6	111.5	1.5
San Joaquin River	—	SW	Stockton, CA	1933	RR	Port of Stockton	39.7	100	3
San Joaquin River	Navy Dr.	SW	Stockton, CA	1940	HWY	U.S. Navy	39.8	100	3
San Joaquin River	—	B	Stockton, CA	1930	RR	BNSF	40.6	98	8
San Joaquin River	SR 4	SW	Stockton, CA	1934	HWY	California	41.6	102	13
San Joaquin River	—	SW	French Camp, CA	1902	HWY	San Joaquin Co.	47.3	103	—
San Joaquin River	—	VL, F	Lathrop, CA	1942	RR	UP	56.1	100	22
San Joaquin River	Manthey Rd.	B, F	Lathrop, CA	1925	HWY	San Joaquin Co.	56.2	115	25
San Joaquin River	SR 120	B, F	Lathrop, CA	1949	HWY	California	56.2	111	25
San Joaquin River	—	SW, F	Lathrop, CA	1928	RR	WP	56.7	69	24
San Joaquin River	—	B	San Joaquin City, CA	1902	HWY	San Joaquin Co.	73.7	120	1.3
San Joaquin River	—	B	Grayson, CA	1893	HWY	Stanislaus Co.	93.4	130	2.9
San Joaquin River	—	RSP	Crows Landing, CA	1949	HWY	Stanislaus Co.	107.2	80	5
San Joaquin River	—	RSP	Patterson, CA	1937	HWY	Stanislaus Co.	108	20	3
San Joaquin River	—	B	Crows Landing, CA	—	HWY	Stanislaus Co.	114.5	110	8
San Joaquin River	—	B	Hills Ferry, CA	1901	HWY	Stanislaus Co.	125.4	100	3.7
San Joaquin River	—	B	Dos Palos, CA	1905	HWY	Merced Co.	181.1	140	5.1
San Joaquin River	—	B	Firebaugh, CA	1885	HWY	Fresno Co.	202.1	70	0.6
San Joaquin River, Burns Cutoff	Jacobs Rd.	SW	Rough and Ready Island, CA	1910	HWY	San Joaquin Co.	1.5	75	—
San Joaquin River, Burns Cutoff	Upper Highway	SW	Rough and Ready Island, CA	1903	HWY	San Joaquin Co.	3	67.5	6.2
San Leandro Bay	—	B	Alameda, CA	1953	HWY	California	0	92	20
San Lorenzo Creek	—	RSP	San Leandro, CA	1984	FTBR	East Bay Regional Park Dist.	0.1	36	7

Sandusky Bay	—	B	Sandusky, OH	1943	RR	NS	7.7	64	5
Sandusky Bay	SR 269	VL	Sandusky, OH	1929	HWY	Ohio	8.2	65	7
Santa Rosa Sound	—	SW	Pensacola, FL	1931	HWY	Escambia Co. & Santa Rosa Co.	2	100	15
Santee River	—	SW	McClellandville, SC	1929	HWY	South Carolina	14	18.75	2.8
Santee River	—	SW	Jamestown, SC	1915	RR	SAL	38	66	1.17
Santee River	—	SW	St. Stephens, SC	1914	RR	ACL	60	45	12.4
Santee River	—	SW	Lanes, SC	1923	HWY	South Carolina	65	66	6
Santee River	—	SW	Rimini, SC	1895	RR	ACL	133	75	15
Santee River	—	SW	Ft. Motte, SC	1916	RR	NS	151	90	0
Sara Bayou	—	SW	Saraland, AL	1927	RR	CSX	0.1	64	3
Sarasota Bay	—	B	Sarasota, FL	1927	HWY	Sarasota Co.	—	55	9.9
Sarasota Bay	—	B	Sarasota, FL	1926	HWY	John Ringling	—	60	8
Sarasota Bay	—	SW	Cortes, FL	1921	HWY	Manatee Co.	—	60	8
Sassafras River	SR 213	B	Georgetown, MD	1918	HWY	Maryland	10	40	4
Satilla River	—	SW	Woodbine, GA	1927	HWY	Georgia	25.7	80	4
Satilla River	—	SW	Woodbine, GA	1945	RR	SCL	25.7	50	5
Satilla River	—	B	Burnt Fort, GA	1929	HWY	Charlton Co.	51.5	80	13.5
Saugatuck River	Saga	B	Saugatuck, CT	1904	RR	Connecticut DOT	1.1	57	13
Saugatuck River	Bridge St.	SW (repl. w. new)	Saugatuck, CT	1884	HWY	Westport	1.3	54	6
Saugus River	SR 1A (Gen. Edwards Br.)	B	Lynn, MA	1936	HWY	Metropolitan Dist. Comm.	1.7	100	27
Saugus River	—	B	Lynn, MA	1912	RR	B&M	2.1	50	7
Saugus River	SR 107 (Foxhall Br.)	B	Saugus, MA	1913	HWY	Massachusetts	2.5	40	6
Savannah River	—	B	Savannah, GA	1908	RR	SAL	13.3	116	8
Savannah River	—	SW	Savannah, GA	1925	HWY	Georgia	18.5	102	1.6
Savannah River	US 17	SW	Port Wentworth, GA	1954	HWY	Georgia	21.6	90	7
Savannah River	—	SW	Hardeeville, GA	1909	RR	ACL	24.4	68	−3.3
Savannah River	—	B	Hardeeville, GA	1969	RR	CSX	27.4	90	7
Savannah River	—	SW	Clyo, GA	1912	RR	SCL	60.9	103	15
Savannah River	US 301	SW	Sylvania, GA	1938	HWY	South Carolina	118.7	100	15

*See key of abbreviations at end.

Waterway[a]	Road Name/ Route Carried	Bridge Type[b]	Location	Const. Date	Traffic Type[c]	Owner*	Miles from Mouth of Waterway	Channel Width/Lateral Clearance (feet)	Clearance Height (feet)
Savannah River	—	B	Augusta, GA	1932	RR	CSX	195.4	100	13
Savannah River	5th St.	SW	Augusta, GA	1932	HWY	Georgia and South Carolina	199.6	75	25
Savannah River	6th St.	B	Augusta, GA	1912	RR	NS	199.9	82	12
Scholfield River	—	R	Reedsport, OR	1929	HWY	Oregon	1	37	16
Schuylkill River	Penrose Ave.	SW (repl. w. fixed)	Philadelphia, PA	1878	HWY	Philadelphia	1.3	183	18.2
Schuylkill River	Passyunk Ave.	B	Philadelphia, PA	1911	HWY	Philadelphia	3.5	200	33
Schuylkill River	Passyunk Ave.	B	Philadelphia, PA	1980	HWY	Philadelphia	3.5	200	50
Schuylkill River	Tasker St.	SW (out)	Philadelphia, PA	1909	RR	CSX	5.1	57	15
Schuylkill River	Grays Ferry	SW	Philadelphia, PA	1902	RR	CR	5.5	65	22
Schuylkill River	Grays Ferry	SW	Philadelphia, PA	1901	HWY	Philadelphia	5.6	75	22.4
Schuylkill River	University Ave.	B	Philadelphia, PA	1929	HWY	Philadelphia	6.2	100	32
Schuylkill River	Christian St.	SW	Philadelphia, PA	1904	RR	CR	6.4	67	26
Schuylkill River	South St.	F, B	Philadelphia, PA	1922	HWY	Philadelphia	6.7	100	36
Scott Creek	—	SW	Portsmouth, VA	1909	RR	ACL	0.5	35	2.9
Scuppernong River	US 64	SW	Columbia, NC	1927	HWY	North Carolina	4.5	41	2
Scuppernong River	US 64	RSP, F	Columbia, NC	1958	HWY	North Carolina	4.6	35	9
Scuppernong River	—	SW	Columbia, NC	1906	RR	NS	5	39.5	5
Scuppernong River	—	SW	Cross Landing Branch, NC	1936	HWY	North Carolina	13.5	42	1.4
Scuppernong River	Spruills Br.	RSP, F	Creswell, NC	1925	HWY	North Carolina	17.5	32	4
Second Narrows	—	VL	Thornton Branch, Canada	1967	RR	CN	—	503	—
Second River	—	SW	Arlington, NJ	—	RR	CR	—	45	36
Seeconk River	India Pt. Br.	SW	Providence, RI	1902	RR	Providence	0.4	84	4
Seeconk River	I 195	B, F	Providence, RI	1928	HWY	Providence	0.6	100	40
Seekonk River	Fox Pt. Blvd.	B	Providence, RI	1930	HWY, TROL	Rhode Island	0.82	100	40.7

Seeconk River	Tunnel Br.	B	Providence, RI	1906	RR	CR	1	92	17
Seekonk River	Waterman St.	SW	Providence, RI	1895	HWY, TROL	Providence	1.56	69.7	13
Severn River	SR 450	B	Annapolis, MD	1925	HWY	Maryland	3	75	12
Severn River	—	SW	Annapolis, MD	1925	RR	BLA	3.6	62	5
Shark River	Ocean Ave.	B	Belmar, NJ	1937	HWY	Monmouth Co.	0.1	90	15
Shark River, South Channel	SR 71	B	Belmar, NJ	1934	HWY	New Jersey	0.8	50	13
Shark River, South Channel	—	B	Belmar, NJ	1937	RR	New Jersey Transit	0.9	50	10
Shark River, South Channel	SR 35	B	Belmar, NJ	1928	HWY	New Jersey	0.92	50	10
Shaws Cove	—	SW	New London, CT	1983	RR	ATK	0	45	3
Sheboygan River	S. 8th Ave.	B	Sheboygan, WI	1995	HWY	Sheboygan	0.7	99	6
Sheboygan River	Pennsylvania Ave.	B	Sheboygan, WI	1910	HWY	Sheboygan	1.14	68.5	14.7
Sheboygan River	13th St.	SW	Sheboygan, WI	1897	RR	CNW	1.57	60	12.5
Sheboygan River	S. 14th St.	B	Sheboygan, WI	1920	HWY	Sheboygan	1.7	54	14
Sheepscot River	Townsend Gut	SW	Boothbay Harbor, ME	1939	HWY	Maine	6	52	10.2
Sheepscot River	—	SW	Barters Island, ME	1931	HWY	Maine	6.8	40	6
Sheepscot River	US 1	SW	Wicasset, ME	1931	HWY	Maine	14	40	10
Sheepscot River	—	B	Wicasset, ME	1915	RR	MEC	15	40	8
Sheepscot River	—	SW	Alna, ME	1920	HWY	Maine	18	30.7	5.6
Sheepscot River	—	R	Dyers River, ME	—	HWY	Newcastle Town	18.1	28	—
Sheepscot River	—	SW	South Newcastle, ME	1921	RR	MEC	20	30	1.9
Shellbank Basin	Nolins Ave.	B	New York, NY	1925	HWY	New York City	0	40	10
Shinnecock Bay	—	B	Hampton Bay, NY	1932	HWY	Suffolk Co.	78	49	10
Shoal Harbor	Compton Creek	SW	Belford, NJ	1928	HWY	Monmouth Co.	0.25	40	4.9
Shrewsbury River	—	SW	Highlands, NJ	1892	RR	CRR (Navesink RR)	1.8	99.5	6
Shrewsbury River	SR 36	B	Highlands, NJ	1933	HWY	New Jersey	1.81	99	35
Shrewsbury River	—	B	Sea Bright, NJ	1946	HWY	Monmouth Co.	4	78	15

*See key of abbreviations at end.

Waterway[a]	Road Name/ Route Carried	Bridge Type[b]	Location	Const. Date	Traffic Type[c]	Owner*	Miles from Mouth of Waterway	Channel Width/Lateral Clearance (feet)	Clearance Height (feet)
Shrewsbury River, North Branch	See Navesink River	B	Oceanic, NJ	1941	HWY	Monmouth Co.	4.5	75.7	22.6
Shrewsbury River, South Branch	See Navesink River	SW	Seabright, NJ	1937	HWY	Monmouth Co.	4.05	75	9
Shrewsbury River, South Branch	See Navesink River	SW	Gooseneck Hwy, NJ	1898	HWY	Monmouth Co.	7.67	66	5.9
Shrewsbury River, Pleasure Bay	See Patten Ave.	SW	Pleasure Bay, NJ	1918	HWY	Monmouth Co.	7.94	77	7.9
Shrewsbury River, Oceanport Creek	—	SW	Oceanport, NJ	1914	RR	NY&LB	8.17	65.1	5
Sicamous Narrows	—	SW	Shuswap, BC, Canada	1982	RR	CP	—	157	—
Siletz River	—	SW	Kernville, OR	1926	HWY	Oregon	2	100	25
Sinepuxent Bay	—	B	Ocean City, MD	1919	HWY	Maryland	0.21	40	4
Sinepuxent Bay	—	B	Ocean City, MD	1940	HWY	Maryland	0.5	70	13
Sinepuxent Bay, Unnamed Gut	—	P	Berlin, MD	—		National Park Service	—	—	—
Siuslaw River	US 101	B	Florence, OR	1936	HWY	Oregon	5	138	18
Siuslaw River	—	SW	Florence, OR	1916	RR	SP	8	100	15
Siuslaw River	—	RSP	Mapleton, OR	1935	HWY	Lane Co.	20	25	21
Siuslaw River, North Fork	—	B	Acme, OR	1926	HWY	Lane Co.	0.5	26.5	15
Siuslaw River, North Fork	—	RSP	Florence, OR	1940	HWY	Lane Co.	2.2	40	7
Six Mile Creek	—	SW	Tampa, FL	1919	RR	Tampa sou	2.3	30	2.2
Skagit River	—	SW	Skagit City, WA	1911	HWY	Skagit Co.	3.5	80	7
Skagit River	—	SW	Fir, WA	1914	HWY	Skagit Co.	5.5	115	10
Skagit River	Division St.	SW	Mt. Vernon, WA	1917	HWY	Skagit Co.	12.25	109	10
Skagit River	—	SW	Mt. Vernon, WA	1954	HWY	Washington	12.5	105	7
Skagit River	US 99	SW	Mt. Vernon, WA	1939	HWY	Washington	17	108	8
Skagit River	—	SW, F	Mt. Vernon, WA	1907	RR	BNSF	17.8	80	5

Skagit River	—	SW	Mt. Vernon, WA	1912	RR	BNSF	22	91	8
Skagit River	—	SW	Sedro–Wooley, WA	1911	HWY	Skagit Co.	22.25	100	10
Skamokawa Creek	US 830	SW, F	Skamokawa, WA	1939	HWY	Washington	0.2	65	12
Skamokawa Creek	—	SW	Skamokawa, WA	1913	HWY	Wahiakum Co.	0.5	61.7	8
Skipanon River	—	SW	Warrenton, OR	1916	RR	BNSF	0.8	33	2
Skipanon River	SH 105	RSP	Warrenton, OR	1916	HWY	Oregon	0.9	37	4
Skipanon River	—	B, F	Warrenton, OR	1929	HWY	Oregon	2.5	24	—
Slaughter Creek	—	SW	Taylors Island, MD	1929	HWY	Dorchester Co.	2.5	26	4.9
Sloop Channel	Meadowbrook Cswy.	B	—	1933	HWY	Long Island State Park Comm.	12.8	75	22
Smacks Bayou	—	B	St. Petersburg, FL	1927	HWY	Shore Acres Property	0.25	23	7.5
Smith Creek	—	SW	Norfolk, VA	1908	HWY	Norfolk	0	49	5.3
Smith River	—	RSP	Reedsport, OR	1949	HWY	Douglas Co.	2.7	55	22
Smith River	—	R	Reedsport, OR	1973	HWY	Douglas Co.	2.7	55	22
Smiths Creek	—	SW	Wilmington, NC	1932	HWY	North Carolina	1.5	45	9
Smyrna River	—	SW	Fleming Landing, DE	1918	HWY	Delaware	4	36	5
Snake River	—	SW	Burbank, WA	1884	RR	NP	0.5	150	15
Snake River	—	VL	Burbank, WA	1972	RR	BNSF	1.5	380	63
Snake River	—	SW	Riparia, WA	1889	RR	OWRR&N	68	158	11
Snake River	US 12	VL	Lewiston, ID	1939	HWY	Washington and Idaho	140	160	45
Snodgrass Slough	—	SW	Walnut Grove, CA	1912	RR	SP	3.5	66	16
Snodgrass Slough	Lambert Rd.	SW	Walnut Grove, CA	1932	HWY	Sacramento Co.	4.4	75	18
Snohomish River	—	SW	Everett, WA	1922	RR	BNSF	3.5	100	10
Snohomish River	US 99 Sb	VL	Everett, WA	1954	HWY	Washington	3.6	105	78
Snohomish River	US 99 Nb	VL	Everett, WA	1927	HWY	Washington	3.6	105	78
Snohomish River	SR 15 Eb	VL	Everett, WA	1940	HWY	Washington	6.9	105	75
Snohomish River	—	SW	Lowell, WA	1922	RR	CMSTP&P	8.75	100	6.25
Snohomish River	—	SW, F	Snohomish, WA	1911	HWY	Washington	15	115	10
Snohomish River	—	SW	Snohomish, WA	1912	RR	NP	15	75	8
Snohomish River	—	SW, F	Snohomish, WA	1893	RR	BNSF	15.5	135	11

*See key of abbreviations at end.

Waterway[a]	Road Name/ Route Carried	Bridge Type[b]	Location	Const. Date	Traffic Type[c]	Owner*	Miles from Mouth of Waterway	Channel Width/Lateral Clearance (feet)	Clearance Height (feet)
Sonoma Creek	—	B	Vallejo, CA	1927	HWY	California	0.14	60	11.5
Sonoma Creek	—	B	Wingo, CA	1921	RR	NWP	5.4	55	3
Sonoma Creek	—	SW	Wingo, CA	1904	RR	S. A. Skaggs	6.65	51.5	2.5
South Bay	—	B	Bay City, WA	1915	HWY	Grays Harbor Co.	0	100	5.7
South River 1	SR 2	SW	Edgewater, MD	1933	HWY	Maryland	5.7	70	13
South River 1	—	SW	Riverview, MD	1923	HWY	Anne Arundel Co.	7	40	8
South River 2	Congress St.	SW	Salem, MA	1917	HWY	Salem	0.83	43	4
South River 3	—	SW	Sayreville, NJ	1918	HWY	Middlesex Co.	2.3	58	6
South River 3	—	SW	South River, NJ	1932	RR	CR	2.8	49	4
South River 4	—	SW	Aurora, NC	1910	HWY	North Carolina	11	40	5.2
South River 4	—	SW	Royal, NC	1907	RR	ACL (Washington and Vandemere)	12.5	35	4.7
South Slough Inlet, Coos Bay	—	SW	Charleston, OR	1934	HWY	Oregon	0.7	60	11
Spa Creek	—	SW	Annapolis, MD	1935	HWY	Anne Arundel Co.	0.3	40	5.1
Spa Creek	SR 181	B	Annapolis, MD	1947	HWY	Maryland	0.4	40	15
Spanish River	—	VL	Mobile, AL	1925	HWY	Alabama	0.1	180	60
Spring Lake	—	SW	Spring Lake, MI	1973	HWY	Ottawa Co.	0.9	53	15
Spring Lake	—	B	Ferrysburg, MI	1921	HWY	Michigan	3.1	66	4.1
Spring Lake	—	SW	Spring Lake, MI	1925	RR	CN	3.2	59	3
S. Thompson River	—	SW	Okanagan, BC, Canada	1927	RR	CN	—	208	—
State Boat Channel	Captree Island	B	Oak Beach, NY	1944	HWY	Long Island Park Comm.	30.7	101	29
State Boat Channel	Robert Moses Cswy.	B	Islip, NY	1968	HWY	Long Island Park Comm.	30.7	100	30

Steamboat Slough, Tributary of Sacramento River	—	B	Grant Island, CA	1925	HWY	California	11.2	200	21
Steamboat Slough, Tributary of Snohomish River	—	SW	Marysville, WA	1908	RR	BNSF	1	87	8
Steamboat Slough, Tributary of Snohomish River	US 99 SB	SW	Marysville, WA	1927	HWY	Washington	1.1	100	10
Steamboat Slough, Tributary of Snohomish River	US 99 NB	SW	Marysville, WA	1954	HWY	Washington	1.2	100	10
Steamboat Slough, Tributary of Sonoma Creek	—	SW	Wingo, CA	1904	PR	S.A. Skaggs	5.2	51	2
Steele Bayou	—	B	Manny, MS	1915	HWY	Issaquena Co.	28	40	−4.3
Steele Bayou	—	B	Scott, MS	1913	HWY	Issaquena Co.	30.4	40	−2.5
Steele Bayou	—	VLS	Issaquena, MS	1920	LOG RD	Issaquena Lumber	36	40	−7.2
Steele Bayou	—	SW	Cary, MS	1917	HWY	Issaquena Co.	37.7	40	−4.7
Steele Bayou	—	VLS	Cary, MS	1921	LOG RD	Belgrade Lumber	41.1	40	−2.8
Steele Bayou	—	SW	Magnolia, MS	1923	HWY	Issaquena Co.	48.9	40	2
Steele Bayou	—	B	Lakeside, MS	1914	HWY	Issaquena Co.	54.1	40	4
Steinhatchee River	SR 358	SW	Stevensville, FL	1949	HWY	Florida	2.3	50	8
Steilacoom Waterways	See Chambers Creek	VL	Steilacoom Creek, WA	1914	RR	NP	0	80	50
Stillaguamish River	—	SW	Stanwood, WA	1909	HWY	Snohomish Co.	1	100	10
Stillaguamish River	—	SW, F	Stanwood, WA	1923	HWY	Snohomish Co.	4.6	86	10
Stono River	SR 700	SW	Charleston, SC	1951	HWY	South Carolina	11	55	8
Stony Creek	SR 173	B	Riviera Beach, MD	1949	HWY	Maryland	0.9	40	18
Straits	Waterway between Core Sound and Beaufort Harbor	SW	—	1941	HWY	North Carolina	—	35	8

*See key of abbreviations at end.

Waterway[a]	Road Name/ Route Carried	Bridge Type[b]	Location	Const. Date	Traffic Type[c]	Owner*	Miles from Mouth of Waterway	Channel Width/Lateral Clearance (feet)	Clearance Height (feet)
Straits	US 70	SW	Harkers Island, NC	1970	HWY	North Carolina	0.6	36	14
Stumpy Bayou	—	RSP	Weeks Island, LA	1948	HWY	Louisiana	1	43	9
Sturgeon Bay	—	B	Sturgeon Bay, WI	1978	HWY	Wisconsin	2.8	160	42
Sturgeon Bay	Michigan St.	B	Sturgeon Bay, WI	1931	HWY	Wisconsin	4.3	139	14
Suisun Bay	See Corquinez Strait	VL	Martinez, CA	1930	RR	SP	0	291.5	135.9
Sulphur River	—	RSP	Sulphur Station, TX	1914	HWY	Texarkana	44.2	40	1
Susquehanna River	—	SW	Havre De Grace, MD	1928	HWY	Maryland	1	97.8	23
Susquehanna River	—	SW	Havre De Grace, MD	1906	RR	ATK	1	100	52
Sutter Slough	—	SW, F	Courtland, CA	1939	HWY	Sacramento Co.	6.4	75	19
Suwannee River	—	SW	Old Town, FL	1935	HWY	Florida	30.5	100	15.5
Suwannee River	—	SW	Old Town, FL	1917	RR	CSX	35	48	5
Suwannee River	—	SW	Luraville, FL	1901	RR	Suwannee River and Florida Rwy.	93.5	100	8
Suwannee River	—	SW	Luraville, FL	1913	HWY	Suwannee Co.	97.5	81	7
Swinomish Slough	—	SW	Whitney, WA	1891	RR	GN	0.25	94	4
Swinomish Slough	—	VL	Anacortes, WA	1937	HWY	Skagit Co.	0.5	100	75
Swinomish Slough	—	SW	La Conner, WA	1915	HWY	Skagit Co.	5	100	6.73
Swinomish Slough	SR 536	VL	Anacortes, WA	1982	HWY	Washington	8.2	160	79
Swinomish Slough	—	SW	Whitemarsh, WA	1953	RR	BNSF	8.4	100	7
Tallahatchie River	—	SW	Greenwood, MS	1911	HWY	Leflore Co.	1.2	102	3
Tallahatchie River	—	SW	Shellmound, MS	1911	HWY	Leflore Co.	19.3	100	5
Tallahatchie River	—	SW	Money, MS	1927	HWY	Leflore Co.	23.8	100	6
Tallahatchie River	—	SW	Minter City, MS	1914	HWY	Leflore Co.	42.2	102	3
Tallahatchie River	—	SW	Philipp, MS	1897	RR	IC	47.3	68.8	8
Tallahatchie River	—	SW	Glendora, MS	1918	HWY	Tallahatchie Co.	65.4	98	5
Tallahatchie River	—	SW	Swan Lake, MS	1900	HWY	Tallahatchie Co.	69.3	85	3
Tallahatchie River	—	B	Jarman Ferry, MS	1917	HWY	Tallahatchie Co.	79.6	70	4

Tallahatchie River	—	VL (abandoned)	Charleston, MS	1921	LOG RD	Lamb Fish Lumber	87		
Tallahatchie River	—	SW	Turners, MS	1906	HWY	Quitman Co.	106.3	70.5	4
Tallahatchie River	—	SW	Near mouth of Coldwater River, MS	1917	HWY	Quitman Co.	110.9	70	5
Tamiami Canal	N.W. S. River Dr.	SW	Miami, FL	—	HWY	Dade Co.	—	40	6
Tante Phine Pass	—	B	Venice, LA	1943	HWY	Plaquemines Pa.	7.6	20	3
Tarpon Bayou	—	B	Tarpon Springs, FL	1928	HWY	Pinellas Co.	0.5	25	8
Taunton Bay	US 1	SW	Hancock, ME	1927	HWY	Maine	1.2	82	10
Taunton River	Slades Ferry	B	Fall River, MA	1938	HWY	Fall River	1.5	100	6.8
Taunton River	US 6 (Brightman St.)	B	Fall River, MA	1906	HWY	Massachusetts	1.8	98	27
Taunton River	—	SW	Somerset, MA	1909	RR	NYNHH	5.5	100	5.4
Taunton River	Center St.	SW	Berkley, MA	1909	HWY	Bristol Co.	10.3	52.8	7
Taunton River	Plain St.	SW	Taunton, MA	1904	HWY	Taunton	14.5	40.5	8.4
Tavernier Creek	US 441	B	Okeechobee, FL	1948	HWY	Florida	0.3	40	9
Tavernier Creek	1st St.	SW	Okeechobee, FL	1929	HWY	Okeechobee	4	40	6
Taylor's Bayou	—	SW	Port Arthur, TX	—	RR	SP	1.9	40	4
Taylor Bayou	—	SW	Seabrook, TX	1930	RR	SP	2	38	5
Taylor's Bayou	—	SW	Port Arthur, TX	1930	HWY	Jefferson Co.	1.9	69	6.5
Taylor Bayou	Gulf Oil Co.	P (temp)	Port Arthur, TX	1973	HWY	U.S. Govt.	6.1	80	—
Taylor's Bayou	—	B	Fannette, TX	1915	HWY	Jefferson Co.	35	20	4
Taylor Creek	—	SW	Okeechobee, FL	1923	HWY	Florida	0.25	50	7
Tchoutacabouffa River	—	SW	Vinnie, MS	1911	HWY	Harrison Co.	5.5	68.5	2
Tchoutacabouffa River	—	SW	Cedar Lake, MS	1973	HWY	Harrison Co.	8	70	5
Teche Bayou	—	P	Patterson, LA	1938	PLNT	Marguerite C. Williams	2.85	112	—
Teche Bayou	US 182	SW	Patterson, LA	1970	HWY	St. Mary Pa.	3.9	60	5

*See key of abbreviations at end.

Waterway[a]	Road Name/ Route Carried	Bridge Type[b]	Location	Const. Date	Traffic Type[c]	Owner*	Miles from Mouth of Waterway	Channel Width/Lateral Clearance (feet)	Clearance Height (feet)
Teche Bayou	—	SW	Patterson, LA	1912	HWY	Shadyside Company	7.5	71	4.4
Teche Bayou	Germania Plantation	SW	Centerville, LA	1906	HWY	W. P. Foster	10.6	71	1.6
Teche Bayou	SR 317	SW	Centerville, LA	1973	HWY	St. Mary Pa.	11.8	60	5
Teche Bayou	Kramer Plantation	SW	Centerville, LA	1934	HWY	W. P. Foster	13.25	59.6	5.2
Teche Bayou	Alice C Plantation	SW	Garden City, LA	1921	HWY	W. P. Foster	14.2	70.2	7.1
Teche Bayou	—	SW	Franklin, LA	1963	HWY	Louisiana	16.3	60	2
Teche Bayou	—	SW	Franklin, LA	1941	HWY	Louisiana	17.1	60	6.5
Teche Bayou	—	SW	Franklin, LA	1971	HWY	Louisiana	17.2	60	4
Teche Bayou	Belleview Plantation	P	Franklin, LA	—	PLNT	Sterling Sugars	20.75	102.2	—
Teche Bayou	—	SW	Franklin, LA	1912	RR	MP (Iberia St. Mary & Eastern)	21.7	60	4.4
Teche Bayou	—	SW	Franklin, LA	1941	HWY	Oaklawn Plantation	22.3	60	6
Teche Bayou	—	P	Baldwin, LA	—	HWY	Sterling Sugars	25	110.2	—
Teche Bayou	Katy Plantation	P	Baldwin, LA	—	HWY	John M. Caffery	26	109.8	—
Teche Bayou	—	SW	Baldwin, LA	1960	HWY	Katy Plantation	27	60	6
Teche Bayou	—	SW	Clarenton, LA	1912	RR	MP (Iberia St. Mary & Eastern)	32	60	6.5
Teche Bayou	—	SW	Clarenton, LA	1941	HWY	Louisiana	32.5	87	7
Teche Bayou	—	SW	Clarenton, LA	1939	HWY	Adeline Plantation	37	61	4
Teche Bayou	—	SW	Jeanerette, LA	1973	HWY	St. Mary Pa.	38.9	60	7
Teche Bayou	—	SW	Jeanerette, LA	1938	HWY	New Orleans	41.25	62.4	8.5

Teche Bayou	—	SW	Jeanerette, LA	1941	HWY	Louisiana	41.75	75	6.3
Teche Bayou	—	SW	Jeanerette, LA	1944	HWY	—	41.8	75	6
Teche Bayou	SR 3128	SW	Jeanerette, LA	1980	HWY	Louisiana	43.5	60	3
Teche Bayou	—	SW	Jeanerette, LA	1929	HWY	U.S. Govt.	46.5	56	5
Teche Bayou	—	SW	Olivier, LA	1962	HWY	Louisiana	48.7	60	4
Teche Bayou	—	SW	Olivier, LA	1927	HWY	Iberia, PA	48.75	60	5.8
Teche Bayou	—	SW	New Iberia, LA	1957	HWY	Louisiana	52.5	60	8
Teche Bayou	—	B	New Iberia, LA	1941	HWY	Louisiana	53	60	6
Teche Bayou	—	SW	New Iberia, LA	1940	HWY	Louisiana	53.3	60	4
Teche Bayou	—	SW	New Iberia, LA	1967	HWY	Louisiana	56.7	50	6
Teche Bayou	—	SW	New Iberia, LA	1939	HWY	Iberia Pa.	56.75	56.7	6.7
Teche Bayou	SR 353	SW	New Iberia, LA	1940	HWY	Iberia Pa.	58	56	4
Teche Bayou	SR 94	SW	Loreauville, LA	1938	HWY	Iberia Pa.	60.7	57	6
Teche Bayou	—	SW	Loreauville, LA	1912	RR	MP	61	60	8
Teche Bayou	SR 344	VL	Loreauville, LA	1965	HWY	Louisiana	62.5	50	50
Teche Bayou	SR 86	SW	Daspit, LA	1967	HWY	Louisiana	69	50	8
Teche Bayou	Used Keystone Lock	SW	St. Martinsville, LA	—	—	—	72.5	36	—
Teche Bayou	—	SW	St. Martinsville, LA	1940	HWY	Louisiana	75.2	40	4
Teche Bayou	—	SW	Levert, LA	1923	RR	SP	77.7	68	8
Teche Bayou	—	VL	Parks, LA	1950	HWY	Louisiana	82	41	50
Teche Bayou	—	SW	Ruth, LA	1934	HWY	Louisiana	87.5	42	8
Teche Bayou	—	VL	Breaux Bridge, LA	1951	HWY	Louisiana	90.5	40	50
Teche Bayou	—	RSP	Breaux Bridge, LA	1962	HWY, RR	SP	91	45	14
Teche Bayou	—	SW	Breaux Bridge, LA	1910	RR	SP	91	60	15.2
Teche Bayou	—	SW	Breaux Bridge, LA	1962	HWY, RR	SP	91.8	45	14
Teche Bayou	I 10	RSP	Poche, LA	1952	HWY	St. Martin Pa.	95.5	46	5
Teche Bayou	—	RSP	Cecelia, LA	1949	HWY	Louisiana	99.5	47	5
Tenean Creek	—	B	Boston, MA	1833	HWY	Boston	0	24.7	4.7
Tennessee River	—	SW	Gilbertsville, KY	1905	RR	IC	22.2	134	0.1
Tennessee River	—	VL	Danville, TN	1932	RR	LN	78.3	275	45.2

*See key of abbreviations at end.

Waterway[a]	Road Name/ Route Carried	Bridge Type[b]	Location	Const. Date	Traffic Type[c]	Owner*	Miles from Mouth of Waterway	Channel Width/Lateral Clearance (feet)	Clearance Height (feet)
Tennessee River	—	SW	Johnsonville, TN	1901	RR	LN NC&STL	96.5	176	1.3
Tennessee River	—	VL	New Johnsonville, TN	1945	RR	CSX	100.5	352	44
Tennessee River	—	VL	Florence, AL	1960	RR	NS	256.5	350	45
Tennessee River	Wilson Lock East	B	Florence, AL	—	HWY	U.S. Govt.	259.4	60	14
Tennessee River	—	VL	Decatur, AL	1973	RR	NS	304.4	399	53
Tennessee River	US 3 (Keller Memorial Br.)	B	Decatur, AL	1928	HWY	Alabama	305	210	18
Tennessee River	Br 123.1	VL	Bridgeport, AL	1982	RR	CSX	414.4	280	59
Tennessee River	—	SW	Jasper, TN	1905	RR	SOU	434.3	200	—
Tennessee River	I 27 (Chief John Ross Br.)	B	Chattanooga, TN	1917	HWY	Chattanooga	464.1	295	32
Tennessee River	—	VL	Hixon, TN	1971	RR	NS	470.7	307	70
Tennessee Tombigbee Waterway	—	SW	Columbus, MS	1904	RR	ICG	284.1	81	20
Tennessee Tombigbee Waterway	1st Ave.	SW	Columbus, MS	1925	HWY	Mississippi	284.5	87	17
Tennessee Tombigbee Waterway	US 45, 82	VL	Columbus, MS	1965	HWY	Mississippi	285	200	52
Tennessee Tombigbee Waterway	—	SW	Waverly, MS	1888	RR	NS	291	95	12
Tennessee Tombigbee Waterway	—	SW	Amory, MS	—	RR	BNSF	322.8	83	2
Tensas River	—	SW	Clayton, LA	1909	RR	MP	27.2	64	0
Tensas River	SR 15	VL	Clayton, LA	1972	HWY	Louisiana	27.3	125	50
Tensas River	SR 128	SW	New Light, LA	1932	HWY	Louisiana	61	64	7
Tensaw River	—	SW	Hurricane, AL	1963	RR	—	15	131	11
Terrebonne Bayou	—	SW	Montegut, LA	—	HWY	Louisiana	22	33.1	0.1
Terrebonne Bayou	—	VL	Montegut, LA	1974	HWY	Louisiana	22.2	45	48

Terrebonne Bayou	—	VL	Klondyke, LA	1972	HWY	Terrebonne Pa.	27.3	45	50
Terrebonne Bayou	Klondyke Br.	SW	Bourg, LA	1942	HWY	Terrebonne Pa.	27.5	47	7
Terrebonne Bayou	—	SW	Bourg, LA	—	HWY	Terrebonne Pa.	28.3	35	2.9
Terrebonne Bayou	SR 24	SW	Bourg, LA	1964	HWY	Terrebonne Pa.	28.8	40	5
Terrebonne Bayou	—	SW	Bourg, LA	—	HWY	Emile Gaudry	28.9	40.3	2.8
Terrebonne Bayou	—	VL	Bourg, LA	1942	HWY	Louisiana	31.3	45	49
Terrebonne Bayou	SR 3087	VL	Houma, LA	1965	HWY	Louisiana	33.9	40	47
Terrebonne Bayou	—	SW	Houma, LA	1955	HWY	Louisiana	35.5	40	3
Terrebonne Bayou	—	VL	Houma, LA	1954	HWY	Louisiana	36.6	40	52
Terrebonne Bayou	—	B	Houma, LA	1941	HWY	Houma	36.8	41	3
Terrebonne Bayou	—	B	Houma, LA	1941	HWY	Houma	36.9	41	3
Terrebonne Bayou	—	SW	Houma, LA	—	HWY	Terrebonne Pa.	38.4	39	4
Terrebonne Bayou	—	SW	Houma, LA	—	HWY	Terrebonne Pa.	39.5	31	3
Thames River	—	SW	New London, CT	1920	HWY	Connecticut	3	200	30.3
Thames River	—	B	Groton, CT	1919	RR	ATK	3.03	151	30
Thames River	Poquetanuck Cove	B	Ledyard, CT	1913	RR	NYNH&H	0	20	2.8
The Gut	SR 129	SW	South Bristol, ME	1930	HWY	Maine	0.2	26	3
Thorofare Bay–Cedar Bay	—	SW	Atlantic, NC	1955	HWY	North Carolina	0.2	30	8
Three Mile Creek	—	SW	Mobile, AL	1925	RR	CSX	0.3	56	10
Three Mile Creek	—	SW	Mobile, AL	1912	RR	Alabama Docks Corp	0.7	53	4
Three Mile Creek	—	SW	Mobile, AL	1932	HWY	Mobile Co.	1	60	5
Three Mile Creek	—	SW	Mobile, AL	1928	RR	NS	1.1	57	2
Three Mile Slough	SR 160	VL	Rio Vista, CA	1949	HWY	California	0.1	150	105
Thunder Bay River	2nd Ave.	B	Alpena, MI	1939	HWY	Alpena	0.5	80	4
Tickfaw River	—	SW	Romes Ferry, LA	1935	HWY	Louisiana	7.2	44	0
Ticonderoga Creek	—	SW	Montcalm Landing, NY	1895	RR	D&H	0	37.5	10
Tigre Bayou	SR 330	SW	Erath, LA	—	HWY	Louisiana	2.3	60	5
Tigre Bayou	—	B	Erath, LA	—	HWY	Vermillion Pa.	3.2	15	2
Tigre Bayou	—	B	Erath, LA	—	HWY	Vermillion Pa.	4.6	17	3
Tillamook River	—	SW	Tillamook, OR	1928	HWY	Tillamook Co.	0.5	70	16
Tittibawassee River	—	SW	Saginaw, MI	1926	RR	PC	3.5	—	—

*See key of abbreviations at end.

Waterway[a]	Road Name/ Route Carried	Bridge Type[b]	Location	Const. Date	Traffic Type[c]	Owner*	Miles from Mouth of Waterway	Channel Width/Lateral Clearance (feet)	Clearance Height (feet)
Tolay Creek	—	VL	Sears Point, CA	1928	HWY	California	4	52.67	51
Tombigbee River	—	VL	Jackson, AL	1948	RR	NS	44.9	300	52
Tombigbee River	—	SW	Jackson, AL	1888	RR	SOU	54.5	99.3	9.1
Tombigbee River	—	VL	Jackson, AL	1928	HWY	—	57.1	200	52
Tombigbee River	—	VL	Naheola, AL	1927	RR	MBRR	128.6	150	55
Tombigbee River	—	VL	Naheola, AL	1927	RR	Alabama State Bridge	141.5	150	—
Tombigbee River	—	VL	Demopolis, AL	1922	HWY	AGS	171	150	52
Tombigbee River	—	VL	Epes, AL	1928	RR	—	226.8	175	53.26
Tombigbee River	—	SW	Cochrane, AL	1908	RR	AT&N	283.4	95.5	19
Tombigbee River	See Tennessee Tombigbee Waterway	SW	Columbus, MS	1904	RR	M&O	333	81.5	20.2
Tombigbee River	St. Johns St. See Tennessee Tombig-bee Waterway	SW	Columbus, MS	1925	HWY	Loundes Co.	333.6	84.75	5
Tombigbee River	See Tennessee Tombigbee Waterway	SW	Waverly, MS	1888	RR	SOU	345.6	94.7	12.6
Tombigbee River	—	SW	Aberdeen, MS	—	RR	SLSF	384.6	83.6	10.4
Tombigbee River	See Tennessee Tombigbee Waterway	SW	Amory, MS	—	RR	SLSF	421.5	84	2.6
Tombigbee River	—	SW	Amory, MS	—	HWY	Monroe Co.	422	108	4.7
Tonawanda Creek	—	SW	Tonawanda, NY	—	RR	PC	0.1	66	6
Tonawanda Creek	—	B	Tonawanda, NY	—	—	—	0.3	100	11
Tonawanda Harbor	—	SW	Tonawanda, NY	1855	RR	PC	0.2	66	8
Town Creek	—	SW	Towncreek, NC	1925	HWY	North Carolina	4	35	3.5
Townsend Gut	SR 27	SW	Boothbay, ME	—	HWY	Maine	0.7	52	10
Townsend Inlet	—	B	Avalon, NJ	1940	HWY	Cape May Co.	0.3	50	23

Trail Creek	Franklin St.	B	Michigan City, IN	1930	HWY	Indiana	0.5	120	17
Trail Creek	—	SW	Michigan City, IN	1946	RR	ATK	0.9	44	7
Trail Creek	2nd St.	B	Michigan City, IN	1929	HWY	Indiana	0.9	79	12
Trail Creek	6th St.	B	Michigan City, IN	1929	HWY	Indiana	1.2	69	6.3
Trent Canal	—	SW	Bala, ON, Canada	1915	RR	CN	—	246.12	—
Trent Canal	—	SW	Havelock, ON, Canada	1897	RR	CP	—	187	—
Trent River	US 70	SW	New Bern, NC	1955	HWY	North Carolina	0	78	13
Trent River	—	SW	New Bern, NC	—	RR	AEC	0	43	0
Trent River	—	SW	New Bern, NC	1933	HWY	North Carolina	0	41	0.4
Trent River	—	VL, F	Pollocksville, NC	1950	RR	CSX	18	40	27
Trent River	—	SW	Pollocksville, NC	1929	HWY	North Carolina	18.1	40	3.1
Trent River	—	SW	Quaker Bridge, NC	1909	HWY	North Carolina	25.5	44.5	1.7
Trinity River	—	SW	Liberty, TX	1921	RR	Texas and New Orleans	40.4	98.9	3.5
Trinity River	—	SW, F	Liberty, TX	1921	RR	UP	41.4	52	22
Trinity River	—	SW	Kenefick, TX	1939	RR	Beaumont Sour Lake & Western	49	92	2.9
Trinity River	—	SW, F	Kenefick, TX	1955	RR	MP	54.8	89	1
Trinity River	—	SW	Romayor, TX	1950	RR	BNSF	96.2	85	7
Trinity River	—	SW	Romayor, TX	1901	RR	Gulf Col. & Santa Fe	96.3	90.3	7.4
Trinity River	—	SW	Goodrich, TX	1897	RR	SP	117.3	40	3
Trinity River	—	SW, F	Riverside, TX	1914	RR	UP	181	126	3
Trinity River	—	SW	Point Blank, TX	1914	RR	I&GN	182.5	127.7	3.9
Trinity River	—	SW	Long Lake, TX	1914	RR	I&GN	311.9	82	4.2
Trinity River	—	SW	Malloy, TX	1911	HWY	Dallas Co.	473.9	94.2	1.3
Trinity River	—	SW	Wilmer, TX	1911	HWY	Dallas Co.	477.8	95.5	2.6
Trinity River	—	SW	Hutchins, TX	1911	HWY	Dallas Co.	482.9	94.7	3.7
Trinity River	—	SW	Dallas, TX	1914	HWY	Dallas Co.	497.6	93.5	−6.8
Trout River	—	SW	Milldale, FL	1928	RR	ACL	0.23	60	1.5
Trout River	—	B	Panama, FL	1929	HWY	Florida	0.88	53	8.7
Trout River	—	SW	Jacksonville, FL	1938	RR	CSX	1	46	2
Trout River	—	SW	Jacksonville, FL	1913	HWY	Florida	4.98	39	6

*See key of abbreviations at end.

Waterway[a]	Road Name/ Route Carried	Bridge Type[b]	Location	Const. Date	Traffic Type[c]	Owner*	Miles from Mouth of Waterway	Channel Width/Lateral Clearance (feet)	Clearance Height (feet)
Tuckahoe Creek	—	R	Tuckahoe Bridge, MD	1890	HWY	Caroline and Talbot Cos.	1.75	39.6	4.1
Tuckahoe River	—	B	Tuckahoe, NJ	1927	HWY	New Jersey	8	30	8
Tulls Creek	—	SW	Moyock, NC	—	HWY	North Carolina	0.3	28	4
Turner Cut	—	P	Stockton, CA	1954	PRIV RD	E. J. Conner	0.2	88	—
Turner Cut	—	R	Stockton, CA	1980	HWY	Reclamation Dist. 2030	2.3	30	16
Turners Creek	—	SW	Thunderbolt, GA	1921	HWY	Georgia	2	90	6
Turtle River	—	SW	Brunswick, GA	1928	HWY	Georgia	4.5	80	4.5
Twin Rivers East	Washington St.	B	Two Rivers, WI	1950	HWY	Wisconsin	0.3	55	7
Twin Rivers East	17th St.	SW	Two Rivers, WI	1894	HWY	Two Rivers	0.3	68	11.6
Twin Rivers East	22nd St.	B	Two Rivers, WI	1932	HWY	Wisconsin	0.7	70	3
Twin Rivers West	Washington St.	B	Two Rivers, WI	1950	HWY	Wisconsin	0.3	55	7
Twin Rivers West	—	SW	Two Rivers, WI	1930	RR	CNW	0.45	50	8
Tybee River	—	SW	Thunderbolt, GA	1921	HWY	Georgia	6.6	95	6
Umpqua River	—	SW	Reedsport, OR	1936	HWY	Oregon	11.1	195	36
Umpqua River	—	RSP	Reedsport, OR	1951	RR	SP	11.1	38	6
Umpqua River	—	RSP, F	Reedsport, OR	1935	HWY	Oregon	11.1	40	8
Umpqua River	—	RSP	Reedsport, OR	1967	HWY	Douglas Co.	11.2	54	7
Umpqua River	—	RSP	Reedsport, OR	1951	RR	SP	11.2	54	6
Umpqua River	—	SW	Reedsport, OR	1916	RR	SP	11.5	150	16
Union Canal	Fuhrmann Blvd.	B	Buffalo, NY	—	HWY	—	0.7	80	9
Urbanna Creek	—	R	Urbanna, VA	1927	HWY	Middlesex Co.	0.6	30.3	6.4
Venice Bay	—	B	Venice, FL	1930	HWY	Florida	—	40	3.6
Vermillion Bayou	—	SW	Perry, LA	1938	HWY	Louisiana	22.25	50.3	7.9
Vermillion Bayou	—	SW	Abbeville, LA	1902	RR	SP	25.25	79	8.1
Vermillion Bayou	—	P	Frederick, LA	—	HWY	Vemillion Pa.	34.5	64.5	—
Vermillion River (Bayou)	SR 335	VL	Perry, LA	1955	HWY	Louisiana	22.4	60	55
Vermillion River (Bayou)	—	SW	Abbeville, LA	1942	RR	SP	25.2	68	8
Vermillion River (Bayou)	US 90	VL	Abbeville, LA	1938	HWY	Louisiana	25.4	50	57
Vermillion River (Bayou)	SR 14	VL	Abbeville, LA	1964	HWY	Louisiana	26	50	55

Vermillion River (Bayou)	US 167	SW	Frederick, LA	1945	HWY	Vermillion Pa.	34.2	60	13
Vermillion River (Bayou)	US 167	VL	Milton, LA	1948	HWY	Louisiana	37.6	61	50
Vermillion River (Bayou)	US 167 (Broussard Store)	VL	Milton, LA	1952	HWY	Louisiana	41	60	52
Vermillion River (Bayou)	US 167	SW	Lafayette, LA	1963	HWY	Louisiana	44.9	55	16
Vermillion River (Bayou)	Pinhook Rd.	VL	Lafayette, LA	1981	HWY	Louisiana	49	40	50
Vermillion River (Bayou)	—	RSP (temp)	Lafayette, LA	1975	HWY	Louisiana	49	25	8
Victoria Channel	—	VL	Bloomington, TX	1964	RR	UP	29.4	80	50
Victoria Harbour	—	B	Victoria, BC, Canada	1923	RR	CP	—	195	—
Wabash River	—	SW	Mauniel, IL	1902	RR	CSX	30.7	100	0
Wabash River	—	SW	Grayville, IL	1973	RR	ICG	63.3	100	5
Wabash River	—	SW	Mt. Carmel, IL	1912	RR	NS	91.5	100	6
Wabash River	—	SW	Mt. Carmel, IL	1954	RR	NS	94.5	93	2
Wabash River	—	SW	St. Francisville, IL	1925	RR	Strangel Farms Grain Service	118.5	100	5
Wabash River	—	SW	Vincennes, IN	1912	RR	BO	128.2	90	2
Wabash River	—	SW	Riverton, IN	1912	RR	IC	162	100	1.6
Waccamaw River	—	SW	Georgetown, SC	1935	HWY	South Carolina	1	120	12.7
Waccamaw River	—	SW	Conway, SC	1904	RR	ACL	44	60	3.6
Waccamaw River	—	SW	Conway, SC	1927	RR	ACL	44.35	80	0.7
Waccamaw River	—	SW	Conway, SC	1972	RR	CSX	44.4	80	8
Wading River	—	B (repl. w. new)	Wading River, NJ	1928	HWY	Burlington Co.	5	30	5
Walden Creek	—	R	Southport, NC	1927	HWY	North Carolina	1	25	6.5
Walluski River	—	SW	Astoria, OR	1938	HWY	Oregon	1	50	6
Wando River	—	SW	Cainhoy, SC	1939	HWY	South Carolina	10	62	6
Wappinger Creek	—	SW	New Hamburg, NY	1888	HWY	New York	0	48	11.1
Wappinger Creek	—	B	New Hamburg, NY	1908	RR	CR	0	40	1.8
Wards Cove	—	B	Ketchikan, AK	1924	HWY	USDA Bureau of Public Roads	0	18	6.5
Wards Creek	—	Slot	Beaufort, NC	1910	HWY	North Carolina	13	16	5.5
Warehouse Bayou	—	RSP	Weeks Island, LA	1948	HWY	Louisiana	1.5	30	7
Wares Creek	—	B	Bradenton, FL	1915	HWY	Bradenton	0.5	40	4.2
Warrior River	—	VL	Demopolis, AL	1927	RR	STLSF	2.1	176	46

*See key of abbreviations at end.

Waterway[a]	Road Name/ Route Carried	Bridge Type[b]	Location	Const. Date	Traffic Type[c]	Owner*	Miles from Mouth of Waterway	Channel Width/Lateral Clearance (feet)	Clearance Height (feet)
Warrior River	—	VL	Eutaw, AL	1919	RR	SOU AGS	56.9	155	72
Warrior River	—	VL	Tuscaloosa, AL	1923	HWY	Tuscaloosa Co.	130.4	175.5	40.86
Washington Lake	Mercer Island	P	Seattle, WA	1940	HWY	Washington	—	190	—
Wateree River	—	SW	Wateree, SC	1903	RR	SOU	10	50	4
Wateree River	—	SW	Acton, SC	—	RR	ACL	15.7	72	—
Watsons Creek	—	B	Manasquan, NJ	—	HWY	Monmouth Co.	0.3	31	9
Weems Creek	SR 437	SW	Annapolis, MD	1930	HWY	Anne Arundel Co.	0.7	28	5
Weems Creek	SR 70	B	Annapolis, MD	1954	HWY	Maryland	0.8	75	28
Wekiva River	—	HNGL	Palm Springs, FL	1938	RR	Wilson Cypress	10	14	3
Welland Canal	—	B	Grimsby, ON, Canada	1929	RR	CN	—	98.12	—
West Bay	—	B	Osterville, MA	1911	HWY	Barnstable town	1.15	33	8.4
West Bay	Bridge St.	B	Osterville, MA	1946	HWY	Barnstable	1.2	31	15
Westchester Creek	E. 177th St.	B	New York, NY	1918	HWY	New York City	1.5	57.8	14
Westchester Creek	Unionport Br.	B	Bronx, NY	1948	HWY	New York City	—	—	—
Westchester Creek	Eastern Blvd.	B	Bronx, NY	—	HWY	New York City	—	—	—
West Palm Beach Canal	—	VL	Canal Pt., FL	1924	RR	FEC	0	50	29
West Palm Beach Canal	US 441	RSP	Canal Pt., FL	1922	HWY	Florida	0.1	36	8
West Palm Beach Canal	—	RSP	Azucar, FL	1927	—	U.S. Sugar Corp.	1.5	36	6
West Palm Beach Canal	—	RSP	—	—	—	Osceola Farms	5.4	32	5
West Palm Beach Canal	—	RSP	—	—	—	Sugar Cane Farms	6.7	32	4
West Palm Beach Canal	—	RSP	—	—	—	Terry Cattle	10.6	32	4
West Palm Beach Canal	—	RSP	State Correcton Farms, FL	1927	HWY	Florida	15.8	32	4
West Palm Beach Canal	SR 80	SW	20 Mile Bend, FL	1922	HWY	Florida	19.4	50	8
West Palm Beach Canal	Ranch Service Rd.	RSP	—	—	HWY	Ousley Sod	26	32	11
West Palm Beach Canal	Wellington Rd.	RSP	—	—	HWY	Florida	29.4	32	11
West Palm Beach Canal	SR 7	RSP	—	—	HWY	Florida	31	32	12
West Palm Beach Canal	Benoist Farm Rd.	RSP	—	—	HWY	Florida	32.7	32	12

West Palm Beach Canal	SR 809 (Military Trail Rd.)	RSP	—	—	HWY	Florida	36.6	36	12
West Palm Beach Canal	Kirk Rd.	RSP	—	—	HWY	Florida	37.2	36	15
West Palm Beach Canal	SR 807 (Congress Ave.)	RSP	—	—	HWY	Palm Beach Co.	38.1	32	12
West Palm Beach Canal	Forest Hill Blvd.	RSP	—	—	HWY	Florida	40.1	32	6
West Palm Beach Canal	—	SW	West Palm Beach, FL	—	RR	CSX	40.1	60	6
West Palm Beach Canal	—	VL	West Palm Beach, FL	—	RR	FEC	41.3	35	47
West Palm Beach Canal	US 1	RSP	West Palm Beach, FL	—	HWY	Florida	41.4	25	8
West Palm Beach Canal	Federal Hwy.	B	West Palm Beach, FL	—	HWY	Florida	41.7	40	9
West Pearl River	—	SW	Dunbar, LA	1954	RR	CSX	1	87	14
West Pearl River	US 90	VL	Indian Village, LA	1932	HWY	Louisiana	7.9	90	50
West Pearl River	US 90	SW	Pearlington, LA	1932	HWY	Louisiana	10	90	10
West Pearl River	I 10	B, F	Logtown, LA	1969	HWY	Louisiana	15.5	115	73
West Pearl River	—	SW	Pearl River, LA	1956	RR	NS	22.1	90	7
West River	Kimberly Ave.	B	New Haven, CT	1906	HWY	New Haven	0.85	45	7.2
Westport River	—	SW	Westport, MA	1904	HWY	Westport	1	34.8	6
Westport River, East Branch	SR 88 (Westport Point Br.)	B	Westport, MA	1960	HWY	Massachusetts	1.2	50	20
Westport Slough	—	RSP	Woodson, OR	1942	HWY	Columbia Co.	4.2	48	11
Westport Slough	—	VL	Clatskanie, OR	1928	HWY	Columbia Co.	9.5	50	14
Weymouth Back River	Lincoln St.	B	Weymouth, MA	1913	HWY	Plymouth and Suffolk Cos.	2.37	50	10
Weymouth Fore River	Washington St.	B	Quincy, MA	1903	HWY	Massachusetts	3.5	175	33.7
Weymouth Fore River	—	B	East Braintree, MA	1919	HWY	Norfolk Co.	5.62	50	12.5
Whitcomb Bayou	—	B	Tarpon Springs, FL	1928	HWY	Pinellas Co.	0.5	25	5
White River	—	VL	Benzal, AR	1974	RR	MP	7.6	300	52
White River	—	SW	Benzal, AR	1906	RR	MP	11.2	166.5	4.8
White River	—	SW	Clarendon, AR	1913	RR	UP	98.9	155	0
White River	US 70	VL	Devalls Bluff, AR	1925	HWY	Arkansas	121.7	197	46
White River	—	VL	Devalls Bluff, AR	1927	RR	RI	122	175	50
White River	—	SW	Georgetown, AR	1929	RR	M&A RR	172.2	128	0.5

*See key of abbreviations at end.

Waterway[a]	Road Name/ Route Carried	Bridge Type[b]	Location	Const. Date	Traffic Type[c]	Owner*	Miles from Mouth of Waterway	Channel Width/Lateral Clearance (feet)	Clearance Height (feet)
White River	—	SW	Augusta, AR	1913	RR	UP	201.3	118	0
White River	—	SW	Newport, AR	1941	RR	UP	254.8	121	0
White River	—	SW	Batesville, AR	1928	HWY	Arkansas	300.9	125	5.5
White River	—	SW	Cotter, AR	1904	RR	MP	403.2	122	13.1
White River 2	—	SW	Decker, IN	1891	RR	C&EI	22.2	138	1.9
Wicomico River	Main St.	B	Salisbury, MD	1962	HWY	Maryland	22.4	40	4
Wicomico River	Camden Ave.	B	Salisbury, MD	1915	HWY	Wicomico Co.	23	40	1
Wicomico River	Division St.	B	Salisbury, MD	1915	HWY	Wicomico Co.	23.4	40	6.25
Willamette River	—	SW (repl. w. below)	St. Johns, OR	1910	RR	BNSF	6.9	230	39
Willamette River	—	VL	St. Johns, OR	1980	RR	BNSF	6.9	230	39
Willamette River	Broadway Br.	B	Portland, OR	1913	HWY	Multnomah Co.	11.7	250	69
Willamette River	Steel Br.	VL	Portland, OR	1913	RR	UP	12.1	205	140
Willamette River	Burnside Br.	B	Portland, OR	1926	HWY	Multnomah Co.	12.4	205	41
Willamette River	Morrison Br.	B	Portland, OR	1958	HWY	Multnomah Co.	12.8	220	48
Willamette River	Hawthorne Br.	VL	Portland, OR	1911	HWY	Multnomah Co.	13.1	165	143.7
Willamette River	—	VL	Salem, OR	1913	RR	UP	84.3	118	68
Willamette River	—	SW	Albany, OR	1922	RR	SP	119.6	110	13
Willamette River	—	VL, C	Albany, OR	1926	HWY	Oregon	120.4	195	22.37
Willamette River	Van Buren Br.	SW	Corvallis, OR	1913	HWY	Oregon	132.1	102	13
Willamette River	—	VL, C	Harrisburg, OR	1926	HWY	Oregon	163.2	172	9
Willamette River	—	VL, F	Harrisburg, OR	1912	RR	Oregon Electric	164.2	190	14
Willamette River	—	SW	Harrisburg, OR	1906	RR	UP	164.3	135	17
Willapa River	—	SW	Raymond, WA	1924	HWY	Washington	7.6	125	19.2
Willapa River	—	SW	Raymond, WA	1934	HWY	Washington	7.8	125	14
Willapa River, South Fork	—	SW	Raymond, WA	1912	RR	BNSF	0.3	125	9
Wisconsin River	—	RSP	Prairie du Chien, WI	1908	RR	BNSF	1.4	103	1
Wisconsin River	—	SW	Woodman, WI	1895	RR	MILW	20	53	1

Wisconsin River	—	SW (repl. w. fixed)	Byrds, WI	1902	HWY	Byrds	37	75	4
Wisconsin River	—	B (repl. w. fixed)	Muscoda, WI	1930	HWY	Wisconsin	43.4	76	9
Wisconsin River	—	SW	Muscoda, WI	—	HWY	Muscoda Bridge Company	43.5	58	8.7
Wisconsin River	—	SW	Lone Rock, WI	1898	RR	MILW	55	42	3.5
Wisconsin River	—	SW	Lone Rock, WI	1897	HWY	Lone Rock Bridge Company	57.4	61.5	5.6
Wisconsin River	—	SW	Spring Green, WI	1908	HWY	Wyoming and Spring Green	66	71	4.8
Wisconsin River	—	SW	Spring Green, WI	1905	RR	MILW	68.2	86	5
Wisconsin River	—	SW	Sauk City, WI	1912	RR	MILW	87	63	1.8
Wisconsin River	Bryant St.	B	Sauk City, WI	1923	HWY	Wisconsin	87.6	75.5	5.2
Wisconsin River	Washington St.	B	Prairie du Sac, WI	1922	HWY	Prairie du Sac	88.5	69	22.8
Wisconsin River	—	SW	Merrimac, WI	1908	RR	UP	98	81	7.2
Wishkah River	River St.	SW	Aberdeen, WA	1909	RR	BNSF	0.1	125	8
Wishkah River	Heron St.	SW	Aberdeen, WA	1951	HWY	Washington	0.2	75	13
Wishkah River	Heron St.	SW	Aberdeen, WA	1906	HWY	Aberdeen	0.25	125	7
Wishkah River	Wishkah St.	B	Aberdeen, WA	1925	HWY	Washington	0.4	125	10
Wishkah River	Young St.	SW	Aberdeen, WA	1911	HWY	Aberdeen	1	100	6
Withlacoochee River	—	SW	Dunnellon, FL	1908	RR	Standard & Hernando	27.83	50	4
Withlacoochee River	—	SW	Dunnelton, FL	1895	RR	CSX	28.4	53	8
Withlacoochee River	—	B	Elliston, FL	—	HWY	Dunnellon Phosphate	33.6	21	6
Wolf River 1	—	B	Pass Christian, MS	1949	HWY	Harrison Co.	1.3	70	13
Wolf River 1	—	SW	Pass Christian, MS	1950	HWY	Harrison Co.	5.5	70	9
Wolf River 2	—	B	Winneconne, WI	1935	HWY	Wisconsin	2.4	70	2
Wolf River 2	Main St.	B	Fremont, WI	1922	HWY	Waupaca Co.	22.43	70.3	3.1
Wolf River 2	—	SW	Gills Landing, WI	1950	RR	SOO	27.8	56	0
Wolf River 2	—	SW	Northport, WI	1898	HWY	Waupaca Co.	42.73	56	0.1
Wolf River 2	Shawano St.	SW	New London, WI	1913	HWY	New London	46.13	69.9	0.7

*See key of abbreviations at end.

Waterway[a]	Road Name/ Route Carried	Bridge Type[b]	Location	Const. Date	Traffic Type[c]	Owner*	Miles from Mouth of Waterway	Channel Width/Lateral Clearance (feet)	Clearance Height (feet)
Wolf River 2	Warren St.	SW	New London, WI	1892	RR	CNW	47.03	51.8	1.7
Wolf River 2	—	SW	Hortonville, WI	1910	HWY	Hortonia Twp.	57.9	62.9	−1.1
Wolf River 2	—	SW	Stephensville, WI	1895	HWY	Ellington Twp.	62.95	87.4	0.6
Wolf River 2	—	SW	Shiocton, WI	1872	RR	GBW	69.05	39	0.1
Woodbridge Creek	—	B	Sewaren, NJ	1926	HWY	Middlesex Co.	0.5	50	7
Woodbridge Creek	—	B	Sewaren, NJ	1921	RR	CR	0.6	50	5
Woodbury Creek	—	SW	Woodbury, NJ	1910	HWY	Gloucester Co.	1.25	40.5	5.5
Woodbury Creek	—	B	Woodbury, NJ	1919	HWY	New Jersey	2	45	4
Wye Narrows	—	R	Wye Landing, MD	1924	HWY	Queen Annes Co.	2	40	5
Yalobusha River	—	SW	Greenwood, MS	1898	RR	IC	1	55	7
Yalobusha River	—	B	Avalon, MS	1910	HWY	Leflore Co.	11	60	2
Yalobusha River	—	SW	Grenada, MS	1902	HWY	Grenada Co.	40.4	60	0
Yazoo River	—	SW	Redwood, MS	1945	RR	ICG	16.7	125	0
Yazoo River	US 49	VL	Satartia, MS	1971	HWY	Yazoo Co.	53.3	166	6
Yazoo River	—	SW	Yazoo City, MS	1882	HWY	Yazoo Co.	75.4	83.3	3
Yazoo River	US 49, SR 6	B	Yazoo City, MS	1937	HWY	Mississippi	79.3	125	15
Yazoo River	—	B	Yazoo City, MS	1937	HWY	Mississippi	80.9	125	19.5
Yazoo River	—	SW	Home Park, MS	1904	RR	Y&MVRR	86.6	96	−6.5
Yazoo River	US 49, SR 12	SW	Belzoni, MS	1908	HWY	Humphreys Co.	126.4	126	13
Yazoo River	US 49, SR 12	SW	Silent Shade, MS	1928	HWY	Humphreys Co	144.9	126	4
Yazoo River	—	SW	Sheppardtown, MS	1912	HWY	Leflore Co.	161.5	126	3
Yazoo River	US 49, SR 12	F, can be converted to VL	Sheppardtown, MS	1967	HWY	Leflore Co.	163.5	159	13
Yazoo River	US 49, SR 7	SW	Roebuck, MS	1913	HWY	Leflore Co.	171.5	126	0
Yazoo River	US 49	SW	Roebuck, MS	1913	HWY	Leflore Co.	171.5	126	0
Yazoo River	—	SW	Ft. Loring, MS	1932	RR	NS	180	125	0
Yazoo River	—	SW	Ft. Loring, MS	1932	RR	NS	180	125	0

Yazoo River	US 49, 82	SW	Greenwood, MS	1925	HWY	Leflore Co.	185.1	126	5
Yazoo River	US 49	SW	Greenwood, MS	1925	HWY	Leflore Co.	185.1	126	5
Yellow Mill Channel (Yellow Mill Pond)	US 1 (Stratford Ave.)	B	Bridgeport, CT	1929	HWY	Bridgeport	0.3	82	11
Yellowstone River	—	VL, F	Fairview, MT (Br in North Dakota)	1915	RR, HWY	BNSF	8.9	250	16
Yolo Bypass	—	P	Courtland, CA	1952	HWY	Yolo Co.	2.2	80	—
York River 1	—	B	York, ME	1908	HWY	York	0	30	5.4
York River 1	—	B	York, ME	1932	HWY	Maine	3.5	40	7
York River 2	US 17 (G. P. Coleman Br.)	SW (repl. w. new)	Yorktown, VA	1952	HWY	Virginia	7	450	60
Youngs Bay	—	VL	Astoria, OR	1964	HWY	Oregon	0.7	130	72
Youngs Bay	—	SW	Astoria, OR	1896	RR	BNSF	0.8	130	10
Youngs Bay	—	B	Astoria, OR	1921	HWY	Oregon	2.4	120	13

*See key of abbreviations below.

[a]Data for waterways listed in roman type are from the U.S. Coast Guard, *Bridges over Navigable Waters,* latest edition, or other confirmed sources. Data for waterways listed in italic type are from the U.S. Army Corps of Engineers (COE) and other references, some not confirmed. Some information may conflict or be incorrect.

[b]B, bascule; C, convertible; CB, conveyor belt; D, drawbridge; F, fixed; H, hinged; HNDL, hand lift; HNGL, hinged lift; MB, movable barge; NO, not operable; P, pontoon, R, retractile; RL, rolling lift; RSP, removable span; SW, swing; T, temporary; TR, trestle; TRAM, tramway; UC, under completion; VL, vertical lift; VLS, vertical lift—submergible.

[c]CC, cattle crossing; ERR, electric railroad; FTBR, footbridge; HWY, highway; LOG RD, log road; PL, pipeline; PLNT, plantation; PRIV RD, private road; R, retractile; RR, railroad; RWY, railway; ST, street; TRAM, tramway; TROL, trolley.

[d]Abbreviations following INTC, INTCW, and INTCWW are as follows:
ALT, Intracoastal Waterway Alternate Route
F, FL, FFLA, Intracoastal Waterway Florida
FFCL, Intracoastal Waterway Florida/Florida Coast Line Canal
G, GULF, Intracoastal Waterway—Gulf of Mexico
G Appal, Intercoastal Waterway—Gulf of Mexico Apalachicola River
Gla, Glb, Glc, Gld, Gle, Glf, various branches of Intracoastal Waterway in Louisiana
Gt, Texas branch of Intracoastal Waterway
ILW, ILWW, INLWW, Intracoastal Waterway Inland Waterway.

Key to Owner Abbreviations in List of Movable Bridges

Abbreviation	Owner
A&ECRR	Atlantic & East Carolina Railway Co.
A&LM	Arkansas & Louisiana Missouri RR Co
A&W	Ahnapee & Western Railway
AC&SHORE RR	Atlantic City & Shore RR Co.
AC&SRR	Atlantic City & Shore RR Co.
ACL	Atlantic Coast Line RR Co.
ACL (Washington & Vandemere rr)	Atlantic Coast Line RR Co.
AEC	Atlantic & Eastern Carolina
AGS RR	Alabama Great Southern Railroad
AN	Appalachicola Northern RR
ASLRY	Atlantic Shore Line Railway Co.
AT&N RR	Alabama Tennessee & Northern Railroad
ATK	Amtrak
ATSF	Atchison, Topeka & Santa Fe Railway Co.
B&A	Boston & Albany RR
B&A(B&O)	Baltimore & Annapolis RR
B&E	Baltimore & Eastern RR Co.
B&M	Boston & Maine RR
B&O	Baltimore & Ohio RR Co.
B&OCT	Baltimore & Ohio Chicago Terminal RR Co.
B4	Cleveland, Cincinati, Chicago & St. Louis Railway (Big Four Route)
Baltimore Transit	Baltimore Transit Company
BEAUF&MOREH	Beaufort & Morehead Railway
BLA	Baltimore & Annapolis RR
BLE	Bessemer & Lake Erie RR Co.
BMH	Beaufort & Morehead Railroad Co.
BMH	Beaufort & Morehead Railway
BMT	Brooklyn, Manhattan Transit
BN	Burlington Northern
BNSF	Burlington Northern and Santa Fe Railway Co.
BO	Baltimore & Ohio RR Co.
BOCT	Baltimore & Ohio Chicago Terminal RR Co.
Bos Elev Rwy	Boston Elevated Railway Co.
Brazoria Co & MP	Missouri Pacific Railway
BRB&L	Boston, Revere Beach & Lynn RR
BRC	Belt Railway Company of Chicago
C of G	Central of Georgia RR
C of GA Rwy	Central of Georgia RR
C&EI	Chicago & Eastern Illinois Railway Co.
C&GRR	Columbus & Greenville Railway
CB&Q	Chicago, Burlington & Quincy RR Co.
CBQ	Chicago, Burlington & Quincy RR Co.
CG	Central of Georgia RR
CH&N	Charlotte Harbor & Northern
CH&NRWY	Charlotte Harbor & Northern Railway
CIL	Chicago, Indianapolis & Louisville Railway Co.
CIN SOU RWY Trustees	Trustees of Cincinnati Southern Railway Co.
CJRWY	Chicago Junction Railway Co.
CMSTP&P	Chicago, Milwaukee, St. Paul & Pacific RR

Abbreviation	Owner
CN	Canadian National
CN IC	Canadian National
CNJ	Central Railroad of New Jersey
CNW	Chicago & Northwestern Railway Co.
CO	Chesapeake & Ohio
COFG	Central of Georgia RR
CP	Canadian Pacific Railway
CR	Conrail
CR PRR	Conrail (PRR)
CR RDG	Conrail (Reading)
CRI CJ per COE 1941	Chicago River & Indiana RR Co.
CRI&P	Chicago, Rock Island & Pacific RR
CRIP	Chicago, Rock Island & Pacific RR
CRR (Navesink RR)	Central RR of New Jersey
CSSSB	Chicago, South Shore & South Bend Railway Co.
CSX	CSX Transportation
CV	Central Vermont RR
CW (PRR)	Calumet Western Railway
CWI	Chicago & Western Indiana Railroad Co.
D&H	Delaware & Hudson Co.
D&M	Detroit & Mackinac Railway Co.
DC	Delray Connecting Railroad Co.
DEL OTSEGO	Delaware Otsego System
DH	Delaware & Hudson Co.
DLW	Delaware, Lackawanna & Western RR Co.
DM	Detroit & Mackinac Railway Co.
DRINW	Davenport, Rock Island & North Western Railway Co.
DTI	Detroit Toledo & Ironton RR Co.
EJE	Elgin, Joliet & Eastern RR Co.
EL	Erie Lackawanna Railway
ERIE	Erie Railroad
ERIE (NYLE&WRR)	Erie Railroad
FEC	Florida East Coast Railway Co.
FMS	Fort Myers Southern RR Co.
FTMSRR	Fort Myers Southern RR Co.
GBW	Green Bay & Western Railway
GH&SA(SP)RY	Galveston, Harrisburg & San Antonio Railway
GHH/ATSF	Galveston, Houston & Henderson RR Co.
GN	Great Northern Railway Co.
GNRR	Great Northern Railway Co.
GNRWY	Great Northern Railway Co.
GT	Grand Trunk Railway
GT (Bay City Terminal RR)	Grand Trunk Western RR
Gulf Col & Santa Fe	Gulf, Colorado & Santa Fe RR
GWW	Gateway Western RR
HBT	Houston Belt & Terminal Railway

Key to Owner Abbreviations in List of Movable Bridges (*Continued*)

Abbreviation	Owner
I&GN	International & Great Northern RR
IC	Illinois Central RR
ICG	Illinois Central Gulf
ICRR	Illinois Central RR
IGN	International & Great Northern Rwy
IHB	Indiana Harbor Belt Railroad
IN(ATSF)(BNSF)	Illinois Northern Railway
INT-MN	Interstate Transfer Railway
IRT	Interboro Rapid Transit RR
IT	Illinois Terminal RR
K&ITRR	Kentucky & Indiana Terminal RR Co.
KCS	Kansas City Southern
KGB&W	Kewaunee, Green Bay & Western Rwy
L&A	Louisiana & Arkansas RR
L&N	Louisville & Nashville RR
LA SOU RR	Louisiana Southern RR
LARR	Louisiana & Arkansas Railway Co.
LC&N(SP)	Lake Chartles & Northern RR
LIRR	Long Island Rail Road
LISPC	Long Island State Parkway Commission
LN	Louisville & Nashville RR
LN NC&STL	Nashville, Chattanooga & St. Louis Railway Co.
M&A RR	Missouri & Arkansas RR
M&ORR	Mobile & Ohio RR
MBRR	Meridian & Bigbee River Ry Co.
MBT	Mariana & Blountstown
MBTA	Massachusetts Bay Transportation Authority
MC	Michigan Central RR
MEC	Maine Central RR Co.
MI RWY (elect)	Michigan Railway
MILW	Chicago, Milwaukee, St. Paul & Pacific RR
M-IRR	Missouri-Illinois RR Co.
ML&T SPRR	Mississippi, Louisiana & Texas RR & Steamship Co.
ML&TRR&SS	Mississippi, Louisiana & Texas RR & Steamship Co.
MNS	Minneapolis, Northfield & Southern
MO RR	Minnesota & Ontario Bridge Co.
MP	Missouri Pacific Railway
MP (Brownsville & Matamoras)	Missouri Pacific Railway
MP (Iberia St Mary & Eastern)	Missouri Pacific Railway
MSB&PRR(SLSF)	Muscle Shoals, Birmingham & Pensacola RR Co.
MSRR	Mississippi Export RR Co.
MTA	Massachusetts Bay Transportation Authority
N&P Belt Line	Norfolk & Portsmouth Belt Line RR
NB	Norfolk & Portsmouth Belt Line RR
NJT	New Jersey Transit Rail Operations
NKP	New York, Chicago & St. Louis Railway Co.
NO&NE (Sou) R.R.	New Orleans & Northeastern RR

Abbreviation	Owner
NOT&M(MP)	New Orleans, Texas & Mexico RR Co.
NP	Northern Pacific Railway Co.
NPB	Norfolk & Portsmouth Belt Line RR
NS	Norfolk Southern Corporation
N-S	Norfolk-Southern Corporation
NS&G	Naples, Seaboard & Gulf RR
NSS	Newburgh & South Shore Railway Co.
NW	Norfolk & Western Railway Co.
NWP	Northwestern Pacific Railroad Co.
NY&LB	New York & Long Branch RR
NYC	New York Central & Hudson River RR
NYC&HR RR	New York Central & Hudson River RR
NYLE&W	New York, Lake Erie & Western RR
NYNH&H	New York, New Haven & Hartford RR Co.
NYNHH	New York, New Haven & Hartford RR Co.
OG&NE(T&P)	Opelousa, Gulf & Northeastern RR
OR ELECT COM	Oregon Electric Railway
OWRR&N	Oregon-Washington Railroad & Navigation Co.
P.M.	Pere Marquette Railway
PANYNJ	Port Authority Trans-Hudson Corporation
PC	Penn Central Transportation Co.
PCCSTL	Pittsburgh, Cincinnati, Chicago & St. Louis Railway Co.
PE	Pacific Electric Railway
PHD	Port Huron & Detroit RR
Phil Trans Co	Philadelphia Transportation Co.
PM	Pere Marquette Railway
PPU	Peoria & Pekin Union Railway
PRR	Pennsylvania RR
PRR (LB&TRR)	Louisville Bridge & Terminal RR Co.
PRSL	Pennsylvania-Reading Seashiore Lines
RFP	Richmond, Fredericksburg & Potomac RR
RI	Chicago, Rock Island & Pacific RR
RT	River Terminal RR Co.
RUT	Rutland Railroad
SAFRR	Seaboard All Florida RR
SAL	Seaboard Air Line Rwy Co.
SC&SRRWY	South Chicago & Southern Railway Co.
SCL	Seaboard Coast Line Railroad
Seaboard	Seaboard All Florida RR
SI	Spokane International Railroad
SLSF	St. Louis-San Francisco Railway
SLSW	St. Louis Southwestern RR
SN	Sacramento Northern Railway
Solvay	Solvay Process Company
SOO	Soo Line Railroad Co.
SOU	Southern Railway Co.
SOU AGS	Alabama Great Southern RR

Key to Owner Abbreviations in List of Movable Bridges (*Continued*)

Abbreviation	Owner
Sou Rwy	Southern Railway Co. X
SP	Southern Pacific Lines
St L&SF RR	St. Louis-San Francisco Railway
STLB&MRR(MP)	St. Louis, Brownsville & Mexico Railway
STLB&MRY(MP)	St. Louis, Brownsville & Mexico Railway
STLSF	St. Louis-San Francisco Railway
Suwannee Riv & FL RY	Suwanee River & Florida Railway Co.
T&NO	Texas & New Orleans RR
Tampa Sou	Tampa Southern Railway Co.
TCRR	Tennessee Central Railway Co.
Texas & New Orleans	Texas & New Orleans RR
TPW	Toledo, Peoria & Western RR
TS	Tampa Southern Railway Co.
TT	Toledo Terminal Railroad
UP	Union Pacific RR
VGN	Virginian Railway Co.
W&T RWY	Wrightsville & Tennille Railway
WAB	Wabash Railway Co.
WC	Wisconsin Central
WLE	Wheeling & Lake Erie Railway
WMD	Western Maryland RR
WP	Western Pacific Railroad Company
Y&MV IC	Yazoo & Mississippi Valley RR
Y&MVRR	Yazoo & Mississippi Valley RR

Note: Railroad ownership updated to greatest extent practical. Railroad bridge ownership by defunct corporation suggests that bridge is abandoned or may have been removed. Unless otherwise indicated, railroad bridge ownership is as shown in referenced government publications except where indicated.

INDEX

Acceleration, 134, 254, 259, 261, 283–285, 287, 290, 446, 449
  swing span, 228
  vertical lift, 404
Access appliances, 221, 236, 266, 359, 379, 399, 406, 456
  evaluation, 489
  inspection, 458
Addendum, 243–244, 254, 427–430, 434, 436
  chordal, 428–429, 430, 434, 436
Air pressure gauges, 423
Alloys, 180
Alternatives to movable bridges, 12–14
Aluminum, 286
American Association of State Highway and Transportation Officials (AASHTO), 107, 123, 129, 235
American Bridge Co., 119
American Railway Engineering Association (AREA), 471
American Railway Engineering and Maintenance-of-Way Association (AREMA), 91, 123, 176, 471
American Society of Mechanical Engineers (ASME), 179, 307
American Society for Testing and Materials (ASTM), 263
Anchor bolts, 164–165, 262, 318, 410
Annealing, 279
Antifriction bearings, 36, 91, 148, 180–181, 193, 235, 237–239, 246, 252, 283, 304
  evaluation, 478, 483
  inspection, 397, 424, 435, 438, 439–440, 443
Applications of movable bridges, 19
Areas, 106, 142, 165, 171, 179, 181, 228, 231, 250, 317, 347–348, 357, 361, 382, 385–387, 398, 413–414, 438, 441, 474, 484
Asbestos, 367, 452
Automatic transfer switches, 289
Auxiliary drives, 220, 224, 305
  evaluation, 486
  inspection, 371, 442–443
  maintenance, 353
Auxiliary power, 288–295
Axle loading, 415
Axles, 196, 391, 410

Backlash, 97, 284, 390, 421–422, 425, 428, 435–436
Balance testing, 445, 449–451, 469
  bascule bridges, 445–448
  vertical lift bridges, 448–449
Balance wheels, 84, 193–195, 230, 331, 360–361, 410–411
  arrangement, 84, 135
  clearance, 193, 195, 361, 411
  inspection, 410–412
  loads, 135
Balancing
  bascule bridges, 20, 33–34, 38, 48, 104, 154–155, 166–169, 217, 283, 300, 476
    construction, 323–324, 335–336
    maintenance, 349
  swing bridges, 5, 82, 91, 95, 207–208
    construction, 323–324
    inspection, 415

Balancing *(continued)*
vertical lift bridges, 55, 65, 75–76, 104, 155, 187, 189–190, 219, 283, 300
construction, 323–324, 336–337
inspection, 405
maintenance, 349
Strauss–Rall types, 72
Bascule bridge(s), 5, 14, 17–20, 22, 24, 33–54
Abt type, 121
advantages, 50–52
applications, 100–101, 106, 112–113, 115
articulated counterweight, 42–46, 53, 108–109
inspection, 445
overhead or over deck, 43–44
underfloor or under deck, 44–46
Brown type, 41, 50
buffer cylinders, 160–161
inspections, 392
centering devices, 159–160, 162–163
clearance, 388
inspection, 388–389
Chicago type, 144
counterweights, 42, 144
counterweights, 5, 10, 33, 35, 40, 154–156
dead load stresses, 139, 475
deck openings, 150–154
definition, 20
disadvantages, 52–53
double deck, 37, 50
double leaf, 20–21, 34–35, 37–40, 48, 52–54, 79, 94, 100–101, 106, 110, 113
barriers, 52, 106
construction, 318
counterweights, 35
design, 138–139, 163, 164, 209, 275–276, 300
inspection, 381, 385–386, 388–390
live load shoes, 358
reactions, 34, 35, 52–53, 138
shear locks, 101, 157–160
inspection, 386
maintenance, 358
economy, 55
heel trunnion, 5, 35, 38–42, 53–54, 108–109, 111–112, 115
balancing, 138, 445, 447
construction, 318–319
design, 137, 139, 152, 156, 166, 212, 283
evaluation, 475
inspection, 376–377, 445
live load, 38
machinery, 115, 212–216
erection, 329
inspection, 390, 392
sleeve bearings, 382
structure, fatigue, 377
Hopkins frame, machinery erection, 329
length, 112
live load anchors, 53, 165
inspection, 389
live load shoes or bearings, 34, 163–166
inspection, 381, 385–386
live load support, 261, 349
machinery
inspection, 380–381, 389–392
installation, 328
live load damage, 390
maintenance, 357–358
multiple girder, 76
noncounterweighted, 48–50, 166, 288
overhead counterweight , 43, 46, 48, 54, 112, 121, 390
Rall type, 73
live load support, 48
replacement, 119–121
rolling counterweight, 446
rolling lift, 5, 46–48, 100
counterweights, 48
machinery, 115
machinery erection, 329
Scherzer, live load support, 47
simple trunnion, 36–38
counterweights, 34, 110
machinery erection, 328–329
single leaf, live load support, 137, 358
single leaf, reactions, 101, 102, 138
span locks, 101, 139, 156–159
inspection, 386–388
Strauss type, 40
St. Charles Air Line, 41
support and stabilization systems, inspection, 381
tail locks, *see* Tail locks, bascule bridges

traffic, 100
trunnions, 76, 145, 148
evaluation, 491
variations, 35
wind load, 104–105
Beam strength, gear teeth, 248, 478
Bearing inspection, 438
Bearing metals
babbitt, 238–239, 301, 304, 439, 482
bronze, 87, 441
cast iron, 304
steel, 413
Bearing supports, 263, 319
Bearings
antifriction, *see* Antifriction bearings
ball, 252
bronze, 85, 88, 148, 168, 180, 238–239, 252, 357, 382, 398, 439, 483
evaluation, 477–478, 482–483, 491
main pinion, 213, 228–229, 263, 329, 391, 416
roller, 6, 84, 92, 146–147, 180–181, 192, 237–238, 283, 345
Bevel gears, 210–211, 229, 251–252, 434–435
Bid documents, 280, 300, 306–308, 309, 314–315
Bolts, 137, 148, 180, 194, 223, 256, 263–264, 322–323, 332, 347–348, 371, 375–376, 378, 382–385, 397–398, 416, 423, 436, 438, 440, 444, 453, 469, 474, 485
coupling, 405, 438
Brake shoes, 440
clearance, 440
Brake(s)
construction, 326, 329
design, 160, 202, 211–212, 214, 219–220, 229, 255, 268, 276, 284, 296, 300, 303
emergency, 392
evaluation, 476, 488
inspection, 389, 391–392, 404, 417, 422, 437–438, 440, 443, 453–454, 456
machinery, 167, 223, 441, 456
maintenance, 353–356
motor, 257, 269, 273–275, 441, 456
thrusters, 257
types, 257
Bridge operators, 10, 23, 69, 100, 125, 158, 202, 259, 267, 268, 271, 273–274, 277–278, 281–282, 288, 292, 296, 334, 367, 387, 415, 442–443, 458, 492
Bronze, 264, 271. *See also* Bearing metals
Brown, Thomas, 41
Bushings, 441. *See also* Bearings
bronze, 46, 180, 237–238, 260, 382, 391, 398, 441

Cables, 13, 19, 23, 26, 35, 40, 94, 331, 459
electrical, 182, 276–277, 406, 451, 455–457, 462, 488
wire rope, 26
Camber, 322
Carter Road Bridge, 60, 80, 121
Cast iron, 48, 55, 154, 195, 286, 304, 336, 407
Centering devices, *see* centering devices subentries under Bascule bridge(s), Swing bridge(s), and Vertical lift bridge(s)
Center locks, jaw type, 157–158
Center of gravity
bascule, 33–34, 43, 46, 150, 154–155, 166–168, 214, 335–336, 445–448, 450, 476
swing, 5, 207
vertical lift, 170, 186
Center pivot bearings, 31, 86–87, 89, 91–92, 135, 191–193, 228, 300–301, 331
clearance, 412
inspection, 370, 410, 412–414
three-piece, 85, 88, 192
two-piece, 88, 191–192
Centers, disc, 84–85, 88, 191, 413
bronze, 85, 88, 191–192, 413
design, 192–194
load suspended, 191, 412–413
pier not level, 191
Center wedges, 84, 101
construction, 331
design, 141, 193–194, 206–207, 229–230, 259, 270, 275–276, 303
inspection, 410, 412–414, 416
maintenance, 361
Chain drives, 443

Circular pitch, 241–242, 246, 248–249, 428, 430–431, 433
Cleaning, 260, 264, 291
 inspection, 367–368, 369, 395, 399, 426–427, 432, 439–440
 maintenance, 341, 350, 371
Clearance
 counterweight, 44, 110
 shaft, 228, 388, 391, 397–398, 421–422, 436, 438–439, 462, 477
 trunnion, 148, 180
Clearance height
 navigation, 7–8, 12–14, 17, 27–28, 46, 55, 58, 75–77, 84, 94, 100–101, 490
 optimization, 35, 52, 106–107, 113, 128, 132, 138
Clearance indicators, 270–271, 444
Clutch reducers, 223, 225–226, 404–405, 435
Clutches
 disconnect, 289, 292, 293
 friction, 229, 259, 288, 417
Communications, 130, 268, 270–271, 273, 277, 282, 296, 443, 457, 463
Continuous spans, 33, 138, 141, 143, 348, 373
Controls, 125, 223, 258–259, 267–269, 270, 304, 333, 452, 477, 492
 electrical, 269–270
Cooling stresses, 319
Copper, 276, 456
Corrosion, 10, 60, 70, 80, 379
 construction, 326
 design, 132, 137, 154, 160–161, 165, 173–175, 180–181, 226, 239, 264, 277, 285–286, 300
 evaluation, 469, 483–484
 inspection, 375–379, 384–385, 389, 392, 395, 397–399, 401, 405–406, 410, 413, 424, 436–438, 441–442, 444, 450, 453–456, 458–459, 462
 maintenance, 343, 347, 349–350, 352, 357–358
 resistance, 173, 285
Cost estimates, 306, 315, 461, 465, 467
Costs
 construction, 24–26, 43, 50, 77, 83, 91, 95, 106, 112, 117–118, 121, 128–129, 131, 135, 139, 141–142, 147, 149–150, 166, 175, 189, 191, 196, 221, 228, 236, 238–239, 250, 256, 260, 288, 292, 296–297, 314, 339, 434
 life cycle, 149
 maintenance, 23, 126, 189, 264, 282, 293, 343–344, 350, 359, 363, 467, 470
 movable versus fixed, 11–12, 14
 operation, 95, 148, 166, 180, 268, 281–283, 288
 repairs, 56, 87, 117, 299–300, 302, 304, 354, 397, 476, 489, 490–491
Counterweight deterioration, 377–378
Counterweight heavy, 215, 217, 219, 223, 336, 405, 476
Counterweight pits, bascule bridges, 10, 36–38, 53–54, 110, 112
 design, 134, 144, 149, 211
 inspection, 367, 375, 389
 maintenance, 358
Counterweight pockets, 155, 169–170, 208, 324, 337, 377–378, 446
Counterweight ropes, *see* Vertical lift bridge(s)
Counterweight trunnions, 40, 42, 46, 283, 329, 445
Counterweights, 3, 6, 10, 24–26, 35, 43
 construction, 169, 320, 323–324
 design, 129, 166, 284
Couplings, 160, 211, 220, 225, 238, 239–240, 258, 293, 303, 307–308
 adjustable, 223, 405
 chain, 240, 437
 clutch, 226
 construction, 326, 331
 disengaging, 223
 flanged, 223
 flexible, 230, 303, 352
 friction, 223
 gear type, 223, 239–240, 304, 329, 437
 grid type, 239, 437
 inspection, 437–438, 440, 443–445, 454–456
 jaw, 240, 304, 352, 437
  clearance, 437
 lubricant, 352
 maintenance, 352
 Oldham, 240
 rigid, 437

Crack inspection, 375–379, 381, 384–385, 392, 395–396, 398, 413–414, 427, 440
Cracks, 109, 134, 147, 149, 156, 165, 180, 192, 212, 228, 304, 319, 339, 347
  fatigue, 109, 147, 149, 156, 180, 339, 475, 483
    inspection, 376–378, 384, 395–396, 398, 437
  sheave shaft, 395–396, 398
Cranks, 199, 259, 289, 386, 388, 418
  hand, 289, 293
Curved treads, 46–47, 149–150, 165, 218, 329, 385, 389
Cylinders, hydraulic, 31, 49, 94, 192, 211, 218, 237, 257, 259–261, 264, 267, 303
  inspection, 392–393, 407–408, 441, 445

Data tables, 344, 372, 424, 461–462
DB Bridge, 92
Dedendum, 243–244, 478
  undercut, 244, 246–247, 254
Deflection
  arm of swing span, 90, 97, 142, 198–199, 228, 230, 415–416
  bascule leaves, 76, 137, 164–165, 210, 319–320, 381, 390
  double leaf bascule, 35, 138–139, 157–158, 163
  machinery supports, 442
  vertical lift bridge, 140, 403
  vertical lift towers, 330
Deflector sheaves, 63, 220–221, 330–331
Design specification, 123, 235, 255, 285
Desirability of movable bridges, 11–12, 80, 107, 111
Diametral pitch, 241–242, 428, 430–431
Direct engine drives, 287, 289, 292–293, 486
Discs, oil grooves, 192
Dock Bridge, 21, 60
Double swing bridges, 21, 84, 94, 96, 107–108, 113, 142, 192
Drawbridges, ancient, 3, 258
Drum girders, 89–90, 141, 196, 409–410
Drums
  brake, 257–258, 353, 355, 440–441
  operating, 26, 63, 66, 190, 219–221, 303, 330–331, 405

Early types of movable bridges, 30–32
Efficiency
  enclosed gear drives, 235
  helical gearing, 252
  hydraulic drives, 260
  machinery, 236, 261, 285, 303
  screw gearing, 285
  spur gearing, 228, 240, 481–482
  worm gearing, 229, 251–252, 285
Electrical component fabrication, 326–327
Electrical installation, 333–334
Electrical system replacement, 304, 477
Electrical wiring, 230, 276–277, 333
Electric motors and brakes, 255–258
  inspection, 440–441
Emergency brakes, 392
Emergency drives, 220, 288–289, 292–294, 295, 305, 371
  evaluation, 486
  inspection, 442–443
Emergency generators, 288–289
End latches, balanced, 203, 304, 331. *See also* Swing bridges
End lift drives, 10, 229–230, 251–252, 259, 261, 265, 270, 275, 285, 288–289, 294, 303, 323, 491
End lifts, 97, 101, 111, 122, 141–142, 193–195, 197–199, 202, 204, 206, 227–229, 231, 275–276, 287, 323, 331, 348, 379. *See also* Swing bridges
  eccentric roller, 199
  inspection, 413, 414–415
  jack and block, 83, 197, 199, 206, 231
  link and roller, 199
  toggle, 199
  wedge, 26, 83–84
    design, 141, 198–199, 203–204, 206, 230–231, 264, 303
    inspection, 415, 418
    maintenance, 360
End rail operators, swing bridges, inspection, 418–419
End rails, 83. *See also* Miter rails
  bascule bridges, 151
  swing bridges, 200, 227, 231, 417, 418, 420
    actuators and machinery, 200, 231–234, 418
  vertical lift bridges, 187

Engine generator sets, 289–292
Engines
diesel, 267–268, 291
gasoline, 267–268, 291
internal combustion, torques for machinery design, 220
steam, 259, 267–268, 288
Equalizers, counterweight ropes, 175

Fasteners, 262–263, 348, 376–378, 397, 436, 442, 454
Fatigue, 80, 111, 147–148, 179–180, 339, 377, 382, 396–397, 485, 491
bolts, 263
counterweight sheave shafts, 395, 485
gears, 434
live load, 38, 142, 149
structural steel, 53–54, 139, 142, 149, 156, 318, 347–349, 376–378, 475
tread plates, 149, 384, 483
wire rope, 60, 65, 109, 125, 171, 173, 176, 484
Fenders, 135–137
Folding bridges, 26, 30–31, 107
Force fits for trunnions, 381
Friction bearings, 148, 180–181, 228, 301, 351–352, 382–383, 391, 397–398, 443, 477, 483
inspection, 422, 438–439
Friction coefficients, 235, 255, 261, 283–284, 287, 295
of machine elements, 261, 283

Gearboxes, 68, 211, 240, 249–251, 325, 424, 431, 436
Gear drives, 240–255
enclosed, inspection, 438
Gearing
bevel, *see* Bevel gears
efficiency, *see* Efficiency
spur, 211, 240, 246, 251–252, 303, 427, 434, 480
worms and wheels, 207, 229, 251–252, 303
Gear inspection, 424–435
Gears
bevel, *see* Bevel gears
helical, 251
herringbone, 251
outside diameters, 243, 428, 430
root diameter, 243
spiral bevel, 251–252
spur, 211, 240, 246, 251–252, 303, 427, 434, 480
Gear teeth, 97, 223, 229, 240, 241–245, 247–250, 252, 264
evaluation, 477, 481
inspection, 390, 404, 406, 426–427, 430, 432–434, 436, 445
Gear tooth calipers, 385, 390, 421, 424, 430–432
Gear tooth measurement, 421, 424, 430, 431–435
Gear tooth strength, 248, 254, 482
Gear trains, 53, 97, 105, 207, 210–211, 228–229, 239–240, 255, 283–285, 427, 434–435, 447, 449, 451, 482–483

Hand drives, 239, 293–295, 367, 380, 486
Herringbone gears, 251
Highway movable bridge type selection, 100–108
Highway traffic safety, 106
Horsepower for bascule span, 50
Horsepower for vertical lift span, 180
Houghton Hancock Bridge, 73
Hurricane locations, movable bridges for, 105–106, 114–115
Hydraulic cylinders, *see* Cylinders, hydraulic
Hydraulic drive machinery, inspection, 392–393
Hydraulic drives, 26, 55, 58, 192, 212, 218, 230, 258–261, 268, 270, 288, 303
construction, 326, 328
evaluation, 469, 486
inspection, 366, 392–393, 407–408, 441–442
maintenance, 345, 353–354
Hydraulic machinery, 25–26, 94, 187, 212, 237, 258, 261, 270, 326, 328
evaluation, 469, 486
inspection, 366, 342–343, 407–408, 441–442
maintenance, 345, 353, 354
Hydraulic system evaluation, 486

Ice load, 155, 208, 282–284, 353, 443, 445–446, 455
Imbalance testing, 449–451
Impact attenuating barrier gates, 106, 277–279, 458–459
Impact loading, 141, 145, 163, 165, 188, 197, 200, 207, 231, 261, 349, 358, 377–379, 385, 386
Impact, ship, 82, 97–98, 126, 135–137, 374–375, 378, 456, 490
Indicators, *see* Clearance indicators; Position indicators
Inspection
 preparation, 372
 report format, 463
 substructure, 373–375
 superstructure, 375–376
 types, 368–371
Inspection equipment
 electrical, 452–453
 machinery, 421–424
 substructure and superstructure, 379
Insulation resistance, 453, 457, 462, 488
Interlocking, 271–276
Involute curve, 243, 482
Involute gear teeth, 247
 module, 242
 stub, *see* Stub tooth gears
Involute rack teeth, 481
Iron, cast, *see* Cast iron

Jacknife bridges, 26, 30–31
Journal design, 145–146, 180
Journal inspection, 382, 391, 397–398

Keys, 302, 414, 437–438, 440, 456
Keyways, 180, 434, 437
Kinetic energy, ship impact, 136

Latches, center, 203
Lateral load resistance, 89, 186, 196
Lead, 154, 336
Levers, 175, 203–204, 229, 231, 257, 272, 386, 417, 454
Lewis formula, 248, 478
Lewis, Wilfred, 248
Limit switches, 207, 268, 272–273, 276, 296, 354–355, 387, 454
 fully seated, 222, 275
 safety interlock, 295
 skew, 70, 275
 span position, 202
Live load anchors, 53, 165, 389
 clearance, 164, 389
Locations, specific, for movable bridges, 14–19
Lockbar clearance, 157–158, 358, 386–388
Locks, center, 35, 53, 110, 143, 157–160, 358, 381, 387–388
Long span movable railroad bridges, 112
Low elevation railroad movable bridges, 115
Lower Newark Bay Bridge, 77
Lubrication, 236, 240–241, 249
 antifriction bearings, 148, 238, 483
 bronze bearings, 238
 charts, 351
 construction, 333
 couplings, 437
 end latches, 202
 evaluation, 483
 inspection, 382, 388, 391, 393, 408, 438–439, 444, 454, 458
 journals, 148, 371, 398
 maintenance, 341, 343, 352, 357
 pivots, 88, 192, 261, 413–414
 racks, 357–358
 span locks, 159
 trunnions, 46, 148, 180, 238, 261, 357, 383
 vertical lift bridge sheave shafts, 180–181, 238, 397–398
 wedges, 230, 360, 415

Machinery
 arrangement, 209
 bascules
  heel trunnion, 212–217
  Hopkins frames, 212
  rolling lift, 218
  simple trunnion, 210–212
 enclosures, 264–266
  evaluation, 489
  inspection, 457–458
 fabrication, 325–326
 hydraulic, 94, 237, 353, 441
 installation, 327–328, 332–333

Machinery *(continued)*
  maintenance, 221, 351–354
  rehabilitation, 300–302
  resistance, design, 235, 284
  supports, 261–262, 284
    inspection, 424
  swing bridges
    end lift and center wedge, 229–231
    turning, 228–229
  upgrades, 302–304
  vertical lift
    span drive, 219–221
    span-tower drive, 224–226
    tower drive, 221–223
Main bearings, 237–238
Maintenance
  control and power systems, 354–356
  manuals, 123, 305, 306, 341–342, 344, 351, 353–355, 365, 367
Manpower, 293, 295, 345, 486
Michigan Avenue bascule bridge, 38, 50
Miter rails, 231, 276
  swing bridges, 26, 83, 98, 111, 203–204, 231, 233–234, 276, 418
Modern types of movable bridges, 20–24, 46, 53, 56–57, 66, 93–94, 108, 177, 329, 331, 358
Modjeski, Ralph, 113, 218
Modulus of elasticity
  metals, 231
  rope wire, 172
Moment of inertia of a swing span, 416
Motors, 239, 254, 255–256, 258, 261–262, 267, 269, 284–285, 287–289, 293, 355
  air, 293, 295, 443
  electric, 48
    design, 239, 254, 255–256, 258, 261–262, 267, 269, 284, 285, 287–289
    direct current, 269–270, 286, 355, 456
    evaluation, 477, 480, 488
    inspection, 440
    maintenance, 355
    torque, 449, 455, 477, 480, 488
      design, 254–256, 261–262, 269, 284–285, 287
  hydraulic, torque, 261, 303
Movable bridge repair, 339
Movable bridge replacement, 338–339
Movable bridge statistics, 21
Movable bridge type selection, 28, 77, 80, 96, 99, 108, 110, 114, 142, 288, 338

Newark Draw, 86
Nickel steel, 180
Noncounterweighted bascule bridges, 48–50

Oilless bearings, 239
Old London Bridge, 4
Operating manuals, 305–306
Operation time, 101, 285
Operators' houses, 170, 207, 220, 224, 229, 291–292, 295–298, 334, 386
  evaluation, 487, 489
  inspection, 457–458
Overstress, 10, 81, 129, 148, 159, 189, 194, 228, 254, 267, 308, 318, 322–323, 386, 388, 400, 403, 437, 455, 465, 474, 477, 480–481, 483

Packing glands, 352
Parallel shaft reducers, 224
Pedestals, 197, 458
Pinions, 26, 66, 68, 72
  construction, 319, 328–330, 333
  design, 194, 210–213, 218, 220–221, 224, 228–229, 238–239, 244, 247, 251–252, 254–255, 261–264, 284, 295, 300–301, 303, 308–309
  evaluation, 478, 481
  inspection, 385, 389–391, 404–405, 416, 421, 426–428, 430, 434–435, 450–451
  maintenance, 349, 357–358
Pins, 40, 180, 388, 390, 392
Pitch diameter, 176, 242–243, 247, 428, 433
Pitch, circular, *see* Circular pitch
Pitch, diametral, *see* Diametral pitch
Pitch radius, 211, 254, 429, 434
Pivot bearing, 86, 92, 126, 410, 412–413
  design, 92, 191–193
  lubrication, 412
Planning, 6, 14–15, 119, 325, 488
  inspection, 370–371
Point-No-Point Bridge, 86

Pontoon bridge
  retractile, 22, 24–26, 308
  swing, 26
Portal Bridge, 86, 98, 233–234
Position indicators, 202, 240, 296, 431, 444
Power, 5, 105, 114, 282–288
  for bascule span, 48, 50, 148, 154, 166, 210–211, 215, 217–218, 380, 392
  construction, 326–327, 333, 336
  design, 135, 236, 239–240, 242, 249, 252, 255–256, 258–261, 267–269, 272–275, 277, 289–295, 303
  evaluation, 468, 476, 480–481, 486, 488–489
  inspection, 436–437, 441–443, 445, 452, 455–458
  maintenance, 345, 351, 353–355, 359
  for swing span, 202, 207, 229–230, 232–233, 417–418, 420
  for vertical lift span, 70, 180, 182–183, 189–190, 220–221, 404, 406
Prairie Du Chien retractile bridge, 24–25
Pressure
  in bolts, 382
  in brake shoes, 257, 353
  in pivots, 191
Proximity switches, 202, 272, 454

Quality, design documents, 309
Quick-acting highway movable bridges, 100–101

Rack(s), 66
  construction, 319, 325, 328–330, 332–333
  design, 135, 194, 210–213, 218, 221, 228, 243, 247, 254–255, 284, 301
  evaluation, 481
  inspection, 385, 389, 391, 416, 421, 451
  and pinions, 26, 72
    construction, 329, 333
    design, 212, 221, 228, 244, 247, 254–255, 261–262, 264, 284, 300–301, 308–309
    inspection, 385, 389–390, 404, 416, 427–428, 430
    maintenance, 349, 357–358
Rack girders, 212, 328, 389
Rack pinion(s), 66
  construction, 319, 328–329
  design, 194, 210, 211–213, 228–229, 238, 244, 252, 254, 284, 295, 303
  inspection, 385, 389, 391, 416, 421, 451
  shaft(s), 251, 303, 328, 390–391, 450
    bearings, 228, 239, 263, 391
Rail joints
  bascule bridges, 162–163
  swing bridges, 200, 231, 233
  vertical lift bridges, 187–188
Rail lifts, swing bridge, 227
Rail locks, 160, 273
Railroad movable bridge type selection, 108–111
Rall bascule, 47, 48, 54, 73, 108, 160, 165, 283, 388, 391, 447
Rall, Theodor, 47, 72–73
Removable spans, 21, 27–29, 344
Replacement of movable bridge electrical and mechanical equipment, 121–122
Replacement of movable bridge superstructure, 121
Reporting of critical items, inspection, 463
Retardation, 269, 287
  of bascule span, 284
  of swing span, 284
  of vertical lift span, 189
Retractile bridges, 3, 6, 21, 22, 24–25, 30–32, 106–107, 154, 237, 259, 270, 288
Right angle reducers, 224, 229, 251, 303
Rim bearings, 83, 85, 89–92, 135, 195–196, 206, 228, 331, 409, 410
  inspection, 410
Roller bearings, 6, 84, 92, 146–147, 180–181, 192, 237–238, 283, 345
Rollers, 24, 213
  antifriction bearings, 440
  Rall bascule bridges, 48, 109, 283
  Rall vertical lift bridges, 73, 83
  rim bearings, 89–91, 94, 141, 171, 193, 195–196, 228, 287, 301, 331, 361, 379, 409–410, 491
  swing bridge centering devices, 203, 417–418
  swing bridge end lifts, 198–199, 203, 206

Rollers *(continued)*
vertical lift bridges, 105, 171, 184, 220, 331, 402
Rolling lift bridges, 5, 100
Rall, 101, 109, 160, 165, 218, 347
Scherzer, 38, 43, 46–48, 54, 100–101, 108, 112
construction, 317, 320
design, 139, 148–149,158–160, 165, 218, 243, 303
evaluation, 475
inspection, 377, 380, 385–386, 388, 390, 447, 455
maintenance, 347–348
Ropes, wire, 29, 171, 173–174
Rust packing, 379, 385

Safety of highway traffic, 106
Safety, inspection, 366–367, 380, 394, 409, 421, 452
Scherzer, Albert, 46, 329
Scherzer, William, 46, 149, 218, 329
Scherzer Rolling Lift Bridge Company, 101, 107
Secondary electrical service, 289
Seismic design, 133–134, 185–186, 196, 491
Seizing of machine elements, 46, 85, 88, 183, 192, 239, 260, 382, 402, 407, 441, 443
Shaft bearings, 238–239
Shafts, 33–34, 36, 66, 68, 80, 122
bending and torsion, 382
counterweight sheave, see vertical lift bridges
and couplings, inspection, 437–438
design, 145, 148, 207, 210–213, 218, 220–225, 228–231, 240, 250, 257–259, 286, 293, 295, 300–301, 303–304, 307
evaluation, 483
inspection, 380, 386, 394, 409–410, 421, 424
maintenance, 352
seals, 352
torsional capacity, 437
torsional, deflections, 53
Shear pole swing bridges, 5, 25–26, 32, 94, 111
Shippingsport Bridge, 221
Shrink fits, 177, 179, 318, 395
Site evaluation, 126–128
Soil conditions, movable bridges, 107, 131, 133, 299
Span balance, evaluation, 476, 491
Span drive vertical lift, *see* Vertical lift bridges
Span heavy, 140, 160, 214–215, 217, 219, 225, 270, 336, 403, 405, 476
Span length optimization, 131–132
Span-tower drive vertical lift, friction, 71–72
Specifications, 308
ASTM, *see* American Society for Testing and Materials (ASTM)
construction, 162, 178, 180, 193, 241, 250, 258, 262–263, 280, 306, 308–309, 311, 315, 320, 327–328, 336, 359
design, 107, 123, 129, 176, 235, 241, 244, 252, 254–255, 261, 263, 282, 285, 299–300, 302, 304
operating machinery, 236, 254
power equipment, 455
wire rope, 174, 359
Speed reducers, 212, 247, 424
Stable movable bridges, 101–103
Steel
alloy, 145, 171
cast, 162, 177, 195, 233, 240, 395, 398
forged, 85, 195, 413
structural, 109, 124–5, 132, 150, 159, 186, 262, 330, 335, 349, 376–378
Steel Bridge, Portland, OR, 60, 73–74, 103
Strain gauges, 336, 405, 450–451, 469
Strauss bascule bridges, 38, 40, 43, 243, 475
Strauss vertical lift bridges, 66, 68, 72, 221, 404
Strauss, Joseph, 38, 40–41, 43–44, 66, 72, 140, 218, 475
Stresses, 43, 112, 125, 137–138, 142, 148–149, 159–160, 165, 171–173, 179, 204, 230, 248, 250, 284, 318–319, 323, 329, 339, 347, 360, 377, 382, 384–385, 388, 395, 426, 475, 478, 480–483, 491
allowable, 52, 147, 177, 249, 254, 318, 465, 474, 479–480

concentration, 180, 264, 359, 485
dead load, 139, 141, 475
internal, 90, 178, 323
live load, 386
reversal, 53, 109, 138–9, 142, 148, 156, 179, 322–323, 348, 376–377, 485
spur gears, 480, 483
Stub tooth gears, 241, 244–245, 252, 427–428, 430
Substructure, 133
bascule bridges, 134
construction, 317
evaluation, 472–473
maintenance, 347–348
rehabilitation, 299
swing bridges, 134–135
vertical lift bridges, 134
Superstructure
bascule bridges, 137–140
erection, 319–321
inspection, 376–377
erection, 323–324
evaluation, 473–476
fabrication, 318–319
maintenance, 348–350
rehabilitation, 300
swing bridges, 140–143
erection, 322–323
inspection, 378–379
vertical lift bridges, 140
erection, 321–322
inspection, 378
Swing bridge(s), 5, 20–22, 52, 76, 81–83
advantages, 95–96
applications, 106–107, 113
balancing, 93
bobtail, 93–94
Camp LeJeune, N.C., 197
center bearing, 84, 88, 135, 141, 193, 206, 410, 413, 416
live load support, 135
centering devices, 97, 199, 200–206, 228, 230–231, 276, 304
clearance, 206, 418
inspection, 417–418
machinery, 231
center latches, 193, 203, 231, 417
clearance, 417
inspection, 417
center pivots, *see* Pivot bearing
center wedges, 206–207
combined bearing, 92
counterweights, 95, 208
definition, 20
disadvantages, 96–98
double, *see* Double swing bridge
double deck, 82, 87, 91, 95, 104
drive machinery inspection, 416–417
end lifts, 141, 197–199. *See also* End lifts
inspection, 415
machinery inspection, 409, 414
live load support, 10, 84, 96–97, 135, 141–142, 193, 195, 227, 261, 323
inspection, 379
machinery installation, 331–332
maintenance, 360–362
replacement, 118–119
rim bearing, 24, 82, 87, 89, 91–92, 108, 361
construction, 332
design, 135, 141, 193, 196, 206, 308
evaluation, 491
inspection, 379, 409–410
variations, 84–95

Tail locks, bascule bridges, 159
Time of operation, *see* Operation time
Tolerances, 308–309
Tomlinson Bridge, 63, 113
Tower drive vertical lift bridge, drive friction, 70
Traffic control systems
highway movable bridges, 277–280
evaluation, 489
inspection, 458–459
Traffic studies, 128–130
Transporter bridges, 12–13, 29, 31–32
Treads, rolling lift bridges
design, 148–150
evaluation, 483
inspection, 384–385
Trunnion(s), 34, 36, 40, 43, 46, 48, 76, 102
construction, 318–319, 328
design, 137, 139, 140, 144–145, 156, 165–166, 179, 300, 302, 308
evaluation, 491
friction, 283

Trunnion(s) *(continued)*
inspection, 377, 380, 381–385, 389
lubrication, *see* Lubrication of trunnions
unit stresses, 147–148
Turned bolt, 256, 262–263, 329, 371, 440

Ultimate strength of materials, 250
Undesirability of movable bridges, 10–11
Unlimited vertical clearance, movable bridges, 106–107, 113–114
Uplift resistance, 91, 165

Vernier calipers, 421–422, 430, 433–434
Vertical lift bridge(s), 6, 20, 21, 24, 33, 52, 55–63
advantages, 75–76
applications, 101, 103, 105, 112, 114–115
auxiliary counterweights, 181–183, 189–190, 226
construction, 337
inspection, 406–407, 448–449
ropes, 181
buffer cylinders, 187
centering devices, 185–186
inspection, 403
counterweight guides, 105, 134, 186–187, 321
counterweight ropes, 10, 57, 60, 65, 68, 140, 170–176, 283, 359
abrasion, 359
evaluation, 469, 483–485
fabrication, 321
inspection, 398–401
maintenance, 358–359
replacement, 79, 109, 189
counterweights, 10, 55, 58, 68, 76, 170–171
erection, 321, 337
counterweight sheaves, 176–178
evaluation, 485
inspection, 395
counterweight sheave shafts, 179–180
bearings, 140, 180–181, 237–238, 345
inspection, 397–398
evaluation, 485
failures, 79, 111, 491
inspection, 396–397
replacement, 79, 339
dead load stresses, 76, 140, 181, 189, 330, 360
definition, 20
disadvantages, 76–79
double deck, 60, 73, 76
Erie Canal type, 74–75
early type, 74
live load shoes, 57, 71, 140, 186, 188–189, 322, 337, 349, 360
inspection, 378, 401–402, 404–405
live load stresses, 76, 102, 140, 186, 349, 378, 403–404
live load support, 76, 102, 140, 155, 378
machinery installation, 330–331
maintenance, 358–360
noncounterweighted, 72, 166, 189
operating machinery inspection, 403–408
operating ropes, 56–57, 63, 65–66, 68, 72, 93, 109, 140, 183, 190, 219–221, 330–331, 404–405, 449–450
Rall type, 72
replacement, 121
skew, 65, 68, 70, 190, 222–223, 275, 402, 435
skewed, 19
South Halsted Street, 56–57
span and counterweight guides, inspection, 402
span drive, 63–68
span guides, 183–186
clearance, 184–185, 402
span locks, 187, 225
inspection, 403
span-tower drive, 71–72
Strauss types, 72–73
Strobel type, 72–73
support and stabilization systems, inspection, 394–395
tower drive, 68–71
towers, 330, 378
traction restraints, 186
variations, 63–75
Wadell type, 56–57, 63, 65–66, 68, 104, 109, 219–220, 303, 330, 405, 450
Vertical lift span operating machinery, 219–226

Very long span movable bridges, 103–105, 112–113

Waddell, J.A.L., 30–31, 56
Wear
  balance wheels, 195, 360
  bearings, 147–148, 180–181, 238–239, 301, 382–383, 397–398, 438–439, 482–483, 491
  brakes, 355
  buffer cylinders, 161
  centering devices, 164, 388, 389, 417–418
  center pivot bearings, 410, 413–414
  commutators, 269, 355
  counterweight ropes, 10, 60, 80, 171, 173–174, 321, 330, 359, 398–400, 406, 469, 483–484
  drive machinery, 202, 225, 228–229, 241, 247, 249–250, 252, 264, 284, 301, 331, 342, 349, 351–352, 357, 385, 390, 404, 406, 416, 424–426, 431–433
  end rails, 418
  gates, 280
  gearboxes, 436
  live load bearings, 186, 349, 358, 360, 386, 401–402
  motor slip rings, 456
  operating ropes, 57, 190, 221, 405
  rail lifts, 418
  Rall bascules, 160, 388
  rim girders, 196, 228, 361, 410
  span guides, 402, 410
  span locks, 53, 157–158, 358, 387–388
  switchgear, 272, 354, 454
  tread plates, 47, 108–109, 165, 320, 385, 388
Wear resistance, 241
Wedges
  center, *see* Center wedges
  end, *see* End lifts, wedge
  friction of, 231
Weld repairs, 300, 384–385, 402, 474, 485
Weldments, 149, 176–179, 318–319, 323, 395, 440
West Seattle swing bridge, 94, 96, 99, 108, 142, 192
Wheels, balance, *see* Balance wheels
Wind pressure, 404
Wind resistance, 35, 48, 97, 105, 114, 209, 287–288
Wire rope(s)
  bending stresses, 171–173
  construction, 171, 484
  elongation, 175–176, 190, 279, 401, 404, 459
  lubrication, 80, 173, 359, 395, 398, 469
  materials, 171, 174, 359
  specifications, 171, 173, 359
  stiffness, 283
  ultimate strength, 172

Zinc, 331, 401

Lightning Source UK Ltd.
Milton Keynes UK
UKHW050126140219
337194UK00004B/98/P